Lecture Notes in Mechanical Engineering

Lecture Notes in Mechanical Engineering (LNME) publishes the latest developments in Mechanical Engineering—quickly, informally and with high quality. Original research reported in proceedings and post-proceedings represents the core of LNME. Volumes published in LNME embrace all aspects, subfields and new challenges of mechanical engineering. Topics in the series include:

- Engineering Design
- Machinery and Machine Elements
- Mechanical Structures and Stress Analysis
- Automotive Engineering
- Engine Technology
- Aerospace Technology and Astronautics
- Nanotechnology and Microengineering
- Control, Robotics, Mechatronics
- MEMS
- Theoretical and Applied Mechanics
- Dynamical Systems, Control
- Fluid Mechanics
- Engineering Thermodynamics, Heat and Mass Transfer
- Manufacturing
- Precision Engineering, Instrumentation, Measurement
- Materials Engineering
- Tribology and Surface Technology

To submit a proposal or request further information, please contact the Springer Editor of your location:

China: Ms. Ella Zhang at ella.zhang@springer.com
India: Priya Vyas at priya.vyas@springer.com
Rest of Asia, Australia, New Zealand: Swati Meherishi at swati.meherishi@springer.com
All other countries: Dr. Leontina Di Cecco at Leontina.dicecco@springer.com

To submit a proposal for a monograph, please check our Springer Tracts in Mechanical Engineering at https://link.springer.com/bookseries/11693 or contact Leontina.dicecco@springer.com

Indexed by SCOPUS. All books published in the series are submitted for consideration in Web of Science.

Elango Natarajan · S. Vinodh · V. Rajkumar
Editors

Materials, Design and Manufacturing for Sustainable Environment

Select Proceedings of ICMDMSE 2022

Springer

Editors
Elango Natarajan
Department of Mechanical and Mechatronic Engineering
Faculty of Engineering, Technology and Built Environment
UCSI University
Kuala Lumpur, Malaysia

S. Vinodh
Department of Production Engineering
National Institute of Technology Tiruchirappalli
Tiruchirappalli, India

V. Rajkumar
Department of Mechanical Engineering
PSG Institute of Technology and Applied Research
Coimbatore, India

ISSN 2195-4356 ISSN 2195-4364 (electronic)
Lecture Notes in Mechanical Engineering
ISBN 978-981-19-3052-2 ISBN 978-981-19-3053-9 (eBook)
https://doi.org/10.1007/978-981-19-3053-9

This Springer imprint is published by the registered company Springer Nature Singapore Pte Ltd.
The registered company address is: 152 Beach Road, #21-01/04 Gateway East, Singapore 189721, Singapore

Preface

The Second International Conference on Materials, Design and Manufacturing for Sustainable Environment (ICMDMSE 2022) was conducted by PSG Institute of Technology and Applied Research, an institution with enormous potential, poised for growth with special emphasis on research. It indeed received a great attention by researchers and professors of renowned institutions in India and across the globe. The expectation and benchmarks were set by our maiden conference on the same theme: Digital Solutions for Sustainable Earth, conducted in 2020 which had great impact among industries and academia. It was conducted in association with Fluid Control Research Institute (FCRI), Palakkad.

The Second International Conference on Materials, Design and Manufacturing for Sustainable Environment (ICMDMSE 2022) with a theme focusing on "Digital Solutions for Sustainable Earth" successfully conducted in collaboration with UCSI University, Malaysia, is also well received and was a grand success. We sincerely thank the members of National and International Advisory Committee, organizers, reviewers, authors and participants of ICMDMSE 2022. Our special thanks to Springer Nature for publishing selected papers of ICMDMSE 2022 as a special volume. Readers shall send us their feedback about this volume.

Select papers from the conference are being published by Springer in the series—*Lecture Notes in Mechanical Engineering*, in four tracks: Materials for Sustainability, Design of Materials for Sustainability, Manufacturing and Industrial Engineering and Thermal and Fluid Sciences for Sustainability.

After a rigorous review process, about 56 technical papers from academia and industry were accepted for the presentation at the conference. In addition, six keynote talks on advanced technologies such as introduction to data science and artificial intelligence, machine learning algorithms for effective manufacturing, high-speed machining of new age alloys, electric vehicles and contemporary issues, renewable energy and fuel cells were presented.

Volume covers a wide variety of technical papers related to composites, nanomaterials, biodegradable materials, optimization techniques, additive manufacturing, green design and manufacturing, sustainable manufacturing and global manufacturing challenges, lean and agile manufacturing, robotics, renewable energy, heat transfer and fluid flow methods. We hope that researchers as well as practicing engineers will find this volume useful.

Kuala Lumpur, Malaysia — Elango Natarajan
Tiruchirappalli, India — S. Vinodh
Coimbatore, India — V. Rajkumar

About the Conference

It gives us immense pleasure to present the proceedings of Materials, Design and Manufacturing for Sustainable Environment—proceedings of International Conference on Materials, Design and Manufacturing (ICMDMSE 2022).

We wish to express our profound gratitude to all the members of the Organizing Committee of ICMDMSE 2022, authors, reviewers, sponsors, volunteers for the active support and their relevant contributions. We express our sincere thanks to our Chief Patron, Shri. L. Gopalakrishnan, Managing Trustee, PSG Institutions, and patrons Dr. G. Chandramohan, Principal, PSG Institute of Technology and Applied Research, Dr. P. V. Mohanram, Secretary, PSG Institute of Technology and Applied Research, Ir. Ts. Dr. Ang Chun Kit, Dean, Faculty of Engineering, Technology and Built Environment, UCSI University, Kuala Lumpur, Malaysia, for their consistent support in successful conduct of the conference.

Our special thanks to the National and the International Advisory Committee members, Chair(s) Dr. N. Saravanakumar, Professor and Head of Mechanical Engineering, PSG iTech, Dr. Elango Natarajan, Associate Professor, UCSI University, Malaysia, for their valuable support and guidance in all conference activities, and Dr. S. Vinoth, Associate Professor, NIT Trichy, in reviewing and editing the book volume.

Our heartfelt thanks to the keynote speakers (national and international) for rendering mind storming presentations on the importance of sustainability with a note on digital solutions.

High-quality manuscripts have been selected after careful peer review. We are confident that chapters included in the proceedings of ICMDMSE 2022 are interesting and thought-provoking.

Finally, we would like to express our gratitude to all the Advisory Committee members of ICMDMSE 2022 for providing necessary guidance and support.

Organizing Team
ICMDMSE 2022

Contents

About the Editors

Dr. Elango Natarajan is working as an associate professor in Mechanical and Mechatronic Engineering Department, UCSI University, Malaysia. He received his bachelor's degree in mechanical engineering and his master's degree in CAD from University of Madras and Ph.D. from Anna University, India. He received C.Eng from Engineering Council, UK. Dr. Elango worked as B.E. Programme Coordinator in Kolej Universiti Linton Coventry University. Malaysia. His research interests include manufacturing and machining of composites, optimization and artificial intelligence, soft actuator and biomedical applications, finite element analysis (CAE) and engineering design. He has published 66 research articles in Scopus/WoS. He is supervising 8 Ph.D. scholars and has 5 funded projects to a tune of RM 637527. He has given many guest lectures and keynote addresses and is an official reviewer in 27 journals.

Dr. S. Vinodh is an associate professor in the Department of Production Engineering, NIT National Institute of Technology, Tiruchirappalli, Tamil Nadu, India. He completed his Ph.D. degree under AICTE National Doctoral Fellowship scheme from PSG College of Technology, Coimbatore, India. He was a Gold Medallist in his undergraduate study. He has published over 175 papers in International Journals. He received Highly Commended Paper Award from Emerald Publishers for the year 2016. He is the recipient of Innovative Student Project Award 2010 based on his Ph.D. Thesis from Indian National Academy of Engineering (INAE), New Delhi, India. His research interests include Sustainable Manufacturing, Lean Manufacturing, Agile Manufacturing, Rapid Manufacturing, Product Development and Industry 4.0.

Dr. V. Rajkumar is working as associate professor in Mechanical Engineering Department at PSG Institute of Technology and Applied Research, India. He did his undergraduate studies in mechanical engineering at PSG College of Technology and postgraduate studies from School of Industrial and Manufacturing Systems Engineering, University of Windsor, Ontario, Canada. He holds his Ph.D. in Manufacturing from School of Mechanical Engineering, Vellore Institute of Technology, Vellore. His research interests are in metal joining, materials characterization and additive manufacturing. He is a reviewer in more than 10 international journals.

Materials for Sustainability

Development of Sustainable Banana Plantain Fibre Extractor

Deepak Eldho Babu, B. Biju, Eldho Geegy, Joseph K. Malil, and Issac Jeoju

Abstract Effective use of materials plays a major role in developing sustainable products. Identifying useful materials from the waste products is an innovative idea towards developing competitive advantage in an industrial economy. Incorporating triple bottom line into sustainable materials is one of the major challenges faced by researchers of present decade. This paper aims in developing a suitable substitute material in the form of banana stem fibre which can replace the existing plastic and conventional coir fibre rope. Further, the paper also aims at designing and developing a fibre extraction machine which can be used for separating the fibres from banana plantain. Since, banana is one of the largest consumed fruit in the world; the paper will also help the researchers and practitioners to understand the innovative concept to use the waste materials from the banana plantain in a sustainable manner.

Keywords Sustainability · Plantain fibre · Banana stem · Fibre extraction machine

1 Introduction

Sustainability has gained much importance in this decade. The most adopted definition of sustainability would be development that meets the needs of the present without compromising the needs future generations. Sustainability refers to an integration of social, environment and economic responsibilities [1]. In engineering practices, sustainability can be explained as the process of using materials and resources that will not compromise the environment or deplete the materials for future generations [2].

D. E. Babu · B. Biju (✉) · I. Jeoju
Mar Athanasius College of Engineering, Kothamangalam, India
e-mail: bbiju@mace.ac.in

E. Geegy
Technische Hochschule Rosenheim, Rosenheim, Germany

J. K. Malil
Massey University, Palmerston North, New Zealand

E. Natarajan et al. (eds.), *Materials, Design and Manufacturing for Sustainable Environment*, Lecture Notes in Mechanical Engineering,
https://doi.org/10.1007/978-981-19-3053-9_1

Ropes are essential product used for household activities. People in developing countries involve in the production of rope as a source of income and hence improving their standard of living. The common raw material which is used for the making of rope includes coir and plastic nylon. Coir or coconut fibre extraction is a time-consuming process where the coir from the husk of coconut is harvested after about six to twelve months and soaked in slow moving water to swell and soften the fibres, which is called as retting. The coir retting process has influence on quality of water and marine flora and fauna [3]. The degrading fibre binding materials of the husk liberate large quantities of organic matter and chemicals into the environment. Consequently, gases like hydrogen sulphide, phosphate and nitrate contents increase, and dissolved oxygen level in the water goes down [4].

Plastic and other synthetic raw materials used in rope production raise great environmental issues. Most importantly, they are non-biodegradable and accumulate in the environment as microplastic waste. Microplastics are very small pieces of plastic material that pollute the environment. These enter natural ecosystem and may even enter the food chain [5].

This paper discusses about an innovative material that exists in nature and can act as a substitute raw material for manufacturing rope. Plantain fibre obtained from the plantain sheath stem could be an effective raw material which is easily available as waste in banana cultivation [6]. The study focuses on the destinations in the State Kerala, the southern part of India which is one of the largest producers of banana in India. Agricultural products in India are expected to accelerate towards 8 per cent GDP growth in the year 2019. The plantain fibres possess comparatively lower density, less weight, recyclable, biodegradable and also appreciable mechanical properties which make it a suitable raw material [7]. Plantain sheath stem constitutes sixty per cent cellulose and twenty per cent lignin which make it fibrous [8]. Fibre extraction from plantain has started long ago, but the methods used for the extraction such as the retting method involve anaerobic, biological organisms like bacteria and fungi presence in the medium, where the stems are stored for long to decompose. This process takes six to seven months for the fibre to decompose. Hence, mechanical methods of fibre extraction are to be developed.

This paper gives attention to the mechanical techniques in plantain fibre extraction in order to reduce the processing time and eliminate the environmental impacts caused by the traditional methods. The paper discusses an electrically powered mechanism to extract fibre from raw plantain stem. This extracted fibre is twisted to rope by feeding the fibres into a rope spinning machine which is also provided with thin nylon strand for physical property enhancing. Further tensile properties of the banana fibre rope are studied.

2 Literature Review

The interest towards sustainability practices for the corporate has increased in the last two decade. Many researchers have pointed out the need to implement sustainability practices in supply chain to develop competitive advantage. Ageron et al. [9] pointed out that sustainability will gain more attention in future due to the importance given to the sustainability practices globally.

2.1 Sustainability Pillars

Elkington (1999) has mentioned three main pillars of sustainability as social, environment and economic factors. The author also pointed out that "the objective of triple bottom line approach to sustainability is to operationalize sustainability where minimum performance is to be achieved in terms of environment, economic and social factors" (Fig. 1).

In a sustainable environment, an ecosystem should maintain populations, biodiversity and overall functionality over an extended period of time. Production of sustainable fibre rope has a positive impact on the environment; with respect to synthetic ropes, banana fibre ropes are natural, biodegradable and resources vastly available. Economic sustainability means creating economic value out from the intended activities performed [10]. In this paper, economic sustainability is explained by using the waste generated from banana cultivation which results in the development of sustainable rope as the by-product. And hence, they can find an additional source of income.

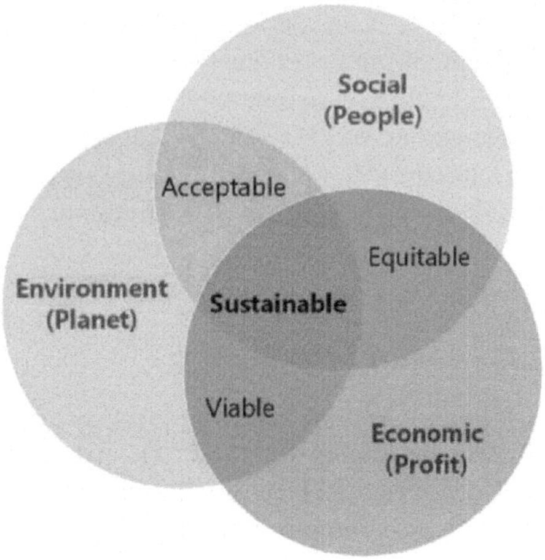

Fig. 1 Pillars of sustainability

2.2 Fibre Extraction

The fibre extraction from coconut involves soaking the dried coconut in running water and drying, then it grinds the coconut to extract the fibre out of the husk and is processed into fine fibres in a raspador machine. Raspador drum consists of blades welded to a shaft which on rotation beat the sheath stem, crushing the stem and yield fibre strands. This whole process consumes more time and manpower; moreover, retting process has its own environmental impact.

On crushing the sheath stem of banana, the pulpy part is removed and fibrous part is obtained. The banana sheath stem passed through rollers only removes some of the pulpy part which is cellulose in the stem, whereas passing the crushed stem through a raspador drum removes all cellulose and yields clean fibre. The machine reduces drudgery and provides with a clean working environment for the workers [11].

Machinery for mass production of these fibres is available, but the quality does not meet the standards. The rotor assembly developed in this paper consists of two discs on which blades are mounted and a shaft design to drive this rotor assembly and the pulley and powering system for the machine.

2.3 Conventional Rope

Nylon plastic rope adds to microplastic waste after its use or after its lifetime. Microplastic is an emerging contaminant of potential concern. These wastes could have a great impact on sustainability by having adverse impact on soil, seawater by contaminating them and could even enter the human food chain.

Remani et al. [3] discussed about pollution due to coir retting and its effect on estuarine flora and fauna. It is visible that the retting process contaminates the water bodies and takes time up to six to twelve month's process in running water.

2.4 Banana Fibre

The banana stem is highly fibrous due to its low lignin high cellulose content, hence so, these fibres can be efficiently extracted using a raspador machine. Kulkarni et al. [7] discuss on the mechanical properties of banana fibres and found that properties such as the Young's modulus (YM), ultimate tensile strength (UTS) and percentage elongation are evaluated, and these properties add to the use of banana fibre as raw material for rope production.

Natural fibres are potential substitute to synthetic fibre in the polymer composites [11]. Natural fibres result in lighter composite materials as compared to synthetic fibre reinforced polymer composites with equivalent mechanical strength. Natural fibres are biodegradable, and their productions are with lower emission compared to the production of synthetic fibre. Also, high natural fibre contents in composites which in turn with polymer expenses result in the economy of energy [12]. The paper explains the scope of increased employments and opportunity when it comes to natural fibres, and the extraction of it will contribute to the emerging countries with rich source of natural fibres.

Joana et al. discusses the Life-Cycle Assessment and Life-Cycle Cost study of Banana (Musa sapientum) fibre Bio composite materials [13]. This article said that biocomposite materials created from banana fibres have a high potential to be eco-friendly and sometimes technically competitive with conventional composites and plastics.

3 Fibre Extraction Machine

Since the conventional retting process has its impact on the environment and manual fibre extraction is time-consuming, a fibre extraction machine was proposed. It was found that the action of raspador drum removes the cellulose content and yields fibre.

In the machine, the pseudo-stem layer is pushed between two rollers, where it is flattened and moves to the rotating raspador drum which is rotating near the third roller. The raspador drum which beats the stem thus making the pulpy cellulose structure loosens from the fibre. Then the stem is pulled, removing the pulpy cellulose and yielding fine fibres. The extracted cellulose gets collected in a tray under the raspador drum. Figure 2 shows the exploded view of the banana fibre extraction machine (Table 1).

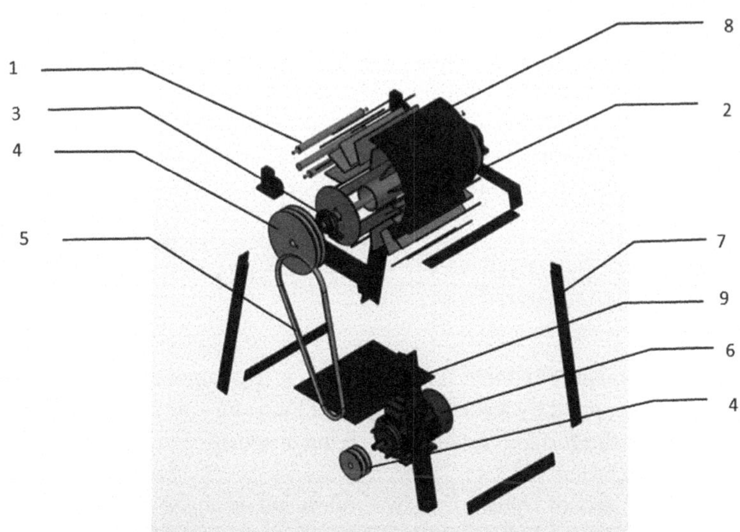

Fig. 2 Fibre extraction machine

Table 1 Specification of machine items

Item No.	Specification
1	Roller setup
2	Raspador/pulping drum
3	UCP bearing/ID 25 mm
4	Pulley 2",5"
5	V belt—B50
6	Motor—1.5HP
7	Frame
8	Drum compartment
9	Waste collection tray

Fig. 3 Raspador drum/pulping drum design

3.1 Raspador Drum/Pulping Drum

The raspador drum consists of blades attached to two circular flanges on a solid shaft. When this drum rotates, it crushes and scraps the banana stem layer. Raspador drum is a driven by an electric motor located under the collecting tray; the drive is transmitted using two pulleys connected by a V belt, the larger one attached to the solid shaft of raspador and the smaller one to the motor shaft (Fig. 3).

3.2 Rollers

The roller setup consists of three rollers, of which two supported by circular ball bearing and one is guided by a slit and supported by spring, allowing vertical movement; this flattens the banana sheath stem. Banana sheath stem enters through the movable and fixed rollers (Fig. 4).

There are three sets of rollers, first two rollers are arranged one on top of other with parallel shaft axes and attached to one pair of metal block. The top roller's shaft is freely suspended without a bearing on the slot cut on the solid metal block; ends

Fig. 4 Roller setup design

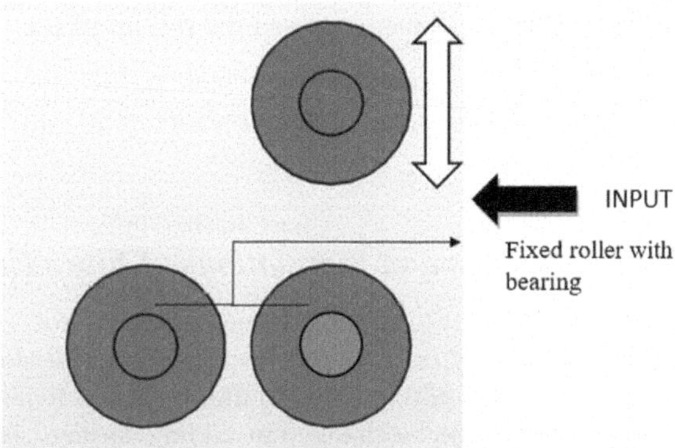

of the shaft are attached to a spring setup which provides tension on the top roller when it lifted.

The third roller is placed near to the drum blades; this roller helps to maintain the clearance of the raspador drum; the banana stem layer will be machined between this roller and the drum. The clearance can be adjusted by moving the third roller.

The following specifications for the plantain fibre machine are considered; plantain width equal to 150 mm maximum; plantain maximum thickness (sliced trunk thickness) equal to 10 mm; driving means equal to electric motor; motor operating voltage equal to 230/415V; type of motor is an induction motor; feeding method is manual; coupling method used is belt drive; belt type used is V belt B50 [14].

Figure 5 is a schematic view of a rolling drum structure showing the radius of the drum and its beating blade; a typical rolling drum carries 14–27 blades.

r is the radius of drum

L is the distance between two blades = length of arc 1 and 2 Number of blades for this design = 15

Distance between two blade, $L = 45$ mm

Drum radius = 50 mm. Drum diameter = 2 × radius = 100 mm Drum thickness = 4 mm

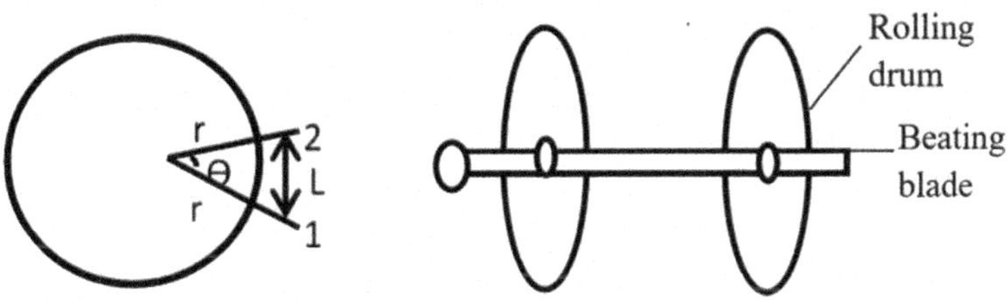

Fig. 5 Rolling drum structure and radius of drum and beating blade

Blade length = distance between the rolling plates = 225 mm Blade thickness = 6 mm flat bar

The 6 mm thickness flat bar has a breadth of 40 mm.

3.3 Drawings and Simulations of Fibre Extraction Machine

Figures 6 and 7 depict the top view and front view of banana fibre extractor, respectively. The height of rollers is positioned such as to allow easy operation. The motor platform height can be changed by adding washers, in case of belt loosening.

Figures 8, 9 and 10 show the Von Misses analysis of the raspador drum and shaft. The points of maximum deflection are shown using ANSYS analysis.

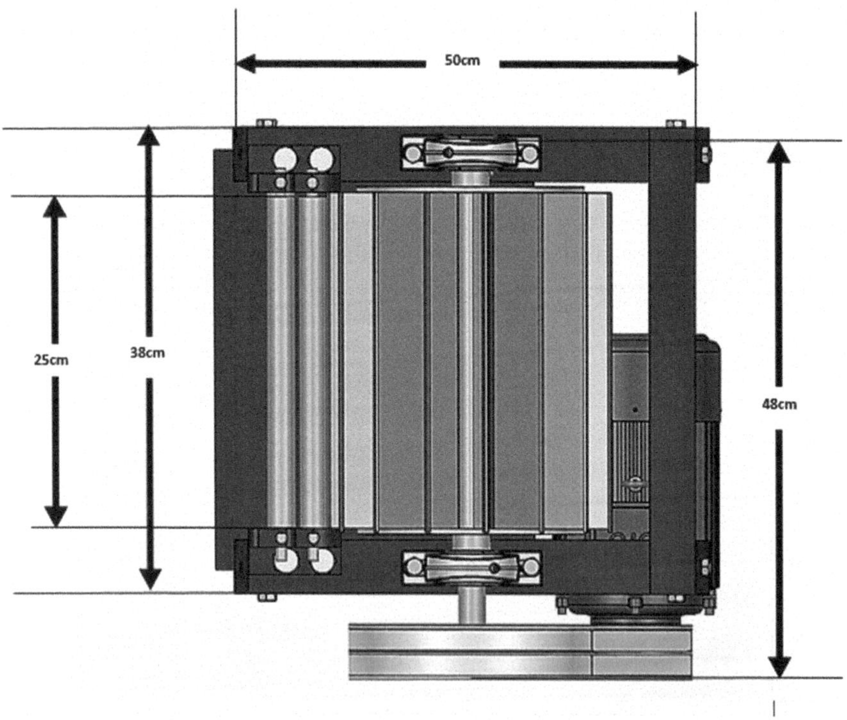

Fig. 6 Top view

Fig. 7 Front view

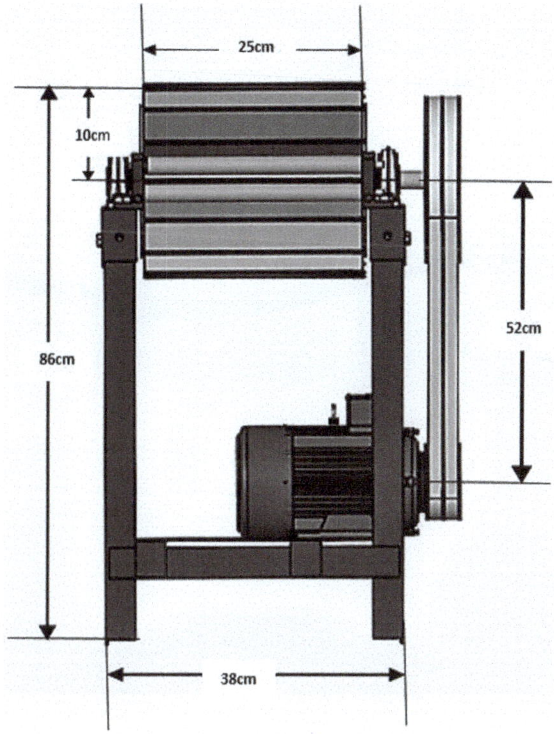

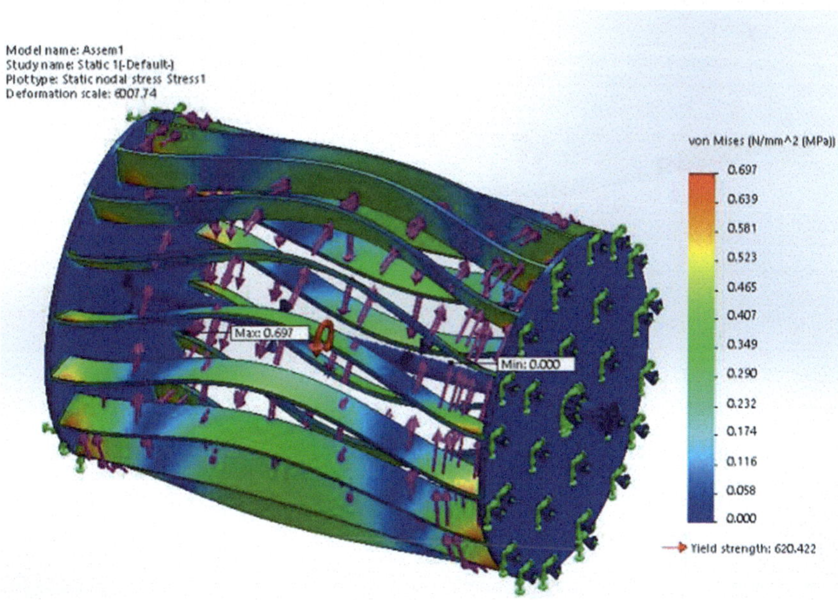

Fig. 8 Von Mises stress analysis of rolling drum under stress–strain conditions

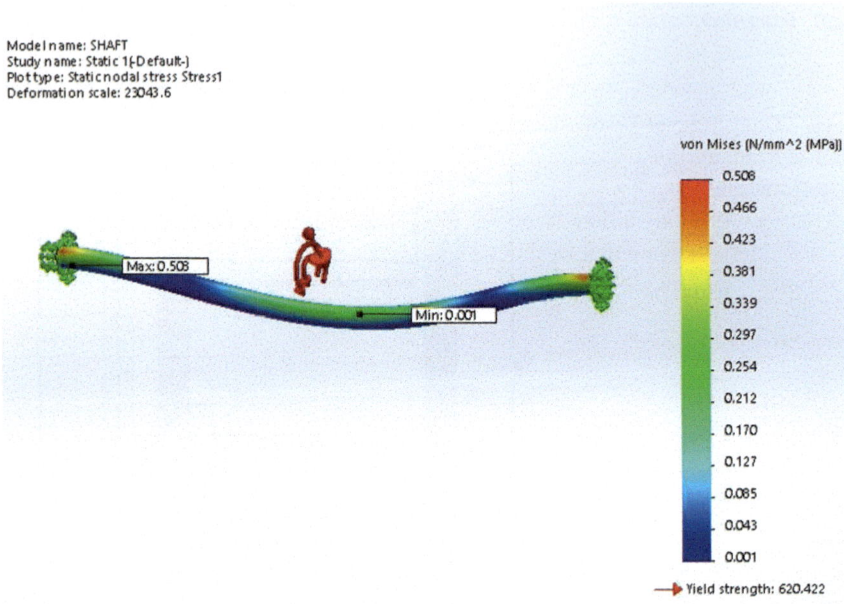

Fig. 9 Von Mises stress–strain analysis of shaft

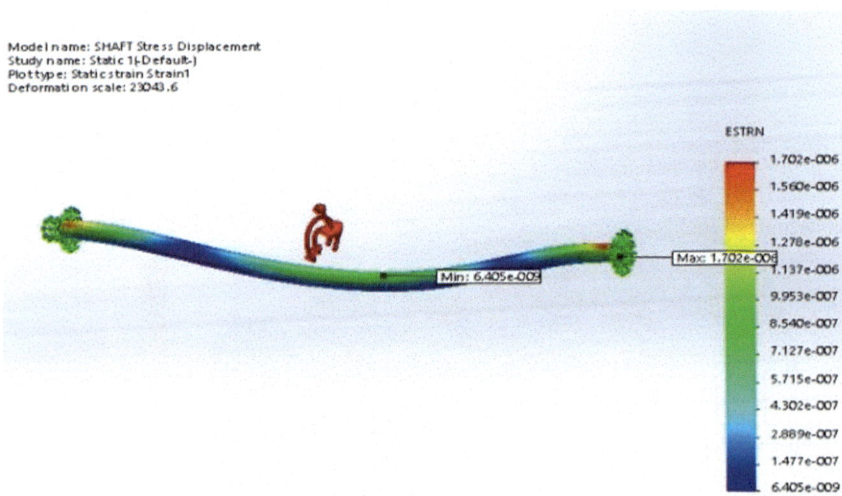

Fig. 10 Von Mises stress–strain analysis of shaft

3.4 *Rope Spinning Machine*

The properties of the processed banana fibres are similar to the properties of the coir fibres. This similarity was used for the rope spinning.

The plantain fibre was dried under normal atmospheric conditions until visible water content was absent. It is then fed into the spinning machine. The machine operates by spinning a thin nylon thread and fibres together to form ropes of small cross section; they are then combined together in the same machine to the required rope size.

4 Workflow

The plantain stem is cut into workable size, and the fibres are extracted in a raspador machine and then it is dried. The dried fibres are then processed into fine fibres. These fine fibres are cut into small size (approx. 10 cm) for smooth machine operation. Fine fibres are spin in a rope spinning machine into rope; also, suitable additives are added to the fibres for strength improvement. Tension test, strain test, moisture test are performed to study the strength and properties of the rope (Table 2).

The extracted fibre strands were dried under normal weather conditions. The fibre extracted using banana fibre extraction machine was found to have very less water content, and hence, drying was quick and easy under normal conditions.

Fibre strands of different length were combined all together in husk form for feeding into rope spinning machine. Thin nylon strands were used to guide the husk into the spinning machine.

Table 2 Fibre extraction test results

Sl. No.	Plantain sliced trunk length (mm)	Width (mm)	Thickness (mm)	Fibre extracting time (s)
1	310	40	4	12
2	480	72	5	18
3	680	130	6	22
4	860	70	7	23
5	590	140	8	24
6	730	145	9	26
7	1240	150	10	31

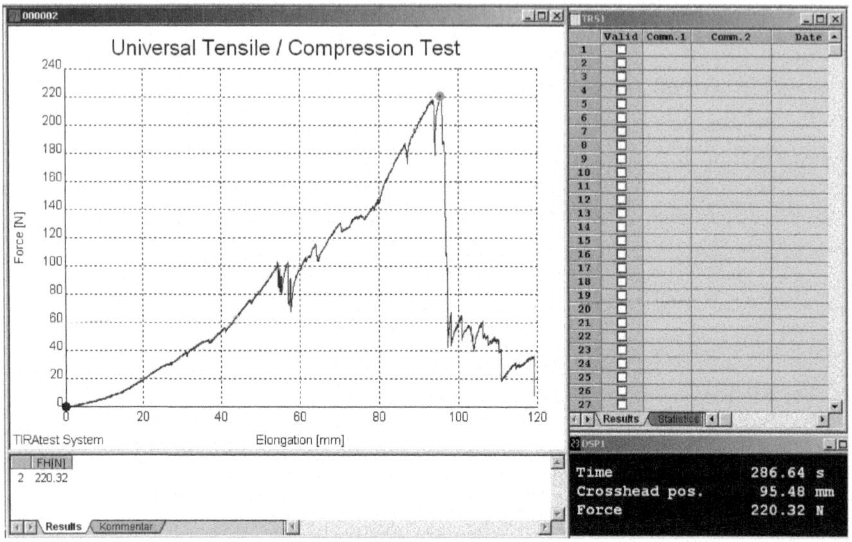

Fig. 11 Tension test result on a 400 mm sample

5 Tension Test

Universal tension test was done for 10 samples with different cross-section area. The graph shows the tension test on a 400 mm length sample. The ASTM standard used is ASTM D3822 which covers the measurement of tensile properties of both natural and man-made single textiles. The ultimate tensile force of the rope was found as 220.32N. Figure 11 shows the plot between force and elongation of the rope, obtained from the tension test.

6 Conclusion

The consumption of natural resources is increasing drastically. Sustainability and sustainable development focus on balancing that fine line between needs which includes our urge to move forward technologically and economically and to protect the environment in which we and our future generations thrive. Hence, it is very important to consider the future, in making our decisions about the present.

In this paper, sustainable banana plantain fibre rope addresses the pillars of sustainability. The plantain farmers can sell their waste as a raw material to the plantain fibre rope making industry; thus, it adds to an additional source of income. The rope is eco-friendly and degradable; thus, it does not affect our environment. Zero wastage, easy and maximum availability of raw material are the goal of any industry [1]. These factors make the rope socially acceptable.

Even though coir rope is sustainable, the higher processing time and the scarcity in availability along with the environment impacts during the processing make banana plantain fibre rope more attractive.

Considering the sustainability aspects and local availability of raw material, we designed a mechanism to convert banana fibre into sustainable rope and further improved its strength by adding composite nylon fibre. Hence, the proposed mechanism to efficiently extract the sustainable plantain fibre rope from the sheath stem is an innovative idea which can develop a competitive advantage in future.

References

1. Deepak EB, Arshinder K (2014) Practices for a sustainable supply chains: a tourism perspective. J Tourism Culture Territorial Develop 3(1):60–61
2. Mishra S (2010) Banana and its by-product utilisation: an overview. J Sci Ind Res 69(5):323–329
3. Nirmala W, Remani KN, Nair SR (1989) Pollution due to coir retting and its effect on estuarine flora and fauna. Int J Environ Stud 32(4):285–295
4. Uragoda CG (1981) Long term exposure to sulphur dioxide during bleaching of coir. Occup Med 31(2):76–78
5. Lithner D, ,Larsson Å, Dave G (2011) Environmental and health hazard ranking and assessment of plastic polymers based on chemical composition. Sci Total Environ 409(18):3309–3324
6. Ramesha M, Palanikumar K, Hemachandra Reddy K (2017) Plant fibre based bio- composites: sustainable and renewable green materials. Asian J Plant 79(1):558–584
7. Kulkarni AG, Satyanarayana KG, Rohatgi P, Vijayan K (1983) Mechanical properties of banana fibres (*Musa sepientum*). J Mater Sci 18(8):2290–2296
8. Shivashankar S (2006) Composition and properties of fibre extracted from pseudostem of banana. J Hortic1(2):95–98
9. Ageron B, Gunasekaran A, Spalanzani A (2012) Sustainable supply management: an empirical study. Int J Prod Econ 140(1):168–182
10. Zaman AU (2015) A comprehensive review of the development of zero waste management: lessons learned and guidelines. City Culture Soc 91(4):12–25
11. Padam BS, Tin HS, Chye FE, Abdullah MI (2014) Banana by-products: an under-utilized renewable food biomass with great potential. J Food Sci Technol 51(2):3527–3545
12. Alvarez-López (2015) Development of self-bonded fibres from plantain leaf: effect of water and organic extractives. Bio Resour 1(11):35–43
13. Rodríguez LJ, Orrego CE, Ribeiro I, Peças P (2018) Life-cycle assessment and life-cycle cost study of banana (Musa sapientum) fiber Biocomposite materials 69:585–590
14. Firbank TC (1970) Mechanics of the belt drive. Int J Mech Sci 12(12):1053–1063
15. Kaebernick H, Kara S, Sun M (2003) Sustainable product development and manufacturing by considering environmental requirements. Robot Comput Integr Manuf 19(6):461–468
16. Dungani R, Karina M, Subyakto, Sulaeman A, Hermawan D, Hadiyane A (2016),' Agricultural waste fibres towards sustainability and advanced utilization: a review. Asian J Plant 15(1):42–55
17. Deepak EB, Arshinder K, Rajendran C (2018) Sustainability practices in tourism supply chain: importance performance analysis. Benchmarking Int J 25(4):1148–1170

Improved Mechanical Properties and Use of Rice Husk-Reinforced Recycled Thermoplastic Composite in Safety Helmets

Ammar A. Al-Talib, Ruey Shan Chen, Elango Natarajan, and Ashley Denise Chai

Abstract Recycled thermoplastics have inferior mechanical properties as compared to virgin thermoplastics. It limits application of recycled thermoplastics in industrial and domestic parts. In this research, abundance of rice husk fibre in the form of agricultural waste was utilised to reinforce and increase mechanical properties of recycled high-density polyethylene and polyethylene terephthalate composite (rHDPE/rPET). Rice husk (RH) fibres (70 wt%)-reinforced recycled thermoplastic composite was fabricated through a two-step extrusion process and hot press moulding technique. The tensile strength and elastic modulus were increased by 4.95% and 162.65% as compared with pure recycled thermoplastic blend. Water absorption and swelling of the composite was analysed by immersing it in water for 4 weeks of time. Furthermore, the applicability of rHDPE/rPET/RH composite in safety helmet was analysed using finite element analysis. The analysis according to EN397 standard reports that RH-reinforced recycled thermoplastic is safer than virgin recycled plastic.

Keywords Recycled plastic · Rice husk · Composite material · Safety helmet · Finite element analysis

1 Introduction

Recycled plastics have weaker physical properties than products made from virgin plastics. This is due to the chemical degradation resulting from heat exposure, ultraviolet light, or oxidation in the recycling process which causes deformation and weakening of the crystal structures, as well as internal polymer structures [1]. This

---

A. A. Al-Talib (✉) · E. Natarajan · A. D. Chai
Faculty of Engineering, Technology and Built Environment, UCSI University, 56000 Kuala Lumpur, Malaysia
e-mail: ammart@ucsiuniversity.edu.my

R. S. Chen
Department of Applied Physics, Faculty of Science and Technology, The National University of Malaysia, 43600 Bangi, Selangor, Malaysia

E. Natarajan et al. (eds.), *Materials, Design and Manufacturing for Sustainable Environment*, Lecture Notes in Mechanical Engineering,
https://doi.org/10.1007/978-981-19-3053-9_2

has resulted with recycled plastics being used as a material only for the fabrication of low-value-added products with very limited applications. With an intention to improve the physical properties of recycled plastics to make it compatible as a permanent replacement of virgin plastics, an eco-friendly solution would be to incorporate a natural fibre into the recycled polymer composition. Natural fibres in simple terms are readily existing fibres in nature in the form of plants or animals and are not manmade or synthetic. Therefore, natural fibres as compared to synthetic fibres have plenty attractive features such as sustainability, biodegradability, renewability, and lightweight, and with it being available in abundance in nature, natural fibres are extremely low cost [2]. For this, the composites shall be custom made by varying reinforcements and matrix phases, to suite the required physical and mechanical properties, in order for the product to fulfil customer's satisfaction, which will then lead to a rise in demand in various industrial sectors.

In regard to enhancing the properties of recycled thermoplastics, focus has been given into natural fibre as they greatly contribute to improving sustainability and environmental impacts. More effort must be placed in trying to replace existing products of pure plastics into natural fibre-reinforced recycled thermoplastics. Reinforced recycled thermoplastics with the use of natural fibres can have superior mechanical properties [3, 4] that are equal or even better than pure virgin plastics. Specifically from agricultural waste such as rice husk (RH) fibre can help to accelerate the biodegradability of polymeric composites [5]. Raw RH consists of 25–35% cellulose, 18–21% hemi cellulose, 26–31% lignin, 15–17% silica, 2–5% soluble, and 7.5% moisture content. These natural polymers offer many advantages for RH fibres; the ability of not releasing a single toxic by-product when it is burnt, its strength and ability to be incorporated with other materials, completely biodegradable, form of renewable source and low cost. Therefore, this has initiated a great interest worldwide in the development of composite materials with the use of agricultural waste like RH. However, the development of such composite material has experienced challenges of poor compatibility between the fibre and base matrix as fibre has hydrophilic properties, and the polymer matrix has hydrophobic properties. The result of the combination between these creates clustering of fibres in the composite or weak distribution of fibres within the composite. A material with weak distribution or cluster of fibres will have reduced and weakened thermal and mechanical properties. As a result, there is a need for the use of coupling agents in the making of natural fibre composite materials to help properly bind the fibre and polymer matrices together. For this research, with reference from the previous studies done on natural fibre polymer composites, maleic anhydride polyethylene (MAPE) was used as the coupling agent [6].

The length of the fibre also plays a significant role in the mechanical properties of NFPCs. As load is transferred to the matrix through fibres, the length of the fibre should be greater than the critical length. Critical length is the length of fibre which bears the maximum tensile load just before fracture [7]. Apart from that the consideration of interfacial bonding between fibres and matrix, fibre dispersion and orientation play a huge role in determining the outcome of mechanical properties of the material. As stress is transmitted to the fibres from the matrix, it is of utmost importance that optimum interfacial bonding is achieved for efficient reinforcement of the composite [8]. Polymer matrices which are hydrophobic in nature create weak bonding with natural fibres that possess cellulose which are the cause of hydrophilic properties in natural fibres. Polymers are known to be water resistant, whilst natural fibres, due to its composition of cellulose, which possess free hydroxyl (OH) molecules that will readily attach with water molecules through hydrogen bonding [9]. Therefore, the main disadvantages of NFPCs are that if interfacial strengths are not chemically enhanced, this leads to undesired risks such as reduced strength and load/stress tolerability, increased moisture and water absorption by fibres which further weakens the bonds, and reduced overall mechanical properties. Thus, optimisation of natural fibres is required by performing chemical treatments on the natural fibres to alter and reduce the hydrophilic chemical properties of the natural fibre, improve wettability and roughness of the fibres, thus resulting with optimum interfacial bonding [10]. Material processing method such as internal mixer, compression or injection moulding, and even counter or co-rotating twin-screw extruders [11] is also important for enhancing the properties.

In regard to fabricating industrial safety helmet with new composite material, the European standard for code of practice may be followed (BS EN 397 standard). This standard covers the specifications on which standardised testing of safety helmets and parameters for the quality of the safety helmet are described [12]. According to EN397 standard, the protective outer layer of the helmet (hard shell) must certify testing requirements in terms of impact/shock absorption, penetration resistance, flame resistance chin strap attachment, etc. [13, 14]. Having understood these, recycled high-density polyethylene (rHDPE) and recycled polyethylene terephthalate (rPET) were used to prepare rHDPE/rPET/RH thermoplastic composite, and the numerical simulation was conducted to ensure the safety of the helmet.

2 Material and Methods

2.1 *Materials and Preparation of Samples*

Recycled high-density polyethylene (rHDPE) with a density of 923 kg/m^3 and a melt flow index of 0.72 g/10 min (190 °C, 2.16 kg) and recycled polyethylene terephthalate (rPET) with an inherent viscosity of 0.68 dL/g were supplied from BioComposites Extrusion Sdn. Bhd., Malaysia. As a suitable agent for the compatibilization of rHDPE/rPET blend, an ethylene-glycidyl-methacrylate (EGMA, which is the trading name of Lotader AX8840) with a melting index of 5 g/10 min (190 °C, 2.16 kg) and 8% glycidyl methacrylate were used. The predetermined quantity of agricultural waste, rice husk (RH) with a particle size of 100 mesh was supplied from BioComposites Extrusion Sdn. Bhd., Malaysia and dried in an oven at 100 °C for 24 h before compounding. As for the coupling agent for the rice husk fibre with the polymer blend matrix, maleic polyethylene anhydride (MAPE) with a melting index

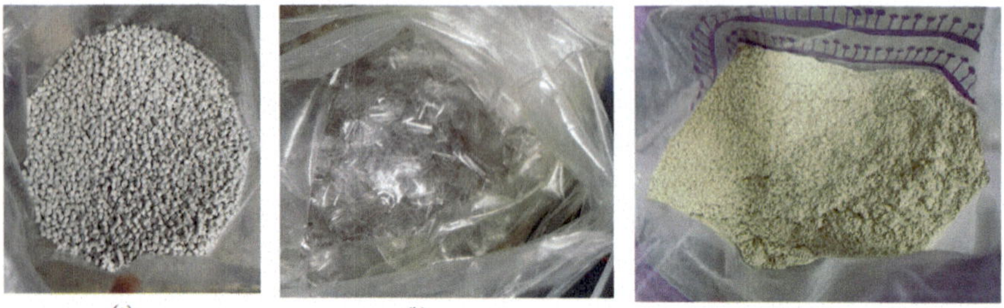

(a) (b) (c)

Fig. 1 **a** Recycled HDPE in the form of pellets, **b** recycled PET in the form of thin layered sheets/pieces, **c** RH fibres in the form of powder

of 5 g/10 min (190 °C, 2.16 kg) and maleic anhydride content of 1 wt% was used (Fig. 1).

A co-rotating twin-screw extruder (Thermo Prism TSE 16 PC, Thermo Electron Company, Germany) with a diameter of 16 mm and a length/diameter ratio of 25 was used to melt the RH-filled recycled thermoplastic mixer. At a screw rotating speed of 30 rpm, a recycled thermoplastic mix of rHDPE and rPET with a weight ratio of 75/25(wt%) and 5% of EGMA (according to the total weight of the rHDPE/rPET mix) was extruded in the first extrusion. The temperature profiles for the barrels were set at 250, 270, 240, and 190 °C, individually starting from feeding to the die zones. The extruded polymer blend composite was then left to cool at room temperature and later granulated with a crusher into pellets. Secondly, dried RH filler powders were melt-compounded thoroughly together with the pre-extruded recycled thermoplastic blend granulates. In order to have enhanced fusing between the fibre and polymer blend matrix, the addition of MAPE was added together into the twin-screw extruder in the second extrusion. During the second extrusion process, the same screw spinning speed of 30 rpm was used but with temperature profiles of 195, 210, 215, and 170 °C, individually starting from feeding to the die zones. The 70 wt% of RH filler was then added into extruder to prepare rHDPE/rPET/RH mix. It was left to cool at room temperature before being pelletized. Finally, rHDPE/rPET/RH thermoplastic composite was compression moulded at 200 °C with a pressure of 6.9 MPa. The timing for preheating, cooling, full pressing, and cold pressing was 3, 2, 5, and 5 min, respectively. After the moulded composite was ready, it was left to cool for 40 h under atmospheric condition, prior to performing any mechanical testing.

2.2 *Tensile Testing*

Samples of rHDPE/rPET/RH composite were prepared according to the ASTM D638-03. They were tested in Testometric M350-10CT tensile testing machine at a crosshead speed of 5 mm/min, at room temperature. The tensile strength and elastic modulus of the eight samples were measured, and mean of the values was considered. Meanwhile, neat, recycled thermoplastic, and rHDPE/rPET samples were also prepared according to ASTM standard and tested in the same condition as discussed above.

2.3 Water Absorption and Swelling Analysis

Test for water absorption was conducted in conjunction with ASTM D570-98 standard. Immersion of rHDPE/rPET/RH composite was done in 200 ml room temperature water. Prior to immersing the samples into the solution, the samples were first oven dried at 105 °C for 24 h to remove any moisture entrapped within the sample. Then, the samples were measured to obtain its oven-dried weight with a high-sensitivity weighing scale and the sample's thickness with vernier calliper. After all measurements were taken, the samples were then immersed into the prepared solution. The weight and thickness of the specimens were measured again after immersion for 2, 24, 48, 72, 96, 120, 168, 336, 504, and 672 h, after being removed from water and wiped dry. The percentages of water absorbed (WA) and thickness swollen (TS) were determined using the following equations:

$$\text{WaterAbsorptionRate} = \frac{(w_t - w_0)}{w_0} \times 100\% \quad (1)$$

where w_0 is the oven-dried weight, w_t is the weight after immersion time (t)

$$\text{Swelling Percentage} = \frac{(T_t - T_0)}{T_0} \times 100\% \quad (2)$$

where T_0 is thickness after over-drying and T_t is thickness after immersion time (t).

2.4 Modelling of Safety Helmets and Impact Analysis

Numerical simulations were conducted on safety helmets made of rHDPE/rPET/RH composite in order to investigate the structural integrity of the developed composite for safety helmet application. With the use of finite element analysis software known as ANSYS, the structural behaviour of safety helmet of rHDPE/rPET/RH composite was analysed.

The testing parameters considered in these simulations are based on the standards that are entitled by the European Safety Helmet EN397 standard:

- Impact/shock absorption test of which the simulation comprises the falling of a 5 kg spherical steel ball onto the top of the helmet, from a height of 1 m. The results of this analysis are to investigate deformation on helmet and stress transformation between helmet and human being.
- Penetration test of which the simulation embodies the falling of a 3 kg structural steel pointed cone onto the top of the helmet, from a height of 1 m. The result of this simulation is to investigate whether pointed impactor can penetrate into the safety helmet.

Firstly, safety helmet model with standard dimension was done using solid works tool and exported to ANSYS in neutral file format (IGES). A spherical impactor was then designed, by simply sketching a 2D semicircle with a radius of 0.04826 m, and then revolving the semicircle by 360° degrees and placing it on the centred position on top of the safety helmet. The assembly of the spherical impactor and the safety helmet is shown in Fig. 2. The respective material properties of both helmet and spherical impactor were input into the model. The spherical impactor was assigned with material properties of steel, whilst helmet was assigned with material properties of rHDPE/rPET/RH composite. From the volume and surface area of impactor, the thickness of line contact between the helmet and impact was computed and input as 0.02176 m.

Meshing was then done with element size of 5.48×10^{-3} m for safety helmet and 8×10^{-3} m for steel impactor. Using ANSYS explicit dynamics module, a drop height of 1 m was assigned to the steel ball, to simulate its fall from that height to impact the safety helmet. With this impact velocity of 4.4287 m/s, testing standard of EN397 for safety helmets is complied. In this way, the deformation and stresses on safety helmet during dynamic impact analysis were observed.

For the penetration test, a pointed cone impactor was designed, by simply sketching a 2D standing rectangle with height of 0.075 m and a width of 0.0382 m. Then, to sketch the pointed end, a line of length 0.025 m was drawn downwards at the bottom left edge of the cylinder. From the same point, a line of length 0.000254 m

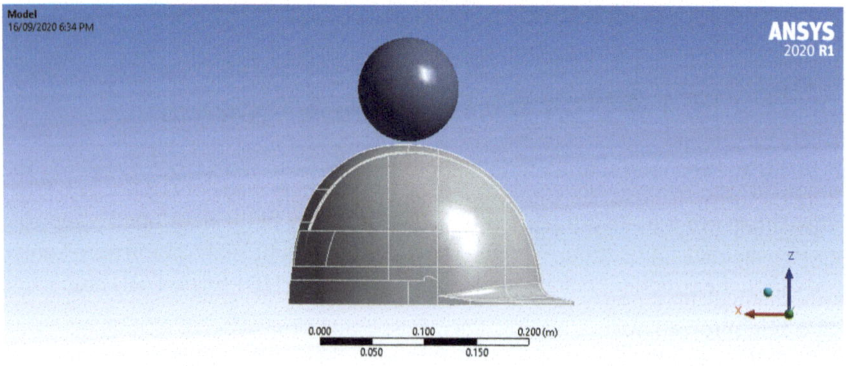

Fig. 2 FE model used in impact absorption test

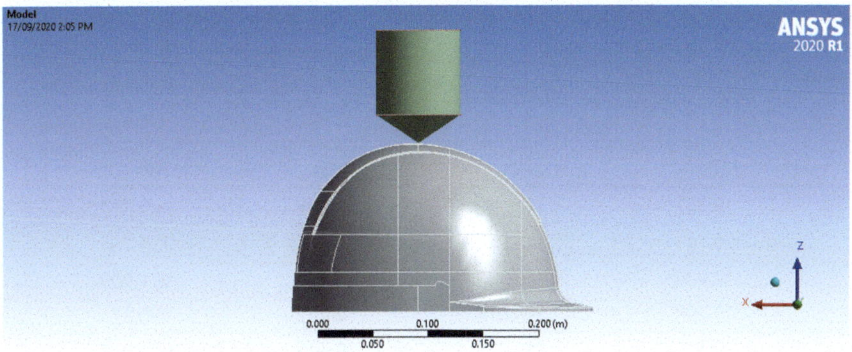

Fig. 3 FE model used in penetration test

was drawn to the right of the point. This final point was then connected to the bottom right edge of the cylinder. The entire drawing was then revolved 360° degrees and was then placed on the centred position on top of the safety helmet. The assembly of the penetrator and the safety helmet is shown in Fig. 3.

The material properties of rHDPE/rPET/RH composite were used for safety helmet, and material properties of steel were used for penetrator. Meshing was then done on both safety helmet and penetrator with element size of 5.48×10^{-3} m and 8×10^{-3}, respectively. The penetrator was dropped at the velocity of 4.4287 m/s from 1 m high from centre of the safety helmet, which complies with testing standard of EN397 for safety helmets. In this way, the deformation and stress on safety helmet during penetration were observed.

3 Results and Discussion

3.1 Tensile Properties of rHDPE/rPET/RH Composite

It is observed that the tensile strength and elastic modulus of recycled thermoplastic rHDPE/rPET blend are 19.667 MPa and 343.438 MPa, whilst, the mean value of tensile strength and elastic modulus of rHDPE/rPET/RH composite is 20.641 MPa and 902.034 MPa, respectively. rHDPE/rPET/RH composite has resulted 4.95% and 162.65% of increase in tensile strength and elastic modulus as compared to rHDPE/rPET blend. As reported in literature, the recycling of plastic waste causes chemical degradation within the polymer's internal and crystal structure and weakens the mechanical properties of recycled thermoplastic. When the recycled thermoplastic is reinforced with RH fibres, its overall tensile strength and stiffness are substantially enhanced because the fibre possesses natural polymers in the composition of 25–35% cellulose, 18–21% hemicellulose, and 26–31% lignin. When these natural polymers are embedded into the recycled polymer matrix, it can improve the

Table 1 The mechanical properties of neat recycled thermoplastic blend and RH-reinforced recycled thermoplastic composite

Sample	Neat recycled thermoplastic, rHDPE/rPET		RH-reinforced recycled thermoplastic, rHDPE/rPET/RH	
	Tensile strength (MPa)	Elastic modulus (MPa)	Tensile strength (MPa)	Elastic modulus (MPa)
1	19.362	331.982	21.801	989.319
2	20.143	386.048	21.362	715.705
3	19.924	346.342	20.422	971.009
4	19.298	328.014	18.991	990.875
5	19.172	350.928	20.406	970.948
6	19.352	326.565	21.628	757.731
7	20.036	357.481	20.182	732.851
8	20.053	320.146	20.335	1087.832
Total	157.34	2747.506	165.127	7216.27
Min	19.172	326.565	18.991	715.705
Mean	19.667	343.438	20.641	902.034
Max	19.924	386.048		

recycled polymer's internal structure by bonding its natural fibre polymers with the polymer chains of the recycled polymer matrix. With great adhesion between the fibre and polymer matrix, the internal structure of the composite will be greatly enhanced. Consequently, when the composite experiences any load or stress, the polymer matrix will be able to absorb the tensile load, and then efficiently disperse and distribute the load throughout the natural fibres length through shear [7]. Therefore, this clearly justifies as to why rHDPE/rPET/RH composite has displayed much superior mechanical strength than rHDPE/rPET blend. Another factor that may contribute to the improved mechanical properties and interfacial bonding within rHDPE/rPET/RH composite is the addition of coupling agent, maleic anhydride polyethylene (MAPE) into the composite during sample preparation. With the presence of a coupling agent, the surface roughness of the RH fibres is modified in order to improve adhesion of the fibre to the polymer matrix. This assists in reducing the existence of voids within the composite and reduces its susceptibility to water absorption [15]. Table 1 shows mechanical properties of the rHDPE/rPET blend and rHDPE/rPET/RH composite samples.

3.2 Absorption Rate and Swelling Test Results

Figure 4 shows the water absorption rate and swelling rate of rHDPE/rPET/RH composite. The results obtained from this immersion comply with Fickian diffusion behaviour theory, of which states that water intake by the sample rapidly increases in the beginning stages and finally over time slowly reaches a state of equilibrium [16]. It is clearly shown that the samples have experienced proportional growth in absorption rate, as the immersion time is increased, and the absorption rate of the samples has reached equilibrium once saturation point is reached. The reason for this is that water absorption in natural fibres-reinforced polymer composites has three methods of which water intake can occur. They are diffusion, capillary, and transport of water molecules. In terms of diffusion, water intake can occur in the micro-cracks that exist in the polymer chains. Meanwhile, water intake by capillary action takes place within the cracks that exist between the assembly of fibre and polymer matrix. Besides that, transport of water molecules into the composite can occur when the composite experiences swelling that creates more micro-cracks within the matrix, thus causing water molecules to transport into these cracks [9]. Therefore, these three mentioned mechanisms justify as to why the samples experience increase in water intake as the immersion time increases. The cause of increase in water intake in samples with higher percentages of RH fibres is due to the availability of free hydroxyl (OH) groups within the natural fibre, causing it to have high-hydrophilic properties. The availability of free hydroxyl molecules within the fibre are due to the fibres having saturated composition of cellulose. The hydrophilic properties of the fibre cause it to readily attach with the water molecules through hydrogen bonding [17]. Therefore, with a larger fibre weight percentage in the composite, there are more available hydroxyl molecules that will bond with water molecules. Thus, explains the increase in water intake by the samples. Though water absorption is higher at inclusion of 70 wt% of RH into recycled thermoplastic, the greater mechanical properties are observed.

It clearly observed that there is a proportional increase in swelling rate in the samples as the immersion time increases. The reason for this is mainly due to the fibre orientation within the matrix. In compression moulding, most of the RH fibres within

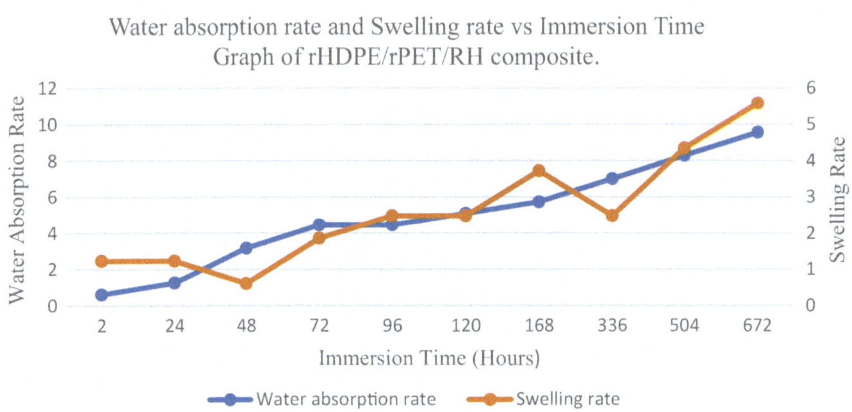

Fig. 4 Water absorption rate and swelling rate versus immersion time graph of rHDPE/rPET/RH composite

the polymer composite were aligned in longitudinal direction of the samples. As swelling of the fibres takes place along its diameter direction, which is perpendicular to the fibres length, this results with the obvious change in thickness of the sample as compared with the width and length [11]. The composite has no resistance protection against water absorption and swelling, mainly due to the hydrophilic properties of the RH fibre within the composite and the concentration of fibre weight percentage within the composite. The mechanical strength of the composite plays a bigger role when considering its suitability for the making of a safety helmet, as its top priority is to protect the user's head from any impact blows.

3.3 Explicit Dynamics Simulation Results

3.3.1 Impact Absorption Results

Figures 5 and 6 show the deformation with respect to time and maximum deformation experienced by the safety helmet of rHDPE/rPET/RH composite material. The maximum deformation is 0.031 m, which is less than 0.05 m that is recommended by EN397 standard for safety helmet. The deformation greater than 0.05 m causes the head injury to the user. The current material is able to absorb the impact energy caused by falling impactor (steel material of 5 kg) without tending to plastic fracture. It is observed that the time taken to absorb and rebound the impactor is much lesser. The reaction force on safety helmet upon falling impactor is 3.088 kN, which lesser than the allowable force (5 kN) mentioned in the specifications of EN397. Figure 7 shows the reaction force on the safety helmet caused by falling weight.

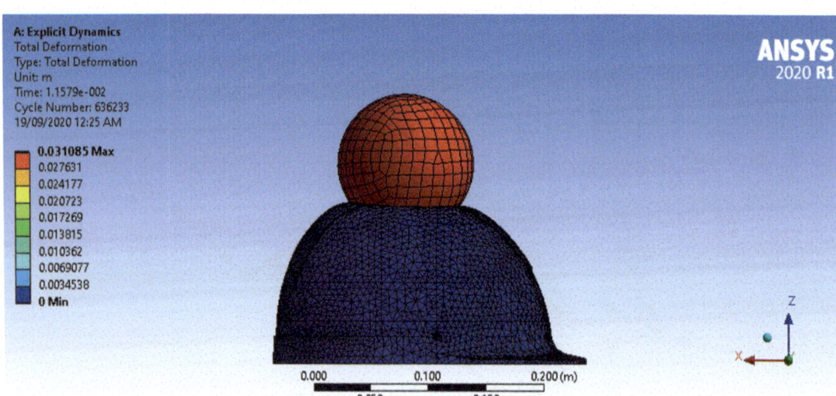

Fig. 5 Deformation caused by falling weight of striker on safety helmet with respect to time

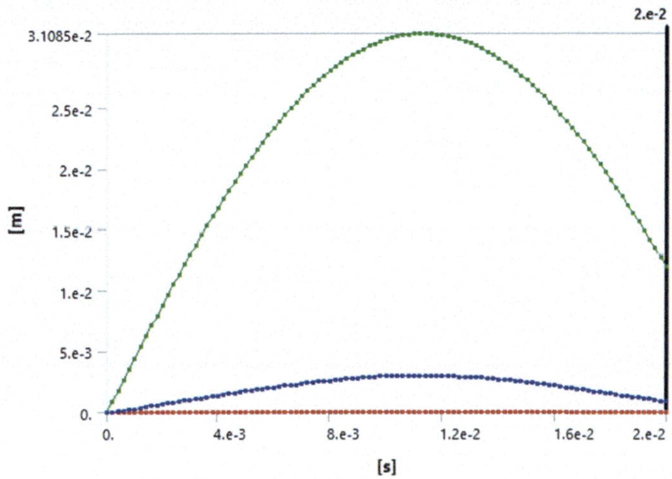

Fig. 6 Maximum deformation on safety helmet of rHDPE/rPET/RH composite upon falling weight impactor

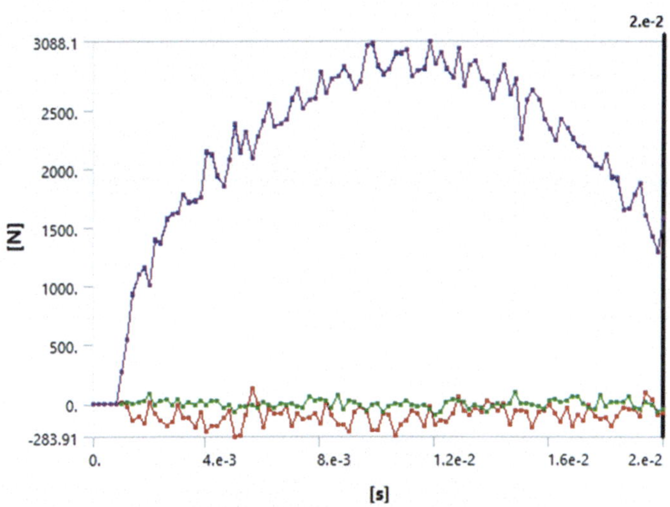

Fig. 7 Reaction force on safety helmet caused by falling weight

3.3.2 Penetration Striker Test Results

The maximum deformation in safety helmet from impact by the striker is 0.025 m. It is less than the allowable deformation (0.05 m) suggested by EN397 standard for safety helmet. Similar to the results in impact absorption test, the time taken to absorb and rebound the impact of the striker is lesser (9.47×10^{-3} s). According to EN397 specification on the requirements of safety helmet, it is necessary that during impact from the pointed striker, the helmet shall not experience any penetration from the striker as this will harm the safety of the user's head. The results from penetration tests are shown in Figs. 8, 9, and 10.

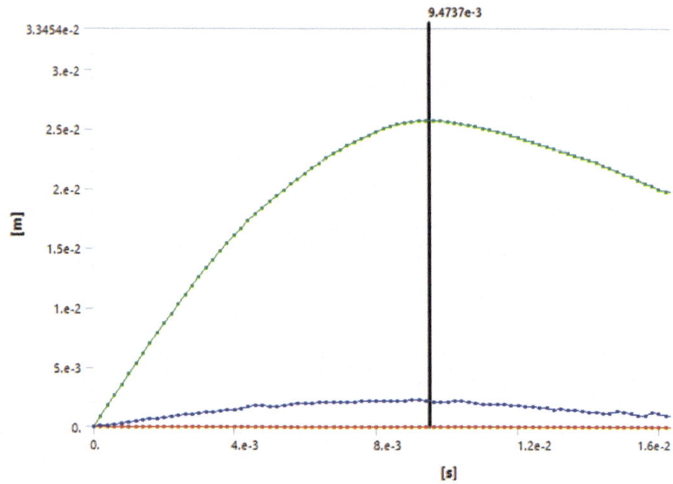

Fig. 8 Deformation on safety helmet with respect to time upon penetration striker

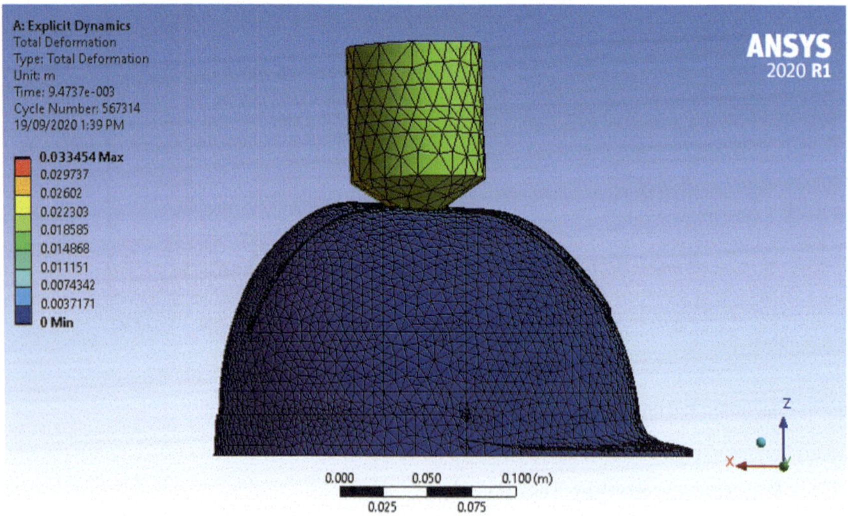

Fig. 9 Maximum deformation on safety helmet of rHDPE/rPET/RH composite upon striking penetration

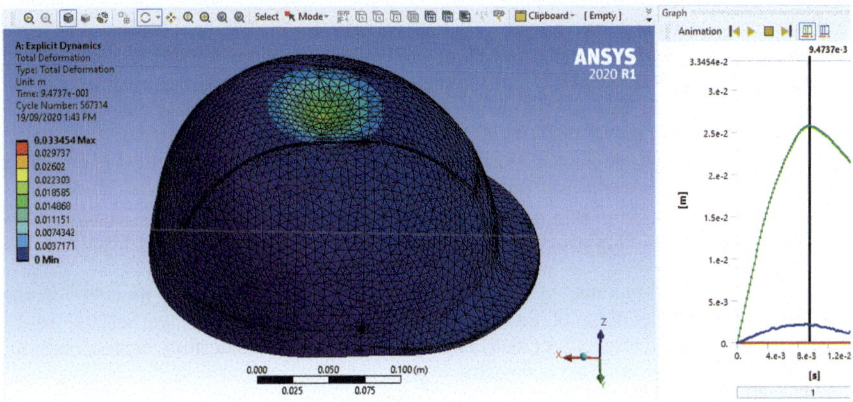

Fig. 10 Depth of deformation on safety helmet

4 Conclusions

Rice husk fibres of 70 wt% were used to reinforce recycled HDPE and recycled PET blend. The tensile test on ASTM D638-03 (type 1) samples has reported that tensile strength and elastic modulus of rHDPE/rPET/RH composite are increased by 4.95% and 162.65% as compared to pure recycled thermoplastic blend. The effective change is caused by lignocellulosic properties of RH fibre and great interfacial bonding produced by MAPE coupling agent. Furthermore, the structural feasibility of rHDPE/rPET/RH composite for safety helmet was investigated by impact analysis. The finite element analysis complying to EN397 safety helmet standard was conducted with falling weight and penetrating striker. The results obtained from these simulations conclude that rHDPE/rPET/RH composite is good to be used in safety helmet. Meanwhile the use of RH-reinforced recycled thermoplastic in safety helmet protects the user from head injury and assists to achieve the sustainability as well.

References

1. Tominaga A, Sekiguchi H, Nakano R, Yao S, Takatori E (2019) Advanced recycling process for waste plastics based on physical degradation theory and its stability. J Mater Cycles Waste Manage 21:116–124
2. Thyavihalli Girijappa YG, Mavinkere Rangappa S, Parameswaranpillai J, Siengchin S Natural Fibers as sustainable and renewable resource for development of eco-friendly composites: a comprehensive review. Front Mater 6(2019):1–14
3. Chen RS, Ahmad S (2016) Characterization of rice husk biofibre-reinforced recycled thermoplastic blend biocomposite. Compos Renew Sustain Mater
4. Palani Kumar K, Keshavan D, Natarajan E, Narayan A, Ashok Kumar K, Deepak M, Freitas LI (2021) Evaluation of mechanical properties of coconut flower cover fibre-reinforced polymer composites for industrial applications. Prog Rubb Plast Recycl Technol 37(1):3–18. https://doi.org/10.1177/1477760619895011

5. Prabhudass JM, Palanikumar K, Natarajan E, Markandan K (2022) Enhanced thermal stability, mechanical properties and structural integrity of MWCNT filled Bamboo/Kenaf hybrid polymer nanocomposites. Materials 15(2). https://doi.org/10.3390/ma15020506
6. Chen RS, Ahmad S, Ghani MHA, Salleh MN (2014) Optimization of high filler loading on tensile properties of recycled HDPE/PET blends filled with rice husk, pp 46–51
7. Pickering KL, Efendy MGA, Le TM (2016) A review of recent developments in natural fibre composites and their mechanical performance. Compos Part A Appl Sci Manuf 83:98–112
8. Kumar R, Ul Haq MI, Raina A, Anand A (2019) Industrial applications of natural fibre-reinforced polymer composites–challenges and opportunities. Int J Sustain Eng 12(2019):212–220
9. Muñoz E, García-Manrique JA (2015) Water absorption behaviour and its effect on the mechanical properties of flax fibre reinforced bioepoxy composites. Int J Polym Sci 2015:16–18
10. Ahmad R, Hamid R, Osman SA (2019) Physical and chemical modifications of plant fibres for reinforcement in cementitious composites
11. Chen RS, Ahmad S, Gan S (2016) Characterization of rice husk-incorporated recycled thermoplastic blend composites, BioResources 11
12. Baszczyński K (2018) Effects of falling weight impact on industrial safety helmets used in conjunction with eye and face protection devices. Int J Occup Saf Ergon 24:171–180
13. Musa A (2018) Dosh-SIRIM ppe approval certification & testing, pp 1–38
14. Sathish Gandhi VC, Kumaravelan R, Ramesh S, Venkatesan M, Ponraj MR (2014) Performance analysis of motor cycle helmetunder static and dynamic loading. Mech Mech Eng 18(2):85–96
15. Chauhan V, Kärki T, Varis J (2019) Review of natural fiber-reinforced engineering plastic composites, their applications in the transportation sector and processing techniques. J Thermoplast Compos Mater
16. Arjmandi R, Hassan A, Majeed K, Zakaria Z (2015) Rice husk filled polymer composites. Int J Polym Sci. https://doi.org/10.1155/2015/501471
17. Shan Chen R, Ahmad S (2019) Compatibilized recycled polymers/clay as a matrix for rice husk composites: Mechanical and physical properties. IOP Conf Ser Earth Environ Sci 268

Thermal, Microstructure, and Hardness Properties of Molybdenum Nanoparticles Added Tin -Bismuth Solder Alloy for Low-Temperature Soldering Application

Amares Singh, Rajkumar Durairaj, Elango Natarajan, Wei-Hong Tan, and Shamini Janasekaran

Abstract The melting, microstructure, and hardness properties of the tin (Sn)-bismuth (Bi) solder alloy added with molybdenum (Mo) nanoparticles were investigated. The results showed that the Sn-Bi solder alloy with Mo nanoparticles fully melts at 148.00 °C, and the melting temperature was not drastically disordered by the Mo nanoparticles additions. The Mo has a high-melting point of 2623 °C, hence could not interfere with the solidification process of the molten Sn-Bi solder alloy. The microstructural morphology of the Sn-Bi added with Mo nanoparticles had much finer and compact lamellar structure compared to that of the unreinforced Sn-Bi solder alloy. Together with the dispersed Mo nanoparticles, there were Bi phases breaking out in the microstructure of the Mo-reinforced Sn-Bi solder alloy. Results from the hardness test showed a leap in the hardness value of 0.4 Hv, for the Mo containing Sn-Bi solder alloy. Presence of Mo nanoparticles and finer microstructure are believed to be the factors behind the increase. Reinforcing small quantity of Mo nanoparticles to the Sn-Bi solder alloy shows improvement to the microstructure and hardness as well as keeping the melting temperature below 150 °C.

A. Singh (✉) · E. Natarajan
School of Engineering, Taylor's University, No 1, Jalan Taylors, 47500 Subang Jaya, Selangor, Malaysia
e-mail: amares.singh@taylors.edu.my

R. Durairaj
Lee Kong Chian Faculty of Engineering and Science, Universiti Tunku Abdul Rahman, Jalan Sungai Long, Bandar Sungai Long, 43000 Kajang, Selangor, Malaysia
e-mail: rajkumar@utar.edu.my

W.-H. Tan
Faculty of Mechanical Engineering Technology, Universiti Malaysia Perlis (UniMAP), Main Campus Pauh Putra, 02600 Arau, Perlis, Malaysia
e-mail: whtan@unimap.edu.my

S. Janasekaran
Faculty of Engineering, Built Environment and Information Technology, SEGi University, Jalan Teknologi, Kota Damansara PJU 5, 47810 Petaling Jaya, Selangor, Malaysia
e-mail: shaminijanasekaran@segi.edu.my

E. Natarajan et al. (eds.), *Materials, Design and Manufacturing for Sustainable Environment*, Lecture Notes in Mechanical Engineering,
https://doi.org/10.1007/978-981-19-3053-9_3

Keywords Sn-Bi solder alloy · Microstructure · Hardness · Low-temperature soldering

1 Introduction

Solder alloys are the medium that provides the bonding between electronic components such as transistors and diodes and the printed circuit board (PCB) in the electronic packaging industry. The bond is achieved by the soldering process. In current soldering technology, the components are usually mounted to the PCB, either by surface mounting technology (SMT) or through-hole technology (THT) [1]. There are many soldering techniques used to mount the components to the board, yet the reflow and wave soldering are two main types of the techniques commonly applied in electronic packaging industries [2]. Reflow soldering process is normally done at reflow temperature between 230 and 250 °C [3]. Soldering process should protect from leaks, besides providing electrical [4], thermal [5], and mechanical continuity to the solder/substrate bond [6, 7].

In general, tin (Sn)–lead (Pb) solder with melting temperature of 183 °C is widely used in the electronics packaging industry [8, 9]. Concern on the usage of the Sn–Pb solder alloy arises upon declaration of lead as among the top 17 hazardous chemicals by the Environmental Protection Agency (EPA) [10]. Lately, lead-free solder alloys are being developed and studied to provide an alternative replacement to the traditional lead solder alloy due to the hazardous exposure [7]. One of the most researched lead-free solder alloys is the tin (Sn)-argentum (Ag)-copper (Cu) [11]. While equipped with competent mechanical properties, SnAgCu solder holds a high-melting temperature of 217 °C which may cause thermal damages to other electronic components such as diodes and transistors during the soldering process. This is because with higher melting temperature, the reflow soldering temperature would need to be increased too. Therefore, the research area shifts toward application of lower temperature solder alloys, among those are Sn-Bi [12], tin (Sn)-indium (In) [13], and tin (Sn)-antimony (Sb) [9]. The Sn-Bi solder alloy gains much attention due to its availability and low temperature (ranging 139–145 °C). However, studies on its mechanical aspects of hardness are scarce. There are some studies that did report the hardness of the Sn-Bi solder such as by Shen et al. [14] where the hardness was 287.3 MPa (29.3 Hv) which then increased almost 6% after adding Al_2O_3 nanoparticles. [15] added Al_2O_3 as well, but the increase in the hardness value was only 0.4 Hv, contradicting too earlier study. Similarly, a minimal increase of 0.3 Hv was reported in the study by Yang et al. [16] after adding 0.5% Al_2O_3 nanoparticles to Sn58Bi. Elsewhere, different weight percentages of graphene nanosheet (GNS) were added to the Sn20Bi solder alloy and found the hardness to drop before increasing with more content of GNS [17]. These mixed trends observed from those studies create gap that has yet to be clearly addressed.

Nanoparticle's additions showed improvement to the microhardness value by altering the microstructural morphology of the solders [18, 19]. The relationship between the microstructure and hardness is inevitable, and both properties are directly related to each other [20]. Saying that, not much of investigation on the nanoparticle additions has been addressed especially for the low-temperature Sn-Bi solder alloy. As an example, a recent study showed it only focuses on the relationship between shear strength and microstructure of the Sn-Bi + Ag and Sn-Bi + porous Cu added with but not to the hardness [21].

In this study, the Mo nanoparticles are added to produce the Sn-Bi composite solder alloy. The effect of adding Mo to the hardness property is studied, and its consequent relationship to the microstructural property is deliberated. The melting temperature of the composite solder is studied as well to provide the relevancy of having a low-temperature soldering.

2 Experimental Procedures

The Sn-Bi solder alloy was developed from melting the tin (Sn) (99.9% pure—Sigma Aldrich) and bismuth (Bi) (99.9% pure—Sigma Aldrich) in a vacuum furnace at 1000 °C. The Sn and Bi were weighed as 8.4 g and 11.6 g, from a total 20 g. This composition is chosen from the eutectic percentages from the phase diagram to obtain the lowest melting temperature at which both Sn and Bi exist as liquid. Prior to the melting, 0.4 g (2%) of Mo (99.9% pure—Sigma Aldrich) was added to the Sn and Bi mixture. After melting for an hour, the Sn-Bi added with 2% Mo nanoparticles was let to solidify in the furnace itself to avoid contamination. The solder alloy was taken out and remelted with stirring to further allow better mixture of the Mo nanoparticles; however, the process was made carefully to minimize contamination. Figure 1a shows the stirring setup. The solder alloys were compressed as shown in Fig. 1b and prepared into round shaped billets with dimensions of 50 × 10 mm dimensions shown in Fig. 1c after the solidification. The reference solder of Sn-Bi (no additions) was also prepared in the similar manner. Using the differentials scanning calorimetry (DSC) machine, the melting properties of the solder alloys were analyzed based on the DSC curve. The heat flow for both reference and Mo added Sn-Bi solder alloys was kept at 20.00 °C/min with temperature from 100 °C to 600 °C under nitrogen (N) atmosphere for this experimental process. The Vickers hardness test was conducted by taking five indentations as in Fig. 1d. The applied load was1 kgf indentation load, and the Vickers hardness value (Hv) can be calculated using (1).

$$\mathrm{HV} = 1.854(F/D) \tag{1}$$

HV = Vickers hardness number, F = Indentation force (1 kgf: 9.81 N), D = Average diagonal diameter (($D1 + D2$)/2).

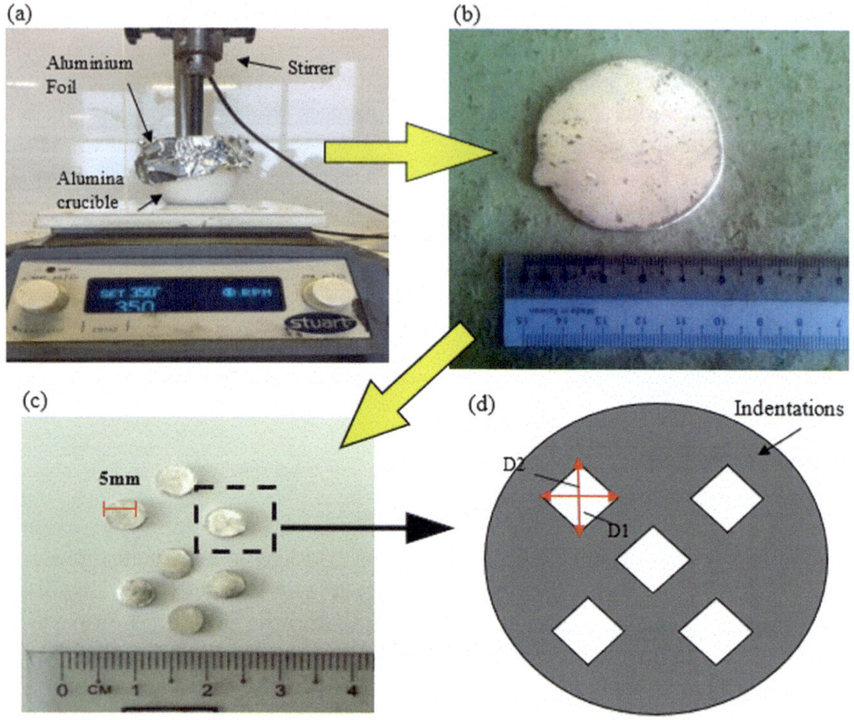

Fig. 1 **a** Stirring process during remelting, **b** compressed of Sn-Bi solder after solidification, **c** solder billets, and **d** indentation on solder billets during hardness test

3 Results and Discussion

3.1 Molybdenum Nanoparticle's Size

Figure 2 shows the SEM imaging of the Mo nanoparticles, which was used to measure the Mo nanoparticles average size. The average size of the nanoparticles was 48.5 nm.

3.2 Melting Properties of the Solder

The melting properties based on the DSC curve of the Sn-Bi and the Sn-Bi added with Mo nanoparticles are shown in Fig. 3a, b. The endothermic reaction of the solder is focused as it relates to the thermal energy absorption during melting. The peak temperature at which the solder fully melts for Sn-Bi added Mo nanoparticles solder is 148.00 °C, 3.17 °C higher than the Sn-Bi solder alloy. The melting temperature (T_L) of the solder was 139.73 °C. The increase here occurs due to the presence of

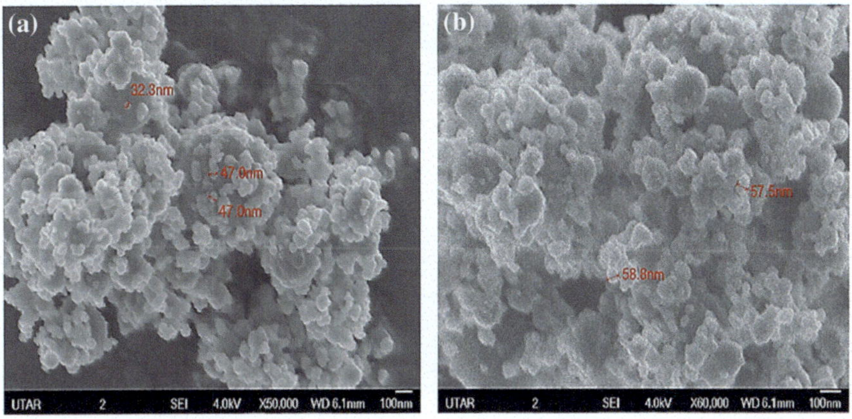

Fig. 2 Mo nanoparticles **a** × 50,000 and **b** × 60,000 magnifications

the Mo nanoparticles which would also absorb the energy during melting. Another possibility to the increase in the melting temperature is the local dissolution of the Mo nanoparticles in the molten Sn-Bi solder.

The pasty range temperature, T_R, is the difference between the solidus (T_S) and liquidus (T_L) temperature and is associated to the microstructure formation during solidification. The T_R values for Sn-Bi and Sn-Bi added with Mo nanoparticles are 4.05 °C and 9.89 °C each. Larger T_R influences in the formation of the lamellar β-Sn and Bi, which is further discussed in Sect. 3.3. Shorter T_R values are expected to benefit in producing smaller grains, but this usually concerned in non-nanoparticles-reinforced solder alloys, whereas, in a nanoparticle-reinforced solder, the nanoparticles would evidently increase the T_R but reduce the grain size of the microstructure by acting as a separate nucleation site for grain's formation [12]. It is interesting to observed that the Mo added Sn-Bi solder alloy starts to melt earlier as can be seen from the lower T_L value but solidifies later compared to the Sn-Bi solder alloy. This indicates the solder loses its surface instability earlier but took longer time to solidify because of the presence of Mo nanoparticles. The Ni nanoparticles additions to the SnSb solder alloy had reported similar statement [9].

3.3 Microstructure Properties of the Solder

Microstructure of the unreinforced Sn-Bi and Mo nanoparticles-reinforced Sn-Bi solder alloy is shown in Fig. 4. Typical lamellar structure containing Bi (white phase) and β-Sn (dark phase) phases were observed and detected from the EDX and XRD analysis. Reinforcing the Mo nanoparticles altered the microstructure by compacting

(a)

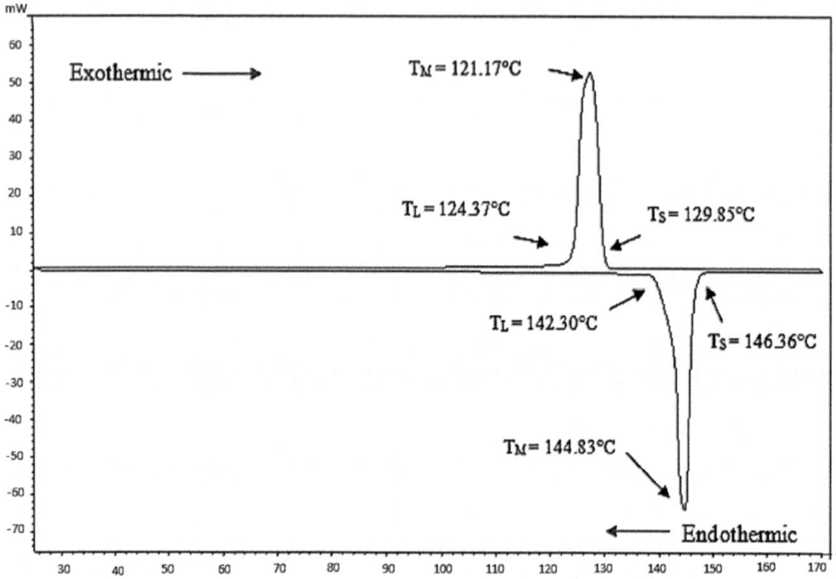

(b)

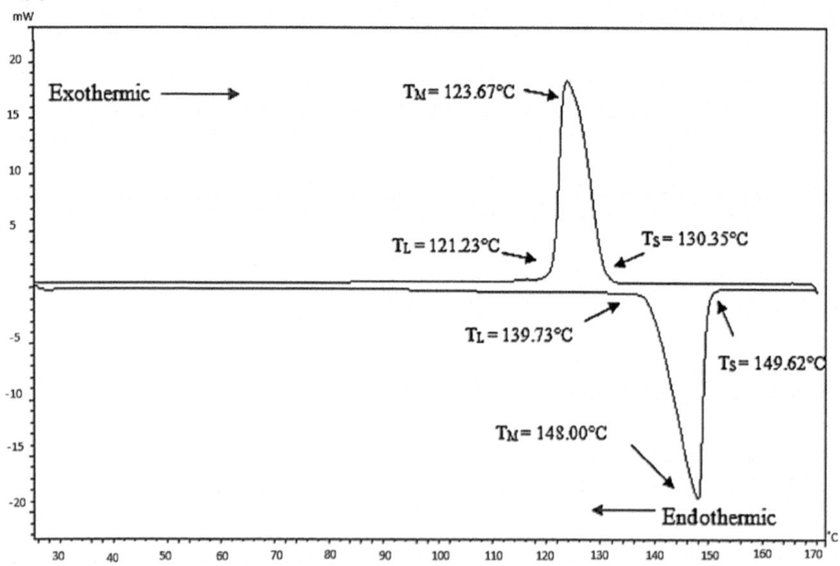

Fig. 3 DSC curve of **a** Sn-Bi and **b** Sn-Bi added with Mo nanoparticles solder alloy

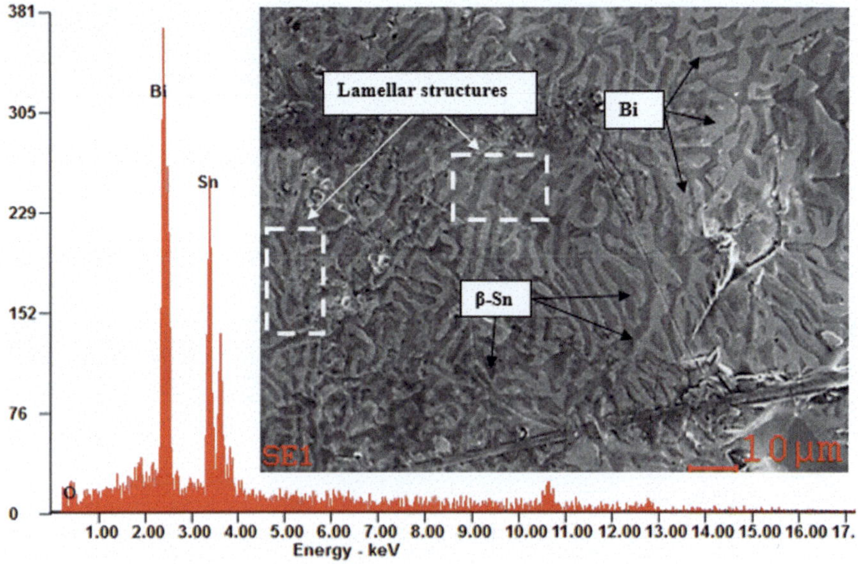

Fig. 4 SEM image and EDX of Sn-Bi solder at 1000 × magnification

and breaking the lamellar structures as highlighted in the red box in Fig. 5. Furthermore, the presence of discrete Mo nanoparticles (white particles) was evident on the microstructure observation.

According to Sn-Bi phase diagram, Bi has low solubility in the Sn, and thence, the Bi would not react with Sn to form any intermetallic compounds in the solder matrix. Hence, the Bi and Sn phases solidify separately and form microstructure as in a lamellar structure. This was too observed in the study by Shen et al. [14] and Silva [22]. Melting point of the Sn-Bi in this study based on the DSC analyzes is 139.73 °C. Linking with the phase diagram (Fig. 6) of Sn-Bi, phases found at the temperature of 139 °C matched to the microstructure phases found in this study.

The Mo nanoparticles have no solubility in the Sn matrix and Bi and does not react with these elements. During the molten state, the molten Sn and Bi compositions are vastly viscous and push the insoluble Mo nanoparticles. Moreover, the lesser dense Mo nanoparticles will be displaced to the surface. This can be observed from Fig. 5, where the Mo nanoparticles seem to be dispersed and appear on top of the lamellar Sn and Bi phases. Since the Mo nanoparticles exist as discrete particles, its presences would change the growth velocity of the phases in the solder alloy. It is predicted that Mo nanoparticles would provide separate site for nucleation during solidification that restricts the surface energy of the molten solder alloy. Although the growth velocity and activation energy are not reported in this paper, the breaking and compacting of the lamellar structure as noticed in Fig. 5 show some relativity with the change in the growth velocity. To further support this claim, study by Haseeb et al. [23] established similar incidence with the reinforcement of Mo nanoparticles in Sn3.8Ag0.7Cu solder alloy, while [24] reported reduction in the growth velocity of

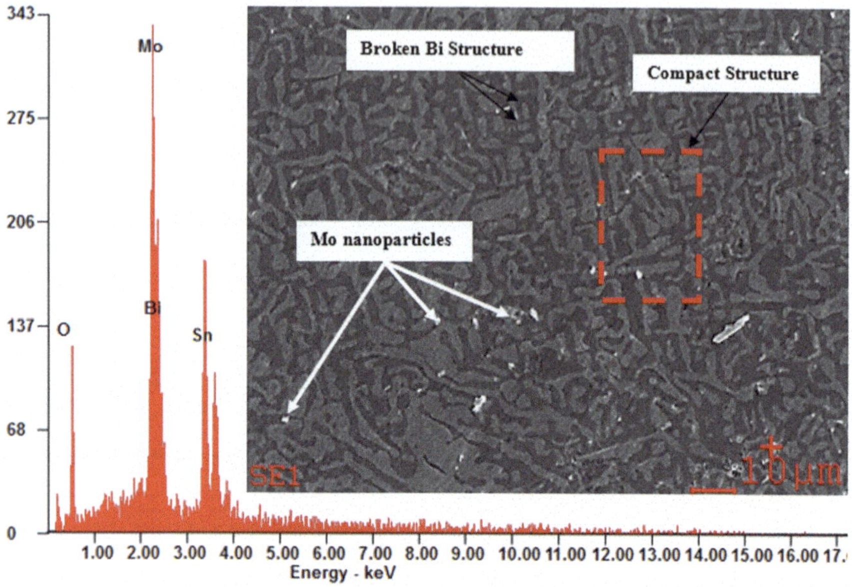

Fig. 5 SEM image and EDX of Mo-reinforced Sn-Bi solder at 1000 × magnification

Fig. 6 Phase diagram of Sn-Bi

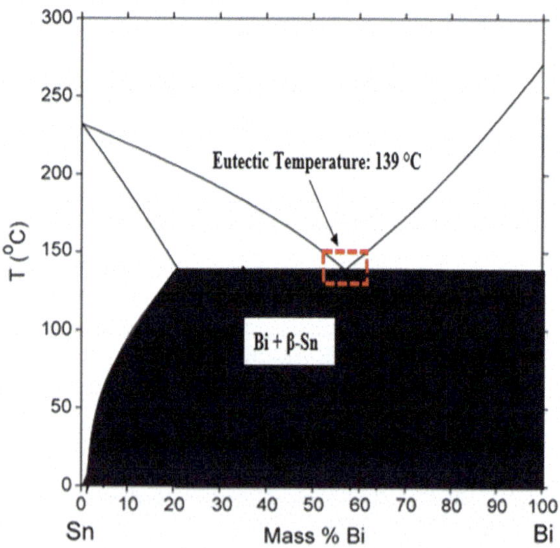

the β-Sn in the Sn-Bi solder with presence of graphene nanoparticles. This explains that the Mo would likewise affect the solidification and grain growth of the β-Sn and Bi [25] as well reported change in the activation energy after adding Ni and Bi to the SnAgCu solder system.

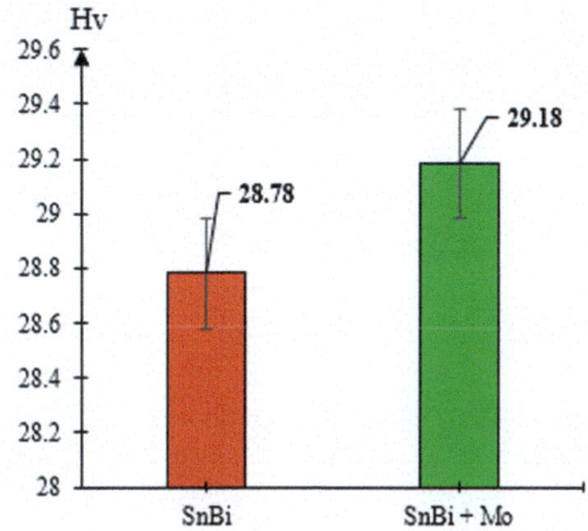

Fig. 7 Vickers microhardness value of Sn-Bi and Sn-Bi added with Mo nanoparticles

The lamellar structures in the Mo added Sn-Bi solder alloy were observed to be more closely packed compared to a more spaced-out structure in the bare Sn-Bi solder alloy. There were also Bi branches breaking to much smaller Bi phases. This justifies the larger pasty range temperature, T_R observed during solidification. Also, this indicates that the Mo disrupts the solidification of the Bi and β-Sn phases, and some Bi phases would sparse out before forming a branch prior solidification. The unreinforced Sn-Bi had T_R value of 4.05 °C.

3.4 Microhardness of the Solder

The average Vickers microhardness value for the pure Sn-Bi and Sn-Bi added with Mo nanoparticles is shown in the Fig. 7, and the values are 28.78 Hv and 29.18 Hv, respectively. A minimal increase in the hardness is directly related to the presence of Mo nanoparticles in the microstructure.

Commonly, the Bi element in the Sn-Bi solder system plays the role as the load inhibitor during the indentation [26]. The existence of dispersed Mo nanoparticles provides additional source of blocking of the load penetration during the indentation. As hard material, the Mo could not be penetrated though by the load, and so, the loads must loop over the particles, and this as consequences increase the resistance ability. Moreover, brittle character of the Bi contributes to the hardness increase of the solder alloy, and as observed, the sparse Bi in the Mo added Sn-Bi solder alloy would offer the additional resistance of deformation. The schematic diagram of the presences of Mo in the Sn-Bi solder alloy is shown in Fig. 8a, and the resistance of the Mo nanoparticles is illustrated in Fig. 8b. In additions to that, looping of the loads will create high-dense dislocations site. This will add to the resistance of penetration during indentation.

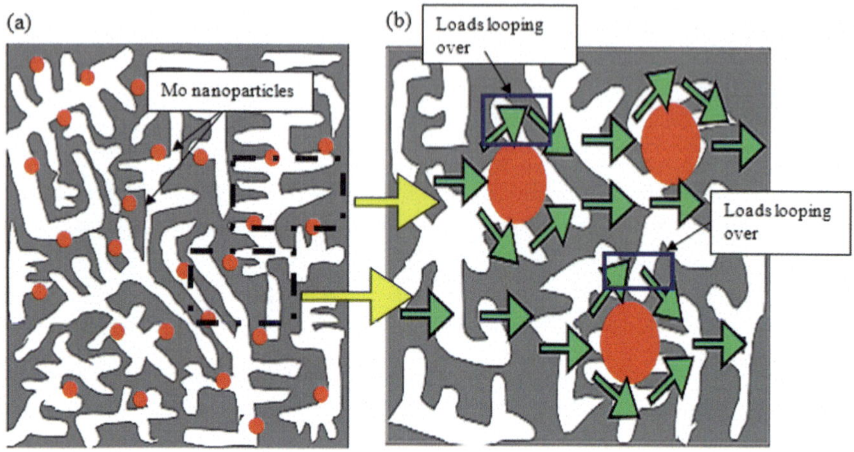

Fig. 8 Schematic drawing of **a** Sn-Bi lamellar microstructure with Mo nanoparticles and **b** load resistance by Mo nanoparticles

The relationship between the microstructure and hardness should be taken to discussion too. Finer and closer alternating lamellar structure contributes to the resistance by providing larger surface area for the load to penetrate through. Kanlayasiri and Meesathien [27] agreed with this claim as the study showed an increase in hardness due to the reduction in the grains sizes of the β-Sn in the ZnO nanoparticles-reinforced SnAgCu solder system. Therefore, the compacted lamellar structure displayed by the Mo-reinforced Sn-Bi solder would act similarly to increase the resistance. It is worth disputing the minimal increase in the hardness, but this is because of a very small percentage of the Mo being added. The main finding here did, however, show an increase to the hardness and the microstructural improvement.

4 Conclusions

This study investigates the effect of adding Mo nanoparticles to the melting, microstructure, and hardness properties of the Sn-Bi solder alloy. As the results shows, the melting temperature increased minimally, 3.17 °C but had no significant effect to the melting properties to the solder alloy. While there was an alteration to the microstructure after adding Mo nanoparticles, the Sn-Bi solder's melting temperature did not elevate drastically, but maintained below 145 °C. This is an important aspect to ensure the soldering could be done at lower temperature and would protect other electronic components from high-temperature damage. Presence of discrete Mo nanoparticles throughout the lamellar β-Sn and Bi was observed and did not diffuse with these elements. The increase in the hardness value witnessed from the results provides an initial perception that the Sn-Bi added Mo nanoparticles solder alloy could be mechanically prominent when being applied in the electronic interconnection. Further studies could be done by adding different and higher weight percentages of Mo nanoparticles to provide sufficient data in gaging this solder as a potential replacement to the traditional SnPb solder alloy.

References

1. Sobolewski M et al (2021) Solder joints reliability of through hole assemblies with various land and hole design. Microelectron Reliab 125:114368
2. Kindesjo U (2001) Phasing out lead in solders, s.l.: s.n
3. Amares S, Durairaj R, Kuan SH (2021) Experimental study on the melting temperature, microstructural and improved mechanical properties of Sn58Bi/Cu solder alloy reinforced with 1%, 2% and 3% zirconia (ZrO_2) nanoparticles. Arch Metall Mater 66(2):407–418
4. Kar A et al (2008) Evolution of mechanical and electrical properties of tin–lead and lead free solder to copper joint interface. Mater Lett 62(1):151–154
5. Billah M, Shorowordi KM, Sharif A (2014) Effect of micron size Ni particle addition in Sn–8Zn–3Bi lead-free solder on the microstructure, thermal and mechanical properties. J Alloy Compd 585C:32–39
6. Zhang Y, Liu Z, Yang L, Xiong Y (2021) Microstructure and shear property of Ni-coated carbon nanotubes reinforced InSn–50Ag composite solder joints prepared by transient liquid phase bonding. J Manuf Process 73:177–182
7. Zhang L, Tu K (2014) Structure and properties of lead-free solders bearing micro and nano particles. Mater Sci Eng 82:1–32
8. Lee J, Jeong H (2014) Fatigue life prediction of solder joints with consideration of frequency, temperature, and cracking energy density. Int J Fatigue 61:264–270
9. Chantaramanee S, Sungkhaphaitoon P (2021) Investigation of microstructure, thermal properties, and mechanical performances of Ni-added Sn–5.0Sb–0.5Cu/Cu solder joints. Microelectron Reliab (17):114421
10. Seelig K, Suraski D (2001) Lead-contamination in lead-free electronic assemblies, s.l.: Technical publication AIM
11. Kuczynska M, Maniar Y, Becker U, Weihe S (2021) Effect of shear and tensile-dominant cyclic loading on failure in SnAgCu solder. Microelectron Reliab 120:114101
12. Yang L, Ma S, Mu G (2021) Improvements of microstructure and hardness of lead-free solders doped with Mo nanoparticles. Mater Lett 304:130654
13. Deshpande M, Chaudhari R, Narayanan PR, Kale H (2021) Evaluation of shear properties of indium solder alloys for cryogenic applications. J Mater Eng Perform 30:7958–7966
14. Shen L, Wu Y, Wang S, Chen Z (2017) Creep behavior of Sn–Bi solder alloys at elevated temperatures studied by nanoindentation. J Mater Sci Mater Electron 28(5):4114–4124
15. Hu T, Li Y, Chan Y-C, Wu F (2015) Effect of nano Al2O3 particles doping on electromigration and mechanical properties of Sn–58Bi solder joints. Microelectron Reliab 55:1226–1233
16. Yang L et al (2017) Electromigration reliability for Al2O3-reinforced Cu/Sn–58Bi/Cu composite solder joints. J Mater Sci Mater Electron 28:3004–3012
17. Yang W et al (2020) Effect of graphene nanosheet addition on the wettability and mechanical properties of Sn-20Bi-xGNS/Cu solder joints. Materials 20:1–14
18. Hu S-H et al (2021) Effects of bismuth additions on mechanical property and microstructure of SAC–Bi solder joint under current stressing. Microelectron Reliab 117:114041
19. Gain AK, Zhang L (2017) Effect of Ag nanoparticles on microstructure, damping property and hardness of low melting point eutectic tin–bismuth solder. J Mater Sci Mater Electron 28:15718–15730
20. Li Ml et al (2021) Materials modification of the lead-free solders incorporated with micro/nano-sized particles: a review. Mater Des (197):109224

21. Liu Y et al (2021) Microstructure and mechanical behavior of SnBi–xAg and SnBi–xAg@P–Cu solder joints during isothermal aging xAg and SnBi–xAg@P–Cu solder joints during isothermal aging. Microelectron Reliab 127:114388
22. Silva BL, Xavier MGC, Garcia A, Spinelli JE (2017) Cu and Ag additions affecting the solidification microstructure and tensile properties of Sn–Bi lead-free solder alloys. Mater Sci Eng A 705:325–334
23. Haseeb ASMA, Arafat MM, Johan MR (2012) Stability of molybdenum nanoparticles in Sn–3.8Ag–0.7Cu solder during multiple reflow and their influence on interfacial intermetallic compounds. Mater Charact 64:27–35
24. Liu Y et al (2019) Interfacial reaction, microstructure, and mechanical properties of Sn58Bi on graphene-coated Cu substrate. Results Phys 13:102256
25. Sivakumar P, O'Donnell K, Cho J (2021) Effects of bismuth and nickel on the microstructure evolution of Sn–Ag–Cu (SAC)-based solders. Mater Today Commun 26:101787
26. Amares S, Durairaj R (2020) Environmental friendly low mass 20g-Sn58Bi/Cu solder alloy as an alternative to lead SnPb and its properties study. IOP Conf Ser Earth Environ Sci 505:012004
27. Kanlayasiri K, Meesathien N (2018) Effects of zinc oxide nanoparticles on properties of SAC0307 lead-free solder paste. Adv Mater Sci Eng

Studies on Elastic–Plastic Behavior of Plasma-Sprayed Ceramic Coatings on TI–6AL–4V Substrate

M. Kalayarasan, P. Dhanabal, and S. Mohanraj

Abstract The elastic–plastic contact between a rigid ball and flat plate with three different types of ceramic coatings was analyzed using the ball indentation test and finite element method (FEM). Micrometer-sized powders of the following compositions were plasma sprayed onto the Ti–6Al–4V substrate: (i). Al_2O_3 (AO), (ii). 8 mol% yttrium stabilized zirconia (8YSZ), and (iii). Al_2O_3–40 wt% 8YSZ (A4Z). The loading and unloading results revealed that the FE model was capable of reproducing the experimental loading and unloading curves for the coated samples. The result from the study suggested that the coating with materials Al_2O_3–40 wt% 8YSZ was the most desirable one for protecting the substrate against yielding since plastic deformation does not initiate until a large plastic zone has been developed in the substrate.

Keywords Coatings · Ball indentation · Experimental · Elastic–plastic · Finite element model

Nomenclature

A	Contact area
E	Young's modulus
R	Radius of the indenter
ω	Interference
P_m	Maximum contact pressure
P_a, p	Average/mean contact pressure
ω_1	Tabor constant
ω_2	Taylor constant
A_e	Contact area
Y	Yield strength

M. Kalayarasan (✉) · P. Dhanabal · S. Mohanraj
Department of Mechanical Engineering, PSG College of Technology, Coimbatore, India
e-mail: mkn.mech@psgtech.ac.in

E. Natarajan et al. (eds.), *Materials, Design and Manufacturing for Sustainable Environment*, Lecture Notes in Mechanical Engineering,
https://doi.org/10.1007/978-981-19-3053-9_4

a_c Contact radius
R Radius of the indenter
ω_c Critical interference

1 Introduction

Instrument indentation testing (IIT) method has peaked its importance in recent years because of its ability for the measurement of material response to the applied load. The ball indentation technique is a non-destructive method for determining the mechanical properties of materials such as strain hardening component, elastic modulus, yield strength, and so on. Metal implants with ceramic coatings are commonly used, and the quantitative study of these coatings might enable us to elucidate the mechanical stability for clinical applications. The interface between the coating and the substrates is very important as it becomes a major cause of interfacial strength failure. Researchers mostly utilized experimental as well as finite element concepts to measure the indentation consequences. The interfacial strength of the coatings was measured experimentally using the pull-out tensile strength, laser spallation technique [1]. Loading and unloading curves can be obtained using the nanoindentation technique, in which a rigid ball was pressed against a coating to predict the fracture mechanism aspects of the coatings [2]. Among the different coating materials, titanium nitride coated on tool steels shows less deformation than that of the other materials [3, 4]. When multilayer is considered, Ti/TiN proves as a promising material to protect the coating against failure [5]. The deposition of multilayer coatings on the surface also helps in improving the tribological behavior of the surfaces. The multilayering and grading process lower the likelihood of cracks which is most common in monolithic materials when indented [6]. With the aid of computational mechanics, numerous technological problems related to contact mechanics were solved owing to its low-operational cost. The material transition from elastic to plastic state was predicted using the finite element method [7–9]. The most common mechanisms of failure include fracture at the surface, plastic deformation, delamination, and interfacial crack which could be observed in the coating substrate system. Some studies dealt with the plastic yielding by considering its locations in a system of sphere with hard coating [10]. The tangent modulus and yield strength of the materials are found to influence the plastic yielding of the materials [11]. The coating thickness and substrate strength have a substantial impact on the material's plastic deformation behavior and load-bearing capacity as well.

Titanium nitride coating finds its application in many engineering fields, because plastic deformation does not initiate until a considerable amount of plastic deformation occurs in the substrate [12]. An analytical model was used to predict the contact friction and found that the indentation profile and load depth curve are affected [13]. Soft metallic multilayers can be introduced to reduce the stress concentration at the coating substrate interface which also increases the fracture resistance [14].

Coating fracture and the methodology for selecting the type of coating process and the coating material can be determined during spherical indentation using numerical analysis technique [14, 15].

It is observed from the above literature that the investigation on the rigid ball and flat plate contact considering the substrate-coating system is minimal. Initially, ball indentation testing was carried out on a Ti–6Al–4V substrate with three different ceramic coatings. Further, this work also focuses on a FE model-based approach for evaluating the elastic–plastic axi-symmetric contact between flat plate and a rigid ball considering three distinct types of ceramic coatings and coating thicknesses. The simulation results are validated with the experimental indentation testing results.

2 Materials and Method

2.1 *Powder Preparation*

An important property that material should possess is to reduce the wear rate and also to minimize friction. Recently, ceramic materials are used as a coating substance to reduce the wear rate. The present study utilizes titanium as the substrate material. Alumina, zirconia, and the composite blend of Al_2O_3–40 wt% ZrO_2 were chosen as the coating material because of the biocompatibility of materials, and they are also used in many clinical applications. Fused and crushed Al_2O_3 powders were commercially purchased and it consists of dense solid particles with angular blocking morphology, and the particle size is in the range of 5–45 μm. In addition, 8 mol% yttria-stabilized ZrO_2 was also obtained with the similar particle size as that of alumina.

In a planetary ball mill, 60% of Al_2O_3 and 40% of 8YSZ were mixed to make composite feedstock powders. The sample was milled for a period of 3 h at 300 rpm with a 30 min settling time for every hour. The blend powders were checked for homogeneous distribution of the particles by using microstructures obtained from SEM analysis shown in Fig. 1.

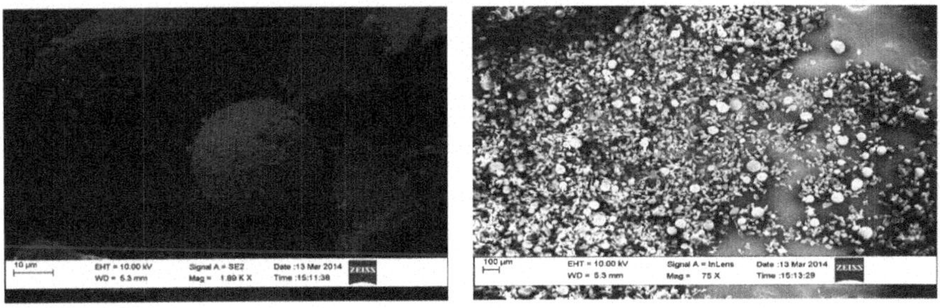

Fig. 1 SEM image of A4Z blend

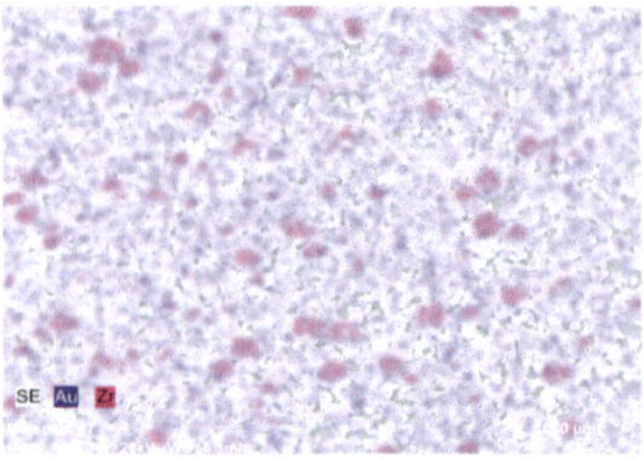

El	AN	Series	unn. C [wt.%]	norm. C [wt.%]	Atom. C [at.%]	Error (1 Sigma) [wt.%]
Al	13	K-series	23.75	60.60	83.87	1.16
Zr	40	L-series	15.44	39.40	16.13	0.62
		Total:	39.18	100.00	100.00	

Fig. 2 Mapping showing the elemental composition

The blended powder contains Al_2O_3 and 8YSZ as it is evident in the series of element mapping images as shown in Fig. 2. It is observed that A4Z powder is composed of stable tetragonal zirconia, and also it consists of α-Al_2O_3 with particle sizes between 10 and 60 μm. In the blend, there was evidence of a high-temperature stable tetragonal ZrO_2 phase as well as α-Al_2O_3. The formation of γ-Al_2O_3 from α-Al_2O_3 is associated with high temperature and rapid splat quenching that takes place during the plasma spraying process.

2.2 *Sample Preparation*

The chosen materials are to be coated on the Ti6Al4V substrate by using the plasma spraying technique [16]. It is a thermal spraying technique where the powder is heated to a temperature of 10,000 K, and the optimized coating parameters for the plasma spraying technique are selected [16]. Initially, alumina is coated on the specimen at a thickness of 500 μm to check the homogeneity of the coatings. The alumina sample had a hardness of about 12.4 GPa, which was in line with published values [16].

The surface roughness of the coated specimens was measured using a surface roughness tester, and the average 'Ra' values taken from the samples were given in Table 1. It has been observed that the results were found to have a close correlation with that of the previous wear studies [16].

Table 1 Roughness measurement for the coated samples

Coating material	Test 1 Ra (μm)	Test 2 Ra (μm)	Test 3 Ra (μm)	Test 4 Ra (μm)	Average Ra (μm)	Literature value [16]
Al_2O_3	3.168	3.228	3.158	3.014	3.142	3.67
A4Z	4.654	4.623	4.635	4.662	4.643	4.75
8YSZ	4.685	4.688	4.681	4.689	4.685	4.52

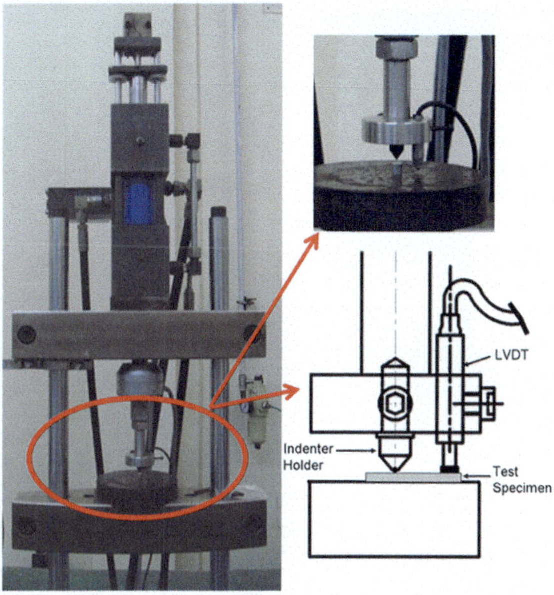

Fig. 3 Ball indentation setup

3 Experimental Analysis

The ball indentation test was used to evaluate the coated samples as depicted in Fig. 3. It consists of a compact servo-hydraulic machine that uses a 1 mm diameter tungsten carbide indenter to perform a loading–unloading cycle. The sample is held on a flat table and a ball with high hardness which is made of tungsten carbide and is made to indent on the sample in the standard sequence of loading steps.

The penetration depth is measured using an LVDT, whereas the load is obtained using a load cell and the LVDT. Figure 4 shows the experimental observations for three different specimens which were obtained while loading and unloading over the sample.

4 Results and Discussion

4.1 Experimental Validation

A 2D axisymmetrical contact model was developed using ANSYS® for a coated substrate indented by a rigid sphere configuration. The indenter was considered to be a rigid sphere of 0.5 mm radius. The substrate was modeled considering a dimension of $3.5 \times 3.5\ \text{mm}^2$ which was similar to the experimental samples considering different thicknesses of coating.

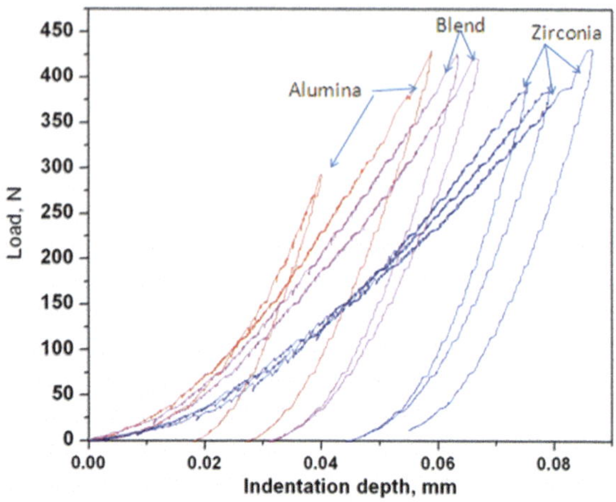

Fig. 4 Variation of load and indentation depth (Al_2O_3, 8YSZ, A4Z)

Constraints were given in the nodes on the axis of symmetry of coating and substrate along the x-direction. The nodes at the bottom of the base substrate were completely constrained in all degrees of freedom. In order to minimize the computational time other than the contact region, the model is meshed with coarse elements as shown in Fig. 5. In order to ensure the contact between the sphere and the coated surface, eight-noded quadrilateral elements (PLANE 183) and three-noded contact elements, namely CONTA172 and target element TARGE169, were used for the contact region. Initially, the finite element results are validated with the experimental ball indentation testing results. The indentation depth during loading and unloading values from FEM results is compared with the obtained experimental results. Figure 6a shows the comparison of results of experimental and numerical simulation for Al_2O_3 coating. A good correlation is found between the numerical and experimental results with a deviation of about 0.69%.

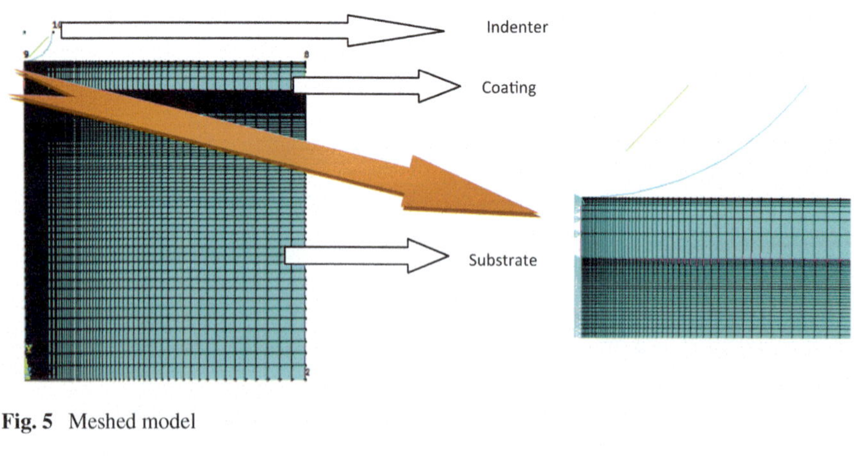

Fig. 5 Meshed model

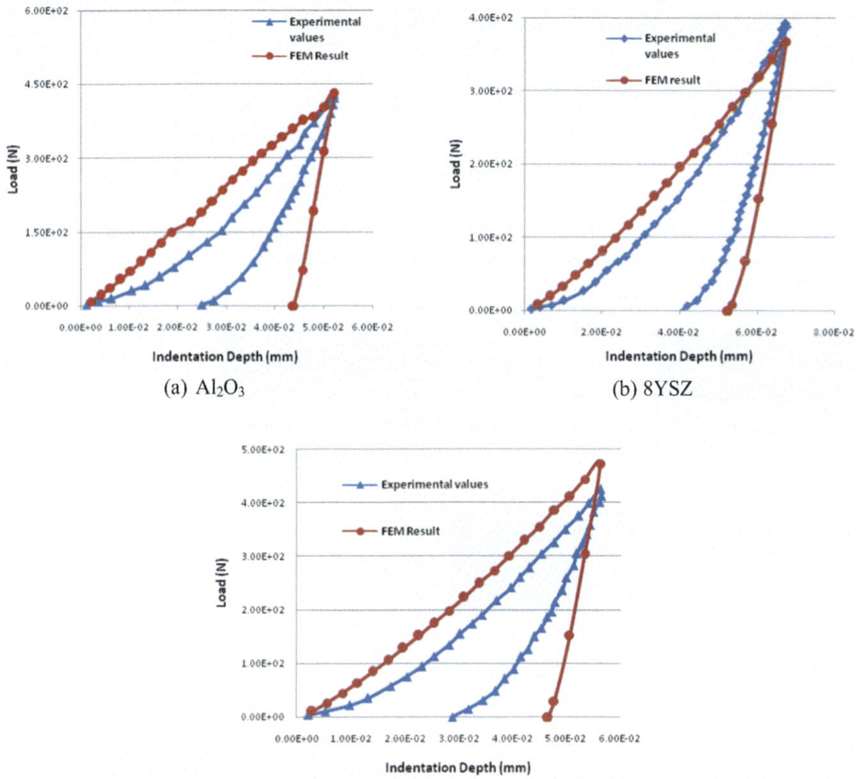

(a) Al_2O_3 (b) 8YSZ

(c) A4Z

Fig. 6 Comparison between the experimental and finite element results of indentation depth with load

The comparison of results for experimental and FE analysis for 8YSZ coating is shown in Fig. 6. The error percentage was found to be 2% in the maximum loading area, and it is uniform throughout. 8YSZ coating shown in Fig. 6c reveals that there is a small variation in the peak load since the coating of 8YSZ of the sample has less adherence to the sample, and thus, it caused the variation. The error percentage was found to be 11.6% for 8YSZ coating.

Subsequently, the validated model is used in order to determine the depth of indentation for varying loads with different coating thicknesses, viz., 0.1 mm, 0.2 mm, 0.3 mm, 0.4 mm, and 0.5 mm, respectively. From the literature, it was found that the coating thickness between 100 μm and 500 μm has a considerable positive effect in the enhancement of the properties along the substrate.

4.2 Microstructural Characterization

The elemental characterization was done using the EDX measurement, in order to find the composition after plasma spraying. The EDX measurements showed that in addition to the coating material, oxygen is also present on the surface of the coating which is a result of the thermal spraying technique. The EDX measurements for Al_2O_3, 8YSZ, and A4Z are shown in Fig. 7a–c. The EDX analysis of the sample reveals Al_2O_3, 8YSZ, and oxygen as significant elements (Fig. 8).

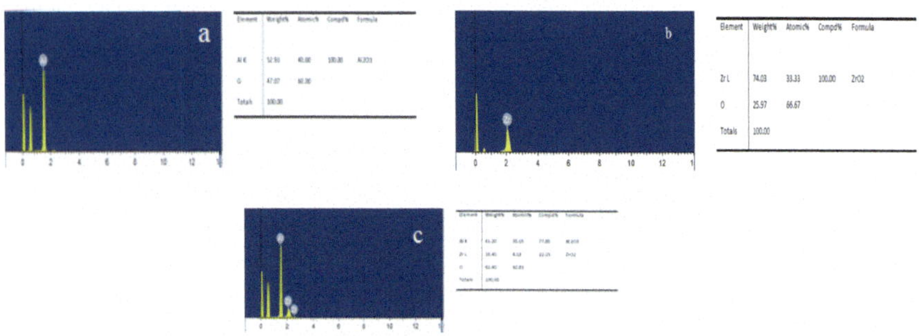

Fig. 7 EDX image of **a** Al_2O_3, **b** 8YSZ, **c** A4Z

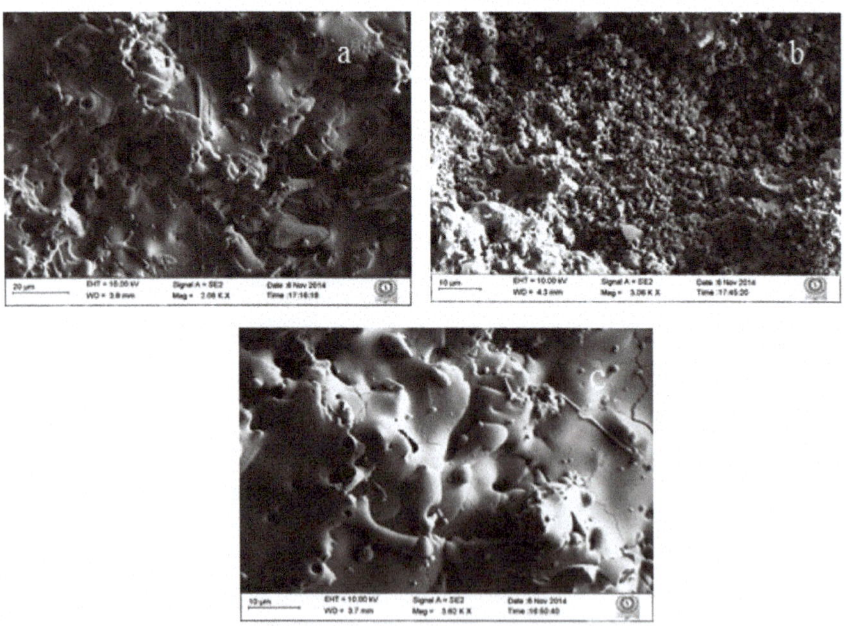

Fig. 8 FESEM images of plasma-sprayed coating of **a** Al_2O_3, **b** 8YSZ, **c** A4Z coatings

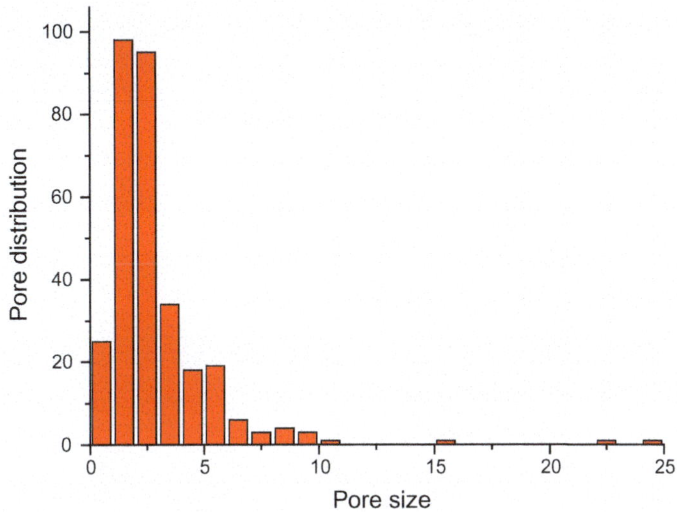

Fig. 9 Pore distribution—Al_2O_3

The particle distribution and pore size (Figs. 9, 10, 11, and 12) distribution after coating were analyzed using FESEM images (Fig. 8a–c) using imageJ software. It was found that for Al_2O_3 coating, the pore size measured was 1.21 μm ± 0.715. The size of the pores influences the hardness of the sample to some extent. The sample with a lesser number of pores had good hardness compared to other samples. It was also found in Al_2O_3 that there are some un-melted particles of size 1.0148 ± 0.7249 μm.

These un-melted particles in the coated sample have contributed to the surface roughness value as given in Table 2. A4Z coating is mainly composed of ultra-fine splats (1–5 μm) and dense fine spheres (~100 nm). In the A4Z composite coating, the ZrO_2 splats were observed to be in the form of flat strips, and these coatings have some micro-cracks and pores perpendicular to the coating/substrate interface. Microfractures were most likely formed by the contraction of rapidly cooled splats onto previously deposited layers, as well as due to differential thermal expansion. Further in the composite coating in A4Z coatings, some partially melted particles in the size range of 10–35 μm are observed. It is evident that the existence of the above said particles have greatly contributed to the surface roughness. A comparison is made between all the three types of coatings, and it is found that the difference between the coefficient of thermal expansion for the composite coating and Ti6Al4V substrate is minimal. This leads to lower residual stress along the interface of the coating substrate and thereby providing better adhesion strength.

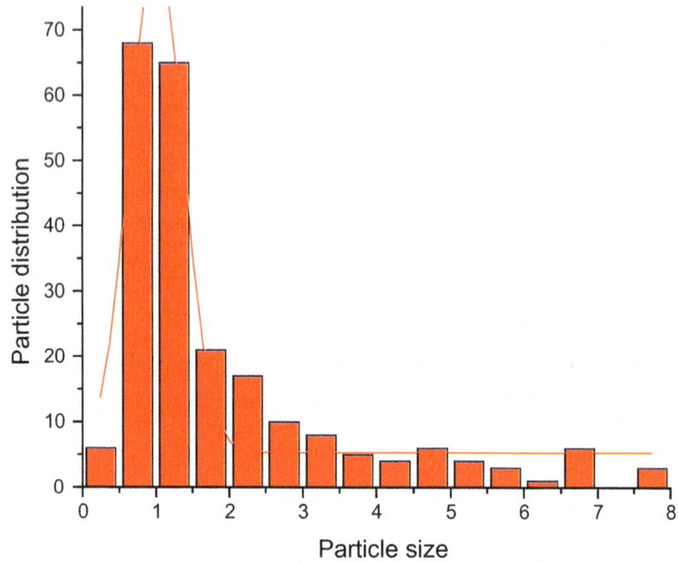

Fig. 10 Particle distribution—Al_2O_3

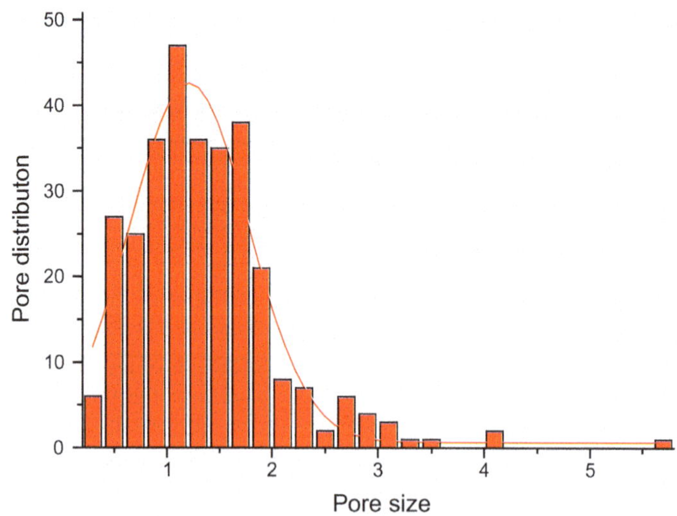

Fig. 11 Pore distribution—A4Z

4.3 Analytical Approach—State of Contact

The nature of contact may be either elastic, elastic–plastic, or fully plastic-type. The state transition level from elastic-to-elastic–plastic and also to fully plastic state is

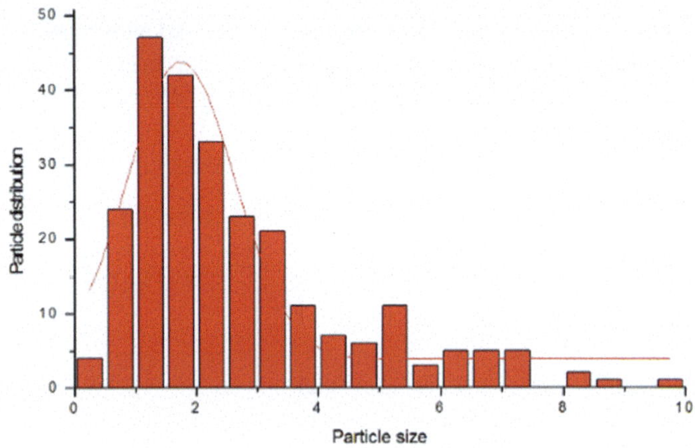

Fig. 12 Particle distribution—A4Z

Table 2 Interference values (ω_1 and ω_2) from the analytical method for different coating samples

S. No.	Sample	ω_1	ω_2 (Tabor)	ω_2 (Taylor)
1	Al_2O_3	0.00098	0.024	0.053
2	8YSZ	0.00507	0.126	0.274
3	A4Z	0.00212	0.053	0.114

to be determined for better performance analysis. When the contact is plastic, the coating behavior may not be predictable. This tends to release many wear debris particles that cause damage which is the major cause for failure. There are both analytical and finite element approaches to find the state of contact. The equations for contact analysis are found from the literature [7], and it is shown below.

$$A_e = \pi^* R^* \omega \tag{1}$$

$$w_e = \left(\frac{4}{3}\right) \mathrm{ER}^{1/2} \omega^{3/2} \tag{2}$$

$$P_m = (2E/\pi)(\omega/R)^{1/2} \tag{3}$$

$$P_a = \frac{2}{3} P_m = \frac{4E}{3\pi} \left(\frac{\omega}{R}\right)^{1/2} \tag{4}$$

$$\omega = \left(\frac{3\pi p_a}{4E}\right)^2 \tag{5}$$

From the above equations, the interference ω is compared with elastic and plastic contact parameters ω_1 and ω_2 as found by Tabor and Taylor. In elastic contact, Tabor [17] found that the material starts yielding when $P_m = 0.6H$ or $P_a = 0.4H$ where H is material hardness.

$$\omega_1 = \left(\frac{3\pi k H}{4E}\right)^2 R \tag{6}$$

$$\omega = \left(\frac{3\pi p_a}{4E}\right)^2 R \tag{7}$$

If $\omega < \omega_1$, the nature of contact is elastic and $\omega \geq \omega_{1,}$ the type of contact is either elastic–plastic or fully plastic. For plastic contact, the conditions were described by Tabor [17] Eq. (8) and Taylor (1985) Eq. (9).

$$\omega_2 > 25\omega_1 \tag{8}$$

$$\omega_2 \geq 54\omega_1 \tag{9}$$

For elastic–plastic contact, the condition is $\omega_1 < \omega < \omega_2$ where ω_1 and ω_2 values are presented in Table 2.

Using the conditions above, the coatings were analyzed for their state of contact, and it is mentioned in Table 3. It is found from Table 3, Al_2O_3 changes its state from elastic-to-elastic plastic at the initial stages of loading and completely plastic when the load is greater than 510.37 N. For 8YSZ coating, the elastic-to-elastic–plastic transformation occurs when the sample is loaded initially, and it does not reach plastic zone for the maximum load considered in this analysis. Similarly, for A4Z coating, the transformations are found to be at initial loads for elastic–plastic and greater than 981.34 N for fully plastic.

It is seen from the approach that Al_2O_3 coating reaches the plastic contact region at higher loads compared to others. 8YSZ coating does not reach the plastic contact for the analyzed depth but it may enter its plastic state with further increase in its depth with their corresponding load. A4Z coating has its capacity between Al_2O_3 and 8YSZ coatings.

Table 3 Variation of interference with load for various coatings using finite element method

Interference ω (mm)	Load (N)		
	Al_2O_3 coating	8YSZ coating	A4Z coating
0	0	0	0
0.01	69.302	33.042	55.648
0.02	163.31	80.98	130.62
0.03	268.15	135.4	216.13
0.04	383.52	192.87	306.23
0.05	510.37	254.39	406.84
0.06	639.39	318.42	511.6
0.07	782.12	388.68	625.39
0.08	926.38	458.21	739.59
0.09	1072.5	531.53	858.63
0.10	1219.9	608.8	981.34
0.09	287.08	209.15	231.99

5 Conclusions

The surface morphological study, experimental, and numerical analysis of three different ceramic coatings- Al_2O_3, 8YSZ, and A4Z on Ti6Al4V substrate lead us to the following conclusions.

Coating proved to be a better alternative to add strength to the substrate to improve its performance. FESEM images of 8YSZ showed a clear insight that the adherence of 8YSZ powders on the Ti6Al4V substrate is less which is indicated by the un-melted and un-adhered particles. The experimental results showed that Al_2O_3 coating can withstand more load comparing others, which is followed by A4Z and 8YSZ coating. From the analytical approach, it is evident that A4Z coating has its capacity between Al_2O_3 and 8YSZ coatings. It does not reach its plastic stage, and therefore, its performance is better than the substrate without coating. This clearly proves that A4Z is the best suitable choice protective layer to enhance the life and reliability of various coating systems.

References

1. Kim H, El-Awady J, Gupta V, Ghoniem N, Sharafat S (2009) Interface strength measurement of tungsten coatings on F82H substrates. J Nucl Mater:386–388
2. Prerez EA, Souza RM (2004) Numerical and experimental analyses on the contact stresses developed during single and successive indentations of coated systems. Surf Coat Technol:572–580
3. Michler J, Blank E (2001) Analysis of coating fracture and substrate plasticity induced by spherical indenters: diamond and diamond-like carbon layers on steel substrates. Thin Solid Films:119–134

4. Kot M, Rakowski W, Lackner JM, Major L (2013) Analysis of spherical indentations of coating-substrate systems: experiments and finite element modelling. Mater Des 43:99–111
5. Cheng YH, Browne T, Heckerman B, Bowman C, Gorokhovsky V, Meletis EI (2010) Mechanical and tribological properties of TiN/Ti multilayer coating. Surf Coat Technol:146–151
6. Yang DK, Wang JT, Fabijanic D, Lu JZ, Hodgson PD (2014) Improved contact load-bearing capacity of ultra-fine grained titanium: multilayering and grading. Mater Des 58:217–225
7. Zhao Y, Maietta DM, Chang L (2000) An Asperity microcontact model incorporating the transition from elastic deformation to fully plastic flow. J Tribol 122:86
8. Kogut L, Etsion I (2002) Elastic-plastic contact analysis of a sphere and a rigid flat. J Appl Mech 69:657
9. Komvopoulos K, Ye N (2001) Three-dimensional contact analysis of elastic-plastic layered media with fractal surface topographies. J Tribol 123:632
10. Goltsberg R, Etsion I, David G (2011) The onset of plastic yielding in a coated sphere compressed by a rigid flat. Wear 271:2968–2977
11. Shankar S, Mayuram MM (2008) Effect of strain hardening in elastic-plastic transition behavior in a hemisphere in contact with a rigid flat. Int J Solids Struct 45:3009–3020
12. Sun Y, Bloyce A, Bell T (1995) Finite element analysis of plastic deformation of various TiN coating/substrate systems under normal contact with a rigid sphere. Thin Solid Films 271:122–131
13. Karthik V, Visweswaran P, Bhushan A, Pawaskar DN, Kasiviswanathan KV (2012) Finite element analysis of spherical indentation to study pile-up/sink-in phenomena in steels and experimental validation. Int J Mech Sci:74–83
14. Kot M (2012) Contact mechanics of coating-substrate systems: monolayer and multilayer coatings. Arch Civ Mech Eng 12:464–470
15. Abdul-Baqi A, Van der Giessen E (2002) Numerical analysis of indentation-induced cracking of brittle coatings on ductile substrates. Int J Solids Struct 39:1427–1442
16. Perumal G, Asokamani GM, R and Alagumurthi N, (2014) Wear studies on plasma sprayed Al_2O_3–40wt% 8YSZ composite ceramic coating on Ti–6Al–4V alloy used for biomedical applications. Wear 311:101–113
17. Tabor D (1951) The hardness of materials. Oxford: Clarendon Press

Biocorrosion Studies on Stainless Steel Implant Material with Different Surface Process Condition

C. Siva Sundaram, K. Gokulraj, N. Hari Vignesh, M. Adam Khan, J. T. Winowlin Jappes, B. Anushraj, and Sankarganesh Arunachalam

Abstract In this study, biocompatible of the implant materials was found to be effective in replacing and supporting fractured natural human bones. In connect to this study, an electrochemical corrosion as the discharge of ions from exposed metal with simulated human body fluid is studied. The biocompatibility is studied on the surface of wire electro spark machined implant, laser pulse shot peened, Co–Cr-coated surface and plain steel. After corrosion studies, the rate of material loss in terms of exposure and its surface quality is evaluated through scanning electron microscope. As a result, the best process in surface protection with simulated body fluid is recommended for implementation.

Keywords Laser · WEDM · Co–Cr · Corrosion SBF

1 Introduction

Implants are used in human anatomy and are replaced to maintain the natural bone mechanics' stability and energy. Self-healing with natural tissues should also be in benefit with mechanical properties. Engineers and orthopedics work together to help a person live a pain-free existence. Stainless steel and titanium-based alloys are wide used as a commercially accessible human implant material. The stainless steel implants are widely used in public under rural, as they are cheap with reasonable advantages. The present essential concerns are mechanical qualities, design in complicated shapes, surface topography, corrosion resistance, and biocompatibility. To tackle the challenges of designing human implant sections such as hip, rib, and

C. Siva Sundaram · K. Gokulraj · N. Hari Vignesh · M. Adam Khan (✉) · J. T. Winowlin Jappes · B. Anushraj
Department of Mechanical Engineering and Centre for Surface Engineering, Kalasalingam Academy of Research and Education, Virudhunagar, Tamilnadu, India
e-mail: adamkhanm@gmail.com

S. Arunachalam
Department of Biotechnology, Kalasalingam Academy of Research and Education, Virudhunagar, Tamilnadu, India

E. Natarajan et al. (eds.), *Materials, Design and Manufacturing for Sustainable Environment*, Lecture Notes in Mechanical Engineering,
https://doi.org/10.1007/978-981-19-3053-9_5

knee joints, highly compatible materials are recommended. Stainless steel, Co–Cr-based alloys, and titanium alloys are often employed. Literatures are available to report on human implant prosthesis development, biocompatibility, deflection, and heat resistance [1–3].

The implant's surface characteristics have a significant impact on tissue growth and response. It will be covered with protein production (naturally) from the body fluid after implantation and then continue to cell/tissue development. Protein and cell growth rates differ depending on the state of a person's body fluids. The surface topography and bone development of an alternative human implant material are readily available for discussion [4–6]. According to the literature, the performance of cast and machined implant materials is evaluated in terms of bone contact and bone filling. It has been discovered that the influence of surface roughness is on par with the impact of bodily fluid. Mozafar [7] attempted to investigate the hydrophilicity and corrosion resistance of orthopedic implants coated with glass/zirconium material.

As a result, bio-components demand a shift from traditional to unconventional processes. Wire electro spark machining is one of the best methods for cutting complex shapes without sacrificing surface quality [8]. The input process parameters are highly influenced to produce surface properties and corrosive nature. On other hand, laser processed implants are investigated with simulated body fluid [9]. It has been identified that the emission of nickel ions from the implant reacts with body fluid to cause severity in human anatomy. The two major elements Ni and Cr are highly reactive with body fluid to cause human effect.

In this paper, the biocompatibility of stainless steel implants material is proposed to report with different surface modifications. Plain stainless steel implant, Co–Cr-coated commercial implant, wire spark machined implant, and laser shot peened implant of stainless steel grade.

2 Experimental Details

Experimentation and procedure for test evaluation are explained in detail, from the material selection, machining process, biocorrosion using simulated body fluid and cell viability have been discussed. The proposed research is to perform experimentation on stainless steel titanium based on implants. On machined surface, biocorrosion studies are performed on using simulated body fluid (SBF) through electrochemical corrosion test. Simulated body fluid (SBF) solution was developed using standard chemical reagents following Kokubo method [10]. A cyclic voltammetry cell (corrosion analyzer) is used to study the behavior of the implants from TAFEL plot. The data extracted from the CV method will provide corrosion rate of the implants in terms of mm/yr. Figure 1 represent the photo of electrochemical test kit used for corrosion analysis. The ACM instrument GillAC, UK consist of three cell electrode which has platinum as counter electrode, saturated calomel electrode as reference electrode, and sample as working electrode. The experiment is carried out in room temperature with the exposed area of 1 cm^2. The test is carried out at room temperature in stimulated body fluid. The measurement is carried out as per ASTM G3-14 Standard. The studies were performed at a sweep rate of −250 mV to 250 mV at a rate of 100 mV/min [11]. The polarization data with TAFEL plot is obtained from the computer attached with the instrument. The morphology of the chemically etched samples was further subjected to characterization studies.

Table 1 represents the list of chemical reagents used for simulated body fluid preparation. To prepare 1000 ml of SBF, in 1000 ml plastic beaker with stirring bar, add 700 ml of ion-exchanged and distilled water. Place the beaker in the magnetic stirrer with water bath setup and cover it with the plastic wrap. Under stirring, heat the water in the container at 36.5 ± 1.5 °C. From the investigation, the machined surface is exposed to scanning electron microscope to examine the metallurgical changes over the exposed surface.

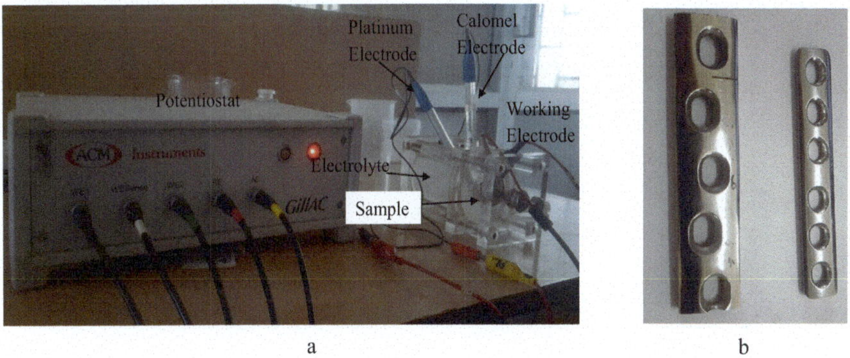

Fig. 1 Photo of **a** electrochemical test kit and **b** impant used for investigation

Table 1 Reagents and their quantity for preparing simulated body fluid

Reagent amount (g/L)	Amount (g/L)
Sodium chloride (NaCl)	6.546
Sodium bicarbonate ($NaHCO_3$)	2.2682
Potassium chloride (KCl)	0.373
Potassium phosphate dibasic trihydrate ($HK_2O_4P{\cdot}3H_2O$)	0.1419
Magnesium chloride hexahydrate ($MgCl_2$. $6H_2O$)	0.3049
1 M hydrochloric acid (1.0 M-HCl)	40 ml
Calcium chloride ($CaCl_2$)	0.3675
Sodium sulfate (Na_2SO_4)	0.071
Tris(hydroxymethyl) aminomethane ($NH_2C(CH_2OH)_3$)	6.057

3 Results and Discussion

3.1 WEDM Surface

The implant with wire electro discharges machined surface was investigated with the simulated body fluid following the ASTM standards. The machined surface has a maximum corrosion rate of 0.01409 mm/year as recorded in Table 2. For this rate, the corrosion potential of the exposed surface found to be −109.2 mV at a current density of 0.001 mA/cm^2. These details are derived from the TAFEL plot arrived on WEDM processed sample (Fig. 2) and the same found similar to the exist literature [2]. The presence of oxide scale one the machined surface has a predominant nature toward surface protection.

Table 2 Electrochemical data recorded for WEDM processed stainless steel implant

Test sample	Ecorr (mV)	Icorr (mA/cm^2)	Corrosion rate (mm/year)
WEDM surface	−109.205	0.0010584	0.0140978

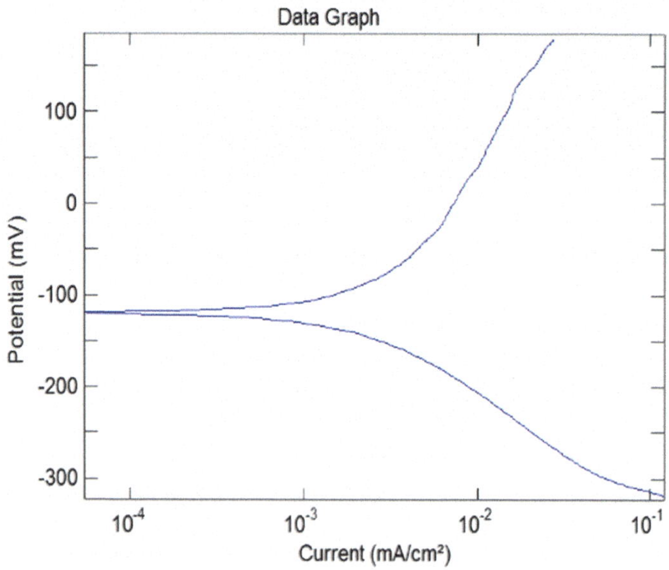

Fig. 2 TAFEL plot recorded for WEDM processed stainless steel implant

3.2 *Co–Cr Coated Surface*

The implant with PVD-coated stainless steel implant was investigated with the simulated body fluid following the same procedure. The coated surface has a maximum corrosion rate of 0.00601 mm/year as recorded in Table 3. For the surface, the corrosion potential of the sample was found to be −142.86 mV at a current density of 0.00045 mA/cm^2. These details are derived from the TAFEL plot arrived on Co–Cr-coated sample and plotted as in Fig. 3.

Table 3 Electrochemical data recorded for Co–Cr coated stainless steel implant

Test sample	Ecorr (mV)	Icorr (mA/cm^2)	Corrosion rate (mm/year)
CoCr coated surface	−142.869	0.0004515	0.0060138

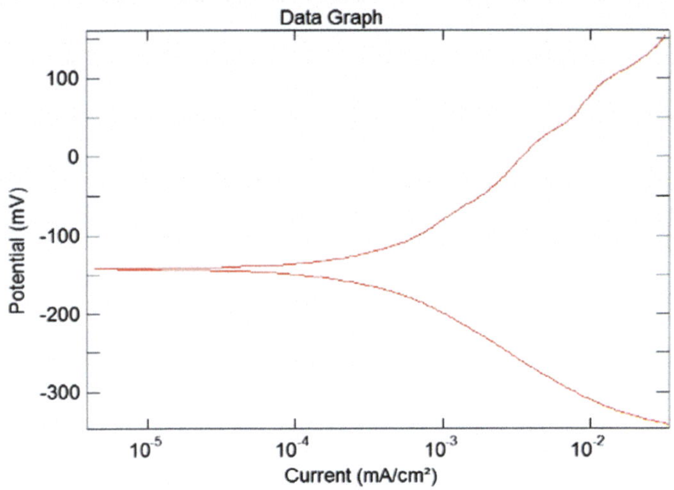

Fig. 3 TAFEL plot recorded for Co–Cr-coated stainless steel implant

3.3 Laser Peened Surface

Metallurgically polished implant was processed with short pulse laser peening process. The exposed surface has a corrosion rate of 0.00354 mm/yr. It has been recorded for a corrosion potential of −245.99 mV with a current density of 0.00026 mA/cm^2. The outcome of the experiment is given in Table 4, and the TAFEL plot as given in Fig. 4.

Table 4 Electrochemical data recorded for short pulsed laser peened stainless steel implant

Test sample	Ecorr (mV)	Icorr (mA/cm^2)	Corrosion rate (mm/year)
Laser peened surface	−245.99	0.0002662	0.0035451

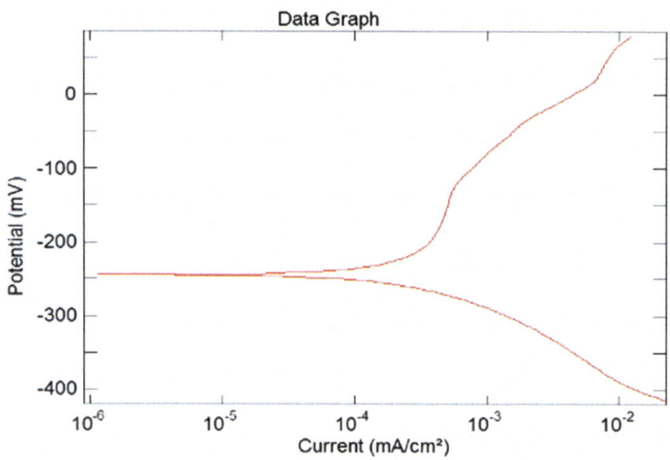

Fig. 4 TAFEL plot recorded for laser peened stainless steel implant

3.4 Plain Stainless Steel Implant Material

The plain implant material is processed through a standard metallurgical procedure for electrochemical corrosion studies. The surface was ensured to metallurgically polished and defect free for investigation. The exposed surface has a corrosion rate of 0.0036 mm/yr. It has been recorded for a corrosion potential of −225.55 mV with a current density of 0.000230 mA/cm^2. The outcome of the experiment is given in Table 5, and the TAFEL plot as given in Fig. 5.

Table 5 Electrochemical data recorded for short pulsed laser peened stainless steel implant

Test sample	Ecorr (mV)	Icorr (mA/cm^2)	Corrosion rate (mm/year)
Plain implant	−225.55	0.000230	0.003665

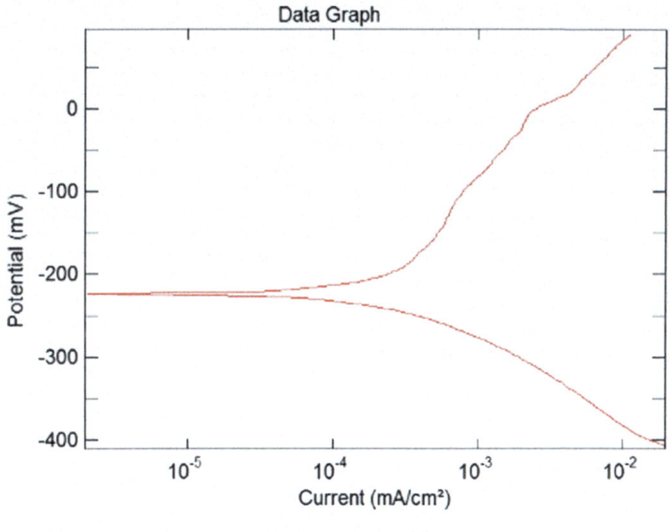

Fig. 5 TAFEL plot recorded for plain stainless steel implant

3.5 SEM Analysis

The exposed surfaces are observed through scanning electron microscope (SEM) at higher magnification (Fig. 6). The WEDM surface has a maximum oxide granule in the form of recast layers. For Co–Cr coated has surface defect due to reaction and found delamination over SBF fluid acceleration. Similarly, the laser peened surface has a spot of thermal effect over exposed surface. It might be forced to protect the material from severe detoxication. Compared to plain steel implant and other processed (WEDM and Co–Cr coated) implant, the surface found better.

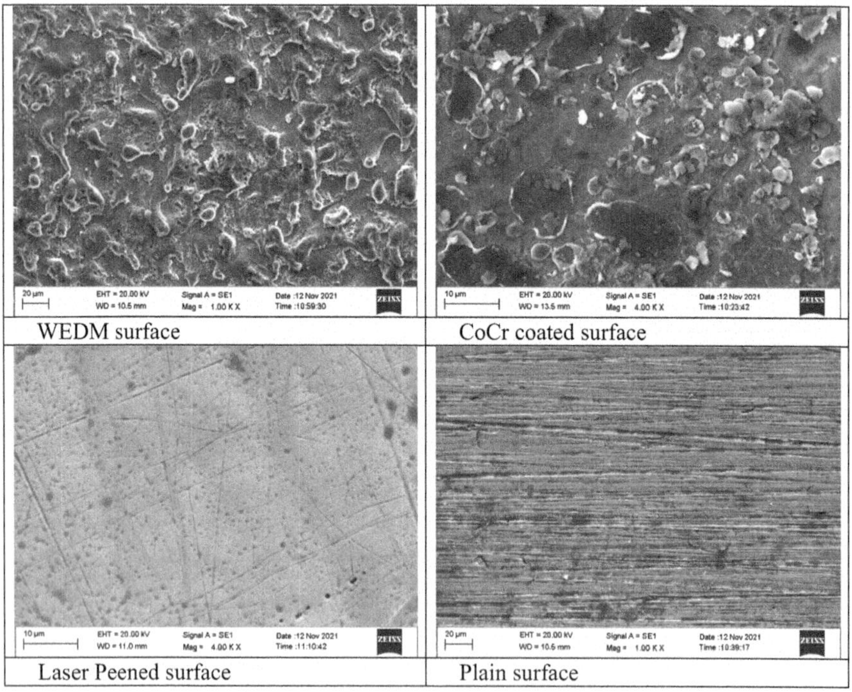

Fig. 6 SEM image of exposed surface

4 Conclusions

From the analysis of austenitic stainless steel under simulated body fluid conditions is deliberated in the discussion. The maximum corrosion is recorded in wire electro discharge machined surface with a rate of 0.014 mm/yr and followed by laser peen (0.0035 mm/yr), plain steel (0.0036 mm/yr), and Co–Cr-coated material is 0.006 mm/yr. The minimum potential is −245.99 mV for laser peened sample and maximum of −109.2 mV for WEDM sample. However, the surface defect found maximum degraded with plain stainless steel surface and found better with WEDM and PVD-coated Co–Cr sample. In a final result, the laser peened surface was superior and good with proposed corrosion studies. As a result, when the surface properties are not compromised, the machined surface can be utilized to replace human implants.

References

1. Alvarado J, Maldonado R, Marxuach J, Otero R (2003) Appl Eng Mech Med 1
2. Sivakumar S, Adam Khan M, Senapathy GJ (2020) Biocompatibility and surface studies on electro spark machined titanium based human implants. J Bio Tribo Corros 6:1–11
3. Sivakumar S, Adam Khan M, Muralidharan B (2020) Studies on surface quality of stainless-steel implant material while machining with WEDM process. Int J Mach Mach Mater. https://doi.org/10.1504/IJMMM.2020.10027931
4. Yazdimamaghani M, Razavi M, Vashaee D, Tayebi L (2015) Surface modification of biodegradable porous mg bone scaffold using polycaprolactone/bioactive glass composite. Mater Sci Eng, C 49:436–444
5. Elias CN, Lima JHC, da Silva MP, Muller CA (2002) Titanium dental implants with different morphologies. Surf Eng 18(1):46–49
6. Dahm KL, Anderson IA, Dearnley PA (1995) Hard coatings for orthopedic implants. Surf Eng 11(2):138–144
7. Mozafari M, Salahinejad E, Sharifi-Asl S, Macdonald DD, Vashaee D, Tayebi L (2014) Innovative surface modification of orthopaedic implants with positive effects on wettability and in vitro anti-corrosion performance. Surf Eng 30(9):688–692
8. Sivakumar S, Adam Khan M, Muralidharan B (2018) Processing of titanium based human implant material using Wire EDM. Mater Manuf Processes 34(6):695–700. https://doi.org/10.1080/10426914.2019.1566609
9. Adam Khan M, Ram Prasad N, Navaneetha Krishnan S, Karthic Raja S, Winowlin Jappes JT, Duraiselvam M (2017) Laser treated austenitic steel and nickel alloy for human implants. Mater Manuf Processes 32(14):1635–1641. https://doi.org/10.1080/10426914.2017.1364746
10. Kokubo T, Ito S, Huang ZT, Hayashi T, Sakka S, Kitsugi T, Yamamuro T, Ca P (1990) J Biomed Mater Res 24
11. Khan MA (2015) Electrochemical polarisation studies on plasma-sprayed nickel-based superalloy. Appl Phys A 120

Experimental Characterization of CNSL-Epoxy Resin Reinforce Natural Fiber

R. Ganapathy Srinivasan, C. Rajaravi, S. Palani, and S. Karthik

Abstract To increase the green content of the matrix material, the investigation is deals with polyester—CNSL resin with jute fiber composite reinforcement. The fiber is used in a long continuous form and has a standard diameter of 2 mm and a length of 20 mm. To improve the adhesiveness and strength of the fibers, different concentrations of alkaline sodium hydroxide (NaOH) are applied to the fibers for varying periods. The hand layup technique is used to prepare composites. The variety of CNSL and epoxy resin concentrations are used. Reinforced fiber composites had the superior mechanical characterization, while composites with the limited porosity had the superior material characterization. Tensile, impact, and flexural tests were used to determine how the properties of jute changed with and without CSNL, as well as with different concentrations of epoxy and CNSL.

Keywords Jute fiber · Cashew nut shell liquid · Synthetic resin · Alkaline sodium hydroxide

1 Introduction

Natural lignocellulose fibers (NLFs) have gained popularity as standard polymer composite material reinforcement in recent decades, replacing non-recyclable synthetic and energy-intensive fibers [1–12]. Of course, NLF-reinforced compounds are investigated for relevance in food packaging [13, 14], civil construction [15],

R. G. Srinivasan (✉) · S. Palani
Department of Mechanical Engineering, Vel Tech Multitech Dr. Rangarajan Dr. Sakunthala Engineering College, Avadi, Chennai, Tamil Nadu, India
e-mail: krgsrini@gmail.com

C. Rajaravi
Department of Mechanical Engineering, Hindusthan College of Engineering and Technology, Coimbatore, Tamil Nadu, India

S. Karthik
Department of Mechanical Engineering, Vel Tech Rangarajan Dr, Sagunthala R&D Institute of Science and Technology, Avadi, Chennai, Tamil Nadu, India

E. Natarajan et al. (eds.), *Materials, Design and Manufacturing for Sustainable Environment*, Lecture Notes in Mechanical Engineering,
https://doi.org/10.1007/978-981-19-3053-9_6

ballistic armor [16, 17], and vehicle parts [18, 19]. Nanocellulose fibrils derived from NLFs have been experimental as a potential reinforcement for proposed bio-nanocomposites with invention of biodegradable plastic films [20, 21], as well as the special properties for medical applications [22–24]. Hemp fiber, a well-known NLF, has been comprehensively reported as a reinforcement of the polymer compound in several researches over the decades and is cited in the majority of research articles [1–12]. Hemp fiber/polypropylene is used as an automatic component [19]. Jute fiber's extraordinary material characterization allows the use of strong thermoset polymeric metrics such as ballistic shield reinforcement-epoxy and polyester [25]. This ballistic application in hemp fiber compounds has yet to be investigated [17].

As a result, the goal of this study is to evaluate the thematic characterizes estimated by curvature and tensile properties of epoxy and polyester composites reinforced with different sizes of hemp fibers for the first time. This will allow for a definition of the compound most suitable for use in a multi-layer armor system based on the matrix and composite volume area. Laminates were composited by the hand layup method. The laminates are done by using epoxy resin and CNSL a natural resin as matrix material. The composites are prepared by varying the concentration of CNSL in epoxy resin concentration and natural fiber. The fabricated component is planned to induce and evaluate mechanical material characterization such as tensile strength, flexural strength, and impact strength. Hand layup was applied to create the laminates. Epoxy resin and CNSL, a natural resin, are produced as matrix materials in the laminates. The composites are made by changing the concentrations of CNSL in epoxy resin and natural fiber. The manufactured component is intended to be induced and evaluated in terms of mechanical qualities like that flexural strength, impact strength, and tensile strength. There are environmental benefits to extracting CNSL from cashew tree fruits nut as a thermosetting resins feedstock. Cashew nut shell liquid is manufacture through processing industry of cashew nut and is used to make resins. The cashew tree produces the cashew nuts and is extensively planted in tropical and subtropical nations' coastal areas. In the previous article [26], we described the major cashew-growing nations and the yearly output of CNSL.

2 Experimental Method

To develop composite material properties, the jute fibers properties were improved using the NaOH treatment method. Fiber contraction during this process has a considerable impact on structure of materials also on mechanical characterization like that modulus of rigidity and tensile properties. The treatment of NaOH of jute yarns (20 min at 25 degrees Celsius in a 25% NaOH solution for 20 min) resulted in a 120 and 150% enhance in tensile characterization and modulus of rigidity. These mechanical property changes are caused by modifying the fiber structure, specifically the crystallization rate, degree of polymerization, and orientation. Chemical binding agents, in general, are particles that activate two occupations. The initial interacts with the cellulose hydroxyl groups, while the next interacts with the matrix's efficient

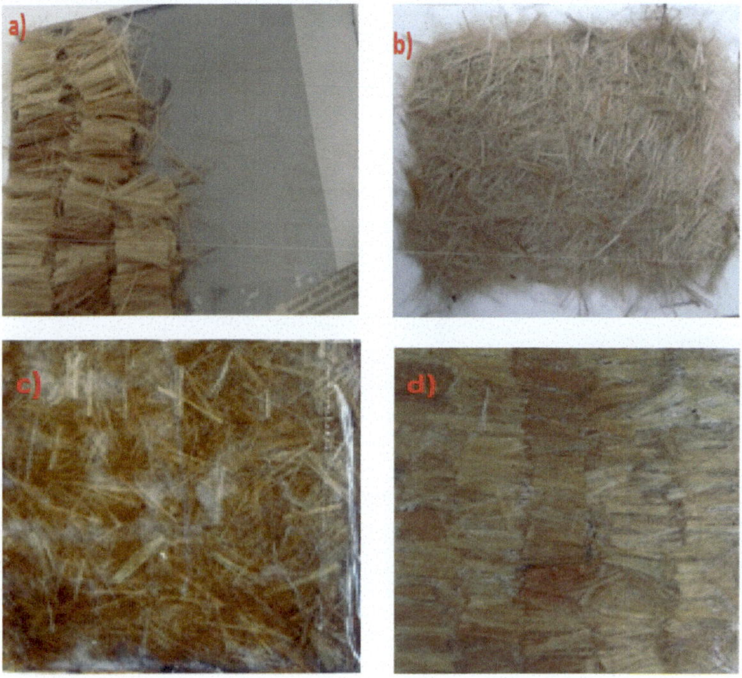

Fig. 1 Sample preparation **a, b** jute fiber, **c** jute with epoxy and CNSL, **d** jute with epoxy

groups. Natural fibers were chemically modified to improve adhesion to the polymer matrix (Figs. 1 and 2).

The proposed jute fibers (200 g) were additional to 10 mL/L of 30% pure H_2O_2 at atmospheric temperature and agitated for 3/4 h before being rinsed with distilled water and dried in a 500 °C oven in anticipation of it achieved a consistent influence. Prior to treatment, the semi-retted fibers be immersed in water for 60 min, dried at 500 °C till constant weight, softly disentangle, and cut to 3 mm length. The process of mercerization involved immersing 200 g of jute fibers in 5% (5 g of sodium hydroxide in 100 ml of water), 10%, and 15% sodium hydroxide solutions for 6 h at 700 °C temperature with stirring and shaking, followed by neutralization with 50% acetic acid to eradicate any absorbed alkali and ultimately thorough washing. The foam sheet is used to create the mold. The dimensions of the foam sheet are 200 × 150 × 5 (mm). Jute fibers were cut to a length of 20 mm and a diameter of 2 mm. It is then aligned horizontally, vertically, and randomly. Tensile specimens are prepared as ASTM D3039 flexural specimens are prepared as ASTM D90 standards, and impact specimens are prepared as ASTM D256 standards (Table 1).

Fig. 2 a 100% epoxy natural fiber, **b** 95% epoxy–5% CNSL natural fiber, **c** 85% epoxy–15% CNSL natural fiber, **d** 75% epoxy–25% CNSL natural fiber

Table 1 Concentration of epoxy and CNSL

% of CNSL in epoxy	Epoxy concentration	Jute fiber %	Resin %
0	100	30	70
5	95	30	70
15	85	30	70
25	75	30	70

3 Result and Discussion

3.1 Tensile Characterization

The specimen is prepared by changing the proportions of CNSL resin and epoxy resin in the mixture. Then, they are machined to meet ASTM specifications of 165*25*5 mm in size. Figure 3 shows the tensile test results obtained according to the ASTM D3039 standard. The data are analyzed to see if the CNSL composition is significant. The value obtained shows that changing the CSNL composition greatly raised the tensile test of composite. As a consequence, 100% epoxy has a tensile strength of roughly 590 N, whereas 5% CNSL composite attributes are as follows: 15% CNSL composite has a tensile strength of 720 N, and 25% composite

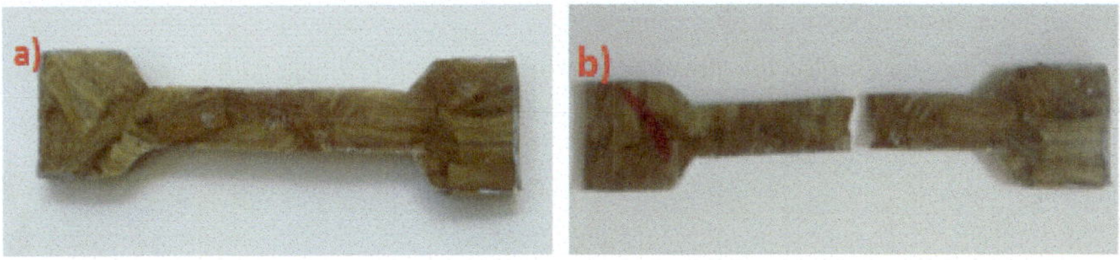

Fig. 3 **a** Tensile specimen before experimentation, **b** tensile specimen after experimentation

has a tensile strength of 810 N. According to the findings, increasing the weight % of CNSL resin enhances tensile strength (Figs. 4, 5; Table 2).

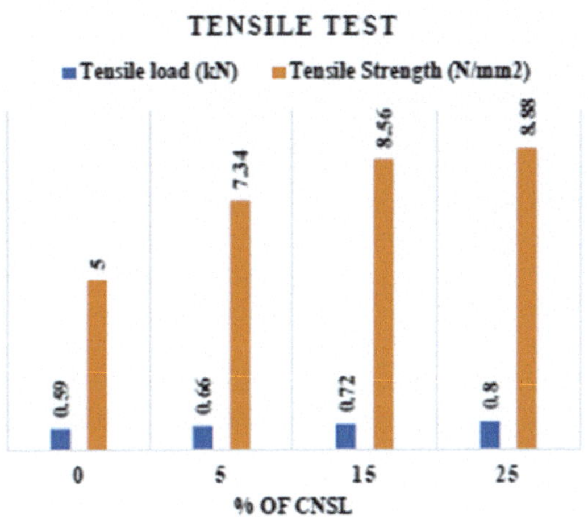

Fig. 4 Tensile test graph with random and combined horizontal and vertical manner

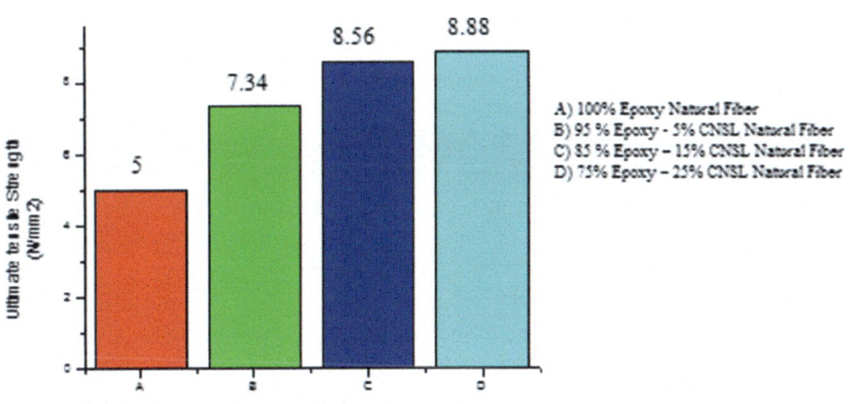

Fig. 5 Tensile characterizations of specimens

Table 2 Experimentation on tensile specimen

Test parameters	Jute fiber with epoxy (horizontal)	Jute fiber with epoxy (random)	Jute fiber with epoxy and 5% CNSL	Jute fiber with epoxy and 15% CNSL	Jute fiber with epoxy and 25% CNSL
Gage width (mm)	17.62	15.21	17.00	17.35	18.23
Gage thickness (mm)	5.20	4.90	5.00	5.85	5.68
Ultimate tensile load (KN)	0.48	0.59	0.66	0.72	0.80
Ultimate tensile strength (N/mm^2)	6.00	5.00	7.34	8.56	8.88

Furthermore, when the tensile strength of random-oriented and horizontal and vertical-oriented composites is compared, random oriented composite has a higher tensile strength of approximately 590 N compared to 480 N. As a result, random-oriented composite has greater tensile strength. Furthermore, when the tensile strength of random-oriented and horizontal and vertical-oriented composites is compared, random oriented composite has a higher tensile strength of approximately 590 N compared to 480 N. As a result, random-oriented composite has greater tensile strength.

3.2 *Impact Characterization*

Charpy test was used to conduct the impact test. The ASTM dimensions for Charpy test specimens were 65 × 15 × 5 mm, and the results were recorded. The impact test was performed with the impact values of both natural fiber composites, and it was discovered to be virtually identical. The specimen is made by varying the composition of CNSL resin combined with epoxy resin. The value is examined in order to determine the significance of CNSL composition. The value obtained demonstrates that the impact test of composite has fluctuated by varying the CSNL composition. The energy absorbed by a random orientation 100% epoxy specimen is 4 J, which is similar to the energy absorbed by a 5% CNSL composite specimen, whereas the energy absorbed by a 15% CNSL composite was found to be increased by about 3 J, followed by a 2.58 J decrease in energy absorbed by a 25% CNSL composite. It is discovered that changing the composition of CNSL causes the energy absorbed by the specimen to fluctuate, i.e., increases and decreases as the composition of CNSL blended epoxy hybrid resin are changed (Figs. 6 and 7).

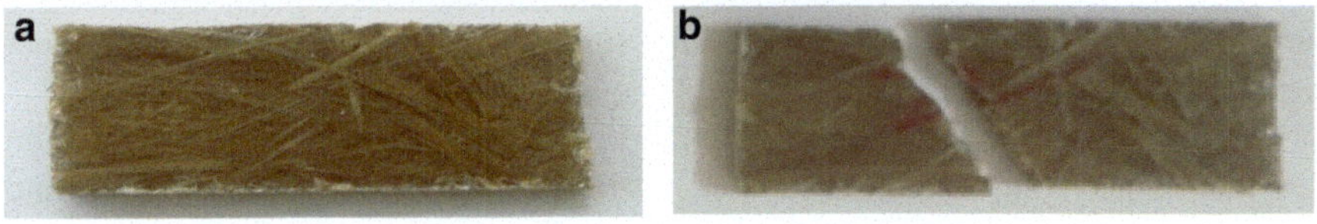

Fig. 6 **a** Impact specimen before experimentation, **b** impact specimen after experimentation

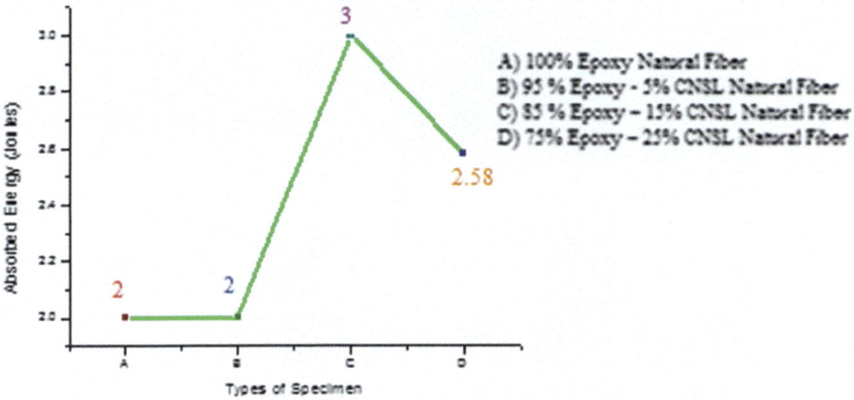

Fig. 7 Impact characterizations of specimens

In addition, when the energy absorbed by horizontal and vertical oriented specimens is compared to random manner specimens, the energy absorbed by horizontal and vertical oriented specimens is greater than random manner specimens. Because fiber content arranged vertically absorbs more energy than fiber content arranged randomly. The specimen energy randomly absorbed is approximately 4 J, which is twice the energy absorbed by the random manner specimen. The random manner specimen energy absorbed is approximately 2 J.

3.3 *Flexural Characterization*

The flexural strength changes for all four CNSL composite variants are depicted. When the CNSL composition is increased, the flexural strength increases. The obtained value demonstrates that by varying the CSNL composition, the tensile test of composite has been significantly increased. As a result, 100% epoxy has a flexural strength of about 30.5 N/mm^2, while 5% CNSL composite has a higher flexural strength of about 32.6 N/mm^2. By comparing the specimens, 5% CNSL composite has low-flexural properties, while 15% CNSL composite has 36.9 N/mm^2 and 25% composite has 39.8 N/mm^2. According to the observations, increasing the weight percentage of CNSL resin increases flexural strength. The obtained results show that the mechanical properties have improved (Figs. 8, 9, and 10).

Furthermore, when the flexural strength of random oriented and horizontal and vertical oriented composites is compared, horizontal and vertical-oriented composite

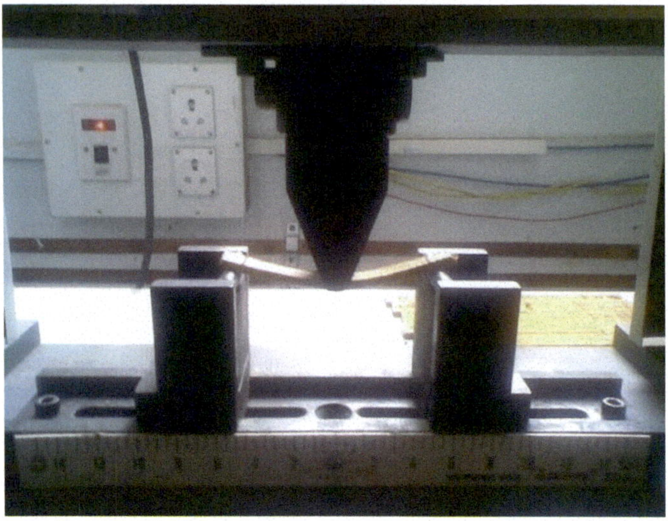

Fig. 8 Flexural testing setup

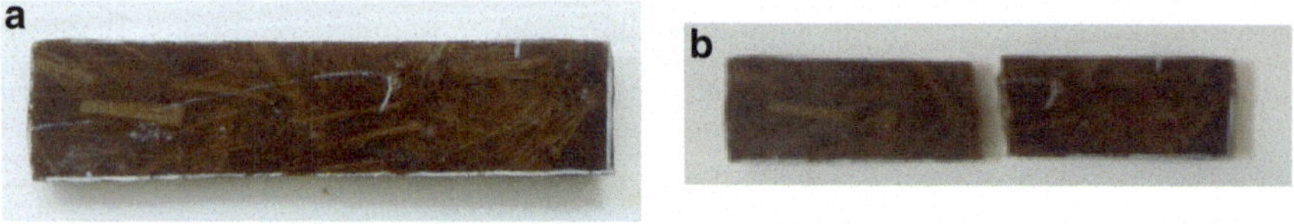

Fig. 9 a Flexural specimen before experimentation. **b** Flexural specimen after experimentation

Fig. 10 Impact characterizations of specimens

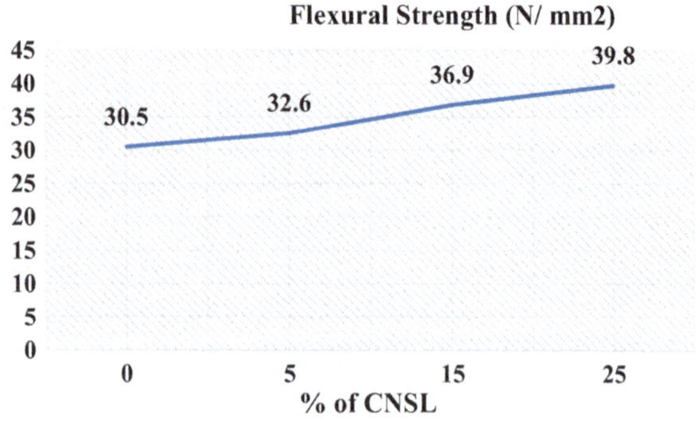

have a higher flexural strength of about 37 N/mm^2 compared to 30 N/mm^2. As a result, horizontal and vertical-oriented composites have higher flexural strength when compared to CNSL resin composites of various compositions. The obtained results are consistent with those obtained recently by other researchers in a similar field of study. The reason for this is an increase in the adhesiveness and compactness between the matrix and the fibers, as well as a decrease in porosity.

4 Conclusions

The mechanical characterization of jute natural fire is investigated by varying the composition of CNSL blended epoxy resin. On various specimens, mechanical characterization such as tensile, flexural, and impact strength is determined. The specimens are 5% CNSL, 15% CNSL, and 25% CNSL hybrid composites that are tested using standard ASTM dimensions.

- This work demonstrates the successful fabrication of jute fiber-reinforced epoxy and CNSL hybrid composites by varying the composition of CNSL for different mechanical property observations.
- It has been exposed that the mechanical characterization of composites, such as flexural strength, tensile strength, and impact strength, is significantly influenced by fiber treatment with different CNSL compositions.
- This research reveals that increasing the CNSL percentage blended epoxy improves the composite flexural strength.
- As the composition of the specimen's CNSL changes, the impact energy absorbed by the specimen varies.

References

1. Monteiro SN, Lopes FP, Barbosa AP, Bevitori AB, Silva IL, Costa LL (2011) Natural lignocellulosic fibers as engineering materials—an overview. Metall Mater Trans A 42(10):2963–2974
2. Faruk O, Bledzki AK, Fink HP, Sain M (2012) Biocomposites reinforced with natural fibers: 2000–2010. Prog Polym Sci 37(11):1552–1596
3. Shah DU (2013) Developing plant fibre composites for structural applications by optimising composite parameters: a critical review. J Mater Sci 48(18):6083–107
4. Mohamed L, Asari MN, Pua G, Jawaid M, Islan MS (2015) A review in natural fiber reinforced polymer composites and its applications. Int J Polym Sci 243947
5. Pickering KL, Efendy MA, Le TM (2016) A review of recent developments in natural fibre composites and their mechanical performance. Compos A Appl Sci Manuf 1(83):98–112
6. Monteiro SN, Lopes FP, Ferreira AS (2009) Nascimento DC. Natural-fiber polymer-matrix composites: cheaper, tougher, and environmentally friendly. Jom 61(1):17–22
7. Ku H, Wang H, Pattarachaiyakoop N, Trada M (2011) A review on the tensile properties of natural fiber reinforced polymer composites. Compos Part B Eng 1:42(4):856–873

8. Muñoz Dominguez E (2016) Fabricación y Caracterización de Green Composites con Bioresina y Tejido de Fibra Natural de Lino mediante Moldeo por Transferencia de Resina (Doctoral dissertation, Universitat Politècnica de València)
9. Thakur VK, Singha AS, Thakur MK (2013) Natural cellulosic polymers as potential reinforcement in composites: physicochemical and mechanical studies. Adv Polym Technol 32(S1):E427–E435
10. Faruk O, Bledzki AK, Fink HP, Sain M (2014) Progress report on natural fiber reinforced composites. Macromol Mater Eng 299(1):9–26
11. Güven O, Monteiro SN, Moura EA, Drelich JW (2016) Re-emerging field of lignocellulosic fiber–polymer composites and ionizing radiation technology in their formulation. Polym Rev 56(4):702–736
12. Sanjay MR, Madhu P, Jawaid M, Senthamaraikannan P, Senthil S, Pradeep S (2018) Characterization and properties of natural fiber polymer composites: a comprehensive review. J Clean Prod 172:566–581
13. Sanyang ML, Ilyas RA, Sapuan SM, Jumaidin R (2018) Sugar palm starch-based composites for packaging applications. In: Bionanocomposites for packaging applications 125–147
14. Ilyas RA, Sapuan SM, Ibrahim R, Atikah MS, Atiqah A, Ansari MN, Norrrahim MN (2019) Production, processes and modification of nanocrystalline cellulose from agro-waste: a review. Nanocrystalline Mater 12:32
15. Islam MS, Ahmed SJ (2018) Influence of jute fiber on concrete properties. Constr Build Mater 189:768–776
16. Naveen J, Jawaid M, Zainudin ES, Sultan MT, Yahaya R (2018) Evaluation of ballistic performance of hybrid Kevlar/cocosnucifera sheath reinforced epoxy composites. J Text Inst 110:1179–1189
17. Benzait Z, Trabzon L (2018) A review of recent research on materials used in polymer–matrix composites for body armor application. J Compos Mater 52(23):3241–3263
18. Atiqah A, Jawaid M, Sapuan SM, Ishak MR, Ansari MNM, Ilyas RA (2019) Physical and thermal properties of treated sugar palm/glass fibre reinforced thermoplastic polyurethane hybrid composites. J Mater Res Technol 8(5):3726–3732
19. Holbery J, Houston D (2006) Natural-fiber-reinforced polymer composites in automotive applications. Jom 58(11):80–86
20. Halimatul MJ, Sapuan SM, Jawaid M, Ishak MR, Ilyas RA (2019) Water absorption and water solubility properties of sago starch biopolymer composite films filled with sugar palm particles. Polimery 64(9):595–603
21. Halimatul MJ, Sapuan SM, Jawaid M, Ishak MR, Ilyas RA (2019) Effect of sago starch and plasticizer content on the properties of thermoplastic films: mechanical testing and cyclic soaking-drying. Polimery 64
22. Ilyas RA, Sapuan SM, Ishak MR, Zainudin ES (2018) Water transport properties of bio-nanocomposites reinforced by sugar palm (*Arenga pinnata*) nanofibrillated cellulose. J Adv Res Fluid Mech Therm Sci 51(2):234–46
23. Abral H, Ariksa J, Mahardika M, Handayani D, Aminah I, Sandrawati N, Pratama AB, Fajri N, Sapuan SM, Ilyas RA (2020) Transparent and antimicrobial cellulose film from ginger nanofiber. Food Hydrocoll 98:105266
24. Ilyas RA, Sapuan SM, Ishak MR, Zainudin ES (2018) Sugar palm nanocrystalline cellulose reinforced sugar palm starch composite: degradation and water-barrier properties. IOP Conf Ser Mater Sci Eng 368(1):012006
25. Pereira AC, Assis FS, Garcia FD, Oliveira MS, Lima ES, Lopera HA, Monteiro SN (2019) Evaluation of the projectile's loss of energy in polyester composite reinforced with fique fiber and fabric. Mater Res 21:22
26. Kini UA, Nayak SY, Shenoy Heckadka S, Thomas LG, Adarsh SP, Gupta S (2018) Borassus and tamarind fruit fibers as reinforcement in cashew nut shell liquid-epoxy composites. J Nat Fibers 15(2):204–218

Development of an Oil Palm Basal Stem Rot Disease Detection Model Via Machine Vision with Optimized Inception-Based Convolutional Neural Network

S. H. Wan, J. C. E. Yong, E. H. Y. Leong, and J. Y. Chan

Abstract The oil palm sector plays an important role in the economic model of Malaysia. As plantations expanded, sustainability becomes critical to ensure the fields pass the relevant environmental, social and economic benchmarks. Oil palm trees are susceptible to basal stem rot disease (BSR) caused by the pathogenic fungus *Ganoderma boninense*. BSR is a fatal and contagious disease which causes a sharp decline in oil palm yields and significantly threatens the production and economic performance of oil palm plantations. Early detection is key to limiting the damage of a BSR infection. Traditional manual detection of BSR is prone to human error and inconsistencies. In response, a number of methods using satellite imagery, spectral imaging and thermal imaging had been introduced with varying degree of success. Alternatively, machine vision (MV) with image processing and feature extraction by deep learning algorithms had been proposed with improved accuracy and feasibility. This paper reviews information on BSR including its stages of infection, symptoms and diagnosis, as well as existing MV models for disease detection in palm-related plants. Additionally, this paper also details the methodology for development of a new MV model for BSR detection based on the Inception convolutional neural network deep learning algorithm.

Keywords Oil palm · Basal stem rot · Disease detection · Machine vision · Convolutional neural network

1 Introduction

Since 2005, palm oil has become the leading commodity in the global oils and fats market. Malaysia, the world's second largest producer and exporter of palm oil behind Indonesia, accounts for 25.8% of total global production and an export volume of 17.37 million tonnes, equivalent to 34.3% of total palm oil trade [1]. Given the high

S. H. Wan (✉) · J. C. E. Yong · E. H. Y. Leong · J. Y. Chan
Department of Mechanical Engineering, Faculty of Engineering, Technology and Built Environment, UCSI University, Jalan Puncak Menara Gading, 56000 Kuala Lumpur, Malaysia
e-mail: 1001542745@ucsiuniversity.edu.my

E. Natarajan et al. (eds.), *Materials, Design and Manufacturing for Sustainable Environment*, Lecture Notes in Mechanical Engineering,
https://doi.org/10.1007/978-981-19-3053-9_7

revenue potential, oil palm plantations in Malaysia experienced a net expansion of 146.6% between 2001 and 2016, growing from 2.59 million hectares to 6.39 million hectares [2]. However, the Malaysian palm oil industry had long been plagued by basal stem rot disease (BSR) originating from the pathogenic fungus *Ganoderma boninense*. BSR is the main cause for premature losses of young and mature oil palm trees in plantations across Malaysia. BSR is incurable, shortening the productive lives of oil palm trees and propagates through the soil undetected. The fatal nature of BSR had led to research in various early detection techniques to limit its spread and subsequent damages inflicted.

This paper is divided into five sections. Section 2 presents a summary of visual symptoms associated with different infection stages of BSR at various oil palm tree sections. In Sect. 3, various techniques developed in previous studies for BSR detection and classification are reviewed. The limitations of the traditional manual inspection method are highlighted, and the concepts of various methods attempting to overcome these limitations are explored. The performance of each method in terms of overall classification accuracy is also presented. In Sect. 4, the methodology for development of a novel machine vision-based deep learning detection model is covered in sequence, from image acquisition to image preprocessing to feature extraction to a new Inception-based neural network algorithm to its training and testing. In Sect. 5, gaps in current BSR research literature are highlighted, and the advantages of the proposed MV model over existing methods are discussed in terms of ease of use, accessibility of data and potential for commercialization.

2 BSR Symptoms

2.1 BSR Stages

BSR is classified based on the severity of infection on each individual oil palm tree. Most studies divided BSR severity into four stages: healthy, mild, moderate and severe. Table 1 gives each stage of infection, the visual symptoms attributed to their classifications and the corresponding images. BSR infections are generally symptomless or display very mild symptoms during early stages. In contrast, symptoms only appear when the plant is moderately or severely infected, where in most cases, it is unable to be saved. BSR impacts oil palm trees of all ages. Young trees with BSR will die within six months, whereas mature trees will die within 12 to 36 months from onset of infection. The fresh fruit bunch production of a tree decreases by 26% under a BSR infection rate of 30%. This rises to 45% as the infection rate reaches 60% [3].

Table 1 Stages of BSR infection according to severity levels

BSR stage	I—healthy	II—mild infection	III—moderate infection	IV—severe infection
Image of tree				
Visual symptoms on stem bark	• No symptoms	• Presence of fungal fruiting bodies (*Basidiomycota* mushroom) • Presence of mycelium • Brittle wood	• Presence of fungal fruiting bodies • Presence of mycelium • Rot onset	• Well-developed fungal fruiting bodies • Rotten stem
Visual symptoms on leaves/foliage	• Healthy leaves	• Yellowing or drying of some leaves	• Necrosis (tissue death) • Declination of older leaves • Collapse of lower leaves	• Widespread necrosis • Total leaf declination • Absence of new leaves
Visual symptoms on canopy structure	• Normal palm canopy	• One or two unopened spears	• Three to five unopened spears	• "Skirt-like" crown shape

3 BSR Detection

3.1 Manual Inspection

Traditionally, BSR was identified by visually inspecting tree-by-tree for foliar symptoms, changes in canopy structure and presence of fungus fruiting bodies (*Basidiomycota* mushroom) on the stem bark. This process was manually carried out by plantation workers such as loose-fruit collectors and sprayers. The recommended interval for BSR status updating was three to six monthly censuses to prevent the spread of disease. However, this method was error-prone due to its reliance on workers' knowledge and experience. The intensity of the job also impacted its final accuracy as workers were fatigued and inconsistent in judgment. Upon detecting a BSR case, workers must isolate the diseased tree by creating a 4 m × 4 m × 75 cm deep isolation trench around it. The soil from the trench would be heaped onto the tree's base to protect the weakened bole from wind and extend its productive life. The impact on fresh fruit bunch yield could be minimized if the infection level is kept below 15% through frequent censuses and quick isolations. The infected bole and root tissues were completely removed during replanting. Further soil treatment involved the creation of a sanitation pit measuring at least 2 m × 2 m × 1 m deep. Each case must also be recorded as part of the Environmental Management Plans (EMPs) imposed on all oil palm plantation owners by the Environmental Impact Assessment (EIA) procedure in Malaysia [4].

3.2 Biochemical Method

A series of biochemical methods were proposed for BSR detection. Polymerase chain reaction (PCR) was used in testing for early detection of BSR [5]. However, PCR is impractical for large-scale field assessments due to its high time cost and limited precision. Bulk screening of samples using an enzyme-linked immunosorbent assay (ELISA) was proposed [6], but the method was found to be susceptible to cross-contamination. Generally, methods involving molecular detection of *Ganoderma* were found to be vulnerable to contamination [7]. DNA-based methods such as nano-sensors and DNA microarrays required further verification and field studies regarding the matter of DNA extraction and purification [8]. Meanwhile, ergosterol quantification was used on field samples and analyzed as a presumptive diagnostic method for BSR detection [9]. Headspace solid-phase micro-extraction (HS-SPME) with gas chromatography-mass spectrometry (GC–MS) was proposed to evaluate the volatile organic compounds (VOCs) released by *Ganoderma boninense* [10]. A significant downregulation in enolase, fructokinase and ATP synthase coupled with a significant upregulation in cysteine synthase and malate dehydrogenase was noted after a BSR infection [11]. The pathogenicity of *Ganoderma boninense* was reported to be associated with a suppression in gene expression of a putative glucan exohydrolase in the root of oil palm seedlings [12]. However, the use of arbitrary cut-off values in the consideration for a gene to be differentially expressed had been replaced with more robust approaches, such as the combined use of fold change and significance testing [13].

3.3 Odor Sampling

A handheld electronic nose sensor system was developed to differentiate healthy and BSR infected oil palm trees by discriminating the odor chemical fingerprint profile of *Ganoderma boninense* and ambient air [14]. Sensors considered include conducting polymer (CP), metal-oxide semiconductor (MOS), quartz crystal microbalance (QCM) and surface acoustic wave (SAW). During data processing, multivariate statistical analysis such as Principal Component Analysis (PCA), Hierarchical Cluster Analysis (HCA) and Linear Discriminant Analysis (LDA) were used to select ten highest contributing sensors in terms of sensitivity, selectivity, stability, robustness, compactness and cost. Table 2 gives the ten selected sensors with their respective target gas or physical quantity measured [14].

An E-nose containing eight MOS gas sensors, a sampling system, a data acquisition unit (DAQ) and a signal processing framework was developed as shown in Fig. 1 [15]. The sampling system was made up of two electronic valves to control airflow from the sample chamber and from ambient atmosphere. The DAQ comprised of 16-bit analog-to-digital converter (ADC) in an Arduino Mega microcontroller. The laboratory assays included pure *Ganoderma* culture compared to other soil fungi (*Trichoderma*, *Aspergillus* and *Omphalina*), and BSR-infected oil palm plants at four infection levels: healthy, early, moderate and severe. The measurements were taken in an environment with temperature 21 ± 2 °C and relative humidity 85 ± 10% [15].

Table 2 The ten selected sensors [14]

Type	Target gas or physical quantity
TGS2201	Gasoline and diesel engine
TGS2600	Air contaminants
TGS826	Ammonia
TGS4161	Carbon dioxide
TGS2620	Solvent vapor
TGS825	Hydrogen sulfide
TGS2611	Methane
TGS2610	Liquefied petroleum gas/propane
TGS2180	Water vapor
SHT77	Temperature and humidity

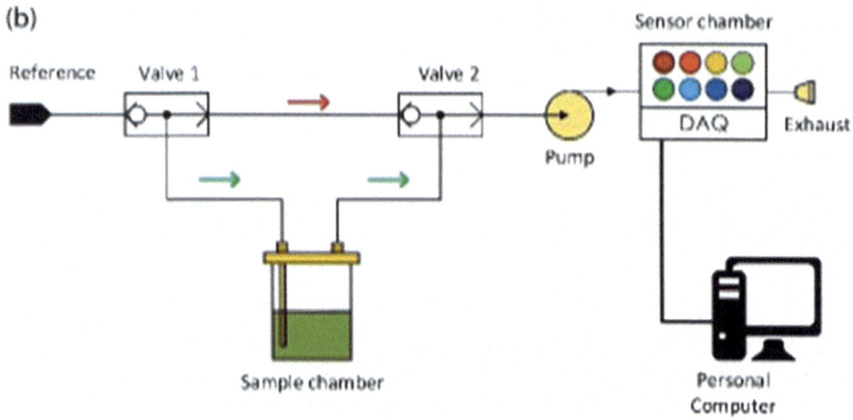

Fig. 1 Schematic diagram of the E-nose device [15]

PCA was used to reduce the dimensionality of dataset to provide a preliminary assessment on E-nose performance. The dataset was normalized through scaling and centering steps. Further, evaluation was conducted using four supervised statistical methods including LDA, Quadratic Discriminant Analysis (QDA) and Support Vector Machine (SVM) with linear and radial kernels. The statistical analysis was conducted using R software, MASS and kernel libraries. Overall, tests for *Ganoderma* mycelium using the unsupervised PCA chemometric system returned significant differences compared with other fungi (*Trichoderma*, *Aspergillus* and *Omphalina*). LDA results show that the pathogenic *Ganoderma* isolates ("Rejosari" and CSB) were clustered largely within the same quadrant, away from the non-pathogenic *Ganoderma lucidum*. SVM was able to differentiate the aroma between the three isolates with 99.64% accuracy. This indicated the ability of the E-nose to differentiate *Ganoderma* from other non-*Ganoderma* fungi. The E-nose also differentiated the four infection levels at the stem, leaves, roots and soil with 97.1%, 89.62%, 90.95% and 87.5% accuracy, respectively [15].

3.4 Hyperspectral Imaging

Hyperspectral imaging uses a field spectroradiometer or scaffoldings to measure the hyperspectral reflectance data from a plant section for disease detection. This technique is based on obtaining the spectrum for every pixel on an image. Hyperspectral imaging differs from multispectral imaging in that it measures across continuous spectral bands rather than spaced spectral bands. Partial least squares discriminant analysis (PLS-DA) was used with preprocessed hyperspectral reflectance data from the canopy of oil palm trees to differentiate several stages of *Ganoderma* fungal infection [16]. The model achieved 98% accuracy when discriminating healthy trees from infected trees, and 94% accuracy when discriminating a four-level disease topology between healthy, light, medium and severe attack. Artificial neural network (ANN) was introduced on raw, first and second derivative spectroradiometer datasets for early detection of *Ganoderma* infection [17]. Compared to previous regression methods, ANN was able to produce more consistent classifications and allows consideration of higher-order interactions among multiple factors using its shared hidden network architecture. The ANN model worked best in the green wavelength of the visible range (540–550 nm), where healthy and mildly infected first-derivative spectral samples were classified with 83.3% and 100% accuracy, respectively.

Time efficiency was improved by automating the detection process using unmanned aerial vehicle (UAV) hyperspectral images (HSI) with multilayer perceptron (MLP) neural network classifier [18]. UAV provided two main advantages: larger scale of detection and simultaneous capture of spatial and spectral information. The UAV captured RGB images for marking ground truth and HSI images covering spectral bands from 510 to 900 nm. The spectral resolution was 1 nm covering 392 bands while raw data resolution was 1024 × 1024 pixel. The samples collected were categorized into Stage A, Stage B, Stage C and Healthy by expert evaluation. The detection model started with a HSI band alignment to counter the misalignment caused by UAV flight instability. This realignment was achieved using Enhanced Correlation Coefficient Maximization [19]. Minimum Noise Fraction (MNF) was used for HSI denoising to generate smoother images without signal degradation. Next, MLP was used for feature extraction from hyperspectral signature. MLP required lower computational cost due to its smaller network size and lesser number of hyperparameters. To reduce overfitting, regularizations with 30% dropout rates were added. The model dataset was generated by randomly selecting pixels from images of all infection stages.

In preliminary testing, MLP achieved 100% accuracy when differentiating infected and healthy plants, and 78.57% accuracy when differentiating Stage A, B and C. However, the model could not differentiate Stage A and B. After removing the dataset of plants with early infection, the model was able to improve its accuracy in differentiating Stage A, B and C to 86.67%, although there seemed to be a lack of spectral signature for Stage A and B. In comparison with vegetation indices, each of the Normalized Difference Vegetation Index (NDVI), Normalized Difference Red Edge (NDRE) and Optimized Soil-Adjusted Vegetation Index (OSAVI) could detect late-stage infection (Stage C) but could not differentiate early-stage infection (Stage A and B) from healthy plants [18].

3.5 Thermal Imaging

Thermal imaging utilizes temperature difference for evaluation and diagnosis. This method works by converting an invisible radiation pattern into a visible form for feature extraction and analysis. For example, the maximum temperature difference (MTD) within a canopy will indicate the healthiness of plant tissues under various environments. Thermal imaging had been used with various classification models on oil palm seedlings for BSR detection at nursery stage [20]. 100 seedlings of 16 months age and two meters height were divided into two categories by Sime Darby experts: healthy and infected. A FLIR E60 handheld thermal camera with temperature range 20–650 °C was used to acquire thermal images of the entire canopy section from three different random angles, each at one meter distance from the seedling. The time of imaging was set from 12 to 2 p.m. Thermal camera parameters were set at 0.98 for emissivity, 20.0 °C for reflected apparent temperature and atmospheric temperature, and 50.0% for relative humidity to ensure consistency.

The thermal images were preprocessed in FLIR Quickreport 1.2, where the temperature scale was standardized to 24–34 °C. This ensured the pixel intensity corresponded to the same temperature representation. The temperature scale was then cropped out in MATLAB. Otsu's thresholding method was used for image segmentation and separating pixels of the seedling from its background. A region of interest (ROI) was created using the segmented image as mask layer. A t-test statistical analysis determined the maximum, minimum and mean pixel intensity values as significant information for extraction from ROI. These information were then fed into the Classification Learner App from Statistics and Machine Learning Toolbox of MATLAB. The classification models used were LDA, QDA, SVM and k-nearest neighbor (KNN). To reduce the dimensionality of dataset, PCA was run on Unscrambler X 10.4. PCA works by projecting each data point onto the first few principal components to obtain lower-dimensional data while retaining data variation. The dataset was split into 70% training and 30% testing. Overall, the SVM model with fine Gaussian kernel was most consistent and achieved the highest accuracy of 80% [20].

3.6 Terrestrial Laser Scanning

Terrestrial laser scanning (TLS) provides accurate information about tree dimensions, morphology and canopy physiology up to the millimeter. TLS was used to study the relationship between oil palm trunk perimeter, diameter at breast height (DBH) and infected canopy area [21]. The differences in crown pixel, frond number and frond angle between healthy and infected oil palm trees were then shown using TLS [22]. Classification models were developed using selected crown strata and top-down view images for BSR detection regardless of infection severity [23]. A FARO laser scanner was used to acquire 3D laser points with repetition of up to 976,000 times per second, yielding a detailed laser point cloud [24]. The laser scanner had a 360° × 305° field of view and was capable of scanning top view of the trees. The point cloud data was stored to a SD card. 40 oil palm trees of 9 years age and between 10 and 11 m height were divided into four categories of severity levels according to Malaysian Palm Oil Board (MPOB) guidelines: T0 for healthy, T1 for mildly infected, T2 for moderately infected and T3 for severely infected. The scanner was mounted on a tripod at 1 m height and 1.5 m distance away from the tree, placed on firm ground and leveled using a bubble balancer. Each tree was scanned from four different locations with 5–8 reference spheres placed around the tree to aid synchronization of laser scan data. All four scans were "registered" in FARO's point cloud processing software, SCENE, to create a complete 3D view of the tree. In AutoCAD, the top-view shape of the frond was drawn using "Polyline" and the distance between each frond in terms of angle was obtained using "Dimension". Crown area was obtained by adding the number of pixels within the crown image. Stratification method was used to segment the point cloud data into a series of horizontal layers (strata) with equal vertical distances according to crown strata density. The five parameters selected for classification models were frond number, frond angle, crown pixel, crown stratum 200 cm from top (C200) and crown stratum 850 cm from top (C850). All five parameters were found to be significant in all infection severity levels with p-values less than 5% in Kruskal-Wallis test run on JMP software. The Pearson correlation calculated by Microsoft Excel between frond angle and frond number was highest at −0.94. It was found that the frond angle increased, while frond number and crown area decreased with increasing infection severity, in agreement with the correlations [24].

3.7 Machine Vision on Detection of Leaf Spots and Blight Spots in Date Palms

The capability of machine vision technology in disease detection of palm plants had been proven in previous studies. In [25], regular images were used with image processing techniques and machine learning to classify leaf spots and blight spots. Leaf spots are characterized by small, and irregular brown to black spots with 3 to 7 mm size on the leaf surface. Blight spots are described as elongated brown to black spots causing cankers on the leaf midrib. The dataset was made up of 35 leaf spot images and 40 blight spot images from Kaggle, plus an additional 50 palm images taken using a Samsung Galaxy A50 mobile camera with 25-megapixel sensor and 26 mm-equivalent f/1.7 lens. The Kaggle images were cropped to an ROI of 224 × 224 pixels to enhance important infection features. Image augmentation techniques such as rotation, flipping and brightness adjustments were used to generate 5250 images for each disease. 80% of the images were allocated for training and 20% for testing. The disruptive noise caused by black backgrounds in some images were solved by converting to a transparent background through the addition of an alpha channel. Convolutional neural network (CNN) classifier was used due to the large dataset size. The CNN model was a VGG16 network with well-established layers and parameters for efficient feature extraction and model learning. Two call backs: Early Stopping and Model Checkpoint were added to layers of the VGG16 model to improve training. The VGG16 model successfully classified leaf spots and blight spots diseases with 97.9% accuracy. The model was also able to detect both diseases correctly from eight regular images taken with different mobile devices at different time of day and distances from the palm tree.

4 Proposal of BSR Detection in Oil Palm by Machine Vision

4.1 Sample Collection and Image Preprocessing

A new MV model is hereby proposed for the detection of BSR in oil palm trees at different stages of infection. The dataset for the proposed MV model will be generated via acquisition of normal images in a FGV oil palm plantation site in Selangor or Perak, Malaysia. Additional details of the image acquisition process are presented in Table 3.

A total of 80 trees of the *Elaeis guineensis* species will be required for this study, where each stage of BSR infection will be represented by 20 trees, respectively. The determination of BSR infection stage will be done according to MPOB experts' evaluation. For each tree, four images will be taken using the Apple iPhone mobile camera from four different random angles, producing 320 images for the entire dataset. The time of imaging will be set from 9 a.m. to 12 p.m. to control the effect of inconsistent natural sunlight intensity on the field. All field images taken will be stored locally in the iPhone prior to transfer to a personal computer via a Lightning-to-USB cable. The entire dataset will be preprocessed before being fed into the MV model for learning purpose. In order to facilitate the extraction of BSR features from

Table 3 Image acquisition details

Sampling parameters	Detail
Oil palm species	*Elaeis guineensis*
Oil palm age	1–20 years old
Oil palm location	FGV plantation in Selangor/Perak
Oil palm section	Fronds (leaves)
Imaging device	Apple iPhone 13 rear wide camera • 12-megapixel sensor • f/1.6 aperture lens • Up to 5 × digital zoom • Red-eye correction • Sensor-shift optical image stabilization • Photo geotagging • HEIF and JPEG image formats
Total number of trees	80 trees
Number of trees per infection stage	20 trees
Number of images per tree	4 images
Number of images per infection stage	80 images (20 trees × 4)
Total number of images	320 images (80 × 4 stages)
Imaging timeslot	9 a.m. to 12 p.m.

normal images, several digital enhancements will be introduced using the Image Processing Toolbox in MATLAB software. A region of interest (ROI) with 256 × 256 pixels size will be created by cropping the raw images to highlight the critical features representative of a certain BSR infection stage [25]. Images containing noisy pixels due to dark backgrounds will be treated with a Gaussian filter to lessen the noise [26]. Additionally, the initial dataset size of 320 images is unlikely to be sufficient for MV model learning to attain a competent level of classification accuracy. Therefore, image augmentation techniques such as shifting, rotation, flipping and brightness adjustment will be used to enlarge the total dataset size to 3,200 images for model training [25]. Each infection stage will be represented by 800 images after augmentation process.

4.2 Feature Extraction

The leaf section of oil palm tree is selected as the main focus in this study. This is to take advantage of the appearance changes manifested on the upper and lower surfaces of the leaves throughout the course of a BSR infection. Visual BSR symptoms such as changes in color, texture and presence of dead tissues on the leaves will serve as primary targets for the proposed MV model. The convolutional neural network (CNN) is selected for its advantage of automatically learning and extracting features from training images. A CNN model consists of convolutional layer, pooling layer and fully connected layer. Feature extraction is carried out in the convolutional layers which are made up of small array of numbers called kernels. These kernels produce feature maps from the input images. In the pooling layers, downsampling operations such as max-pooling, min-pooling and average-pooling are conducted to reduce the dimension of the convolutional feature map. This is important to keep the data size and its corresponding computational cost at a feasible level. The following layer is a fully connected layer which is responsible for transforming the output feature maps into one-dimensional vectors connecting every input to every output by a set weight [27].

4.3 Deep Learning Algorithm

The proposed CNN classifier model for this study is based on the Inception architecture. The basic Inception module which is elemental to any Inception-based neural networks is shown in Fig. 2 (left). In order to reduce the number of parameters for the sake of computational cost saving and depthwise separable convolution will be introduced to replace the standard 3 × 3 and 5 × 5 convolutions in the filters of the basic Inception module to yield a new module shown in Fig. 2 (right). Depthwise separable convolution is a new method in which the input channels are kept separated during convolution, as opposed to performing the convolution over multiple

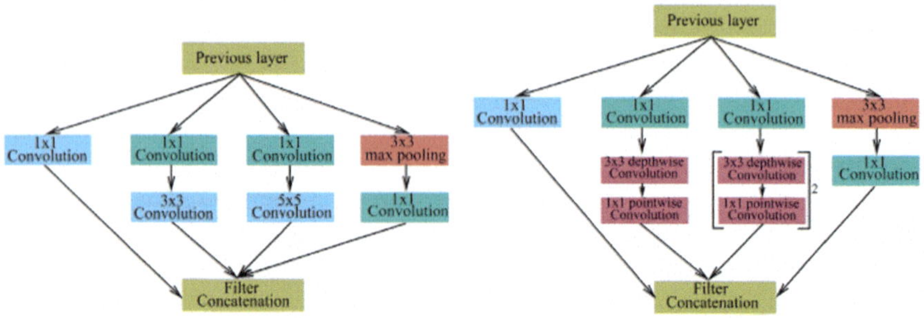

Fig. 2 Original basic Inception module (left) and modified Inception module with depthwise separable convolution (right)

or all input channels. Depthwise separable convolution is made up of two stages: depthwise convolution and pointwise convolution. During depthwise convolution, filters with depth 1 will be split into different channels and each filter will be applied on one channel, generating an output with the same depth as the input image. This is followed by pointwise convolution, where 1×1 filters with the same depth as the input image are iterated through every single point to generate an output with depth 1. The number of 1×1 filters can be increased to produce outputs with greater depth. The output from each channel is then stacked together in concatenation to yield an output for the entire input tensor.

The proposed new Inception deep learning network is shown in Fig. 3. The modifications include the addition of batch normalization and ReLU activation function after every standard convolutional layer and Inception module. The deep network may encounter internal covariate shift leading to lower learning rates and stricter parameter initialization. In response, batch normalization standardizes the activations of prior layer to a mean of zero and a standard deviation of one, while still allowing the relationships between units and nonlinear statistics of single units to change. As a result, weight updating across multiple layers becomes more coordinated and generalization errors are reduced. Batch normalization also enables the use of higher learning rates for faster network convergence. ReLU rectifies the vanishing gradient problem and provides nonlinearity to the model. A 50% dropout rate is applied to reduce overfitting by randomly dropping out nodes during training to approximate the training of different architectures. The dropout layer also prevents co-adaptations between nodes by making the training process noisy and forcing individual nodes to adjust for any given input. Besides, a fully connected layer is added to learn the nonlinear combinations of high-level features in the feature map. Softmax activation function classifies images into four class labels of Healthy, Mild, Moderate and Severe Infection. The layer also calculates the error between expected and predicted results using a loss function which is then backpropagated to update the model. An epoch size of five is selected for model training.

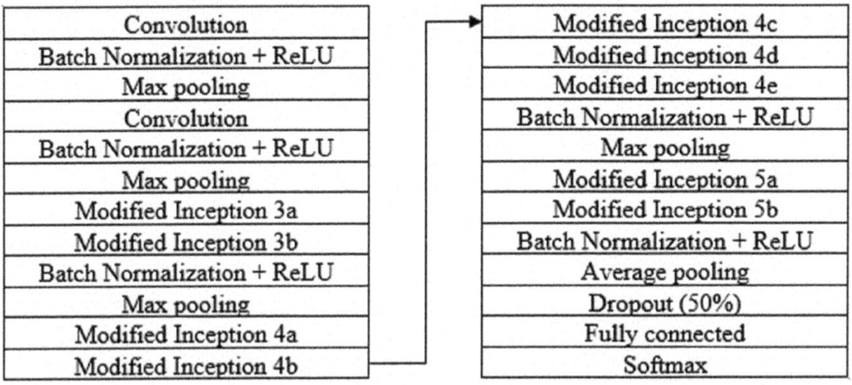

Fig. 3 Proposed network for MV model

The performance of the model will be evaluated based on several criteria including accuracy, precision, recall and F1-score. The model is targeted to achieve not lower than 90% for all four criteria across all four classes of infection severity.

4.4 Pseudo-Code

4.4.1 Model Training

```
Initialize epoch and dataset size
Initialize weight parameters
Set counter and image counter = 1
While counter ≤ epoch size
   While image counter ≤ dataset size
      Input one image into first convolutional layer
      Feedforward the image through the entire network
    Classification into one of four class labels by softmax function
      Calculate error between expected and predicted result
      Update model weights by backpropagation
      Image counter + 1
   Counter + 1
```

4.4.2 Model Testing

```
Initialize epoch and dataset size
Initialize weight parameters
Set counter and image counter = 1
While counter ≤ epoch size
   While image counter ≤ dataset size
      Input one image into first convolutional layer
```

```
        Feedforward the image through the entire network
      Classification into one of four class labels by softmax function
          Tabulate the amount of true positive, true negative, false
positive and false negative results
        Image counter + 1
    Counter + 1
Calculate the model's accuracy, precision, recall and F1-score
```

5 Research Gap

Although classification accuracies of over 90% had been reported from various BSR classifier models, there remains one general problem underlying the majority of existing detection techniques. The lack of obvious visual symptoms at the early stages of BSR remains an obstacle that must be tackled by future deep learning models for more effective differentiation between healthy and mildly infected samples. This is critical because early detection can drastically limit the damage inflicted on surrounding trees through the spread of *Ganoderma* pathogen. Novel techniques in image acquisition, image preprocessing and training algorithms of deep learning models must be continually explored to close this gap and enhance the feasibility of these models as a complete unit for BSR disease detection.

Comparatively, the proposed MV model offers several key advantages. Firstly, it comprises a process flow which is more user-friendly. Since image acquisition in MV only requires the use of normal cameras which could be that of a smartphone, a tablet or a digital camera and the learning curve in terms of image shooting is more straightforward and can be picked up by plantation workers with minimal training. Secondly, the MV model uses normal imaging which is a more accessible and readily available data source. This helps plantation owners to avoid the cost and scarcity associated with third-party resources such as satellite data from aerospace agencies. Thirdly, the MV model is designed with commercialization in mind. In contrast to previous methods, the MV technique runs on and takes advantage of the computational powers offered by modern everyday gadgets such as smartphones, tablets and laptops. This would increase the appeal of the MV model to stakeholders within the palm oil industry and make its adoption a smoother, if not seamless transition.

Acknowledgements This work is supported by the Faculty of Engineering, Technology and Built Environment, UCSI University under the grant number REIG-FETBE-2020/011.

References

1. Parveez GKA, Tarmizi AHA, Sundram S, Loh SK, Ong-Abdullah M, Palam KDP, Salleh KM, Ishak SM, Idris Z (2021) Oil palm economic performance in Malaysia and R&D progress in 2020. J Oil Palm Res 33(2): 181–214 (Lembaga Minyak Sawit Malaysia). https://doi.org/10.21894/jopr.2021.0026
2. Xu Y, Yu L, Li W, Ciais P, Cheng Y, Gong P (2020) Annual oil palm plantation maps in Malaysia and Indonesia from 2001 to 2016. Earth Syst Sci Data 12(2):847–867. https://doi.org/10.5194/essd-12-847-2020
3. Hashim IC, Shariff ARM, Bejo SK, Muharam FM, Ahmad K (2021) Machine learning approach using SAR data for the classification of oil palm trees that are non-infected and infected with the basal stem rot disease. Agronomy 11(3):532
4. Lim KH, Lim SS, Parish F, Suharto R (2012) RSPO manual on best management practices (BMPs) for existing oil palm cultivation on peat (First). Roundtable Sustain Palm Oil
5. Mandal PK, Babu MK, Jayanthi M, Satyavani V (2014) PCR based early detection of *Ganoderma* sp. causing basal stem rot of oil palm in India. J Plantation Crops 42(3):392–394
6. Utomo C, Niepold F (2000) Development of diagnostic methods for detecting *Ganoderma*-infected oil palms. J Phytopathol 148(9–10):507–514
7. Hushiarian R, Yusof NA, Dutse SW (2013) Detection and control of *Ganoderma boninense*: strategies and perspectives. Springerplus 2(1):1–12
8. Dutse SW, Yusof NA, Ahmad H, Hussein MZ, Hushiarian R (2013) DNA-based biosensor for detection of *Ganoderma boninense*, an oil palm pathogen utilizing newly synthesized ruthenium complex [Ru (phen) 2 (qtpy)] 2 based on a PEDOT-PSS/Ag nanoparticles modified electrode. Int J Electrochem Sci 8(9):11048–11057
9. Muniroh MS, Sariah M, Abidin MZ, Lima N, Paterson RRM (2014) Rapid detection of *Ganoderma*-infected oil palms by microwave ergosterol extraction with HPLC and TLC. J Microbiol Methods 100:143–147
10. Zainol Hilmi NH, Idris AS, Mohd Azmil MN (2019) Headspace solid-phase microextraction gas chromatography–mass spectrometry for the detection of volatile organic compounds released from *Ganoderma boninense* and oil palm wood. Forest Pathol 49(4):e12531
11. Al-Obaidi JR, Mohd-Yusuf Y, Razali N, Jayapalan, JJ, Tey CC, Md-Noh N, Junit SM, Othman RY, Hashim OH (2014) Identification of proteins of altered abundance in oil palm infected with *Ganoderma boninense*. Int J Molecular Sci 15(3):5175–5192
12. Yeoh KA, Othman A, Meon S, Abdullah F, Ho CL (2012) Sequence analysis and gene expression of putative exo-and endo-glucanases from oil palm (*Elaeis guineensis*) during fungal infection. J Plant Physiol 169(15):1565–1570
13. Siddiqui Y, Surendran A, Paterson RRM, Ali A, Ahmad K (2021) Current strategies and perspectives in detection and control of basal stem rot of oil palm. Saudi J Biol Sci 28(5):2840–2849. https://doi.org/10.1016/j.sjbs.2021.02.016
14. Abdullah AH, Adom AH, Shakaff AYM, Ahmad MN, Zakaria A, Saad FSA, Isa C MNC, Masnan MJ, Kamarudin LM (2012, February). Hand-held electronic nose sensor selection system for basal stamp rot (BSR) disease detection. In: 2012 third international conference on intelligent systems modelling and simulation. IEEE, pp 737–742
15. Kresnawaty I, Mulyatni AS, Eris DD, Prakoso HT, Triyana K, Widiastuti H (2020, March) Electronic nose for early detection of basal stem rot caused by *Ganoderma* in oil palm. In: IOP conference series: earth and environmental science, vol 468, No 1. IOP Publishing, p 012029
16. Lelong CC, Roger JM, Brégand S, Dubertret F, Lanore M, Sitorus NA, Raharjo DA, Caliman JP (2010) Evaluation of oil-palm fungal disease infestation with canopy hyperspectral reflectance data. Sensors 10(1):734–747
17. Ahmadi P, Muharam FM, Ahmad K, Mansor S, Abu Seman I (2017) Early detection of *Ganoderma* basal stem rot of oil palms using artificial neural network spectral analysis. Plant Dis 101(6):1009–1016

18. Lee CC, Koo VC, Lim TS, Lee YP, Abidin H (2021) A multi-layer perceptron-based approach for early detection of BSR disease in oil palm tree using hyperspectral image. Available at SSRN 3936335
19. Evangelidis GD, Psarakis EZ (2008) Parametric image alignment using enhanced correlation coefficient maximization. IEEE Trans Pattern Anal Mach Intell 30(10):1858–1865
20. Johari SN, Bejo SK, Lajis GA, DaimDai LD, Keat NB, Ci YY, Ithnin N (2021) Detecting BSR-infected oil palm seedlings using thermal imaging technique. Basrah J Agric Sci 34:73–80
21. Khairunniza-Bejo S, Vong CN (2014) Detection of basal stem rot (BSR) infected oil palm tree using laser scanning data. Agric Agric Sci Procedia 2:156–164. https://doi.org/10.1016/j.aaspro.2014.11.023
22. Azuan NH, Khairunniza-Bejo S, Abdullah AF, Kassim MSM, Ahmad D (2019) Analysis of changes in oil palm canopy architecture from basal stem rot using terrestrial laser scanner. Plant Dis 103:3218–3225. https://doi.org/10.1094/PDIS-10-18-1721-RE
23. Husin NA, Khairunniza-Bejo S, Abdullah AF, Kassim MS, Ahmad D, Azmi AN (2020) Application of ground-based LiDAR for analysing oil palm canopy properties on the occurrence of basal stem rot (BSR) disease. Sci Rep 10:1–16. https://doi.org/10.1038/s41598-020-62275-6
24. Husin NA, Bejo SK, Abdullah AF, Kassim MS, Ahmad D (2021) Relationship of oil palm crown features extracted using terrestrial laser scanning for basal stem rot disease classification. Basrah J Agric Sci 34:1–10
25. Alaa H, Waleed K, Samir M, Tarek M, Sobeah H, Salam MA (2020) An intelligent approach for detecting palm trees diseases using image processing and machine learning. Int J Adv Comput Sci Appl 11(7):434–441
26. Darwin B, Dharmaraj P, Prince S, Popescu DE, Hemanth DJ (2021) Recognition of bloom/yield in crop images using deep learning models for smart agriculture: a review. Agronomy 11(4):646
27. Hassan SM, Maji AK (2022) Plant disease identification using a novel convolutional neural network. IEEE Access 10:5390–5401. https://doi.org/10.1109/ACCESS.2022.3141371

Effect of the Hardness in Al/TiB_2 MMC with Sand Mould and Permanent Mould

C. Rajaravi, B. Gobalakrishnan, Ganapathi Srinivasan, S. Palani, and Karthik

Abstract Fabrication of Aluminium matrix reinforced with in-situ formed TiB_2 MMCs. It has higher strength and stiffness than Al/TiB_2 MMCs. Al/TiB_2 MMCs can also be made using the vortex process. These features make Al/TiB_2 MMCs very desirable. Typically, experimental techniques are employed to characterize as-cast Aluminium (356), Al/TiB_2 MMCs. The goal of this research is to determine the effectiveness of various moulds, such as sand moulds and permanent moulds, in improving the hardness qualities of Al-TiB_2 MMC. The basic idea behind this research is that as the TiB_2 reinforcement particle size increased, the hardness value decreased, resulting in higher hardness with reduced particle size growth.

Keywords Al/TiB_2 · Hardness · Sand mould · Permanent mould

C. Rajaravi (✉)
Department of Mechnical Engineering, Hindusthan College of Engineering and Technology, Coimbatore, Tamil Nadu, India
e-mail: rajaravi.mech@gmail.com

B. Gobalakrishnan
Department of Mechnical Engineering, Care College of Engineering, Tiruchirapalli, Tamil Nadu, India
e-mail: pgkrish.mech@gmail.com

G. Srinivasan · S. Palani
Department of Mechanical Engineering, Veltech Multitech, Chennai, Tamil Nadu, India
e-mail: krgsrini@gmail.com

S. Palani
e-mail: planiraji.s@gmail.com

Karthik
Department of Mechnical Engineering, Veltech Rangarajan Dr. Sakunthala R&D Institute of Science and Technology, Chennai, Tamil Nadu, India

E. Natarajan et al. (eds.), *Materials, Design and Manufacturing for Sustainable Environment*, Lecture Notes in Mechanical Engineering,
https://doi.org/10.1007/978-981-19-3053-9_8

1 Introduction

Aluminium alloy matrix composites are best suited to temperatures below 400 °C. At higher temperatures, titanium-based alloys show promise as matrices. Fibre reinforced superalloys are utilized for applications exceeding 900 °C (1650°F), with tungsten alloy wires being the most promising reinforcement. Such composites, despite their high density, are being evaluated for aviation and rocket motor turbine blades and vanes. In terms of creep and fatigue strength, as well as thermal conductivity, they outperform unreinforced superalloys. Cast Aluminium (356), a component, is utilized in a variety of automotive and aerospace applications. Cast Aluminium is more extensively used than cast iron because to its light weight, low melting temperature, and ease of manufacture. Aluminium castings can also be made thinner than cast iron castings. Aluminium is also a better conductor of heat and electricity. Aluminium is more cost-effective than cast iron for all of these reasons. Aluminium's success over magnesium can be attributed to its design flexibility, high wettability, and strong bonding at surfaces [1, 2].

TiB_2 has shown to be an excellent reinforcement. This is due to TiB_2's rigidity and hardness, as well as the fact that it will not react with Al to form reaction product at matrix-reinforcement interface. Melt-stirring (vortex casting) is a popular liquid state processing technology due to its low cost and availability of a wide range of metal materials, which compensates for the increased transverse and fabrication conditions. Many aircraft components, sports automobiles, worn parts, and piston rings and seals require lighter materials with higher strength and lower weight, such as aluminium and magnesium. When compared to Al/SiCp MMCs, cast Aluminium with TiB_2 reinforcement generated in-situ has greater strength and stiffness. The vortex process, which is similar to that of the Al/SiCp MMC stated above, can also be used to quickly create Al/TiB_2 MMC. These characteristics make Al/TiB_2 MMCs particularly appealing. Al/TiB_2 MMC has the potential to be competitive with magnesium, a currently utilized material whose melting and manufacture are extremely difficult [3, 4].

Sand mould castings are made in foundries that specialize in this type of casting. Through metal casting, 70% of the casting procedure will be produced in sand casting. Even for steel foundries, sand casting is reasonably inexpensive and suitably refractory. An appropriate bonding material usually clay is required in addition to the sand is added occurs when sand is used. The dough has been moistened, generally but not with water occasionally in combination with other substances, to increase the clay's flexibility and strength, as well as to make the aggregate ideal for moulding [5].

Permanent mould casting, also known as gravity die casting, is a method in which a metal mould made up of two or more components is used repeatedly to produce several castings of the same form. Gravity draws the liquid metal into the mould. Metal is commonly used for simple removable cores, although sand or plaster is used for more sophisticated cores. The procedure is known as semi-permanent mould casting when sand or plaster cores are employed. Permanent mould casting is ideal for high-volume manufacturing of castings with relatively consistent wall thickness, limited undercuts, and precise internal coring [6].

The procedure can also be used to make intricate castings, but the quantity of castings produced must be sufficient to justify the cost of the moulds. Brinell hardness testing entails using a 10 mm diameter hardened steel or carbide ball to imprint the test material and then exposing it to a 3000 kg load. To minimize severe indentation, the weight can be lowered and softer materials could be up to 1500 or 500 kg. In the case of iron and steel, a whole weight is applied for 10 to 15 s, and for at least 30 s in the case of other metals. A low-powered microscope is used to measure the depression left in the material for the test diameter.

The Brinell harness number is calculated by multiplying the applied load by the surface area of the indentation. The diameter of the impression is the average of two readings taken at right angles, and using a Brinell hardness number table can make determining the Brinell hardness easier. A well-structured Brinell hardness value, such as "75 HB 10/500/30," implies a Brinell hardness of 75 was achieved using a 10 mm diameter hardened steel and a 500 kg load for 30 s. A tungsten carbide ball is used instead of a steel ball in tests of exceptionally hard metals. Comparatively, various hardness test procedures, because the Brinell ball creates the deepest and broadest depression, the difficulty of the test is averaged over a greater area quantity of material, more precisely accounting for various structure of grains and faults in the homogeneity of the substance [7]. The aim of this study fabricates both sand mould and permanent mould Al/TiB_2 MMCs. The Brinell hardness tester is used to determine the hardness value of both sand mould and permanent mould.

2 Experimental Work

As a base metal, cast Aluminium was employed. To produce and construct TiB_2 reinforcement, two salts were used: Potassium Tetra Fluoro Borate (KBF4) and Potassium Hexa Fluoro Titanate (K2TiF6). Aluminium was melted and kept at varied temperatures of 820 °C, and two types of salts were prepared and kept at 250 °C for around 30 min before being slowly stirred into the molten aluminium for 15 min with a graphite rod.

The fabrication technique of Al/TiB_2 MMC used in this work is self-explanatory, as shown in Fig. 1. Chips of aluminium ingot are put into the stir casting furnace for melting in Fig. 2a. Pre-weighted titanium and boron salts are shown in Fig. 2b, ready to be filled into tiny crucibles and loaded into a muffle furnace for preheating. The snapshot of the experimental setup is shown in Fig. 3 [8, 9].

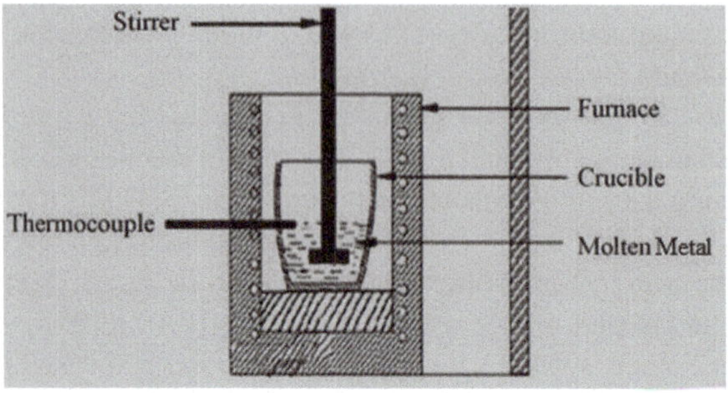

Fig. 1 Schematic diagram of fabrication process of Al/TiB_2 MMC

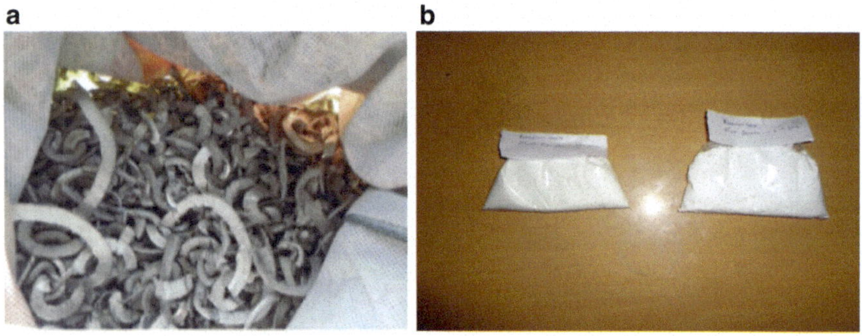

Fig. 2 **a** Chipped aluminium ingot and **b** Ti and B salts

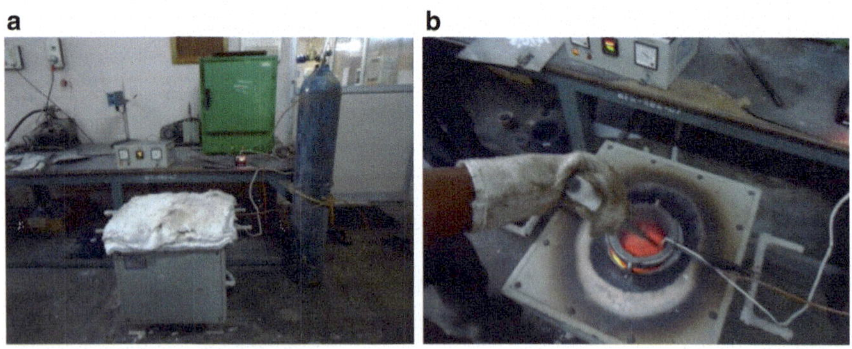

Fig. 3 **a** and **b** Stir casting furnace

2.1 Permanent Mould

Permanent moulds are usually made of metal, and they can be reused and fed by gravity. However, a vacuum or gas pressure can also be employed. Aluminium,

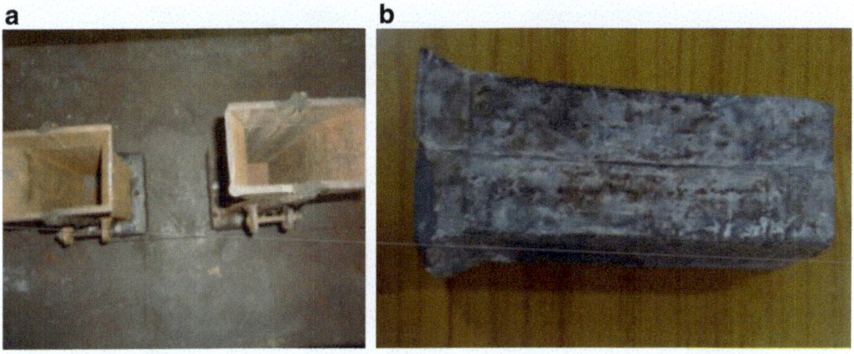

Fig. 4 **a** Permanent mould and **b** ingot

magnesium, and copper alloys are common casting metals. The permanent mould is formed of mild 60 steel sheets that are 5 mm thick and measures 70 mm × 55 mm × 180 mm, with top gating, and the metal solidifies in the mould before being removed as casting. Figure 4 depicts the permanent mould utilized in this investigation as well as a typical cast ingot produced.

2.2 *Sand Mould*

Air set sand moulds were used to cast the ingots. The sand mould is made by combining 15 kg of dry sand with 0.562 kg of air set resin, then pouring and shoving it into a detachable wooden mould box while keeping the pattern in place. The sand in the mould box is left to air set for 1 h before removing the detachable wooden panels of the wooden mould box and the pattern to reveal the air set sand mould ready for pouring using the top gate. Figure 5 shows a shot of the air set sand mould, which has been prepared for pouring, as well as the cast ingot.

2.3 *Specimen for Microstructure*

The specimens were polished with emery sheets of 200, 400, 600, 1000, and 1500 grade, respectively. After that, Alumina was used to polish the disc. Keller's solution was used to etch the specimens after they were cleansed in water. The microstructures were examined using an optical microscope at various magnifications, as well as a scanning electron microscope.

Fig. 5 Sand mould and ingot

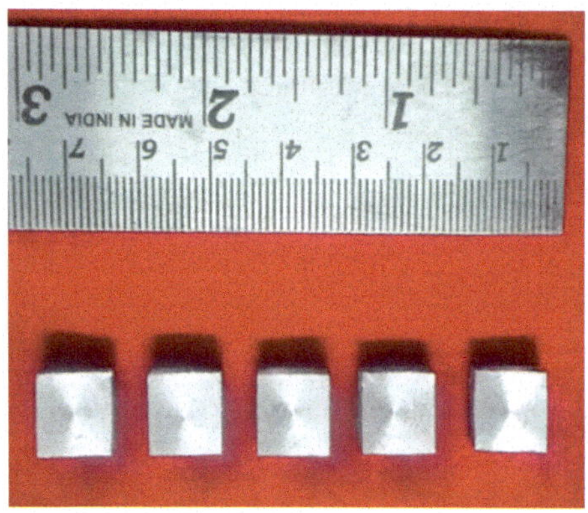

Fig. 6 XRD test specimens

2.4 XRD Specimens

XRD examination was performed on samples with total dimensions of 5 mm × 10 mm × 10 mm was performed to detect the phases present in the produced Al/TiB_2 MMC castings. Figure 6 depicts the test specimens for XRD analysis. The findings were matched to data from the Joint Committee of Powder Diffraction Standards.

2.5 Hardness Test

The cast specimens' hardness of the material was determined using a Brinell testing equipment with a 100 kgf load and a 15 s dwell period, as illustrated in Fig. 7. The readings were obtained at several points, and the average hardness value for each set of samples was determined. The hardness tests were replicated three times using the Brinell hardness testing equipment, and the hardness values of the samples were measured using a steel ball with a 1.58 diameter and a load of 100 kgf on the (10 × 10) mm square samples as per ASTM E10-18 standard. The Brinell hardness machine assesses the material's overall hardness values and is essentially unaffected by localized effects (5). Figure 8 depicts the hardness test specimens.

Fig. 7 Brinell harness testing machine

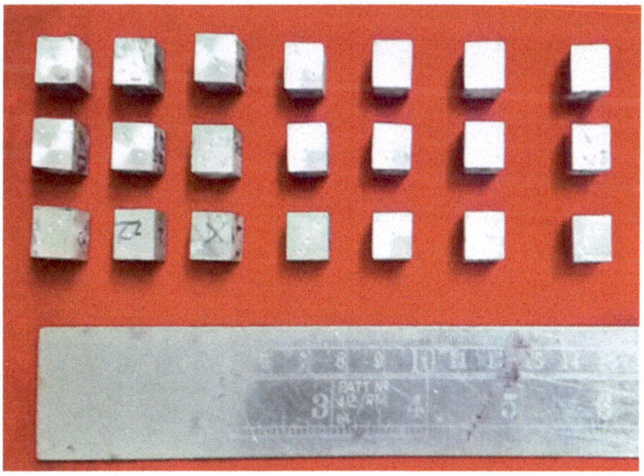

Fig. 8 Hardness test specimens

3 Results and Discussion

3.1 Micro Structural Investigation of Permanent Mould and Sand Mould Castings of Al/TiB$_2$

Optical micrographs of Al/TiB$_2$ MMC cast in a permanent mould at a pouring temperature of 820 °C are shown in Fig. 9. Figure 9a, b is optical micrograph, and it show the Al/TiB$_2$ particles present sand and permanent mould at a temperature of 820 °C pouring. The TiB$_2$ particles in the Al/TiB$_2$ MMC permanent mould are smaller than those in the sand mould. This is because TiB$_2$ particles produced at higher pouring temperatures expand in size [10, 11].

Figure 10a, b shows Al/TiB$_2$ MMC under SEM at a magnification of 1500X. The Al/TiB$_2$ MMC cast in both micrographs was made under the same conditions. In both micrographs, there is an equitable dispersion of TiB$_2$ particles. The molten state of the sand mould and the Al/TiB$_2$ MMC lasts longer before solidification because the sand moulds are insulated.

The XRD patterns of Al/TiB$_2$ MMCs cast in permanent and sand mounds, as shown in Fig. 11a and Fig. 11.b, confirm the production of TiB$_2$ particles in-situ. The evidence of Al3Ti, a brittle phase, is further confirmed by the patterns [12, 13]. By keeping the melt temperature between 700 and 800 degrees Celsius, A3Ti can be prevented.

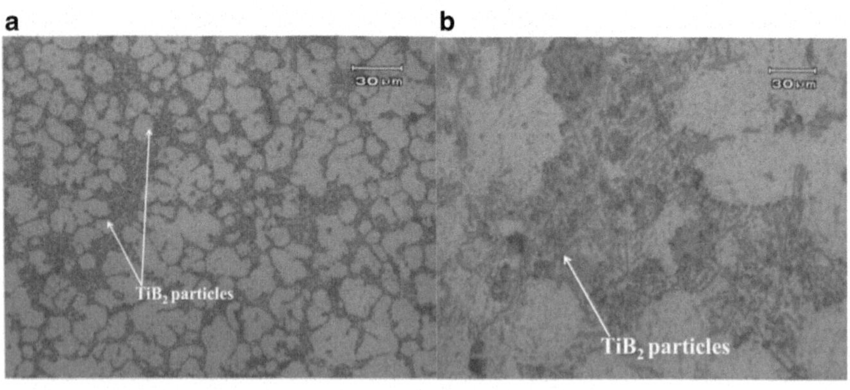

Fig. 9 **a, b** Optical micrographs of Al/TiB$_2$ MMCs in sand and permanent mould

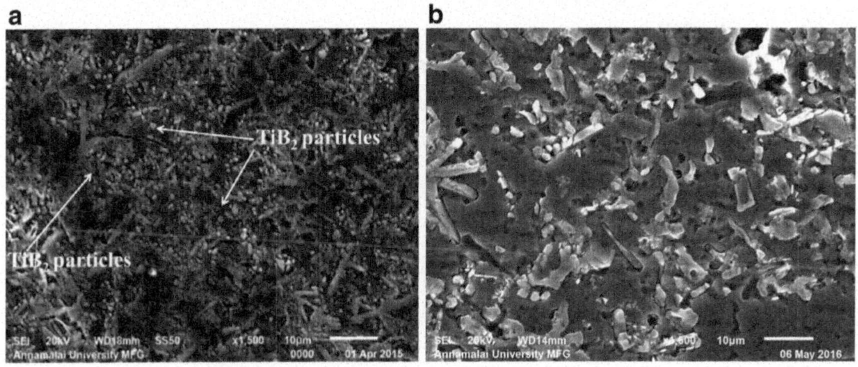

Fig. 10 **a, b** SEM of Al/TiB$_2$ MMCs cast in permanent mould and sand mould

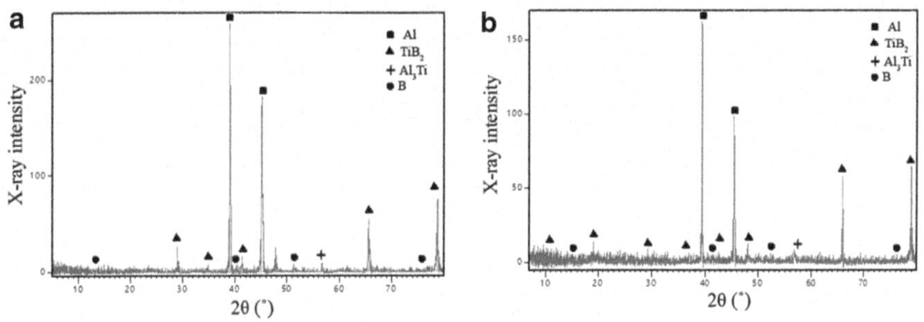

Fig. 11 **a, b** XRD of pattern for Al/TiB$_2$ MMC cast in permanent mould and sand mould

3.2 Comparison of Hardness of Al/TiB$_2$ MMC Cast in Sand and Permanent Moulds

Figure 12 shows the results of the Brinell hardness test, and the hardness values for sand mould and permanent mould conditions are compared. The hardness of the range examined increases as the temperature rises in permanent and sand mould settings. Permanent mould have higher hardness ratings than sand moulds due to the finer grains created in permanent moulds and the faster cooling rates [14]. Due to TiB$_2$ reinforcement particle size increased at that time decrease the hardness value of the sand mould conditions, longer as the sand mould insulated, in permeant mould As a result, the particle size grows smaller but the hardness increases.

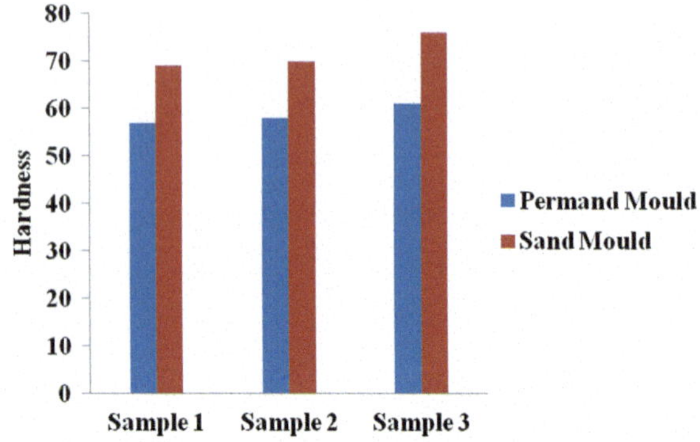

Fig. 12 Comparison of hardness values in sand mould and permeant mould

4 Conclusions

The in-situ formed Al/TiB_2 metal matric composite cast through permanent and sand moulds is created by melting A356 Aluminium alloy and combining estimated quantities of KBF4 and K2TiF6 to produce a maximum of 6% TiB_2. The Brinell hardness of the final material is determined experimentally both permeant mould and sand mould. Microstructural examination reveals the presence of TiB_2, and it is evenly distributed in the metal matrix. The XRD graphs shown confirm the presence of TiB_2 particles. The hardness values are high in the permeant mould than a sand mould at a pouring temperature of 820 °C during fabrication. Due to TiB_2 reinforcement, particle size increased at that time decrease the hardness value of the sand mould conditions, longer as the sand mould insulated. In permeant mould, resulting in increased hardness with decrease the growth of the particle size.

References

1. Ramanathan A, Krishnan PK, Muraliraja R (2019) A review on the production of metal matrix composites through stir casting Furnace design, properties, challenges, and research opportunities. J Manuf Process 42:213–245
2. Kumar S, Singh R, Hashmi MSJ (2020) Metal matrix composite: a methodological review. Adv Mater Process Technol 6(1):13–24
3. Gopalakrishnan S, Murugan N (2012) Production and wear characterisation of AA 6061 matrix titanium carbide particulate reinforced composite by enhanced stir casting method. Compos B Eng 43(2):302–308
4. Mandal A, Maiti R, Chakraborty M, Murty BS (2004) Effect of TiB_2 particles on aging response of Al–4Cu alloy. Mater Sci Eng A 386(1):296–300
5. Rajaravi C, Lakshminarayanan PR (2016) Analysis on mechanical properties of Al/TiB_2 MMCs and validated for FEA with different mould condition. Indian J Eng Int J 13:634–645
6. Tee KL, Lu L, Lai MO (2001) In situ stir cast Al–TiB_2 composite: processing and mechanical properties. Mater Sci Technol 17(2):201–206
7. Rajaravi C (2018) Comparative study of aluminium with SiC and TiB_2 as reinforce particulate and with FEA. Mater Res Express 2(5):2053–2091

8. Christy TV, Murugan N, Kumar S (2010) A comparative study on the microstructures and mechanical properties of Al 6061 alloy and the MMC Al 6061/TiB_2/12p. J Minera Mater Charact Eng 9(1):57
9. Gobalakrishnan B, Rajaravi C, Udhayakumar G, Lakshminarayanan PR (2021) Effect of ceramic particulate addition on aluminium based ex-situ and in-situ formed metal matrix composites. Metals Mater Int 26(11):3695–3708
10. Xie L, Wang L, Jiang C, Lu W (2014) The variations of microstructures and hardness of titanium matrix composite (TiB+ TiC)/Ti–6Al–4V after shot peening. Surf Coat Technol 244:69–77
11. Rajaravi C, Niranjan K, Lakshminarayanan PR (2015) Comparative analysis of Al/TiB_2 Metal matrix composites in different mould conditions. J Adv Microsc Res 10(4):260–264
12. Xiong HH, Zhang HN, Dong JH (2017) Adhesion strength and stability of TiB_2/TiC interface in composite coatings by first principles calculation. Comput Mater Sci 127:244–250
13. Niranjan K, Lakshminarayanan PR (2013) Dry sliding behavior of in situ Al/TiB_2 composites. Mater Des 47:167–173
14. Rajaravi C, Gobalakrishnan B, Lakshminarayanan PR (2019) Effect of pouring temperature on cast Al/SiCp and Al/TiB_2 metal matrix composites. J Mech Behav Mater 28:162–168

Studies of Effects of Pollutants M Sand, Wood Ash, Rice Husk Ash and Graphene on Mechanical Properties of Recycled Aluminium

C. Bhagyanathan, P. Karuppuswamy, S. Sathish, and D. Elangovan

Abstract The present study focuses on the influence of pollutants that affects the microstructural and mechanical characteristics of recycled Al 6063 grade aluminium alloy and affect its inherent characteristics. As the recycled scraps shown to have more contaminants, known pollutants like carbon, ash, graphene powder and sand were introduced in a control percentage to the recycled aluminium. The samples were fabricated with a common 4wt% addition of the materials. The material characteristics considered for analysis are hardness and tensile strength. Further, the properties were correlated with the microstructure which was analysed through the microstructural analysis of the samples. The Brinell hardness of the recycled alloy samples that were subjected to the addition of wood ash and graphene increased by 6% and 15%, respectively, due to the modified SDAS in the samples. The tensile strength of the samples with the added impurities reduced by 7%.

Keywords Recycled aluminium · Al6063 alloy · Pollutants · Graphene · SDAS

1 Introduction

Recycled aluminium has been in use for several decades and has been meeting the continuous demand by compensating the energy consumption for several years. There are certain areas of concerns for the aluminium recyclers. The key problem that is facing the recycling sector as per the survey from multiple literature articles is that undesirable components are accumulating in all recycled material streams. The list of

C. Bhagyanathan (✉) · P. Karuppuswamy · S. Sathish
Department of Mechanical Engineering, Sri Ramakrishna Engineering College, Coimbatore, Tamil Nadu, India
e-mail: bhagyanathan@srec.ac.in

S. Sathish
e-mail: sathish.selvaraj@srec.ac.in

D. Elangovan
Department of Mechanical Engineering, PSG Institute of Technology and Applied Research, Coimbatore, Tamil Nadu, India

E. Natarajan et al. (eds.), *Materials, Design and Manufacturing for Sustainable Environment*, Lecture Notes in Mechanical Engineering,
https://doi.org/10.1007/978-981-19-3053-9_9

troublesome impurities in aluminium is rather long, and includes, but is not limited to, Fe, Si, Ni, Pb, C and Ca. Owing to the superior mechanical and tribological qualities, aluminium alloys are employed in sophisticated technical appliances and industries such as the automotive and aerospace sectors, as well as low-density applications, to meet rising demand of the industries. Al6063 alloys have increased hardness and tensile strength. According to Su and Teng, fly ash was added to aluminium melts to form composites through the sintering techniques. The volume proportion of the particles was shown to improve the material's resistance to indentation, wear resistance and modulus of the samples with fly ash particles. Tiwari, S. and Pradhan used friction stir processing to produce and characterize AA6061/18vol. percent rice husk ash in AMC. The interfacial connection between the cracked particles and the aluminium matrix was found to be strong. The addition of rice husk ash particles to the composite increased its tensile strength [1]. Investigation on the microstructural and mechanical characteristics of Al6061 materials subjected to the addition of SiC and Al7075 alloy subjected to the addition of Al_2O_3. The totting of the particles to the alloys enhanced the micro hardness, density and tensile strength [2]. Idusuyi et al. reviewed the behaviour of Al7075 alloy with 10% volume fractions of SiC, Al_2O_3 and B_4C through stir casting process. In comparison to other particle-reinforced composites, they found Al7075-B_4C samples to show enhanced behaviour to mechanical testing. Investigations on the wear and mechanical characteristics of aluminium alloy 7075 with Al_2O_3 and graphite additions were carried by Zhang et al. [3]. It was discovered that raising the weight % of reinforced particles improves the mechanical characteristics. Further, the investigations of Yang et al. analysed the behaviour of alumina combined with rice husk ash and graphite particles [4]. The outcomes revealed that the hardness reduced as the percent of graphite and rice husk ash increased in the samples. The presence of graphite had a substantial impact on wear resistance, although increasing the percent of graphite in the alloy lowered mechanical qualities. According to Xies et al., a material that is alloyed with impurities/composites outweighs pure aluminium in terms of mechanical characteristics. The stir casting process was utilized to create the hybrid composite from Al 6061 with varying percentages of SiC and B_4C. Because of the number of carbides contained in the composites, the hybrid composite had greater tensile, flexural and hardness values. Graphene's presence is the most important factor impacting the yield strength of stir cast samples. Similarly, melting temperature was shown to be the most influential parameter on yield strength in the squeeze cast specimen.

The presence of impurities tends to affect the properties of the materials, and most sorting techniques are not able to eliminate the impurities from the molten material. According to Gustad et al. (2012), removing the impurities from the molten aluminium is not a simple procedure. The sources of such impurities vary according to the applications that the particular aluminium alloys served. Therefore, it is necessary to understand the mechanism by which the material properties are affected before making amendments to reduce the presence of such impurities. This means that not only is it difficult to manage impurity levels, but it's also tough to get the desired alloy composition. The melt process, which is based on energy considerations, determines the removal of undesired elements and contaminants in the scrap flux [5].

In this study, the effects of pollutants that are mixed with the scrap aluminium during improper sorting processes were analysed by adding them in constant quantities. Analysis was done based on the observations made from the microstructure and mechanical properties of recycled Al6063 scraps. For this purpose, four different materials, namely graphene, M sand, wood ash and rice husk ash were selected as pollutant materials.

2 Materials and Methods

The sorted scarps are degreased and cleaned with water to remove its pollutants like oil, sand, paint, grease, etc., and preheated at 200 °C in an industrial oven to dry. Then 1.5 kg of scarp is loaded in a crucible and kept in a small stir casting furnace at 750 °C. The slags are removed after melting and the first sample was poured out onto the dies which were preheated in the industrial oven for 200 °C. Appropriate pollutants like wood ash, M sand, rice husk carbon and graphene powder were added to the melt of 4%. This percentages affecting the mechanical and microstructural characteristics of 6063 grade aluminium alloy. The melt is then transferred to a preheated die and specific samples were made to test for its characteristics.

It was made sure that the added pollutants were in uniform sizes so the particles were sieved. The particles were stirred to the molten alloy after being initially preheated up to 300 °C for 2 h in small graphite crucibles. The metal was melted in a graphite crucible located inside a stir casting furnace. The scrap added to the crucible was 1.5 kg for all trials. The addition of the various particles in separate experimental trials was done as shown in Table 1. The melt temperature was maintained at 780 °C before mixing, and the temperature was reduced to 720 °C during mixing and poured at 720 °C. The samples were allowed to cool and they were removed from the dies for analysis.

The elementary composition of the five samples was analysed with the spectromaxx optical emission spectrometer and the tensile testing was carried out in a FIE make UTM (UTES 40) on the samples that were sectioned as per ASTM E8 standards. The samples were standardized to a 50 mm gauge length to evaluate percent elongation, yield strength and ultimate tensile strength. Using a standard Brinell hardness test machine [6], hardness measurements were taken on the samples with

Table 1 Description of experiments

Sample	Material added	% addition	Stir speed (RPM)	Stir time (min)
C0	None	–	45	15
C1	M Sand	4	45	15
C2	Wood ash	4	45	15
C3	Rise husk ash	4	45	15
C4	Graphene	4	45	15

Table 2 Chemical composition of the samples

Sample	Al%	Si%	Cu%	Fe%	Mg%	Ca%	C%	Zn%
C0	98.7	0.55	0.005	0.22	0.40	0.02	0.01	0.006
C1	98.8	0.75	0.001	0.20	0.37	0.02	0.01	0.001
C2	98.7	0.56	0.001	0.23	0.37	0.02	0.10	0.005
C3	98.8	0.53	0.001	0.18	0.37	0.15	0.07	0.001
C4	98.7	0.58	0.001	0.21	0.39	0.02	0.23	0.001

and without the addition of the pollutants. BHN analysis was carried out to scrutinize the effect of the various pollutants on the hardness [7]. A 10 kg was the load set for the experiments on the hardness tester which had a square-based diamond pyramid indenter. The microstructure of the material was examined using a metallurgical microscope. After the hot mounting process, the samples were etched with 0.5 percent HF solution to improve the surface for the micro structural examination.

3 Results and Discussion

3.1 Elementary Composition

The below Table 2 shows the elementary composition of as cast samples that were determined by (OES) Optical Emission Spectrometry. It was observed that the addition of these pollutants did not affect the basic chemical composition of the recycled Al 6063 scraps, although there was a slight change in the silicon content after the addition of the M sand and the slight increase in the presence of carbon in the samples with rice husk ash, wood ash and graphene addition.

3.2 Tensile and Hardness Properties

The UTM analysis was done and the values of the tensile strength, yield strength and the elongation of the materials were observed and are mentioned in Table 3.

The various types of pollutants had a negative impact on melt quality. M sand and rice husk ash have almost the same effect on mechanical properties and degrade the mechanical properties of Al6063 recycled scarp [8], whereas graphene and wood ash, which are derived forms of carbon powders commonly used to coast dies, do not appear to have any negative impacts and have similar mechanical properties.

Table 3 Tensile and hardness results of the samples

Sample	TENSILE strength (MPa)	Yield strength (MPa)	Elongation %	Hardness HB
C0	125.4	75.3	10.9	73
C1	116.9	67.9	9.3	61.7
C2	121.3	79.2	11.7	75.9
C3	116.4	70.5	9.6	67.7
C4	121.6	72.6	12.3	84.5

3.3 Microstructural Analysis

The microstructural analysis shown in Fig. 1 exhibits minor variations in the grain structure of the observed samples and that is due to the addition of the pollutant material, and it can be observed that the porosity with the presence of rice husk particles. The grain structure became more fine with the addition of graphene which enhanced the eutectic silicon formation in the samples as indicated in Table 4, but similar results were not found in the samples with the addition of rice husk ash [9]. The sample with the rice husk ash addition showed the presence of porosity on the surface which could possibly be the effect of the carbon that settled along the grain boundries, improper wetting of the particles to the melt or the poured temperature.

The results that are obtained for the samples with and without the pollutants subjected to various mechanical properties and the influence of the secondary dendrite arm spacing (SDAS) and avg. length of the eutectic silicon are shown in the Fig. 2.

The results of the comparative analysis of the mechanical and microstructural properties as seen from Fig. 2 show a decreasing trend of hardness with increase

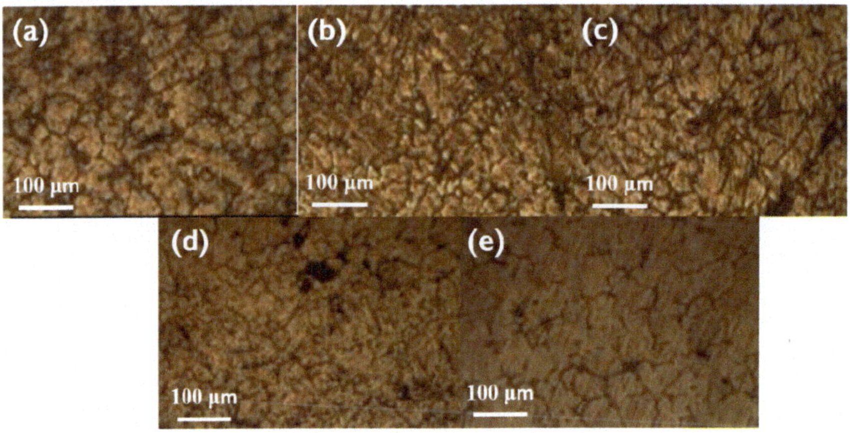

Fig. 1 OM images of **a** recycled 6063 alloy; **b** with M sand addition; **c** with wood ash addition; **d** with rice husk ash; **e** with graphene

Table 4 Properties derived from the microstructural evaluation

Sample	SDAS, μm	Avg. length of eutectic silicon, μm	Avg. thickness of eutectic silicon, μm	Aspect ratio of eutectic silicon
C0	45	102	3.175	3.77
C1	34	125	2.98	3.74
C2	16	135	3.125	3.81
C3	25	123	2.562	3.85
C4	37	134	2.26	3.36

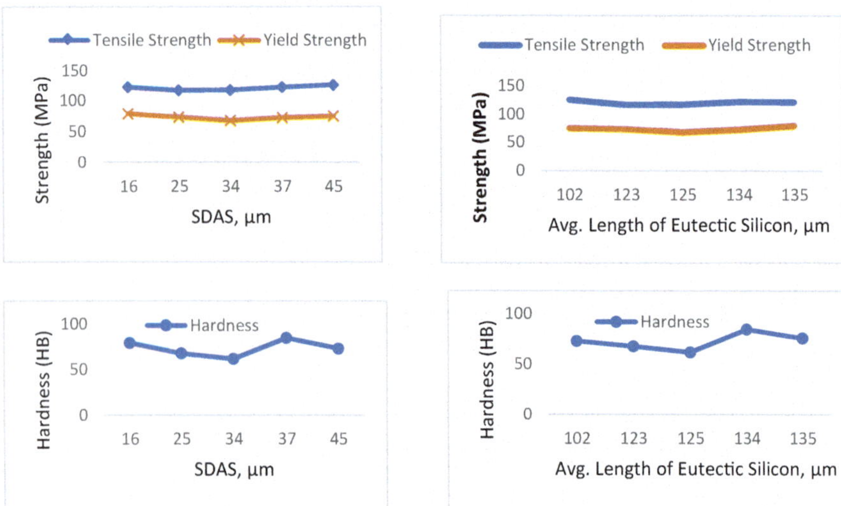

Fig. 2 Effect of SDAS and average length of eutectic silicon on the tensile and hardness behaviour of the samples

in the average length of the eutectic silicon and with lower SDAS values. Thus, the modifications that are done to the microstructure of the recycled scrap due to the presence of rice husk ash and granulated M sand affect the surface properties of the alloy being produced and also affects the indentation. The presence of rice husk ash and M sand in the recycled alloy diminishes the dislocation density which does not resist the deformation when it is subjected to strain. When the presence of these particles is above 4 wt %, hardness decreases from 73 to 60 BHN which was in contrast to the findings of Subrahmanyam et al. [10].

The properties of the materials were reduced as the modifications to the microstructure and the chemical composition had occurred due to the presence of the impurities in the melt [11] and David and Kopac [5]. The properties of the samples with the addition of rice husk ash also did not show any enhancement of the mechanical properties that were obtained with the experimental analysis carried out by Subrahmanyam et al. [12], Tiwari et al. [13].

4 Conclusions

The experimental trials to determine the influence of the pollutants were carried out. The results from the conducted trials describes the negative influence of foreign particles in the recycled aluminium scrap and these were the observations that were made from the experimental works.

- The tensile behavior of the samples is affected with the presence of wood ash, graphene, rice husk ash and M Sand reduced by 10% in the recycled Al6063 alloy.
- The recycled alloy in the presence of Wood ash and graphene was not affected but showed slight improvement in the hardness by 6% and 15% respectively.
- Modifications were observed in the SDAS and eutectic silicon values with the addition of the pollutants [14].
- Improved hardness properties were observed with lower SDAS values at the expense of the tensile properties.
- The experimental studies highlight that the presence of impurities like M Sand in the scraps tend to affect the properties of the Al–Si–Mg alloys by modifying the SDAS and the average eutectic silicon length of the alloy.

References

1. Parveen A, Chauhan NR, Suhaib M (2019) Study of Si3N4 reinforcement on the morphological and tribo-mechanical behaviour of aluminium matrix composites. Mater Res Express 6(4):042001
2. Subburaj A, Antony Joseph Decruz AMM, Chandra Moorthy VA, Durairaj R (2022) Mechanical characterization and micro-structural analysis on AA2024 hybrid composites reinforced with WC and graphene nanoparticles. Transactions of the Indian Institute of Metals, pp 1–10
3. Zhang X, Wang S, Tu J, Zhang G, Li S, Tian D, Jiao S (2018) Flower-like vanadium suflide/reduced graphene oxide composite: an energy storage material for aluminum-ion batteries. Chemsuschem 11(4):709–715
4. Subramaniam B, Natarajan B, Kaliyaperumal B, Chelladurai SJS (2018) Investigation on mechanical properties of aluminium 7075-boron carbide-coconut shell fly ash reinforced hybrid metal matrix composites. China Foundry 15(6):449–456
5. David E, Kopac J (2015) Use of separation and impurity removal methods to improve aluminium waste recycling process. Mater Today Proceed 2(10):5071–5079
6. Iyandurai N, Duraisamy P, Boopathi MM, Muniyappan M (2020) Effect of graphite and molybdenum disulfide on AA 2024 reinforced with slag and calcium carbonate hybrid metal matrix composites. Mater Today Proceed 26:3615–3622
7. Kordijazi A, Behera SK, Akbarzadeh O, Povolo M, Rohatgi P (2020) A statistical analysis to study the effect of silicon content, surface roughness, droplet size and elapsed time on wettability of hypoeutectic cast aluminum–silicon alloys. In: Light metals 2020. Springer, Cham, pp 185–193
8. Raj R, Bhardwaj K, Sharma S, Kumar N, Srivastava P (2021) Fabrication of hybrid material (Al-SiC-Fly Ash) for industrial application. In: Advances in engineering materials. Springer, Singapore, pp 461–471

9. Kaplan Y, Tan E, Ada H, Aksöz S (2019) Comparison of the effects of B_4C and SiC reinforcement in Al-Si matrix alloys produced via PM method. In: Light metals 2019. Springer, Cham, pp 129–134
10. Capuzzi S, Timelli G (2018) Preparation and melting of scrap in aluminum recycling: a review. Metals 8(4):249
11. Gaustad G, Olivetti E, Kirchain R (2012) Improving aluminum recycling: a survey of sorting and impurity removal technologies. Resour Conserv Recycl 58:79–87
12. Subrahmanyam APSVR, Narsaraju G, Rao BS (2015) Effect of rice husk ash and fly ash reinforcements on microstructure and mechanical properties of aluminium alloy (AlSi10Mg) matrix composites. Int J Adv Sci Technol 76:1–8
13. Tiwari S, Pradhan MK (2017) Effect of rice husk ash on properties of aluminium alloys: a review. Mater Today Proceed 4(2):486–495
14. Yajjala RK, Inampudi NM, Jinugu BR (2020) Correlation between SDAS and mechanical properties of Al–Si alloy made in Sand and Slag moulds. J Market Res 9(3):6257–6267

Comparisonal Analysis of "V" Punches with Various Radii and Its Impact on the Steel Material Used for Sheet Metal Operations

P. Karuppuswamy, C. Bhagyanathan, S. Sathish, and D. Elangovan

Abstract Accurate production is a critical aspect of the forming process in sheet metal businesses. Die clearance, Spring back, press load and tool radius are the most common reasons for uneven sheet metal parts during the bending process. The experimental investigation was conducted on a thick plate made of EN-3B Steel material with a thickness of 8 mm. A striking issue in sheet metal forming operation in both theory and engineering has been the bending of the steel material. Predicting this event and accounting for spring back throughout the process of bending remains a significant task today, the trials on this issue are limited. The three-point bending method is employed in this project work for the 8 mm thick plate bending process. ANN was used to analyse the spring back prediction model. It was observed that the matching loading stroke can potentially negate the deflection of the plate after spring back. Numerical simulations utilising the finite element method were run for punches with a radius of R8, R12 and R16 to explore the impact on spring back. Validation of the finite element model was done as per results.

Keywords Bending · Fe-410WC sheet metal · Numerical simulations · Spring back

1 Introduction

Nowadays, bending in thick plates (i.e., eight mm and above eight mm thick) is gaining popularity in the industrial field due to its use in structural components

P. Karuppuswamy · C. Bhagyanathan (✉) · S. Sathish
Department of Mechanical Engineering, Sri Ramakrishna Engineering College, Coimbatore, Tamil Nadu, India
e-mail: bhagyanathan@srec.ac.in

S. Sathish
e-mail: sathish.selvaraj@srec.ac.in

D. Elangovan
Department of Mechanical Engineering, PSG Institute of Technology and Applied Research, Coimbatore, Tamil Nadu, India

E. Natarajan et al. (eds.), *Materials, Design and Manufacturing for Sustainable Environment*, Lecture Notes in Mechanical Engineering,
https://doi.org/10.1007/978-981-19-3053-9_10

and as an alternative for casting supporting components. These applications make thick plate bending a critical one, despite the challenges we face in its bending. The spring-back effect is primarily case-dependent. Thus, with sheet metals (i.e., less than six mm thick), even though the spring back is high, the correction stroke is easily accessible by the technician and sufficient important research has been undertaken on those. However, even if the spring back is less in this thick plate bending, the problem we confront in this bending is very high.

A large deformation algorithm based on TEIP was used to model a typical sheet metal bending process. The impact of load on spring back when the die thickness and radius are adjusted was analysed. With an acceptable degree of precision, the numerical findings accord with the experimental data [1]. Elastic recovery is not only associated with various tool settings due to the complicated material deformation behaviour, but it is also difficult to predict correctly. The Lagrangian thermo-elastoplastic finite element approach was used to investigate a plane-strain sheet metal U-bending process [2]. The metal shell crimping assembly method was utilised to join a crimp-nut to a sheet metal structure to get the correct stress-strain curve for the already work-hardened low carbon steel. The cold forging technique is proposed as a feasible option since it provides for the fabrication of pieces with superior surface quality and mechanical qualities.

An inelastic buckling mode of deformation was developed as the magnitude of pre-strain increased, with the peak buckling load increasing and the minimum load dropping. It was established that pre-strain effects in bending of AA5754 sheet must be assessed in terms of present sheet thickness and material hardening [3]. The Voce-MA material model demonstrated capable of modelling pre-strain and strain rate effects for all instances of AA5754 sheet bending investigated in this work. For investigation of the T-section beam bending process, the spring back prediction model is created using an artificial neural network technique. Extensive use demonstrates that the suggested approach may achieve an allowed straightness error [4].

Advanced high strength steels (AHSS) are limited by their modest formability and changed failure behaviour as compared to mild steel grades. Bendability is very important in automobile applications [5]. Stringer sheets are sheet metal structures with cross-sectional bifurcations that strengthen them. They can be utilised as sheet metal hydroforming preforms to create spatially curved lightweight structures. The stiffness of sheets modified by stringers is expected to influence hydroforming outcomes. Particular attention is paid to the impact of geometric inhomogeneity on the amount and orientation of spring back. The spring back law and spring back prediction are always issues in the field of profile stretch–bending processing. Spring back is a common problem in the forming process of AHSS. The (P-HGS) is a meta-model-assisted optimization method for spring back minimization. The proposed method's efficiency has been greatly improved [6].

The evolution of the spring back and neutral layer for AZ31B magnesium alloy sheet was studied, and the results revealed that the neutral layer shifts to the sheets' tension zone. The coefficient of the neutral layer (k-value) drops as temperature and punch radii rise. This is due to the asymmetry of the outer tension layer and the interior compression layer during bending [7]. Bending and spring back of a stainless-steel clad aluminium sheet during V-shaped air bending [8]. The sheet-set condition (Alin/SSout or SSin/Alout) influences the bending phenomena significantly [9]. The constituent high-strength materials were stretched beyond the intended fracture strain limit in these testing, resulting in better formability. Multi-layered steel sheets were effectively formed as a result of this phenomenon in V-bending and hemming testing.

Bending is the most critical step in production, and achieving accuracy in that process is not as straightforward as it is in other machining operations. The key issue identified is the accuracy of this procedure as a result of its spring back. In truth, it was only via trial and error that it was conquered. In previous research, theoretical analysis of this spring back has been performed, and our numerical technique and attempt to fix this has been completed. However, presently, advanced software packages analyse spring back and material properties and a valid attempt to overcome this problem is made and succeeded with 80% success [3, 4, 10, 11].

Air bending of a "V" form for low thickness is accomplished with more accuracy and less spring back. Several studies have been conducted on spring back and the effect of bending with low thickness low carbon steel sheets. When bending is necessary for a plate-like model and the thickness of the plate is greater than 6.4 mm, it is difficult to execute the bend without side bulging and exact load to bend operation values [5].

The main problem observed is the accuracy of this operation, due to its spring back. In reality, it has been overcome just by the trial-and-error method. In the past studies, theoretical analysis of this spring back has been done, by our numerical approach and attempt to rectify this is done. But in nowadays studies the spring back and material properties are well studied by the advanced software packages and a valid attempt to overcome this problem is done and achieved with eighty per cent success. Also, in the latest studies optimization of these processes is done well-advanced approaches and nearly the exact and greatest solution is got to these spring back problems.

Nowadays, bending in thick plates (i.e., 8 mm and above 8 mm thick) is gaining popularity in the industrial field due to its use in structural components and as an alternative for foundry casting supporting components; this type of use makes thick plate bending a critical one, despite the challenges we face in its bending. The spring-back effect is primarily case-dependent. Thus, with sheet metals (i.e., less than 6 mm thick), even though the spring back is high, the correction stroke is easily accessible by the technician and sufficient important research has been undertaken on those. However, even if the spring back is less in this thick plate bending, the problem we confront in this bending is very high. Aside from their bending problems, their increasing demand as supporting components in heavy machines and all types of structures makes the study of this type of thick plate bending more unavoidable, necessitating additional research work in this area. Moreover, solving these types of challenges will be significantly more difficult. Because the interface pressures are low (well below the flow stress of the workpiece material), contact is limited to the asperities, and the coefficient of friction can be quite high [12]. If the die/workpiece combination is prone to adhesion, pickup develops. If the workpiece is covered with hard abrasives, die wear occurs. In either case, the workpiece becomes scored, and lubrication would be needed. Moreover, by bending this thick plate material in air bending, the technique becomes more versatile and may be done for any size of the plate. The need for separate presses is eliminated, making the process less expensive and more cost-effective.

2 Materials and Methods

This grade is also known as EN-3B. The dimensions of the selected material were 150 × 30x8mm. The minimum tensile strength of the material is 410 MPa, and the minimum yield stress is 250 MPa [5] (Table 1).

Using the Autodyne programme, a proper explicit model of low carbon steel is chosen for the sheet model, and tool steel is used for the die and punch. For the material model, proper hardening law is applied; all models are obtained by the software from various research papers given in the introduction section.

The output of a perceptron was created by feeding that sum via an activation function. The activation function decides whether or not the perceptron "fires" in the case of a simple binary output. An LED is attached to the output signal; if it lights up, the light goes on; if it doesn't, it stays off. The activation functions were closely watched (Fig. 1).

Table 1 Elementary composition of EN-3B steel

C%	Si%	Mn%	Ni%	Cr	Mo	P	S
0.25	0.35	1.0	–	–	–	0.06	0.06

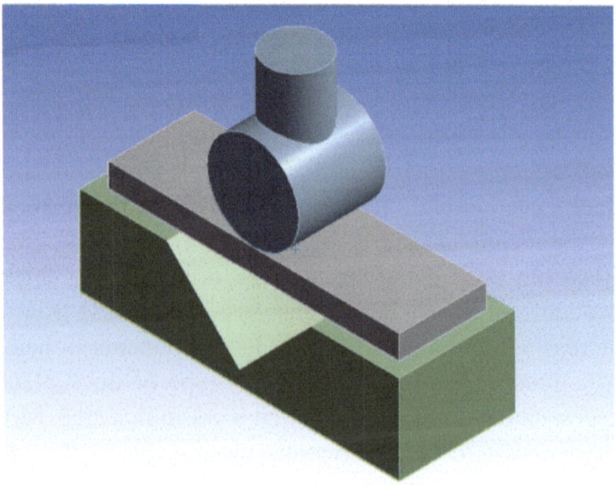

Fig. 1 Design of the die and punch materials

2.1 Analysis Setup

The dimensions of the EN-3B material considered for the research was 150 × 30 × 10 mm. The boundary was considered to be a simply supported structure.

The load on the punch is converted to Newton's employing pressure applied on the punch and the diameter of the hydraulic cylinder piston. Three types of punch are used for experimental and the same is used for simulation with calculated loads. The pressure for the three punches R8, R12 and R16 were calculated to be 6, 8 and 10 kg/cm^2 based on the observations from [13].

2.2 Experimental Setup

The hydraulic press with 10 T capacity was the machine that was used. 'V' type die was used for the analysis and the experiments were conducted with three different punches namely R8, R12 and R16. The punch material was tool steel and a total of 120 experiments were considered keeping 40 experimental analyses per punch (Fig. 2).

3 Results and Discussion

The three-point data analysis was carried out for 40 experiments on the three punches and this data is used for performing neural network simulation and predicting the desired result without real-time experiments.

From the graph shown in Fig. 3 the load range is from 8 to 10 kg/cm^2 and the 'V' forming angle is from 89° to 90.45° and for the R12 Punch the load range is from 8 to 9 kg/cm^2 and the 'V' forming angle is from 89° to 91° and for the R16 Punch the load range is from 6 to 7 kg/cm^2 and the 'V' forming angle is from 89° to 91°.

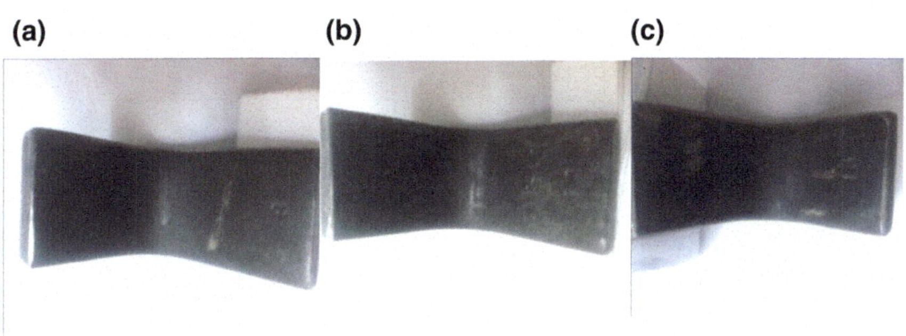

Fig. 2 Material subjected to bending with **a** R8 punch; **b** R12 punch; **c** R16 punch

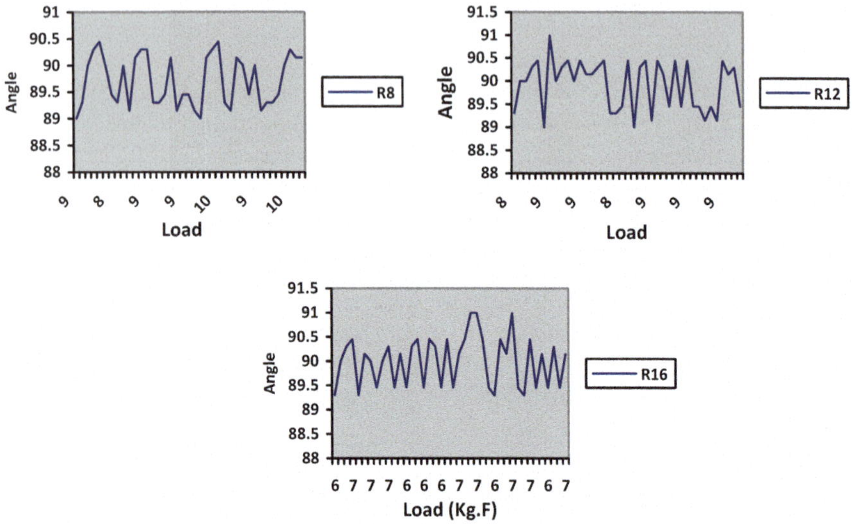

Fig. 3 Load versus angle **a** R8 punch; **b** R12 punch; **c** R16 punch

As the punch die radius increases load range decreases for the same amount of bend angle (i.e. bend angle varies between 89° and 91°). The graph range shows the optimum load range of EN-3B materials of plate thickness 8 mm and the validation of trained neurons is confirmed in the performance curve as shown in Fig. 4 its error-free validation is carried out by using an error histogram and its error is minimum.

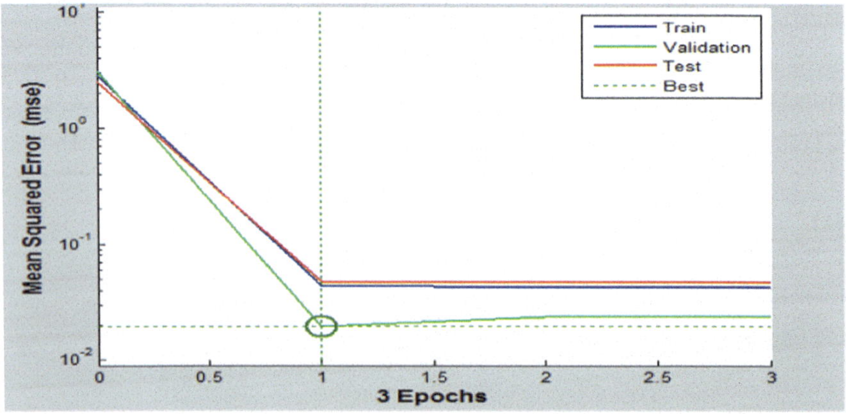

Fig. 4 Performance curve

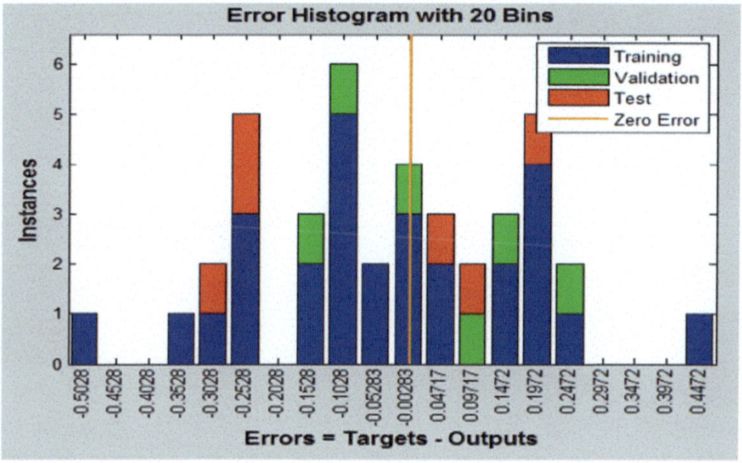

Fig. 5 Error histogram

3.1 *MATLAB—Neural Network Simulation*

The MATLAB coding of neural network training and prediction for three-point 'V' bending based on load and 'V' forming angles. The neural network needs to be trained on the given inputs and targets to predict the expected results for the user given values. The experimental data is divided into two parts, one as load and another the angle, neural network requires both data for prediction. Figures 4, 5, 6 shows the simulation and its results on performance, error histogram and regression curve for the data.

It was found that the best performance was observed at 0.0196 Mean squared error and 1 epoch based on the analysis from Fig. 4.

From Fig. 6 it was determined that the best fit was found for die angles 89°–89.5°.

4 Conclusions

The experimental investigation of "V" die air bending with three different radius punches was carried out successfully. A single die setup is used for the experiment; it has been deduced that:

- For this single die setup, which has no clearance, the load to bend the specimen increases as the punch radius increases [14];
- It has also been noted that the spring-back angle is less for R8 punch and significantly more for R16 punch (load) for each of the three radii [15].

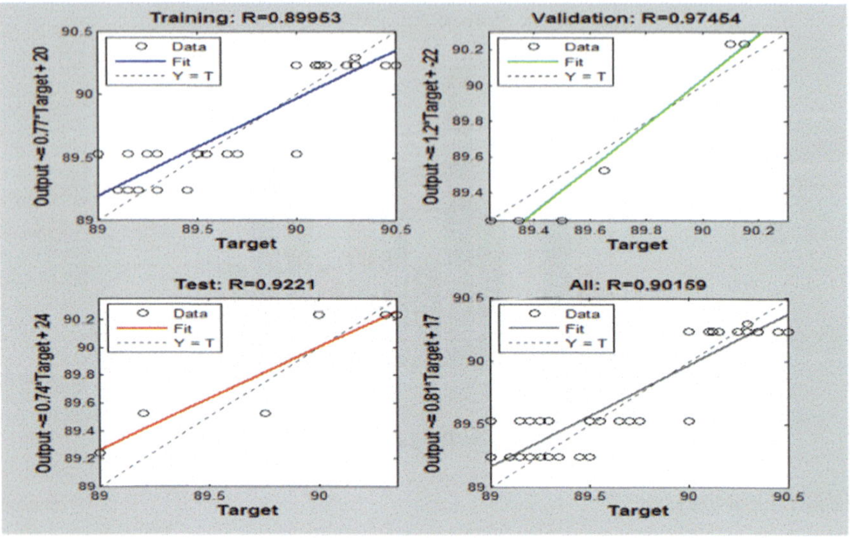

Fig. 6 Regression plot

- Each test specimen was examined, and it was found that, while the spring back is significantly less, the more side bulging in the test piece leads to the conclusion that the R8 punch radius cannot be used to form this 8 mm thick plate.
- Although the spring back is quite high, the load applied to bend the specimen is also quite high in the other two radii.

A study can be conducted on this to overcome the spring back and reduce the load required to bend the specimen by using a 'V' die with larger clearances. As a result, the punch radiuses R12 and R16 can be used for further research.

The experimental results were optimised using a neural network in MATLAB, and the same was given as output and used in subsequent experiments. The R 12 and R 16 punch results were primarily used in the real-time process. Future research will also be beneficial in fine-tuning and improving the process. And this will be the project's scope in the future.

References

1. Panthi SK, Ramakrishnan N, Pathak KK, Chouhan JS (2007) An analysis of springback in sheet metal bending using finite element method (FEM). J Mater Process Technol 186(1–3):120–124
2. Cho JR, Moon SJ, Moon YH, Kang SS (2003) Finite element investigation on spring-back characteristics in sheet metal U-bending process. J Mater Process Technol 141(1):109–116
3. Wowk D, Pilkey K (2013) An experimental and numerical study of prestrained AA5754 sheet in bending. J Mater Process Technol 213(1):1–10

4. Song Y, Yu Z (2013) Springback prediction in T-section beam bending process using neural networks and finite element method. Arch Civ Mech Eng 13(2):229–241
5. Kaupper M, Merklein M (2013) Bendability of advanced high strength steels—a new evaluation procedure. CIRP Ann 62(1):247–250
6. Zhao J, Zhai R, Qian Z, Ma R (2013) A study on springback of profile plane stretch–bending in the loading method of pretension and moment. Int J Mech Sci 75:45–54
7. Wang L, Huang G, Zhang H, Wang Y, Yin L (2013) Evolution of springback and neutral layer of AZ31B magnesium alloy V-bending under warm forming conditions. J Mater Process Technol 213(6):844–850
8. Serban FM, Bâlc N, Achimas G, Ciprian C (2013) Research concerning the springback prediction in the bending operations. In: Advanced engineering forum, vol 8. Trans Tech Publications Ltd., pp 490–499
9. Yilamu K, Hino R, Hamasaki H, Yoshida F (2010) Air bending and springback of stainless steel clad aluminum sheet. J Mater Process Technol 210(2):272–278
10. Kodli BS, Vijayanand BK () Leak proof sheet metal crimp simulation and stress analysis using FEM. STRESS 240:0–0034
11. Roque CMOL, Button ST (2000) Application of the finite element method in cold forging processes. J Braz Soc Mech Sci 22:189–202
12. Groche P, Bäcker F (2013) Springback in stringer sheet stretch forming. CIRP Ann 62(1):275–278
13. Gopal M (2021) Experimental investigation of duplex stainless steel using RSM and multi-objective genetic algorithm (MOGA). In: Materials, design, and manufacturing for sustainable environment. Springer, Singapore, pp 813–834
14. Balaji SV, Karuppuswamy P, Karuppusamy S (2014) Experimental analysis of 'V' die air bending of Fe 410W-C steel plate. Int J Curr Eng Technol. 2277–4106
15. Ahmed GS, Ahmed H, Mohiuddin MV, Sajid SMS (2014) Experimental evaluation of springback in mild steel and its validation using LS-DYNA. Procedia Mater Sci 6:1376–1385

Study of Improvement in Mechanical Properties of Chemically Treated Hybrid Fibre-Reinforced Polymer Composites

C. Boopathi, V. Vadivel Vivek, N. Natarajan, and R. Siva Balaganesh

Abstract Natural fibres such as banana, jute, coir, hemp, sisal, and bamboo are utilised as reinforcements for composite materials at random because of their inherent features such as low-cost, low-density, and low-energy consumption when compared to man-made fibres. Furthermore, they are both renewable and biodegradable, and a wide range of natural fibres are accessible. In this study, mechanical properties such as tensile, flexural, compression, and impact behaviour of a hybrid banana–jute fibre-reinforced polymer composite were examined using compression moulding techniques. Jute and banana fibre were used as reinforcements, and polyester resin was used as the matrix in this investigation. The jute and banana fibres were chemically treated for 48 h with potassium hydroxide (KOH) solution at 5% solution concentration. Fibre composites were made using the compression moulding technique. Mechanical and physical evaluations revealed the impact of surface chemical treatments on reinforcements. An optical microscope (SEM) was used to investigate the impact of chemical alteration on composites. Thermogravimetric analysis was used to determine the thermal stability of hybrid composites. Finally, it is discovered that the hybridization of jute and banana fibres might be highly useful in the production of composites.

Keywords Hybrid fibre · Banana · Jute · Mechanical behaviour · Thermal stability · Surface treatment

1 Introduction

Material selection is a key step in ensuring that a product can be used in a variety of applications. Synthetic fibres such as glass, Kevlar, aramid, and carbon fibres

---

C. Boopathi · V. Vadivel Vivek · R. S. Balaganesh (✉)
Department of Mechanical Engineering, Bannari Amman Institute of Technology, Sathyamangalam, India
e-mail: sivabalaganeshr@gmail.com

N. Natarajan
Department of Mechanical Engineering, Excel Engineering College, Komarapalayam, India

E. Natarajan et al. (eds.), *Materials, Design and Manufacturing for Sustainable Environment*, Lecture Notes in Mechanical Engineering,
https://doi.org/10.1007/978-981-19-3053-9_11

dominate in all dimensions, however, they come at a high cost, require extensive processing, are not environmentally friendly, and are potentially dangerous to humans. Natural fibre-reinforced polymer matrix composites have the ability to compete with synthetic fibre in this regard [1–5].

The binding ability of fibre and matrix determines the attributes of fibre-reinforced composites. Fibre chemical composition, hydrophobicity, and hydrophilicity are all essential criteria that influence fibre binding ability. Natural fibres, in general, have a water-absorbing (hydrophilic) character, which will have an impact on the mechanical properties of composites [6–8]. Chemical surface modification is a promising strategy for converting the hydrophilic nature of fibre to the hydrophobic nature. A thorough examination of various chemical treatments on natural fibre for natural fibre composites was conducted [9–11]. A study was conducted on potassium hydroxide (KOH) treated areca fibres, and it was discovered that KOH treated areca fibres had a considerable change in flexural strength [12–16].

In addition to increased physical and mechanical qualities, hybrid glass/jute/epoxy composites have improved mechanical properties due to the hybrid reinforcements [17, 18). After surface treatments, the flexural and impact strength of the hybrid kenaf/banana composites improved, as did the mechanical properties of the woven fabric hybrid composite [19]. In general, natural fibre hybrid composites give increased qualities by incorporating at least one synthetic fibre, such as glass or Kevlar fibres, into the composites [20–22].

Furthermore, the goal of this research is to create natural fibre hybrid polymer composites. Jute and banana fibres were used as reinforcements, whilst polyester resin was used as the matrix. The surface treatment was completed with a KOH chemical solution. The influence of KOH chemical treatment on the mechanical, physical, and thermal behaviour of natural fibre hybrid composites was investigated experimentally.

2 Materials and Method

2.1 Materials

The hybrid reinforcements of banana and jute fibres, as well as polyester resin as a matrix phase, make up the composite. Jute and banana reinforcements are readily available in India's local region. The chemical composition of banana and jute fibres was examined using FTIR techniques SITRA Coimbatore, and the well extracted banana and jute fibres were purchased in the local market for Coimbatore. Sheenu & Co. Coimbatore also supplied the matrix phase of polyester resin as well as its chemicals (Fig. 1).

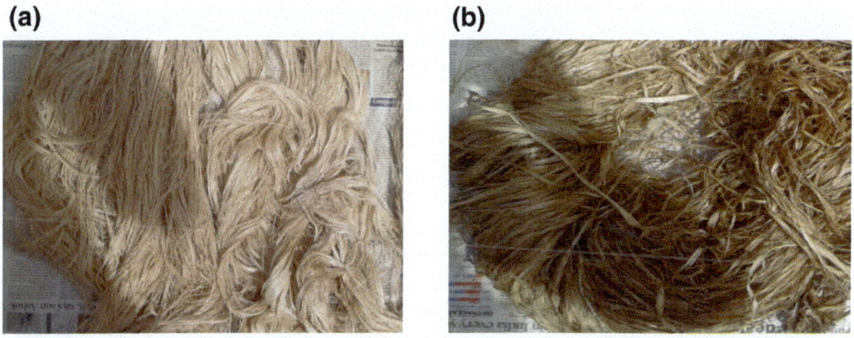

Fig. 1 **a** Jute fibres. **b** Banana fibres

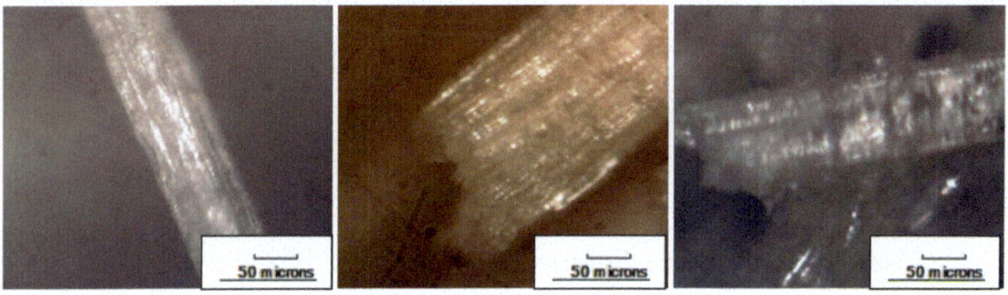

Fig. 2 Microscope analysis of untreated fibre surface

2.1.1 Microscope Analysis

The optical microscope clearly revealed the morphology examination of fibre and composites. For treated and untreated fibres, two micrograph analyses were performed, as well as fractography study for mechanically tested composites. The validity of micrographs is thoroughly tested using a large number of microscope images (Fig. 2).

2.2 Methods

2.2.1 Chemical Treatment

Natural fibres are made up of cellulose, hemicellulose, lignin, pectin, ash, and other elements, with pectin, ash, and other components physically decreasing the fibres' capacity to attach to the matrix. To eliminate undesired contaminations from natural fibre surfaces, a number of chemical treatments are accessible; in this case, chemical processing of alkaline treatment plays a critical role. The chemical treatment of potassium hydroxide (KOH) treatment was absorbed for this experiment in order

Fig. 3 Chemical treatments of hybrid fibre

to evaluate the influence of KOH chemical treatments on fibre surface. For this treatment, the fibres were individually treated for 49 h in a 5% solution concentration. The influence of chemical treatments on fibre surface was investigated using an optical microscope (Fig. 3).

2.2.2 Processing Methods—Compression Moulding Techniques

One of the most essential parameters is composite processing methods; in this study, compression moulding was used to create hybrid-reinforced polymer composites. In general, compression moulding has numerous advantages, including proper matrix material distribution, the avoidance of air holes, the achievement of desired dimensions, and the arrangement of fibres. For varying concentrations of hybrid reinforcements, the 300 mm * 300 mm * 5 mm plates were manufactured using a compression moulding machine. Compression moulding machines have a temperature-controlled zone for properly curing resin and improving fibre-to-matrix bonding (Fig. 4).

3 Mechanical Properties of Hybrid Composites

Tensile, flexural, compression, and impact tests were performed on compression moulded composites that were sized according to ASTM standards. The composites' tensile (ASTM D638) and flexural (ASTM) strength were tested using an Instron 50 k universal testing machine (UTM). The grippers were attached to the correctly dimensioned composite examples, and the testing process began. The impact characteristics of manufactured composites were measured experimentally using a Tinius Olsen Izad impact tester (ASTM D256). An average value obtained by testing five specimens in each unique composite combination for documentation purposes. Compression testing was also performed on all composite specimens to determine their capacity to withstand compression loads. Finally, all fractured specimens were subjected

Fig. 4 Compression moulding machine with plate

to fractography examination to determine the impact of chemical treatments on reinforcements.

4 Thermogravimetric Analysis (TGA)

Thermogravimetric analysis is an essential experimental technique for determining composite thermal stability. The TGA approach was used to characterise the untreated and chemically treated fibres. The untreated and chemically treated fibres were first ground to convert particles, which were then stored in a temperature-controlled zone. The temperature-controlled chamber constantly displays temperature variations and composite particle weight loss.

5 Results and Discussion

5.1 Effect of Fibre Surface Treatment

This section is mostly concerned with the effects of fibre treatments (KOH) on composites. For preliminary experiments, the obtained banana and jute fibres were treated with KOH chemical at a 5% solution concentration for 48 h. Natural fibres contain cellulose, hemicellulose, lignin, pectin, wax, and other compounds; nevertheless, the highest concentrations of wax and other compounds are highly detrimental to the binding capacity of fibres with matrix. To begin chemical treatment, the natural fibres were rinsed in distilled water and dried at room temperature. The fibres were

Fig. 5 Microscope images of chemically treated fibres

treated with a 5% solution concentration for around 48 h as part of the KOH treatment. To remove the continuing influence of chemicals on the fibre surface, the treated fibres were rinsed with distilled water and dried at room temperature. Chemical treatment was characterised using an optical microscope, and the impact of chemical treatment on fibre surface was graphically analysed.

Chemical treatments are designed to remove undesirable wax and other contaminants from fibre surfaces, but in actuality, they also remove all necessary chemical compositions such as cellulose, hemicellulose, lignin, and pectin. In this regard, the single filament test demonstrates that the strength of treated fibre was sufficient for use as a perfect reinforcement in composites even after chemical treatment. In another perspective, the surface nature of the fibres is the most crucial parameter in terms of binding ability. Fibre surfaces can be classed as crystalline or amorphous. The crystalline fibre surfaces are smooth in nature, with no flaws in the atoms arrangement, whereas the amorphous fibre surfaces are rough or uneven, with atom imperfections present. The binding ability of amorphous nature fibres will be greatest in this aspect due to their rough surfaces. Overall, the goal of this chemical treatment was to change the crystalline character of fibre surfaces to amorphous for better binding.

The microscope images are graphically exposed the fibre surfaces before and after chemical treatments in Fig. 5. As discussed earlier, the treated fibre surfaces are highly influenced by chemicals and fibre surfaces become rough (amorphous nature) intentionally to improve binding ability.

5.2 *Effect of Treated Hybrid Reinforcement on Mechanical Properties of Composites*

Chemically treated banana and jute fibres were reinforced with polyester matrix, and mechanical properties such as tensile, flexural, compression, and impact were studied. The composites were made with varied concentrations of hybrid fibre, and a total of 12 composite specimens were made using a compression moulding machine, six of which were untreated and six of which were made with KOH treated fibres. The tensile, flexural, compression, and impact properties of untreated KOH treated fibres are shown in Tables 1 and 2, respectively.

Table 1 Mechanical properties of untreated composites

Sample No.	Tensile strength	Flexural strength	Compression strength	Impact strength	Hardness
	(MPa)	(MPa)	(MPa)	(kJ/m^2)	(RHC)
A1	24.5	46	39	27	83.5
A2	26.5	46.5	40	28.5	84
A3	28	47	39.5	31	84.5
A4	23	43	30	24.5	81.5
A5	22.5	43.5	29.5	23.5	80
A6	22	42	26	16	78

Table 2 Mechanical properties of KOH treated composites

Sample No.	Tensile strength	Flexural strength	Compression strength	Impact strength	Hardness
	(MPa)	(MPa)	(MPa)	(kJ/m^2)	(RHC)
C1	33.5	52.5	41	33	91.5
C2	34	54	42.5	34.5	93.5
C3	35	55.5	44	38	94
C4	31.5	49	36.5	35	91
C5	29.5	48.5	31	32.5	90.5
C6	28	43.5	29.5	31	86

6 Graphical Representations

The mechanical characteristics for varied concentrations of hybrid reinforcements differed when comparing the results from Table 1. The differences in mechanical characteristics of by various peaks are shown in the graphs, where the tensile and flexural properties were raised by around 25% when the fibre concentration was increased (10% banana and 15% jute). Similarly, at the same 25% fibre concentration, the impact and compression properties were increased (5% of banana and 20% jute) (Fig. 6).

7 Effect of Temperature Difference on Composites

The thermal stability of natural fibres and composites is the most significant issue to consider when considering the use of composites in all applications. The fibres (banana and jute) were kept in a temperature-controlled environment to evaluate their performance over a wide range of temperatures. TGA/TA instruments were used to heat fine grained fibres in weight percentages ranging from 4 mg (2 mg banana and

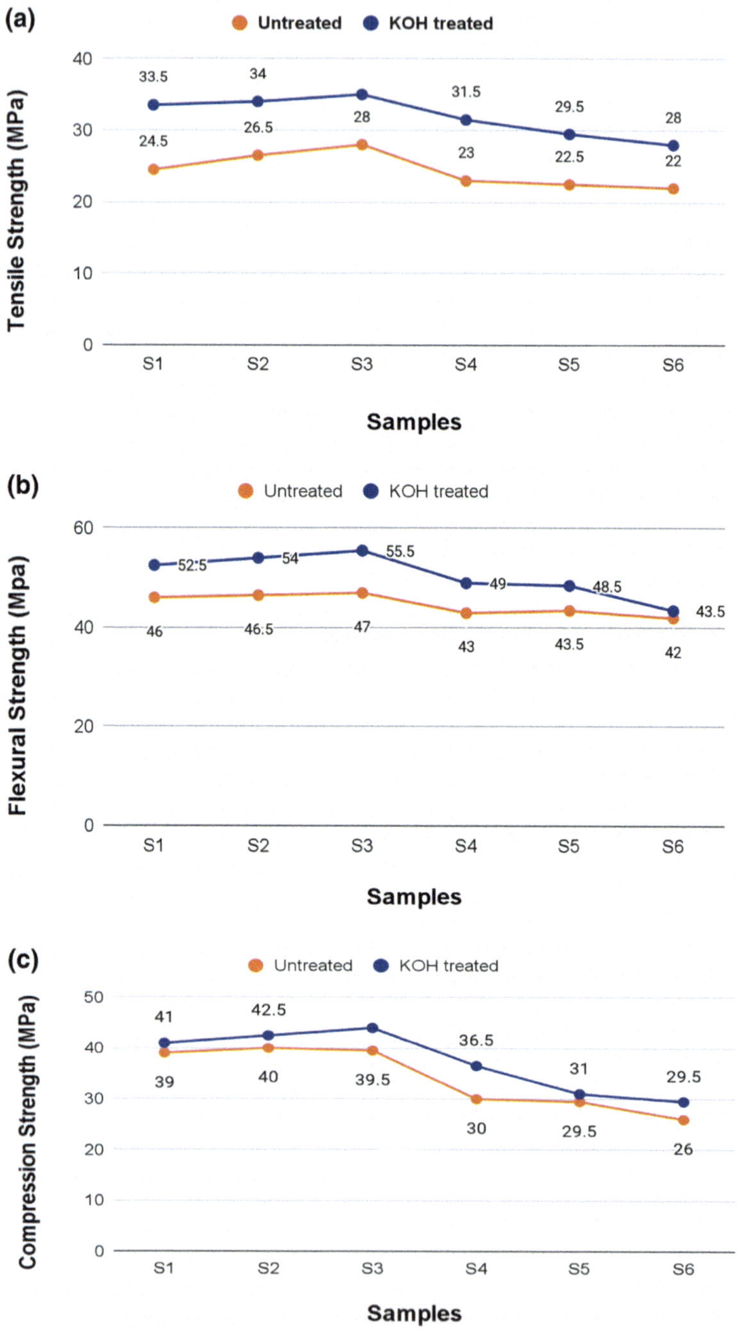

Fig. 6 **a** Graph representing tensile strength of treated and untreated samples. **b** Graph representing flexural strength of treated and untreated samples. **c** Graph representing compression strength of treated and untreated samples. **d** Graph representing impact strength of treated and untreated samples. **e** Graph representing hardness of treated and untreated samples

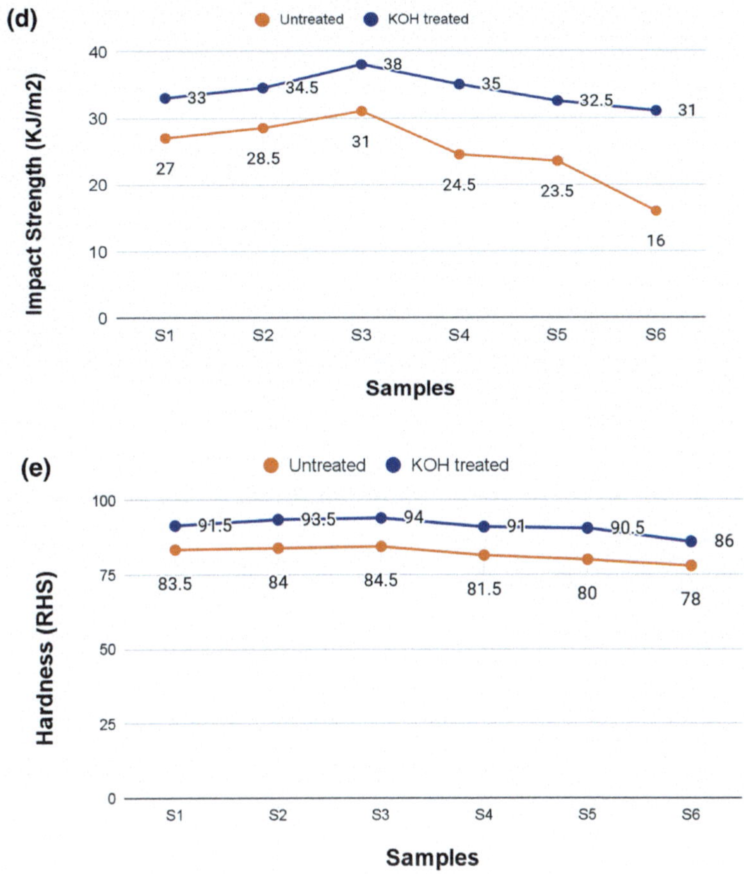

Fig. 6 (continued)

2 mg jute fibres) from 30 to 800 °C at a rate of 10 °C per minute. TGA/TA instruments were used to heat fine grained fibres in weight percentages ranging from 4 mg (2 mg banana and 2 mg jute fibres) from 30 to 800 °C at a rate of 10 °C per minute. The Fig. 6 expressed the reaction of fibres for various temperature ranges, from this we may conclude the fibres could sustain up to 550 °C above 550 °C, the fibres became carbon. The chemical composition of fibres was burned in different stages of temperatures. The results show that the KOH treatment resulted in the removal of chemical composition on the fibre surface, implying that chemically treating fibres reduced thermal stability (Fig. 7).

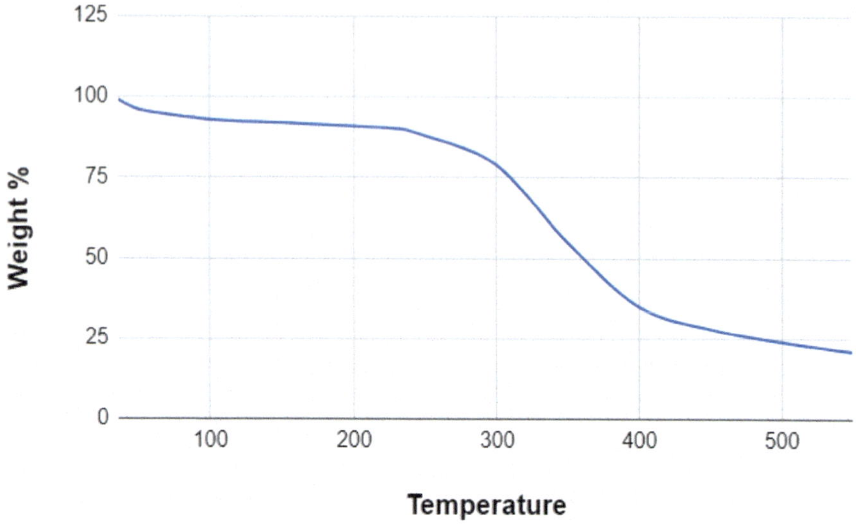

Fig. 7 TGA analysis of graph

8 Conclusion

Chemically treated hybrid fibre-reinforced polymer composites were tested in terms of mechanical and thermal behaviour. The fibre surface was modified using a KOH chemical solution, and the banana and jute hybrid fibres were chemically treated adequately in terms of mechanical qualities. The composite's thermal stability was also calculated, and the results show that the composite properties were very gradual as the jute fibre volume concentrations increased.

- The jute and banana hybrid fibre-reinforced composite show better mechanical and physical property than jute fibre-reinforced composite and banana fibre-reinforced composite. From this, it is concluded that the hybridization of jute and banana fibre exposes good agreement between them for hybrid composite.
- There are two chemical surface modification process carried out for both jute and banana fibres by KOH treatments. As comparatively untreated hybrid composite, the treated fibre composite exhibits better properties in all aspects.

References

1. Al-Oqla FM, Sapuan SM (2014) J Cleaner Prod 66:347
2. Al-Oqla FM, Sapuan SM, Ishak MR, Nuraini AA (2014) J Compos Mater 0:1
3. Ahmad F, Choi HS, Park MK (2015) Macromol Materi Eng 10:300
4. Madsen B, Thygesen A, Lilholt H (2007) Part A: Appl Sci Manufac 67:1584
5. Yao J, Ma L, Lu W, Tan H (2015) Bioresour 10:697

6. Zampaloni M, Pourboghrat F, Yankovich SA, Rodgers BN, Moore J, Drzal LT, Mohanty AK, Misra M (2007) Part A: Appl Sci Manuf 38:1569
7. Srinivasan VS, Somasundaram S, Ravindran D, Manikandan V, Narayanasamy R (2011) Mater Des 32:453
8. Saba N, Paridah MT, Jawaid M (2015) Constr Build Mater 76:87
9. Puniyamoorthy R, Sampath Kumar D, Srinivasa CV, Bennehalli B (2015) Bioresour 10:697
10. Li X, Tabil LG, Panigrahi S (2007) J Polym Environ 15:25
11. Srinivasa CV, Bennehalli B, Kenchappa MG, Ranganagowda RPG (2015) Bioresour 10:697
12. Srinivasa CVB, Bharath KN (2013) Int J Chem Nucl Metall Mater Eng 7:4
13. Van de Weyenberg I, Truong TC, Vangrimde B, Verpoest I (2006) Part A: Appl Sci Manuf 37:1368
14. Gassan J, Andrzej Bledzki K (1999) Compos Sci Technol 59:1303
15. Threepopnatkul P, Kaerkitcha N, Athipongarporn N (2009) Compos: Part B 40:628
16. Phan NH, Rio S, Faur C, Le Coq L, Le Cloirec P, Nguyen TH (2006) Carbon 44:2569
17. Rahman MR, Huque MM, Islam MZ, Hasan M (2008) Compos: Part A 39:1739
18. Mohamed Jalaludeen S, Malayalamurthi R, Jayabal S, J Mech Sci Technol
19. Alavudeen, Ranjani N, Karthikeyan S, Thiruchitrambalam M, Venkateshwaren N, Mater Des
20. Birat KC, Panthapulakkal S, Kronka A, Agnelli JA, Tjong J, Sain M, J Appl Polym Sci
21. Venkateshwaran N, Elayaperumal A, Sathiya GK (2012) Compos: Part B 43:793
22. Thwe MM, Liao K (2002) Compos: Part A: Appl Sci Manuf 33:43

Comparative Study of Mechanical Strength and Piezoelectric Coefficient of Post-processed Polyvinylidene Fluoride Nanofibrous Films

M. Satthiyaraju, K. Ananthakumar, R. Shankar, and C. K. Arvinda Pandian

Abstract The polyvinylidene fluoride electrospun nanofibers were fabricated through the electrospinning technique. The electrospun PVDF nanofbrous films were post-processed as annealed and mechanically stretched for enhancing its mechanical strength and piezoelectric coefficient to utilize for the different piezoelectric applications. The FESEM images are shown beadles and randomly oriented nanofibers. After post-processing, the mechanical strength and elongation due to break were significantly enhanced. The annealed and stretched sample P-A-S is shown increment up to 22.74 and 35.15% due to the interconnection of nanofibers. Moreover, the piezoelectric coefficient D33 is enhanced to the significant level up to 23.94% of increment of P-A-S nanofiber films. These research woks will assist the future work to obtain the various interesting findings to enhance the piezoelectric properties of the piezoelectric polymers which can be used for wearable electronics, energy harvesting, biomedical, sensors, and actuators applications.

Keywords PVDF · Nanofiber · Post-processing · Mechanical strength · Piezoelectric coefficient

---

M. Satthiyaraju
Department of Mechanical Engineering, Kathir College of Engineering, Coimbatore 641062, India

K. Ananthakumar
Department of Mechanical Engineering, Karpagam College of Engineering, Coimbatore 641032, India

R. Shankar
Department of Mechanical Engineering, Kings College of Engineering, Pudukkottai 613303, India

C. K. Arvinda Pandian (✉)
Department of Automobile Engineering, School of Mechanical Sciences, B. S. Abdur Rahman Crescent Institute of Science and Technology, Chennai 600048, India
e-mail: doctratearvind@gmail.com

E. Natarajan et al. (eds.), *Materials, Design and Manufacturing for Sustainable Environment*, Lecture Notes in Mechanical Engineering,
https://doi.org/10.1007/978-981-19-3053-9_12

1 Introduction

The unlimited need of piezoelectric materials applications directs the current research which focuses on the multiple development for fulfilling the ever growing needs. The energy harvesting, sensors and actuators, drug delivery systems, piezoelectric nanogenerators, and wearable electronics are the focused applications for the piezoelectric polymers, ceramics, and their nanocomposites [1–5]. The piezoelectric polymer polyvinylidene fluoride is most probably famous for its unique advantages such as good mechanical and thermal stability, nontoxic, highly flexible compared to the piezoceramic materials.

Among all the piezoelectric materials, the PVDF is semi-crystalline polymer which possesses the main five polymorphic phases such as alpha (α), beta (β), gamma (γ), delta (δ), and epsilon (ε) [6]. The main monomer of the in the PVDF is $-CH_2CF_2-$. Among the polymorphic crystalline phases, the β and γ phases are mainly responsible for the piezoelectric responses. Also, not only the PVDF is having piezoelectric property, also it is having the pyroelectric property. There are different methods to form the nanofibers as solvent casting, phase inversion, and electrospinning. Moreover, the electrospinning technique shows the mechanical stretching and thermal heating of the polymeric nanofibers which enhances the piezoelectric phases [7].

The electrospinning method is cost effective, easy, and simple to produce the nanofibers compared to the other techniques. The dimensions from micro-meters to nanometers are possible to produce the polymeric fibers using the electrospinning. Based on the dimensions, the nanolevel fibers can be shown good piezoelectric properties due to high-fiber surface to its volume ratio which leads to many biomedical and electronics applications [8]. Also these structures can be classified into the one dimensional structures. Based on the literature, the heat treatment and mechanical stretching can lead to enhance the piezoelectric properties and mechanical stability of the PVDF nanofibrous films.

In this research work, the PVDF polymeric nanofibers were prepared through the electrospinning method. There are no other works have been published related to the comparative analysis of post-processed piezoelectric nanofibers. The post-processing methods are thermal annealing and mechanical stretching. Here, both methods are used and compared to analyze the morphology, mechanical properties, ad piezoelectric responses of the nanofibers. The additional heating and additional mechanical drawing lead to the molecular alignment of the polymorphic piezoelectric phases.

Table 1 Sample details of post-processed PVDF films

S. No.	Sample code	Post-processing treatment	Temperature (°C)	Strain rate (mm/min)
1	P	As-spun	–	–
2	P-S	Stretching	–	1
3	P-A	Annealing	100	–
4	P-A-S	Annealing and stretching	100	1

2 Experimental Methods

2.1 *Materials*

Polyvinylidene fluoride (PVDF) has been purchased from Sigma-Aldrich with molecular weight of 275,000 g/mol. Dimethyl formamide (DMF) and acetone were used as a solvent for obtaining homogeneous polymeric solution.

2.2 *Preparation of PVDF Nanofibers*

The PVDF nanofibers were prepared using the electrospinning method. The homogeneous solution is the key for obtaining proper nanofibrous structures of PVDF polymers. The PVDF pellets were added into the DMF: acetone mixture. Here, the acetone is helped to evaporate the solvents quicker from the polymeric solution. The solution is filled with the 5 ml metallic syringe. In which, the nanofibers have been formed due to the electrostatic effect in between the needle tip and static collector. The optimal parameters were followed in this work is applied voltage 17 kV, distance to collector 18 cm and PVDF 20 wt%.

2.3 *Post-processing Details of PVDF Nanofibers*

The PVDF nanofibers have been heated to the different temperatures under the category of thermal annealing. Some set of samples were proposed to various strain rates as a post-processing treatment of mechanical stretching. Hence, the post-processing treatment was mainly applied to these PVDF samples to realign the crystalline structure of polymorphic phases. The post-processed films were named as PVDF, PVDF-A, PVDF-S, and PVDF-A-S, respectively. The details of post processing treatment on PVDF nanofibers are given in the Table 1.

2.4 *Characterization*

The morphological studies of the PVDF nanofibers were examined using the scanning electron microscope (SEM). These studies are most important to get the nanofiber diameters and alignment of the nanofibers. The minimal layer of gold helps to obtaining the clear morphology. The mechanical properties of post-processed films were analyzed through the stress strain curves with the proper dimensions as per the D882 ASTM Standard. The D33 piezometer was used to find out the D33 piezoelectric coefficient under the mechanical load applied on the surface of the nanofibrous films. All the tests were repeated three times to get the accurate values from the experiments.

3 Results and Discussion

3.1 *Scanning Electron Microscope Studies*

The surface morphology of the nanofibrous mat is examined through the FESEM micrographs which is shown in the Fig. 1. The micrographs of the PVDF mat have not shown any beads formation and accumulation of nanofibers which leads to confirm that the nanofibers have formed properly after evaporation of all the solvent content while forming the fiber jet from the polymeric solution. The FESEM images have shown the neat and randomly oriented nanofibers without any residues. Also, it shows high amount of interconnected electrospun nanofibers and the range of porosity. The main reason behind the no bead formation and agglomeration of nanofibers is the solvents of DMF and acetone were mixed and evaporated properly.

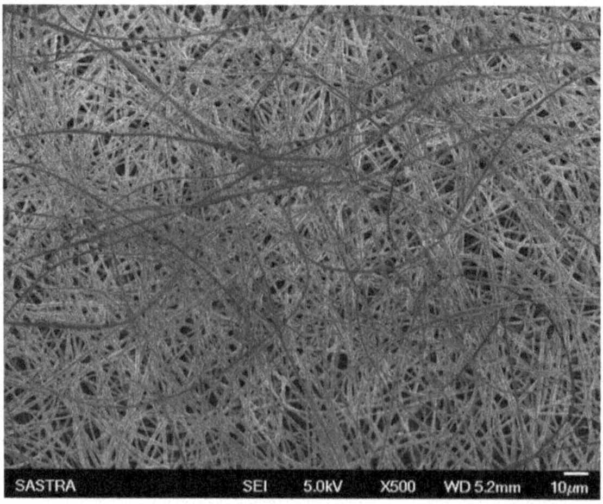

Fig. 1 SEM micrographs of the prepared PVDF nanofibrous films

3.2 Mechanical Studies

The mechanical responses of the post-processed films are obtained through the uniaxial tensile test in which the stress–strain curves were obtained and illustrated in Fig. 2. The mechanical tensile strength is increased to certain level based on the different treatments after the fabrication of the PVDF nanofibers. Due to the annealing treatment, the mechanical strength of the sample P-A is increased to 17.78%. Moreover, the stretched films (P-S) were shown 5.54% decrement due to the pre-stretching effect. Hence, the mechanically stretched nanofibers after the heat treatment (P-A-S) are shown good mechanical response as 23.92% increment compared to the as-spun PVDF nanofibers.

The elongation of post-processed nanofibrous films also was obtained from the mechanical test. Where, the percentage of the elongation is decreased to 16.15, 23.75, and 35.16% compared to the as-spun PVDF nanofibers. This effect of decrement shows that the pre-stretching effect and annealing could be reduced the percentage of the elongation. The fusion among the nanofibers in different spots is lead to the enhancement of the mechanical strength [9]. Hence, the intermolecular relation and the nature of crystallinity direct to the good mechanical properties of the electrospun nanofibers.

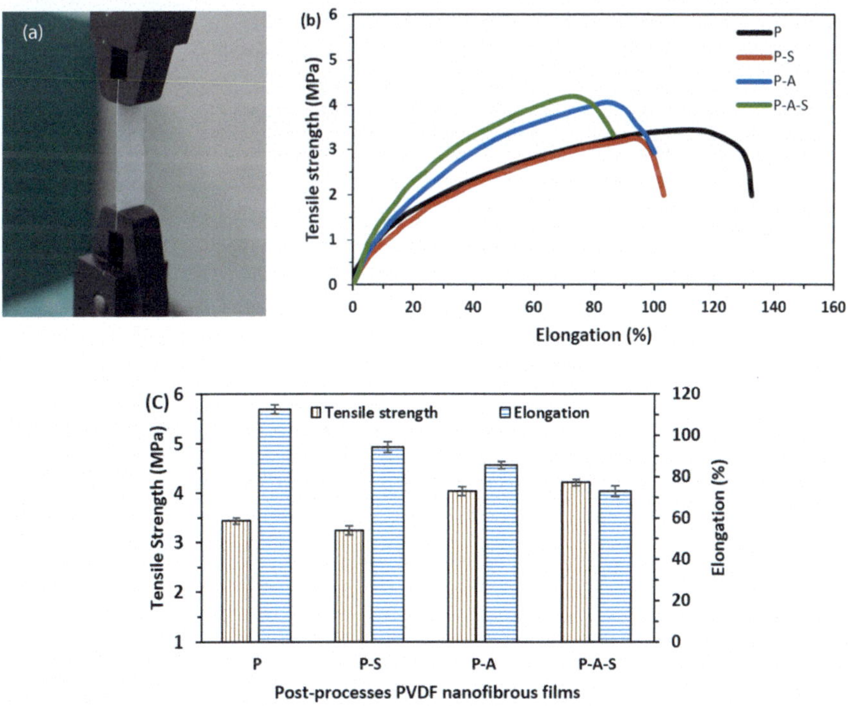

Fig. 2 **a** Tensile testing of PVDF nanofibrous films, **b** stress–strain plots of post-processed PVDF films, **c** mechanical strength and elongation from the tensile tests of PVDF films

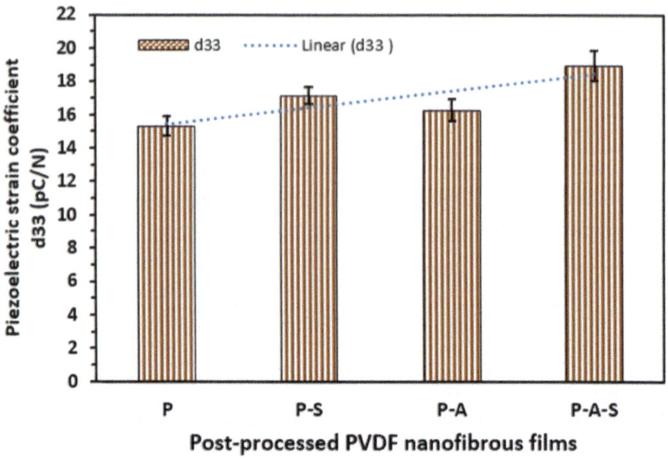

Fig. 3 Piezoelectric D33 piezometer responses on the post-processed PVDF films

3.3 Piezoelectric Studies

Among the polymorphic phases in the PVDF, the β phase could be enhanced due to the high voltage and mechanical pulling of the nanofibers. Here, the electrostatic force is the main reason for the molecular rearrangement which is the possible cause for the enhancement of the β phase in the nanofibers.

The piezoelectric coefficient, D33, is highly related to the piezoelectric phase of the PVDF where it is enhanced over the post-processing treatment. Even the fiber diameters can be influenced the piezoelectric effect of the nanofibers. The electrified jets of the polymeric solution are highly elongated in the electrospinning process. The measured piezoelectric coefficients have been shown in the Fig. 3. Hence, the piezoelectric coefficient of the annealed nanofibers films is increased to (17.12 pC/N) 11.92%, and the mechanically post stretched films were increased up to 16.25 pC/N as 6.2% increment. The D33 coefficient of both annealed and stretched nanofibers is explored as 23.92% of increment as 18.96 pC/N compared to the as-spun PVDF. This piezoelectric coefficient is directly related to the piezoelectric effect of the nanofibrous material [10]. Conclusively, these post-processed fibers can be used for the various piezoelectric applications.

4 Conclusion

In this study, the response of the post-processing treatment on the mechanical strength and piezoelectric effect of the PVDF nanofibers were investigated. The conclusive points are following as,

- The PVDF nanofibrous films were fabricated through the electrospinning method
- The nanofibers were processed under the annealing and mechanical stretching as post-processing treatment
- The FESEM micrographs were shown beadless, non-accumulative, and randomly oriented nanofibers
- The mechanical strength of the nanofibers was shown 22.74% increment for the P-A-S which is annealed and stretched specimen. Also the percentage of elongation of the nanofibers were decreased to percentage of decrement of 35.15% for the P-A-S.
- The piezoelectric coefficient is enhanced due to the annealing and mechanical stretching up to 18.96 pC/N.

The post-processed PVDF nanofibers can be used for the wearable electronics, energy harvesting, and sensors and actuator applications.

References

1. Dias JC, Correia DC, Lopes AC et al (2016) Development of poly(vinylidene fluoride)/ionic liquid electrospun fibers for tissue engineering applications. J Mater Sci 51:4442–4450
2. Satthiyaraju M, Ramesh T (2019) Nanomechanical, mechanical responses and characterization of piezoelectric nanoparticle-modified electrospun PVDF nanofibrous films. Arab J Sci Eng 44(6):5697–5709
3. Xie Y, Wang J, Yu Y et al (2018) Enhancing breakdown strength and energy storage performance of PVDF-based nanocomposites by adding exfoliated boron nitride. Appl Surf Sci 440:1150–1158
4. Satthiyaraju M, Ramesh T, Jagatheswaran K (2019) Annealing and ZnO doping effects on hydrophilicity and mechanical strength of PVDF nanocomposite thin films. In: Advances in manufacturing technology, pp 463–471
5. Satthiyaraju M, Ramesh T (2019) Effect of annealing treatment on PVDF nanofibers for mechanical energy harvesting applications. Mater Res Express 6(10):105366
6. Martins P, Lopes AC, Lanceros-Mendez S (2014) Electroactive phases of poly(vinylidene fluoride): determination, processing and applications. Prog Polym Sci 39:683–706
7. Chen HJ, Han S, Liu C et al (2016) Investigation of PVDF-TrFE composite with nanofillers for sensitivity improvement. Sens Actuators, A 245:135–139
8. Liang CL, Mai ZH, Xie Q et al (2014) Induced formation of dominating polar phases of poly(vinylidene fluoride): positive ion-CF_2 dipole or negative ion-CH_2 dipole interaction. J Phys Chem B 118:9104–9111
9. El Mohajir BE, Heymans N (2001) Changes in structural and mechanical behaviour of PVDF with processing or thermal treatment. 2. Evolution of mechanical behaviour. Polymer (Guildf) 42:7017–7023
10. Roji AMM, Jiji G, Raj ABT (2017) A retrospect on the role of piezoelectric nanogenerators in the development of the green world. RSC Adv 7:33642–33670

Bidirectional Jute-Reinforced Polyester Composites: Influence of Sodium Bicarbonate Treatment on Static Mechanical Properties

P. Ravikumar, G. Rajeshkumar, K. C. Nagaraja, S. Rajanna, and M. Karthick

Abstract The present work involved the fabrication of jute fiber-reinforced polyester composites by employing compression molding method and determined their tensile, flexural, and impact properties. The neat polyester (NP), untreated jute fiber (UTJF)/polyester composites, and sodium bicarbonate-treated jute fiber (SBTJF)/polyester composites were subjected to the above-mentioned mechanical testings. The test results indicate that incorporating jute fiber into a polyester matrix improves the mechanical properties. This improvement is attributed to stiffness of the composites increased with the incorporation of jute fibers. Additionally, an attempt was made to improve the mechanical properties of the polyester by reinforcing it with SBTJF. The fibers' surface treatment had a beneficial effect on their mechanical qualities. This improvement in properties is attributed to the interfacial bonding at fiber-matrix interface, and it was evidenced in the morphological analysis.

Keywords Polyester composites · Sodium bicarbonate · Jute fiber · Mechanical properties · Interfacial bonding

---

P. Ravikumar
Department of Mechanical Engineering, Adithya Institute of Technology, Coimbatore, Tamil Nadu, India

G. Rajeshkumar (✉) · M. Karthick
Department of Mechanical Engineering, PSG Institute of Technology and Applied Research, Coimbatore, Tamil Nadu, India
e-mail: rajesh@psgitech.ac.in

K. C. Nagaraja
Department of Mechanical Engineering, Acharya Institute of Technology, Bengaluru, Karnataka, India

S. Rajanna
Department of Mechanical Engineering, Government Engineering College, Hassan, Karnataka, India

E. Natarajan et al. (eds.), *Materials, Design and Manufacturing for Sustainable Environment*, Lecture Notes in Mechanical Engineering,
https://doi.org/10.1007/978-981-19-3053-9_13

1 Introduction

Polymer composite materials are used to make low-weight structural elements for automobiles, airplanes, and other commercial applications because of their outstanding mechanical qualities, high strength-to-weight ratio, biodegradable nature, lower density, repeatability, and minor abrasiveness to processing equipment [1–3]. In the goal of environmental safety and protection, the scientific community and industries have recently been more fascinated in natural fiber-based products for civil constructions, car and construction interiors, consumer, and athletic products among other applications [4]. Polymer composites may be made using a variety of production processes, including manual lay-up, extrusion, sealing, and resin transfer molding, and each process has its influence on the finished composite's qualities. Compression molding is the most extensively utilized fabrication process for creating randomly oriented natural fiber-reinforced composites since it accommodates over 70% of the reinforcements [5–7].

Jute is a well-known vegetable fiber that is mostly farmed in India, Bangladesh, China, Nepal, and Thailand. They collectively generate over 95% of the world's jute fibers [8]. In this aspect, jute fibers may find use in vehicle parts, construction, and sports. Many research works reported the improvement of various characteristics of polymers after reinforcing them with jute fibers. Raghavendra et al. [9] from his results conveyed that the jute fiber-reinforced polymer composites offer 55 and 61% tensile and flexural strengths, respectively, of glass fiber composites. Mishra and Biswas [10] found that incorporating bidirectional jute fibers into epoxy greatly improved the composites' mechanical characteristics. Composites containing 48 wt.% jute fibers, in particular, exhibit superior overall characteristics.

Though the jute fiber addition in polymers offers many advantages, it also has a major drawback of incompatibility because of hydrophilic and hydrophobic natures of jute fiber and polymers, respectively. Hence, the surface of fibers is treated with physical and chemical procedures to ensure compatibility and interfacial bonding at the fiber-matrix interface [11, 12]. Among other approaches, the majority of researchers use chemical treatments to alter the structure and surface of the fibers to increase the compatibility of composite constituents [13, 14]. Chemical treatment opens up the cellulose content and removes superfluous elements, making fibers smooth, easy to bind, durable, and flexible. It also has a long-term effect on natural fiber mechanical characteristics, particularly strength and stiffness [15]. In this aspect, Ramakrishnan et al. [16] explored the viscoelastic and free vibration characteristics of treated jute/epoxy composites and found that composites containing 5% NaOH-treated jute fiber performed better. Similarly, Wang et al. [15] perceived the improvement in mechanical qualities of composites containing treated jute fibers.

Very recently, researchers have stated using sodium bicarbonate (SBC) solution for treating natural fibers due to its environmental friendliness and cost-effectiveness. Literature indicated that the fibers' surface altered with 10 wt% SBC solution will offer excellent results [17–19]. Santos et al. [20] evaluated the consequence of SBC treatment (10 wt%) on coir fiber-based epoxy and polyester composites. The

Table 1 Characteristics of constituent materials [22]

Properties	Polyester	Jute fibers
Density (g/cm^3)	1.16	1.3
Diameter (mm)	–	0.015–0.002
Tensile strength (MPa)	8–19	393–773
Tensile modulus (GPa)	0.58	26.5
Strain at break (%)	1.6	1.5–1.8

outcomes indicated that fibers treated for more than 96 h exhibited superior characteristics when incorporated into polymer matrices. In another work, Zindani et al. [21] explored the mechanical qualities of 10 wt% SBC treated Punica Granatum-reinforced bio-epoxy composites. The results disclosed that the composites added with 5 days (120 h) treated fibers perform better. To the author's knowledge, no study has been conducted on the influence of SBC treatment on the mechanical performance of bidirectional jute/polyester composites. Therefore, in the present work, mechanical characterization of neat polyester (NP), untreated jute-based polyester composites and SBC (10 wt%) treated (120 h) jute fiber-based polyester composites was carried out in order to find the effect of fiber treatment. Additionally, a morphological analysis was conducted to better understand the bonding properties of the jute fiber and polyester matrix using scanning electron microscope (SEM).

2 Materials and Method

2.1 Materials and Their Properties

For the present work, the bidirectional woven jute fibers were used as reinforcement to polyester composites. The bidirectional jute fibers, polyester, and hardener were procured from local retailers, Coimbatore, India. The physical and mechanical characteristics of constituent materials are presented in Table 1.

2.2 Sodium Bicarbonate Treatment

In ambient circumstances, the bidirectional weaved jute reinforcements were manually washed and dried for 48 h. After that, the reinforcements were soaked in a 10 wt% of SBC solution for the time duration of 120 h and then rinsed with distilled water. Following that, the treated reinforcements were dried for 24 h at 40 °C.

2.3 Preparation of Composites

The composites were made using the compression molding procedure, which involved reinforcing raw and sodium bicarbonate-treated bidirectional woven jute fibers into the polyester matrix. The silicone spray was first sprayed on the mold's inside surface to facilitate the removal of the composite laminate once it had cured. The matrix solution of polyester and hardener in a 10:1 combination was poured over the bidirectional woven jute fibers placed on bottom part of the mold cavity. To obtain a composite laminate with a constant thickness, the mold was closed and held under 0.5 MPa pressure for 6 h. Finally, these laminates were post-cured for 4 h at 60 °C and trimmed according to ASTM standards for mechanical testings.

2.4 Mechanical Testings

Universal testing equipment (8801-INSTRON) was used to assess the composites' tensile and flexural characteristics. The tensile test was performed following ASTM 3039 (specimen dimension: $250 \times 25 \times 3\ mm^3$) using a gage length of 150 mm and a crosshead speed of 2.5 mm/min. With a span length of 40 mm and a crosshead speed of 2.5 mm/min, the flexural test was conducted according to ASTM D790 (specimen size: $140 \times 12.7 \times 2\ mm^3$). The impact test was conducted following ASTM D256 utilizing Izod impact testing equipment (Tinius Olsen IT 503, UK).The impact hammer weighing 1.30 kg was made to impact the specimen ($63.5 \times 12.5 \times 3\ mm^3$) loaded on the machine. Mechanical testing was performed on five samples from each category, and the average value was recorded.

2.5 Morphological Analysis

SEM (Hitachi S 3500) was used to analyze the eminence of fiber to matrix interfacial bonding in UTJF and SBTJF-reinforced polyester composites at a 10 kV accelerated voltage.

3 Experimental Results and Discussion

3.1 Tensile Property

Tensile characteristics of NP, UTJF, and SBTJF composites are shown in Fig. 1. SBTJF composites have greater tensile strength and modulus than NP and UTJF

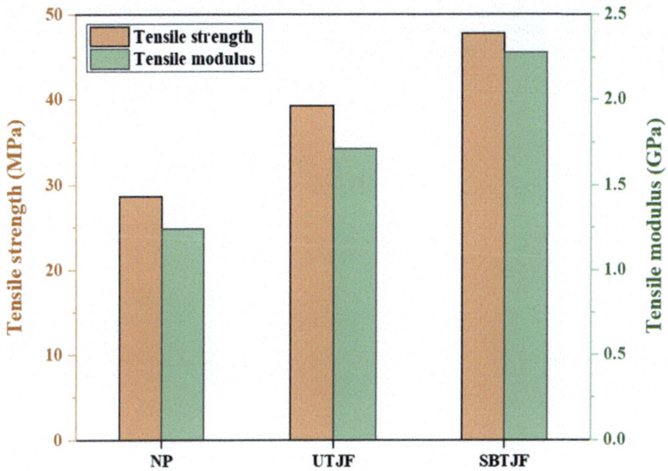

Fig. 1 Tensile properties

polyester composites. This can be ascribed to the increase of stiffness after the addition of treated jute fiber into the polyester matrix. Compared to NP, 37.20 and 67.30% increase in the tensile strength was noted for UTJF and SBTJF composites, respectively. Furthermore, the surface treatment helps the composites to reduce the porosity as well as micro-cracks present on the surface of the fiber thereby increase in tensile strength could be achieved. Similarly, the UTJF and SBTJF composites showed 37.9 and 83.87% improvement in tensile modulus, respectively, compared to NP. Apart from the stiffness, the interfacial bonding between the constituents contributes significantly toward the improvement of tensile properties.

3.2 *Flexural Property*

Figure 2 depicts the impact of fiber treatment on the composites' flexural strength and modulus. When flexural characteristics of surface-treated jute fiber were compared to UTJF and NP, it was discovered that surface-treated jute fiber had the best results. This could be because the surface treatment created a rough surface on the fiber, which aided in the attachment of jute fiber to polyester matrix, resulting in an improvement in flexural characteristics. Furthermore, the flexural strength of UTJF and SBTJF composites was 10.54% and 23.44%, respectively, greater than NP. Similarly, SBTJF composites had flexural moduli that were 101.45 and 20.28% greater than UTJF composites and NP. This demonstrates that the jute fiber improves the flexural characteristics of the polyester matrix. Furthermore, the improved interfacial bonding and considerable load transmission between the jute fiber and polyester can be attributed to this improvement [23–25].

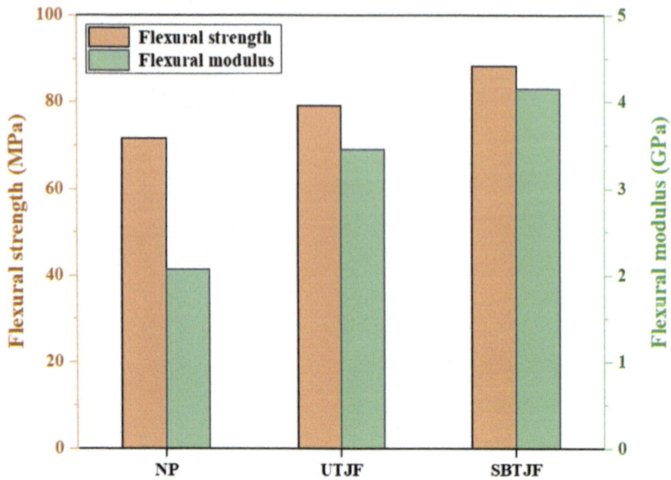

Fig. 2 Flexural properties

3.3 Impact Property

Impact strength refers to a material's ability to absorb and disperse energy when subjected to impact loading. The impact strength of NP, UTJF, and SBTJF composites is depicted in Fig. 3. SBTJF composites were shown to have a 21.94 and 67.3% greater impact strength than UTJF composites and NP, respectively. This revealed that jute fibers are capable of absorbing and dissipating shock loads. Matrix fracture, fiber pull-out, and composite debonding are the key elements that influence impact failure [20, 26, 27]. These failure mechanisms were identified in SEM micrographs of the composites, which enable composites to withstand greater impact loads.

3.4 Morphological Analysis

Figure 4 shows SEM micrographs of broken surfaces of UTJF and SBTJF composites. Due to insufficient interfacial bonding between the elements, the UTJF composites exhibited fiber pull-outs and micro-cracks at the interface. On the contrary, the SBTJF composites exhibited excellent bonding between the composite constituents. As a result, treated fiber composites absorb more energy upon failure, resulting in improved mechanical properties. The surface treatment with SBC would also help in increasing the stiffness of the jute fiber, which would enhance the mechanical properties [25, 28].

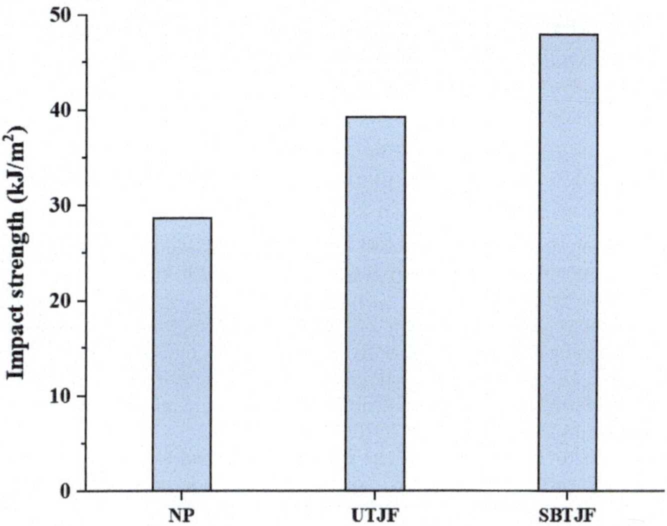

Fig. 3 Impact strength

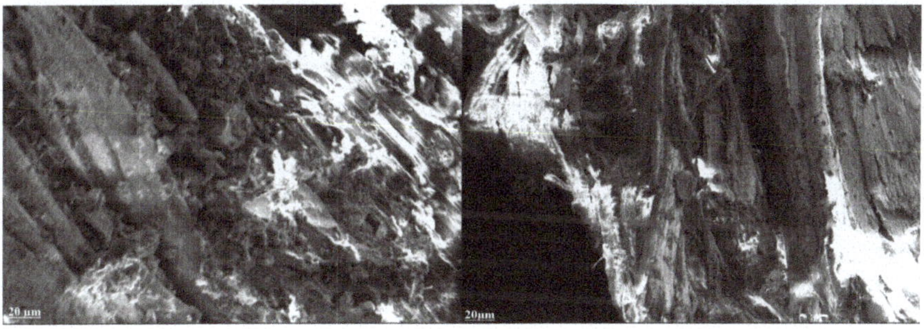

Fig. 4 SEM images of UTJF and SBTJF composites

4 Conclusions

The polyester-based natural composites were manufactured by reinforcing untreated and sodium bicarbonate-treated jute fibers and evaluated their static mechanical performance. In terms of characteristics, all of the produced composites outperformed the plain polyester, demonstrating the use of jute fiber as a major reinforcement in the fabrication of polyester-based composites. Furthermore, mechanical testing findings show that sodium bicarbonate treatment has a beneficial impact on them due to its compatibility, greater fiber-matrix interfacial bonding, mechanical interlocking, and increased stiffness. SEM micrographs evidenced that the treated jute fiber composites had stronger interfacial bonding. Finally, it can be concluded that the polyester-based composites fabricated by reinforcing sodium bicarbonate-treated jute fibers perform satisfactorily and could be a better fit for lightweight structural components.

References

1. Ravikumar P, Suresh AR, Rajeshkumar G (2020) An investigation into the tribological properties of bidirectional jute/carbon fiber reinforced polyester hybrid composites. J Nat Fibers 1–11. https://doi.org/10.1080/15440478.2020.1764444
2. Zhan J, Li J, Wang G, et al (2020) Review on the performances, foaming and injection molding simulation of natural fiber composites. Polym Compos 1–20. https://doi.org/10.1002/pc.25902
3. Rajeshkumar G, Devnani GL, Maran JP et al (2021) Characterization of novel natural cellulosic fibers from purple bauhinia for potential reinforcement in polymer composites. Cellulose 28:5373–5385. https://doi.org/10.1007/s10570-021-03919-2
4. Sumesh KR (2020) An investigation into the mechanical and wear characteristics of hybrid composites: influence of different types and content of biodegradable reinforcements. J Nat Fibers 1–13. https://doi.org/10.1080/15440478.2020.1821297
5. Lotfi A, Li H, Dao DV, Prusty G (2019) Natural fiber-reinforced composites: a review on material, manufacturing, and machinability. J Thermoplast Compos Mater. https://doi.org/10.1177/0892705719844546
6. Jaafar J, Parlaungan J, Salwani S et al (2019) Important considerations in manufacturing of natural fiber composites: a review. Int J Precis Eng Manuf Technol. https://doi.org/10.1007/s40684-019-00097-2
7. Nagarjun J, Kanchana J, Kumar GR (2020) Improvement of mechanical properties of coir/epoxy composites through hybridization with Sisal and Palmyra palm fibers. J Nat Fibers 1–10. https://doi.org/10.1080/15440478.2020.1745126
8. Alves C, Ferrão PMC, Silva AJ et al (2010) Ecodesign of automotive components making use of natural jute fiber composites. J Clean Prod 18:313–327. https://doi.org/10.1016/j.jclepro.2009.10.022
9. Raghavendra G, Ojha S, Acharya SK, Pal SK (2014) Jute fiber reinforced epoxy composites and comparison with the glass and neat epoxy composites. J Compos Mater 48:2537–2547. https://doi.org/10.1177/0021998313499955
10. Mishra V, Biswas S (2013) Physical and mechanical properties of bi-directional jute fiber epoxy composites. Procedia Eng 51:561–566. https://doi.org/10.1016/j.proeng.2013.01.079
11. Nagarajan KJ (2021) A comprehensive review on cellulose nanocrystals and cellulose nanofibers: pretreatment, preparation, and characterization
12. Madhu KNBP, Gowda TGY, Vinod MRS, Suchart K (2020) Alkaline effect on characterization of discarded waste of Moringa oleifera fiber as a potential eco-friendly reinforcement for biocomposites. J Polym Environ. https://doi.org/10.1007/s10924-020-01818-4
13. Madhu P, Sanjay MR, Senthamaraikannan P et al (2018) Effect of various chemical treatments of Prosopis juliflora Fibers as composite reinforcement: physicochemical, thermal, mechanical, and morphological properties. J Nat Fibers 1–12. https://doi.org/10.1080/15440478.2018.1534191
14. Mukhtar I, Leman Z, Zainudin ES et al (2018) Effectiveness of alkali and sodium bicarbonate treatments on sugar palm fiber: mechanical, thermal, and chemical investigations. Effectiveness of alkali and sodium bicarbonate treatments on Sugar. J Nat Fibers 1–13. https://doi.org/10.1080/15440478.2018.1537872
15. Wang H, Memon H, Hassan EAM et al (2019) Effect of jute fiber modification on mechanical properties of jute fiber composite. Materials (Basel) 12. https://doi.org/10.3390/ma12081226
16. Ramakrishnan S, Krishnamurthy K, Rajeshkumar G, Asim M (2021) Dynamic mechanical properties and free vibration characteristics of surface modified jute fiber/nano-clay reinforced

epoxy composites. J Polym Environ 29:1076–1088. https://doi.org/10.1007/s10924-020-01945-y

17. Fiore V, Sanfilippo C, Calabrese L (2019) Influence of sodium bicarbonate treatment on the aging resistance of natural fiber reinforced polymer composites under marine environment. Polym Test 80. https://doi.org/10.1016/j.polymertesting.2019.106100
18. Zindani D, Kumar S, Ranjan S, Sumit M (2020) Mechanical characterization of bio-epoxy green composites derived from sodium bicarbonate treated Punica Granatum short fiber agro-waste. J Polym Environ. https://doi.org/10.1007/s10924-020-01868-8
19. Belaadi A (2019) Effect of eco-friendly chemical sodium bicarbonate treatment on the mechanical properties of flax fibers: Weibull statistics
20. do Santos JC, de Oliveira LÁ, Gomes Vieira LM et al (2019) Eco-friendly sodium bicarbonate treatment and its effect on epoxy and polyester coir fiber composites. Constr Build Mater 211:427–436. https://doi.org/10.1016/j.conbuildmat.2019.03.284
21. Zindani D, Kumar S, Maity SR, Bhowmik S (2021) Mechanical characterization of bio-epoxy green composites derived from sodium bicarbonate treated Punica Granatum short fiber agro-waste. J Polym Environ 29:143–155. https://doi.org/10.1007/s10924-020-01868-8
22. Ravikumar P, Rajeshkumar G, Maran JP et al (2021) Evaluation of mechanical and water absorption behaviors of Jute/Carbon fiber reinforced polyester hybrid composites. J Nat Fibers 1–13. https://doi.org/10.1080/15440478.2021.1924339
23. Nagaraja KC, Rajanna S, Prakash GS et al (2020) Studying the effect of different carbon and glass fabric stacking sequence on mechanical properties of epoxy hybrid composite laminates. Compos Commun 21:100425. https://doi.org/10.1016/j.coco.2020.100425
24. Fiore V, Scalici T, Nicoletti F et al (2016) A new eco-friendly chemical treatment of natural fibres: effect of sodium bicarbonate on properties of sisal fibre and its epoxy composites. Compos Part B 85:150–160. https://doi.org/10.1016/j.compositesb.2015.09.028
25. Nagaraja KC, Rajanna S, Prakash GS, Rajeshkumar G (2020) Mechanical properties of polymer matrix composites: effect of hybridization. Mater Today Proc 2019–2021. https://doi.org/10.1016/j.matpr.2020.03.108
26. Gupta PSMK (2019) Lowering in water absorption capacity and mechanical degradation of sisal/epoxy composite by sodium bicarbonate treatment and PLA coating. Polym Compos 1–14. https://doi.org/10.1002/pc.25397
27. Biswas S, Kindo S, Patnaik A (2011) Effect of fiber length on mechanical behavior of coir fiber reinforced epoxy composites. Fibers Polym 12:73–78. https://doi.org/10.1007/s12221-011-0073-9
28. Rajeshkumar G, Arvindh Seshadri S, Devnani GL et al (2021) Environment friendly, renewable and sustainable poly lactic acid (PLA) based natural fiber reinforced composites—a comprehensive review. J Clean Prod 127483. https://doi.org/10.1016/j.jclepro.2021.127483

Design of Materials for Sustainability

Investigation on Pull-Out Strength and Stripping Torque of Joint Produced by Ultrasonic Insertion and Pre-moulding: A Comparative Study

K. Anand, S. Elangovan, S. Pratheesh Kumar, and S. Hari Chealvan

Abstract This paper investigates the performance of joints produced by ultrasonic insertion and plastic injection moulding processes. The metal insert and thermoplastic test specimen, which need to be joined, are prepared as per standards. Ultrasonic insertion process is optimized through response surface methodology (RSM). The experiments are conducted, and joints are produced for optimum inserting conditions. Similarly, the metal insert and thermoplastic part are joined using plastic injection moulding process. The pull-out strength and stripping torque of joints obtained from pre, post-moulding processes are measured using tensile testing and torque testing facilities then compared. The comparative results show that ultrasonically inserted joint yields better outcomes than joints produced by pre-moulding process.

Keywords Ultrasonic insertion · Pre-moulding · Pull-out strength · Stripping torque · RSM

1 Introduction

As the usage of plastic parts rapidly increases in industries, it is necessary to identify a technique to join the plastic part with metal parts. Generally, the plastic components are fastened by bolt and self-tapping screws. When the plastic parts are joined by these fasteners, it may be spoiled due to the stripping of threads. In these situations, a technique is required to join the plastic parts, also it has to enhance the joint performance. Pre-moulding, cold pressing, thermal insertion and ultrasonic insertion are such techniques, which are used to join the plastic parts with the help of metal insert. In this study, two methods, namely pre-moulding and ultrasonic insertion, are investigated with respect to joint performance. Metal inserts are fasteners which are being used for joining the plastic components with other mating parts. Metal inserts

K. Anand (✉) · S. Elangovan · S. Pratheesh Kumar · S. Hari Chealvan
Department of Production Engineering, PSG College of Technology, Coimbatore, Tamil Nadu, India
e-mail: mechanand@gmail.com

E. Natarajan et al. (eds.), *Materials, Design and Manufacturing for Sustainable Environment*, Lecture Notes in Mechanical Engineering,
https://doi.org/10.1007/978-981-19-3053-9_14

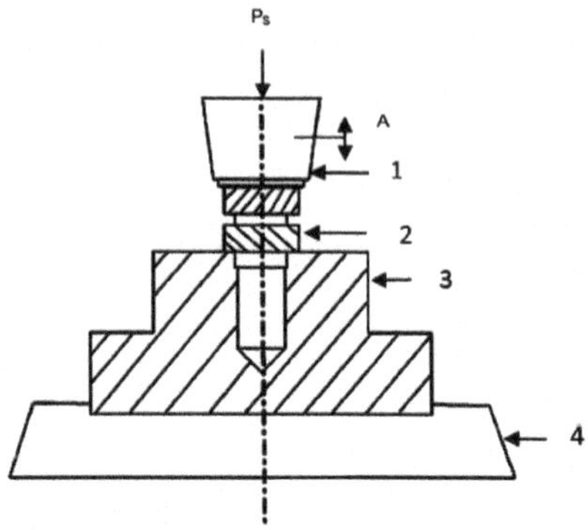

Fig. 1 Schematic representation of ultrasonic insertion [4]. 1 Solid horn, 2 metal insert, 3 thermoplastic component 4 Anvil, A—displacement amplitude and P_S—static pressure

improve joint performance (pull-out strength and stripping torque) by having knurls, grooves and undercuts on its exterior [1].

Metal inserts can be pre-moulded (plastic injection moulding) or post-moulded (ultrasonic insertion) into the plastic component. In pre-moulding, metal inserts are placed on guide pins in the mould cavity and softened plastic is injected around the exterior of metal insert. The inserts are positioned when the plastic cools off evenly around the metal insert.

In ultrasonic insertion process, a metal insert is embedded into thermoplastic component using ultrasonic energy and static pressure [2]. In this method, a metal insert is placed on mounting hole which is smaller in diameter. The ultrasonic energy generated by horn is transferred to the metal insert. Frictional force developed at the interface of metal insert and thermoplastic component, produces heat which makes the thermoplastic material to soften and flow around exterior of metal insert to produce joint [3]. The schematic representation ultrasonic insertion is given in Fig. 1.

Poor joint quality (axial pull-out resistance and torque resistance) is the major issue in joining of metal insert with thermoplastic part. Improper joint affects the performance of the product and leads to damage. This occurs because of the inappropriate selection of joining process. The performance of joint also depends on the design of metal insert and thermoplastic part. For obtaining better resistance against pull-out and torsional loads, sufficient volume of thermoplastic has to flow around exterior of metal insert to encapsulate it. To deal with these problems, ultrasonic insertion and pre-moulding processes need to be investigated and compared with respect to joint performance.

2 Materials and Methods

In this work, a brass insert is prepared as per design standards with grooves and helical knurls to improve both pull-out strength and stripping torque [5]. The specification of brass insert with internally threaded hole is pictorially represented in Fig. 2.

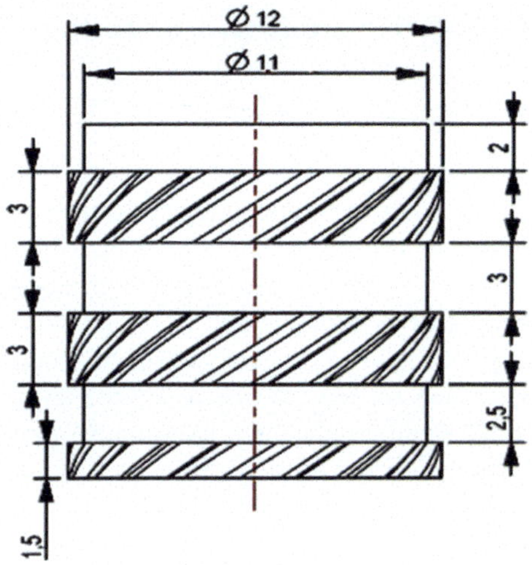

Fig. 2 Dimensions of brass insert. All dimensions are in mm

The thermoplastic mounting hole, in which brass insert to be embedded needs to be prepared in such a way that softened thermoplastic to be moved around the exterior of metal insert to confirm the functional requirements of produced joint. The joint performance can be improved while boss diameter of thermoplastic mounting hole is twice the diameter of brass insert. The diameter of mounting hole is usually smaller than (0.38–0.51 mm) brass insert. The depth of thermoplastic mounting hole needs to be slightly longer than height of brass insert to avoid "back out" of metal insert [5]. The dimensions of ABS thermoplastic mounting hole are shown in Fig. 3.

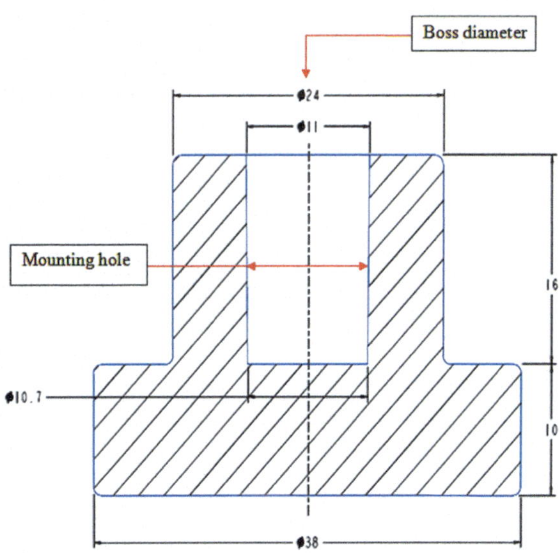

Fig. 3 Dimensions of ABS thermoplastic mounting hole. All dimensions are in mm

3 Experimentation and Testing

3.1 Ultrasonic Insertion of Metal Insert

The insertion experiments are performed with ultrasonic insertion machine (1500 W, 20 kHz). The ultrasonic horn made of aluminium alloy (AA63511) was used in this study because of its excellent acoustical properties. The maximum amplitude developed at vibrating end of the horn is 60 μm without any of load. The experimental facility is shown in Fig. 4. Ultrasonic insertion experiments are planned according to three factors and three-level face-centred central composite designs [6, 7]. In this study, three insertion parameters, namely pressure, inserting time and holding time, are identified through preliminary experiments for brass insert and ABS thermoplastic component. The insertion parameters and their ranges are shown in Table 1.

Fig. 4 Ultrasonic insertion machine. 1 Ultrasonic horn, 2 Anvil, 3 Control panel

Table 1 Insertion parameters and their levels [4]

Parameter	Notation	Units	Factor level		
			−1	0	1
Pressure (x_1)	P	Bar	2.5	2.75	3
Inserting time (x_2)	IT	Seconds	2	2.25	2.5
Holding time (x_3)	HT	Seconds	3	3.25	3.5

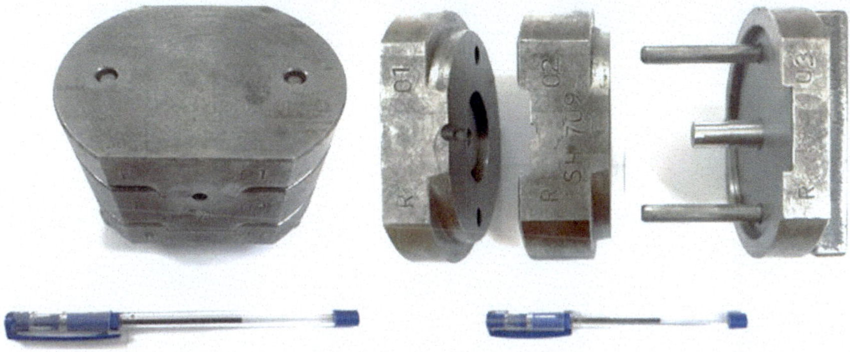

Fig. 5 Assembled and exploded views of injection moulding die

3.2 *Plastic Injection Moulding of Metal Insert*

Three plate injection moulding die is designed and manufactured using mild steel for producing joint between brass insert and ABS thermoplastic mounting hole. The ABS thermoplastic granules are used in injection moulding machine to produce joints. The assembled and exploded views of injection moulding die are shown in Fig. 5.

3.3 *Pull-Out Strength and Stripping Torque Measurement*

The pull-out strength of joints is measured using tensile testing facility with strain rate of 15 mm/min. A fixture is designed and fabricated to hold the thermoplastic test specimen while measuring pull-out strength. Testing facility used for measuring pull-out strength is shown in Fig. 6.

The maximum torque at which metal insert is stripped out from mounting hole is known as stripping torque. Drilling machine with drill tool dynamometer is used to measure stripping torque joint. The M10 thread is produced on the shaft, and it is mounted on spindle of drilling machine. The ABS thermoplastic component is fixed on the three-jaw chuck which is wired with control unit of drilling dynamometer. When the spindle rotates, M10 thread engages with metal insert, and it is stripped out at maximum torque [8]. The maximum torque is recorded by the control unit. The experimental setup for measuring stripping torque is shown in Fig. 7.

Fig. 6 Computerized tensile testing machine with initial setup

Fig. 7 Setup for measuring stripping torque of joint

4 Results and Discussion

The polynomial models have been developed to establish the relationship between inserting parameters and response parameters (pull-out strength and stripping torque) using experimental data. The mathematical models for three factors considered in this study are given in Eq. (1) [9, 10].

$$y_i = \beta_0 + \sum_{j=1}^{3} \beta_j x_j + \sum_{j=1}^{3} \beta_{jj} x_j^2 + \sum_{i=1}^{2} j \sum_{j=i+1}^{3} \beta_{ij} x_i x_j \tag{1}$$

where y_i represents pull-out strength/stripping torque; x_j represents inserting parameters; β_0, β_j, β_{jj} and β_{ij} represent the constant, linear, quadratic and interaction terms, respectively.

A set of 20 combination of inserting parameters are obtained from central composite design, and experiments are conducted as per design matrix. The photograph of joints produced from ultrasonic insertion is shown in Fig. 8.

Pull-out strength and stripping torque are measured using tensile testing and torque testing facilities. The experimental results are used to develop nonlinear models for joint performance which are given Eqs. (2) and (3).

$$\begin{aligned} Y_{\text{Pullout}} = & -86.6 - 68.1X_1 + 228.0X_2 - 39.0X_3 + 11.17X_1^2 - 38.03X_2^2 \\ & + 13.57X_3^2 + 3.80X_1X_2 - 21.56X_2X_3 \end{aligned} \tag{2}$$

$$\begin{aligned} Y_{\text{Torque}} = & -745.4 - 54.4X_1 + 139.66X_2 - 419.5X_3 + 40.00X_1^2 - 38.53X_3^2 \\ & - 15.30X_1X_2 - 39.98X_1X_3 - 28.26X_2X_3 \end{aligned} \tag{3}$$

The experimental and predicted pull-out strength and stripping torque are compared in Table 2. From Table 2, it is observed that the deviation between experimental and theoretical results is minimum.

Table 3 shows the influence of individual inserting parameter on pull-out strength. From the table, it is noticed that inserting time is the most influential parameter. The main effect plot for pull-out strength is shown in Fig. 9. From the main effect plot

Fig. 8 Thermoplastic components with insert assembly (before testing)

Table 2 Comparison of experimental and predicted joint performance of ABS test specimen

S. No.	Pressure (bar)	Inserting time (s)	Holding time (s)	Pull-out strength (MPa)		Stripping torque (Nm)	
				Experimental	Predicted	Experimental	Predicted
1	2.75	2.25	3.25	13.25	13.51	19.42	19.50
2	2.75	2.50	3.25	10.43	10.44	21.30	20.93
3	3.00	2.00	3.50	14.81	15.12	17.76	17.57
4	2.75	2.25	3.00	14.08	14.18	17.99	18.21
5	3.00	2.50	3.00	13.60	13.87	25.80	25.76
6	3.00	2.25	3.25	14.47	14.68	22.10	22.31
7	2.75	2.25	3.25	13.65	13.51	19.82	19.50
8	3.00	2.50	3.50	11.52	11.53	14.91	14.99
9	2.50	2.50	3.00	12.41	12.46	21.97	22.06
10	2.75	2.25	3.25	13.28	13.51	19.91	19.50
11	2.75	2.25	3.25	13.30	13.51	19.33	19.50
12	2.50	2.00	3.50	14.70	14.66	20.10	20.03
13	2.75	2.25	3.25	12.99	13.51	19.62	19.50
14	2.50	2.25	3.25	13.81	13.74	22.07	21.69
15	2.50	2.00	3.00	11.26	11.61	13.93	13.74
16	2.75	2.25	3.25	13.55	13.51	19.72	19.50
17	3.00	2.00	3.00	12.21	12.08	21.62	21.27
18	2.50	2.50	3.50	9.75	10.11	21.04	21.28
19	2.75	2.25	3.50	14.50	14.53	16.37	15.97
20	2.75	2.00	3.25	11.70	11.82	17.27	18.06

for pull-out strength, it is found that optimum inserting condition is P (3 bar), IT (2.25 s), HT (3.5 s). The experiments are conducted for optimum inserting condition, and pull-out strength is measured which is 15.88 MPa.

From Fig. 9, it is observed that maximum pull-out strength is obtained for higher values of pressure. The reason is high pressure induces frictional force on plastic material and plastic softens then moves around the exterior of metal insert. This improves more pull-out strength for the joint. Also, it is noted that pull-out strength has higher values at mid-level of inserting time. The reason is that plastic material

Table 3 Average pull-out strength at various levels of parameter

Insertion parameter	Levels			Difference	Rank	Optimum level
	1	2	3			
Pressure (bar)	12.386	13.073	13.322	0.936	2	P3
Inserting time (s)	12.936	13.688	11.542	2.146	1	IT2
Holding time (s)	12.712	13.043	13.056	0.344	3	HT3

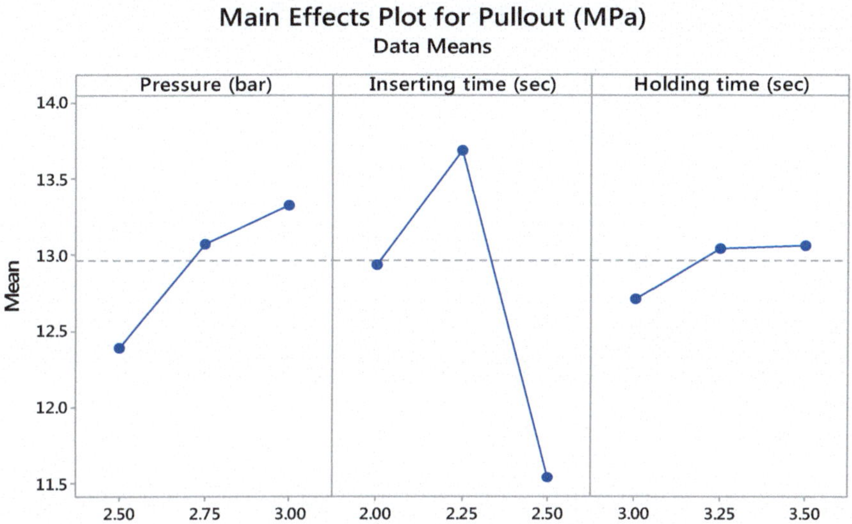

Fig. 9 Mean values of pull-out strength at various levels of parameters

has adequate time to soften and flow around the exterior of the insert [11]. This gives good retention between insert and mounting hole. Beyond 2.25 s, the pull-out strength again starts decreasing. This is due to excessive inserting time deteriorate the bonding between the metal insert and plastic component.

Table 4 shows the effect of ultrasonic inserting parameter on stripping torque. From the table, it is noticed that inserting time is the most influential parameter. The main effect plot for stripping torque is shown in Fig. 10. From the main effect plot for torque, it is noticed that optimum inserting condition is P (3 bar), IT (2.5 s), HT (3 s). The experiments are conducted for optimum condition and stripping torque measured which is 21.63 Nm.

From Fig. 10, it is observed that stripping torque increases for higher values of pressure. The reason may be higher pressure induce frictional force on plastic material and plastic softens then flows around the exterior of the insert. This gives more torque withstanding capability for the fastened joint. Also, it shows that stripping torque increases with the increase of inserting time. The reason may be, when the insertion

Table 4 Average stripping torque at various levels of parameter

Insertion parameters	Levels			Difference	Rank	Optimum level
	1	2	3			
Pressure	19.822	19.075	20.438	1.363	3	P3
Inserting time	18.136	19.635	21.004	2.868	1	IT3
Holding time	20.262	20.056	18.036	2.226	2	HT1

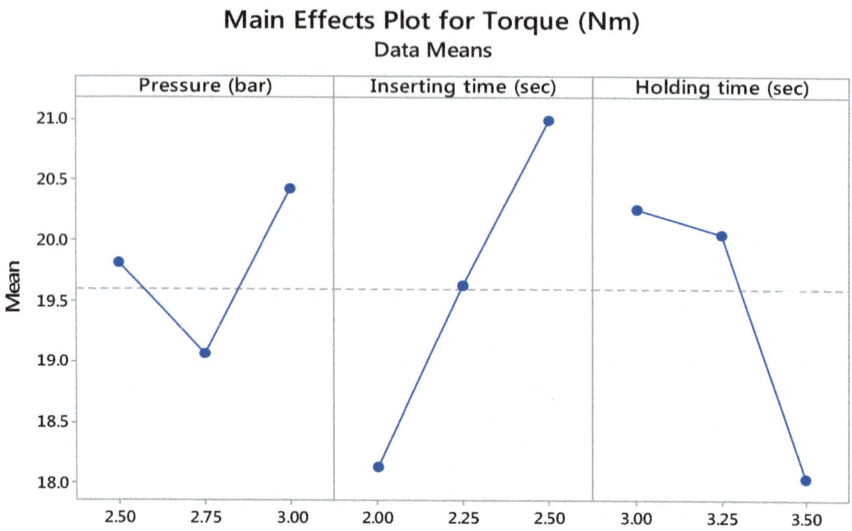

Fig. 10 Mean values of stripping torque at various levels of parameters

time is increased, it gives more time for the softened plastic material to flow and settle around the external features of the metal insert.

In pre-moulding process, brass insert is kept on the guide pins in the mould cavity. Guide pins hold the brass inserts firmly and ABS thermoplastic is injected into the mould cavity to flow around the exterior of brass insert. Brass insert is locked in the respective position in the thermoplastic part then polymer cools off and joint is produced. The joints produced from plastic moulding process are shown in Fig. 11 and tested using tensile testing and torque testing facilities. The results obtained from pre-moulding and post-moulding processes are compared in Table 5.

From Table 5, it is observed that joints produced from ultrasonic insertion yields better results than pre-moulded joint. The reason is localized heating at joint. The frictional force applied at joint by ultrasonic vibrations softens the ABS thermoplastic. The softened plastic flows around the exterior of metal insert which enhances the pull-out strength and stripping torque as well. In the case of pre-moulding process, molten plastic is not that much penetrated into the exterior features of metal insert and results in poor joint performance.

5 Conclusions

In this study, processes like ultrasonic insertion and plastic moulding were used to produce joint between brass insert and ABS thermoplastic material. The pull-out strength and torque withstanding capability of joints produced from pre- and post-moulding processes were compared. The conclusions obtained are as follows:

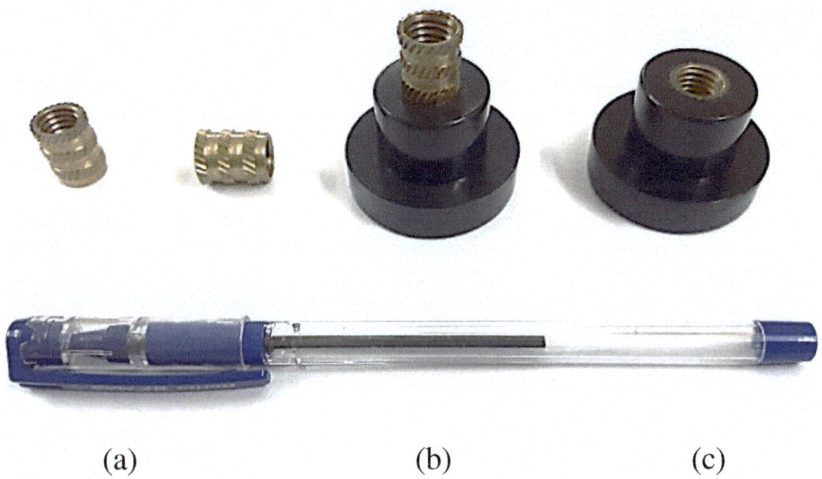

Fig. 11 Joints produced from pre-moulding. **a** Helically knurled brass insert with two grooves, **b** ABS thermoplastic test specimen with brass insert (before assembling), **c** ABS thermoplastic test specimen with insert assembly

Table 5 Comparison of responses obtained from ultrasonic insertion and pre-moulding

Ultrasonic insertion		Pre-moulding	
Pull-out strength (MPa)	Stripping torque (Nm)	Pull-out strength (MPa)	Stripping torque (Nm)
15.88	21.63	10.40	12.50

(1) Inserting parameters which affect the performance of joint were studied, and experimental trials were carried out to find the working range of inserting parameters.
(2) Polynomial models were developed for ultrasonic joints to improve the pull-out strength and stripping torque using experimental results which are planned and conducted as per face-centred central composite design of RSM.
(3) The mean effect plot shows that maximum pull-out strength corresponds to $P = 3$ bar, $IT = 2.25$ s and $HT = 3.5$ s, and the maximum stripping torque corresponds to $P = 3$ bar, $IT = 2.5$ s and $HT = 3.0$ s.
(4) The pull-out strength and stripping torque obtained from pre-moulding process are less compared with post-moulding process. The reason is in ultrasonic insertion that produces localized heat at joint.

References

1. Jansen JA (2016) Case studies of plastics failure associated with metal fasteners. Publishes and presented at ANTEC. https://www.madisongroup.com/publications/ANTEC2016JansenPlasticwithMetal%20Fasteners.pdf. Accessed 19 Jan 2021
2. Benatar A, Cheng Z (1989) Ultrasonic welding of thermoplastics in the far field. Polym Eng Sci 29:1699–1704
3. Pausan DV, Popovici V (2009) Research on installing metal inserts using ultrasonic oscillation. In: The annual symposium of the institute of solid mechanics SISOM, Bucharest, 28–29 May, pp 339–344
4. Anand K, Elangovan S (2017) Optimizing the ultrasonic inserting parameters to achieve maximum pull-out strength using response surface methodology and genetic algorithm integration technique. Measurement 99:145–154
5. Michael Troughton J (2008) Handbook of plastics joining a practical guide. PDI publications, New York
6. Krishnaiah K, Shahabudeen P (2012) Applied design of experiments and Taguchi method. PHI Learning Private Limited, New Delhi
7. Mathews PG (2005) Design of experiments with MINITAB. ASQ Quality Press, Milwaukee, Wisconsin
8. McKeen LW (2010) Fatigue and tribological properties of plastics and elastomers. William Andrew Inc, Norwich
9. Elangovan S (2011) Experimental and theoretical investigation on ultrasonic metal welding and optimization of process parameters to achieve quality welds. Ph.D. Thesis, Anna University, Chennai
10. Elangovan S, Anand K, Prakasan K (2012) Parametric optimization of ultrasonic metal welding using response surface methodology and genetic algorithm. Int J Adv Manuf Technol 63:561–572
11. Chuah YK, Chien Chang BC, Liu (2000) Effects of the shape of the energy director on far field ultrasonic welding of thermoplastics. Polym Eng Sci 40:157–167

Investigation on Machinability of EN8 Steel Through Taguchi Method, ANOVA and Genetic Algorithm

K. Anand, S. Pratheesh Kumar, and S. Hari Chealvan

Abstract This work studies the machinability of EN8 steel by employing Taguchi method, analysis of variance (ANOVA) and genetic algorithm (GA). The CNC turning experiments are planned and conducted as per L27 orthogonal array and polynomial models for surface roughness and circularity error of EN8 steel shaft are developed. Reliability of polynomial models and significance of turning parameters are tested using ANOVA. The polynomial models are integrated with GA as fitness function to find the optimal turning conditions which has to minimize the surface roughness and circularity error as well. Further, the optimum turning conditions obtained from GA are validated by confirmation experiments. From the results, it was noted that turning conditions obtained from GA correlate well with experimental results. This shows that Taguchi method, ANOVA and GA can be used for minimizing the surface roughness and circularity error in the turning of EN8 steel.

Keywords CNC turning · Surface roughness · Circularity error · Taguchi DOE · ANOVA · GA

1 Introduction

In the field of machining, the quality of machined component is much important. The quality of machining is improved by proper selection of cutting tool, work material combination and optimal machining conditions. The cutting tool improves the quality of the shaft and ensures the low-cost manufacturing. In machining, the surface and geometric qualities are the major requirements of customers. As the demand of high-precision components increases, surface finish and geometric quality of machined components play important role in modern manufacturing processes. EN8 is an unalloyed medium carbon steel, and it is used for producing shafts, studs, keys, general purpose axles, etc.

K. Anand (✉) · S. Pratheesh Kumar · S. Hari Chealvan
Department of Production Engineering, PSG College of Technology, Coimbatore, Tamil Nadu, India
e-mail: mechanand@gmail.com

E. Natarajan et al. (eds.), *Materials, Design and Manufacturing for Sustainable Environment*, Lecture Notes in Mechanical Engineering,
https://doi.org/10.1007/978-981-19-3053-9_15

Kumar et al. [1] have studied the influence of input conditions in CNC turning of carbon alloy steels. The input variables feed rate and spindle speed are varied to study their influence on surface texture. Various carbon alloy steels used are SAE8620, EN8, EN19, EN24 and EN47. The study reveals that the surface finish is influenced by feed rate and speed. Sharma et al. [2] optimized the surface finish of mild steel in CNC turning through Taguchi design of experiments (DOE). The results show that input parameters like feed rate, speed and depth of cut play a major role in improving the surface finish. Prashant et al. [3] have optimized the surface finish of EN8 steel in turning process using Taguchi DOE and analysis of variance (ANOVA). Shet et al. [4] have optimized surface roughness of EN1A steel in CNC turning process without coolant by Taguchi DOE. Krishna et al. [5] investigated the surface finish in turning of OHNS high-carbon steel with DNMG carbide insert using Taguchi analysis and ANOVA. The results show that cutting velocity is the most influential turning parameter on surface finish. Bhiksha et al. [6] have studied the effect of turning conditions in machining of EN8 steel using Taguchi DOE and ANOVA. The input variables considered for this study are speed, feed rate and depth of cut. The output variables are surface finish (Ra) and material removal rate (MRR). The results show that the most significant parameter for MRR is feed rate and for Ra are speed and depth of cut. Neelesh and Atul [7] have performed multi-objective optimization for improving machinability in turning process using RSM, teaching learning-based optimization (TLBO), JAYA and genetic algorithms. From the results of their study, it was concluded that TLBO and JAYA algorithm yield better results than GA. Manjunath et al. [8] have modelled, analyzed and optimized turning parameters for improving the machinability of Al 7075 alloy. The experiments are planned, conducted and analyzed using RSM. Variance analysis was carried out to check the significance of machining parameters, and the same were optimized using JAYA algorithm.

From the literature review, it is understood that surface finish of EN8 steel has been improved by optimizing the turning parameters using Taguchi DOE. Also, it was observed that the circularity error of EN8 steel shaft is not investigated in turning process. It is also noted that both surface finish and circularity error are not optimized using multi-objective optimization technique. So, theoretical and experimental investigations on machinability of EN8 steel using Taguchi DoE, ANOVA and GA have been tried in this study.

2 Materials and Method

EN8 is a medium carbon steel used for producing axles, shafts, gears, bolts and studs. The chemical composition of EN8 steel is given in Table 1. The DNMG insert is adopted to machine EN8 steel for improving surface and geometrical qualities. DNMG inserts are diamond or 'D' shaped turning inserts with a 55° angle.

Design of experiments is a statistical method used for investigating both input and response variables. The experimental conditions are created with the help of design

Table 1 Chemical composition of EN8 steel

Carbon (%)	Silicon (%)	Manganese (%)	Sulphur (%)	Phosphorous (%)
0.36–0.44	0.10–0.40	0.60–1.00	0.050 (max)	0.050 (max)

Table 2 CNC turning parameters and their levels

Parameter	Units	Factor level		
		−1	0	1
Speed (X_1)	rpm	800	900	1000
Feed rate (X_2)	mm/rev	0.2	0.25	0.3
Depth of cut (X_3)	mm	0.5	1.0	1.5

matrix, in which each parameter has an equal number of test conditions [9]. The turning process involves with several input variables which may affect the response variable. From preliminary experiments, three turning parameters such as speed, feed rate and depth of cut are considered for this work. The factors with their ranges for turning EN8 steel are given in Table 2. In this work, turning experiments are planned as per Taguchi's L27 orthogonal array which is given in Table 3.

3 Experimentation and Testing

3.1 CNC Horizontal Turning Centre

The EN8 shaft with length of 150 mm and diameter of 50 mm is mounted on head stock the CNC lathe. The CNC programme for turning is generated manually. The experiments are carried out with various combination of turning parameters within the range provided. The facility used for turning of EN8 steel is shown in Fig. 1, and the specifications of CNC–HTC are displayed in the Table 4.

3.2 Measurement of Surface Roughness

The surface finish of turned EN8 steel shaft is measured using surface roughness tester. The facility used for measuring surface roughness is shown in Fig. 2, and Table 5 gives the specifications of surface roughness tester. The circularity error is measured using dial gauge.

Table 3 L27 orthogonal array for CNC turning

S. No.	Speed (rpm)	Feed rate (mm/rev)	Depth of cut (mm)
1	800	0.2	0.5
2	800	0.2	1.0
3	800	0.2	1.5
4	800	0.25	0.5
5	800	0.25	1.0
6	800	0.25	1.5
7	800	0.3	0.5
8	800	0.3	1.0
9	800	0.3	1.5
10	900	0.2	0.5
11	900	0.2	1.0
12	900	0.2	1.5
13	900	0.25	0.5
14	900	0.25	1.0
15	900	0.25	1.5
16	900	0.3	0.5
17	900	0.3	1.0
18	900	0.3	1.5
19	1000	0.2	0.5
20	1000	0.2	1.0
21	1000	0.2	1.5
22	1000	0.25	0.5
23	1000	0.25	1.0
24	1000	0.25	1.5
25	1000	0.3	0.5
26	1000	0.3	1.0
27	1000	0.3	1.5

4 Results and Discussion

4.1 Development of Surface Roughness and Circularity Error Models

The nonlinear models have been developed to establish the association between turning parameters and response parameters (surface roughness and circularity error) using experimental data. The nonlinear equation for three factors considered for this study is given in Eq. (1) [8, 10].

Fig. 1 CNC horizontal turning centre

Table 4 Specifications of CNC horizontal turning centre

Capacity	Maximum chuck diameter	165 mm
	Maximum turning diameter	200 mm
Spindle	Hole through the spindle	53 mm
	Spindle speed	4500 rpm
	Spindle motor power	5.5/7.5 KW
Feed	X axis = 105 mm, Z axis = 320 mm	
Turret	Number of stations	8
	Total shank size	20 × 20
Tail stock	Quill diameter	65 mm
	Quill stroke	80 mm
	Quill taper	(MT4)

$$y_i = \beta_0 + \sum_{j=1}^{3} \beta_j x_j + \sum_{j=1}^{3} \beta_{jj} x_j^2 + \sum_{i=1}^{2} j \sum_{j=i+1}^{3} \beta_{ij} x_i x_j \quad (1)$$

where y_i represents, i.e. surface roughness/circularity error; x_j represents turning parameters; β_0, β_j, β_{jj} and β_{ij} represent the constant, linear, quadratic and interaction terms, respectively.

The surface roughness and circularity error measured through experiments for various turning conditions are considered as input in Minitab software, and polynomial models are developed to predict surface roughness and circularity error. The mathematical models developed are given in Eqs. (2) and (3).

Fig. 2 Surface roughness tester

Table 5 Specifications of surface roughness tester

Drive unit	
Measuring speed	0.25 mm/s; 0.5 mm/s; 0.75 mm/s
Traverse	17.5 mm, S-type 5.6 mm
Detector	
Measuring method	Differential inductive
Measuring range	360 μm
Stylus	Diamond tip
Skid radius	40 mm
Mass	500 g

$$Y_{\text{Surface roughness}} = -16.53 + 0.03177X_1 + 47.13X_2 + 4.254X_3 - 0.000019X_1^2 - 120.67X_2^2 - 0.18X_3^2 + 0.01433X_1X_2 - 0.005417X_1X_3 + 2.00X_2X_3 \quad (2)$$

$$Y_{\text{Circularity error}} = 5.718 - 0.010294X_1 - 10.878X_2 + 0.7600X_3 + 0.000005X_1^2 + 11.778X_2^2 - 0.32889X_3^2 + 0.005667X_1X_2 - 0.02333X_2X_3 \quad (3)$$

The experimental and predicted responses of turning process are compared in Table 6. From the Table 6, it can be noted that the deviation between experimental and theoretical results is minimum.

Table 6 Comparison of experimental and predicted responses

S. No.	Speed (rpm)	Feed rate (mm/rev)	Depth of cut (mm)	Surface roughness (μm)		Circularity error (mm)	
				Experimental	Predicted	Experimental	Predicted
1	800	0.2	0.5	3.96	3.73	0.12	0.14
2	800	0.2	1.0	3.99	3.76	0.22	0.25
3	800	0.2	1.5	3.89	3.69	0.16	0.26
4	800	0.25	0.5	4.21	4.00	0.06	0.07
5	800	0.25	1.0	4.31	4.07	0.15	0.18
6	800	0.25	1.5	4.29	4.06	0.09	0.20
7	800	0.3	0.5	3.83	3.66	0.06	0.07
8	800	0.3	1.0	4.00	3.78	0.15	0.17
9	800	0.3	1.5	4.05	3.82	0.08	0.22
10	900	0.2	0.5	3.99	3.70	0.04	0.07
11	900	0.2	1.0	3.71	3.45	0.14	0.18
12	900	0.2	1.5	3.33	3.11	0.08	0.19
13	900	0.25	0.5	4.32	4.03	0.01	0.03
14	900	0.25	1.0	4.13	3.84	0.11	0.14
15	900	0.25	1.5	3.81	3.55	0.04	0.16
16	900	0.3	0.5	4.04	3.77	0.04	0.06
17	900	0.3	1.0	3.91	3.62	0.13	0.16
18	900	0.3	1.5	3.67	3.38	0.06	0.21
19	1000	0.2	0.5	3.59	3.28	0.06	0.10
20	1000	0.2	1.0	3.04	2.76	0.17	0.21
21	1000	0.2	1.5	2.62	2.16	0.10	0.22
22	1000	0.25	0.5	4.04	3.69	0.06	0.10
23	1000	0.25	1.0	3.55	3.22	0.16	0.20
24	1000	0.25	1.5	2.94	2.66	0.09	0.22
25	1000	0.3	0.5	3.84	3.49	0.12	0.15
26	1000	0.3	1.0	3.42	3.07	0.21	0.25
27	1000	0.3	1.5	2.89	2.57	0.13	0.32

4.2 *Analysis of Variance (ANOVA) and Model Fitment Test*

Variance analysis is a technique of partitioning total variation into accountable sources of variation in an experiment. The variance analysis test is performed to check the reliability of developed nonlinear models and significance of turning variables [11]. In this study, the confidence level considered is 95%. If the model is adequate, then the P-Value is less than 0.05 for desired level of confidence. Table 7 represents the variances analysis results surface roughness.

Table 7 ANOVA table for surface roughness

Source	Degree of freedom	Adj sum of square	Adj mean square	*F* value	*P* value	Comments
Regression	9	5.33073	0.592303	298.47	0.000	Significant
Speed	1	0.17869	0.17869	90.05	0.000	Significant
Feed rate	1	0.23330	0.23330	117.56	0.000	Significant
Depth of cut	1	0.39149	0.39149	197.28	0.000	Significant
Speed^2	1	0.20907	0.20907	105.35	0.000	Significant
Feed rate^2	1	0.54602	0.54602	275.14	0.000	Significant
Depth of cut^2	1	0.01215	0.01215	6.12	0.024	Significant
Speed × feed rate	1	0.06163	0.06163	31.06	0.000	Significant
Speed × depth of cut	1	0.88021	0.88021	443.55	0.000	Significant
Feed rate × depth of cut	1	0.03000	0.03000	15.12	0.001	Significant
Error	17	0.03374	0.001984			
Total	26	5.36447				

From Table 7, it is clearly understood that turning parameters are significant, and developed polynomial model is adequate to predict the surface finish. The correlation coefficient 'R^2' value is 99.04%. This shows that the developed polynomial model is adequate to predict the surface finish. Table 8 shows the influence of turning parameters on surface roughness. From the table, it is noticed that feed rate is most influential parameter. The main effect plot for surface roughness is shown in Fig. 3. From the main effect plot for surface roughness, it is found that optimum turning condition is speed (1000 rpm), feed rate (0.2 mm/rev) and depth of cut (1.5 mm). Table 8 represents the variances analysis results of circularity error.

From Table 9, it is clearly understood that turning parameters are significant, and developed polynomial model is adequate to predict the circularity error. The correlation coefficient 'R^2' value is 99.68%. This shows that the developed polynomial model is adequate to predict the circularity error. Table 10 shows the influence of individual turning parameter on circularity error. From the table, it is noticed that feed rate is most influential parameter. The main effect plot for circularity error

Table 8 Average surface roughness at various levels of parameter

Turning parameter	Levels			Difference	Rank	Optimum level
	1	2	3			
Speed (rpm)	4.05	3.87	3.32	0.73	3	S 3
Feed rate (mm/rev)	3.56	3.95	3.73	0.17	1	FR 1
Depth of cut (mm)	3.98	3.78	3.49	0.49	2	DOC 3

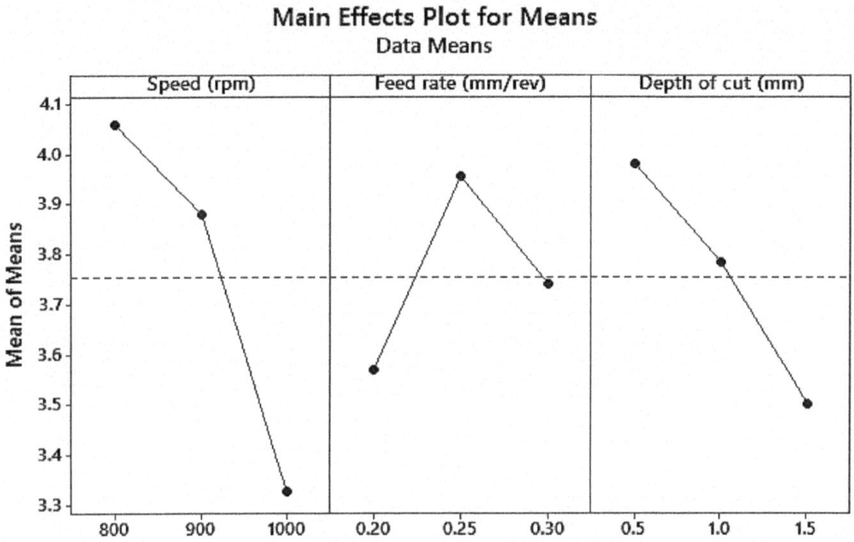

Fig. 3 Mean values of surface roughness at various levels of parameters

is shown in Fig. 4. From the main effect plot for circularity error, it is found that optimum turning condition is speed (900 rpm), feed rate (0.25 mm/rev) and depth of cut (0.5 mm).

Table 9 ANOVA table for circularity error

Source	Degree of freedom	Adj sum of square	Adj mean square	*F* value	*P* value	Comments
Regression	9	0.074917	0.008324	899.00	0.000	Significant
Speed	1	0.018766	0.018766	2026.72	0.000	Significant
Feed rate	1	0.012426	0.012426	1342.04	0.000	Significant
Depth of cut	1	0.012496	0.012496	1349.58	0.000	Significant
Speed^2	1	0.014669	0.014669	1584.20	0.000	Significant
Feed rate^2	1	0.005202	0.005202	561.80	0.000	Significant
Depth of cut^2	1	0.040563	0.040563	4380.80	0.000	Significant
Speed × feed rate	1	0.009633	0.009633	1040.40	0.000	Significant
Speed × depth of cut	1	0.000008	0.000008	0.90	0.356	Insignificant
Feed rate × depth of cut	1	0.000408	0.000408	44.10	0.000	Significant
Error	17	0.000157	0.000009			
Total	26	0.075074				

Table 10 Average circularity error at various levels of parameter

Turning parameter	Levels			Difference	Rank	Optimum level
	1	2	3			
Speed (rpm)	0.121	0.070	0.122	0.052	2	S 2
Feed rate (mm/rev)	0.121	0.080	0.108	0.041	1	FR 2
Depth of cut (mm)	0.060	0.160	0.090	0.1	3	DOC 1

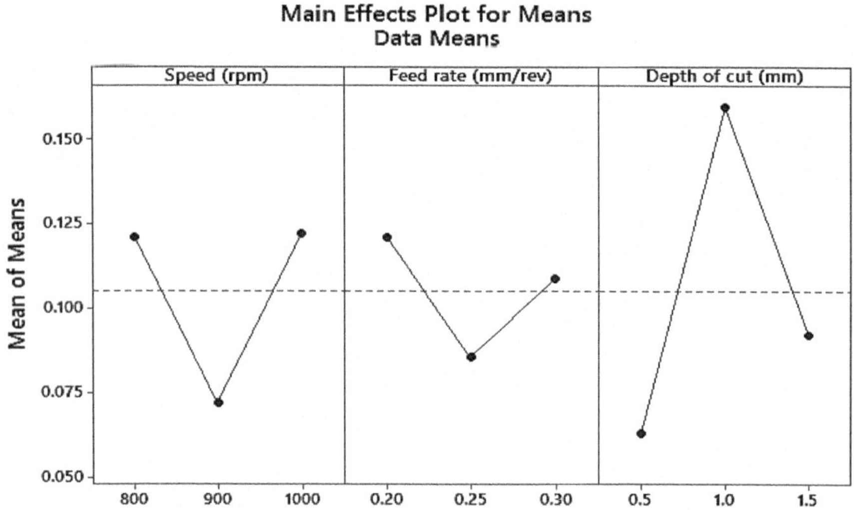

Fig. 4 Mean values of circularity error at various levels of parameters

From ANOVA tables and main effect plots, it is clearly understood that combination of speed, feed rate and depth of cut that minimizes the surface roughness may not be applied for reducing circularity error. So, it requires a multi-objective optimization tool for minimizing the surface roughness and circularity error simultaneously.

4.3 Multi-objective Optimization of CNC Turning Parameters Using Genetic Algorithm

Genetic algorithm (GA) is non-traditional optimization method and works based on the principles of natural selection and genetics. GA yields better results than traditional optimization techniques in irregular experimental regions, because GA works with a string – coded variables instead variables. So, multi-objective optimization of turning parameters is performed using GA.

Using the Eqs. (2), (3) and turning conditions given in Table 2, parametric optimization of turning of EN8 steel was attempted to minimize the surface roughness

Table 11 Pareto-optimal turning parameters

S. No.	Turning parameters			Response parameters	
	Speed (rpm)	Feed rate (mm/rev)	Depth of cut (mm)	Surface roughness (μm)	Circularity error (mm)
1	961	0.25	0.6	3.83	0.11
2	900	0.25	0.5	4.03	0.06
3	990	0.25	1.5	2.84	0.17
4	918	0.25	0.6	3.97	0.09
5	939	0.25	0.5	3.92	0.09
6	936	0.25	1.5	3.32	0.13
7	981	0.25	1.5	2.90	0.16
8	920	0.25	1.5	3.43	0.12
9	930	0.25	1.5	3.36	0.12
10	970	0.25	1.5	3.02	0.15
11	976	0.25	1.5	2.97	0.16
12	939	0.23	1.5	3.19	0.13
13	990	0.25	1.5	2.84	0.17
14	945	0.25	0.6	3.89	0.10
15	911	0.25	0.5	4.00	0.08
17	946	0.22	1.5	3.03	0.14
18	900	0.25	0.5	4.03	0.06

and circularity error as well. For this, GA tool box of MATLAB is used. Pareto-optimal solution obtained from GA tool is given in Table 11. Figure 5 shows the formation of pareto-optimal solution for the problem proposed. The results shown in the Table 11 give 18 optimal turning conditions. From the 18 optimal conditions, any optimal condition can be selected based on the functional requirement, and it may be recommended for machining.

Based on optimum turning conditions obtained from genetic algorithm (Table 11), two conditions (Trial No. 5 and 9) are randomly selected to verify the prediction of responses (surface roughness and circularity error). Comparative results of predictive responses and experimental validation are shown in Table 12. The results show that experimental values are in good agreement with predicted values of responses (approximately less than 10% of error).

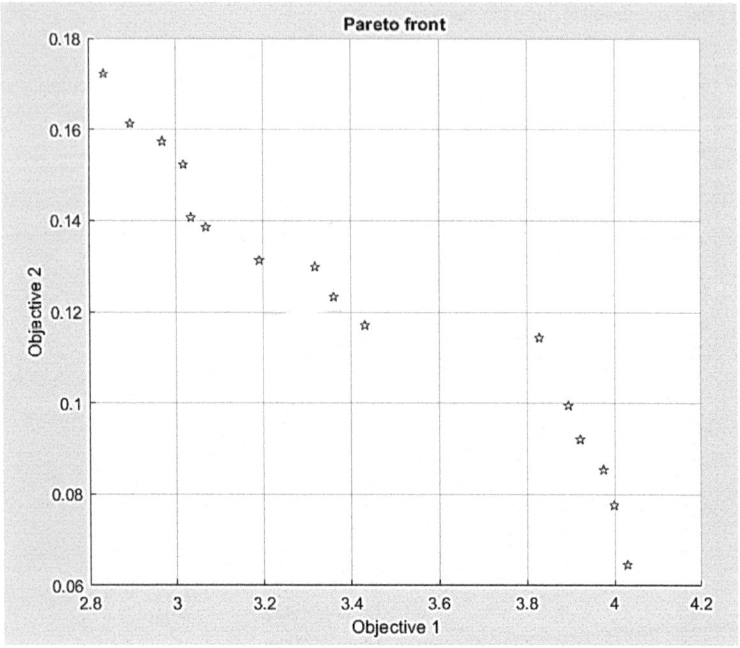

Fig. 5 Pareto-optimal front surface roughness and circularity error

Table 12 Comparison of predicted responses with experimental validation

Optimal turning conditions	Predicted from GA		Experimental		% Error	
	Surface finish (μm)	Circularity error (mm)	Surface finish (μm)	Circularity error (mm)	Surface finish	Circularity error
S—939 rpm FR—0.25 mm/rev DOC—0.5 mm (Trial No. 5)	3.92	0.09	4.02	0.08	2.48	11.12
S—930 rpm FR—0.25 mm/rev DOC—0.5 mm (Trial No. 9)	3.36	0.12	3.51	0.11	4.27	8.33

5 Conclusions

In this study, techniques like Taguchi DOE, ANOVA and GA were integrated to improve the surface finish and geometrical accuracy of EN8 steel shaft. The conclusions obtained are as follows.

(1) Turning parameters which affect the machinability of EN8 were studied, and experimental trials were conducted to find the working range of turning conditions.
(2) Polynomial models were developed to improve the surface finish and circularity error using experimental results which are planned and conducted as per L27 design matrix.
(3) Variance analysis results show that the developed polynomial models are reliable to predict as R^2 values are close to 1. The linear, quadratic and interaction terms of regression models are significant. Also, all turning parameters are significant.
(4) The mean effect plot shows that minimum surface roughness corresponds to speed = 1000 rpm; feed rate = 0.2 mm/rev; depth of cut = 1.5 mm; and the minimum circularity error corresponds to speed = 900 rpm; feed rate = 0.25 mm/rev; depth of cut = 0.5 mm.
(5) The developed regression models were connected with GA, and multi-objective optimization was carried out, and 18 optimal turning conditions were obtained. The optimal conditions were randomly selected and experimentally verified.

References

1. Satheesh Kumar N, Shetty A, Shetty A, Ananth K, Shetty H (2012) Effect of spindle speed and feed rate on surface roughness of carbon steels in CNC turning. Procedia Eng 38:691–697
2. Sushil P, Sharma K, Kumar ES (2014) Optimization of surface roughness in CNC turning of mild steel (1018) using Taguchi method. Int J Eng Res Appl 3:2928–2932
3. Nikam KG, Kadam SS (2014) Optimization of surface roughness of EN8 steel by changing cutting parameters and insert geometry in turning process. Int J Sci Res 3:1331–1335
4. Shet GT, Lashmana Swamy N, Somashekar H (2014) Optimization of surface roughness parameters in turning EN1A steel on a CNC lathe without coolant. Int J Eng Res Technol 3:1648–1657
5. Krishna DG, Kumar MV (2017) Taguchi analysis on surface roughness in turning OHNS high carbon steel with DNMG carbide insert. Int J Innov Res Sci Technol 4:1–7
6. Gugulothu B, Kumsa DK, Kassa MB (2017) Effect of process parameters on MRR and surface roughness in turning process of EN8. Mater Today Proc (Published in Online)
7. Sahu NK, Andhare AB (2019) Multiobjective optimization for improving machinability of Ti-6Al-4V using RSM and advanced algorithms. J Comput Des Eng 6:1–12
8. Babbar A, Prakash C, Singh S, Gupta MK, Mia M, Pruncu CI (2020) Application of hybrid nature-inspired algorithm: single and bi-objective constrained optimization of magnetic abrasive finishing process parameters. J Mater Res Technol 9:7961–7974
9. Mathews PG (2005) Design of experiments with MINITAB. ASQ Quality Press, Milwaukee, Wisconsin
10. Anand K, Elangovan S (2017) Optimizing the ultrasonic inserting parameters to achieve maximum pull-out strength using response surface methodology and genetic algorithm integration technique. Measurement 99:145–154
11. Krishnaiah K, Shahabudeen P (2012) Applied design of experiments and Taguchi method. PHI Learning Private Limited, New Delhi

Prediction of Strength and Durability Characteristics of Rice Husk Ash Concrete Using Artificial Neural Network (ANN)

V. Rajkumar, M. Kabeerhasan, R. Mirdula, and D. Suji

Abstract Rice husk ash (RHA) is an agro-based waste used as a sustainable supplement in concrete. The RHA produced by controlled incineration completely blends in concrete mix by increasing the pozzolanic property since it holds silica without compromising on cement properties. After replacing RHA partially in cement, a fair refinement in porous structure increases strength and durability characteristics. The present paper investigates the application of statistical models to predict the characteristics using MATLAB software by ANN tool with networks like FFNN, LRNN, CFNN and ENN. The network performance characteristics such as RMSE, MAE, MRE, prediction accuracy percentage and computational time are used to find the optimal network. The dependent variables are 28th day compressive strength, ultrasonic pulse velocity test results and water absorption percentage. Two hundred and eleven mix design samples of RHA concrete were collected from the various reputed journals published within a decade. Water to binder ratio, cement, RHA, water, fine, coarse aggregate and super plasticizer were used as input parameters to develop the models and ultimately to predict strength and durability characteristics of RHA concrete. The comparison results of the various prediction showed that all the four networks performed roughly same, but based on overall performance characteristics, the developed CFNN model is identified as the optimal ANN, used for predictions in the future.

Keywords Rice husk ash · Artificial neural network · Pozzolanic property · Strength and durability characteristics · Ultrasonic pulse velocity test

V. Rajkumar
Department of Mechanical Engineering, PSG Institute of Technology and Applied Research, Coimbatore, India
e-mail: rajkumarv@psgitech.ac.in

M. Kabeerhasan · D. Suji (✉)
Department of Civil Engineering, PSG College of Technology, Coimbatore, India
e-mail: dsu.civil@psgtech.ac.in

R. Mirdula
Department of Civil Engineering, PSG Institute of Technology and Applied Research, Coimbatore, India

E. Natarajan et al. (eds.), *Materials, Design and Manufacturing for Sustainable Environment*, Lecture Notes in Mechanical Engineering,
https://doi.org/10.1007/978-981-19-3053-9_16

1 Introduction

With the tremendous development of construction, the considerations for sustainable development such as protection of natural resources play a significant role in construction industry. There are various alternative substitutes from industrial and agricultural by-products which exhibits fair pozzolanic properties [1]. RHA, a supplementary material in concrete which is an alternative approach toward sustainable environmental solution as it reduces the cement consumption and CO_2 generation, reduces heat evolution and increases strength for a long period of time [2–5]. Every year, approximately 4.4 million tonnes of rice husk ash are generated from 20 million tonnes of paddy production in India. Most of them are dumped into the land which alters the pH level of the soil. The rice husk is an agro-based waste obtained during the process of milling paddy. Rice husk ash is composed into amorphous silica after the process of controlled incineration or burning which contains about 80–90% of silica [6]. This pozzolanic material advantageously blends in the mixture of concrete, thereby improving the interfacial bond between the aggregates and the cement paste [7]. It significantly increases the durability and compressive strength.

Artificial neural network (ANN) is a computational model which is neuro-biologically inspired paradigm that works similar to the neurons in the brain. The ANN detects the input data based on the inputs and attempts to perform a comparative study of the data network [8]. Artificial neural network does not need any specific equations for predictions. It needs only sufficient input and output data related to the prediction. In ANN, the prediction values are highly dependent on the input and output data. In the process of ANN prediction, testing and training process is very much important. After the continuous retrain of new data set on ANN, adopt the new data set for its network algorithm [9].

Back propagation, a function that trains networks, is used in modeling process. It is reported that ANN network uses three layers for development in which input layer and the output layer consist of five neurons and one neuron, respectively. The meta-heuristics is done in two phases, namely exploration and exploitation, which generates algorithm successfully on the optimization problem. The solutions are searched as separate fragments [10].

Three types of ANN (feed-forward back propagation, cascade-forward back propagation and layer recurrent back propagation) were analyzed to predict the dam displacement values and noticed an increase in accuracy with the increase in neighboring object points and monitoring series [11]. For effective prediction, both historical and forecasted data along with the dam displacement data are considered.

From the prediction results, performance of CFNN is better than FFNN for all concrete mixes. It was investigated using Elman neural network which identified and forecasted the data feasibly and the prediction error is below the average absolute relative error [12]. The flexural and compressive strengths of concrete are predicted using a recurrent neural network (RNN) [13]. From the simulation results, much more desired results were obtained from RNN, and during the validation, the least values of root mean square error (RMSE) and mean average error (MAE) values were obtained.

The main objective of this study is to collect the theoretical and mathematical specifications about the various network types and training function in ANN, to develop a prediction model for strength and durability characteristics of RHA concrete with different types of networks and training function using collected data sets in MATLAB software and to compare and identify the optimal network model based on network performance characteristics of prediction results.

2 Methodology

In general, there are five basics steps for constructing and analyzing the ANN model: (1) collecting data, (2) processing data, (3) building the network, (4) training and (5) test performance of ANN model.

2.1 Data Collection

The mix designs of RHA concrete data, namely water to binder ratio, cement content, RHA content, water content, fine aggregate content, coarse aggregate content and super plasticizer content, were collected from the recently published reputed journals. And the same was used for training and testing ANN model which will be used for the prediction of strength and durability characteristics of RHA concrete, namely 28th day compressive strength, water absorption percentage and ultrasonic pulse velocity test results. For 28th day compressive strength 129 data sets, for water absorption percentage 42 data sets and for ultrasonic pulse velocity test results 40 data sets were collected (Tables 1 and 2).

Table 1 Compressive strength data set on day 28

Parameters	Minimum value	Maximum value
W/B	0.23	0.55
C (kg/m^3)	277	783
RHA (kg/m^3)	0	171
W (kg/m^3)	132.4	233.75
FA (kg/m^3)	344	910
CA (kg/m^3)	590	1419
SP (kg/m^3)	0	72.6
28-day compressive strength (N/mm^2)	8.6	92.21

Table 2 Water absorption % data set

Parameters	Minimum value	Maximum value
W/B	0.35	0.65
C (kg/m^3)	291.43	474
RHA (kg/m^3)	0	108
W (kg/m^3)	148.75	233.75
FA (kg/m^3)	433.72	750
CA (kg/m^3)	840	1419
S (kg/m^3)	0	21.228
Water absorption %	1.2	8.5

Table 3 Ultrasonic pulse velocity test result data set

Parameters	Minimum value	Maximum value
W/B	0.23	0.96
C (kg/m^3)	90	783
RHA (kg/m^3)	0	168
W (kg/m^3)	168	224
FA (kg/m^3)	344	978
CA (kg/m^3)	933	1196
SP (kg/m^3)	0	9.02
UPVT results (km/s)	3.5	4.95

2.2 *Preprocessing Data*

After data collection, the chosen data sets were randomly divided into two: the training group, corresponding to 77.5% of the patterns in 28th day compressive strength, 76% of the patterns in water absorption percentage and 75% of the patterns in ultrasonic pulse velocity test results and the test corresponding to 22.5% of the patterns in 28th day compressive strength, 24% of the patterns in water absorption percentage and 25% of the patterns in ultrasonic pulse velocity test results (Table 3).

2.3 *Building Network*

A three-layered structure with one input, one hidden and one output layer was selected with tan sigmoidal (tansig) transfer function for hidden and output layers. The gradient descent momentum (GDM) and adaptive learning rate (ALR) MATLAB neural network toolbox algorithms were employed in training process. The ANN is characterized with weights and bias values. In architecture selection, several experiments with different architectures were carried out like training and testing process.

Better architectures were selected depending on results as a validation set using various training sessions. The architecture consists of seven nodes in the input layer, ten hidden nodes in the second layer and one in output layer for all 12 networks.

2.4 Testing Network

The ANN model outputs of strength and durability characteristics of RHA concrete, namely 28th day compressive strength, water absorption percentage and ultrasonic pulse velocity test results, were used to evaluate the prediction process of these models. The prediction accuracies were evaluated by network performance characteristics such as RMSE, MAE, MRE, prediction accuracy percentage and computational time. The ANN model simulations were done using MATLAB software. Finally, the optimal network gets identified based on the comparison of network performance characteristics.

3 Modeling in ANN

To predict the data sets, modeling of data is very much important. In the artificial neural network modeling, two types of variables are mainly used (input variables and target variables). In our work, input variables are water to binder ratio, cement, RHA, FA, CA contents, water and super plasticizers content. Target variables are selected based on the prediction characteristics. In the ANN modeling, the available data sets are separated into three types using Excel sheets. The three types of data sets are training data, testing data and target data. From the literature survey, references 2/3 times of data sets are used as training data and 1/3 times of data sets are used as testing data.

3.1 Specifications of the ANN Model

Implementing artificial neural network in MATLAB software, the prediction models are developed. Each prediction model consists of seven input variables and one target variable. Input variables are water to binder ratio, cement, RHA, water content, fine aggregate content, coarse aggregate content and super plasticizers content. The target variables are 28th day compressive strength of concrete, water absorption percentage and UPVT results.

Table 4 Developed ANN models

Model name	Network type	Purpose
FFNN 1	Feed-forward neural network	To predict 28th day compressive strength of RHA concrete
CFNN 1	Cascade-forward neural network	
LRNN 1	Layer recurrent neural network	
ENN 1	Elman neural network	
FFNN 2	Feed-forward neural network	To predict water absorption percentage of RHA concrete
CFNN 2	Cascade-forward neural network	
LRNN 2	Layer recurrent neural network	
ENN 2	Elman neural network	
FFNN 3	Feed-forward neural network	To predict ultrasonic pulse velocity test results of RHA concrete
CFNN 3	Cascade-forward neural network	
LRNN 3	Layer recurrent neural network	
ENN 3	Elman neural network	

3.2 Modeling

In the MATLAB software, the artificial neural network is analyzed using NNTOOL command. Initially, the available input data sets were separated into training data and testing data. For 28th day compressive strength, 100 numbers of training data and 29 numbers of testing data sets were taken; for water absorption percentage, 32 numbers of training data and ten numbers of testing data sets were taken, and finally for ultrasonic pulse velocity test results, 30 numbers of training data and ten numbers of testing data sets were taken. A number of target data sets are same as number of training data sets for all the three models. These data sets are initially tabulated in the Excel sheet, and these all are exported to the MATLAB workspace. For the modeling purpose, the rows of the data sets are placed as columns and columns of the data sets are placed as rows in the MATLAB workspace (Table 4).

3.3 Network Performance Characteristics

To find the optimal network in prediction process, the networks were compared based on their network performance characteristics. Some of the network performance characteristics which we used in this work are root mean square error (RMSE), mean absolute error (MAE), mean relative error (MRE), prediction accuracy percentage and computational time.

4 Results and Discussion

Strength and durability characteristics of RHA concrete were predicted by using ANN with four different network types. Every individual network analysis contains testing and training process, so the results are also like that only. To compare and find the optimal network in prediction of strength and durability characteristics of RHA concrete, every network was constructed in a same way and the given input parameters for each network were also same.

4.1 Prediction of 28th Day Compressive Strength

Totally, 129 data sets were collected and used in prediction process of evaluating 28th day compressive strength for RHA concrete. Out of 129 data sets, 100 data sets were used for training data and 29 sets for testing data. In the prediction process, four types of networks were used. The outputs and results of this process are shown below.

Relationship plot between observed and predicted values of 28th day compressive strength prediction results is shown in Fig. 1. From the relationship plot, we can identify the prediction accuracy of each network type, and based on the network performance characteristics, optimal network gets identified. Network performance characteristics results are given in Table 5.

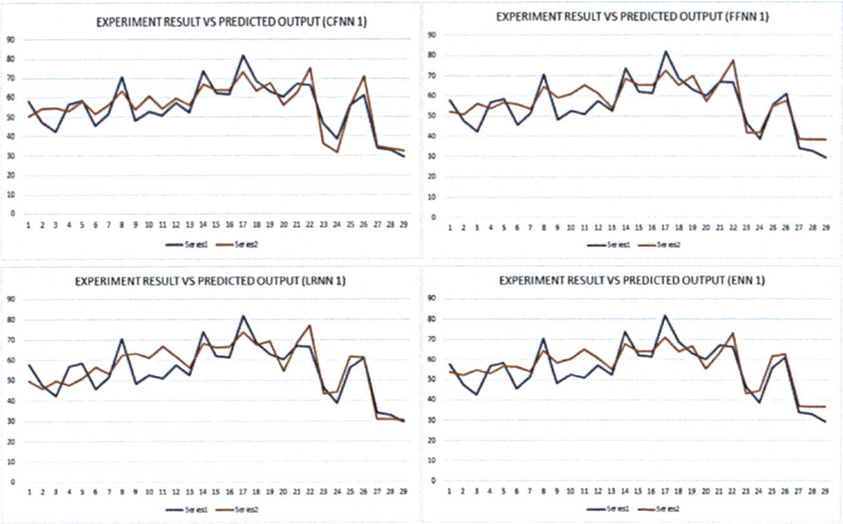

Fig. 1 Comparison plot of 28th day compressive strength prediction results

Table 5 Comparison of networks in 28th day compressive strength results

28th CS	Purpose	FFNN	CFNN	LRNN	ENN
RMSE	Training	6.4440	6.0979	8.8705	8.2606
MAE	Training	5.00924	4.4531	6.1492	5.3511
MRE	Training	0.11245	0.0932	0.1450	0.1182
RMSE	Testing	6.74464	6.0011	7.0966	6.4067
MAE	Testing	5.6055	5.1406	5.9033	5.5097
MRE	Testing	0.11182	0.0956	0.1101	0.1077
Time (s)		2	2	16	15
Accuracy (%)		88.817	90.434	88.987	89.229

Table 6 Comparison of networks in water absorption % results

WA%	Purpose	FFNN	CFNN	LRNN	ENN
RMSE	Training	0.1399	0.2190	0.2266	0.1452
MAE	Training	0.0861	0.1730	0.1186	0.0802
MRE	Training	0.0197	0.0466	0.0309	0.0219
RMSE	Testing	0.6806	0.4709	0.3801	0.5922
MAE	Testing	0.3921	0.3352	0.2798	0.4124
MRE	Testing	0.0622	0.0589	0.0479	0.0731
Time (s)		2	2	17	16
Accuracy (%)		93.779	94.105	95.201	92.686

4.2 Prediction of Percentage Water Absorption

Totally, 42 data sets were collected and used for the prediction process of water absorption % of RHA concrete. In that 42 data sets, 32 were used for training data and 10 were used as testing data. In the prediction process, four types of networks were used. The outputs and results of this process are given in Table 6.

Relationship plot between observed and predicted values of water absorption % prediction results is shown in Fig. 2. From the relationship plot, we can identify the prediction accuracy of each network type, and based on the network performance characteristics, optimal network gets identified. Network performance characteristics results are given in Table 6.

4.3 Prediction of UPVT Results

Totally, 40 data sets were collected and used for the prediction process of ultrasonic pulse velocity test result of RHA concrete. In that 40 data sets, 30 sets were employed

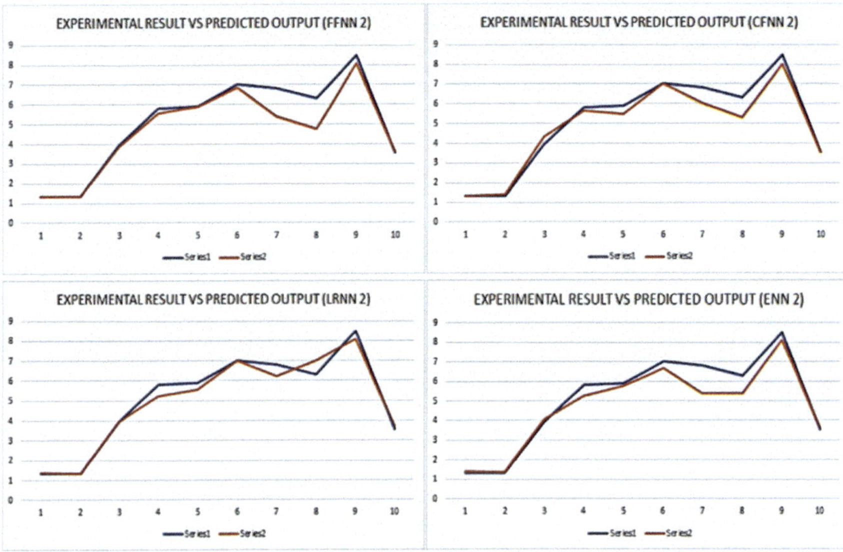

Fig. 2 Comparison plot of water absorption % prediction results

for training data and 10 sets of data were taken as testing data. In the prediction process, four types of networks were used. The outputs and results of this process are given in Table 7.

Relationship plot between observed and predicted values of UPVT results is shown in Fig. 3. From the relationship plot, we can identify the prediction accuracy of each network type, and based on the network performance characteristics, optimal network gets identified. Network performance characteristics results are given in Table 7. Training and testing process of ANN is mainly affected by the quality of the data set such as accuracy, flexibility and precision. In our case, durability characteristics show more accuracy than the strength characteristics of RHA concrete. This is because of variation in the data sets only. Based on the accuracy percentage only, we cannot

Table 7 Comparison of networks in UPVT results

UPVT	Purpose	FFNN	CFNN	LRNN	ENN
RMSE	Training	0.1578	0.1439	0.1883	0.1744
MAE	Training	0.1127	0.1101	0.1256	0.1220
MRE	Training	0.0259	0.0259	0.0296	0.0287
RMSE	Testing	0.2295	0.2085	0.2017	1.4151
MAE	Testing	0.2038	0.1886	0.1733	0.5626
MRE	Testing	0.0448	0.0418	0.0376	0.0332
Time (s)		2	2	13	20
Accuracy (%)		95.512	95.813	96.232	96.671

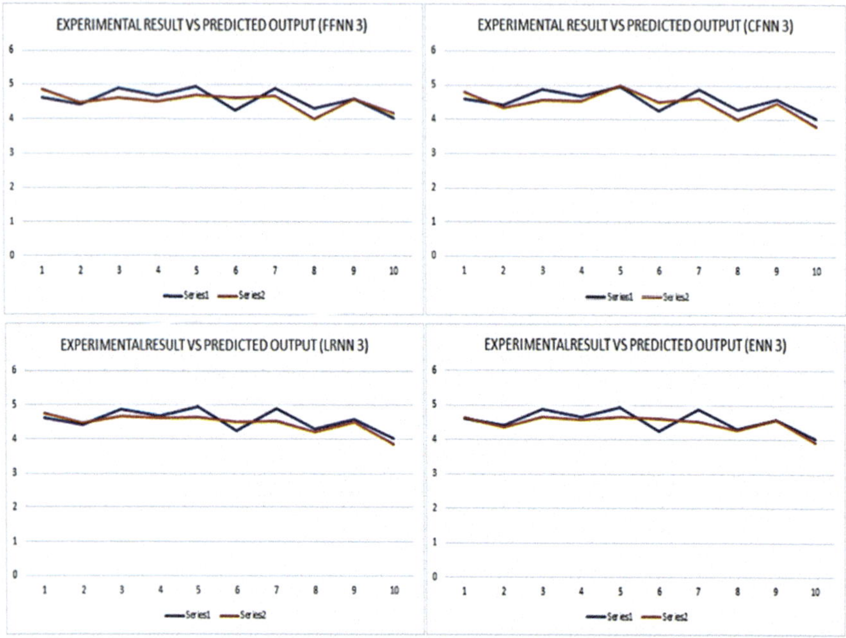

Fig. 3 Comparison plot of UPVT prediction results

finalize the optimal network. Each and every parameter is as important as the other one, so based on all the eight parameters only, the optimal network should be defined.

5 Conclusions

From the analysis and results, the following conclusions can be made.

- Using ANN in predicting strength durability characteristics of RHA concrete reduces computational time and effort.
- From the results, all the four networks predict more or less similar to the other networks. Difference in prediction accuracy is also very less.
- Comparing to strength characteristics prediction results, prediction of durability characteristics results shows better results. This is because of a smaller number of data variations and the quality of data set.
- The main limitation of using ANN in RHA concrete prediction process is that it required large numbers of input data for the training process. Also, regression graphs are not available for LRNN and ENN.
- As per the literature study, 1/3 times of total data sets are used as testing data and the remaining 2/3 times of data sets are used as training data. This ratio of data separation is suitable for this work.

- The main advantage of this neural network modeling is to predict the new sets of data in the future without any alterations. If changes are required in the network, model editing is also possible in MATLAB software.
- As per the results, the difference between maximum and minimum value in a data set has a higher amount of influence in the prediction accuracy percentage.
- From the results comparing all the eight network performance characteristics, cascade-forward neural network (CFNN) shows better results than the other networks in both strength and durability characteristics of RHA concrete.

References

1. Chao-Lung H, Le Anh-Tuan B, Chun-Tsun C (2011) Effect of rice husk ash on the strength and durability characteristics of concrete. Constr Build Mater 25(9):3768–3772
2. Islam MN, Mohd Zain MF, Jamil M (2012) Prediction of strength and slump of rice husk ash incorporated high-performance concrete. J Civ Eng Manage 18(3):310–317
3. Memon SA, Shaikh MA, Akbar H (2011) Utilization of rice husk ash as viscosity modifying agent in self compacting concrete. Constr Build Mater 25(2):1044–1048
4. Madandoust R, Ranjbar MM, Moghadam HA, Mousavi SY (2011) Mechanical properties and durability assessment of rice husk ash concrete. Biosyst Eng 110(2):144–152
5. Gautam A, Batra R, Singh N (2019) A study on use of rice husk ash in concrete. Eng Heritage J 01–04
6. Ismail MS, Waliuddin AM (1996) Effect of rice husk ash on high strength concrete. Constr Build Mater 10(7):521–526
7. Chopra D, Siddique R (2015) Strength, permeability and microstructure of self- compacting concrete containing rice husk ash. Biosys Eng 130:72–80
8. Badde DS, Gupta AK, Patki VK (2013) Cascade and feed forward back propagation artificial neural network models for prediction of compressive strength of readymix concrete. IOSR J Mech Civ Eng 3(1):1–6
9. Hemeida AM, Hassan SA, Mohamed AAA, Alkhalaf S, Mahmoud MM, Senjyu T, El-Din AB (2020) Nature-inspired algorithms for feed-forward neural network classifiers: a survey of one decade of research. Ain Shams Eng J
10. Hammoudi A, Moussaceb K, Belebchouche C, Dahmoune F (2019) Comparison of artificial neural network (ANN) and response surface methodology (RSM) prediction in compressive strength of recycled concrete aggregates. Constr Build Mater 209:425–436
11. Hamzic A, Avdagic Z (2016) Multilevel prediction of missing time series dam displacements data based on artificial neural networks voting evaluation. In: 2016 IEEE International conference on systems, man, and cybernetics (SMC). IEEE, pp 002391–002396
12. Ren L, Liu Y, Rui Z, Li H, Feng R (2009) Application of Elman neural network and MATLAB to load forecasting. In: 2009 International conference on information technology and computer science, vol 1. IEEE, pp 55–59
13. Gupta T, Sachdeva SN (2020) Recurrent neural network-based prediction of compressive and flexural strength of steel slag mixed concrete. Neural Comput Appl 1–13

Experimental Study on Thrust Force and Wall Angle in Single Point Incremental Forming of Ti6Al4V

S. Pratheesh Kumar, K. Anand, R. Mohanraj, and R. Arun Srinivasan

Abstract Incremental sheet forming (ISF) is a process where three dimensional objects can be fabricated without the application of die. The process deforms the material in small increments to obtain the final part geometry. Incremental forming is a reliable, flexible and high quality method of sheet metal products. In this study, Ti-6Al-4V is subjected to experimentation by the model obtained by central composite design. The input parameter considered are tool rotational speed (rpm), step depth (mm), feed rate (mm/min), and the output parameter measured are forming force (kg) and wall angle (°). The process is performed using a hemispherical and forming tool with the custom specific fixture for incremental forming process. This study aids in understanding the influence of the process parameters in obtaining the required part geometry. The outputs of this study can thereby use in producing custom specific products in real-time production.

Keywords Incremental sheet metal forming · Response surface methodology · Central composite design · Vertical machining centre · Ti-6Al-4v alloy

1 Introduction

Incremental sheet forming (ISF) is useful for prototyping and small batch production. Create component in computer aided design (CAD), then generate programme in computer aided manufacturing (CAM) software to import into computer numerical control (CNC) vertical machining centre (VMC) to create component using forming tool in CNC vertical machining centre. The sheet is held in place by a fixture while the forming tool shapes it according to the programme. The material is deformed locally to shape it. This method uses a simple general tool to make a variety of shapes without requiring a specific die setup. So, it is only good for prototypes and small runs. Researchers are working hard to fully grasp this process's benefits.

S. Pratheesh Kumar (✉) · K. Anand · R. Mohanraj · R. Arun Srinivasan
Department of Production Engineering, PSG College of Technology, Coimbatore, Tamil Nadu, India
e-mail: spratheeshkumarth@gmail.com

E. Natarajan et al. (eds.), *Materials, Design and Manufacturing for Sustainable Environment*, Lecture Notes in Mechanical Engineering,
https://doi.org/10.1007/978-981-19-3053-9_17

Experimenting with incremental sheet metal forming (ISMF) would provide not only basic parameter information but also a process understanding. Incremental sheet forming is deformation of sheet material in layer-by-layer process by applying force using a single point tool. Any changes required in the part can be done with minimum cost compared to other sheet forming process. The formability of the process depends on the radius of the tool and speed of the tool.

The need of this experiment is to increase the strength of the part and forming ability of the process. This works on finding the nominal value of feed rate, speed and tool movement to get preferred wall angle and improving quality of the part. This research depicts the basic concepts, experimental setup and final result of the incremental sheet forming of Ti6Al4V. The research also explains about approach shown by various scientists towards the improvisation of process with different process parameters.

2 Literature Review

In recent years, several academics have worked to improve incremental sheet forming technology. Many variables have been studied to improve process robustness and formability. We have previously discussed the evolution of incremental sheet formation. Kim and Park [1] used a variety of ball and hemispherical tools also used to assess formability. Lower input rate improved process formability. Formability varies with rolling tool movement. Vanhove et al. [2] used an AA 5182-O with feed rates up to 600 mm/min. High feed rates increased the maximum forming angle to 65° by improving material ductility.

Al 5052 alloy was studied by Reddy et al. [3] using the Box-Behnken method. For all depths, surface roughness is said to decrease with tool diameter. Up to a point, surface roughness increases with depth. Increased wall angle reduces surface roughness. Petek et al. [4] investigated process-induced deformation and forming force. The forming force increases with vertical step depth and tool diameter, the study found. With minor oxidation, Fan et al. [5] used electric hot incremental forming to work the Ti-6Al-4V alloy to form hard to form sheet metals. The surface quality of Ti-6Al-4AV is poor at high temperature and to enhance the surface quality proper lubricant is required at different processing temperature. The effect of sheets is compared by using various lubricants to find the best forming temperature. Ti-6Al-4V titanium workpiece was formed with high surface quality while using the lubricant film of nickel matrix with MoS_2 self-lubricating material.

It was Park and Ji [6] who made the magnesium sheets. The finite element analysis was used to investigate higher temperature forming. In both axisymmetric and plane-strain experiments, formability increased with temperature. Progressive forming was developed to overcome forming constraints by increasing the forming angle. Ambrogio et al. [7] used Joule's influence on aeronautical alloys. This methodology was supported by a workability window for the selected materials, which specifies the nominal conditions for wall angle and specific energy consumptions. The results show that as specific energy increases, surface quality decreases while heating improves formability rather than cold forming.

Araghi et al. [8] built a hemispheric with a groove using incremental sheet metal forming and stretch forming. Comparable to two-point incremental shaping, SPIF combined with stretch forming reduces sheet thinning to a greater extend. In both cases, finite element simulation was performed where results show that the combined technique also reduces forming time and produced uniform wall thickness. These frustums were made by Minitoli et al. [9]. The pyramid is 35 mm high and 70 mm in diameter. Conical shapes had more wall angles than pyramids. LS-DYNA numerical simulations showed that free surfaces can be generated for different strain values in the most stressed zone. For the greatest wall angle generated by incremental forming, Suresh and Srinivasa [10] experimented with varying wall angles. 110 mm top base and 40°–80° wall angle are compared. The theoretical model predicted thickness distribution better than the finite element model. Hamilton and Jeswiet [11] found that increasing feed rate to 8.89 m/min reduces surface roughness. One could only shape so much before the "orange peel" effect set in. Yu et al. [12] studied about laser metal deposition Ti6Al4V parts. LMD produced Ti6Al4V products which have good yield and ultimate tensile strength compared to annealed parts. The experiment was performed using a 7 kW IPG fibre laser with a 600 μm diameter firing fibre. The laser spot had a top hat energy distribution, and the powder was continuously transported through a coaxial nozzle with a stream of argon inert gas. The mechanical properties and microstructure of fabricated laser metal deposited Ti6Al4V parts were investigated. The microstructure is consistent between layers and tracks and total depth of the dilution, and heat affected zone is lowered.

Improvisation of one parameter affects the quality of other parameter, such that altering the value of other parameter also affects the quality of the product. Response surface methodology helps in finding the correlation and the interaction effect of all parameters to make better quality parts. Thereby, this methodology is used in this study. The quality of the part is improved by parameter optimisation in which wall angle and thrust force are analysed to the required limit.

Since ISF is cost effective for low volume production, the process is used in wide range of application. Parts made by ISF process helps in medical field such as in attachments of broken bones and other prosthesis. Moreover, Ti6Al4V founds wider application in aerospace structural members. This study uses Ti-6Al-4V alloy sheet metal, which can only be formed into simple shapes using standard forming methods. Since this study considers thrust force and wall angle as output responses considering industry relevant process parameter ranges. The output of this study results can be directly applied in custom specific part production in biomedical and aerospace applications, where the novelty of this study lies in. The primary goal of this study is to predict the forming force and maximum deformation wall angle that could be produced in incremental sheet metal forming.

3 Experimental Works

The experiments used a 150 mm × 150 mm Ti-6Al-4V alloy 0.5 mm thick. Table 1 shows the material's chemistry. In experimental 3-axis, CNC vertical machining centre 8000 rpm spindle, 5.5 kW drive motor and 450 × 350 × 350 mm maximum stroke length, the machine receives the CAM tool path. Experiment setup for single point incremental formation is shown in Fig. 1. The shaping tool (Fig. 2) is made of HSS M2 and TiN coated for durability.

A CMM measures the manufactured object's wall angle. The component was initially clamped perpendicular to its surface. The cloud points' geometric properties

Table 1 Chemical composition of Ti-6Al-4V alloy [12]

Element	Weight per cent
Carbon	<0.08%
Iron	<0.25%
Nitrogen	<0.05%
Oxygen	<0.2%
Nickel	0.05
Hydrogen	0.015
Yttrium	0.005
Aluminium	5.5–6.76%
Vanadium	3.5–4.5%
Titanium	Balance

Fig. 1 ISMF experimental setup

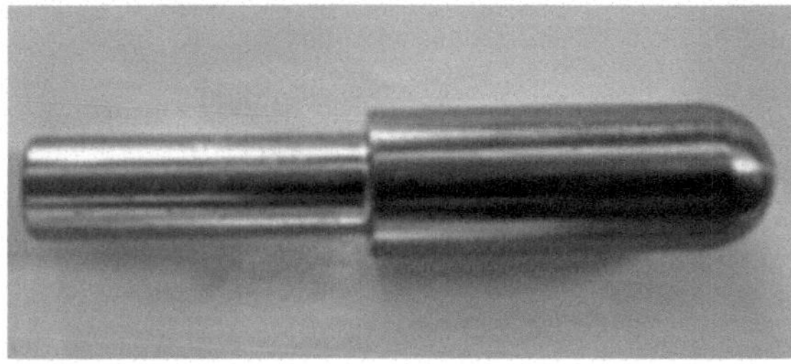

Fig. 2 Hemispherical form tool of diameter 15 mm

Table 2 Design parameters and their levels

Levels	1	2	3
Tool rotational speed (rpm)	100	150	200
Incremental depth (mm)	0.1	0.2	0.3
Feed rate (mm/min)	1000	1500	2000

were retrieved. A CNC machine tool dynamometer measures forces during forming. A three factor three level factorial design was used to optimise tool rotation speed, feed rate and incremental depth. The experiment was designed using a response surface method (central composite design). Table 2 summarises the levels of the factors.

4 Results and Discussion

These include tool rotation speed, incremental depth and feed rate. A CMM measures profile accuracy. The best ideal process parameters were predicted using response surface methodology using wall angle and force as responses. A wall angle and force are shown in Table 3. Figure 3 is an ISMF test subject.

4.1 Forming Force

Incremental sheet forming is a process in which forming force is applied incrementally using round tipped tool. Forming force varies with the tool diameter and higher spindle speeds. The forming force values change with respect to bending conditions corresponding to part geometry. The increase of forming force is mainly due to friction and load applied between tool and blank specific to spindle speed. Other factor contributing effects include the forming force which include rotational spindle speed, step size, feed rate and tool diameter. Forming force was calculated using data from each run's milling tool dynamometer. It helps visualise parameter effects. To the contrary, increasing feed rate from 1000 to 2000 mm/min has no effect on forming

Table 3 Experimental design and response values

Run	Tool rotational speed (rpm)	Step depth (mm)	Feed rate (mm/min)	Forming force (kg)	Max wall angle (degree)
1	100	0.3	1000	214	27.45
2	200	0.3	1000	219	27.92
3	150	0.3	1500	210	27.67
4	200	0.1	1000	156	29.74
5	150	0.2	1500	192	28.78
6	150	0.2	1500	196	28.32
7	100	0.2	1500	188	28.2
8	150	0.2	1000	202	28.85
9	150	0.1	1500	177	29.61
10	150	0.2	1500	192	28.45
11	150	0.2	1500	190	28.7
12	100	0.1	1000	180	29.39
13	100	0.1	2000	174	29.1
14	200	0.2	1500	198	28.88
15	200	0.3	2000	208	27.52
16	150	0.2	1500	192	28.65
17	150	0.2	1500	194	28.43
18	200	0.1	2000	172	29.63
19	150	0.2	2000	185	28.38
20	100	0.3	2000	204	27.16

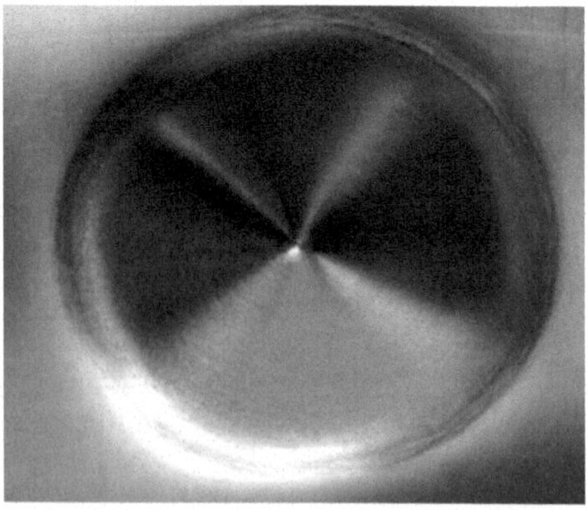

Fig. 3 Experimental specimen of ISMF

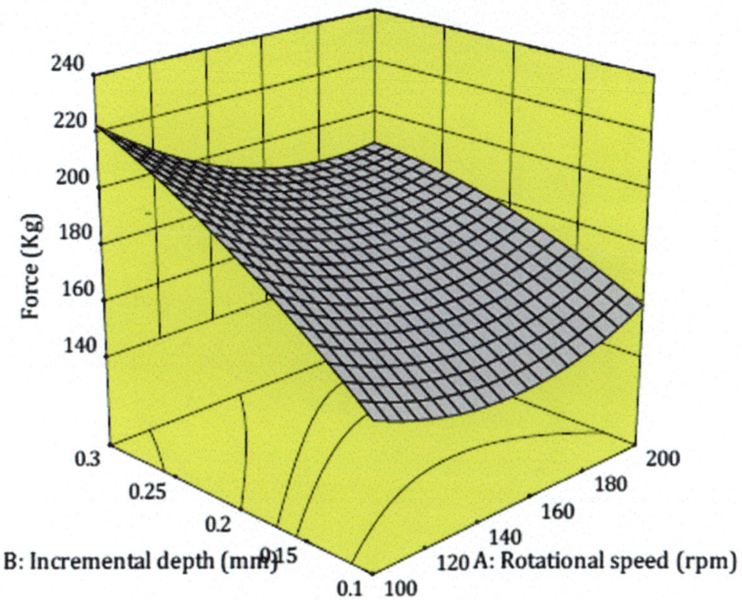

Fig. 4 Effect of tool rotational speed and incremental depth on forming force

force. Figures 4, 5 and 6 show the variation in rotational speed and feed rate on forming force. The forming force increases from 10 to 23 N with depth. The reaction change is dictated by the incremental depth. More material to distort for major steps down more rapid tool rotation reduces forming force. By increasing tool sheet friction and localising heat, high rotation speed reduces forming forces incremental depth of 0.1 mm, form force 156 kg at 200 rpm.

4.1.1 Wall Angle

Wall angle is subjected to vary with respect to the part geometry and is directly related to the extent to which material is formed without failure. In real time manufacturing product produced have varying wall angles and therefore the influence of process parameters also differs over a wider extent. There is also considerable increments on tool geometry with respect to wall angle of the component. Springback in die-formed parts causes non-conformance but storing elastic energy reduces springback. Figures 7, 8 and 9 show a minimum feed rate and step depth. Feed rate affects forming time. You can release elastic energy from the springback region before the forming is finished. As step depth increases, so does the amount of localised deformation. The wall angle increases with tool speed. So, a faster spindle speeds up forming. With a 200 rpm spindle, 1000 mm/min feed rate, and 0.01 mm incremental depth, the maximum wall angle is 29.74°. The 29.74° wall angle is within 0.26° of the conventional 30° wall angle.

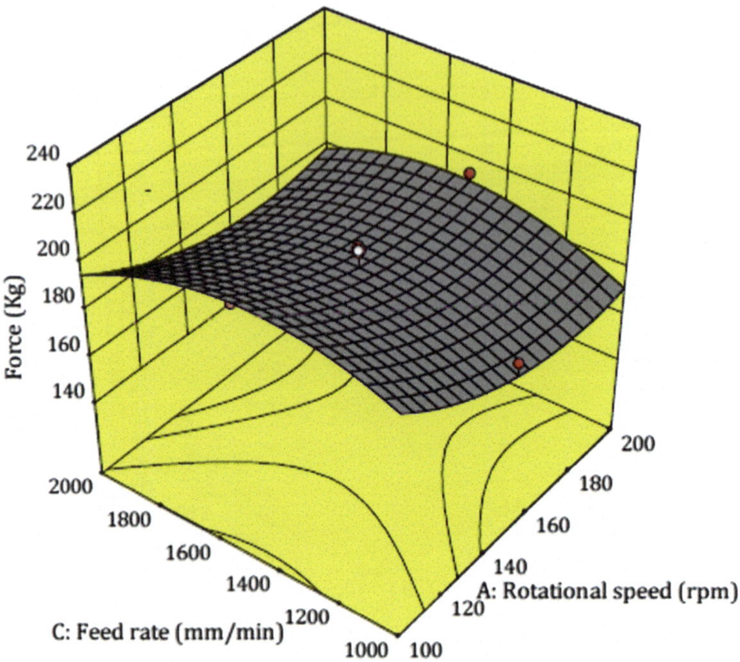

Fig. 5 Effect of tool rotational speed and feed rate on forming force

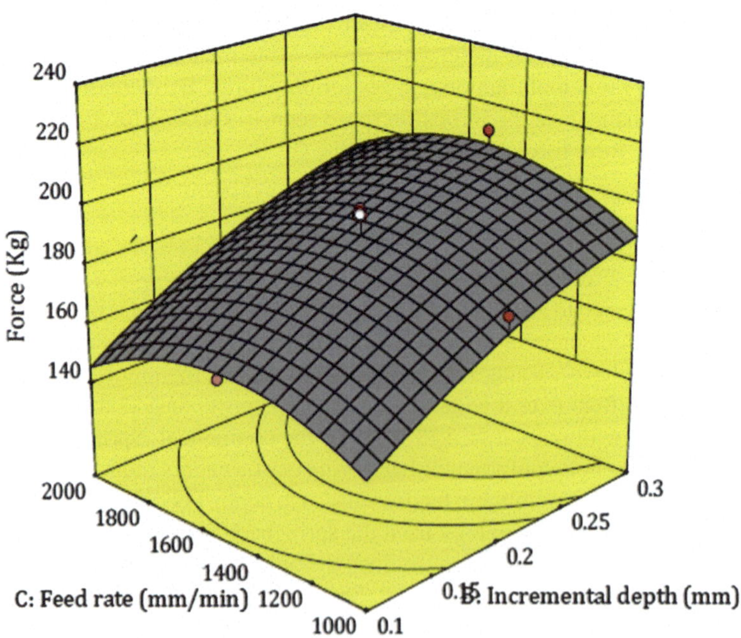

Fig. 6 Effect of incremental depth and feed rate on forming force

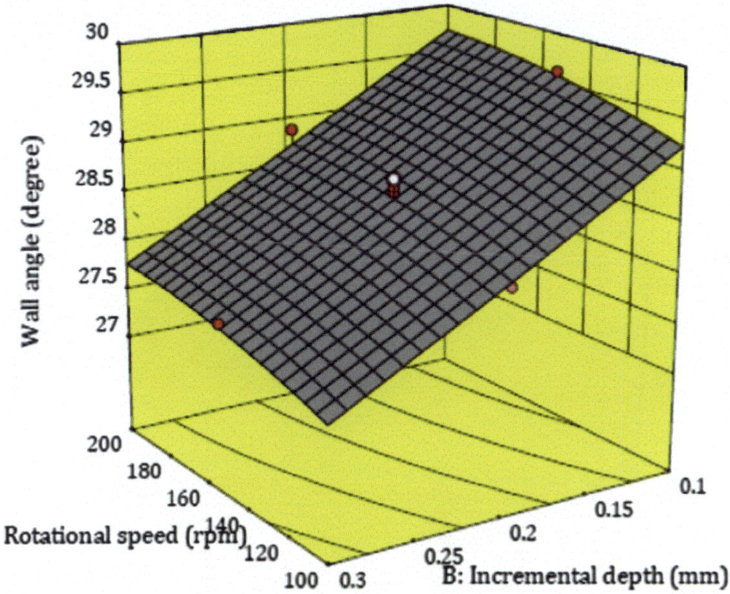

Fig. 7 Effect of tool rotational speed and incremental depth on wall angle

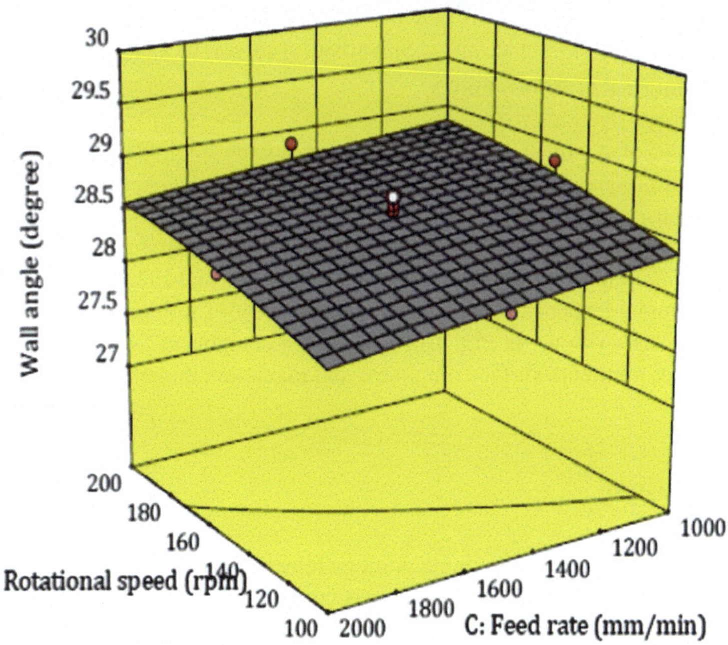

Fig. 8 Effect of tool rotational speed and feed rate on wall angle

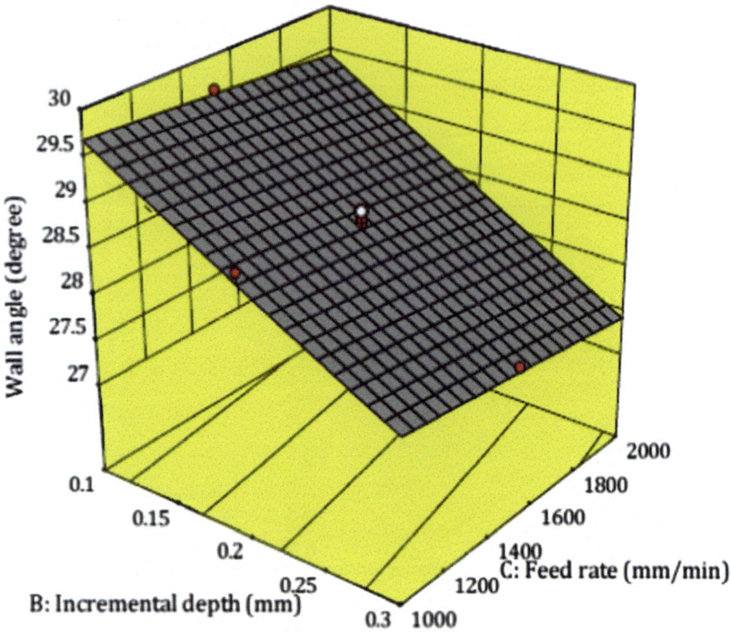

Fig. 9 Effect of incremental depth and feed rate on wall angle

4.1.2 Numerical Optimisation

By minimising or maximising optimization functions, the required response is chosen from the resulting regression equations. Desirability measures optimality of the output responses. Springback always influences wall angle, emphasising the importance of determining optimal process parameters. With input parameter combinations of 1044.09 mm/min feed rate, 0.1 mm incremental depth, and 192.298 rpm tool rotation, the nominal responses obtained were 155.637 kg of forming force and 29.76 degree wall angle (shown in Fig. 10). Prepare the component for optimal forming conditions for minimum surface roughness and maximum thinning.

5 Conclusion

With truncated part geometry, single point incremental forming of Ti-6Al-4V alloy sheet metal was studied. This work necessitates an end tool fixture design. Wall angle and forming force are the parameters studied in this section. This research study highlights that forming force changes due to increase in step depth but not due to feed rate, and it can be reduced by high tool rotation rate. Future, scope of this

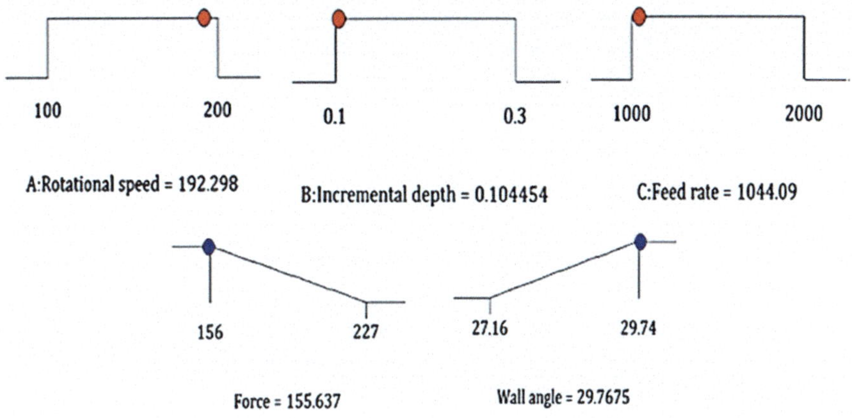

Fig. 10 Optimisation results

study defines that thickness deviation, material property changes, form tool influence in thickness of the sheet can be defined in future through this study.

1. The wall angle is greatly affected by tool speed, and 30° wall angle has good geometric accuracy.
2. Although increases in spindle speed decrease forming force due to tool-blank friction, increases in incremental depth increase forming force.
3. The spring back region has less elastic energy stored and can release it before forming.
4. Inferred divergence between measured and actual wall angles is 97.80%.
5. On average, 192.3 rpm spindle speed, 1044.09 mm/min feed rate and 0.1 mm incremental depth produce 155.637 kg and 29.76° on average.

References

1. Kim YH, Park JJ (2002) Effect of process parameters on formability in incremental forming of sheet metal. J Mater Process Technol 130:42–46
2. Vanhove H, Mohammadi A, Guo YS, Duflou JR (2014) High-speed single point incremental forming of an automotive aluminium alloy. In: Key engineering materials, vol 622. Trans Tech Publications Ltd., pp 433–439
3. Bhattacharya A, Maneesh K, Venkata Reddy N, Cao J (2011) Formability and surface finish studies in single point incremental forming. J Manuf Sci Eng 133(6)
4. Petek A, Kuzman K, Kopač J (2009) Deformations and forces analysis of single point incremental sheet metal forming. Arch Mater Sci Eng 35(3):107–116
5. Fan G, Sun F, Meng X, Gao L, Tong G (2010) Electric hot incremental forming of Ti-6Al-4V titanium sheet. Int J Adv Manuf Technol 49(9):941–947
6. Ji YH, Park JJ (2008) Formability of magnesium AZ31 sheet in the incremental forming at warm temperature. J Mater Process Technol 201(1–3):354–358
7. Ambrogio G, Filice L, Gagliardi F (2012) Formability of lightweight alloys by hot incremental sheet forming. Mater Des 34:501–508

8. Araghi BT, Manco GL, Bambach M, Hirt G (2009) Investigation into a new hybrid forming process: Incremental sheet forming combined with stretch forming. CIRP Ann 58(1):225–228
9. Minutolo FC, Durante M, Formisano A, Langella A (2007) Evaluation of the maximum slope angle of simple geometries carried out by incremental forming process. J Mater Process Technol 194(1–3):145–150
10. Kurra S, Regalla SP (2014) Experimental and numerical studies on formability of extra-deep drawing steel in incremental sheet metal forming. J Market Res 3(2):158–171
11. Hamilton K, Jeswiet J (2010) Single point incremental forming at high feed rates and rotational speeds: surface and structural consequences. CIRP Ann 59(1):311–314
12. Yu J, Rombouts M, Maes G, Motmans F (2012) Material properties of Ti6Al4 v parts produced by laser metal deposition. Phys Procedia 39:416–424

Study and Development of Remote Control Appliances in DailyLife

S. Pratheesh Kumar, R. Dinesh, V. Raja, and S. Karthikeyan

Abstract As the world progresses, new technology becomes more and more commonplace in our daily life. Controlling a television, radio or other electrical item via an infrared remote control. As far as anyone can tell, the infrared photons don't do anything at all. Excessive exposure, on the other hand, may damage the retina. The basic function of an infrared (IR) remote control is as a source and receiver of transmission signals in devices that have IR transmitter and receiver circuits. In order to control some home appliances, we designed a circuit that connects an infrared receiver circuit to the appliance. The appliance may then be operated with a regular infrared remote control. This method makes use of infrared photons. Pushing the command button on the IR transmitter circuit generates infrared rays, which are then sent to the IR receiver circuit, which is connected to the appliance that will be controlled by the remote. This remote control can be used when the transmitter and receiver circuits are 10 m apart. It uses a TSOP 1738 receiver module and operates at a frequency of 38 kHz. Freshly made appliances are the subject of this investigation. This circuit can be used to supply power to already-existing home electronics that are compliant.

Keywords Remote control appliances · Technology · Internet of things · Smart appliances · Infrared

---

S. Pratheesh Kumar (✉) · R. Dinesh · V. Raja · S. Karthikeyan
Department of Production Engineering, PSG College of Technology, Coimbatore, Tamil Nadu, India
e-mail: spratheeshkumarth@gmail.com

E. Natarajan et al. (eds.), *Materials, Design and Manufacturing for Sustainable Environment*, Lecture Notes in Mechanical Engineering,
https://doi.org/10.1007/978-981-19-3053-9_18

1 Introduction

It is possible to change the characteristics of a gadget using a remote control that sends digitally-coded pulses. These gadgets are often small wireless hand-held devices having a selection of customisable television station, music number and volume control buttons. Remote manipulation has all feature controls for numerous devices, whereas managed tools have few or no number one controls. IR, RF, bluetooth, movement sensors and voice manipulation have all been added to remote controls as they've become more advanced over time [1, 2]. Air conditioners, lighting and satellite receivers and televisions are amongst the many things that may be controlled using an IR remote control that is commonly seen in offices.

Most homes have a range of electric and electronic home appliances, such as a television, a ceiling fan, an air conditioner, light bulbs, DVD players, a radio and so on. All of these home appliances have traditionally required a switch to turn them on and off. When performing routine switching operations, physically challenged or elderly people, as well as younger and busier people, may find it inconvenient to use this guided switching technique on any domestic device. Thus, a simplified switching strategy is developed to update this guide switching technique, which makes use of a sophisticated switching technique for digital home equipment [3].

Early tethered or tripped remotes allowed users to control their home appliances from a distance, but they also restricted their mobility. If the buyer ran the wire of the far-off mange under the rugs and furnishings, the tool would be anchored to a single location in the room, limiting its intended freedom of movement [4]. If the wire is disregarded and left in the open, customers may trip over it. The Wi-Fi remote manager solved most of the problems of the stressed-out remote manager and effectively takes control of the home. Wi-Fi remote management. In order to re-establish order in the household, distance is used. This changes the way people live; it simplifies lives that have many different characteristics and platforms [5]. As time has progressed, new technologies have emerged, such as radio waves for information management, electromagnetic waves for executing instructions and a battery-operated low-frequency radio transmitter remote control system for client electronics. The design of a digital remote control that used a light beam to control a photoelectric cell to operate wirelessly.

The most frequent form of light in residential remote controllers is infrared (IR) [6]. While infrared light cannot be seen with the naked eye, it may be seen with a digital camera, video camera, or phone camera when sent between a remote control handset and the instrument it controls. The transmitter inside the remote control handset sends out a sequence of moderate infrared pulses when the user presses a button. A transmitter is mainly a light-emitting diode (LED) integrated inside the pointing stop of the distant manipulator handset. The tool's receiver recognises the sample and gives instructions on how to respond. Use a TV remote control to turn on and off any connected home equipment (such as a lamp, fan or radio).

The circuit can be started up to ten metres away. The remote control's 38 kHz infrared (IR) rays are received by the circuit's IR receiver module TSOP 1738. R5 connects pin 2 of the TSOP 1738 to the power supply, and pin 1 is grounded. The output of the TSOP 1738 comes from pi3. Transistor T1 boosts the signal at the output (BC558). Relay pole and neutral terminal are connected together in order to control an appliance. When the relay is energised, a normally open (N/O) contact connects it to the live terminal of the alternating current mains.

A remote control that sends out digitally encoded pulses can be used to change the characteristics of the gadget. These devices are often small wireless handhelds with a selection of customisable TV stations, music numbers and volume buttons. When performing routine switching tasks, people with disabilities, the elderly, young and busy people may find it inconvenient to use this controlled switching method on all their home devices. The most common form of light in home remote controls is infrared (IR). Infrared is invisible to the naked eye, but visible to a digital camera when infrared is transmitted between the remote control handset and the device it controls. These mechanisms must be controlled by the landlord from anywhere, as the users may be away from work or travel elsewhere on the weekend. Therefore, a remote control and management system is essential.

1.1 Circuit Diagram Description

The 38 kHz infrared rays from the remote control are picked up by the IR receiver module TSOP 1738 in the circuit, as shown in Fig. 1. There are three connections on the TSOP 1738: a ground pin, a power pin and an output pin. Q1 boosts the signal at the end. Pin 14 of the decade counter ICCD 4017 clock receives the increased signal (IC1). When the appliance is turned off, the LED on pin 3 (red) illuminates, which indicates that pin 8 (Ground) of IC 1 is grounded. Pin 2 of IC 1 provides the device's output. D3, which is connected to pin 2, indicates if the appliance is turned on. Pin 2 of IC 1 is used by Q2 to control the relay RL1. D1 is a diode that may be moved about freely. Relay pole and neutral terminal are connected together in order to control an appliance. When the relay is energised, a normally open (N/O) contact connects it to the live terminal of the alternating current mains. This relay can be used with a regulated DC 12-V power source, as long as the circuit voltage does not go above DC 5 V.

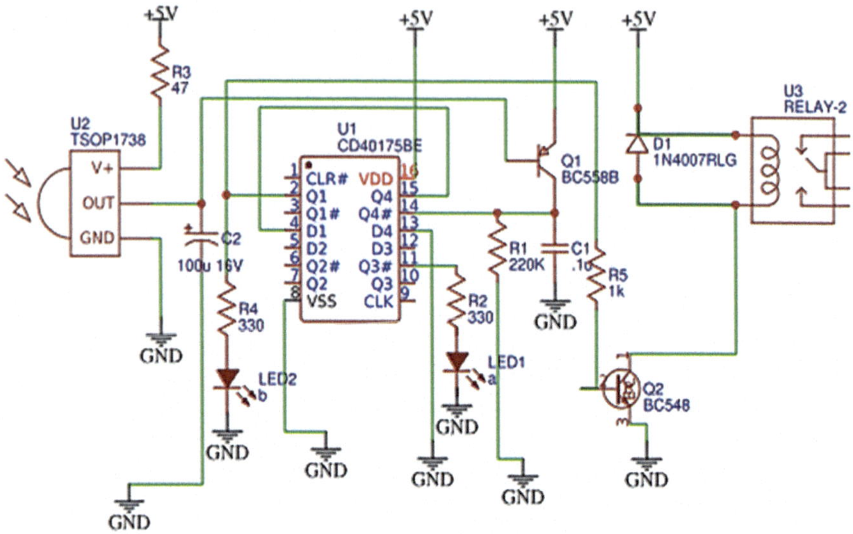

Fig. 1 IR circuit diagram

1.2 Components Used

1.2.1 IC4017

It's possible to use the IC 4017 in many different situations thanks to its flexibility in the CMOS family. On the interior, there's a decade counter/divider with 10 stages. When a clock pulse is applied to it, the outputs become sequential logic 'hi' and 'lo' (positive and negative) (one after the other). It can be used in a variety of ways. The IC 4017 will have a value from 0 to 10, and the circuit will use less board space and designing time by using this IC. There are striking similarities between this decade counter and Johnson's 10 degree decade counter. It is common to use this CMOS IC in the building of ten LED-primarily based complete circuits to power the light for newbies [7]. Because it counts up to ten and has ten separate outputs, it's one of the most potent counters around. In addition to the decoder, this integrated circuit features a counter.

1.2.2 Pin Configuration

A 16-pin dual in line packaging IC, the 4017 in Fig. 2, can be seen in action. On printed-side ICs, pin 1 begins at the left side of the semicircle notch. If you do not see a depressed circle there, it's easy to miss. As seen in Fig. 3, the IC's outputs are represented by pins 1–7, 8, 9 and 10. Pin 16 provides power, whereas pin 8 serves as a ground. Pin 15 is the IC reset pin. Connecting this pin to ground or to logic 0 turns

Fig. 2 IC4017

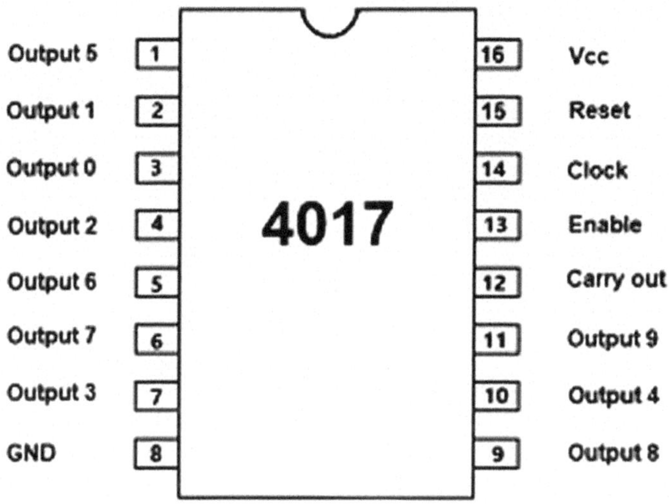

Fig. 3 IC 4017 pin configuration

on the IC and makes it operational. In this case, pressing the logic "1" will stop the procedure and start it over. There is no logic "hi" in any of the other outputs of the IC 4017 in this state. Pin 14 of the IC 4017 serves as the clock input. An external clock signal will cause logic '1' to continue from pin 3 to pin 11, if applied here.

The results of a quick literature review were summarised below to help better understand remote appliance technology. To run home appliances, Narendra Reddy and Sukumar [8] use an ARM board and a universal remote controller to develop a circuit for the Internet of Things (IoT), Wi-Fi and bluetooth (BLE). By utilising the concept of multiple data fusion, the circuit is built with decreased complexity, security in the regulating, and low power consumption. Using IoT, Wi-Fi, and bluetooth technologies, data can be viewed on a mobile phone, and a smart global remote control system can be developed. Customers can control linked home appliances (TV, VCR, camera, CD speller, etc.) via a mobile device such a mobile phone or a Web pad using a system described by Nikolova et al. [9]. Examples of typical packages are the following: remote mobile video cassette recorder (VCR) programming, remote mobile heating thermostat manipulation, remote mobile safety camera tracking and others. Because we're interested in middleware, we created a widely used domestic-cell community gateway architecture. Increasing real-time networking, data collection and analytics capabilities are all part of what Aheleroff et al. [10] call the "Internet of Things" (IoT). To put it another way, statistics are now used as a way to charge clients for stressful but low-price SPS that is more personalised than ever. When Internet of Things (IoT) became popular in the early 2010s, the home equipment industry was one of the first to adopt cutting-edge technologies like IoT and Cloud. IoT provides continuous collaboration for sensing, actuation, and real-time data sharing.

According to Mbunwe [11], a remote control switching tool for the home is a domestic instrument used to remotely control the switching of family home appliances. Its goal is to make it easier for people of all ages and abilities to switch between household appliances, whether they are physically able or not. This research develops a remote control system based on the radio frequency era, comprising multiplexers, demultiplexers, encoders, decoders, and a radio frequency module, as well as an evaluation of several technologies that can be used to produce a remote control system. As long as you utilise the remote control safely, the gadget can be used up to roughly 10 m away. Because of that, a transmitter device is designed and built that receives a signal when a button is depressed, and a receiver device that receives and processes the signal received from a transmitter device before activating or turning off home equipment.

A safety device is also included, which only permits signal transmission when certain conditions are met. Without using a microprocessor, this study built a low-cost, dependable and powerful remote control system for residential use. Using a Raspberry Pi and relay circuits, Ashraf's [12] study shows how to design and build a home automation system that doesn't require expensive IP-based devices and instead leverages conventional electrical household equipment for remote control. Standard lighting, heating, ventilation and air conditioning systems, as well as a variety of digital gadgets can all be used with this system. Customers can use a drag-and-drop smartphone software to customise their smart household to match their actual home. Both the local community and distant access can use the machine to control the appliances. Facts in the event of a gateway failure, logging to the Microsoft Azure cloud database supports machine healing while documenting facts for lateral use.

Users can also benefit from periodic notifications by making better use of home equipment. It is hoped that the gadget developed by Vijayalakshmi et al. [13] will provide a low-cost solution for controlling domestic appliances remotely and protecting the home from intruders while the owner is away. In order to offer remote access to the home equipment and allow domestic security, a low-cost control system for home equipment was envisioned. Appliances for the home and business use power. Keeping an eye on and turning on and off these gadgets should be a top priority. For the most part, everything was done manually. In addition to the literature review discussed, most suitable literature is compared.

More effective and efficient device control is now required at all times and from anywhere. On this technology, we plan to expand a remote control-based home and office equipment [24, 25]. So, we design a circuit that includes an IR module, a CD4017 IC, LEDs to show reception of IR radiations, otherwise signalling the ON/OFF country, a relay and other parts.

A demo model was developed and tested to demonstrate how a luminaire can be controlled from a distance. In an atmosphere where people's movements are restricted and daily tasks are handled by remote control, products are of promising future. Ideal for making home appliances easier to manage for patients, people with disabilities and the elderly. It's good to conserve energy in unused conditions. It uses IR to turn appliances on and off without actually accessing the switchboard.

This is complicated by the fact that the user may need to control the device's (on/off) for necessary reasons, or may need to do so on a daily basis to save energy bills and costs incurred on a regular basis. If users can conveniently operate multiple home appliances while sitting in one place when needed, it not only provides a lot of convenience depending on the situation, but also encourages them to invest and use the time saved.

The main thing of remote control units is that both time and energy consumption are reduced. Relevance in today's global market. When you focus on your current task, you don't have to worry about leaving it in the middle. The range of the remote control is determined by the frequency used and the effectiveness of the circuit. The advantage is remote control. For example, it can be used in a variety of devices and applications such as TVs, VCRs, cameras, CD recorders and radios, and even for simple tasks such as unlocking a door. A single remote control can be programmed to operate on different frequencies, each corresponding to a specific purpose the device must fulfil, such as watching TV, for example. Circuits can be easily assembled using readily available and inexpensive parts. It is also easy to keep clean.

2 Features and Testing

2.1 Features

(1) Photo detector and preamplifier in one package
(2) Internal filter for PCM frequency
(3) Improved shielding against electrical field disturbance
(4) TTL and CMOS compatibility
(5) Output active low
(6) Low power consumption
(7) High immunity against ambient light
(8) Continuous data transmission possible (up to 2400 bps)
(9) Suitable burst length $\geq$10 cycles/burst.

2.2 Infrared Remote Controls

Using light-emitting diodes (LEDs) to convey coded pulses of infrared light to a receiving device has become the norm for remote controllers in today's world. Infrared remote controllers have the disadvantage of requiring an unobstructed line of sight to work; however, you can bounce a signal off the ceiling or a wall and they can operate up to 20 m away. Infrared (IR), commonly referred to as "heat," is the most popular remote control technology in the home [26]. Adding an IR device assumes that sunlight can be used to keep signals flowing between a remote control and the device it's controlling. Aspects of magnetic attraction that are invisible to the unaided eye include a form of radiation known as adonic radiation.

Home appliance infrared remote controls use comparable technology to a television remote. Actinic radiation is the part of the electromagnetic spectrum that lies between microwaves and visible light. In infrared communication, a diode sends off non-visible light bursts as the infrared signal. To recover the data stored in the image, a photodiode or photoreceptor is used to detect and capture the sunlight pulses. The pulses of actinic radiation emitted by an infrared device transmitter indicate the unique binary codes. Binaries such as adore Power ON/OFF are used to represent commands.

The receiver picks up the signal and sends it to the microchip, which subsequently executes the appropriate command. The remote control's incoming signal is received by the device's infrared receiver, which is on the device's front. Infrared remote controls use a microcircuit to detect when a switch is pressed and deliver a binary command to the remote's front-facing diode.

2.3 Controlling the Controls

In recent years, the number of remote controllers in each home has expanded substantially. Thanks to universal remotes, this problem has been solved. Universal remotes can pick up commands from other remotes.

2.4 Test Working

Check that the batteries' supply voltage and polarity are correct, and that all connections are in the proper position.

2.5 Soldering Process

Connect the soldering iron to the wall socket and the stand. Once the iron gets there, it'll take a few minutes to do the work. Allow the damp sponge to air dry for a few minutes. As a result, the tip will be thoroughly cleansed. Wipe the damp sponge's tip. This will make the journey much more understandable. Use a soldering iron to melt a little amount of solder.

3 Working

You may use a remote control to link your regular home appliances to this appliance remote control. To use the circuit, all you need to do is connect it to a light, fan, radio or other common household item. A frequency-controlled remote control can now be used to operate the appliances. It is possible to start the circuit from up to 10 m away. The IR receiver module TSOP1738 in the circuit receives the remote control's 38 kHz infrared beams. There is no need for networking or an expensive controller with this device, and it can be operated using only a TV remote control (Fig. 4).

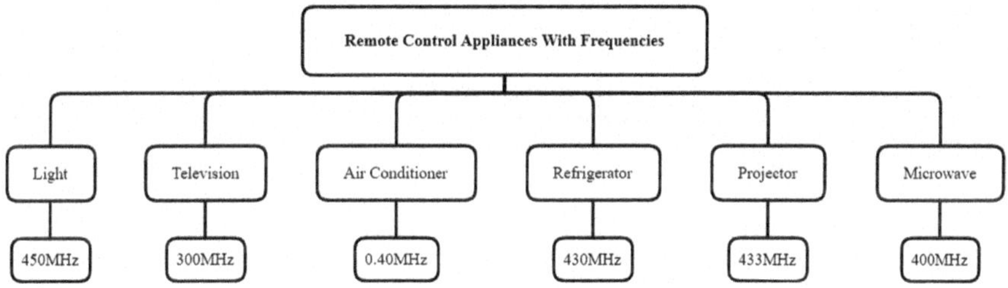

Fig. 4 Remote control appliances with different frequencies

4 Modelling

4.1 CAD Model

The cad model and concept representations of the model is shown in Figs. 5 and 6.

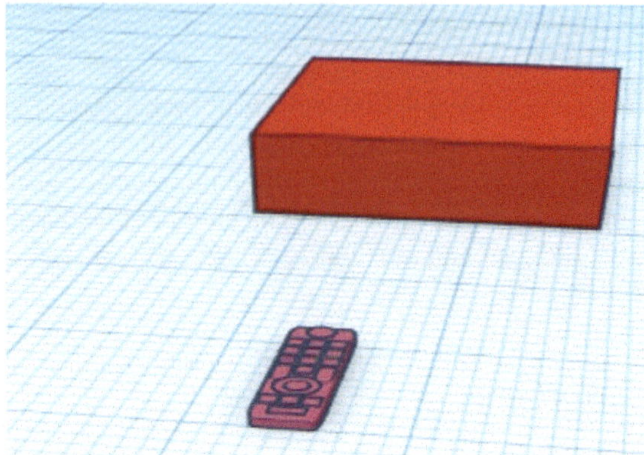

Fig. 5 Cad model

4.2 AutoCAD Model

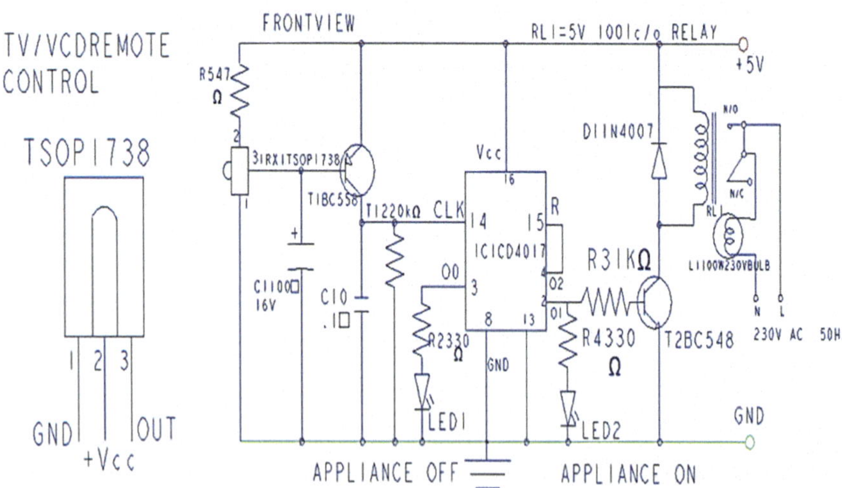

Fig. 6 Auto Cad model

Fig. 7 Fabricated model

4.3 Fabricated Model

See Fig. 7.

5 Cost Estimation

Many products are available on the market, but they are unable to meet the demands of the elderly, the poor and those with physical or mental disabilities. To achieve our primary objective, we must ensure that our customers are satisfied and that they are able to operate the home appliance in their preferred posture on their own. When we compared our estimations with those of the other companies, ours was lower, as shown in Fig. 8 and Tables 1, 2.

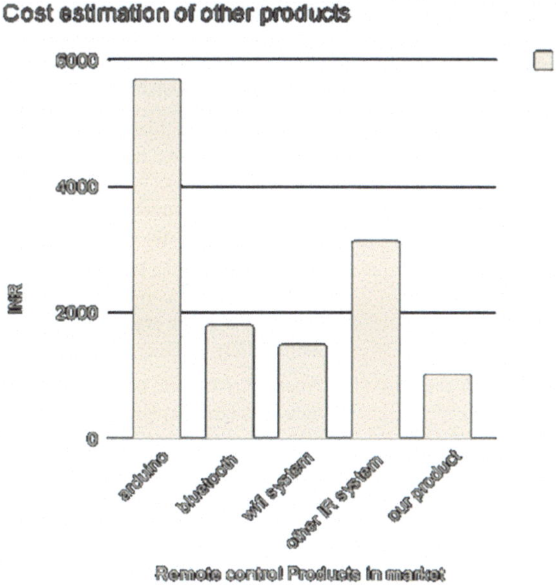

Fig. 8 Cost comparison

Table 1 Comparison of literature review with existing literatures

S. No.	Authors	Description	References
1	Hasan et al.	Introducing an external device switching circuit for an automated home with a microcontroller capable of controlling the device. Circuit is incorpated with smartphone applications by using bluetooth where it sends command to microcontroller	[14]
2	Maiti and Nandi	Describes the design and implementation of an infrared (IR) remote control signal decoder for use in a variety of home control applications	[15]
3	Shinde et al.	By introducing a home automation system using IR remote control, bluetooth and GSM, AC devices are controlled with an Android app that is easier to use than the existing switching method	[16]
4	Bajaj et al.	To give users better control over their home appliances, a home automation system has been proposed that uses an android application to initiate the on/off and turn on angles of installed electrical appliances	[17]
5	Roy et al.	Advanced universal remote control and control lighting, fans and all home appliances controlled by IR such as TVs, home theatres and air conditioners. He tells how to make a remote and how to turn a smartphone into a universal remote	[18]
6	Krishna et al.	Describes how to control in remote and automating the offices and homes using microcontrollers units, relays, water level sensor and motor drive	[19]
7	Myers	Describes an overview of features being developed for portable wireless devices such as cell phones and palm-sized computers. Not just for cell phone use, the device can also be used as a remote control device for computers and home and office supplies	[20]
8	Richmond et al.	Applying various linear and nonlinear models, it can be seen that connected homes can use more electricity and use much more electricity to power four categories of appliances: lighting, refrigerator, entertainment and others by taking survey in some states	[21]
9	Hsu et al.	He developed some techniques and some skills related to 3C which means computer, communication and content to make the home appliances more intelligent. He done with the refrigerator	[22]
10	Zhou et al.	Potential risk could be IoT devices that support IR remote control exploit sensitive leaks from the airgap network using malicious IR hardware	[23]

Table 2 List of components

S. No.	Components	Quantity	Cost
1	TV remote	1	300
2	Bulb	1	80
3	Soldering	-	400
4	Resistor	6	24
5	Capacitor	10	20
6	Transistor	9	18
7	IR Sensor	1	40
8	LED	2	3
9	5 V relay	1	15
10	Diode	1	1
11	PCB board	1	120
12	Total	–	1021

6 Applications

The primary benefit of a remote controlled gadget is the reduction of both time and energy consumption. Its relevance in the global marketplace today. They don't have to worry about leaving it in the middle if they utilise it while they're focused on their current task. The remote control's range is determined by the frequency used and the circuit's efficiency. It has the benefit of remote control operation. For example, it can be utilised in a wide range of gadgets and applications like TVs and VCRs and cameras and CD spellers and radios and even simple tasks like unlocking doors. It is possible to programme a single remote control to operate at various frequencies, each of which corresponds to a certain purpose that the device is supposed to perform, such as watching TV. Using generally available and inexpensive components, the circuit is straightforward to build. Keeping it clean is a breeze as well.

Controlling the devices using cell phone's remote option such appliances are television, air conditioner, light bulbs, fans, DVD players, projectors, set up boxes and cameras. Smart lighting makes to control wall switches, blinds and lamps. You can schedule lights to turn on and off, determine which rooms should be lit at specific times, and choose the level of light emitted. And other applications are smart fridges let you experiment grocery save receipts and preserve a stock of your items, and indicators you if an object is set to expire. The smart oven synchronises with your smartphone and automatically heats up to the right temperature according to the recipe selected from the database. For security purpose, it offers many features including door and window sensors, motion detectors, video cameras and recording mechanisms. All connected to your mobile device and accessed via the cloud, giving you access to real-time information about the security posture of your home.

7 Conclusion

In this study, a demo model was developed and tested to show how a light fixture can be controlled from a distance. In an atmosphere where people's movements are limited and their daily tasks are performed by remote control, the product is a promising future. The implementation of the proposed control system is currently limited to sensors. You need to define more stateful devices. Therefore, controlling devices with more accurate targeting mechanisms and supporting automatic detection mechanisms for state dependencies are two possible directions for future research. The key outcomes of the study are listed below,

1. The wireless remote control eliminates the need for operators to directly touch the machine in operation.
2. Users, control several electrical appliances by just sitting at one place, whenever needed.

There are two sensors IR1 and IR2 on the door for remote monitoring and control of household appliances. The IR sensors are arranged in such a way that when a person enters, the IR1 sensor is triggered first, then the IR2 sensor is triggered, and the output is vice versa. Whenever a person enters, it signals the ESP8266 to turn on light 1 and whenever a person leaves the door it signals the ESP8266 to turn the light off. The IR signal from the ESP8266 is sent. Therefore, these devices can also be controlled remotely. This method also enables users to control multiple devices and monitor the health of their devices from anywhere.

References

1. Lin YN, Wang SK, Yang CY, Shen VR, Juang TTY, Hung WH (2020) Development and verification of a smart remote-control system for home appliances. Comput Electr Eng 88:106889
2. Khiyal MSH, Khan A, Shehzadi E (2009) SMS based wireless home appliance control system (HACS) for automating appliances and security. Issues Inform Sci Inf Technol 6
3. ShahriyarR HE, Sohan SM, Naim I, Akbar MM, Khan MK (2008) Remote controlling of home appliances using mobile telephony. Int J Smart Home 2(3):37–54
4. Van Nguyen T, Lee DG, Seol YH, Yu MH, Choi D (2007) Ubiquitous access to home appliance control system using infrared ray and power line communication. In: 2007 3rd IEEE/IFIP international conference in Central Asia on Internet. IEEE, pp 1–4
5. Mishra RK, Bankar S, Shinde S (2020) Design & implement infrared (IR) controlled home appliances
6. Kaphungkui N K (2015) Rf based remote control for home electrical appliances. Int J Innov Res Electr, Electron, Instru Control Eng
7. Das CK, Sanaullah M, Sarower HMG, Hassan MM (2009) Development of a cell phone based remote control system: an effective switching system for controlling home and office appliances. Int J Electr Comput Sci IJECS 9(10):37–43
8. Narendra Reddy K, Sukumar P (2017) Controlling home appliances by using universal remote control system (IoT and Bluetooth)

9. Nikolova M, Meijs F, Voorwinden P (2003) Remote mobile control of home appliances. IEEE Trans Consum Electron 49(1):123–127
10. Aheleroff S, Xu, X, Lu, Y, Aristizabal M, Velásquez J P, Joa B, Valencia Y (2020) IoT-enabled smart appliances under industry 4.0: A case study. Adv Eng Inform 43:101043
11. Mbunwe MJ (2017) Design and Construction of a remote control switching device for household appliances application
12. Ashraf I, Umer M, Majeed R, Mehmood A, Aslam W, Yasir MN, Choi GS (2020) Home automation using general purpose household electric appliances with Raspberry Pi and commercial smartphone. Plos One 15(9):e0238480
13. Vijayalakshmi K, Sahoo S, Shah D, Goyal S, Chaurasiya P (2018) Journal of IR remote controlled home automation using arduino. J Netw Commun Emerg Technol 8(4)
14. Hasan M, Anik MH, Chowdhury S, Chowdhury SA, Bilash TI, Islam S (2019) Low-cost appliance switching circuit for discarding technical issues of microcontroller controlled smart home. Int J Sens Sens Netw 7(2):16–22
15. MaitiS NPK (2014) IR remote control signal decoder for home automation. Int J Eng Sci Innov Technol 3(3):262–268
16. Shinde A, Kanade S, Jugale N, Gurav A, Vatti R A, Patwardhan M M (2017) Smart Home automation system using IR, bluetooth, GSM and android. In: 2017 Fourth international conference on image information processing (ICIIP), pp 1–6
17. Bajaj D, Kumar A, Sahu J, Charaya N, Prakash V, Scholar UG (2016) Home automation by android application based remote control. Int J Eng Sci 6053
18. Roy J, Roy JK (2014) Design of smart universal remote using mobile for home automation. IOSR J Comput Eng 16(5):73–80
19. Krishna AR, Devi CS, Sneha PR (2016) Home automation using remote control system. Int J Eng Res Sci (IJOER) 2:84–91
20. Myers BA (2005) Using handhelds for wireless remote control of PCs and appliances. Interact Comput 17(3):251–264
21. Richmond J, Agrawal S, Urpelainen J (2020) Drivers of household appliance usage: evidence from rural India. Energy Sustain Dev 57:69–80
22. Hsu CL, Yang SY, Wu WB (2010) 3C intelligent home appliance control system–example with refrigerator. Expert Syst Appl 37(6):4337–4349
23. Zhou Z, Zhang W, Li S, Yu N (2019) Potential risk of IoT device supporting IR remote control. Comput Netw 148:307–317
24. Madan V, Reddy SRN (2012) GSM-Bluetooth based remote monitoring and control system with automatic light controller. Int J Comput Appl 46(1):20–28
25. Al Mghawish AJ (2013) A practical approach for mobile-based remote control. Eur Sci J 9(18)
26. Dai F, Ouyang Y, Qin Y, Bian C, Wei B, Chang S, Liu B (2018) Development of integral smart home appliances. J Robot Netw Artif Life 4(4):291–294

Formability Assessment and Process Parameter Optimization in Single Point Incremental Forming of AA2024

S. Pratheesh Kumar, S. Elangovan, S. Boopathi, and R. Ramanathan

Abstract Incremental forming is a novel forming process that provides a clear advantage in complex multi-feature final products. This decade's desire for new goods, facelifts, and strategic product implementations tends to increase customisation and flexibility. With incremental depth, tool rotation, and traverse speed, this article's objective issue is formability assessment. A superior profile precision and supportive action are enhanced by numerical optimization technique. The output response was thinning, average-surface roughness, and geometric precision. Based on these findings, the meridional influence of rotational tool speed on marginal forming limit propagates localized streamline fracture initiation was studied in a series of studies. So as to gain relevant ideas to further research aspirant, correlation between all components of analysis established with relation and validated. The study results show that with reduced speed rate and incremental depth most suitable part quality can be obtained. In addition, spindle speed and feed rate show a significant contribution on forming force temperature. Which is crucial in reducing the internal stress concentration is deformed.

Keywords Incremental forming · Mathematical modelling · Deformation analysis

1 Introduction

Metal forming technologies that can match market demands and flexible for new product development have been increasingly important in recent decades. The most common sheet metal forming processes are die moulding, drawing, and roll bending. These processes were created for mass production. The design of punches and dies is critical here. The industrial growth is transforming mass production into quality production based on small batch manufacture. Conventional forming procedures are inadequate for producing bespoke pieces. Designing unique punches and dies for

S. Pratheesh Kumar (✉) · S. Elangovan · S. Boopathi · R. Ramanathan
Department of Production Engineering, PSG College of Technology, Coimbatore, Tamil Nadu, India
e-mail: spratheeshkumarth@gmail.com

E. Natarajan et al. (eds.), *Materials, Design and Manufacturing for Sustainable Environment*, Lecture Notes in Mechanical Engineering,
https://doi.org/10.1007/978-981-19-3053-9_19

prototyping and low volume component production takes effort. With the exception of material and time consumed for die and punch fabrication, incremental sheet metal forming (ISMF) was developed. Incremental forming (IF) is a modern manufacturing technology used to build complicated three-dimensional items. Through a sequence of minor incremental deformations, a sheet is used to make the final component. Formed components are then fed into computer numerical control (CNC) vertical machining centres to be machined using forming tools. As a result of a predefined CNC programme, the sheet metal is clamped tightly in an ISMF fixture. With this method, simple generic tools are used to generate localized deformation in the sheet, allowing it to form a broad range of shapes. Its adaptability improves forecasting and planning. Aviation, automotive, bio-medical, architectural, customisation, and research and development are a few engineering areas that invest in incremental form. Contact sites, hybrid techniques, and physical variables all affect incremental formation. The graphic below depicted incremental forming classifications (Fig. 1).

During its initial prior decades, traditional incremental sheet metal forming (ISMF) comprises a tool and sheet with predefined contact points with specified tool-paths able to form the sheet to desired geometry. Last two decades, ISMF revolutionized with different strategies to overwhelm the complexities in incremental forming such as single point incremental forming (SPIF) with hydroforming, ISMF with positive or negative fixtures and ISMF with desired wooden or metallic support. These

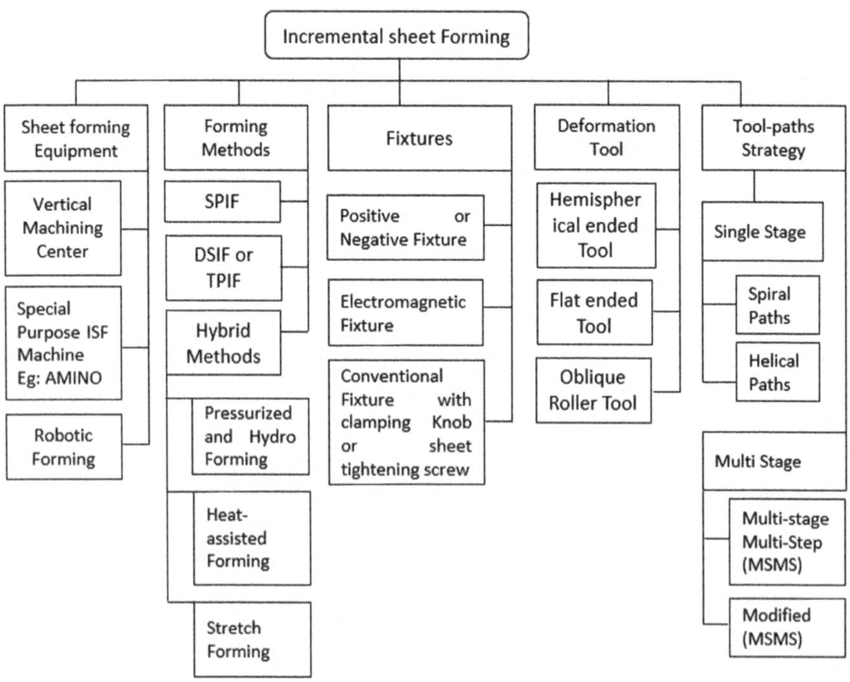

Fig. 1 Classification of incremental forming

strategies have significant advantages over initial ISMF techniques like geometrical accuracy, formability, sophisticated thinning characteristics, flexibility in implementation and reduction in forming force. SPIF is a dieless incremental forming method which is used for manufacturing a wide range of customized components, and it is highly efficient in material formability. Hence, this paper mainly focuses on the formability assessment as the key aspect in forming of AA2024.

Systematic literature review is carried out in order to better understand about ISMF process and modern techniques involved and to identify the potential research areas in the field of SPIF. Nirala and Agrawal [1] illustrates the effect of infeed force and process parameter that has a wide range of complex applicability for the industries. Here, the author mainly contributed over the fixture and forming force, the fracture initiation is compared with simulation and experimental approach. The material chosen here is SS 430 and deforms the sheet metal incrementally by D-shell and frustum of a square pyramid has the forming depth of 7 and 25 mm, respectively. The sheet material is annealed for better formable. A newly developed novel electromagnetic camping has significantly reduced the forming force about 20% for D-shell components and for frustum of a square pyramid is observed forming force reduced to be 16.67% and increased the machining lead time, and the compared fracture initiation result shows the experimental approach fracture initiated at 15 mm but in simulation resulting failure occurs at 22 mm. Ai and Long [2] conducted a review on fracture mechanism in incremental sheet metal forming. This paper is mainly reviewed on various formability enhancement techniques in ISF and proposes the different type fracture behaviour occurring in the experimental investigation. In fracture mechanism, the empirical way of fracture analysis is forming limit curve that is mainly discussed with plane strain condition; the progressive tool paths on ISF mainly form the nonlinear loading strain paths. The sine law is approximately related to the sheet thickness distribution and helps to measure the thickness with various forming angles. The two different fractures like fracture initiated and necking initiated failure are more observed in SPIF and are challenging. The DSIF has very limited in occurring fracture but expected is more in-depth analysis of fracture behaviour occurring in the material. There is a need for a damage model with more focusing on complicated loading conditions to predict the ISF material fracture.

Torsakul and Kuptasthien [3] this paper is mainly focused on the effect of process parameters including material type, forming angle, and revolution speed for SPIF. All these parameters have various aspects resulting over the influence of forming force. As the results show more favour to brass sheet, the high tool revolution speed resulting in low forming force and low wall inclination angle resulting in high forming force with replication of high coefficient of friction. The experimental values revealed that the influence of resultant force involved in different sheet material from the value brass had the maximum force of 165.155 N. The optimized value shows the high wall inclination angle and tool revolution speed indicate the reduction in forming force.

Bansal et al. [4] this paper is based on the prediction of thickness, forming forces, and contact area during the single and multi-stage toolpaths for deformation of sheet metal. Using strategies from literature to alter the analytical model so that it can forecast the parameter more precisely and with less processing time. The experimental investigation and analytical model are compared, and the result shows maximum error for predicting the forming force is decreased from 18.1 to 7.8% at 60° wall inclination angle and for thickness distribution about maximum error is 3.85% without bended region. Diabb et al. [5] this paper illustrated the experimental study of lubrication and wear characteristics in single point incremental forming for AA 6061 in addition with lubrication as sunflower and corn oils with weight per cent of 0.0125, 0.025, 0.05, and 0.1% of sio2 nanoparticles prepared as the nanofluids. The experimental results show the 0.025% weighs concentration of nanofluids has better lubrication and low surface roughness. The Stribeck curves help for friction effect; the curve shows before SPIF has hydrodynamic fluid film lubrication and during the process is mixed fluid film lubrication. The normal forces decrease with the bending moment also decrease only with magnitude.

Durante et al. [6] this paper reported the experimental investigation of tool-polycarbonate sheet interface and toolpaths strategies. The variation in toolpaths increases the formability and minimum twisting as compared to unidirectional toolpaths. The alternate toolpaths with twisting angle of 0.5° have the high geometrical accuracy, and elastic springback is not influenced by toolpaths. The thickness measured from sample and prediction of thickness using cosine law in good agreement; the value of prediction is 0.95 mm and measured sample with mean value of 0.978 mm. Gupta and Jeswiet [7] this article conducted the research on the effect of temperature variation on formability for new geometry using a flat ended tool using an AA5754-H32 sheet. The high relative motion between the tool-sheet interfaces does not affect the surface roughness but the generated temperature distribute the shear stress throughout the sheet wall thickness. Geometrical error occurs at bending near to the backing plate and lobe interaction due to the lack of support and unavailability of die when tool passes at high wall angle. Whereas, higher feed rate and tool rotation result in friction stir heating in sheet metal forming. A continuous zig-zag fracture occurs at 9.6 mm deformation depth.

Kumar and Kumar [8] this paper conducted the hybrid process by combining the ISF and SHF process to achieve the higher wall angle and reduce the failure of the component. The MSMS forming strategy is able to form the component up to 78.75°, and the primary issue of MSMS is the failure of components due to the thinning of sheet. The modified MSMS forming strategy and hydroforming technique enable higher forming angles of even 90°. The analytical model is developed and based on the principle of constant volume; this analytical model is able to predict the thickness up to forming angle at 90°. The developed model has the two cases to predict the vertical thinning and bottom bend thickness. Gupta et al. [9] illustrated the vertical wall component was subjected to multiple passes causing a residual strain distribution and decrease in failure with high material strain. Here, the sheet metal is deformed to C-channel for aerospace fuselage application with deployment of SPIF. The sheet metal was annealed for reducing the geometrical issue and has no constraints on the backing plate. The results show the maximum under forming and over-forming which occur at the corner and base, respectively. The sheet metal underwent multiple passes in plastic deformation and was clearly visible in microstructure as high as variation in orientation, grain size and fragmentation between the locations. This variation in plastic deformation might cause the inconsistency in material properties at different locations can be avoided through the post material processing of annealing.

McAnulty et al. [10] the detailed review on limitation of forming force, process capability and influence of forming temperature, sheet thickness, and tool shape. Here, the elevated temperature achieves through three strategies like laser assisted, induction assisted, and electric hot assisted ISF processes. In laser assisted, the sheet metal was heated on strategic location of sheet and tool interface, the result observed for forming force was decreased with increase in the temperature. In case of electrically assisted, the current flowing through the tool to sheet increases the ductility of material at contact face. When the sheet thickness increases proportionally axial peak forces increase. The roller and oblique roller ball tool were improving the surface quality but maximum wall angle is expected one parameter. The hemispherical and flat end tool was deployment of deform the sheet metal results in lower peak force achieved over the hemispherical end tool. In simulation, the adaptive re-meshing technique improves with near finished component utmost result achieved without compromising the accuracy of parameters and quality of mesh. Najafabady and Ghaei [11] performed the EHIF process with titanium alloy sheet material to deform the different profile strategies for investigation of geometrical accuracy and surface roughness. Moreover, three different geometries are preferred such as CWACF, VWACF and PF. In the EHIF process, it has a lack of heat transfer and high friction affects the surface quality. For that molybdenum di-sulphide powder was sprayed over the sheet metal before the sheet and tool interface. These powders act as a self-lubricant, and result shows nearly 64% reduction in average Ra value. Then, the components are scanned and extracted into cloud points for deviation analysis.

Pacheco and Silveira [12]. justified the relation between EHISF and EHISF with preheating through numerical simulation with Johnson–Cook's material modelling and simplified CAD model for reducing the simulation time. The investigation is based on plastic strain and von-Mises stress to identify the thermal stress and heat distribution on sheet and tool. Further, data analysis for the above two processes compared with original CAD data to obtain the geometric error. EHISF with preheating shows good geometric accuracy with an average value of 4.7%. After performing the process, there is unlikely to observe the residual stress grows due to room temperature cooling and there need to be concentrate. Further, it was claimed that above 45% wall angle and proportionally geometrical error increases.

Barnwal et al. [13] illustrated for AA 6061 on behalf of behaviour of forming characteristics and evaluating the microstructural deformation on SPIF experimentation. Before experimentation all mechanical properties were examined and determined the tensile test data acquisition under ASTM-E8 standard, and r-bar test was conducted and then experimentation performed with lesser spindle rotation to ensure lesser tool wear, constraints, etc. Digital-image-correlation was done after post forming to evaluate surface strains. The three main constituents such as major together with minor strain and thickness strain are illustrated from DIC surface strains techniques. Where major-true strain ranging from 0.30 to 0.35 observed on deformed sheets. Nevertheless, deformed wall region has appropriate constant thickness together absence of strain localization. In FE simulation, coarse–fine mesh with 1 mm utilized for constraints and overcoming the computational difficulties. However, both the test data acquisition shows better correlation in agreement. Due to maximal Taylor factor causes maximum resistance on deformation throughout ID direction.

Chang and Chen [14] developed the predicting model for average-surface roughness and validated with various sheets for effective output responses while investigating the surface morphologies. The developed analytical model has derived based upon localized-plastic deformation and elastically deflected and tendency to describes effect on tool diameter, vertical angle, sheet metal thickness, incremental depth, etc. Two consecutive paths, i.e. scallop height constitutes to uneven sheet thinning distribution over meridional direction. The ratio of Ra and Rz ends-up with 0.268 ratio through experiments series and reveals the critical effectual parameters are spindle tool speed, friction mode, and transverse speed. Unappropriated spindle speed together with feed rate and unsuitable lubricating condition resulted in plough grooves. ISMF is a free-form forming technique to manufacture the complex contour end products. The conventional press-forming technique demands for capital and dedicated die and control systems, while ISMF is exceptions of dedication dies. Numerous researchers pointed out formability and overwhelming technique, but still, it researches progress despite of highest industrial requirements and norms. Only a few studies have been done on aircraft components with sophisticated profile geometry. The mathematical model has been constructed, but acquiring the coefficient and parameters is difficult. This lacks the availability of primal process variable for demanding sheet metal applications.

Shim et al. [15] in their research work used a forming tool with a freely rotating ball to test the formability of Al 1050 sheet. The tool was tested with various strains paths. The forming limit diagram of the sheet showed that its formability is dependent on the strain path. It was suggested that the forming limit curve with an incremental increase in tool depth should be used.

Filice et al. [16] have researched on the material formability in incremental forming. Several experiments were devised with the goal of achieving various straining situations in the material and, as a result, determining forming limit diagrams for progressive forming procedures. The study discusses the characteristics and applications of a FLD.

Kim and Park [17] have discussed how the sheet metal appears to be more formable in incremental forming than in conventional forming. Experiments and FEM studies were used to explore the effect of process parameters on formability, including tool type, tool size, feed rate, friction at the tool-sheet interface, and sheet plane anisotropy. It was discovered that using a ball tool of a specific size with a slow feed rate, and little friction improves formability. The formability varies depending on the tool movement direction due to planar anisotropy.

Bhattacharya et al. [18] have worked in the formability investigation, the Box–Behnken technique is utilized, and for the surface finish study, the complete factorial approach is used. The formability of incremental forming reduces as the tool diameter increases, according to the findings of the experiments. With incremental depth, the formable angle increases and subsequently drops, and it is also noted that the variation in the formable angle is not substantial in the range of incremental depths regarded to generate satisfactory surface finishes during the current investigation. The stress levels during incremental sheet metal forming are estimated using a basic analytical model assuming that the deformation occurs largely under plane strain conditions. In ISMF, a stress-based criteria is employed in conjunction with the previously described analysis to predict formability.

The sheet metal has capabilities to deform the complex contour geometry without intending to geometrical inaccuracies known as formability. Formability is associated with secondary attributes like vertical angle, twisting angle, positive and negative deviation due to bending, tool deflection, and higher axial force, etc. The traditional ISF process does not meet the feasible solution for minimizing the error in desired sheet components. Then, achieving the precision contour of hard to deform materials such as titanium alloy, Inconel and stainless steel are more complicated and result in adverse effects. This effect can resolve with optimized physical variables and compensation methods. In the SPIF process, the sheet undergoes localized straining and incremental contact area causing the geometrical inaccuracies. By optimizing the incremental pitch and toolpaths result in higher plane strain mode and maximum vertical angle. The current study aimed to improve the formability and geometric accuracy of AA2024 sheet material.

Even though there are many research findings for different materials formability assessments and process parameter optimization under SPIF technique but AA2024 aluminium alloy has not been concentrated or focused in any research work. So, we have taken AA2024 alloy as the major point of focus in this paper. AA2024 alloy has very good mechanical properties and workability with light weight. Hence it is a important alloy to be studied in order to enhance it's SPIF process capabilities.

We have selected SPIF as the experimental setup for this research work as it is a die less sheet metal process the elimination of the die in the manufacturing process reduces the per piece cost and increases turnaround time for low production runs, and this research work has more significance as this material is highly used in two main industries which is aviation and automobile, it has to been given more importance and the process should evolve more. In this paper, we can get the extent to which the alloy can be formed and how its process parameters can influence its efficient output. Aviation and automotive are the main industries which can benefited by this work and bio-medical, architectural, and customisation are the other areas in which the research findings can be used.

2 Experimental Methods and Materials

The flowchart below represents the process of the incremental forming and overview of the current research work. This includes findings of various output responses and inference of output responses. Finally, comparative validation was determined

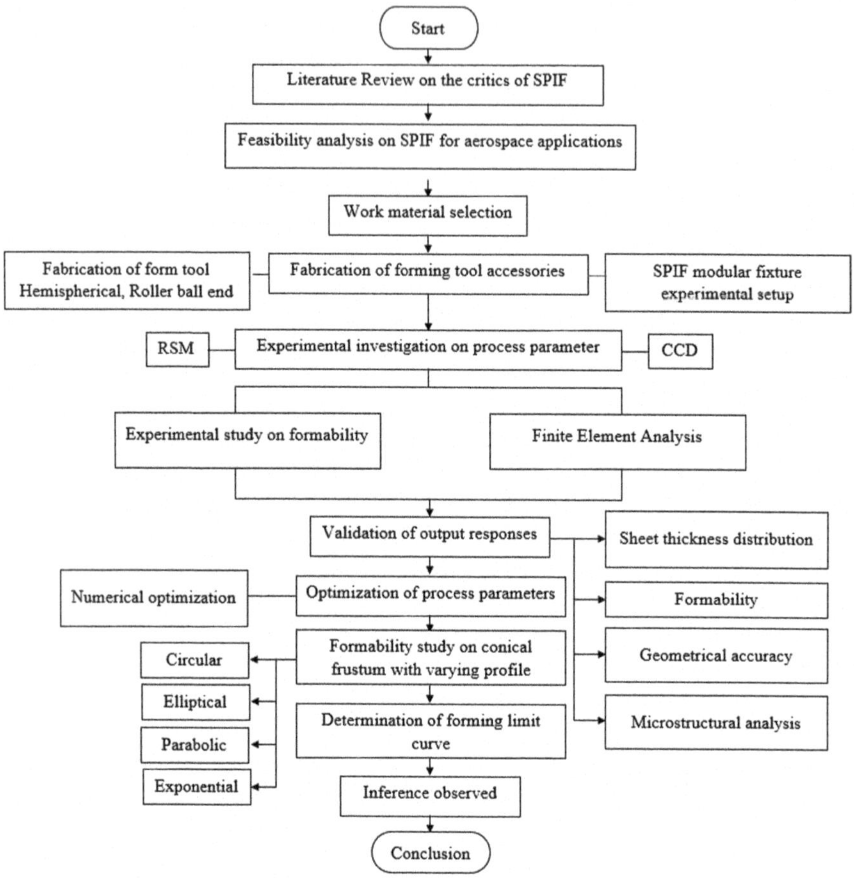

Fig. 2 Methodology

between theoretical and experimental values to establish the appropriate suggestions (Fig. 2).

2.1 CNC Vertical Machining Centre

The ISF experimentation was carried on a vertical machining centre, presented in Fig. 3. The CNC machine has a spindle speed of 8000 rpm and power output of 7.5 KW. The experimental work area or guideway travel is constrained over 550 mm × 400 mm × 350 mm in *x*, *y*, z directions. The helical toolpaths were constructed using the MASTERCAM package and interface with the CNC machine.

Fig. 3 SPIF-experimental setup

2.2 *Fabrication of Tool Shape and Type*

The hemispherical-ended forming tool of 10 mm diameter was fabricated. In order to maintain optimum responses, forming tool has tendency to withstand the forming force and load together with adhesive wear and well-polished chrome alloy steel 52100 (65 HRC) is used as form tool material and servo 68 lubricant oil (kinematic viscosity at 40 °C, 68 mm^2/s) is employed to reduce friction between tool-sheet interface.

2.3 *Applications of Sheet Metal Components*

2.3.1 Aluminium Alloy 2024—Aircraft Structural Applications

Pre-processing of sheet metal: Sheet metal has high hardness and relatively low ductility and toughness, which tends to consume more energy, higher forming force, tool adhesion, and wear. Therefore, the heat treatment has required more elongation, ductility, and also toughness to enhance the required formability. However, with reference to literature to process, the sheet metal with annealing at 450 °C and soaked for 120 min and furnace cooling to obtain the respective factors.

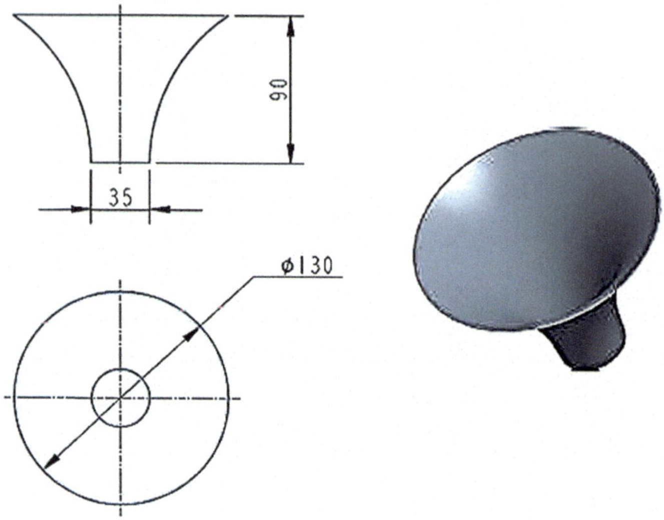

Fig. 4 Component design

2.4 Modelling of Desired Design

This experiment uses a 130 mm base diameter and 90 mm depth with variable wall angle frustum cone as a geometric profile (Fig. 4).

2.5 Process Parameter Selection

After conducting a series of trails, the most influencing parameters were found from varying all the parameters affecting the SPIF, feed rate, spindle speed, and incremental depth (Table 1).

Table 1 Major influencing parameters and their corresponding values

S. No.	Parameter (units)	Low	Medium	High
1	Feed rate (mm/min)	1000	2000	3000
2	Spindle speed (rpm)	100	300	500
3	Incremental depth (mm)	0.3	0.6	0.9

2.6 *Toolpaths Generation*

Tool path influences component dimensional accuracy, processing time, and surface quality. The CNC programme determines and controls tool path trajectory. The CAD model is transferred into the CAM software to create NC data. The final stage is to pick the machine type, tool diameter, tool path generation, surface parameter, and rough contour parameter. An inter-spiral tool route has been built to lower the circle diameter for the developed model (Fig. 5).

3 Results and Discussions

The experiments were performed as per the run order of the design matrix and the corresponding output responses such as average-surface roughness (R_a), thinning (t_f), maximum height formed without failure (h_{max}), and forming force (F_a) were measured (Table 2). The research objectivity is to explore the best combination of forming parameters to form the sheet metal without any defects and with excellent surface finish.

The quadratic model explains the empirical relationship well since it incorporates all major components determining the output responses. So RSM is necessary on this pivot job. The design matrix considered two input elements, tool transverse and rotational speed, on three levels, and one category factor, tool geometry, and on five levels.

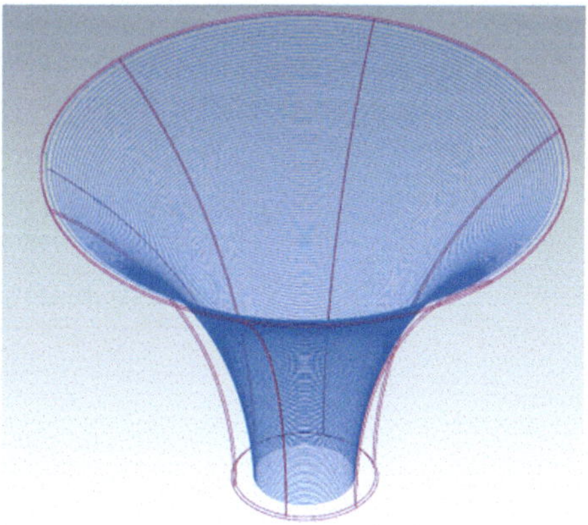

Fig. 5 Toolpaths generated

Table 2 Central composite design layout and output responses

Run order	Feed rate (mm/min)	Spindle speed (rpm)	Incremental depth (mm)	Maximum height (mm)	Final thickness (mm)	Surface roughness (μm)	Forming force (N)	Temperature (°C)
1	3000	300	0.6	13.939	0.503	1.4	3119.58	60.1
2	1000	100	0.3	12.486	0.507	2.64	2903.76	52.7
3	1000	300	0.6	15.14	0.5034	2.11	2913.57	54
4	2000	300	0.6	11.5	0.515	2.43	2766.42	55.8
5	2000	300	0.6	11.5	0.515	2.43	2766.42	55.8
6	3000	100	0.9	10.246	0.5155	1.91	3570.84	58.7
7	1000	500	0.9	19.848	0.4776	1.49	3050.91	60.4
8	2000	300	0.6	11.5	0.515	2.43	2766.42	55.8
9	2000	100	0.6	11.55	0.5098	1.36	3404.07	60.2
10	2000	500	0.6	10.358	0.5042	1.77	2217.06	56.4
11	2000	300	0.3	10.389	0.513	2.52	2236.68	53.2
12	2000	300	0.9	13.596	0.4994	1.91	3384.45	60.8
13	2000	300	0.6	11.5	0.515	2.43	2766.42	55.8
14	3000	500	0.3	14.5	0.5015	1.33	2501.55	57.5
15	2000	300	0.6	11.5	0.515	2.43	2766.42	55.8

3.1 Mathematical Modelling

3.1.1 Instantaneous Thickness

The given equation is derived from Chang and Chen [14] et al., the equation which validate the end thickness of wall region under the plane strain compression and bi-axial stretching condition.

$$T_{\mathrm{f}} = ti(\cos\alpha + (1 - \cos\alpha))\frac{2\gamma}{\alpha + \beta} \tag{1}$$

$$\gamma = \frac{\Delta Z - \text{Scallop height}}{\Delta Z}.\sin\frac{\Delta Z}{2tr\sin\alpha} \tag{2}$$

$$\text{Scallop height} = \frac{2 * \Delta Z}{4d\sin\alpha} \tag{3}$$

$$\beta = \cos\left(\frac{tr - \text{scallop height}}{tr}\right) \tag{4}$$

where t_{i}—initial thickness, α—wall angle, t_{r}—tool radius, d—tool diameter, and ΔZ—incremental depth.

3.2 Height of Wall

$$H = R(\cos\theta_1 - \cos\theta_2) \tag{5}$$

H = Total height of varying wall angle.
R = Radius of wall quarter curvature.
θ_1, θ_2—Two different wall angle.

3.3 Forming Depth up to Fracture Limit

The given equation is derived from Nirala [1] et al., this equal established under the relation between cosine law and modified Nirala for asymmetric component thickness prediction equation.

$$(H - H_f) = (ud - ti.\cos\alpha.tr) \tag{6}$$

3.4 Flow Stress

The given equation is derived from power hardening law.

$$\sigma = \frac{K(\varepsilon p^{n+1})}{n+1}(ud - ti.\cos\alpha.tr) \tag{7}$$

Effective plastic strain

$$\varepsilon p = \frac{2}{\sqrt{3}}\ln\left(\frac{ti(\cos\alpha + (1-\cos\alpha))\frac{\gamma}{\alpha+\beta}}{ti\cos\alpha}\right) \tag{8}$$

where K-strength coefficient.
n-strain hardening exponent.

3.5 Axial Forming Force

$$Fa = -\frac{2}{\sqrt{3}}\frac{p.ud}{4d\sin\alpha}.\left(\frac{ti}{tr+ti}\right).\cos\left(\frac{\alpha-\gamma}{2}\right) \tag{9}$$

3.6 Influence of Ease-to-Form Material with Concerning Output Responses

The interaction between feed rate and spindle speed has been plotted on 2D surfaces in Fig. 6 and illustrates the role of output responses under insignificance or stabilized manner. Forming height is majorly influential on feed rate, thereby describing that excessive strain at permissible limit throughout the meridional together with forming direction. Final thickness has higher contribution towards rise in feed rate and minimal tool rotation. Each iterative successive path has maximal contact region; this intends to localize plastic strain and elastically deflect at each deformation cycle resulting in a superlative threshold. The average-surface roughness perceived low degree of peak and valley with manifested wall surface and substantiates minimal distance of scallop height with enough material deformation, where both spindle speed along with feed rate should be higher as far. Forming force has substantial effect on the higher feed rate and little influence towards lower spindle speed and these effects clearly governed by flow stress and stretching mode effect. Forming temperature has increased by both factors.

3.7 Numerical Optimization of Process Parameters

Optimization is a minimal or maximal set of data-driven variables with limiting the factors based upon DOE. Numerical or analytical optimization is a multiple segment of variables optimized with optimization technique through regression equations, which can express the predicted appropriate combination of factors that intend to propagate the final output responses. Wall inclination angle has been described as vertical angle far from nominal geometrical profile, thus resulting in geometrical inaccuracies. This effect could dominate based upon region where might deviate on positive or negative springback because of bending, meridional stretching and twisting angle and compensated through closed loop or iterative compensating with inline optimal process variables. Average-surface roughness is the effect of the deepest of two consecutive traced step paths or scallop height, tool-sheet contact region, variation in rotational spindle speed together feed rate impact texture of vertical surface, normally Ra value should be minimal as to best fitted factor. For that minimum value is considered an optimized variable. Final thickness and maximum height intend to be maximum and avoid forming fracture. Forming force has inherently constituted many factors such as contact pressure distribution, energy consumption, rigid-body displacement, and forming strategy and therefore, much literature concentrate on minimizing the force component; probably minimal force is an optimal parameter. Temperature has to be marginally minimal, and therefore, geometrical accuracy with maximum sheet thinning could be constraints on numerical optimization. From the six output responses, better fitted combination of output factors extracted data in Fig. 7 and nominal factors achieved through 2.03719 μm, 0.498224 mm,

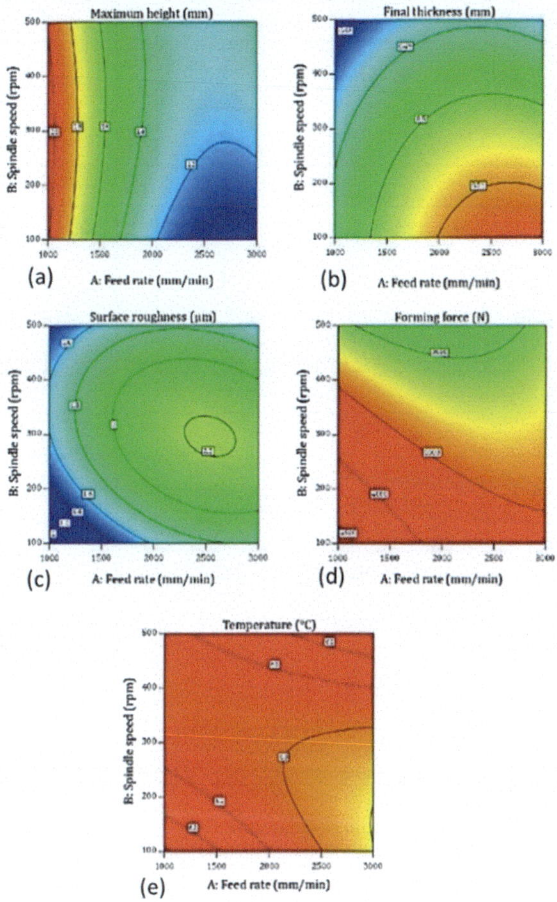

Fig. 6 Influence on output responses (a) maximum height, (b) final thickness, (c) surface roughness, (d) forming force, and (e) temperature

14.4102 mm, 2252.4 N, and 53.1528 °C along with feed rate of 1000 mm/min, spindle speed 500 rpm and step depth of 0.662768 mm. Three data sets of experiments with optimal process parameters progressed, and results showed better agreement of 98% as regards predicting the output responses.

4 Conclusion

This article expressed for aluminium alloy 2024, this selected sheet has wide range of automotive and aerospace applications concerning cold or hot forming as major advantages, and ISMF has relatively dependent upon material stretching these aspects examined with both the experimental and analytical way-off analysis. The central composite design is considered for DOE with 15 experimental runs with three sets

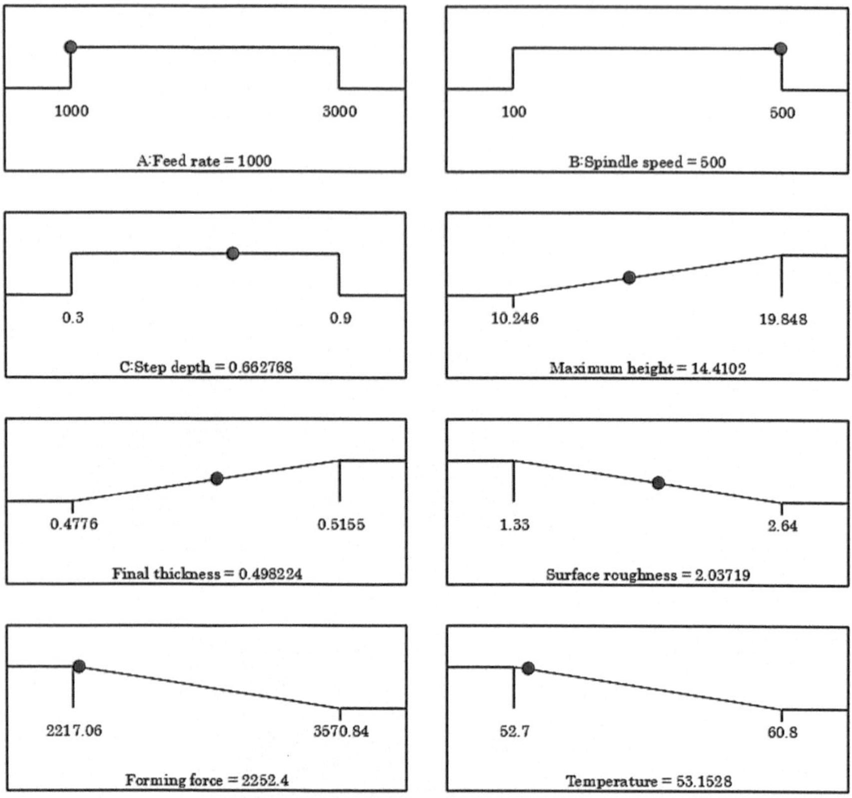

Fig. 7 Numerical optimisation results

of process parameters to examine the tool geometry contributing towards various output responses and optimistic process parameters observation. The key outcomes of this study is listed below,

- The numerically optimized process parameter has validated and brings out important consideration of better fitted combination of output factors achieved through 2.03719 μm, 0.498224 mm, 14.4102 mm, 2252.4 N, and 53.1528 °C along with feed rate of 1000 mm/min, spindle speed 500 rpm and step depth of 0.662768 mm.
- The mathematical modelling is evaluated with input parameters to meet the appropriate output responses, both the data comparatively examined observed better agreement in predicting the output responses.

References

1. Nirala HK, Agrawal A (2019) A novel electromagnetic fixture for incremental sheet metal forming. J Manuf Sci Eng 141(3)
2. Ai S, Long H (2019) A review on material fracture mechanism in incremental sheet forming. Int J Adv Manuf Technol 104(1):33–61
3. Torsakul S, Kuptasthien N (2019) Effects of three parameters on forming force of the single point incremental forming process. J Mech Sci Technol 33(6):2817–2823
4. Bansal A, Lingam R, Yadav SK, Reddy NV (2017) Prediction of forming forces in single point incremental forming. J Manuf Process 28:486–493
5. Diabb J, Rodríguez CA, Mamidi N, Sandoval JA, Taha-Tijerina J, Martínez-Romero O, Elías-Zúñiga A (2017) Study of lubrication and wear in single point incremental sheet forming (SPIF) process using vegetable oil nanolubricants. Wear 376:777–785
6. Chang Z, Chen J (2020) Mechanism of the twisting in incremental sheet forming process. J Mater Process Technol 276:116396
7. Gupta P, Jeswiet J (2018) Effect of temperatures during forming in single point incremental forming. Int J Adv Manuf Technol 95(9):3693–3706
8. Kumar Y, Kumar S (2019) Experimental and analytical evaluation of incremental sheet hydro-forming strategies to produce high forming angle sheets. Heliyon 5(6):e01801
9. Gupta P, Szekeres A, Jeswiet J (2019) Design and development of an aerospace component with single-point incremental forming. Int J Adv Manuf Technol 103(9):3683–3702
10. McAnulty T, Jeswiet J, Doolan M (2017) Formability in single point incremental forming: a comparative analysis of the state of the art. CIRP J Manuf Sci Technol 16:43–54
11. Najafabady SA, Ghaei A (2016) An experimental study on dimensional accuracy, surface quality, and hardness of Ti-6Al-4 V titanium alloy sheet in hot incremental forming. Int J Adv Manuf Technol 87(9):3579–3588
12. Pacheco PAP, Silveira ME (2018) Numerical simulation of electric hot incremental sheet forming of 1050 aluminum with and without preheating. Int J Adv Manuf Technol 94(9):3097–3108
13. Barnwal VK, Chakrabarty S, Tewari A, Narasimhan K, Mishra SK (2018) Forming behavior and microstructural evolution during single point incremental forming process of AA-6061 aluminum alloy sheet. Int J Adv Manuf Technol 95(1):921–935
14. Chang Z, Chen J (2019) Analytical model and experimental validation of surface roughness for incremental sheet metal forming parts. Int J Mach Tools Manuf 146:103453
15. Shim MS, Park JJ (2001) The formability of aluminum sheet in incremental forming. J Mater Process Technol 113(1–3):654–658
16. Filice L, Fratini L, Micari F (2002) Analysis of material formability in incremental forming. CIRP Ann 51(1):199–202
17. Kim YH, Park JJ (2002) Effect of process parameters on formability in incremental forming of sheet metal. J Mater Process Technol 130:42–46
18. Bhattacharya A, Maneesh K, Venkata Reddy N, Cao J (2011). Formability and surface finish studies in single point incremental forming. J Manuf Sci Eng 133(6)

Bio-inspired Design and Optimization of Motor Body of an Electronic Lock Using Taguchi DOE-FEA-GA Integration Technique

S. Ashruth Gukan, K. Sivaranjani, Avinash V. Nair, K. Nandha, and K. Anand

Abstract Injection moulding is the most used manufacturing process for fabrication of plastic parts. A wide range of products are manufactured using injection moulding. The motor body of an electronic lock system is the identified component for this study. In a motor body cover, when cyclic loading is applied, it was found to fail at specific part on the cover. Failure of the components affects the functionality of the product. In this study, the bio-inspired design approach is adopted for improving the mechanical properties of the motor body cover. Motor body is designed based on the principles of bio-mimicry and finite element analysis (FEA). Then design was optimized using genetic algorithm (GA). The optimized design of motor body can be fabricated to validate the design experimentally. Fabrication of the component can be done using injection moulding process and tested using ASTM D7791 testing standards.

Keywords Bio-inspired design · Taguchi DOE · Finite element analysis · Genetic algorithm

1 Introduction

Injection moulding is most frequently used manufacturing process for fabrication of plastic components. A wide variety of products are manufactured using injection moulding, which varies in size, complexity, and application.

Bio-mimicry is an approach of applying design from nature to make existing design more sustainable. With bio-mimicry, new products can be developed or improve existing design concepts. It helps to change the thinkers' approach towards design problems and uncover new solutions. Imitating bio-inspired structures from animals and plants is a very promising for sector replacing conventional structures in future. In recent years, the use of the bio-mimicry concept in energy absorption applications has been inclining.

S. Ashruth Gukan (✉) · K. Sivaranjani · A. V. Nair · K. Nandha · K. Anand
Department of Production Engineering, PSG College of Technology, Coimbatore 641004, India
e-mail: ashruth.gukan3101@gmail.com

E. Natarajan et al. (eds.), *Materials, Design and Manufacturing for Sustainable Environment*, Lecture Notes in Mechanical Engineering,
https://doi.org/10.1007/978-981-19-3053-9_20

A genetic algorithm (GA) is a computerized search and optimization algorithm based on the concept of natural genetics and natural selection. GA is different from traditional optimization methods used. Because of its simplicity, fewer prerequisites, and global perspective, GA has been widely suggested for a variety of problems.

The literature survey was conducted for better understanding of theoretical background of the bio-inspired design. The literature reviewed were from research papers published between 1983–2020 which includes articles on theoretical, experimental investigations, design characteristics of finite element analysis, Taguchi design of experiments. The major literatures that were reviewed are described below.

Burns et al. [1] have conducted a study which primarily focuses to utilize the design strategies of structural design concepts found in nature to create new concepts for hybrid composite joints. Sysaykeo et al. [2] have discussed the methodology to design a bio-inspired bearing to replace bush bearing in industrial machines. Zhang et al. [3] discusses about a bio-inspired generative design and optimization method for creating a tree-type support structure using a plastic additive manufacturing process. It explains the entire design and optimization process through six steps starting from identifying support areas to post-processing the optimum solution. Akrami et al. [4] have tried to introduce a novel bio-inspired design planning concept based on the optimized topology of joint of a bird's bone to improve the strength-to-weight ratio and damage tolerance of composite joints.

Burns et al. [5] have designed carbon/epoxy T-joint with improved structural properties for lightweight structures. The optimization procedure used in this research paper minimizes the interlaminar stress concentration in the T-joint radius bend and improves strength. Alfano et al. [6] focus on the fracture analysis of adhesive-bonded double cantilever beam (DCB) specimens with 3D printed nylon substrates. The result from the DCB fracture test shows that in comparison with bulk samples, bio-inspired subsurface channels cause a considerable increase in total dissipated energy. Takayama et al. [7] have discussed about the dependency on fossil fuels and minimisation of CO_2, technological development of materials produced from plants (e.g., polylactic acid (PLA), and the like). Sigmund et al. [8] have written a detailed study on multi-scale structures, such as those found in nature (such as bone and bamboo), have the potential to provide greater performance while being lightweight, resilient, and multi-functional.

Kikuchi et al. [9] spoke about social behaviours of living organisms found in nature, like food searching, environment fitness, and survival are inspirations for meta-heuristic mathematical models. Audibert et al. [10] have discussed how nowadays, additive manufacturing processes greatly simplify the production of openwork workpieces providing new opportunities for workpieces design. Altan et al. [11] have spoken about all mass-produced discrete parts which are made utilizing production procedures that involve dies, the design and manufacturing of dies, and moulds represent a significant link in the entire production chain. Design and manufacturing of dies and moulds represent a significant link in the entire production chain because nearly all mass-produced discrete parts are formed using production processes that employ dies and moulds. Ryan et al. [12] have discussed how bio-inspired design has been applied in sustainable design (e.g. lightweight structures) to learn from nature and support material structure functionalities. Panneerselvam et al. [13] described a new method of selection of process parameters for obtaining optimal weld tensile strength. A genetic algorithm was used to optimize the parameters for the process.

From the above-mentioned literature review, it was understood that mechanical structures with T and L-type joints made of composite materials experienced fatigue failure in the respective joint regions. Also, it was observed that the motor body was not designed based on bio-mimicry and non-traditional optimization techniques. So, design and optimization on motor body were attempted using bio-mimicry, Taguchi DOE, FEA, and GA.

2 Materials and Methods

The methodology, which is being followed for the completion of the project is shown below

- Identification of component for which mechanical properties has to be improved.
- Based on constraints of an existing design develop alternate designs based on the principle of bio-mimicry.
- Conduct FEA to predict the mechanical properties and develop design parameters for the selected structures.
- Develop a Taguchi L27 DOE for the selected parameters within the constraints.
- Based on results from Taguchi DOE, develop a CAD model and analyse it using FEA.
- Conduct variance analysis to check the significance of design parameters.
- Optimization the design using genetic algorithm.

2.1 Part Dimensions and Assembly

The CAD model of the motor body is shown in Fig. 1.

2.2 Material Properties

The material properties of ABS plastic are mentioned in Table 1.

2.3 Identification of Failure

Figure 2 shows fatigue failure of the component which has resulted in the breakage at a part in the component. The component is subjected to cyclic loading of both bending

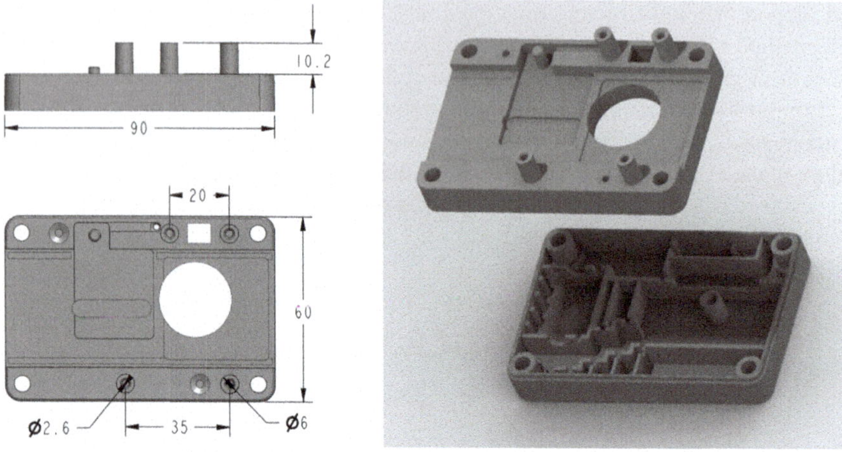

Fig. 1 CAD model of the motor body

Table 1 Material properties

Material	ABS plastic
Density	1030 kg m^{-3}
Young's modulus	1.628 gpa
Bulk modulus	2.9784 gpa
Poisson's ratio	0.4089
Shear modulus	5.7776×10^8 Pa
Yield strength	2.741×10^7 Pa
Ultimate tensile strength	3.612×10^7 Pa

and torsional. The reason for the failure is due to poor design selection which was not capable of withstanding the required loading conditions in the industry.

Fig. 2 Failure identification

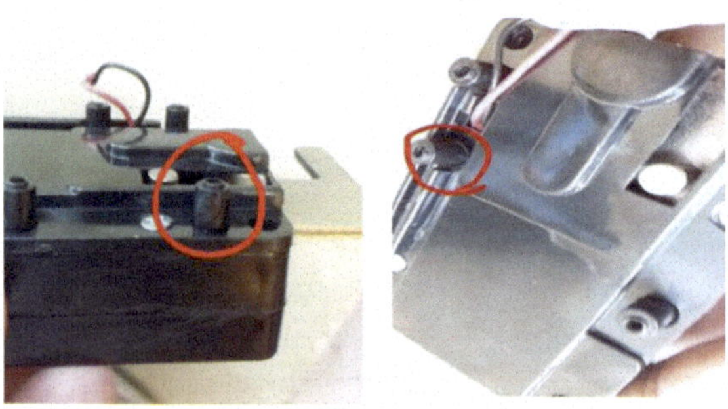

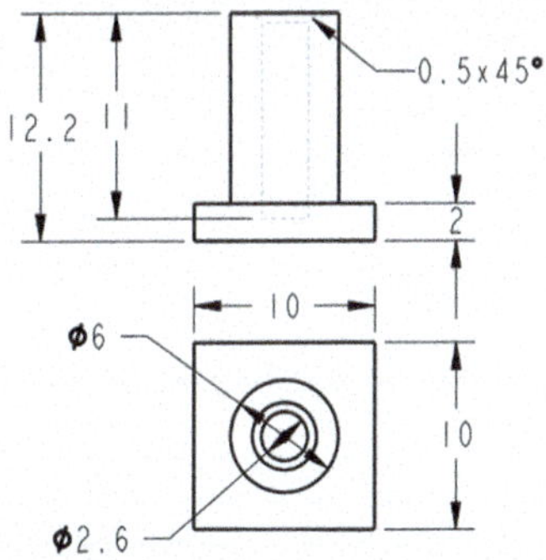

Fig. 3 ABS thermoplastic test specimen

2.4 Test Specimen

Test specimen shown in Fig. 3 was designed to analyse the failure.

2.5 Loads Acting

The test specimen considered is subjected to cyclic loading with the combination of bending load of 50 N and torsional load of 50 Nmm about its central axis, as shown in Fig. 4.

Load Acting
Cyclic loading

1. Bending load—50 N
2. Torsional load—50 Nmm.

Fig. 4 Load acting

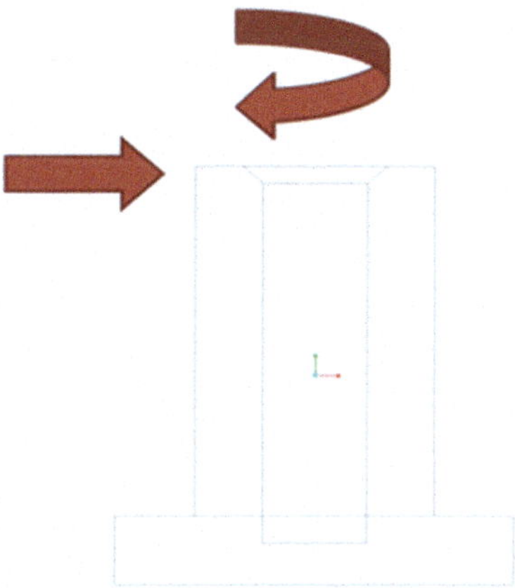

2.6 *Stress Analysis*

Stress analysis is a very important process for the particular component. It can help to determine the stress induced in the component. Figures 5 and 6 show results obtained from stress induced and factor of safety obtained from and finite element analysis. The results are

Equivalent alternating stress = 24.47 M Pa, factor of safety = 0.85, and weight = 0.44 g.

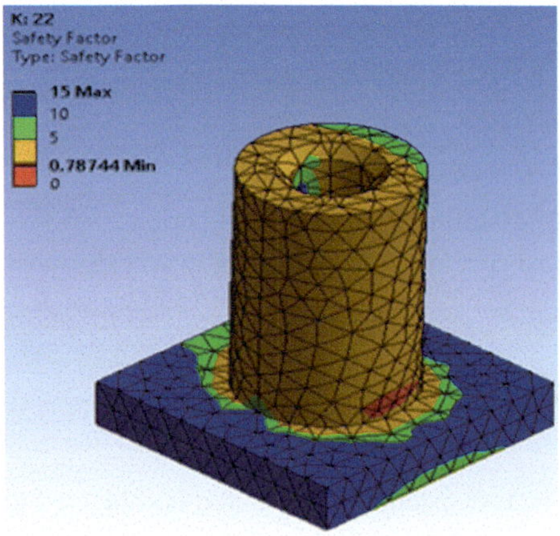

Fig. 5 Stress analysis

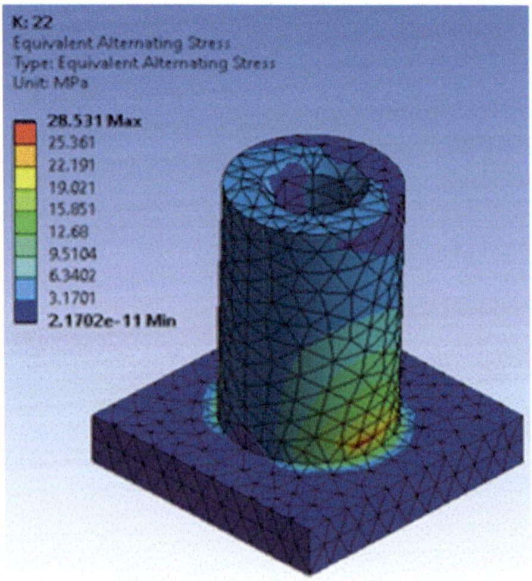

Fig. 6 Stress analysis

2.7 *Bio-inspired Designs*

2.7.1 Tree Root Spread

This design has been inspired from the tree root spread. Root structure is generally described by four regions or zones. These regions, starting from the end of the root, are the root cap, region of cell division, region of elongation, and region of maturation. Figures 7 and 8 show the CAD model and actual root spread of tree.

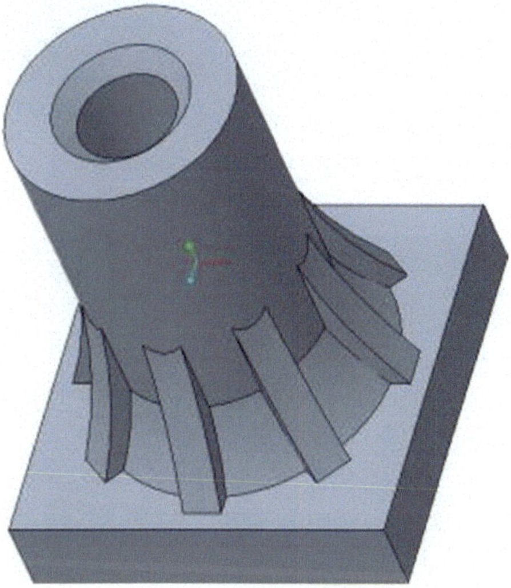

Fig. 7 CAD model of tree root spread

Fig. 8 Tree root spread

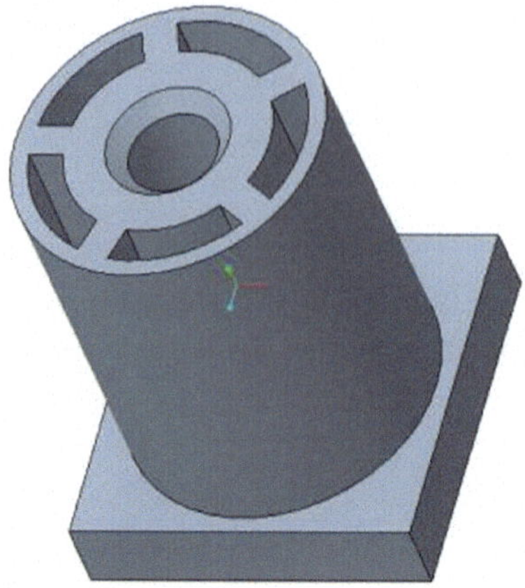

Fig. 9 CAD model bamboo cross-section

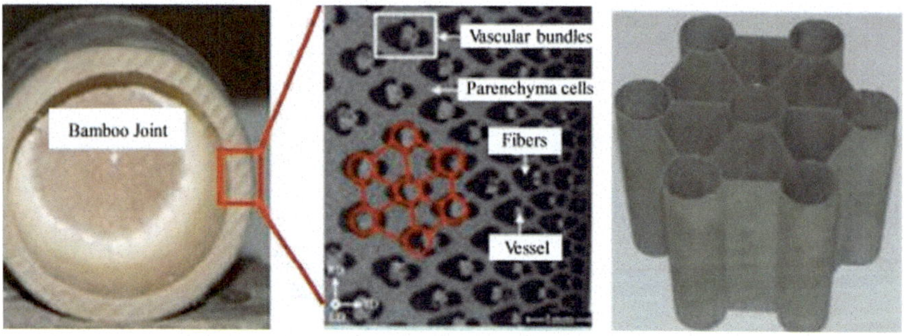

Fig. 10 Cross-section of bamboo

2.7.2 Cross-Section of a Bamboo

Bamboo is a lightweight biological composite material with higher stiffness–mass ratio than some metallic materials such as steel and aluminium. Bamboo can withstand extreme loads such as bending loading from the wind and axial loading from its weight. Figures 9 and 10 show the CAD model and actual cross-section of bamboo.

2.7.3 Leg of an Ostrich

The bones of an ostrich are potential nominees for designing energy absorbers. Simulation results showed that ostrich leg had superior energy absorption ability and a stable collision process to those of the original. The CAD model and actual leg of Ostrich is shown in Figs. 11 and 12.

Fig. 11 CAD model ostrich leg

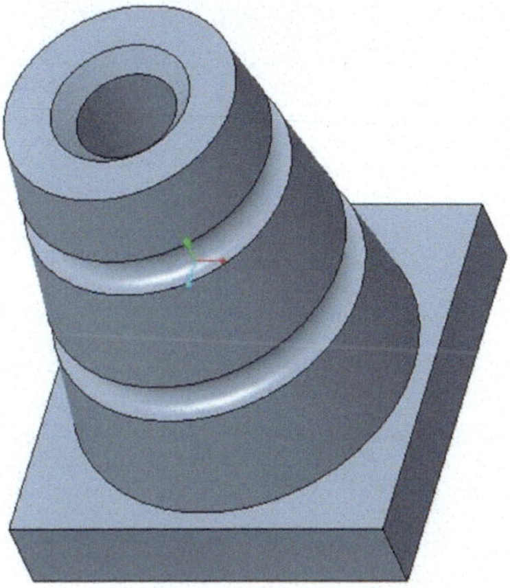

Fig. 12 Leg of an ostrich

2.7.4 Lotus Stem

The unit cell of the lotus stem core is composed of two hexagonal frustums. The energy absorption capacity of unique lotus root-filled tube is greater than that of the conventional tube under the condition of the same mass. Based on validation and ease of manufacturing, Lotus stem structure was finalized. Figures 13 and 14 show the CAD model and actual root of lotus stem.

2.7.5 Comparison of Strength

The alternating designs are subjected to the same static structural loading conditions, and the factor of safety and weight values are compared. From the comparison, the factor of safety of the alternate design 4 was found to be the highest. The strength comparison of four selected alternate designs are shown in Figs. 15, 16, 17, 18, and Table 2.

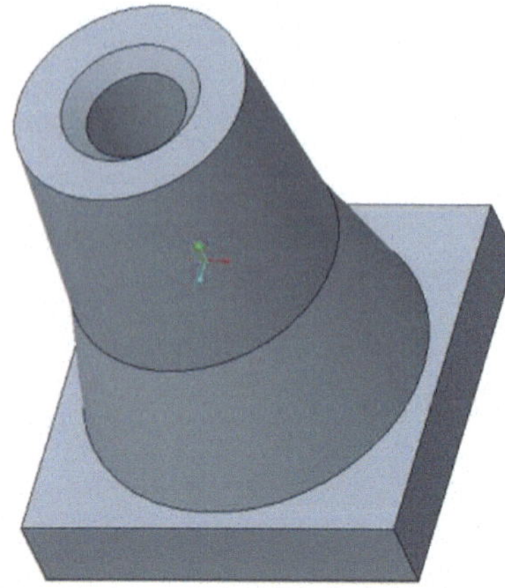

Fig. 13 CAD model of lotus stem

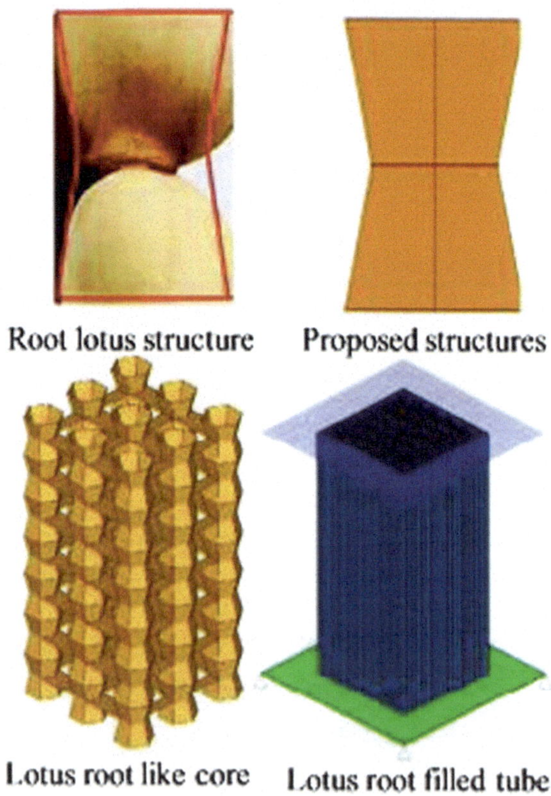

Fig. 14 Lotus stem

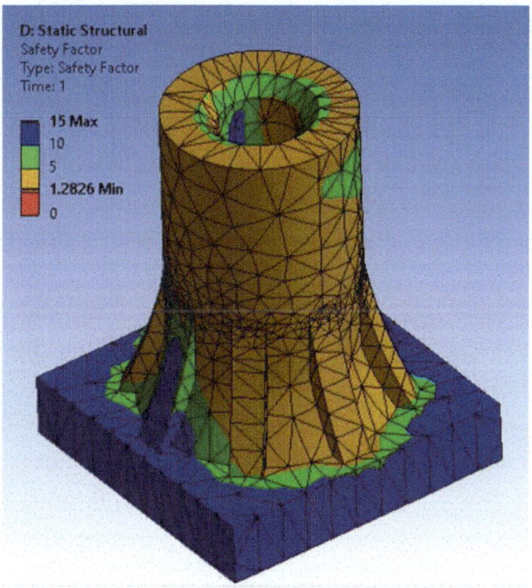

Fig. 15 Tree root spread

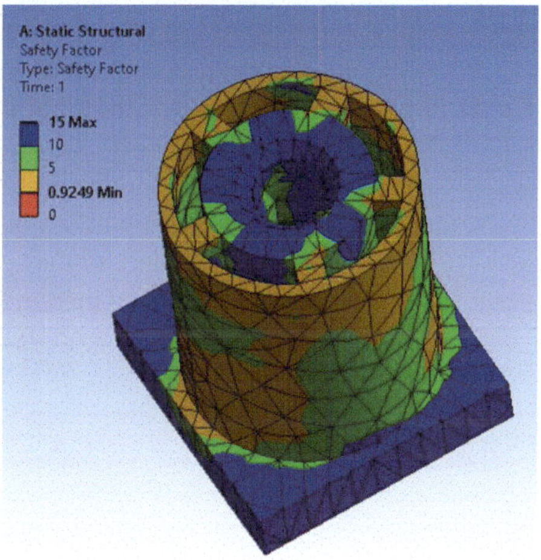

Fig. 16 Cross-section of bamboo

3 Finite Element Analysis of Designs

3.1 Development of Regression Model for Factor of Safety

The design factors and their levels for design optimization are given in Table 3. In this work, motor body is designed as per L27 orthogonal array which is given in Table 4.

Fig. 17 Ostrich's leg

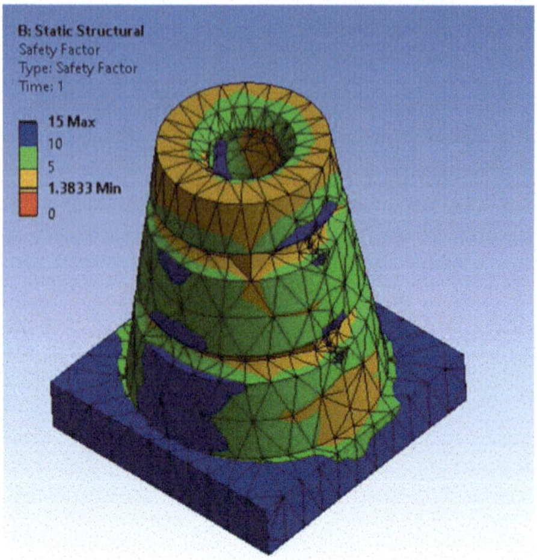

Fig. 18 Lotus stem

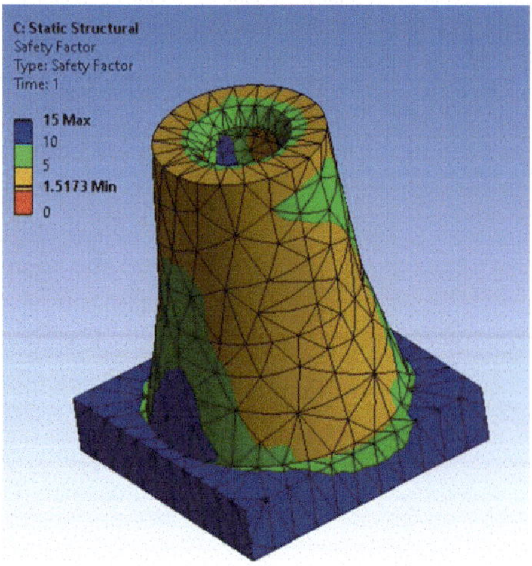

Table 2 Design outputs

S. No.	Design name	Factor of safety	Weight (in g)
1	Tree root spread	1.28	0.48
2	Cross-section of a Bamboo	0.92	0.64
3	Leg of an ostrich	1.98	0.59
4	Lotus	1.51	0.57

Table 3 Design parameters

Notation	Design parameter	Level 1	Level 2	Level 3
P1	Top diameter (mm)	4	5	6
P2	Middle diameter (mm)	5	6	7
P3	Bottom diameter (mm)	6	7	8
P4	Step height (mm)	4	6	8

Table 4 L27 Orthogonal array

Exp. No.	P1	P2	P3	P4	Factor of safety	Equivalent alternating stress (MPa)	Weight (g)
1	4	5	6	4	0.40	2158	0.34
2	4	5	7	6	0.41	742	0.39
3	4	5	8	8	0.47	155	0.46
4	4	6	6	6	0.44	229	0.41
5	4	6	7	8	0.52	85.7	0.46
6	4	6	8	4	0.41	861.5	0.43
7	4	7	6	8	0.51	99.5	0.47
8	4	7	7	4	0.45	194	0.46
9	4	7	8	6	0.45	200	0.52
10	5	5	6	4	0.78	28.9	0.37
11	5	5	7	6	1.05	16.96	0.41
12	5	5	8	8	1.10	15.7	0.47
13	5	6	6	6	0.86	23.7	0.42
14	5	6	7	8	1.22	13.38	0.47
15	5	6	8	4	1.23	13.27	0.46
16	5	7	6	8	0.89	22.09	0.48
17	5	7	7	4	1.39	11.08	0.49
18	5	7	8	6	1.41	10.88	0.54
19	6	5	6	4	0.78	29	0.39
20	6	5	7	6	1.06	16.54	0.42
21	6	5	8	8	1.21	13.55	0.48
22	6	6	6	6	0.78	21.53	0.44
23	6	6	7	8	1.25	12.87	0.48
24	6	6	8	4	1.30	12.2	0.48
25	6	7	6	8	0.88	22.78	0.49
26	6	7	7	4	1.4	10.94	0.52
27	6	7	8	6	1.81	7.81	0.56

3.2 Structural Analysis of Designs

Stress analysis was conducted for all 27 combination of design variables using FEA. The results obtained for few combinations of design variables are shown in Figs. 19, 20, and 21.

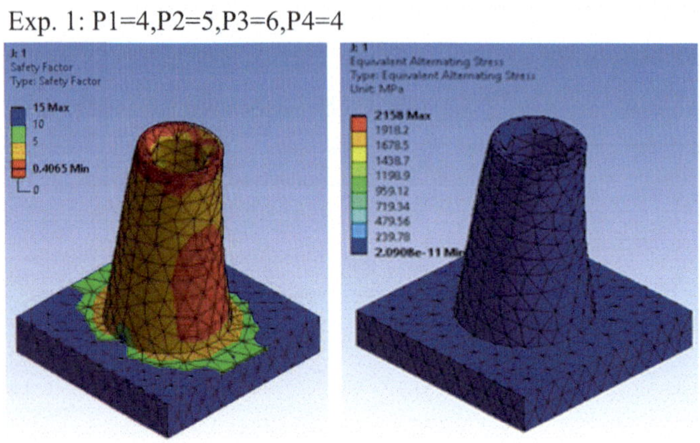

Fig. 19 Stress analysis of Exp. 1

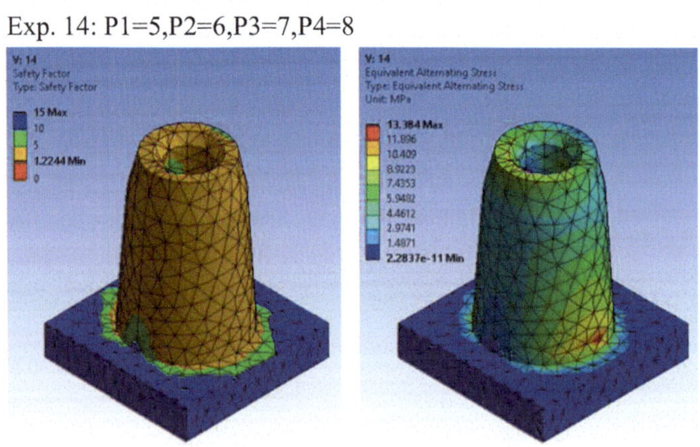

Fig. 20 Stress analysis of Exp. 14

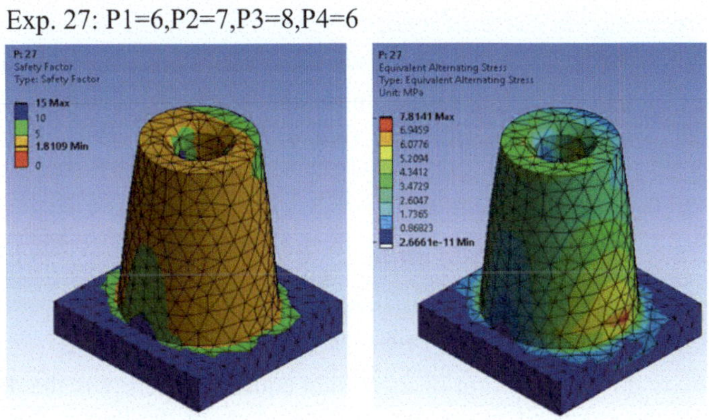

Fig. 21 Stress analysis of Exp. 27

3.3 Optimization of Design

Product weight is one of the most important properties for an injection-moulded part. The determination of process parameters for obtaining reduction in the weight of the component is essential. From the Taguchi DOE, the optimized parameters are calculated. Figure 22 shows the main effects plot for FOS.

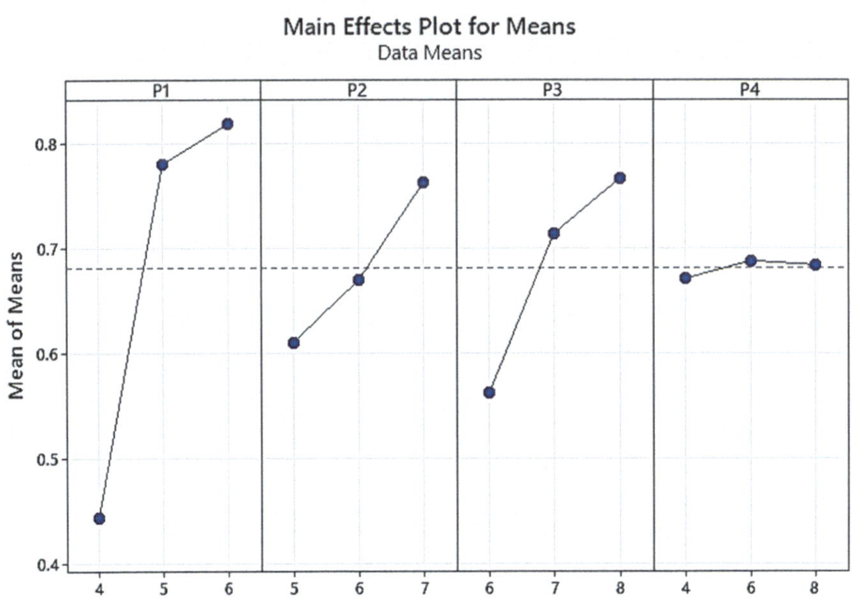

Fig. 22 Main effects plot for FOS

Table 5 ANOVA table

Source	DF	Adj SS	Adj MS	F-value	P-value
Regression	13	4.03016	0.310012	40.83	0.000
Error	13	0.09870	0.007592		
Total	26	4.12885			

3.4 Variance Analysis of Nonlinear Model of FOS

$Y = a + bx$ is the general form of regression equation where y is the dependent variable, x is the independent variable, b is the slope of the line, and a is the y-intercept. The obtained regression equation is given in the following equation, and variance analysis of FOS nonlinear model is given in Table 5.

Model Summary

S	R-sq	R-sq (adj)	R-sq (pred)
0.0871318	97.81%	95.22%	87.75%

$$
\begin{aligned}
Y_{\mathrm{FoS}} = {} & -1.93 + 1.849x1 - 1.123x2 - 0.116x3 + 0.133x4 \\
& - 0.2961x1 * x1 + 0.0239x2 * x2 \\
& - 0.0794x3 * x3 + 0.0074x4 * x4 + 0.0758x1 * x2 \\
& + 0.1583x1 * x3 - 0.0158x1 * x4 \\
& + 0.0978x2 * x3 - 0.0200x2 * x4
\end{aligned}
$$

3.5 Design Optimization Through Genetic Algorithm

The equation developed using FEA results was used as a fitness function in genetic algorithm. The optimized design was obtained and given below.

- P1—6 mm [Top diameter]
- P2—7 mm [Middle diameter]
- P3—8 mm [Bottom diameter]
- P4—4 mm [Step height].

The optimized structure can be seen in Fig. 23.

Fig. 23 CAD model of optimized structure

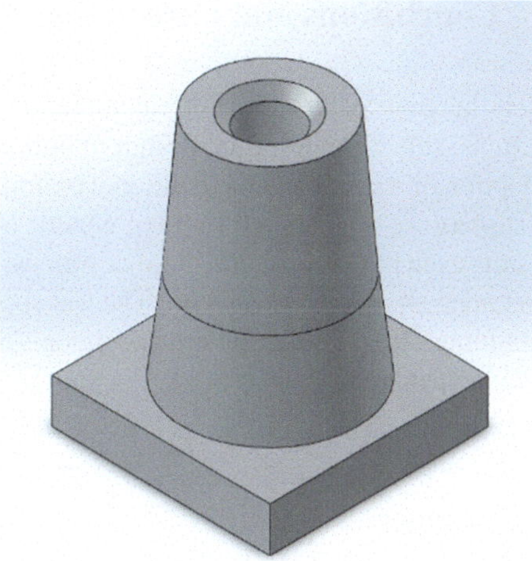

4 Results and Discussion

- The selected motor body of an electronic lock has undergone fatigue failure which has resulted in the breakage at a part in the component. The component is subjected to cyclic loading of both bending and torsional. The reason for the failure is due to poor design which was not capable of withstanding the required loading conditions in the industry. To solve this problem, alternative designs were developed. These alternative designs were inspired from nature itself. The four designs generated were tree root structure, cross-section of a Bamboo, an Ostrich's leg, and Lotus stem structure. Finite element analysis were carried out for all the four designs. The Taguchi design of experiments (L27) was formed with reference to the selected structure. The analysis was conducted from which 27 distinct sets of design parameters were obtained. The CAD models were generated for all the 27 distinct sets.
- Using the CAD models, FEA was conducted for all 27 combination of design parameters. From finite element analysis, the factor of safety, equivalent alternating stress, and weight were obtained. Based on FEA results, variance analysis was performed and non-linear model for factor of safety was developed. From the ANOVA table , it can be inferred that the parameters are significant, the R square value was found as 97%, which proves that the developed non-linear regression model is significant. Since the non-linear regression model is significant, it can be used as a fitness function for the genetic algorithm technique. The role of the genetic algorithm is to optimize the design even further, using which the fittest design parameters can be obtained.

5 Conclusions and Future Work

Identification of component, identification of failure, generation of alternative designs, selection of a final design to optimize, selection of design parameters, and adoption of a suitable non-traditional optimization technique were carried out in the first phase of this study . Further, genetic algorithm has to be used to optimize the results even better. From the final design parameters obtained from GA results, a test specimen should be developed. The test specimen can be fabricated using injection moulding die . Finally, the test specimen should be tested according ASTM D7791 testing standards.

References

1. Burns LA, Mouritz AP, Pook D, Feih S (2012) Bio-inspired design of aerospace composite joints for improved damage tolerance. Compos Struct 94(3):995–1004. https://doi.org/10.1016/J.COMPSTRUCT.2011.11.005
2. Sysaykeo D, Mermoz E, Thouveny T (2020) Clearance and design optimization of bio-inspired bearings under off-center load. CIRP Ann 69(1):121–124. https://doi.org/10.1016/J.CIRP.2020.03.006
3. Zhang Y, Wang Z, Zhang Y, Gomes S, Bernard A (2020) Bio-inspired generative design for support structure generation and optimization in additive manufacturing (AM). CIRP Ann 69(1):117–120. https://doi.org/10.1016/J.CIRP.2020.04.091
4. Akrami R, Fotouhi S, Fotouhi M, Bodaghi M, Clamp J, Bolouri A (2019) High-performance bio-inspired composite T-joints. Compos Sci Technol 184. https://doi.org/10.1016/J.COMPSCITECH.2019.107840
5. Burns L, Mouritz AP, Pook D, Feih S (2015) Bio-inspired hierarchical design of composite T-joints with improved structural properties. Compos Part B Eng 69:222–231. https://doi.org/10.1016/J.COMPOSITESB.2014.09.041
6. Alfano M, Morano C, Bruno L, Muzzupappa M, Pagnotta L (2018) Analysis of debonding in bio-inspired interfaces obtained by additive manufacturing. Procedia Struct Integr 8:604–609. https://doi.org/10.1016/J.PROSTR.2017.12.059
7. Takayama T, Komabayasi K, Itou M, Miyake Y (2009) Development of bio-based plastics for injection moulding. SAE Int J Mater Manuf 2:12–17. https://doi.org/10.4271/2009-01-0019
8. Wu J, Sigmund O, Groen JP (2021) Topology optimization of multi-scale structures: a review. Struct Multidiscip Optim 63(3):1455–1480. https://doi.org/10.1007/s00158-021-02881-8
9. Carreira P, Alves N, Ramos C, Bártolo P (2013) Bio inspired algorithms for injection moulding optimization. Adv Mater Res 683:771. https://doi.org/10.4028/www.scientific.net/AMR.683.771
10. Audibert C, Chaves-Jacob J, Linares JM, Lopez QA (2018) Bio-inspired method based on bone architecture to optimize the structure of mechanical workspieces. Mater Des 160:708–717. https://doi.org/10.1016/J.MATDES.2018.10.013
11. Altan T, Lilly B, Yen YC (2001) Manufacturing of dies and moulds. CIRP Ann 50(2):404–422. https://doi.org/10.1016/S0007-8506(07)62988-6
12. Ryan-Johnson W, Byron C, Nagel J, Zhang H (2021) A systems approach of topology optimization for bioinspired material structures design using additive manufacturing. Sustainability 13:8013. https://doi.org/10.3390/su13148013
13. Panneerselvam K, Aravindan S, Noorul Haq A (2009) Hybrid of ANN with genetic algorithm for optimization of frictional vibration joining process of plastics. Int J Adv Manuf Technol 42(7):669–677. https://doi.org/10.1007/s00170-008-1641-z

Machinability and Surface Integrity Characteristics in Hard Turning of High Hardened Steels Using Different Types of Inserts

G. Muthu Krishnan and J. Pradeep Kumar

Abstract Hard turning is performed by manufacturing industries for its applications to achieve the required machining surface characteristics by eliminating finishing processes such as cylindrical grinding. Hard turning is widely used in the fields of automotive and aerospace industries. The objective is to investigate the surface characteristics of different hardened steels AISI 1018 (35 ± 1 HRC), AISI 4140 (43 ± 1 HRC), AISI D2 (50 ± 1 HRC), and AISI H13 (60 ± 1 HRC) steel using the wiper carbide insert with designation (WNMG060408MW), cemented carbide insert (TiCN/Al_2O_3) with designation (CNMG120412MR), ceramics insert with designation (CNGA433T0820), and carbide insert with designation (CNMG120404B) under the dry turning conditions. The hardness of selected material is tested using the Rockwell Hardness Tester. The experiment was conducted on the CNC Lathe Galaxy Midas-6 model. The significant parameters are cutting speed, hardness of workpieces, feed, depth-of-cut, and type of inserts are considered as input parameters, and surface roughness, tool wear, tool wear rate, MRR "material removal rate", and cutting temperature are considered as output response parameters. DOE using Taguchi's L16 orthogonal array with five factors and four levels was selected. The experimental results reveal that the type of insert, cutting speed, and workpiece hardness are the important significant parameters in the surface characteristics of the workpiece along with the feed rate and depth-of-cut. The workpiece hardness and depth-of-cut play a significant role in tool wear followed by the type of insert, cutting speed, and feed. The workpiece's hardness, insert type, and time to complete the hard turning operations play a major role in cutting temperature. The optimization result reveals that the VIKOR optimization is best suitable for multi-criteria decision making optimization.

Keywords Hard turning · VIKOR optimization · Dry turning · Tool wear · MRR—"material removal rate" · Tool wear rate · Surface roughness (Ra)

G. Muthu Krishnan (✉) · J. Pradeep Kumar
Department of Production Engineering, PSG College of Technology, Coimbatore, India
e-mail: gmuthukrish95@gmail.com

J. Pradeep Kumar
e-mail: jpk.prod@psgtech.ac.in

E. Natarajan et al. (eds.), *Materials, Design and Manufacturing for Sustainable Environment*, Lecture Notes in Mechanical Engineering,
https://doi.org/10.1007/978-981-19-3053-9_21

1 Introduction

Hard turning is an advanced turning operation performed on hardened materials with a hardness value above 45 HRC. Hard turning is performed by manufacturing industries for various applications to achieve the required machining surface characteristics by eliminating finishing processes such as cylindrical grinding. Hard turning is widely used for many applications in the fields of the aerospace industry and automotive industry. The different cutting tools like carbide, wiper carbide, ceramics, and CBN are very useful in hard turning operations. This work in hard turning of AISI4340 helps to study influence of input parameters on the response variables like tool wear and Ra. The workpiece selected was AISI 4340 with 36 mm dia and 150 mm in length. The inserts selected for the hard turning operations were cemented carbide. It is found that feed is an important parameter for the surface finish in dry turning conditions. The lower tool wear and surface roughness were achieved by the optimal parameters [1]. The workpiece material selected was AISI 420 with a workpiece hardness of 48 HRC. The Kennametal's coated carbide insert was used for the hard turning process. The experimental results showed that wiper carbide inserts performed well even in dry turning conditions. The optimum parameters for tool life are achieved with lower cutting speed and lower feed [2]. The investigation in turning of AISID2 steel with ceramic wiper inserts. This study focus on the effects of input parameters on the response variables using the ceramic insert with multi-radii. Wiper insert can achieve good surface finish values when used in dry conditions. Result reveals that the tool life was higher in lesser feed and low cutting speed. The parameters were optimally controlled to achieve the low tool wear and good surface finish values [3]. Investigated AISIH13 in dry condition with ceramic insert. The cutting performances of ceramic insert in dry condition of AISIH13 were studied. From result found that lower feed rate, cutting speed, and depth-of-cut improved the tool life. The parameters are optimally controlled in order to obtain lower tool wear and a better surface finish [4]. The experiment on AISI 4340 and AISI D2 steel. The workpiece hardness is 58 HRC, the work was divided into two phases with two workpieces and performed the hard turning operations, and the insert used for the operations is mixed alumina inserts. It is noted that the tool wear was higher when the speed and depth-of-cut were higher. The optimum combinations were determined for the lower tool wear [5]. This work using wiper and conventional ceramic inserts, the workpiece selected for the experiment was AISI 4140 with a hardness of 60 HRC. The work was conducted in dry turning conditions with the wiper and ceramic inserts. It is found the wiper is capable of producing a good surface roughness value when compared with the other inserts. The optimum parameters were determined for the lower tool wear and Ra value [6]. This investigation on AISIH13 Steel. The workpiece selected was AISI H13 with 150 mm length and dia 50 mm. The workpieces are heat treated with different hardness values of 45, 50, and 55 HRC. The CBN insert was used for the turning of H13 steel in dry conditions. It is found the important factor for the temperature during the hard turning are hardness of the workpiece, speed, and feed. To obtain the lesser tool wear and Ra, the parameters are optimally controlled [7].

2 Experimental Setup

AISI 1018, AISI 4140, AISI D2, and AISI H13 steels were selected as the workpiece materials in this current investigation for their applications in manufacturing industry, automobile industry, and even in aerospace industry. The workpiece dimensions are 35 mm in diameter and 150 mm in length. The workpiece is shown in Fig. 1. The initial hardness of the workpiece materials was AISI 1018 (15 ± 1) HRC, AISI 4140 (26 ± 1) HRC, AISI H13 (28 ± 1) HRC, and AISI D2 (24 ± 1) HRC. The workpieces were heat treated to attain the required hardness value (HRC). The required level of hardness levels is AISI 1018(35 ± 1) HRC, AISI 4140 (43 ± 1) HRC, AISI H13 (50 ± 1) HRC, AISI D2 (60 ± 1) HRC. The hardness was measured using the Rockwell Hardness Tester. Hardness of every work material were tested again with the Rockwell Hardness Tester post-heat treatment process. Before starting the experimental work, facing operations was done on both side of the workpieces and countersink operations were performed in the workpieces to attain proper support to the tailstock while performing hard turning operations in the CNC turning center machine. Hard turning of length 100 mm was fixed as a constant for every workpiece and performed the turning operations. The images of the workpiece and inserts are shown in Fig. 1a. (insert workpieces in order (1.AISI 1018, 2.AISI 4140, 3.AISI H13, and 4.AISI D2) inserts in order (1. carbide, 2. emented carbide, 3. ceramics, 4. wiper carbide).

3 Machine Setup

Hard turning operations are carried out in a CNC Galaxy Midas 6 model with tailstock support. This machine has speed of 4000 RPM with a max power rating of 15 KVA with a FANUC OI MATE T CNC system. To perform the hard turning operations, different types of inserts were selected from the Kennametal manufacture, namely, carbide insert with designation CNMG120404B, cemented carbide insert (coated with TiCN/Al_2O_3) with designation CNMG120412MR, ceramic insert with designation CNGA433T0820, and wiper carbide insert with designation WNMG060408MW. The surface roughness was measured with "Mitutoyo Surface Roughness Tester Model: SJ-201" with a cut-off length of 12.5 mm was selected to find the surface roughness of (Ra) arithmetic surface roughness. The cutting parameters and levels are shown in Table 1. The experiment was planned with five factors four levels, Taguchi L16 orthogonal array was selected with the 16 runs to minimize the overall cost of the experiment shown in Table 2. The experimental layout plan is shown in Table 3.

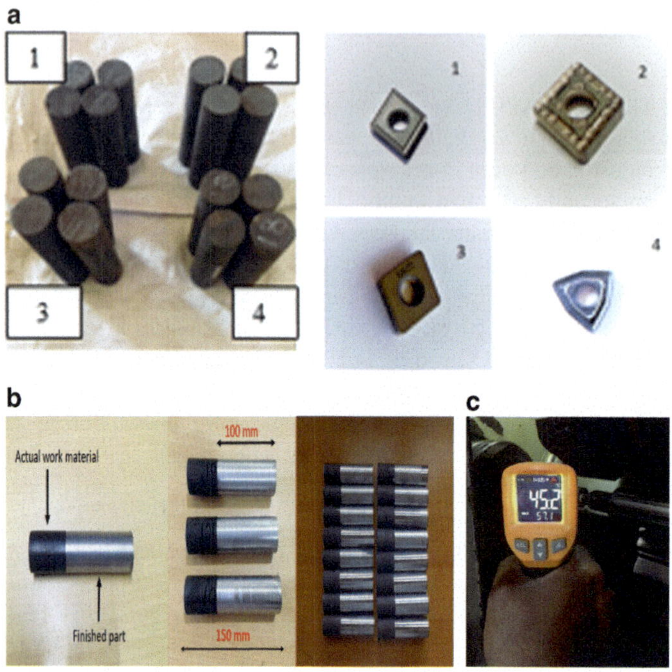

Fig. 1 **a** Workpieces and inserts, **b** finished workpieces, **c** temperature measurement

Table 1 Parameters and selected levels

		Levels			
Parameters	Units	1	2	3	4
Speed	m/min	100	120	140	160
Feed	mm/rev	0.05	0.2	0.35	0.5
Depth-of-cut	mm	0.1	0.2	0.3	0.4
Hardness	HRC	35	43	50	60
Inserts		WC	CC	CE	CA

Where WC—wiper carbide, CC—cemented carbide, CE—ceramic, CA—carbide

4 Result and Discussions

The experimental results are shown in Table 3. ANOVA—"Analysis of Variance" was performed to analyze influence of the cutting parameters like speed, hardness of workpiece, feed, depth-of-cut, and type of inserts on the output parameters like surface roughness (Ra), tool wear rate, and MRR. The MINITAB software helped find out the performance of the models. The performance of models was tested using ANOVA under 95% confidence level. The weight of the workpieces before the hard

Table 2 DOE

Run	Workpiece	Hardness (HRC)	Inserts	Speed (m/min)	Feed (mm/rev)	Doc (mm)
1	1018	35	WC	100	0.05	0.1
2	4140	43	CC	100	0.2	0.2
3	H13	60	CE	100	0.35	0.3
4	D2	50	CA	100	0.5	0.4
5	H13	60	CA	120	0.05	0.2
6	D2	50	CE	120	0.2	0.1
7	1018	35	CC	120	0.35	0.4
8	4140	43	WC	120	0.5	0.3
9	D2	50	CC	140	0.05	0.3
10	H13	60	WC	140	0.2	0.4
11	4140	43	CA	140	0.35	0.1
12	1018	35	CE	140	0.5	0.2
13	4140	43	CE	160	0.05	0.4
14	1018	35	CA	160	0.2	0.3
15	D2	50	WC	160	0.35	0.2
16	H13	60	CC	160	0.5	0.1

Table 3 Experimental result

Run	TEMP (°C)	TWR (g/min)	MRR (g/min)	Ra (μm)
1	28.63	0.00049	1.367	0.8
2	35	0.0000191	1.565	2.0125
3	34.53	0.00000236	2.24	1.14
4	48.11	0.000528	10.965	2.345
5	61.69	0.0000575	0.691	1.995
6	39.48	0.0000898	1.134	0.8875
7	28.85	0.00035	2.983	2.4825
8	32.93	0.00000158	3.699	3.895
9	34.39	0.0000636	0.387	0.8675
10	39.82	0.0000402	1.813	0.945
11	34.58	0.0000417	1.936	2.0367
12	33.66	0.000207	2.214	1.875
13	48.12	0.0000116	0.648	0.7675
14	32.43	0.000328	2.21	2.315
15	38.3	0.000247	0.387	0.55
16	36.9	0.00002198	3.087	0.8825

Table 4 ANOVA for (Ra)

Source	DF	Seq SS	Adj SS	Adj MS	F-value	P-value
Speed (m/min)	3	3.048	3.048	1.0159	1.47	0.380
Feed (mm/rev)	3	2.978	2.678	0.8927	1.29	0.420
DoC (mm)	3	1.633	1.633	0.5442	0.79	0.576
Hardness (HRC)	3	2.903	2.903	0.9678	1.40	0.395
Error	3	2.076	2.076	0.6919		
Total	15	12.337				

turning operations and after the hard turning operations was calculated to find the MRR. The weight of the inserts was also measured before and after hard turning operations to find out the tool wear rate.

4.1 Arithmetic Mean Roughness (Ra)

The good surface finish can be achieved with the help of higher cutting speed, types of cutting inserts, and geometry of the cutting inserts followed by the feed rate and depth-of-cut. ANOVA for (Ra) is obtained. ANOVA for SN ratios (Ra) versus speed, feed, depth-of-cut, hardness, and inserts was plotted and the result reveals that the cutting insert with (24.84%) is the significant input parameter followed by the speed (23.67%), hardness of the workpieces (22.66%), feed (18.67%), and depth-of-cut (10.15%). ANOVA table is shown in Table 4. The Ra was measured four times around the workpiece as the workpiece is a round bar, and the average of the four readings is considered as final surface roughness value. The surface plot and normal probability plot are shown in Fig. 2. Actual versus predicted values for (Ra) is shown in Fig. 2c.

4.2 Tool Wear

SEM "Scanning Electron Microscope" was helpful to find out the tool wear. The images obtained from the SEM are shown in Fig. 3. Tool wear was lesser in wiper carbide inserts when compared with the other inserts during the hard turning operations. (SEM) images were taken to study and analyze the tool wear. In this study, SEM analysis was done on different inserts, namely, wiper carbide insert, ceramic, carbide and cemented carbide. The images of each inserts are shown in Figs. 3, 4, 5 and 6, respectively. When compared with the other cutting inserts, wiper carbide inserts showed lesser tool wear and are able to produce good surface finish even when the hardness of the workpieces is 60 HRC. From the SEM images, it is known that only micro cracks were obtained during the hard turning process.

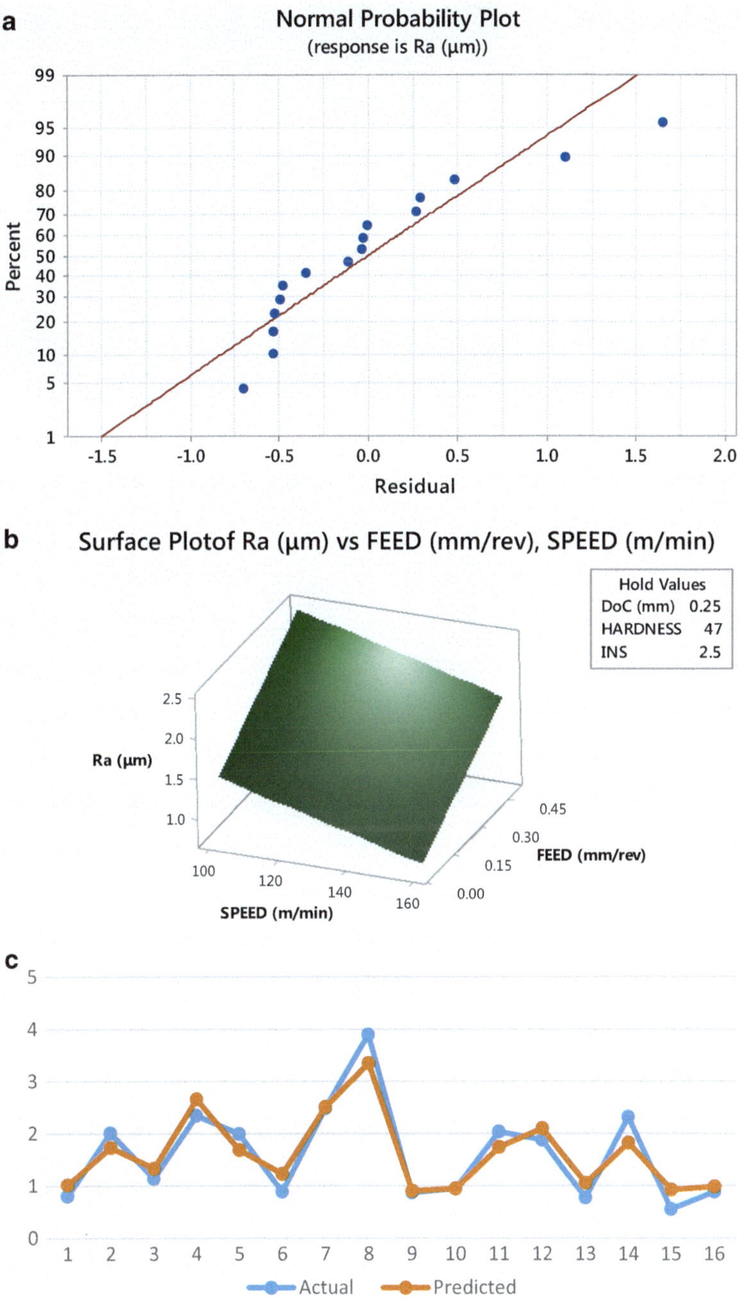

Fig. 2 **a** Normal probability plot for Ra, **b** surface plot, **c** actual versus predicted values for surface roughness (Ra)

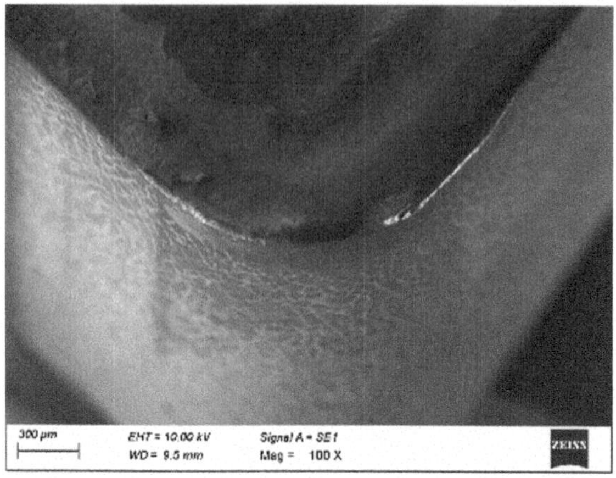

Fig. 3 SEM images of WC

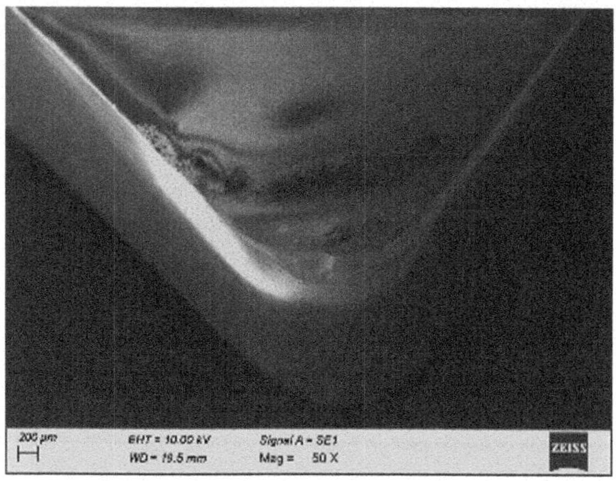

Fig. 4 SEM images of CE

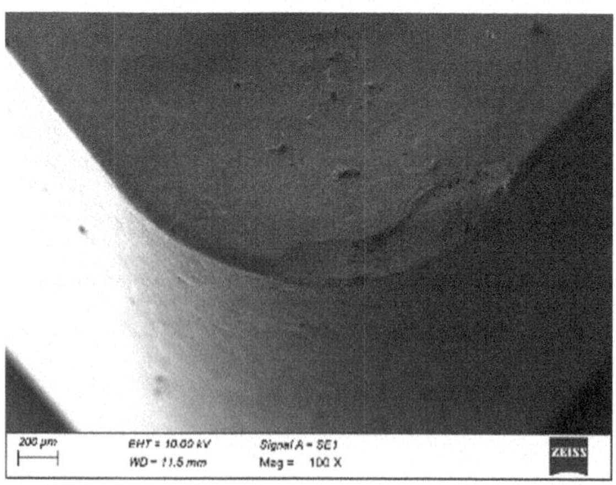

Fig. 5 SEM images of CA

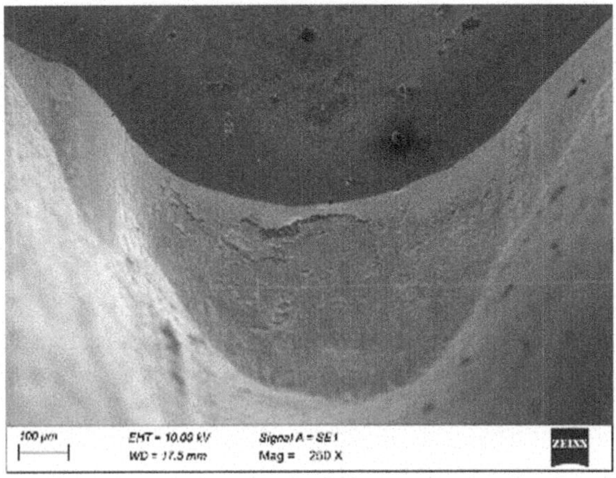

Fig. 6 SEM images of CC

4.3 Tool Wear Rate and Cutting Temperature

ANOVA for SN ratios, tool wear rate versus speed, feed, depth-of-cut, hardness, and inserts was plotted. The result reveals that the hardness of the workpiece is the significant input parameter for tool wear rate with the contribution of 68.26% followed by the depth-of-cut (ap) 17.30%, types of insert 11.91%, speed 1.41%, and feed 1.11%.The surface plot and normal probability plot for tool wear rate and cutting temperature are shown in Figs. 7 and 8. The actual vs predicted values for the tool wear rate are shown in Fig. 9. The cutting temperature increases when the hardness of the workpieces, depth-of-cut, feed rate, and speed increases. It is found that from the ANOAV results, the hardness is the most significant parameter followed by depth-of-cut, feed, and speed. ANOVA is shown in Table 5.

4.4 MRR

ANOVA for SN ratios tool wear rate vs speed, depth-of-cut, feed, hardness, and inserts was plotted shown in Table 6. The surface plot and normal probability plot for MRR are shown in Figs. 10 and 11. The result showed that the feed is the significant input parameter for material removal rate with a highest contribution of 55.60% followed by the depth-of-cut 15.65%, speed with 15.06%, insert with 7.28% and hardness of the workpiece with 6.41%. The actual vs predicted values for MRR shown in Fig. 12.

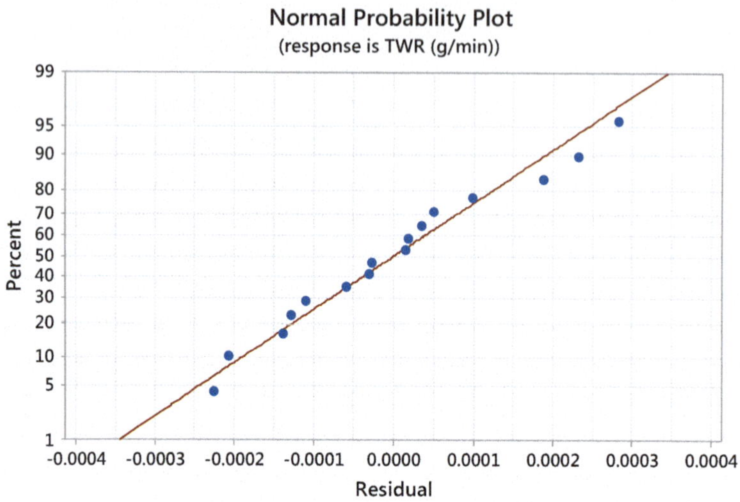

Fig. 7 Normal probability plot for Ra

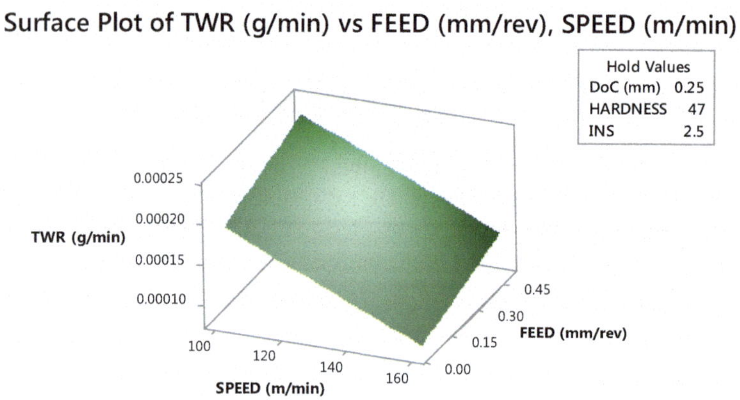

Fig. 8 Surface plot

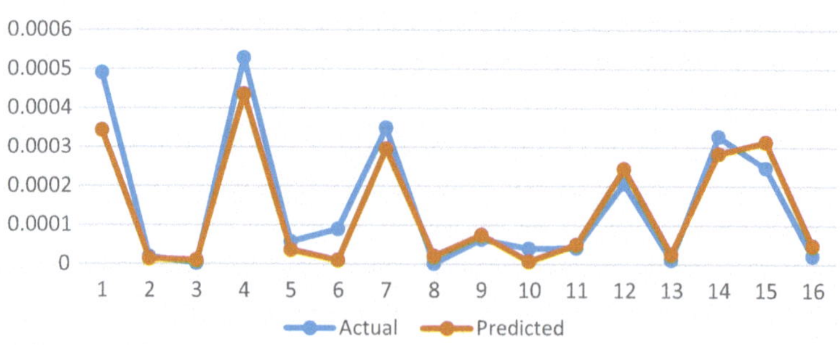

Fig. 9 Actual versus predicted values for tool wear rate

Table 5 ANOVA for cutting temperature

Source	DF	Seq SS	Adj SS	Adj MS	*F*-value	*P*-value
SPEED (m/min)	3	64.48	64.48	21.49	0.24	0.863
FEED (mm/rev)	3	177.37	177.37	59.12	0.67	0.627
DoC (mm)	3	227.89	227.89	75.96	0.86	0.550
HARDNESS	3	329.27	329.27	109.76	1.24	0.433
Error	3	266.46	266.46	88.82		
Total	15	1065.48				

Table 6 ANOVA for MRR

Source	DF	Seq SS	Adj SS	Adj MS	*F*-value	*P*-value
Speed (m/min)	3	16.220	16.220	5.407	1.13	0.461
Feed (mm/rev)	3	40.497	40.497	13.499	2.82	0.209
DoC (mm)	3	18.503	18.503	6.168	1.29	0.420
Hardness (HRC)	3	4.326	4.326	1.442	0.30	0.825
Error	3	14.358	14.358	4.786		
Total	15	93.904				

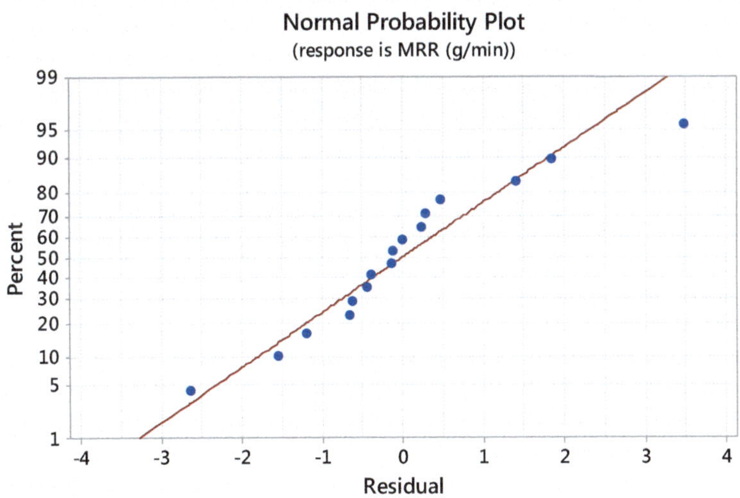

Fig. 10 Normal probability plot for MRR

5 Optimization

The main objective of this work is to reduce tool wear and surface roughness. The VIKOR optimization is useful in finding out the optimal solutions. Based on the

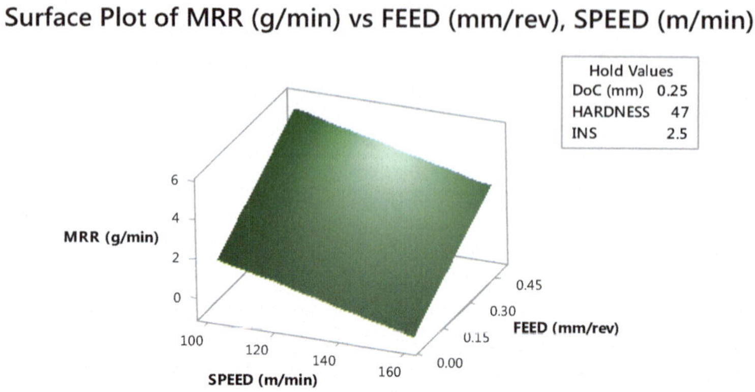

Fig. 11 Surface plot

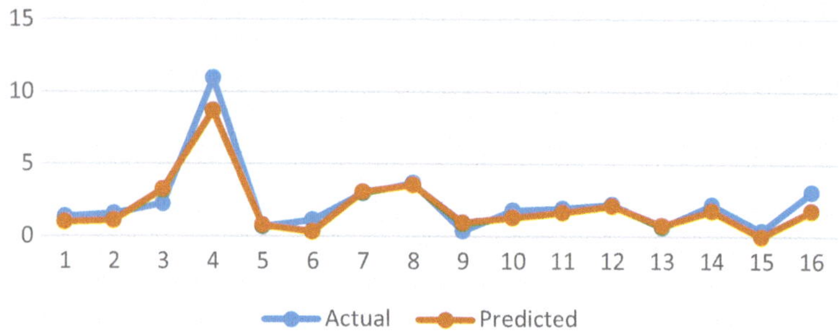

Fig. 12 Actual versus predicted values for MRR

VIKOR equations, Step 1: Assign the weights to response variables from (0–1) and calculate the unity measure (si). Step 2: Find out beneficiary and non-beneficiary and identify best and worst values for response variables then calculate individual regret (Ri). Step 3: Calculate (Qi). The ranking of the values based on Qi, Si, and Ri is calculated. The ranking based on the VIKOR optimization is shown in Table 7. The optimal solutions obtained with this method are speed 100 (m/min), feed (0.05 mm/rev), depth-of-cut 0.1(mm), cutting insert wiper carbide, and hardness of 35 HRC will result in minimum (Ra) value of 0.55 and 0.8(μm).

6 Conclusions

In this investigation, the hard turning of high hardened workpieces using different types of inserts was studied. The main process parameters like speed, feed, depth-of-cut, inserts, and hardness are optimally controlled to obtain a lower tool wear and Ra value. The following conclusions are made from this research work

Table 7 VIKOR ranking

Experiment run	Rank Qi	Rank Si	Rank Ri
1	1	1	6
2	9	9	7
3	14	14	12
4	16	16	16
5	3	4	4
6	8	11	5
7	7	3	13
8	13	7	15
9	5	8	1
10	12	13	8
11	11	10	9
12	10	6	11
13	6	12	3
14	4	2	10
15	2	5	1
16	15	15	14

- ANOVA result reveals that for the surface roughness cutting speed, is the main significant input parameter followed by hardness of workpieces, feed, and depth-of-cut.
- Feed is the most significant input parameter for material removal rate with a highest contribution of 43.13% followed by the depth-of-cut with 19.70%, cutting speed with 17.20%, and hardness of the workpiece with 4.61%.
- For tool wear rate, the hardness of the workpiece is the significant input parameter for tool wear rate followed by speed, depth-of-cut with, and feed.
- The hardness of the workpieces is the most significant parameter for cutting temperature followed by depth-of-cut, feed, and speed.
- Wiper carbide insert is capable of performing better with a lesser tool wear rate in dry turning conditions even when the hardness of the workpiece is (60 ± 1) HRC.
- Wiper carbide inserts are capable of producing the good surface roughness values of 0.5–0.8 (μm).
- This experimental work confirms the potential and effectiveness of using wiper carbide inserts in the hard turning process under dry turning conditions. Wiper carbide inserts are able to achieve the required machining surface characteristics and servers as a cost-effective option.

References

1. Pradeep Kumar J, Kishore KP, Ranjith Kumar M, Saran Karthick KR, Vishnu Gowtham S (2018) Experimental investigation on the hard turning of AISI 4340 steel using cemented coated carbide insert. Mater Sci and Manuf 316
2. Kurniawan D, Yusof NM, Sharif S (2010) Hard machining of stainless steel using wiper coated carbide tool life and surface integrity. Mater Manuf Process 25:370–377
3. Ozel T, Karpat Y, Luis Figueria J, Davim P (2007) Modelling of surface finishand tool flank wear in turning of AISI D2 steel with ceramic wiper inserts. J Mater Process Technol 189(2007):192–198
4. Jinpeng S, Jiaojiao G, Ahmad R, Ming L (2020) Cutting performances of TiCN–HfC and TiCN–HfC–WC ceramic tools in dry turning hardened AISI H13. Adv Appl Ceram 119:380–386
5. Lima JG, Avila RF, Abrao AM, Faustino M, Paulo Davim J (2005) Hard turning AISI 4340 high strength low alloy steel and AISID2 cold work tool steel. J Mater Process Technol 169(2005):388–395
6. Elbah M, Yallese MA, Aouici H, Mabrouki T, Rigal J-F (2013) Comparative assessment of wiper and conventional wiper ceramic tools on surface roughness in hard turning AISI 4140 steels. Measurements 46:3041–3056
7. Kumar P, Chauhan SR (2015) Machinability study on finish turning of AISI H13 hot working die tool steel with CBN cutting tool inserts using RSM. Arab J Sci Eng 40:1471–1485

Analysis of Impact Strength on Polymer Matrix Composites Using Projectile Parameters for Combat Vehicle Application

S. Palani, S. Karthik, C. Rajaravi, and R. Ganapathy Srinivasan

Abstract Our research objective recognizes projectile parameters effect for impact strength of carbon fiber reinforced epoxy sap composite. For this reason, epoxy carbon unidirectional 3D finite element analysis (FEA) model has been demonstrated by using ANSYS (ACP)/Explicit. High velocity/ ballistics impact characteristics such has deformation, stress and strain induced on the epoxy carbon unidirectional has been stimulated. The damage becomes asymmetric and increases with the increase in the projectile material (Titanium alloy), size (120 mm) and speed (700, 750, 800 and 850 m/s) as a consequence of change in stress field profile. The FE strategy is used to model the specimen in high speed/ballistic impact situation. The projectile and the composite layers are modeled utilizing ANSYS/ACP. The FEA model contains two particular parts such as projectile and composite layers. At first, projectile has modeled as a rigid body, properties of material are imported from the ANSYS material library, then underlying speed of the projectile is set in m/sec and the range of the projectile is set in similar manner. At that point so as to show a composite layer ANSYS/ACP is used, at first, a fabric is made which incorporates the directional properties of the composite. At that point, a stack is made where various fabrics could be included with shifting edges. The present model has fabric with thickness of 5 mm and a stack comprising four fabrics at angles [0/45/90/-45]. The mesh sensitivity, maximum stress and maximum strain induced in the test sample, and the projectile materials are found and the parameters that influence the impact strength of the epoxy carbon composites are identified.

S. Palani (✉) · R. Ganapathy Srinivasan
Department of Mechanical Engineering, Vel Tech Multitech, Avadi, Chennai, Tamil Nadu, India
e-mail: palaniraji.s@gmail.com

S. Karthik
Department of Mechanical Engineering, Vel Tech Rangarajan Dr. Sagunthala R&D Institute of Science and Technology, Avadi, Chennai, Tamil Nadu, India

C. Rajaravi
Department of Mechanical Engineering, Park College of Engineering and Technology, Coimbatore, Tamil Nadu, India

E. Natarajan et al. (eds.), *Materials, Design and Manufacturing for Sustainable Environment*, Lecture Notes in Mechanical Engineering,
https://doi.org/10.1007/978-981-19-3053-9_22

Keywords Ballistics · Projectile · Impact characteristics · Composite materials · FEA-ANSYS

1 Introduction

Most recent researches contribute the materials from composites for applying different structural in light, more explicit firmness as well as quality proportion, particularly structural with aviation anywhere scarcely noticeable effect harm be a basic issue. Causes of different harms on composite laminated structural when low-speed impact with example of breakage of fiber, splitting of metal matrix, failure due to shear also delamination [1]. The damages like delamination caused critical decreases of basic trustworthiness as well as compressive strength at residual [2]. Contrasted and flawless shield and impact performance of scarf fixed cover are progressively convoluted because of the presentation of scarf joint [3].

There are two broadly winning methodologies for considering impact issues with HC structural by sandwich composite particularly exploratory system as well as numerical strategy [4]. Exploratory examination was given effect with direct vision also typical damage in addition to along these lines advantageous and viable for getting the fundamental assets for additional examination. An extensive variety of effect vitality extending static to dynamic effect setting actualized on graphite and epoxy skins by means of various thicknesses [5]. Sort along with degree of damage in the face sheets of graphite and epoxy through HC centers contrast with froth centers beneath low-speed impact along with outlined and there were two different types of harm so as to happen among expanding impact vitality [6].

Glass fiber affectability rate strengthened and epoxy skins among the aluminum center. Also, analyses showed so as to the episode vitality parcel was fundamentally affected through effect shot geometry [7]. The trials of harm reaction in the HC structured sandwich composite beneath sideways as minimum impact speed [8]. Technique of arithmetical for examination turns out to be progressively appealing because of its focal points in limiting the measure of testing, accordingly sparing structure cost and time [9, 10]. The model of an arithmetical considering enormous avoidances, plastic disfigurement of aluminum HC center also layer of composite in plane debasement, in favor of contemplating damage conduct beneath impact with minimum speed [4]. For getting failure reaction of HC structured composite sandwich through 3D arithmetical model in addition to relate among test outcome [11], a model of arithmetical representing impact with maximum speed on structured composite sandwich by means of center with aluminum HC, furthermore, decided energy assimilations comparable for relating the mechanism of failure [12].

The express business code along with the client characterized of material subroutine for considering failure model as of in plane breaking for delamination beneath of impact with minimum speed. In addition, contrasted their numerical outcomes and the test bends of damage model to confirm the built up [13]. The model of damage disappointment criteria of composite with face sheet as well as versatile plastic intended for HC center impact with minimum speed on the sandwich composite bars [11]. The arithmetical strategy for recreating effect on composites texture strengthened covered and the materials of nonlinearity under the damage reactions. For example, fiber break, intra-laminar also, between laminar breaking of network and interface, were mulled over [14]. The numerical models with respect to gum-based glass fiber strengthened plastics sandwich boards with Nomex HC centers based on adjustment with the trials, and furthermore examined and talked about the strain-rate-subordinate dynamic properties [15].

All the above works for the most part focused on the low-speed impact quality. However, for the battle vehicle, the effect is because of the high speed of the shots which is especially more prominent than 700 m/s. Subsequently, to withstand such a high impact, it is important to examine the segment for high-speed impact which has been gathered in our proposed research work.

2 ANSYS Modeling

For testing of planning as well as assessment to obtain outcomes of structural composites ANSYS Prep Post composite dealing as specific device. It fills in an extra module used by the professional of ANSYS and completely incorporated in the stage of work bench. The blend of this permits compelling planning of individual composite parts as well as multifaceted gatherings which are the part of composite joint through different part of materials as like the shell of tank by means of steel ribs. Additionally, conceivable to effectively consolidate various sorts of investigation like static, explicit, dynamic and temperature together with different physical examination. The composite ANSYS Prep Post would be incorporated with parametric improvement through the design of ANSYS Xplorer device. Rosettes are facilitating frameworks that used to set the reference direction of oriented selection sets. Rosettes characterize the 0° bearing for the composite lay-up. Arrange frameworks characterized in mechanical are imported naturally. Extra Rosettes could be characterized in ACP. There are five unique kinds of Rosettes. The root and headings of Rosettes are given in the worldwide arrange framework. All the time, fiber headings follow the external structure of a plan. To utilize these characterizing fiber headings, the ANSYS composite Prep Post could be characterized the bearings dependent on edges in the model. The application (draping) on doubly bended surfaces changes the hypothetical fiber directions. Much of the time, the impact is little and would be dismissed. On the opposite side, it is essential to realize how enormous this impact and on the off chance that it must be considered. ACP permits to assess the draping fiber headings. Selection rules permit to choose components through geometrical tasks. For demonstrating a composite in ACP, pre portion of the essential parameter is shown in Fig. 1.

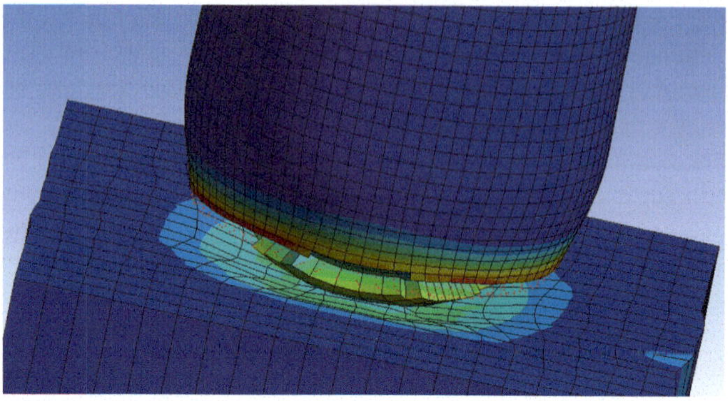

Fig. 1 Composite model generated with the help of ANSYS/ACP

These choices could be joined with Oriented Selection Sets or Modeling Plies to characterize and utilizes of discretionary shape. This component could be utilized to characterize nearby fortifications (fixes) or faltering.

3 Composite Damage Model for Impact

Impact with high velocity damage is having two categorize. One is the failure of intra-laminar caused due to matrix tensile as well as compressive failure on fiber that occurs within a ply and another one is the inter-laminar failure which caused between different plies. The impact damage caused in the composite stimulated through damage with progressive model includes the failure model of initiation criteria, damage evolution as well as continuum.

3.1 Intra-Laminar Failure

The list of criteria with failure upon most popular one by Hashin's [16] in favor of anticipating failure commenced with single layer overlays composite which fiercely utilized together scholarly as well as the field of designing. Our proposed research work executed Hashin criterion with 3-dimentionalwhich incorporates matrix failure component of normal stress. The failure mode with 4-diverse is given below:

Failure with tensile on fiber: $(\sigma_{11} \geq 0)$

$$F_{\text{ft}} = \left(\frac{\sigma_{11}}{X_{\text{T}}}\right)^2 + \alpha\left(\frac{\sigma_{12}}{S_{12}}\right)^2 + \alpha\left(\frac{\sigma_{13}}{S_{13}}\right)^2 \geq 1 \tag{1}$$

Fiber compressive failure:$(\sigma_{11} < 0)$

$$F_{\text{ft}} = \left(\frac{\sigma_{11}}{X_{\text{C}}}\right)^2 \geq 1 \tag{2}$$

$$\text{Matrix tensile failure}: (\sigma_{22} + \sigma_{33} \geq 0) \tag{3}$$

$$F_{\text{mt}} = \left(\frac{\sigma_{11} + \sigma_{33}}{Y_T}\right)^2 + \frac{\left(\sigma_{23}^2 + \sigma_{22}\sigma_{33}\right)}{S_{23}^2} + \left(\frac{\sigma_{12}}{S_{12}}\right)^2 + \left(\frac{\sigma_{13}}{S_{13}}\right)^2 \geq 1 \tag{4}$$

Matrix compressive failure: $(\sigma_{22} + \sigma_{33} < 0)$

$$F_{\text{mt}} = \left(\frac{\sigma_{22} + \sigma_{33}}{2S_{23}}\right)^2 + \frac{\sigma_{22} + \sigma_{33}}{Y_{\text{C}}}\left[\left(\frac{Y_{\text{C}}}{2S_{23}}\right)^2 - 1\right] + \frac{\left(\sigma_{23}^2 + \sigma_{22}\sigma_{33}\right)}{S_{23}^2} + \left(\frac{\sigma_{12}}{S_{12}}\right)^2 + \left(\frac{\sigma_{13}}{S_{13}}\right)^2 \geq 1 \tag{5}$$

where

σ is the effective stress.

X_{T} and X_{C} are the fiber direction of tensile as well as compression, respectively.

Y_{T} and Y_{C} are the transverse direction of tensile as well as compression, respectively.

S_{12}, S_{13} and S_{23} are shear strength.

α is the shear failure coefficient ($\alpha = 1$).

3.2 *Damage Evolution Model*

The model for damage evolution has been defined after the satisfaction of the criterion of failure. In order to do this, Lapczyk [17], Fang [18] and also Chao [19] implemented using model with crack bandcreated by Bažant [20].

$$G_{\text{t}} = \frac{1}{2}\sigma_{\text{eq}}^f \epsilon_{\text{eq}}^f l_c \tag{6}$$

where

G_{t} = fracture energy density.

3.2.1 Inter-Laminar Damage

The relationship with separation bilinear traction is utilized for stimulating with delamination of inter-laminar in cohesive zone element between the adjacent piles of laminate composite. Here taken mutually criterion with quadratic failure as well as criterion with Benzeggagh and Kenane (B-K) [21] for initial predicting as well as the damage with delaminate propagation.

$$\frac{t_n^2}{N^2} + \frac{t_s^2}{S^2} + \frac{t_t^2}{T^2} = 1 \tag{7}$$

Here
t_n, t_s and t_t are the tractions of normal as well as shear, respectively.
N, S and T are the interface strength of normal as well as shear.

4 Results and Discussion

From the mesh sensitivity test, the impact of mesh size on the finite element modeling for at first explored for the composite model of epoxy carbon unidirectional with four stacks at angles 0°, 45°, 90° and −45° and each stack of thickness 4 mm and subsequently the absolute composite thickness of 16 mm. The projectile speed is set at 700 m/s. Maximum stress is estimation obtained for various mesh size keeping every other variable steady. The estimation of maximum stress and the computational time are arranged. This shows that the estimation of most maximum stress changes by 1.6% when the mesh size is diminished from 1 to 0.4 mm; it possibly changes 0.13%, if the mesh size is lessen to 1 mm. Figure 2 shows the different mesh size used for this study.

Anyway utilizing a mesh size of 1 mm requires 30% more computational time than utilizing a mesh size of 0.4 mm. In this manner, it is recommended to force a base edge size of 0.4 mm for tetrahedral component produces finite element solution with great precision and a reasonable computational time. This mesh size is utilized in all the incitement.

From the delamination, we observed incited on the test model for the different speeds of the projectile. In the graph, X-rotate shows the thickness of the various model, and the Y-center point shows the delamination impelled on the material. The graph includes four twists exhibiting the four one of a kind paces of the projectile material. The outline obviously shows that the delamination increases with the extension in the projectile speed and the thickness factor accept a basic activity in confining the delamination instigated. Figure 3a shows the damaged projectile after the impact, and Fig. 3b shows the delamination.

All the four curves bit by bit increase and gets collected at the delamination factor of 1.5. From this, it very well may be unmistakably expressed that the most extreme delamination caused is around 1.5 independent of the projectile speed. Figure 4 shows the thickness versus delamination impelled on the material.

From the maximum stress consideration, the Fig. 5 shows the most extreme stress prompted on the test example and the projectile for different speeds of the projectile.

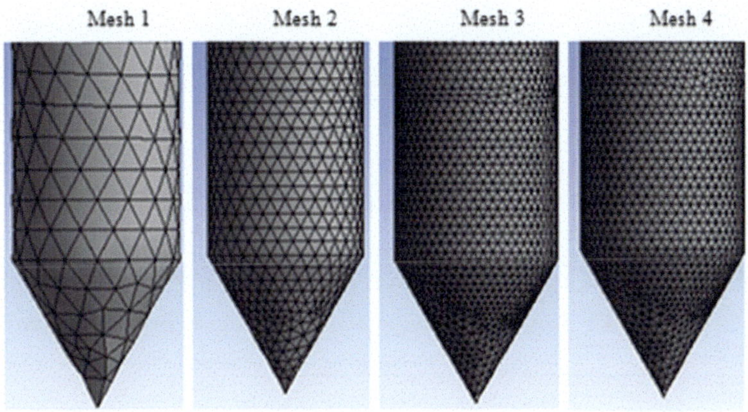

	Mesh 1	Mesh 2	Mesh 3	Mesh 4
Minimum element size(mm)	1	0.5	0.4	0.1
Number of tetrahedral elements	10660	42720	47167	137007
CPU Time(mins)	15	30	35	50
Maximum Stress(MPa)	1960.83	1930.91	1930.31	1927.7

Fig. 2 Different size of mesh used for analysis

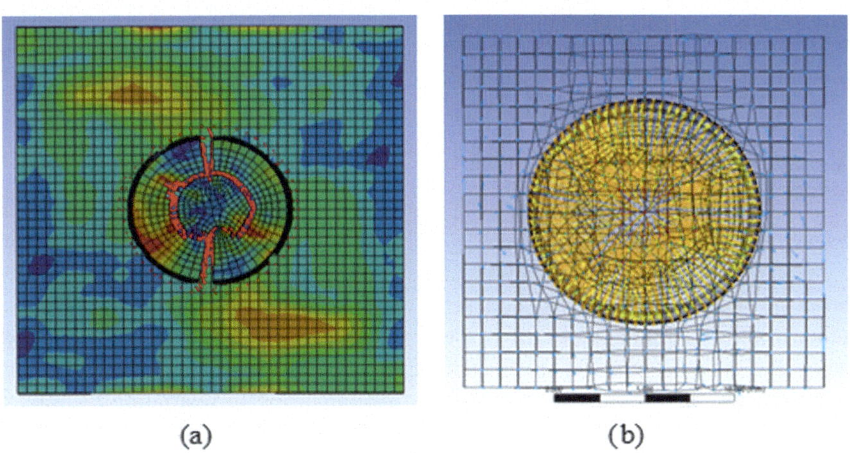

Fig. 3 **a** Projectile destroyed after impact and **b** delamination caused in the fiber

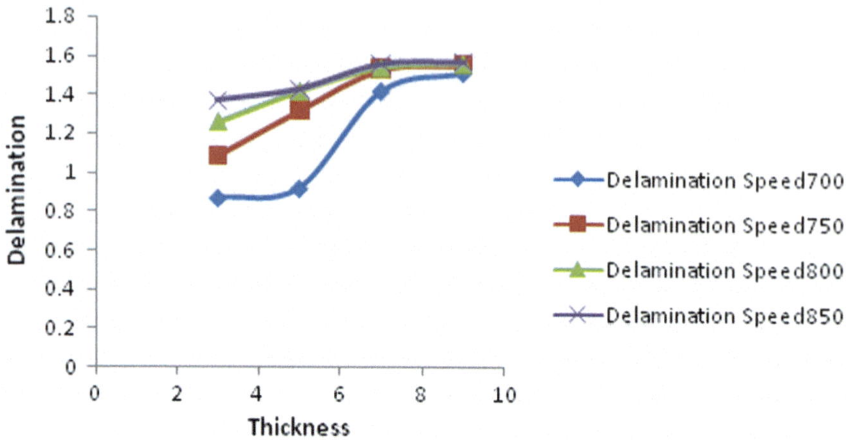

Fig. 4 Thickness versus delamination

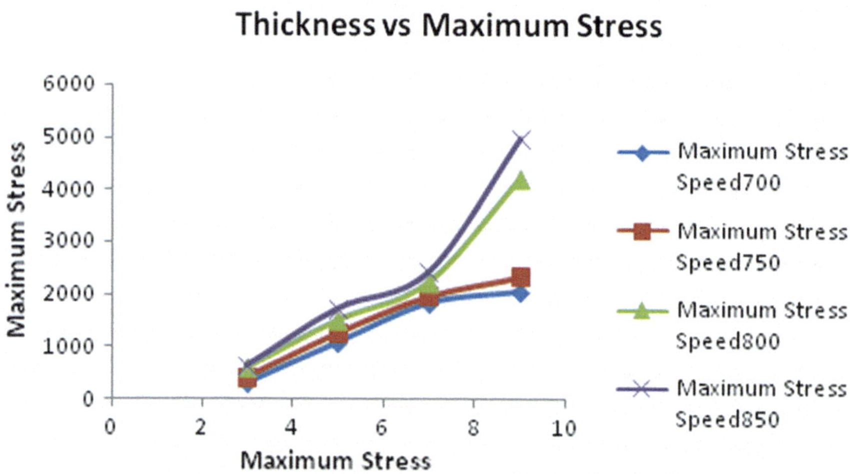

Fig. 5 Most extreme stress prompted on the test example and the projectile for different speeds of the projectile

In the chart, X-pivot shows the thickness of the diverse example, and the Y-hub shows the maximum stress instigated on the material. The chart comprises four bends demonstrating the four unique rates of the projectile material. The chart unmistakably shows that the most extreme stress increments with the expansion in the projectile speed and the thickness factor assume an imperative job in restricting the stress incited. All the four bends step by step increment till the thickness is 7 mm and afterward it increments quickly. Consequently composite material with a thickness 7 mm is practical to restrict most extreme stress initiated. In the diagram, X-rotate shows the thickness of the different model, and the Y-center shows the maximum

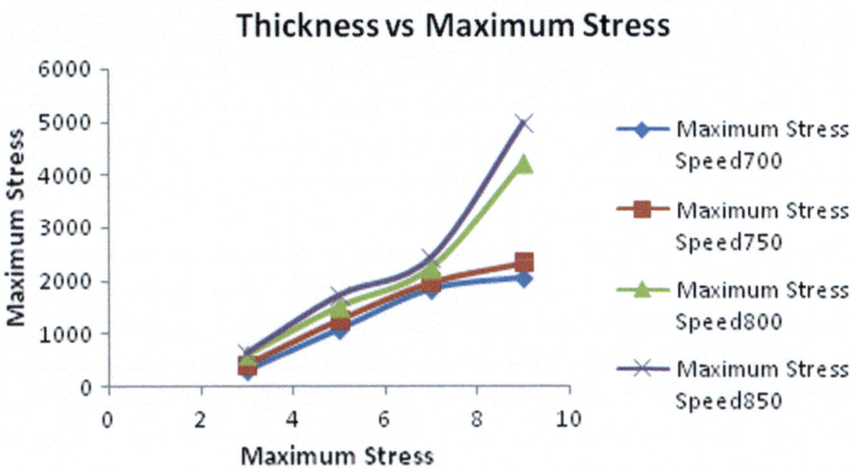

Fig. 6 Most extraordinary strain provoked on the test model and the projectile for different projectile speeds

strain incited on the material. The outline contains four twists exhibiting the four one of a kind paces of the projectile material. The outline undeniably shows that the most outrageous strain augments exponentially with the extension in the projectile speed and the thickness factor expect a basic activity in limiting the strain actuated. All the four twists increase exponentially till the thickness is 7 mm and a while later it augments rapidly. From the maximum strain consideration, Fig. 6 shows the most extraordinary strain provoked on the test model and the projectile for different projectile speeds.

Thus composite material with a thickness of 7 mm is common sense to confine most outrageous strain started.

5 Conclusions

From this research work, we concluded that, a large portion of the battle vehicles are made with steel and titanium so as to withstand the high impact of the shot; however, in the long run, it builds the heaviness of the vehicle in the way of diminishing the vehicle versatility. Furthermore, it is shown that, the battle vehicle protective layer thickness is around 100–120 mm; however, the equivalent strength able to be accomplished by assembling the layer defensive with epoxy carbon composite at a thickness of 7–9 mm which diminishes the heaviness of the vehicle by expanding the portability of the vehicle.

Acknowledgements We wish to express our sincere thanks to almighty and the people who extended their help during the course of our work. We are greatly and profoundly thankful to our honorable Chairman, Col. Prof. Vel. Shri Dr. R. Rangarajan B.E. (ELEC), B.E. (MECH), M.S. (AUTO), D.Sc., and Vice Chairman, Dr. Mrs. Sakunthala Rangarajan MBBS., for facilitating us with this opportunity. We take this opportunity to extend our gratefulness to our respectable Chairperson and Managing Trustee Smt. Mrs. Rangarajan Mahalakshmi Kishore B.E., M.B.A., for her continuous encouragement. Our special thanks to our cherishable Vice President Mr. K.V.D. Kishore Kumar B.E., M.B.A., for his attention toward students community.

References

1. Anderson T, Madenci E (2000) Experimental investigation of low-velocity impact characteristics of sandwich composites. Compos struct 50:239–2347
2. Ghasemnejad H, Soroush V, Mason P, Weager B (2012) To improve impact damage response of single and multi-delaminated FRP composites using natural Flax yarn. Mater Des 36:865–873
3. Zhang J, Cheng X, Guo X, Hu R (2013) Study of tensile failure mechanisms in scarf repaired CFRP laminates 38(6):10–14
4. Besant T, Davies G, Hitchings D (2001) Finite element modeling of low velocity impact of composite sandwich panels. Compos Part Aappl Sci Manuf 32:1189–1196
5. Herup E, Palazotto (1998) Low-velocity impact damage initiation in graphite/epoxy/nomex honeycomb-sandwich plates. Compos Sci Technol 57:1581–1598
6. Dear J, Lee H, Brown S (2005) Impact damage processes in composite sheet and sandwich honeycomb materials. Int J Impact Eng 32:130–154
7. Hazizan M, Cantwell W (2003) The low velocity impact response of an aluminium honeycomb sandwich structure. Compos Part A Appl Eng 34:679–687
8. Ivañez I, Moure M, Garcia-Castillo S, Sanchez-Saez S (2015) The oblique impact response of composite sandwich plates. Compos Struct 133:1127–1136
9. Hou S, Ren L, Dong D, Han X (2012) Crashworthiness optimization design of honeycomb sandwich panel based on factor screening. J Sandw Struct Mater 14(6):655–678
10. Hou S, Shu C, Zhao S, Liu T, Han X, Li Q (2015) Experimental and numerical studies on multi-layered corrugated sandwich panels under crushing loading. Compos Struct 126:371–385
11. Foo C, Chai G, Seah L (2008) A model to predict low-velocity impact response and damage in sandwich composites. Compos Sci Technol 68:1348–1356
12. Buitrago B, Santiuste C, Sánchez-Sáez S, Barbero E, Navarro C (2010) Modelling of composite sandwich structures with honeycomb core subjected to high-velocity impact. Compos Struct 92:2090–2096
13. Ivañez I, Sanchez-Saez S (2013) Numerical modelling of the low-velocity impact response of composite sandwich beams with honeycomb core. Compos Struct 106:716–723
14. Schwab M, Todt M, Wolfahrt M, Pettermann H (2016) Failure mechanism based modelling of impact on fabric reinforced composite laminates based on shell elements. Compos Sci Technol 128:131–137
15. Menna C, Manes Zinno A, Asprone D, Prota A (2013) Numerical assessment of the impact behavior of honeycomb sandwich structures. Compos Struct 106:326–339
16. Hashin Z (1980) Failure criteria for unidirectional fiber composites. J Appl Mech 47(2):329–334
17. Lapczyk I, Hurtado J (2007) Progressive damage modeling in fiber-reinforced materials. Compos Part A: Appl Sci Manuf 38(11):2333–2341
18. Guo-dong F, Jun L, Bao-lai W (2009) Progressive damage and nonlinear analysis of 3d four-directional braided composites under unidirectional tension. Compos Struct 89(1):126–133
19. Zhang C, Li N, Wang W, Binienda W, Fang H (2015) Progressive damage simulation of triaxially braided composite using a 3D meso-scale finite element model. Compos Struct 125:104–116

20. Bažant Z, Oh B (1983) Crack band theory for fracture of concrete. Matér Constr 16(3):155–177
21. Benzeggagh M, Kenane M (1996) Measurement of mixed-mode delamination fracture toughness of unidirectional glass/epoxy composites with mixed-mode bending apparatus. Compos Sci Technol 56(4):439–449

Design and Development of Portable Blender Cum Heater

N. Muthuram, C. Ajay Krishna, S. Dharshn, G. Moorthi, and C. Rithik Kumar

Abstract Numerous industries in India produce portable blenders with new designs to attract customers all over the world. The efforts are always being made to satisfy their growing demands. The goal of this project is to design and analyze a portable blender equipment upgrade. Material selection and design enhancement are critical for producing a portable blender that is user-friendly. In order to serve the customer a dual purpose of blending and heating installed in a single unit, the design is made and a prototype is developed in a scientific approach. The CAD modeling of the prototype is made using the PTC CREO 6.0 software. Thermal analysis is carried out using ANSYS 16.2 software based on the heat and power required to operate the unit. From the numerical analysis, it is found that the heat developed during the operation of the unit is approximately equal to that of the calculated value and also it indicates the safety of the product during the service.

Keywords Portable blender · Heating · Blending · Thermal analysis · Static structural analysis

---

N. Muthuram (✉) · C. Ajay Krishna · S. Dharshn · G. Moorthi · C. Rithik Kumar
Department of Production Engineering, PSG College of Technology, Coimbatore, Tamil Nadu, India
e-mail: muthuramprod@gmail.com

C. Ajay Krishna
e-mail: 18p602@psgtech.ac.in

S. Dharshn
e-mail: 18p613@psgtech.ac.in

G. Moorthi
e-mail: 18p625@psgtech.ac.in

C. Rithik Kumar
e-mail: 18p640@psgtech.ac.in

E. Natarajan et al. (eds.), *Materials, Design and Manufacturing for Sustainable Environment*, Lecture Notes in Mechanical Engineering,
https://doi.org/10.1007/978-981-19-3053-9_23

1 Introduction

The solicitation of new processes, the quick implementation of new practices, or the establishment of novel perceptions to create new value restores and keeps a field, a product, or a resource up to date. Administrations can improve their efficiency in one of the mentioned two ways: by following pricing guidelines or by developing a parting strategy. Creativity is required in both situations. In the first part, designers were challenged to come up with several unique concepts based on an existing product. As prompted, they come up with a variety of inventive concepts and narrowed down a few options based on factors such as portability, the convenience of use, cost-effectiveness, and the least skill level required to handle the product. Blending has become more of an hour-by-hour operation as the demand for juices, milkshakes, and other beverages has risen dramatically because of growing consumer demand and the fast rising temperature. Heating, on the other hand, is critical in cold climates where hot beverages are a must. In this situation, a product combines blending and heating, with the added benefits are desirable. The beauty of this product is that it is inexpensive, so nearly anybody can buy it, and it eliminates the need to go to the store to quench thirst by carrying a blender, the raw material to be blended, and the liquid to be heated.

To establish a new design of the portable blender cum heater, various literatures have been studied to understand the scientific requirements to build the product. Sahil et al. [1] introduced to the market a new low-cost insulating flask. On the model of the flask, a test was conducted. The flask was filled with 373 K boiling water and left to cool for a day at 298 K ambient temperature. As a result, one may have a low-cost insulated flask (approximately Rs. 350 INR) that can be used for a variety of purposes such as thermos, icebox, insulated lunch box, and so on. Syahmi et al. [2] designed and developed a compact shaker for powder metallurgical research. Aluminum plate, steel shaft, AC motor, electrical circuit, nuts and bolts, and the main platform make up the majority of their project. With minor errors below 30 rpm in speedometer readings, the machine was usable and capable of delivering the expected results.

Effendi et al. [3] examined the possibility of enhancing sustainable design by reducing the amount of energy required. The DFMA and the Life Cycle Analysis (LCA) were combined in their work. The product was made up of lot of different parts that the customer does not need. The redesign's purpose was to merge some of the old product's features while maintaining its functioning. Farizuan et al. [4] developed and tested a novel material for usage in portable blenders. The approach that aids in data collection for client demands contains data assembly, devising, design idea, design selection, and detailed design. In comparison with the previous version, the redesigned product was more compact. As the number of product components has been reduced, design efficiency has gradually grown, as measured by the design for manufacturing assembly approach.

Mehta et al. [5] used a lumped capacitance approach to model the transient heat transfer of a thermo-flask container, with the premise that there was no temperature differential between the container's wall and the fluid. Because the pressure rise in the thermo-flask was modest, the work done by the venting fluid was ignored. Formulating an overall energy balance on the thin wall fluid container yields the transient temperature response. The temperature can be changed by changing the values of the parameter (hA/mCP) for various electro-magnetic intensity (mV/m) values.

Raju et al. [6] described steady-state thermal analysis on a variety of fins, including rectangular, rounded, triangular, and random rectangular fins. They computed the total heat flux, temperature distribution, and directional heat flux. Rectangular, circular, triangular, and interrupted rectangular fins were included in a three-dimensional model of the heat sink. The fins are all thought to be the same size. The material, the base temperature, and the thermodynamic conductivity are all the same. The simulated system is split down into finite components, and the mesh was generated in the following phase. Total heat flow was greater for interrupted rectangular fins than for other rectangular fins, circular fins, and triangular fins, according to the thermal analysis grades. Schmidt et al. [7] investigated the cutting forces for frozen wood based on tool geometry. Special buzz saw blade designs were developed based on these findings to improve the long-term viability of industrial sawing operations for cutting frozen wood. As a result, positive rake angles greater than 30° are more effective than smaller ones at generating lower cutting forces F_c. In addition, the clearance angle affects the cutting force F_c significantly. When milling frozen spruce at −30 °C, the impact of clearing angle and rake angle on cutting forces F_c was investigated.

Sudhir et al. [8] designed a prototype that can do simultaneous heating and cooling as well as electricity generation for mobile charging. The bottle can be operated manually or automatically. The temperature of the bottle was read using a temperature sensor in automated mode. In their work, authors created a device that transfers heat from one side of the body to the other, causing one side of the body to become extremely cold (ice formation may occur depending on the current given and the room temperature) while the other side of the body becomes extremely hot. This method was handier because it was smaller, lighter, and uses less electricity. This bottle is suitable for use in both summer and winter.

Since the heating and blending functions are generally present in different machines, it is understood that there is a scope for developing a product that combines both operations in a single compact machine, which is portable and occupies less space. This uses Lithium-ion battery rather than a regular battery because it has a faster charging and discharging rate. The objective of this product development is to incorporate both blending and heating functions in a portable container unit. Here, the product can be used for home and commercial applications.

2 Methodology

Based on the need of the customers and the market survey the blender combined with heater unit is selected for designing. Several variants of the mixed units were identified and conceptual sketches were made. To choose the best design among the different variants of a blender or heater, a market survey is conducted and it is given ranking based on the technical parameters. A concept scoring matrix is significantly more useful in determining the best design among them. Based on importance of every variant, each criterion is given a percentage weight as given in Table 1. The design with the highest score, based on the created criteria, is the best option among the four candidate designs.

For the four variants of the design, scores were allotted based on their feasibility and technical compatibility, and the best one was selected (i.e., Concept A: Portable blender cum heater).

Table 1 Concept scoring matrix

S. No.	Criteria	Concept A: portable blender cum heater	Concept B: compact cup heater	Concept C: USB charged heater	Concept D: high rpm blender
1	Strategic alignment	4	4	4	3
2	Market size: how big is the product opportunity?	4	4	4	3
3	Market growth forecast: how much is the market expected to grow by this idea?	5	4	3	3
4	'Wow' factor/value: how much will it positively impact the customer?	5	4	3	3
5	Technical feasibility: how feasible will it be to get the technology?	5	3	3	2
6	Manufacturing feasibility: how feasible will it be to produce at scale?	4	4	3	3
7	Delivery feasibility: how feasible will it be to distribute?	4	3	3	3

Table 2 Material properties

S. No.	Component	Material	Coefficient of thermal expansion (/°C)	Thermal conductivity (W/(mK))
1	Heating coil	Copper	16.7	386
2	Blade	Stainless steel	17.3	25
3	Outer casting	LDPE	10	0.3175

2.1 Material Selection by Ashby Chart

The material selection technique is critical in every product design for each application. This selection process may be explained using application concepts, possible materials, and physical ideology [9]. The properties considered for the selection of portable blender cum heater are listed in Table 2.

By accounting for the above-mentioned properties of the material from the Ashby chart, suitable materials like copper and stainless steel were selected for the fabrication. The above materials were selected from a region where all the candidate materials were identified based on the properties requirements. The region is identified at the intersection of the vertical axis (the coefficient of thermal expansion) and the horizontal axis (thermal conductivity) in the Ashby chart [9].

2.2 CAD Modeling

CREO 6.0 has a collection of computer-aided design (CAD) programs developed by PTC to aid discrete manufacturers with product design. The built environment is made up of multiple apps, and each with its own set of functionalities tailored to a certain user role in product development. Creo 6.0 is a Microsoft Windows-based 3D-CAD parametric feature that has solid modeling, 3D direct modeling, 2D orthographic views, FEA and simulation, graphic design, procedural drawings, and observing and visualization tools [10]. The Creo 6.0 software was used to model the prototype's parts as well as to assemble the prototype as shown in Fig. 1.

2.3 Analysis

In the thermal analysis, transient and structural type analysis is performed to study the behavior of the product. The finite element analysis by ANSYS 16.2 (or structural

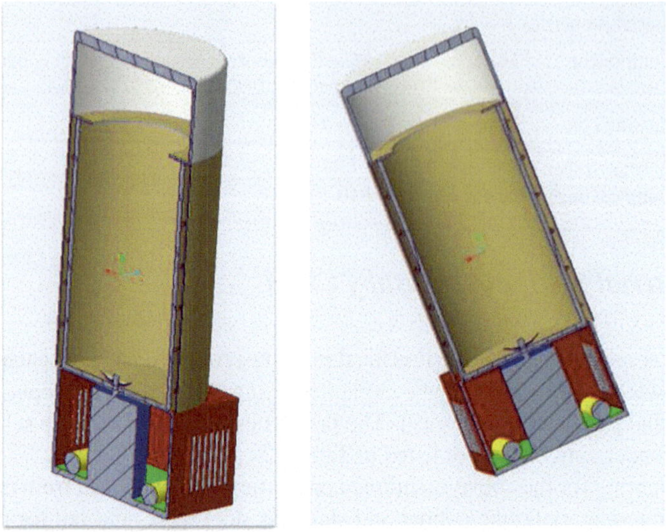

Fig. 1 Rechargeable blender cum heater—CAD Model

analysis) software allows us to digitally test and anticipates the behavior of structures under static and dynamic loading situations, as well as solve difficult structural engineering issues.

Meshing Condition

Coarse type—10 mm.

Steady-State Thermal Analysis

- Boundary conditions:

  Minimum temperature −120 °C.
  Maximum temperature −120 °C.

- Heat applied—On coil.

The thermal equilibrium of a system in which the temperature stays consistent after sometime is assessed by steady-state thermal analysis. As such, steady-state thermal analysis is concerned with deciding analysis, a system's equilibrium state under steady heat loads and ambient factors. Linear steady-state analysis is the most basic type of steady-state analysis in which input characteristics like material qualities are treated as independent variables. Nonlinear steady-state behavior is observed in real systems. When the input parameters are inter-dependent, it is called nonlinear steady-state analysis. Systems having temperature-dependent thermal conductivity, radiation, or natural convection are examples of nonlinear features.

Steady stable-state thermal analysis can be used to design and assess systems in multiple simulations, even at the end of the design phase. For example, when modeling a server's thermal state, which is normally consistent throughout time, steady-state thermal analysis can be employed. When utilized to drive early design and prototype efforts with quick feedback for simple models, steady-state thermal analysis is also advantageous. It aids in the design of typical operating situations and serves as a baseline for transient control analysis and failure analysis.

Transient Thermal Analysis

- Boundary conditions:

  Minimum temperature −25 °C.
  Maximum temperature − 120 °C.

- Heat applied—Coil.

The assessment of how a system reacts to fixed and changing boundary conditions on heat input all through time is known as transient thermal analysis. A measure to achieve a steady-state temperature, just as how long working conditions might be supported before arriving at a threshold temperature can be determined for fixed boundary conditions. Transient analysis can show the system's thermal reaction as the boundary conditions change after sometime. Transient analysis is often used to look at functional condition transitions before the system restores to steady-state. In such cases, it is normal to recognize the steady-state answer for one set of working conditions before playing out a transient analysis with an alternate set. Transient analysis can be utilized to assess control algorithms that utilize time and temperature as contributions to control which yields power and fan speed. The importance of steady-state analysis is in transient thermal analysis of electronics modeling. The most crucial operations are transient, whether testing control algorithms or cooling problems [11]. It illustrates about a system's time-dependent reaction to changing conditions via transient thermal analysis.

Structural Analysis

- Type of force—Shear force.
- Boundary condition: Applied load—25 N and displacements at one end are arrested.

Structural analysis is a thorough examination that ensures that the deformations caused by the load in a structure are acceptable and within permissible limits and that the structure will never fail. It is a way of determining how a structure reacts to varying load conditions. The process of defining a structure's response or behavior under certain loads is known as structural analysis [12]. The strength, stiffness, stability, and vibration of a structure are all factors to consider when analyzing it. A structural engineer must first establish the varying loads, forces, and moments that a particular element and the structure as a whole must sustain before proceeding with the design.

2.4 Power Requirement

2.4.1 Maximum Heat of the Container

Mass m, specific temperature C_p, temperature change T, and Temperature 'Q' are related to the calculation [13]:

$$Q = mC_p\Delta T \tag{1}$$

Certain temperatures depend on the factors and the category of the object.
ΔT—Change in temperature $P(H_2O) = 0.92$ g/ml
$M = 300$ ml * 0.92 g/ml
$M = 276$ g of ice
$\Delta T = 120\ °C - 20.445\ °C = 9.555\ °C$.
Cp (H_2O) = 2111.4 J/kg K, $m = 0.276$ kg

$$Q = 0.276 * 2111.4 * 99.555$$
$$Q = 58,015.3\,J \approx 58\,\text{kJ}$$

2.4.2 Power Used to Heat

P_t—Thermal power (in kWh) can be defined as [13],

$$P_t = (4.2 * L * T)/3600 \tag{2}$$

where
L—Liters and T—Difference in temperature
In our case,
$L = 0.750$ ml, T = Final – Initial = 120 – 23 °C = 97 °C

$$P_t = 4.2 * 0.750 * 97/3600 = 0.0848\,\text{kW}$$

2.5 *Failure Mode and Effects Analysis (FMEA)*

FMEA is a process for detecting possible failures in the design of the product or process. A process' failure modes are the various ways in which it might fail. Effects are the ways in which these errors might result in unused, flaws, or undesirable results for the user. The FMEA approach is used to identify, prioritize, and restrict failure modes [14].

3 Results and Discussions

The material for each component of this prototype was decided using the Ashby chart. The required values of coefficient of thermal expansion and thermal conductivity were plotted in the chart, and the appropriate material was chosen for the development of the product. The innovative product has been prototyped based on the design requirements. Based on the analysis made, it was found that the blending efficiency and temperature induced in the assembled prototype are in control and safe region. The prototype is made with the use of the electrical and mechanical assembly of components and thus it performs the expected work output as required. The initial temperature of the liquid (say, water) was noted down to be 98.5 °F using a clinical thermometer. After a heating time of 2 min, the temperature of the liquid experienced an increase and was noted down to be approximately above 99.5 °F. The blender was also efficient in its operation to blend solids at around 15,000 rpm.

In steady-state thermal analysis, the input temperature is fixed as 120 °C, and the heat is applied to the copper coil. As a result, the total heat flux of the coil is obtained, and it shows the heat distribution throughout the container. From the result, it is observed that the maximum heat flux is 1.638×10^{-9} W/cm^2 (Shown in Fig. 2).

From a minimum temperature of 25 °C to a maximum temperature of 120 °C, the transient thermal analysis (shown in Fig. 3) was carried out. The various temperature zones mentioned inside the Fig. 3 indicate the temperature distribution at different zones of the container. It is noticed that the maximum temperature occurs on the periphery of the container which are in the proximity of the heating coil. The maximum temperature observed is 80 °C.

In order to identify how the system responds to deformation, static structural analysis is performed (shown in Fig. 4). The input load is 25 N is applied on the top of the container and the container responded by withstanding that load with minimum stress concentration. The maximum deformation for the applied load is observed to be 7.57×10^{-9} cm.

Also, the causes for failure of the model designed are analyzed, and their weight factors are listed in Table 3 based on their severity, occurrence, and detection.

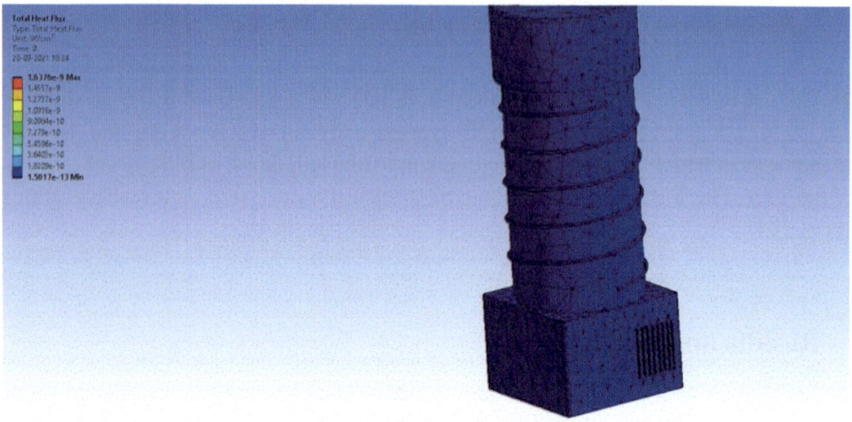

Fig. 2 Steady-state thermal analysis

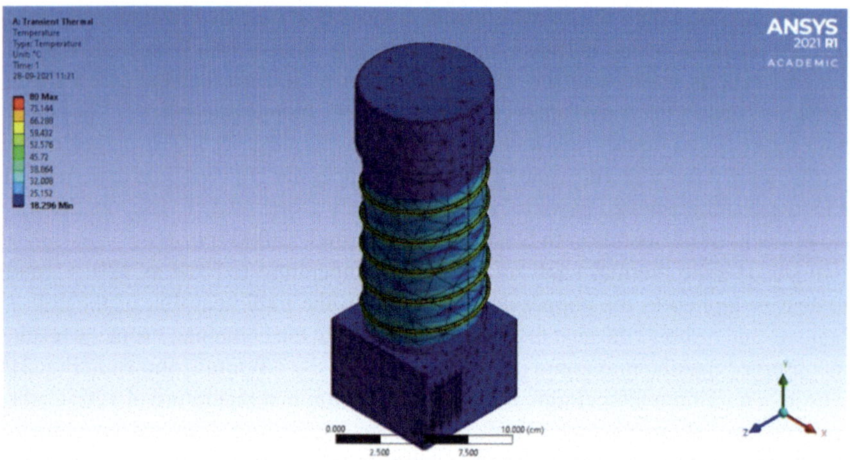

Fig. 3 Transient thermal analysis

Model Calculation

Consider the case of components and function of a motor,

Severity factor = 8, Occurrence rating = 6, Detection rating = 8.

This implies, Risk Priority Number = 8 * 8 * 6 = 384.

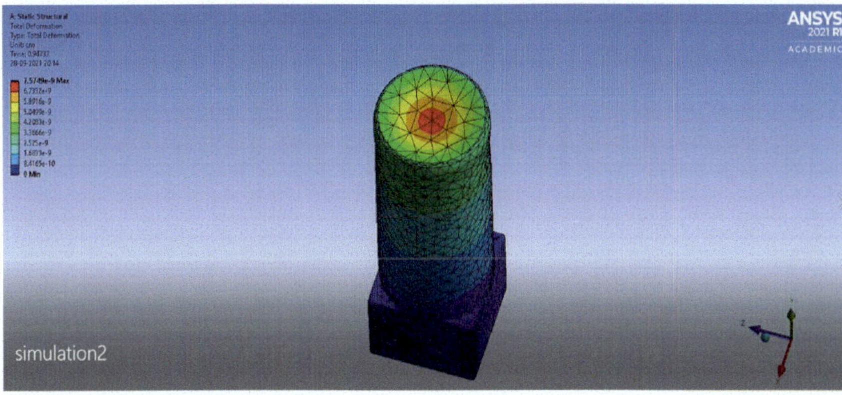

Fig. 4 Static structural analysis

4 Conclusions

Based on these results, the following conclusions can be made. Steady-state thermal analysis is performed to aid the design for typical operating situations and serves as a baseline for transient control analysis and failure analysis. FMEA analysis was formulated based on all possible occurrences of failure and corresponding measures to address them. Using FMEA and thermal analysis methodologies, an environmentally friendly product can be created. This method is effective for implementing sustainability into the industry on a bigger scale while also guaranteeing that the basics and practicalities of a sustainable design are addressed. The new design is less harmful to the environment. In terms of assembly time and material used, it also gradually lowers manufacturing costs to 40%. As the prototype is made with the use of the electrical and mechanical assembly of components, it performs the expected work output as required. Based on the analysis made, it was found that the blending is 20% more efficient, and the temperature induced in the assembled prototype is in control and is in a safe region.

Table 3 Failure modes and effects analysis

Components	Failure mode	Effects of failure	Severity factor	Cause of failure	Occurrence	Detection	Risk priority number	Recommended actions
Motor: powers the blade to spin	Motor overheating	Motor will not able to turn the blades and would render blender useless	8	Motor could overheat if left on too long when trying to blend foods of a harder consistency	6	8	384	Incorporate a switch that would shut the blender off if the torque approaches its failure point
Blade: spins at high speed to both grind up and smooth food	Blade edge dulls	It will take longer to mix and smooth as well as making it more difficult to break	3	Overuse and attempting to break apart dense and hard items	5	2	30	Find an appropriate material, thickness and angle that will be least likely to dull
Pitcher keeps all items to be blended	Cracks	Items being blended will leak outside the blender	4	Dropped from a high altitude	7	3	84	Test different shapes and wall thickness to find a container that is least likely to break
Handle allows user To carry pitcher	Breaks off	No handle to carry Blender	3	Dropped from a high altitude	7	2	42	Test other ways to fit parts together to find most effective solution

(continued)

Table 3 (continued)

Components	Failure mode	Effects of failure	Severity factor	Cause of failure	Occurrence	Detection	Risk priority number	Recommended actions
Motor coupling connects motor shaft to blade that will be Turning	Uncouples	Blade will not spin even when motor does	8	Motor spins in incorrect direction	2	8	128	Find a controller or switch that could deactivate motor if it were to spin in a wrong direction
Wiring allows current to flow to motor	Wire becomes disconnect ed	Current will be unable to power Motor or similar function	8	Wire tangled with moving parts	2	3	48	Test different positions for wire to be fastened to reduce tangling
cap unscrew cap	Cannot remove cap from bottle	Cannot drink the contents	8	Cap on Bottle too Tight with damaged threads	4	6	192	Ensure screw on cap correctly after each use
Copper winding allows the container to get heated	Winding failure/shortage	Uneven heating	8	Wear	3	10	240	Ensure proper winding inside the casting

References

1. Padha ES, Anthal MT, Sharma MR, Singh MG, Wani MM (2017) Enhancement in thermos flask a review. Global Res Dev J Eng
2. Syahmi M, Anwar MA, Ashraf MA, Fariz M (2016) Design and development of mini shaker mechanical system design. UMP Pekan Malaysia
3. Effendi MSM, Shayfull Z, Radhwan H, Ahmad SAS, Fai KK, Nain LT, Zulkeply SZ (2021) The improvement of a portable blender design using DFMA and environmental impact assessment. In: AIP conference proceedings, vol. 2347(1). AIP Publishing LLC, p 020255
4. Farizuan RM, Irfan AR, Radhwan H, Ahmad SAS, Fai KK, Jing KX, Zulkeply SZ (2021) Study of material selection and design improvement for a portable blender by using finite element analysis. In: AIP conference proceedings, vol 2339(1). AIP Publishing LLC, , p 020216
5. Mehta RC, Sch J Eng Tech (2018) Transient thermal analysis of a thermo-flask. A Unit of Scholars Academic and Scientific Society, India
6. Raju VN, KL SKPN, Mallikarjun K, Krishna G (2017) Steady-state thermal analysis of heat sink with fins of different geometry. Int J Mech Eng Technol 8(5):196–206
7. Divekar SN, Patil VN, Puri SB, Pawar RR, Shinde SA, Ujagare MR (2020) Portable water bottle heating and cooling system
8. Schmidt C, Westermann HH, Steinhilper R (2019) An investigation of buzz saw blade cutting forces depending on tool geometry for cutting frozen wood. Procedia Manuf 33:778–785
9. Shercliff HR, Ashby MF (2016) Reference module in materials science and materials engineering
10. Groover M, Zimmers E (2005) CAD-CAM: computer aided design and manufacturing
11. Choudhary RB (2020) Introduction to ANSYS 16.0 2nd edn.
12. Boni L, Quarta AA, Mengali G (2019) Thermal-structural analysis of a square solar sail. Acta Astronaut 154:268–277
13. Chien LH, Chen DC, Liu YJ, Yan WM, Ghalambaz M (2021) Heat and mass transfer of evaporative cooler with elliptic tube heat exchangers-an experimental study. Int Commun Heat Mass Transf 127:105502
14. Stamatis DH (2003) Failure mode and effect analysis, 2nd edn.

Design and Performance Optimization of Electric Resistance Furnace Using ANSYS Simulation Software

Addisu Boshe, Endalkachew Mosisa Gutema, and Mahesh Gopal

Abstract The use of renewable energy in resistance electric furnaces avoids the need for gas flow minimizes energy and impurity related to chemical contamination of the metal. In an electric resistance furnace, the electrical power translates thermal energy using convection and eradiation law. In resistance furnaces, metals' heating operations and melting are below 1200°C. The maximum temperature is limited to electric heating elements and inefficient furnace design due to the electrical heating elements' frequent burnout due to overheating and surface oxidation, resulting in more significant downtime. This research addresses these challenges and other performance-related issues through optimization techniques. ANSYS simulation software was used for optimization purposes. Performed tests under two different design alternatives of actual conditions for melting three-light metals: lead, aluminum, and brass, as sample materials have low, intermediate, and high-temperature behaviors. Simulation results were compared, and validation was performed. The results show that it is possible to use electrical resistance furnaces for melting a greater quantity of metal and achieve rapid heating rates than the currently available designs. The thermal efficiency ranges from 64, 88, and 89%, and the volumetric efficiency of 23, 45, and 67%. The achieved results are best suitable for small, medium, and large furnaces.

Keywords Resistance electric furnace · ANSYS simulation software · Lead · Aluminum · Brass

1 Introduction

Fuel-fired furnaces use the combustion of chemical fuels such as coal, oil, or gas to generate thermal energy. The forced flow of the fumes is transferred around the melting chamber. Currently, electrical resistance melting furnaces are limited to minimal (lift-out crucible type) furnaces, which encounter frequent wire burnouts

A. Boshe · E. M. Gutema (✉) · M. Gopal
Department of Mechanical Engineering, College of Engineering and Technology, Wollega University, Post Box No 395, Nekemte, Ethiopia
e-mail: endalkem5@gmail.com

E. Natarajan et al. (eds.), *Materials, Design and Manufacturing for Sustainable Environment*, Lecture Notes in Mechanical Engineering,
https://doi.org/10.1007/978-981-19-3053-9_24

due to oxidation by air. Another problem restricting resistance furnaces only to small capacity devices is inefficient design. Lift-out crucible design necessitates a sufficient air gap between the heating wire and the crucible to avoid electric shocks and short circuits, which increases the overall volume of the furnace body. The furnace body absorbed the heat energy rather than the molten metal, wastage of power, and lesser. The experiment used an electric resistance furnace to reuse heavy metals as a resource by incorporating a smelting process for zinc and lead in the heavy metal. The authors [1] suggested that the furnace provides high performance in separating heavy metals. The researchers [2] conducted the physical modeling experiment to study the behavior of first-order linear, bilinear, and direction-dependent models by using LabVIEW 7 software. The result emphasizes that the first-order system is the dominating factor. Mg–Al alloys were melted using an electric resistance furnace under SF6 and CO_2 gases. A thermocouple was used to monitor the temperature. The CaO addition takes place significantly at ignition results AES depth profile. Without the addition of CaO, the formation of a thick and porous MgO oxide layer on the surface [3]. The model was developed to identify the electrical-related parameter based on timing in electrical resistance furnace to produce silicon carbide [4]. The researchers [5] carried out backtracking optimization techniques in an electric resistance furnace. The graph theory algorithm was used to achieve process output results. The result emphasizes that implementing a newly designed furnace reduces the losses essentially. The researcher experimented with controlling the temperature using the fuzzy proportional–integral–derivative controller. To link object and object control are controlled using MATLAB software. The results show that fuzzy PID controller performance is better than conventional PID controller to control temperature [6]. The authors [7] developed a model to calculate the temperature distribution inside the electric resistance furnace. The simulation model is created Simulink in MATLAB to compare the dynamic characteristics of irregularity influences in distribution. Experimental work was conducted on an electric resistance furnace at a maximum temperature of 1800 °C to measure the vacuum inside the furnace. To calculate the thermal conductivity of the thermal insulation, the authors [8] created the zero-dimensional and one-dimensional model. The researcher [9] developed a fuzzy controller to measure the electric resistance chamber furnace temperature using MATLAB. A mathematical model was developed to calculate the life hours of the resistance furnaces under temperature and pressure conditions [10]. An algorithm was designed to degrade the vacuum furnaces of pure refractory metals such as tungsten or molybdenum. Using CFD software FLUENT, the authors proposed a method to analyze temperature simulation of the super alloy in a resistance furnace. The developed model was satisfactory [11]. The experimental investigation to study the furnace's dimensions and theoretical analysis were performed to investigate the loss of temperature in different insulating materials [12]. The temperature analysis was conducted in the furnace to study the processes of thermal effects to measure the heat flow and influencing parameters in the design of the furnace [13]. The author [14] calculated the transfer function based on the modeling process in electric resistance furnace using simulation and suggested that machine learning is a very effective method for analyzing the system's behavior.

2 Design Procedure and Simulation

ANSYS simulation software developed a virtual model under the transient thermal model conditions. Effects of variations in selected dimensions such as diameter, thickness, and surface area were optimized at rapid melting using minimum power consumption for different volume capacities. The design consists of two cylindrical vessels: shell crucible and external casing. The resistor and refractory electrical insulator are fixed at the bottom of the inner vessel, and the refractory material was used to fill the inner and outer containers gap. The finite element analysis was performed using ANSYS mathematical modeling and simulation as per the design features, as shown in Fig. 1, based on geometric dimensions and thermal conditions. The selected parameters are temperature, chamber volume, crucible thickness, and insulation layer thickness based on material, volumetric capacity, and power supply.

To maximize desirable values and minimize undesirable values of output parameters for the given iteration time of 1800s (30 min) are the objective to conduct a simulation. UL safety standard recommends 10% factor of safety for small systems, 20% for intermediate applications, and 20–35% for large systems. In other words, the global maximum temperature should be at 10% below the melting point temperature of the heating element for small, 20% for intermediate, and 20–35% for large heating applications. The recommended temperature to calculate the global maximum temperature: maximum pouring temperature of brass was 1040 °C, and melting point temperature of Ni—chrome heating element was 1450 °C as shown as a benchmark in Eq. (1).

$$\mathrm{SF\%} = \frac{T_{\text{melt(heating element)}} - T_{\text{global max}}}{T_{\text{melt(heating element)}}} \times 100$$

$$T_{\text{globalmax}} = \left(1 - \left(\frac{\mathrm{SF\%}}{100}\right)\right) \times T_{\text{melt(heatingelement)}} \tag{1}$$

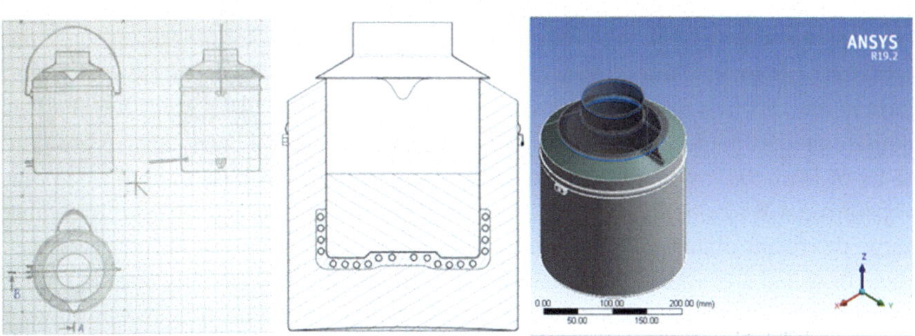

Fig. 1 Design of furnace

Table 1 Criteria for analysis on ANSYS Workbench

Description	Objective	Lower bound (°C)	Upper bound (°C)
Final average melt temperature	Maximize	940	105
Global maximum temperature	Minimize	–	1200
Final external surface temperature	Minimize	–	70

The global maximum temperature should be below 1305 °C for small, 1160 °C for intermediate, and 942.5–1160 °C for large heating applications. Criteria for analysis on ANSYS workbench are shown in Table 1.

Bond strength is the resisting force that exerts during the separation of concrete from the reinforcing bars. Figure 1 gives the flow of bond components in bond degradation [2].

3 Simulation

3.1 Mesh Convergence (Mesh Sensitivity) and Number of Iteration and Duration

A universal mesh element size and a control body size were selected for the shell crucible considering its relative thinness. Mesh element size, mesh body size, and final average melt temperature were parameterized to achieve an optimal precision with an acceptable error margin and minimum run time. The test was performed by varying the mesh element size and body size from 2 to 30 mm and observing changes in the output parameter (final average melt temperature). Progressively reducing the mesh element size indicates the error margin as an asymptote to the perfect result. The acceptable error margin for simulation is less than five percent variation compared to the best solution (i.e., less than 5% deviation from the most acceptable mesh solution). Transient thermal analysis was used to calculate the furnace's temperature and total heat flux over time. Time steps are temporal equivalents of mesh elements, i.e., subdivisions through time iterations are updated. The entire time input for analysis was estimated by manual calculation of the ideal heat transfer. That means, to simulate one kg of lead, the total input time is set to 240 s (4 min), because the melting point temperature was attained within this time duration. The iteration time is set to a value double the result obtained by manual calculation considering delays by resistances to heat flow and absorption by furnace body. If the result shows that the melting point temperature of the metal is not achieved within the inputted time, then the simulation is repeated with more extended time intervals.

$$E = \frac{\text{Result obtained by finest mesh size} - \text{Result by desired mesh size}}{\text{The result obtained by finest mesh size}} \times 100 \tag{2}$$

$$t = \frac{m\left[C_{p,s}(T_m - T_i) + H_f + C_{p,l}(T_P - T_m)\right]}{Q}$$

$$= \frac{1\text{kg} \times \left[\frac{129\text{J}}{\text{kg}^\circ\text{C}} \times (327.5^\circ\text{C} - 25^\circ\text{C}) + \frac{23000\text{J}}{\text{kg}} + \frac{140\text{J}}{\text{kg}^\circ\text{C}(400^\circ\text{C} - 327.5^\circ\text{C})}\right]}{500\text{J}/\text{S}}$$

$$t = 144\text{s} \tag{3}$$

Properly adequate subdivisions were made to obtain good enough results to an acceptable error margin (E < 5%).

3.2 Consistency Test to Minimize Redundant Computations

The consistency test validates the correctness of the result comparing the partial (revolved cross-section body) solution with the complete 3D model by checking the degree of variations on essential variables. This approach reduces redundancies by omitting bodies based on similarities and symmetric objects. Revolved cross-section that yields result with a tolerable error margin (E < 5%) was selected for the parametric optimization. By introducing a custom coordinate system and setting the slicing surfaces as boundary planes, it is possible to obtain projected solutions for 3600 bodies from the partial solution.

4 Experimentation

The experimental test was performed on electric resistance melting furnaces. Two resistance melting furnaces were developed to a power range of 5 kW to melt light metals such as zinc and aluminum and heavier metals such as brass and copper. The materials include stainless steel crucible, brick, gypsum, magnesia, wire heater, tube heater, and wirings. The thermostats and temperature regulators are fixed to prevent wire burnout. Shell-type crucible design consists of an inner cylindrical container, the shell crucible, and a large external container. The furnace is 0.33 mm wall thickness, and the heater tube is bent spirally, providing V-slot in the inner bucket. Five tube heaters, each having around 54Ω of resistance, make up 12Ω in parallel, generating about 5 kW heat when connected to the 240 V terminal. The gap at the top rim was sealed by gypsum powder (aka plaster of Paris), later sealed by refractory mortar (cement plasticizer) to avoid gypsum powder inclusion. Figure 2 shows the construction of electrical resistance furnaces to perform physical tests. Physical testing was performed by melting lead, aluminum, and brass.

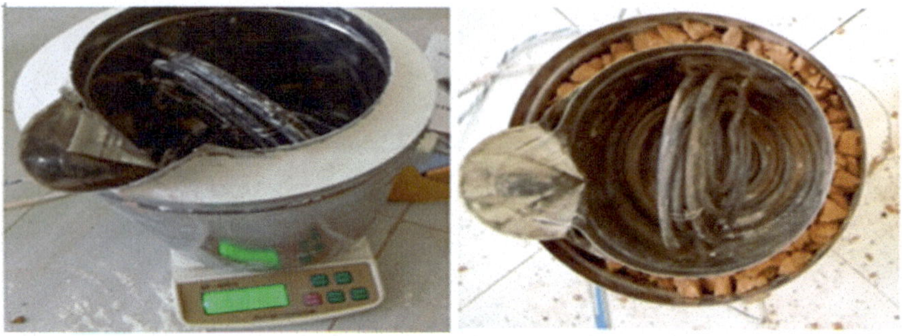

Fig. 2 The outer and inner casing of electrical resistance furnace

5 Results and Discussion

When the mesh element size is varied from the finest (2 mm) mesh to the coarsest (30 mm) mesh. The final result (Average Melt Temperature) initially shows slight deviations, as shown in Figs. 3 and 4. Still, it becomes more visible as it increases further. However, reducing the mesh below a specific critical size will not significantly improve the result; it creates unnecessary load on the computer and requires significant running time. Therefore, the necessary mesh element size was determined by observing the converging behavior of the result on the parametric chart aiming for deviation, not more than 5%. Two time division techniques were utilized; program controlled (default) iterations and manually input iterations. The results were compared, and the manual input with even divisions produced the most convenient data for plotting graphs with sufficient precision. A total time of 1800s was provided with iteration substeps divided from 120 to 2400 for running simulation on brass 60/40. Total duration of 1800s evenly subdivided into 360 iteration steps produces results with only 0.02% error while maintaining processing speed of the computer (Figs. 5 and 6).

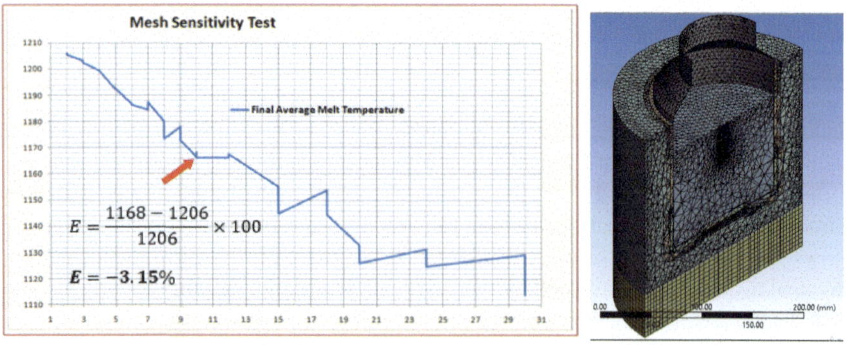

Fig. 3 Mesh sensitivity test result ANSYS parametric plot

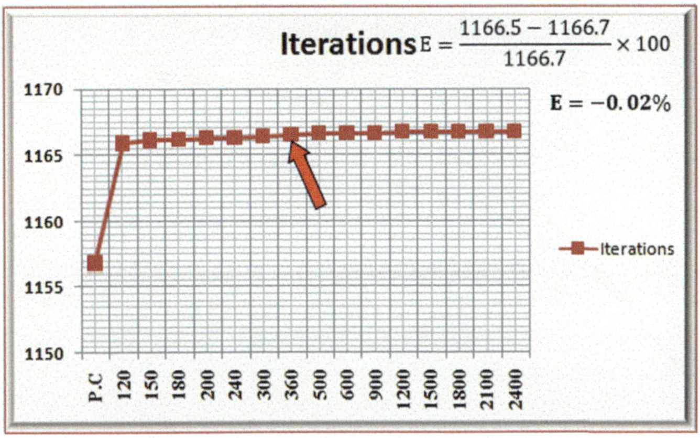

Fig. 4 Effect of iteration substeps on final results

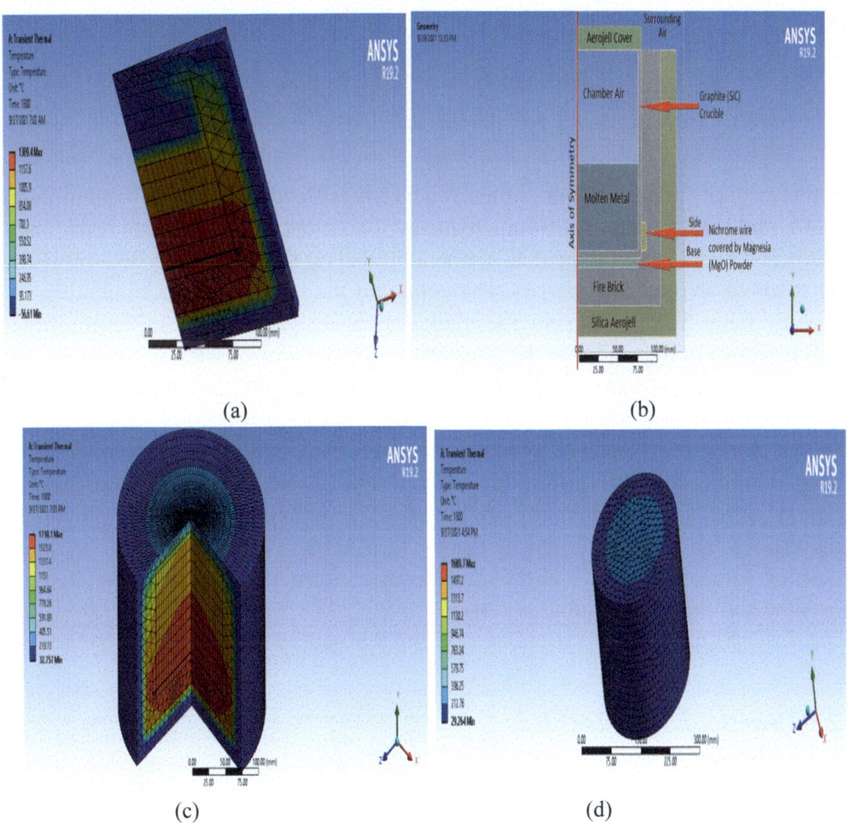

Fig. 5 Reducing redundant solutions **a** 3° partial slice solution (3 min 23 s) **b** and **c** 3° with boundary conditions (beta) projection (42 min 27 s) **d** 360° complete solutions (7 h 19 min)

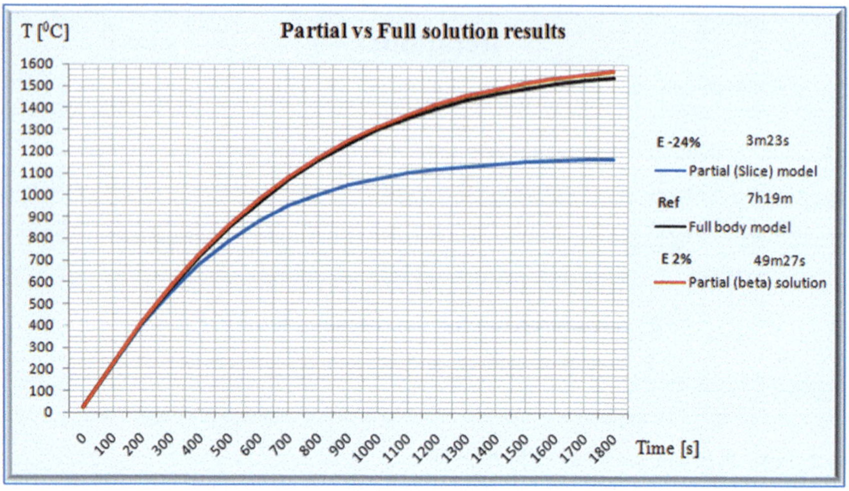

Fig. 6 Graph of partial versus complete solution

5.1 Simulation Results for Validation by Physical Tests

The three sample materials, low antimony acid battery lead (0.3–0.9% Sb), aluminum (A380), and brass 6040, were simulated using two models (gypsum and brick insulation). The results were plotted as temperature versus time graphs to determine the time required to melt the charges. The details of the test model for simulation are shown in Table 2. Results from 18 simulations were summarized in the multiplot graph is shown in Fig. 7. From the temperature versus time charts, time values were sorted into the mass versus time table to relate the amount of molten metal to the time required to fuse completely. The final pouring temperature is assumed to be 1000 °C above the melting point temperature for each case. As observed from the graphs, doubling the melt quantity does not result in twice the melting time; instead, the time is less, and the same amount of metal was to be melted in several batches. This observation was confirmed by the physical tests, which consistently increase efficiency when the quantity of the melt increases from 1 to 2 kg and 5 kg. Table 3 shows the summary of simulation tests.

5.2 Practical Test Results

The two furnaces are used for melting A380 window frame metals and low antimony lead in measured quantities (1, 2, and 5 kg batches), as shown in Table 3. The resistance of heating elements was checked using a digital multi-meter before each melting cycle (it read 12Ω ± 2%). The current was monitored using the digital clamp-on meter (it read 18A ± 10%).

Table 2 Details for simulation of the test model

Object	Material	Variable	Dimensions
Crucible	Stainless steel	Chamber radius	70 mm
		Thickness	0.33 mm
		Height	160 mm
Melt	Aluminum	Height	25 mm (1 kg)
			50 mm (2 kg)
			125 mm (5 kg)
	Lead	Height	6.2 mm (1 kg)
			12.5 mm (2 kg)
			31.5 mm (5 kg)
	Brass	Height	8 mm (1 kg)
			16 mm (2 kg)
			40 mm (5 kg)
Heating element	Nichrome resistor covered in MgO powder	Volume	108,941 mm^3
		Watt density	0.036717 W/mm^3
Refractory	Gypsum	Thickness	25 mm
	Brick	Thickness	25 mm
Cover	Gypsum	Thickness	10 mm
Container	Stainless steel	Diameter	190 mm
		Thickness	0.33 mm
		Height	180 mm

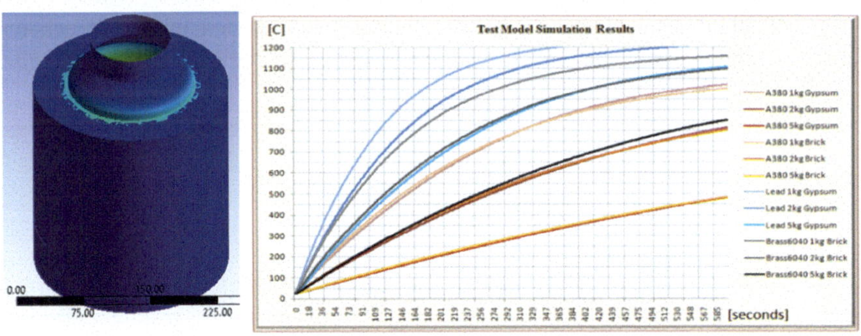

Fig. 7 Customized model for validation test [maximum power < 5 kW]; temperature versus time multiplot graph of 18 simulations

Table 3 Summary of the simulation: melting time as a function of the mass of molten metal

Metal type	Melt quantity [kg]	Pouring time [s]	
		Gypsum wall	Brick wall
A380	1	290	287
	2	520	525
	5	> 600	> 600
Lead	1	50	48
	2	62	59
	5	108	95
Brass 6040	1	325	317
	2	480	467
	5	> 600	> 600

Table 4 Comparison of average physical test results t_2 to corresponding simulation t_1

Details	A380		Lead		
	1 kg	2 kg	1 kg	2 kg	5 kg
Gypsum wall (t_1/t_2)	0.88	0.92	0.51	0.46	0.62
Brick wall (t_1/t_2)	0.85	0.93	0.46	0.45	0.53

$$Q_{\text{Supply}} = \text{Electrical Power} = I^2 R \tag{4}$$

These results conclude the validation tasks obtaining 45% minimum and 93% full closure of actual physical tests subjected to human error and manufacture precision Table 4. Comparison of average physical test results t_2 to corresponding simulation $t_{1.}$

5.3 *Parametric Optimization Results*

Table 5 shows the materials and their attributes for candidate selection. Quick (partial solution) simulations were performed on each material type, and the results confirmed the validity of manual selection. Brass 60/40 is the set of highest temperature limits to determine the upper limit of melting capacity for light metals. The initial optimization was performed on Brass 60/40, and the other metals were tested based on the optimized design. Composite furnace wall was designed using fire brick wall covered by silica aerogel lining to take advantage of brick's high melting point temperature and the aerogel's excellent thermal and physical properties at the exterior (less extreme) zones. Magnesia powder was used as electrical insulation. However, it was modeled as an internal heat generation body since its weight and thermal effects dominate the heating element. Graphite is the best candidate for crucible material because it has outstanding properties that produce desirable results. Direct optimization on the resistance furnace was performed for three capacity ranges (10, 100, and 1000 kW). The 390 samples (design points) for each optimization process were simulated using four input parameters (thirteen variable) parameterized to optimize three output parameters. Optimization for small-scale (range 1) furnaces with a maximum of 10 kW, medium scale (range 2) 100 kW, and large scale (range 3) 1000 kW power supply was performed by limiting the range input variables within table size dimensions (portable size). Table 6 shows the parametric relation and the range of input variables. A total of 390 sample design points were generated automatically by varying the input variables within each specified range. All the 390 design samples were solved in one simulation. Figure 8 Optimization range 1 multi plot graph of candidate design points. Figure 9 Optimization range 2 multi plot graphs of candidate design points. Figure 10 optimization range 3 multi plot graphs of candidate design points. Range 1: With the outlined dimensions, the volume of the composite side and bottom heater models make up 185,000 mm^3. The recommended operating watt density of 0.062 W/mm^3 of tube heater with Nichrome wire core, total power of 11,500 W (or 11.5 kW) was found. **Range 2**: Volume of the composite side and bottom heater models makeup 749,290 mm^3. The recommended operating watt density of 0.062 W/mm3 of tube heater with Nichrome wire core, power of 46,456 W (46.5 kW) was found. **Range 3**: The volume of the composite side and bottom heater models is 27,317,300 mm^3. Taking the recommended operating watt density of 0.062 W/mm3 of tube heater with Nichrome wire core, total power of 1,693,673 W (1694 kW) was found (Figs. 11 and 12).

Table 5 Material library

Table 5. Material library

Details	Material	Density [kg/m^3]	Melting point [^{0}C]	Conductivity [W/m^{0}C]	Heat capacity [J/kg^{0}C]	Latent Heat [kJ/kg]
Melt	Brass 60/40	8800	940	96	380$^{(s)}$/490$^{(l)}$	168
	Aluminum (A380)	2770	660	237	921$^{(s)}$/1180$^{(l)}$	38.8
	Lead (low Sb)	11290	327.5	33.8	129$^{(s)}$/140$^{(l)}$	23
Refractory	**Aerogel (Silica)**	100	1200	0.017	730$^{(s)}$	N/A
	Brick	1922	2500	1.09	800$^{(s)}$	N/A
	Gypsum	2960	1460	0.5	1000$^{(s)}$	N/A
	Magnesia	3580	2852	30	800$^{(s)}$	N/A
	Sand	1220	3000	2.05	730$^{(s)}$	N/A
Crucible	Graphite (SiC)	1800	3500	140	710	N/A
	Stainless steel	8000	1510	25	466	N/A
Heaters	Ni-chrome	8415	1450	11.3	450	N/A
	SiC graphite	1800	3500	140	710	N/A

Table 6 Parametric relationships (input range of variables)

Details		Variables	Lower bound			Upper bound			Optimum		
			Range 1	Range 2	Range 3	Range 1	Range 2	Range 3	Range 1	Range 2	Range 3
Input variables	Crucible dimension [mm]	Side thickness	0.3	0.3	0.5	10	20	20	8.28	11.37	16.3
		Base thickness	0.3	0.3	0.5	10	20	20	9.84	10.11	12.0
		Chamber radius	30	100	600	100	200	800	77.6	147.9	715
		Height	90	200	980	150	300	1100	130.4	249.8	1055
	Melt Quantity [h, mm]		10	100	100	90	980	980	56.5	189.5	447
	Heating element	Side length	6	30	100	50	180	800	10.8	44.7	690
		Base thickness	5.4	5.4	5.4	12	18	12	6.3	6.29	5.9
		Side thickness	5.4	5.4	5.4	12	18	12	10.7	6.93	5.6
	Refractory thickness [mm]	Brick base	10	10	10	70	70	70	28.9	63.71	47.4
		Brick wall	10	10	10	70	70	70	60.2	52.22	17.7
		Aerogel base	2	2	5	30	50	50	22.1	16.28	14.0
		Aerogel wall	2	2	5	30	50	50	14.3	35.00	29.5
		Cover	5	3	5	30	30	30	26.5	20.06	12.3

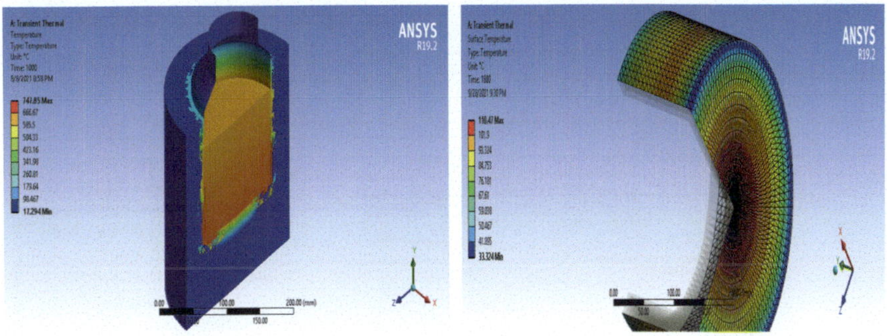

Fig. 8 Temperature distribution across and over an external surface

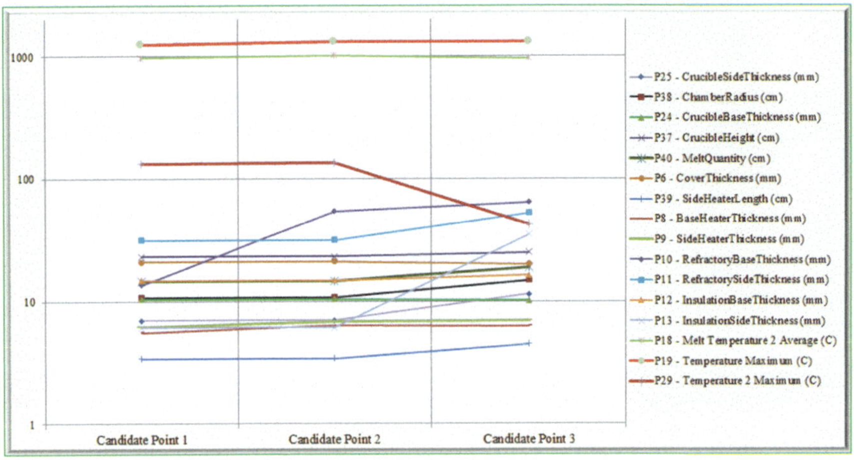

Fig. 9 Optimization range 1 multiplot graph of candidate design points

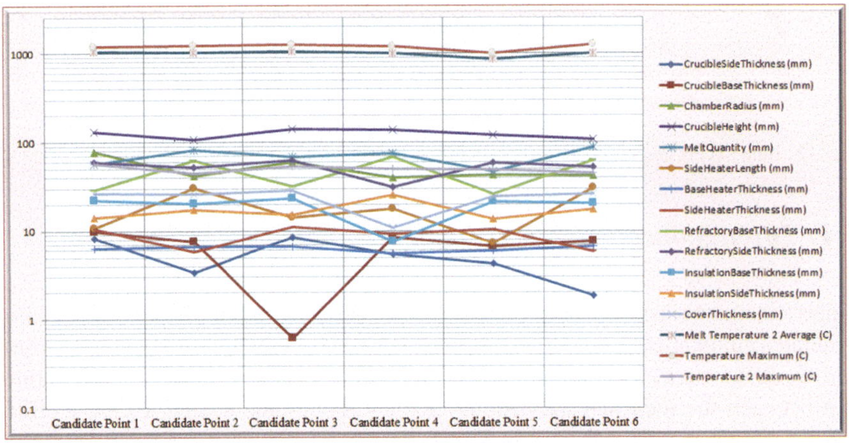

Fig. 10 Optimization range 2 multiplot graphs of candidate design points

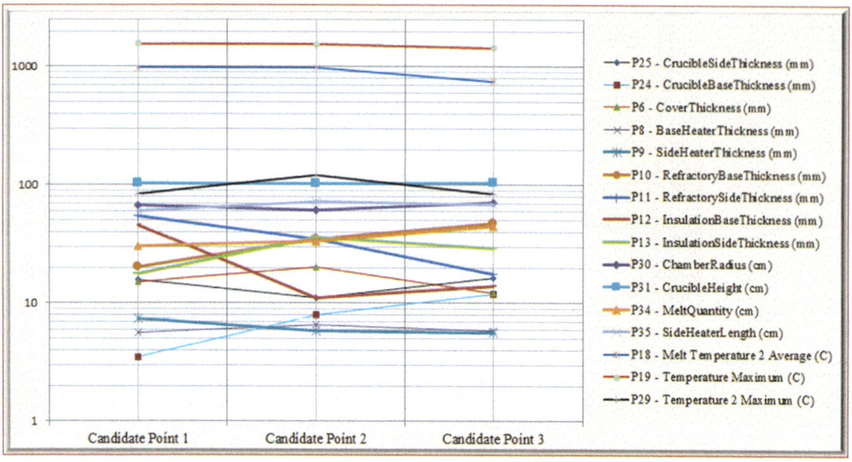

Fig. 11 Optimization range 3 multiplot graphs of candidate design points

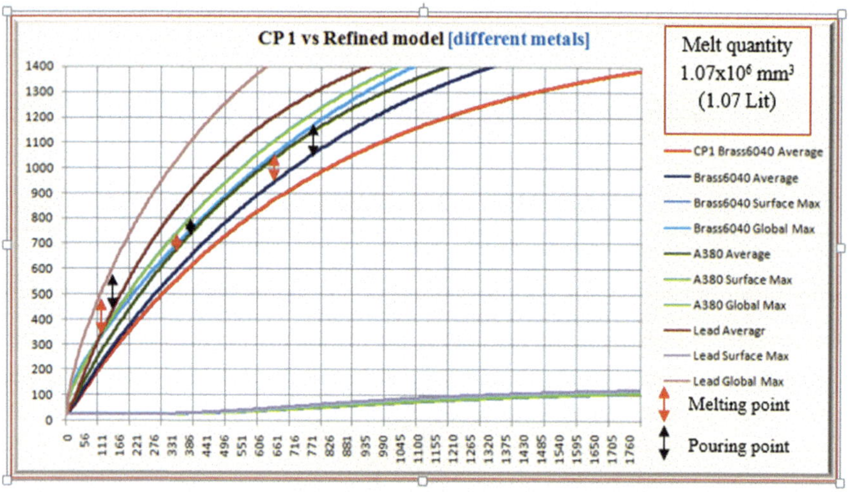

Fig. 12 Beta axis symmetric solution graph of CP 1 versus refined model (different metals)

5.4 *Discussions*

Final refinement was performed on the optimized model by modifying the dimensions generated automatically by the software to match horizontal and vertical dimensions and facilitate space, use, and handling. This refinement minimizes the temperature gap between the melt and the heating element. Nichrome heating element (10% safety factor for small-scale stems) was chosen as the minimum design criteria. The maximum and average melt temperature gap must be minimized to ensure the safety limit up to the pouring stage. The refined design still shows a better combination of results than DP1. The table below summarizes solutions based on the progressive model for three cycles of melting brass 6040, A380, and lead with the half-filled chamber (1 L melt). Based on the dimensions specified in the above table, the composite (side and bottom) heating elements make up a volume of 170,808 mm^3. The recommended operating watt density of 0.062 W/mm^3 of tube heater with Nichrome wire core, total power of 10,590 W (or 10.6 kW) was found. The maximum temperature does not exceed 1305°C. The above graphs observed that it went side-by-side with the average melt temperature by an excess of around 100°C up until the pouring stage. The furnace has a maximum melting capacity of 2.3 L, and it occupies 9.7 L of space that was fitted in a box with 204 mm × 204 mm × 120 mm overall dimensions. About a quarter (23.4%) of the total volume can be utilized for melting at maximum capacity. Essential details about the refined small-scale furnace are presented in the table (Tables 7 and 8).

Table 7 As-symmetric dimensions to the nearest digits

Variables	Range 1	Range 2	Range 3
Crucible side thickness (mm)	4	10	12
Crucible base thickness (mm)	5	10	12
Chamber radius (mm)	78	150	700
Crucible height (mm)	130	400	120
Melt quantity (h, mm)	56.56	189.5	447
Side heater length (mm)	18	120	300
Base heater thickness (mm)	6	6	6
Side heater thickness (mm)	6	6	6
Brick lining (mm)	25	20	20
Aerogel lining (mm)	15	15	15
Cover thickness (mm)	15	15	15
Final average melt T°	**1548.6**	**1269.7**	**656.1**
Global maximum T°	**1685.8**	**1634.8**	**1473.8**
Outer surface final T°	**110.47**	**111.7**	**120**

Table 8 Important volumetric details - Maximum melting capacity of 2.3 Litres (ANSYS)

Metal type	Volume (mm^3)					Mass (kg)		
	Total region	Surrounding air	Maximum melt	Heating element	Total furnace	Total	Melt	Net Furnace
Brass6040	12,975,000	3,278,500	**2,268,400 (2.3 L)**	170,808	**9,696,500 (9.7 L)**	27.255	**19.962**	**7.3**
A380						13.576	**6.2834**	
Lead						32.903	**25.61**	

The refined design was selected to lower the maximum power supply to 100 kW and reduce the global maximum temperature below 1160°C. However, such temperature was exceeded before achieving pouring point temperature when melting Brass 6040. Since the time it takes to melt large quantities is long enough to cause overheating of the heating element, it is necessary to incorporate temperature control techniques in the model. The graph below summarizes solutions based on the refined model for three cycles of melting brass 6040, A380, and lead with the half-filled chamber (13.4 L melt) with a temperature regulation device to prevent overheating. The composite (side and bottom) heating elements make up a volume of 1,161,510 mm^3 based on the dimensions specified in the above table. Taking the recommended operating watt density of 0.062 W/mm^3 of tube heater with Nichrome wire core, total power of 72,014 W (or 72 kW) was found. The global maximum temperature can be regulated and maintained by temperature control circuits below 1160°C. In this case, the initial heating rate (°C/min) was used to describe the furnace's performance. The maximum safe heating rate was obtained by running full power (72 kW) on the brass 6040 sample for 800 s (until the upper-temperature limit is achieved). The refined furnace has a maximum melting capacity of 24.7 L, and it occupies 55.1 L space that can be fitted in a 42 cm × 42 cm × 45 mm box. Nearly half (45%) of the total volume can be utilized for melting at maximum capacity. Essential details about the refined small-scale furnace are presented in Table 9. The refined design was selected to lower the maximum power supply to 1000 kW and reduce the global maximum temperature below 1160ºC. However, temperature was exceeded before the pouring point temperature for melting such a large quantity of metal (shown in the above graph). Since the time it takes to melt large amounts is long enough to cause overheating of the heating element, it is necessary to incorporate temperature control techniques in the model. The graph below summarizes solutions based on the refined model for three cycles of melting brass 6040, A380, and lead with the half-filled chamber (688 L melt) with a temperature regulation device to prevent overheating. The composite (side and bottom) heating elements make up 17,322,800 mm^3 based on the dimensions specified in the above table. Taking the recommended operating watt density of 0.062 W/mm^3 of tube heater with Nichrome wire core, total power of 1,080,336 W (or 1080 kW) was found. Temperature control circuits below 1160ºC maintain the global maximum temperature. In this case, the initial heating rate (ºC/min) was used to describe the furnace's performance. The maximum safe heating rate was obtained by running full power (1000 kW) on the brass 6040 sample for 1000 s (until the upper-temperature limit). The refined furnace has a maximum melting capacity of 1500 L, and it occupies 2320 L space that can be fitted in a 1.6 m × 1.6 m × 1.25 mm box. Two third (67%) of the total volume can be utilized for melting at maximum capacity. Essential details about the refined small-scale furnace are presented in the Table 10 (Figs. 13 and 14).

Table 9 Important volumetric details - Maximum melting capacity of 24.7 Litres (ANSYS)

Metal type	Volume (mm^3)					Mass (kg)		
	Total region	Surrounding air	Maximum melt	Heating element	Total furnace	Total	Melt	Net furnace
Brass 6040	64,164,000	9,082,000	**24,740,000 (24.7 L)**	1,161,510	**55,082,000 (55.1 L)**	251.1	**217.7**	**33.4**
A380						101.9	**68.5**	
Lead						312.7	**279.3**	

Table 10 Important volumetric details (obtained from ANSYS Mechanical)

Metal type	Volume (mm^3)					Mass (kg)		
	Total region	Surrounding air	Maximum melt	Heating element	Total furnace	Total	Melt	Net furnace
Brass 6040	2.32e +9	98,325,000	**1.5086e + 9 (1500 L)**	17,322,800	**2.2205e + 9 (2220 L)**	13,763	**13,276**	**489**
A380						4665	**4179**	
Lead						1752	**1703**	

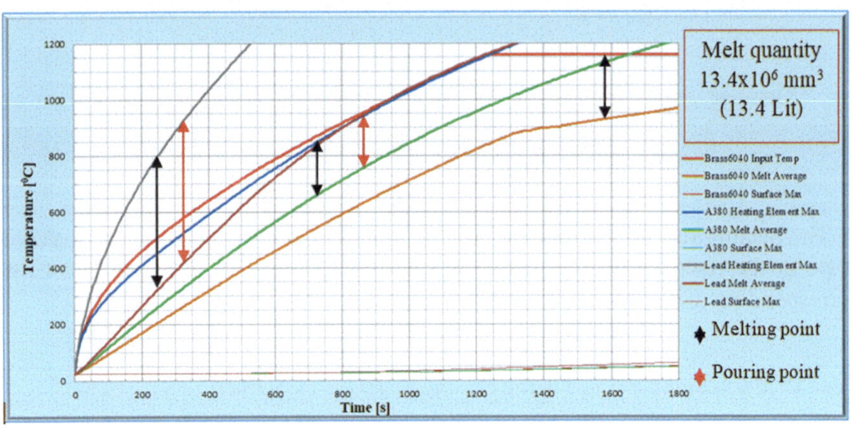

Fig. 13 Beta axis symmetric solution graph of CP 1 versus refined model

5.5 *Computing Furnace Efficiency*

The simulation results for all three optimization ranges are used to calculate the thermal efficiency of the electrical resistance furnace in terms of energy. By using the energy relation given by Eqs. (5) and (6), and Table 11, considering the pouring temperature for each metal as 100°C above its melting temperature, furnace efficiency was computed as follows:

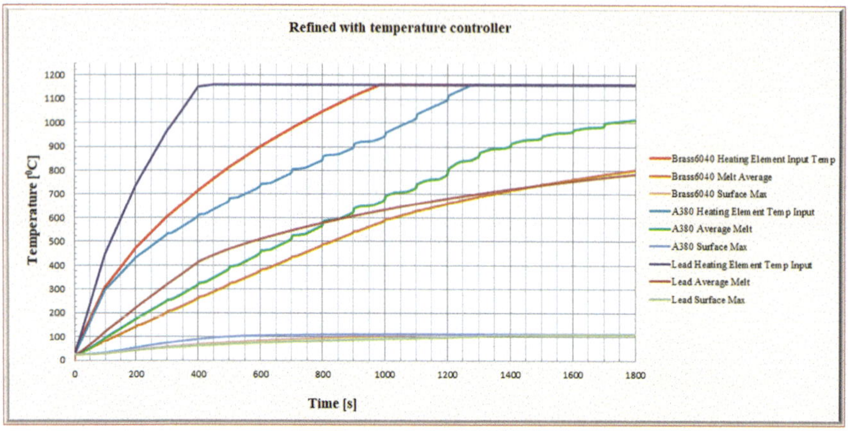

Fig. 14 Result with maximum heating element temperature maintained below 1160 °C

$$\text{Efficiency} = \frac{\text{Energyabsorbedbymoltenmetal}}{\text{Energy Supplied to Heating element}} \times 100 \tag{5}$$

$$Q_{\text{absorbed}} = \rho V \frac{C_{p,s}(T_m - T_i) + H_f + C_{p,l}(T_p - T_m)}{t} \tag{6}$$

$$Q_{\text{absorbed}} = t \times Q_{\text{absorbed}} = \rho V\left(C_{p,s}(T_m - 25) + H_f + C_{p,l}(100\,^{\circ}\text{C})\right)$$

$$Q_{\text{Supply}} = t \times Q_{\text{Supply}}$$

The heating rate of the furnaces for each metal type was taken as the average slope of the temperature–time curve below the melting point temperature of the corresponding metal.

Table 11 Furnace efficiency of different capacity ranges and metal types

Details	Optimization range 1			Optimization range 2			Optimization range 3		
	Lead	A380	Brass	Lead	A380	Brass	Lead	A380	Brass
Q_{absorb} (kJ)	917.6	2198.1	5317.2	11,491	27,528	66,589	572,854	1,372,292	1,731,189
Time (s)	150	380	780	238	432	1050	580	1550	1800
Q_{Supply} (kJ)	1590	4028	8268	17,136	31,104	75,600	626,400	1,674,000	1,944,000
Heating rate (°C/min)	164	119	84	69	77	73	35	34	25
Efficiency (%)	57.7	54.6	64.3	67.1	88.5	88.1	91.4	82.0%	89.1%

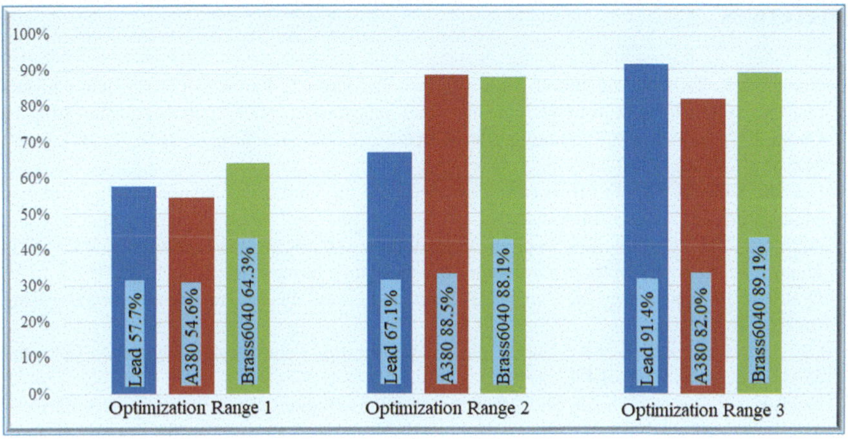

Fig. 15 Furnace efficiency of different metals

The total period to raise the temperature of the metal from room temperature (25 °C) to pouring temperature (taken as $T_{\text{melting point}} + 100$ °C) is shown in Fig. 15. The following table summarizes all the details and calculated values of efficiency for each material type under each optimization range.

6 Conclusion

The findings show that the electrical resistance furnace can melt light metals. Medium- and large-scale furnaces require temperature-controlled input power to maintain the temperature of the heating element below recommended safe values. The optimization was carried out based on the least safe criteria of the heating element (Nichrome wire heater). So, better heating elements such as ceramic carbide or graphite resistors can improve the service life of heating elements. Optimizing the furnace with less reliable material (NiCr) used in the heating element guarantees better safety and durability of the equipment. Molten metal production was achieved while maintaining the temperature gap between the heating element and the molten metal low as recommended by safety standards. The external surface temperature was also kept below 100ºC by natural cooling. The conclusion from the physical experiment is applicable to construct functional electrical resistance melting furnaces with appreciable performance. The small-scale furnace can be safely operated with full power until pouring point temperatures of light metals are attained. The temperature control system is unnecessary unless the operation demands extra safety or constant holding time.

References

1. Nakahara K, Tada M, Yamamoto H, Shinagawa T, Akashi T, Tsurui M (2001) Heavy metals recovery system for dust from NKK electric-resistance furnace. NKK Tech Rev (Japan) 84:8–15
2. Chook KC, Tan AH (2007) Identification of an electric resistance furnace. IEEE Trans Instrum Meas 56(6):2262–2270
3. Lee JK, Kim SK (2011) Effect of CaO addition on the ignition resistance of Mg-Al alloys. Mater Trans 52(7):1483–1488
4. Alkhimov EA, Dovgal AN, Ivanova OB, Burtsev AG, Kaplya VI, Nosenko VA (2011) Identification of the electrical parameters of a resistance furnace. Meas Tech 54(7):808–812
5. Dimitrov B, Yordanova M, Nenov H Optimization of electric resistance furnace using backtracking algorithm
6. Cao J, Ye Q, Li P (2015) Resistance furnace temperature control system based on OPC and MATLAB. Meas Control 48(2):60–64
7. Rubtsov VP, Goryachikh EV, Shcherbakov AV (2015) Studying the influence of irregularity of an electric resistance furnace as a control object. Russ Electr Eng 86(7):403–406
8. Litvintsev KY, Finnikov KA, Kharlamov EB (2017) Features of a mathematical model of heat transfer in a vacuum resistance furnace. In: Journal of physics: conference series, IOP publishing. 891(1):012108
9. Jevtić M, Đorđević N, Radaković Z (2018) Experimental tests of fuzzy logical temperature controller of an electric resistance chamber furnace. IJEEC-Int J Electr Eng Comput 2(1)
10. Pogrebisskiy MY, Bodarev AI, Salmanova EF, Bulgakov AS (2018) Modelling of degradation processes for refractory metallic heating elements of vacuum resistance furnace. In: IOP conference series: materials science and engineering. 313(1): 012011
11. Fu Z, Yu X, Shang H, Wang Z, Zhang Z (2019) A new modelling method for superalloy heating in resistance furnace using FLUENT. Int J Heat Mass Transf 128:679–687
12. Tsvetkova N (2020) Coordination of the thermal insulation in an electric resistance furnace for art glass melting. In: 2020 21st international symposium on electrical apparatus and technologies (SIELA), pp 1–4
13. Naumkin AS, Borisov BV, Syrodoy SV Malishev D Y (2020) Temperature analysis in the heated region of the chamber electric furnace of resistance. In: AIP conference proceedings. 2212(1):020042
14. Dev P, Jain S, Kumar H. (2021) Simulation driven approach for deriving the PID parameters for the control of an electric resistance furnace (ERF). In: IOP conference series: materials science and engineering. 1104 (1):012016

Numerical Studies on VMC Base Made of Epoxy Granite

C. Shanmugam, P. R. Thyla, and P. Dhanabal

Abstract Advanced machine tools play an important role in the modern manufacturing system. To achieve a higher quality of workpieces, higher damping and dynamic stability of machine tools are necessary. Machine tool structures are conventionally made from cast iron because of good compressive strength, rigidity and damping properties but the dynamic stability is limited by higher inertia and less vibration damping at higher operating conditions. Hence, there is a need for developing an alternate material which will meet the desired functions. Polymer composites have been shown to possess good strength and better damping properties than cast iron. In this work, an attempt is made to investigate the behavior of polymer composite base made of epoxy granite as an alternate for CI base through numerical simulation. The static rigidity of the EG base is enhanced by reinforcing with steel reinforcement in the form of stirrup arrangements. The steel-reinforced hybrid EG base with better static and dynamic behavior is proposed as an alternate to the existing base.

Keywords Alternate materials · Machine tools · Finite element analysis and epoxy granite

1 Introduction

High stiffness and damping are essential properties for machine tools in order to operate them at high cutting speeds, which improve productivity and product quality. The materials with which the machine tools structures are built contribute to the static and dynamic stability of the machine tools [1]. Cast iron, one of the traditional materials, has lower Young's modulus and higher damping. Steel, on the other hand, has higher Young's modulus and lower damping coefficient. Hence, traditional machine tool materials have a high static stiffness but low damping or vice versa [2].

C. Shanmugam (✉) · P. R. Thyla · P. Dhanabal
Department of Mechanical Engineering, PSG College of Technology, Coimbatore, India
e-mail: csn.mech@psgtech.ac.in

E. Natarajan et al. (eds.), *Materials, Design and Manufacturing for Sustainable Environment*, Lecture Notes in Mechanical Engineering,
https://doi.org/10.1007/978-981-19-3053-9_25

Epoxy granite is a special type of particulate composite with very good dynamic properties, and hence, it can be an alternate to cast iron in machine tool structures [3]. Experimental investigations on EG composite have shown that it possesses excellent material damping properties in the order of 3–10 times than that of CI while the compressive strength was 90–120 MPa, which is higher as compared to other polymer concrete composites. Epoxy granite can dampen the vibration of the structure better than that of cast iron but the volume of the structure for the same stiffness of cast iron has to be increased [2, 3]. In order to replace CI structure with EG structure, several design iterations have to be carried out to arrive at a suitable design of EG considering the static rigidity of CI structure as an equivalence.

Since the elastic modulus of EG is one-third that of CI for the equal static rigidity, EG structure must be cast to three times larger than that of CI structure which is not viable, since it may affect the mounting and assembly constraints of other subassemblies of VMC. A better way to improve the rigidity of EG structure is by incorporating reinforcements made of material with high elastic constant. Steel rods of small diameters can be reinforced with EG to improve the stiffness. The best suited EG steel reinforcement is adopted for the VMC base.

The finite element method (FEM) is a powerful method that will be helpful in carrying out various analyses through design iterations to arrive at an optimized design of structures. In this work, an attempt has been made to use EG as structural material in vertical machining center (VMC) base as an alternate to cast iron. Initially, a finite element model of the VMC base made of CI is developed and numerically analyzed under the action of unit force at tool center point (TCP) to predict the structural behavior. Modal analysis is performed to study the modal parameters of the existing VMC base. The deflection of CI base, modal frequencies and mode shapes were used as benchmark data for the development of VMC base made of EG. Free body diagram concept is used to predict the static deflection of the VMC base, and the FE model is validated using the hybrid modeling method (HMM) [4]. In HMM, the whole VMC is considered for analysis in which the base alone is treated as a deformable member with assigned material properties of CI, while the other structures are assumed as rigid and assigned with the elastic modulus of several orders higher than that of the actual material properties. These rigid structures do not deform under the action of forces but they transfer the resultant forces and moments to the base. Predicted deflection of VMC base using HMM is used to validate the results of free body diagram approach, i.e. single module modeling (SMM) approach. The validated SMM approach is used to carry out the design iterations of the base made of EG. Finally, hybrid EG base reinforced with steel which offers better static and dynamic characteristics is proposed.

2 Numerical Studies on Existing VMC Base

2.1 Structural Analysis of Existing Base

A vertical machining center (VMC) is a versatile computer numerically controlled (CNC) machine tool which performs a variety of machining operations to produce workpieces of required form and surface finish without restrictions on the shape and size. The base, bed and column are the structural elements which assure the geometric configuration and rigidity of VMC [5].

The approximate outer dimensions of the existing VMC were measured, and the geometric model is developed using modeling software. The finite element model is obtained by meshing using higher-order elements and is shown in Fig. 1.

A unit force at the TCP and corresponding reaction force on the workpiece is applied; the base is constrained at the bottom face of foundation pads. Static structural analysis is performed by assigning CI material properties to the base and the corresponding total deflection on the base is predicted which is found to be 4 μm as is evident from the total deformation plot shown in Fig. 2.

Instead of modeling the full model, handling the base alone simplifies the FE model reduces the computations, storage requirement and computational time. Hence, in this work, the behavior of the base is considered under static and dynamic conditions. The base is separated from VMC, numerically studied using structural and modal analysis.

In order to predict the structural behavior of the base, a free body diagram concept is used [2]. For the unit force acting at TCP, the corresponding reaction forces and moments were applied at the interface areas of the cross-slide and column with the base. The results of static structural analysis of the base under this condition show a total deflection of 4.2 μm with a deviation of 6% to that of deflection predicted

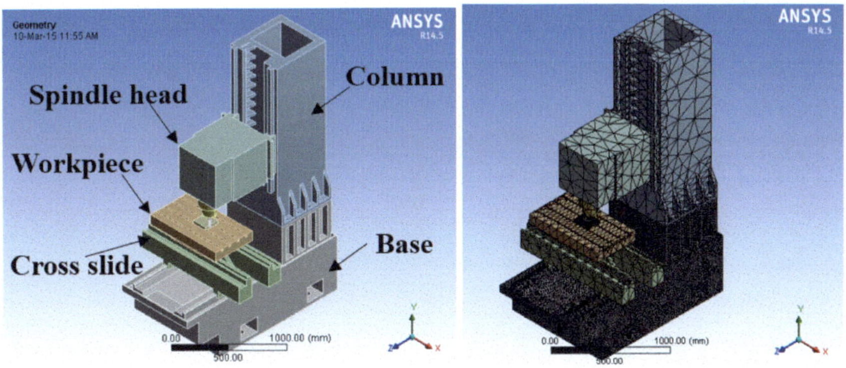

Fig. 1 Structural components of VMC and finite element model

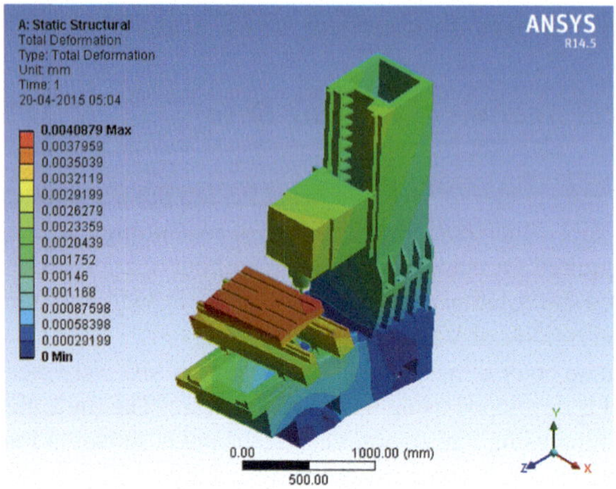

Fig. 2 Total deformation plot of the base in HMM

using HMM. Since both the methods predict the same structural behavior of base, the free body diagram approach is adopted for further numerical studies.

In order to predict the structural behavior of the existing base undercutting conditions, the reaction forces were transferred from TCP to the interfaces of the base with column (B, C, D and E) and cross-slide (F, G, H and J) as shown in Fig. 3. The resultant of cutting force reactions transferred from TCP at locations B, C, D and E was 2424 N, while the same at F, G, H and J was 1487 N, respectively. In addition, a cutting reaction force of 2452.5 N from workpiece is also applied at I. The base is constrained at four mounting locations at A.

The base is analyzed to predict its deflection under the action of cutting forces, and the result of static structural analysis of base is shown in Fig. 4 as total deformation plot with a maximum total deflection of 43.7 μm.

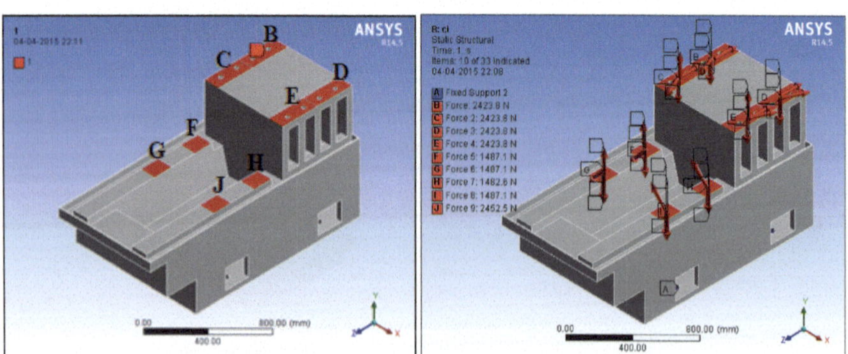

Fig. 3 Loading and boundary conditions of existing base

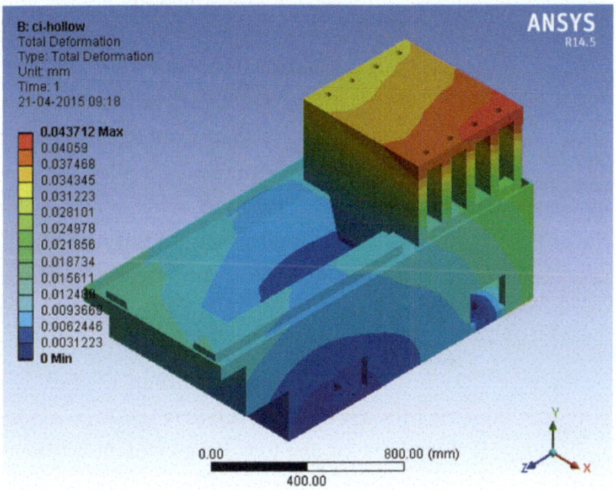

Fig. 4 Maximum deformation plot of existing base

It is inferred from Fig. 4 that the base deflects more at the interface areas of the column with base due to cantilever actions. The base is further numerically analyzed to predict the dynamic modal parameters.

2.2 *Modal Analysis of Existing Base*

The goal of modal analysis is to determine the natural mode shapes and frequencies of the structure during free vibration. If the structure is operated at the natural frequency unduly large amplitude of vibration will result which will affect the product quality. Hence to avoid resonance, any structure has to be operated far away from the zone of natural frequency. FE models will be helpful to predict the modal behavior of any structure, and the results of which can be validated through experiments. The numerical modal analysis of the exiting base is performed under free-free vibrating conditions, and the mode shapes of the first four fundamental modes are shown in Fig. 5

The base is found to vibrate under longitudinal mode in '*X*–*Z*' plane with maximum deflection of 2 mm in the first mode of 103 Hz. The 2nd and 3rd modes are observed to be in '*Y*–*Z*' plane at 116 and 208 Hz, respectively. The 4th mode of twisting is observed about '*X*' axis at 234 Hz with maximum twisting of 2.76 mm at the interface of the column with base. These data pertaining to the static and dynamic modal behavior of the existing base are used as benchmark data to design the base made of epoxy granite.

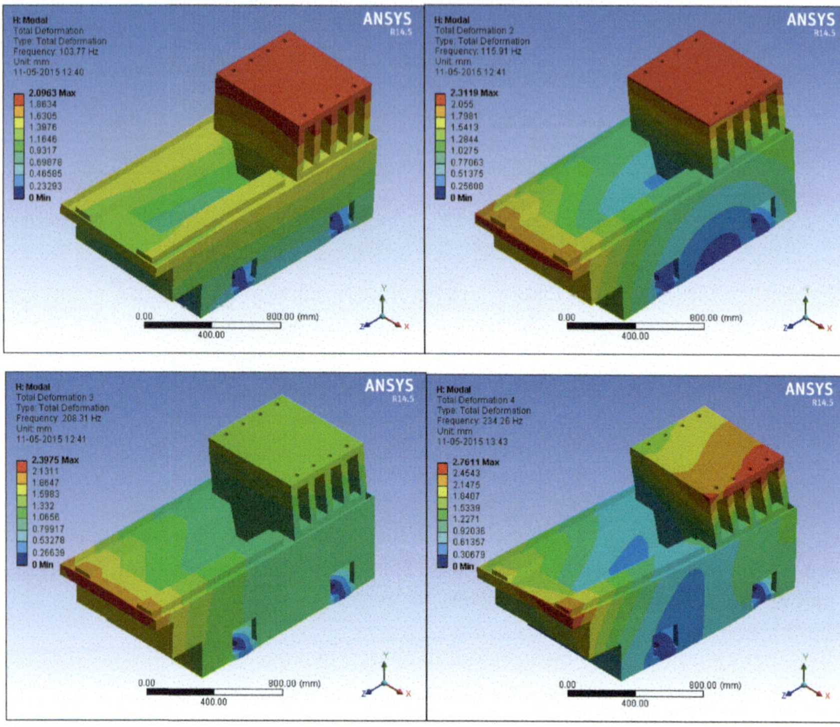

Fig. 5 Mode shapes of existing base

3 Numerical Studies on Epoxy Granite VMC Base

3.1 Structural Analysis of EG Base

Epoxy granite (EG) is a particulate-based polymer composite which is composed of granite as an aggregate system bonded with thermoset-based polymeric resins such as epoxy. These materials are found to be one of the favorable alternatives to conventional materials in machine tool structures owing to their superior damping characteristics [1, 6]. For the static structural analysis of base made of EG undercutting conditions, the same geometry of the existing CI base is retained. The FEM of the base is assigned with EG material properties (Density: 2100 kg/m^3, Poisson's ratio: 0.25 and elastic modulus: 30 GPa) and analyzed with the same boundary conditions as that of existing CI base.

Since the elastic modulus of EG is one-third of that of CI, the deflection of the EG base will be approximately 3.5 times than that of existing CI base and is evident as shown in Fig. 6. Moreover, the maximum deformation in the base is observed at the column-based interface due to induced reaction forces and moments from TCP

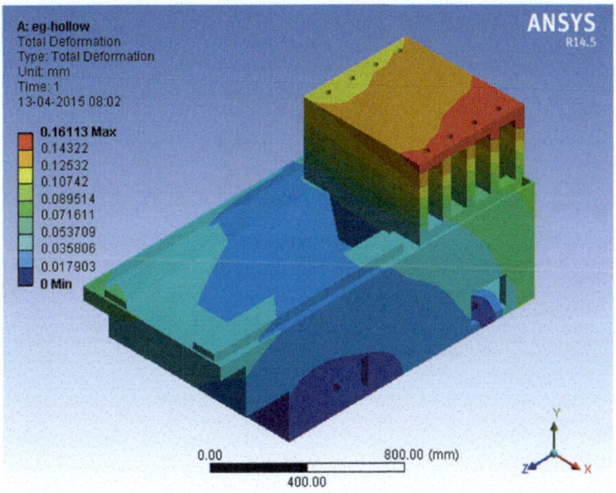

Fig. 6 Maximum deformation plot of EG base

through the column and the overhung portion of the base due to cantilever action. Further, numerical modal analysis is performed to study the dynamic characteristics of the EG base.

3.2 *Modal Analysis of EG Base*

In order to find the fundamental natural frequencies and understand the modal behavior of EG base, the numerical modal analysis of EG base is performed under free-free boundary conditions, which is similar to that of CI base. The numerically extracted mode shapes of the EG base are shown in Fig. 7.

Figure 7 depicts the fundamental mode of 107 Hz in '*X*–*Z*' axis, followed by 120 Hz, 216 and 240 Hz at higher modes, and the mode shapes are similar to those obtained for CI base. The results of the modal analysis revealed that a marginal shift in the observed natural frequencies in EG base which in turn helps to enhance the dynamic characteristics of the machine tool. Further, the results of static analysis of EG base revealed that inferior in rigidity compared to that of CI base. This necessitates to explore the possibility of enhancing the rigidity of the EG base. The rigidity of EG structure can be improved by form design modifications, introducing stiffening features and material with higher elastic modulus. In the present study, the rigidity of the EG base is enhanced by reinforcing with structural steel. So that, the beneficial damping property of EG can be utilized in machine tool structures [7, 8].

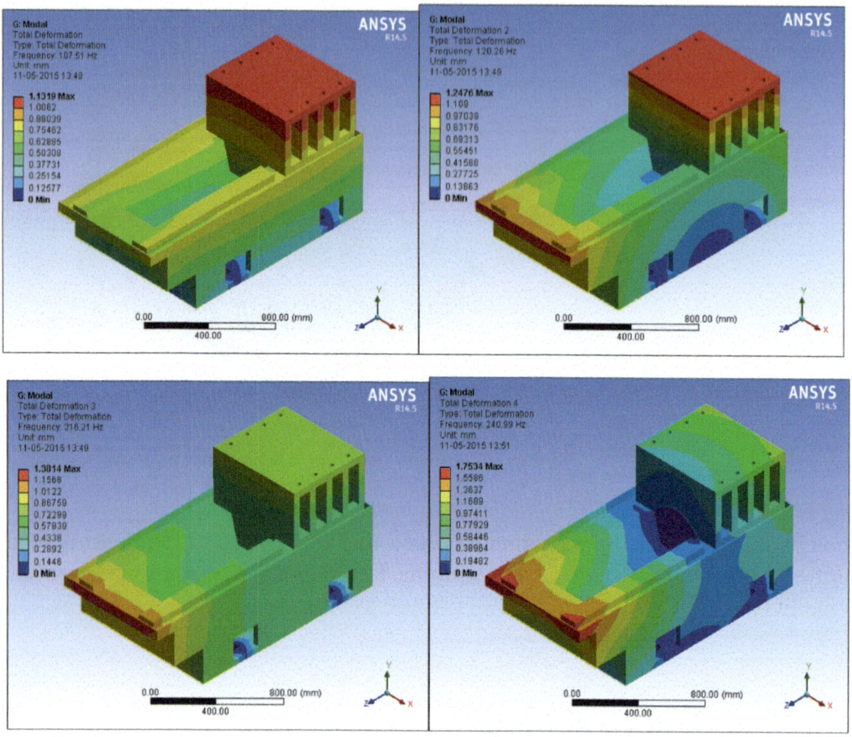

Fig. 7 Mode shapes of EG base

4 Numerical Studies on Steel-Reinforced Epoxy Granite VMC Base

The static rigidity of the EG base is enhanced by reinforcing with steel rods of diameter 6 mm. Different steel reinforcements in the form of stirrups for EG base were numerically analyzed against unit force, and the design which offered better resistance against deflection is selected [9, 10]. Hybrid EG base is designed by reinforcing with finalized reinforcement as shown in Fig. 8

Static structural analysis on hybrid EG base is carried out undercutting conditions same as that of existing CI base. The maximum total deflection of hybrid EG base is found to be 49 μm as shown in deformation plot Fig. 9, thus the static behavior of hybrid EG base is found to be similar to that of existing base made of CI.

The dynamic behavior of hybrid EG base is studied by performing modal analysis, and the boundary conditions were similar to that applied on CI base. The hybrid EG base is characterized by fundamental natural frequency of 115 Hz in '*X*–*Z*' plane, followed by 118, 186 and 278 Hz at higher modes as shown in Fig. 10.

The 2nd natural frequency of 118 Hz is found to be bending mode in '*Y*–*Z*' plane, and the edges of base at the tail end portion are subjected to maximum bending effect

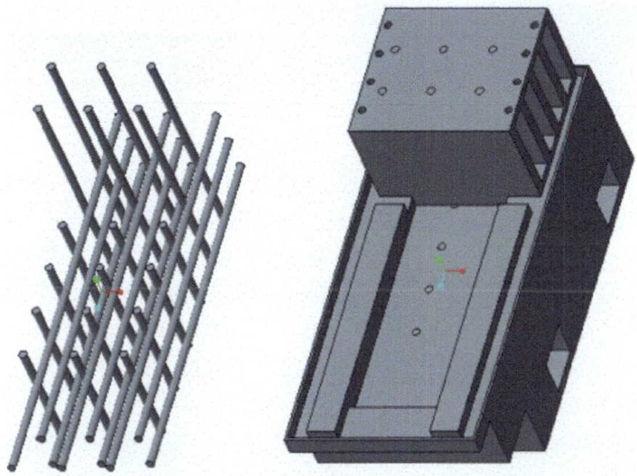

Fig. 8 Steel reinforcement and hybrid EG base

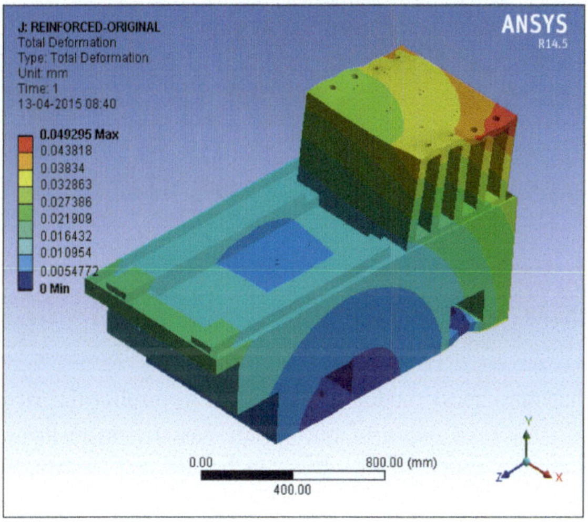

Fig. 9 Total deformation plot of hybrid EG base

in this mode. The 2nd and 3rd bending modes were observed at 186 and 278 Hz in '*Y*–*Z*' plane, and the interface portions of base with column are subjected to maximum deflection in these modes.

The static structural analysis of hybrid EG base revealed equal static rigidity as compared to that of CI base under similar loading conditions. The modal analysis revealed that there is a considerable shift in the natural frequency of 15–20% in the observed modes for the hybrid EG base compared to that of existing CI base. Numerical studies on machine tool structures made of ferrocement and composites have been reported in the existing research works [11, 12]. This work extends the possibility of using EG composite in machine tools, particularly for bed and base kinds of structures through numerical studies. A numerical model of EG base is developed by assuming homogeneous material properties. Further, the study confirms

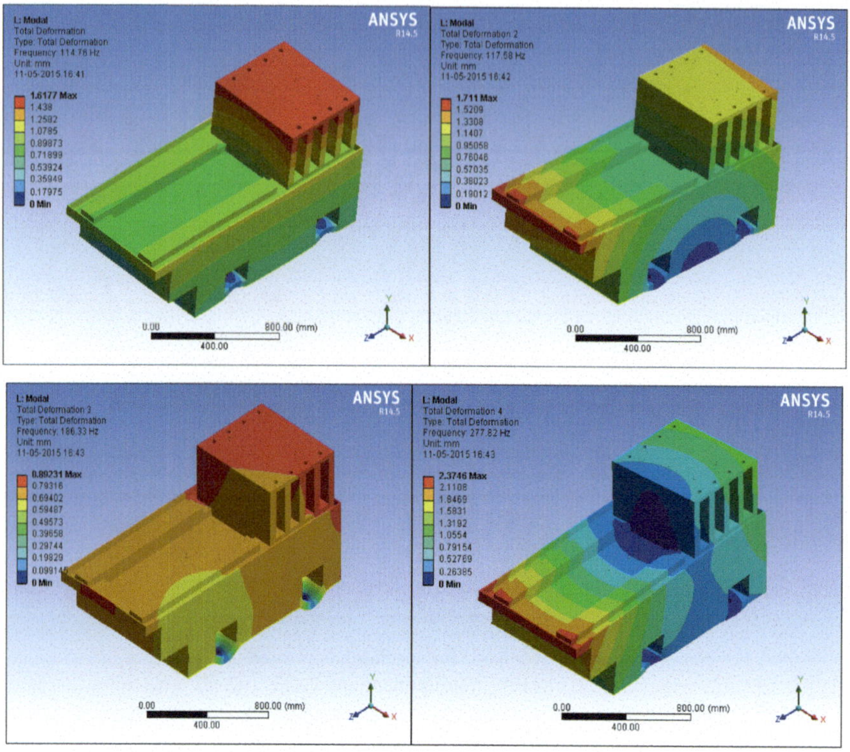

Fig. 10 Mode shapes of hybrid EG base

that numerical models can be used to predict the structural and dynamic behavior of structures made of particulate-based composites. Thus, the enhanced material damping properties of EG can be utilized in structures to improve the dynamic stability of machine tools with equal or greater rigidity as compared with existing structures. The numerical studies of this work pave a way to analyze the rigidity and dynamic behavior of machine tools made of EG structures.

5 Conclusions

In this work, an attempt has been made to numerically study the suitability of EG as a structural material in precision machine tool structures. A VMC base is considered for the study. Initially, a finite element model of the existing base made of cast iron is developed, and its behavior is studied using static structural and modal analysis. The deflection and modal natural frequencies of the existing base were used as benchmark data for the design of proposed base made of epoxy granite. Initially, the geometry of the existing base is retained, and the base is numerically analyzed with EG material properties. In spite of better damping properties, the static rigidity of EG base is found to be lesser than that of CI base owing to the low elastic modulus of EG material (1/3rd of that of CI).

In order to enhance the static rigidity of EG, a hybrid EG base reinforced with steel reinforcement is designed by retaining the same footprints of CI base and is numerically analyzed. Through the numerical studies, it is concluded that the hybrid EG base offers equal rigidity under similar loading conditions. The positive shift in natural frequency of 15–20% will help to enhance the dynamic performance of machine tools. Widened frequency ranges will help to operate machine tools at high cutting speeds. Beneficial material damping characteristics (three to ten times higher than that of cast iron) of EG will help machine tools to carry out machining operations at higher depth of cut with better stability against chatter. To sum up, the EG helps to improve the performance of machine tools and hence the product quality will be improved in terms of geometric accuracy and surface finish with enhancement in productivity.

References

1. Heisel U, Gringel M (1996) Machine tool design requirements for high-speed machining. CIRP Ann 45(1):389–392
2. Rahman M, Mansur A, Karim B (2001) Non-conventional materials for machine tool structures. JSME Int J Ser C 44(1):1–11
3. Selvakumar A, Mohanram PV (2012) Analysis of alternative composite material for highspeed precision machine tool structures. Ann Fac Eng Hunedoara Int J Eng Tome X. Fascicule 2. ISSN 1584–2665
4. Huang DT-Y, Lee J-J (2001) On obtaining machine tool stiffness by CAE techniques. Int J Mach Tools Manuf 41:1149–1163
5. Malleswarsa Swami B, Sunil Ratna Kumar SK (2012) Design and structural analysis of CNC vertical milling machine bed. Int J Adv Eng Technol 3(4):97–100
6. Ranganna H, Ashok Kumar M, Ramesh A, Development and behavior of mechanical properties of graphite/granite filled epoxy hybrid composites. Int J Mod Trends Eng Res 02(08)
7. Kushnir EF, Patel MR, Sheehan TM, Hardinge Inc, Elmira (2001) Material considerations in optimization of machine tool structure. In: Proceedings of ASME international mechanical engineering congress and exposition, 11–16 Nov, New York
8. Suh JD, Lee DG (2008) Design and manufacture of hybrid polymer concrete bed for high-speed CNC milling machine. Int J Mech Mater Des 4(2):113–121
9. Schulz H, Nicklau RG (1983) Designing machine tool structures in polymer concrete. Int J Cement Compos Lightweight Concr 5(3):203–207
10. Zaw MM (2008) Strength analysis of bed ways for CNC lathe machine. In: GMSARN international conference on sustainable development: issue and prospects for the GMS 12-14 Nov 2008
11. Rahman M, Mansur MA, Ambrose WD, Chua KH (1987) Design, fabrication and performance of a ferrocement machine tool bed. Int J Mach Tools Manuf 27(4):431–442
12. Hadree RH, Tariq M, Saleh F, Dynamic stability of milling machine bed using natural fibre composite. Int J Adv Res Eng Technol (IJARET). ISSN 0976–6480(Print), ISSN 0976–6499(Online) 5(4)

Experimental Investigation on the Influence of Cutting Parameters During Dry Machining of Ti–6Al–4V

S. Sudhagar, A. Ajay Sivaraman, R. Bibeye Jahaziel, B. Geetha Priyadarshini, and V. Krishnaraj

Abstract Titanium alloys have its applications in aerospace industries because of its mechanical properties like high-specific strength, high hardness, and high resistance to corrosion. Due to lower ability to conduct heat, high heat generated at the tool chip contact area which causes high-cutting force and increased tool wear during turning. In this study, the dry turning of titanium alloy was carried out using carbide inserts. Under response surface methodology (RSM), Box-Behnken design of experiments was designed with 3 factors (cutting speed, feed, depth of cut) and 3 levels of total 15 experiments. The optimization has been carried out using design expert analysis software, and process parameters are related to response cutting force as a quadratic regression equation with goodness of fit 93.8%. The contour and surface plots were interpreted for minimum cutting force requirement. The predicted and actual cutting force values have been compared. The analysis of variance was implemented to get the highly influential process parameters on the cutting force and its contribution. It was found that the depth of cut and feed are the most influential parameters on cutting force. The cutting speed has insignificant effect on the output cutting force. The optimum machining parameters were identified.

Keywords Cutting force · Machining · Ti–6Al–4V · RSM · Optimization

1 Introduction

Titanium alloys find an extensive application in aerospace, biomedical industries, and many other engineering fields because of its high strength at high-temperature, high-specific strength, and excellent corrosion resistant [1]. The low-thermal conductivity (7.1 W/m k) of titanium alloys has become a huge problem in machining of these

S. Sudhagar · A. Ajay Sivaraman · R. Bibeye Jahaziel
Department of Production Engineering, PSG College of Technology, Coimbatore, India

R. Bibeye Jahaziel · B. Geetha Priyadarshini · V. Krishnaraj (✉)
PSG Institute of Advanced Studies, PSG College of Technology, Coimbatore, India
e-mail: vkr.prod@psgtech.ac.in

E. Natarajan et al. (eds.), *Materials, Design and Manufacturing for Sustainable Environment*, Lecture Notes in Mechanical Engineering,
https://doi.org/10.1007/978-981-19-3053-9_26

alloys which make it as a difficult to machine material. This leads to increased heat generation at tool chip interface, high-cutting force, and adversely affects the tool life [2, 3]. In recent years, high-speed turning where the cutting speed greater than 54 m/min has gained a great attention to reduce the machining time, cutting force, tool wear [4, 5]. In general, for the machining of grade 5 Ti–6Al–4V, low-cutting speed and high feed with lubricant will be used at stated in [6]. Sutter and List performed high-speed machining of Ti–6Al–4V alloy and stated that the major influence was from the cutting speed on the chip formation and cutting force [7]. The machining of titanium was performed by Chen et al. [8] and Wang et al. [9] and found that reduction in tool wear and improved tool life was achieved with cutting speed less than 2 m/s. Study on high-speed turning of titanium alloys by Shaw [10] and Trent [10] becomes a great solution to reduce the machining time with increased productivity. It also leads to increased generation of heat at the tool chip interface during machining. Boujelbene [11] modelled the tangential force during turning of titanium alloy, and it was reported that the cutting speed and feed rate were the most impactful parameters on cutting force. The tangential force minimization was achieved when the cutting speed and feed rate were set at low levels. Krishnaraj et al. [12] have done a dry milling of Ti–6Al–4V where high temperature was generated and analyzed the influence of cutting speed, feed, and depth of cut on the cutting force and temperature. It was observed that the feed has major impact on cutting force as well as temperature. Thermodynamic chip formation study was done during titanium machining at different cutting speeds by Cook [13]. The dry machining has got attention nowadays because of the environmental protection. The use of metal working fluid will reduce the temperature at the cutting zone. However, it also imparts health effects to the operator [14]. The disposal of these fluids as well as the recycling will be a tedious task for an industry which increases the production cost [15]. Shahebrahimi and Dadvand [16] investigated on the turning of Ti–6Al–4V, and the influence of cutting speed, feed, and depth of cut was analyzed on surface roughness. The ANOVA analysis revealed that the depth of cut and feed have high influence on surface roughness, whereas cutting speed has lesser influence on surface roughness. Rangaraju and Vijayan [17] have performed a milling operation on Ti–6Al–4V, and the optimization by using RSM (central composite design of experiments) was done. The optimum machining parameters for minimum surface roughness were found, and it was stated that the minimum value of surface roughness was obtained with low feed, depth of cut, and high-cutting speed. Vijay and Krishnaraj [18] carried out an optimization of end milling operation on Ti–6Al–4V for minimum cutting force requirement and to minimize the value of surface roughness. The depth of cut and feed per tool were the significant factors that impact the cutting force. The cutting speed has less significance on cutting force. Ansari et al. [19] performed a face milling operation on Ti–6Al–4V, and the effect of feed rate on cutting force component was analyzed. Karkolos et al. [20] have studied the impact of process parameters on surface roughness by performing milling of titanium alloys. The optimum machining parameters were found by comparing the RSM results with developed ANN model. The low-feed led to low-surface roughness and improved finish.

From the literatures, it could be observed that numerous investigations have been carried out in the optimization of machining parameters for Ti–6Al–4V with the aid of cutting fluids. In view of attaining a cleaner production environment, recent literatures have shifted towards the machining investigation under dry machining environments. The present work focusses on the effect of process parameters on cutting force. This in turn could establish a relationship on the cutting mechanics that determine the prolonged tool life in turning of titanium alloy. A comparison of the optimal process parameters has been compared with that of RSM and ANOVA.

2 Materials and Method

2.1 Experimental Setup

The Kennametal TCMT16T304LF KC5025 PVD TiAlN coated cutting tool was used for turning Ti–6Al–4V alloys. The holder used was STGCR2020K16. The geometry of insert used was orthogonal with rake angle of 5°, corner radius of 0.4 mm, clearance angle of 7°. Cutting speed, feed, and depth of cut were used as the input parameters, and cutting force was taken as output response. The workpiece length and diameter are 250 mm and 60 mm, respectively. Tables 1 and 2 detail about the chemical composition and mechanical properties of Ti–6Al–4V Alloy, respectively.

The experiments were carried out using a Pinacho auto feed lathe. The measurement of cutting force was done using Kistler dynamometer which was attached to the tool post of the lathe machine. All the three cutting force components were recorded on the digital monitor screen of dynamometer. The setup for carrying out the experiment and the dynamometer setup are shown in Figs. 1 and 2, respectively.

Whilst performing the machining of Ti–6Al–4V alloy, three cutting force Fz was measured through the dynamometer. The cutting force directions were shown in Fig. 3.

Table 1 Chemical composition of Ti–6Al–4V

	Mass percentage				
Ti–6Al–4V	Al	C	V	Fe	Ti
	5.96	0.022	4.073	0.138	89.578

Table 2 Mechanical properties of Ti–6Al–4V

Ti–6Al–4V	Strength (MPa)	Yield point (MPa)	Ductility (%)	Hardness (HB)	Modulus of elasticity (GPa)	Density (kg/m^3)	Thermal conductivity (W/m K)
	940	864	9.65	343	114	4424	7.50

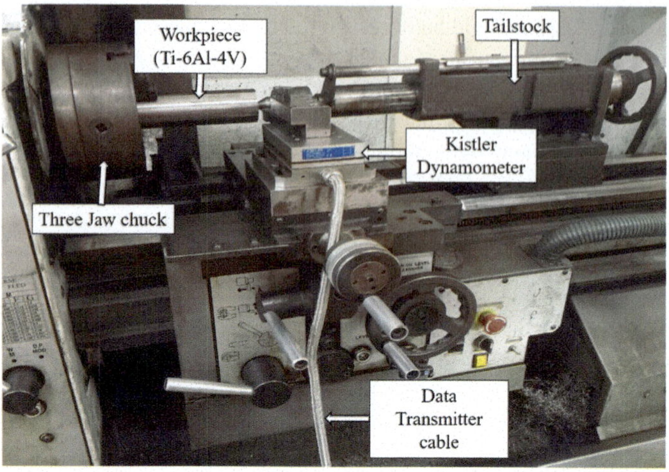

Fig. 1 Pinacho lathe with dynamometer

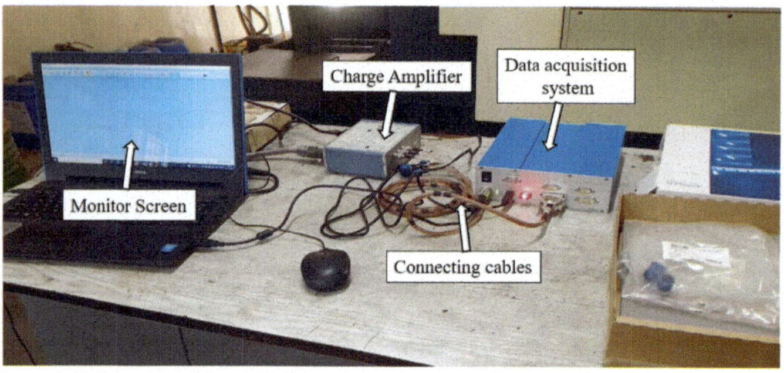

Fig. 2 Dynamometer setup

2.2 Design of Experiments

The response surface methodology technique was used for optimization. The RSM optimization has a capability to frame a relation between input and output variables efficiently. It also helps in studying the interaction of different parameters on the output. The 3D surface and contour plots facilitate easy interpretation of results visually. It also reduces number of experiment trials which will ultimately reduce the experimentation cost. Under RSM, Box-Behnken design with 15 experiments was selected by setting 3 factors at three levels. The factors are cutting speed (m/min), feed (mm/rev), and depth of cut (mm) which are used for experimentation. The different factors with its associated levels along with DoE were shown in Tables 3 and 4, respectively.

Fig. 3 Cutting force direction

Table 3 Factors and levels of DoE

Level	Parameters		
	cutting speed (m/min)	Feed (mm/rev)	Depth of cut (mm)
−1 (low)	60	0.1	0.4
0 (medium)	75	0.15	0.6
1 (high)	90	0.2	0.8

3 Results and Discussion

The experimentations were performed as per the order in Table 4. The cutting force Fz was noted for each experimentation as an output. The result of experiments is shown in Table 5.

3.1 Optimization Using RSM

ANOVA Analysis

The experimental data were analyzed to find the impact of cutting speed, feed, and depth of cut on cutting force component Fz. The analysis of variance (ANOVA) carried out and the same shown in Table 6. The analysis assumed with a 95% of confidence level and 5% significance level.

Table 4 DoE table

Run	Cutting speed (m/min)	Depth of cut (mm)	Feed (mm/rev)
1	75	0.4	0.2
2	75	0.4	0.1
3	90	0.6	0.2
4	90	0.4	0.15
5	75	0.6	0.15
6	90	0.6	0.1
7	60	0	0.1
8	75	0.6	0.15
9	75	0.8	0.1
10	60	0.4	0.15
11	60	0.6	0.2
12	75	0.6	0.15
13	60	0.8	0.15
14	90	0.8	0.15
15	75	0.8	0.2

Table 5 Experimental results

Run	Cutting speed (m/min)	Depth of cut (mm)	Feed (mm/rev)	Fz (N)
1	75	0.4	0.2	58.42
2	75	0.4	0.1	54.58
3	90	0.6	0.2	132.3
4	90	0.4	0.15	95.05
5	75	0.6	0.15	99.2
6	90	0.6	0.1	98.21
7	60	0.6	0.1	87.62
8	75	0.6	0.15	111.8
9	75	0.8	0.1	129
10	60	0.4	0.15	97.61
11	60	0.6	0.2	151.4
12	75	0.6	0.15	154.4
13	60	0.8	0.15	171.5
14	90	0.8	0.15	223.9
15	75	0.8	0.2	242.9

Table 6 ANOVA for cutting force Fz

Source	Sum of squares	df	Mean	*F*-value	*p*-value	% contribution
Model	38,853.59347	9	4317.065941	8.489118635	0.014897342	
A-cutting speed	213.5211125	1	213.5211125	0.419869902	0.545576837	0.516
B-depth of cut	26,638.9362	1	26,638.9362	52.38305201	0.00078589	64.351
C-feed	5810.959013	1	5810.959013	11.42672387	0.019666826	14.037
AB	755.1504	1	755.1504	1.484934773	0.277357899	1.824
AC	220.374025	1	220.374025	0.433345533	0.539450336	0.532
BC	3028.3009	1	3028.3009	5.954879067	0.058634788	7.315
AA^2	421.6465442	1	421.6465442	0.829129688	0.404283723	1.019
BA^2	779.3892058	1	779.3892058	1.532598186	0.270679695	1.883
CA^2	842.3012827	1	842.3012827	1.65630908	0.254459853	2.035
Pure error	1673.52	2	836.76			
Cor total	41,396.29909	14				

From Table 8, the impact of depth of cut on cutting force Fz was very high with a contribution of 64.315% (*p*-value 0.000785 which is less than 0.05). Next to depth of cut, the feed has significance of 14.03% on cutting force with (*p*-value 0.01966 which is less than 0.05). Interactions do not have any influence and significant impact on cutting force Fz.

Regression Equations and Validation

The quadratic regression equations were established to form the relation between input and output responses. The regression model for cutting force Fz was shown in Eq. 3. The equation satisfies for its goodness of fit with Rsq values.

$$\begin{aligned} \text{Fz} = {} & 446.379 + -8.04325\text{Vc} + -903.563d \\ & + 1442.82f + 4.58\text{Vc} \times d + -9.89667\text{Vc} \times f + 2751.5d \\ & \times f + 0.0474944\text{Vc}^2 + 363.219d^2 + -6041.5f^2(\text{Rsq} = 93.86\%) \end{aligned} \quad (3)$$

The regression model was verified for its satisfaction of assumptions. The residuals have to follow normal distribution via a straight line as shown in Fig. 4. The residual versus run order plot should not follow any pattern, and it should be a random as shown in Fig. 4b. Also, the predicted and actual values were also been plotted in Fig. 4c.

Mean Effect Plot

The mean effect plot was generated, and the effect of each parameters was identified as shown in Fig. 5. From the mean effects plot, it could be observed that for the

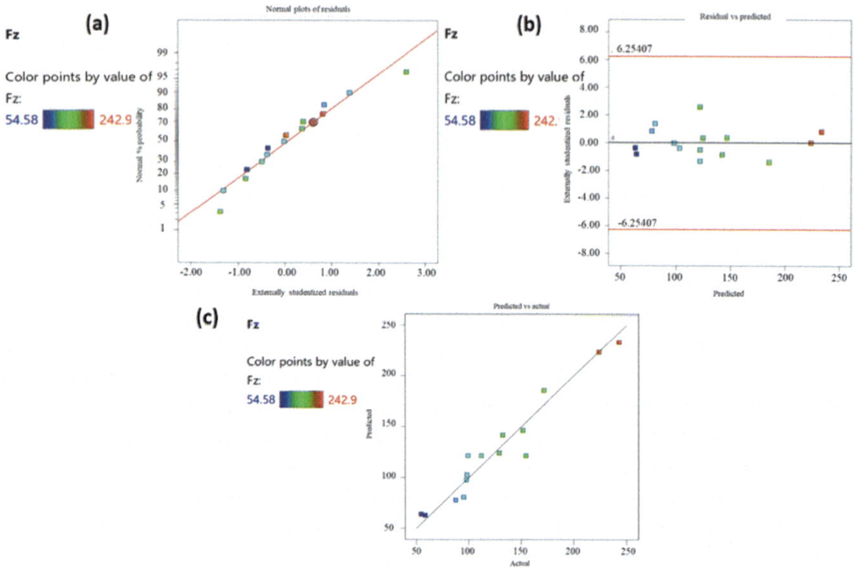

Fig. 4 **a** Normal plot of residuals for Fz. **b** Residual versus predicted plot for Fz. **c** Predicted versus actual plot for Fz

selected range of process parameters, depth of cut has been a vital factor followed by feed. More amount of force is required to machine materials at higher depth and hence an increase in the cutting force. With increase in the depth of cut, the cutting force increases. A similar trend was also observed for feed. At higher feeds, the cutting force increases significantly to a certain extent. The lease significant factor in the present research work was identified to be cutting speed. The cutting force was higher at lower speeds and further increase in cutting speed decreased the cutting force. The cutting force increased with higher cutting speed of 90 m/min and this in turn results in thermal softening phenomenon.

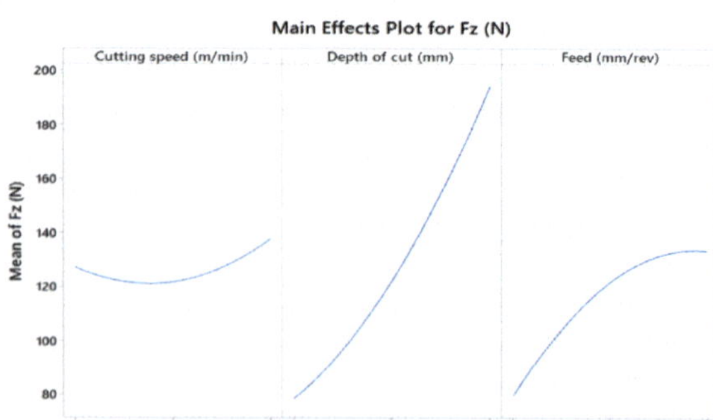

Fig. 5 Mean effect plot for cutting force

Contour and Surface Plots

To facilitate a better understanding about the influence of the selected process parameters, contour plots were plotted against process parameters as shown in Fig. 6. In the plot between cutting force and depth of cut (6a), it was noted that lower cutting force was generated at lower speeds and minimum depth of cut. Increase in the depth of cut leads to higher cutting forces inspite of the cutting speed being low. At higher speeds and higher depth of cut, the cutting force was observed to be the maximum. In a contour plot between cutting speed and feed rate, it was observed that the cutting force was higher at higher speeds. The cutting forces were moderate even at lower cutting speed and higher feed rate. From Fig. 6b, with the increase in feed, the cutting force also increases up to 242 N. The cutting speed shows a minimum contribution on cutting force after 70 m/min.

The contour plot for depth versus feed showed that the highest cutting force was observed at maximum feed and depth of cut. It was also observed that at lower depth of cut, with increase in the feed rate, the cutting forces were observed to be minimal. From Fig. 6, as the depth of cut crosses 0.55 mm, there is a significant increase in cutting force. The feed has no effect on the force generated. From the contour plot analysis, the optimal machining conditions that could yield lower cutting force was arrived at and is shown in Table 7.

Similarly, the surface plots have also been generated for understanding the effect of each parameter on cutting force with better visuals are shown in Fig. 7. The contour plot of cutting speed versus depth showed that a minimal change in cutting force was observed as cutting speed increased from a minimum of 60 to maximum of 90 m/min. High depth of cut and low-cutting speed resulted in higher cutting force. A maximum

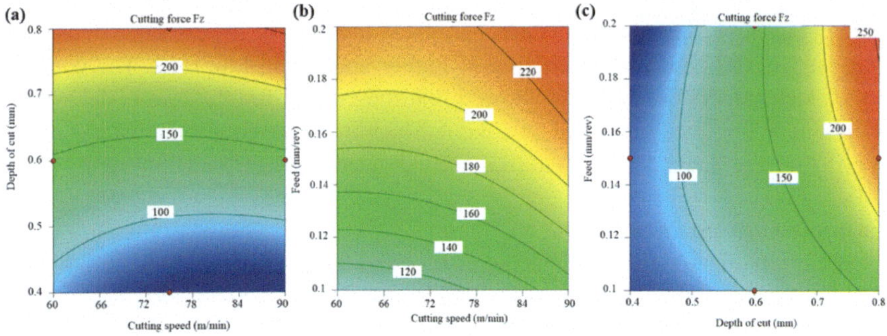

Fig. 6 Contour plots of cutting force with respect to **a** cutting speed and depth of cut, **b** cutting speed and feed, **c** depth of cut and feed

Table 7 Parameters for minimum force condition

Force component	Parameters	Value
Fz	Vc (m/min)	60–90
	d (mm)	0.4–0.51
	f (mm/rev)	0.1–0.12

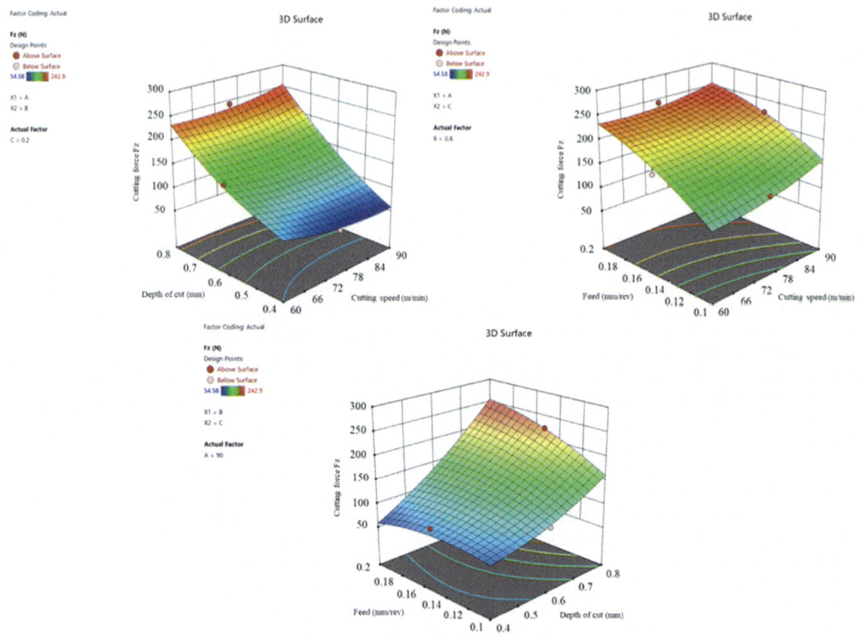

Fig. 7 **a** Response of cutting speed, depth of cut on cutting force Fz. **b** Response of cutting speed, feed on cutting force Fz. **c** Response of feed and depth of cut on cutting force Fz

cutting force of ~ 250 N was noted at a condition of high-cutting speed and depth of cut. In contour plot between cutting speed versus feed rate, it revealed that the range of cutting force values were comparatively higher. The lowest value of cutting force was ~ 150 N whilst machining at higher speed and low feed. The contour plot of feed rate vs depth of cut showed that the cutting forces were lesser whilst machining at low-feed rate and depth of cut. Machining trials done at maximum feed rate (0.2 mm/rev) and depth of cut (0.2 mm) showed the maximum cutting force.

Optimum Cutting Parameters

The optimum machining parameters were investigated during turning of Ti–6Al–4V alloy in dry condition. The goal was to minimize the cutting force for the selected cutting parameters, namely cutting speed, feed, and depth of cut. The optimized values are shown in Table 8.

After the analysis, the optimum values were identified based on the desirability function. The value of desirability ranges from 0 to 1 (from least 0 to most favourable 1). The optimum parameters of Table 8 have a desirability value of 1.00. Based on

Table 8 Optimized values of process parameters for minimum cutting force

Parameter	Optimum values
Fz	Vc 86.246 m/min, *d* 0.400 mm, *f* 0.200 mm/rev

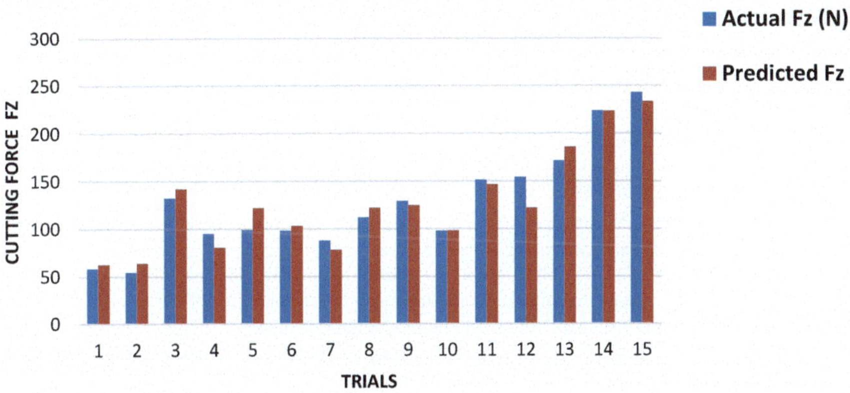

Fig. 8 Predicted versus actual cutting force Fz

the desirability function values and the objective of minimizing cutting force, the process parameters shown in Table 8 can be implemented in order to attain the goals set.

Based on the design of experiments, the dry machining trials of Ti–6Al–4V were carried out. Based on the RSM technique DoE and the regression analysis, the cutting forces were theoretically calculated. The theoretical cutting forces from regression equations were compared with the experimental cutting forces measured using a dynamometer are shown in Fig. 8.

4 Conclusion

Dry turning of Ti–6Al–4V was experimentally performed and its cutting forces analyzed. The experimental observations were compared with the theoretical analysis, and the influence of various process parameters was investigated. Based on the observations, the following conclusions have been made.

- The quadratic regression model was established to understand the machining behaviour during dry turning, and the accuracy of model was determined by coefficient of determination of 93.86%.
- The ANOVA analysis was performed for finding the most influential cutting parameters, its contribution on output, and its interactions on cutting force.
- Depth of cut and feed were the parameters which have high influence on cutting force Fz with contribution of 64.351% and 14.037%, respectively.
- The cutting speed has less impact on cutting force with the contribution of 0.516%.
- The optimum cutting conditions were identified from design expert optimization tool and summarized from this research work.

References

1. Balaji JH, Krishnaraj V, Yogeswaraj S (2013) Investigation on high-speed turning of titanium alloys. Procedia Eng 64:926–935. https://doi.org/10.1016/j.proeng.2013.09.169
2. Pradhan S, Singh S, Prakash C, Krolczyk G, Pramanik A, Pruncu CI (2019) Investigation of machining characteristics of hard-to-machine Ti-6Al-4V-ELI alloy for biomedical applications. J Mater Res Technol 8(5):4849–4862. https://doi.org/10.1016/j.jmrt.2019.08.033
3. Ezugwu EO, Bonney J, Yamane Y (2003) An overview of the machinability of aero-engine alloys. J Mater Process Technol 134:233–253
4. Schalz H, Moriwaki T (1992) High-speed machining. Ann CIRP 41(2):637–645
5. Vaughn RL (1968) Faster machining of titanium. Am Machinist 6(3):98–99
6. President Titanium (1999) Machining and technical data, Hanson
7. Sutter G, List G (2013) Very high-speed cutting of Ti-6Al-4V titanium alloy-change in morphology and mechanism of chip formation. Int J Mach Tools Manuf 66:37–43. https://doi.org/10.1016/j.ijmachtools.2012.11.004
8. Chen P (1991) Characteristics of self-propelled rotary tools in machining high-performance materials. Jpn Soc Precis Eng 25(4):267–272
9. Wang ZM, Ezugwu EO, Gupta A (1998) Evaluation of a self-propelled rotary tool in the machining of aerospace materials. Tribol Trans 41(2):289–295
10. Shaw MC (2005) Metal cutting principles. Oxford University Press, London
11. Boujelbene M (2018) Investigation and modelling of the tangential cutting force of the Titanium alloy Ti-6Al-4V in the orthogonal turning process. Procedia Manuf 20:571–577. https://doi.org/10.1016/j.promfg.2018.02.085
12. Samsudeensadham S, Mohan A, Krishnaraj V (2020) A research on machining parameters during dry machining of Ti-6Al- 4V alloy. Mater Today Proc. https://doi.org/10.1016/j.matpr.2020.02.821
13. Cook NH (1953) Chip formation in machining titanium. In: Proceedings of the symposium on machine grind. Titanium, Watertown Arsenal, pp 1–7
14. Klocke F, Eisenblätter G (1997) Dry Cutting. CIRP Ann Manuf Technol 46:519–526
15. Sreejith PS, Ngoi BKA (2000) Dry machining: machining of the future. J Mater Process Tech 101:287–291
16. Shahebrahimi SP, Dadvand A (203) Optimization of cutting parameters for turning operation of titanium alloy Ti-6Al-4V material workpiece using the Taguchi Method, vol 685, pp 57–62. https://doi.org/10.4028/www.scientific.net/AMR.685.57
17. Rangaraju A, Vijayan K (2017) Study of process parameters on surface roughness during end milling of Titanium alloy (Ti-6Al-4V), ICRIPE
18. Vijay S, Krishnaraj V (2013) Machining parameters optimization in end milling of Ti-6A1-4V. Procedia Eng. 64:1079–1088. https://doi.org/10.1016/j.proeng.2013.09.186
19. Ali MH, Khidir BA, Ansari MNM, Mohamed B (2013) FEM to predict the effect of feed rate on surface roughness with cutting force during face milling of titanium alloy, HBRC, pp 263–269, vol 9
20. Karkolos NE, Galanis NI, Markopoulos AP (2016) Surface roughness prediction for the milling of Ti–6Al–4V ELI alloy with the use of statistical and soft computing techniques. Measurement, pp 25–65

Experimental Investigation on EN 19 Substrate Weld Cladded with Austenitic Stainless Steel for Improvement of Material Properties

R. Paullinga Prakash, S. Palani, S. Babu, and M. Selvam

Abstract The microstructure and other material properties like hardness, wear resistance and corrosion resistance are studied and compared between non-cladded substrate and cladded plate to come up with better option for cladding the material for suitable applications in the respective areas where it could be used. The tests are carried out on the cladded stainless steel of austenitic in the plates of steel with medium carbon as well as low alloy. The cladded plate has better results in corrosion and wear resistance which makes it suitable for applications, where that factor is majorly considerable. The materials are cladded using MIG welding, and the specimen undergoes various testing and microstructure study to determine the correlations and inference between cladded and non-cladded material.

Keywords MIG welding · Corrosion · Wear resistance · Cladding · EN 19 substrate · Austenitic stainless steel

1 Introduction

In order to reduce the rate of corrosion as well as erosion, a material of corrosive resistance is deposited on the base material and that process is called cladding of weld. This process is also improving the service life span of the component of engineering even an aggressive conditional environment. In the present research scenario, more number of researchers is applying this process in multiple industries like power plant which is operated by steam, industries for processing food fertilizing companies' also chemical oriental industries and so on. Especially, the cladding is utilized for mainly resisting of corrosiveness.

R. Paullinga Prakash · S. Palani · M. Selvam
Department of Mechanical Engineering, Vel Tech Multitech, Chennai 600062, India
e-mail: paullingaprakash@veltechmultitech.org

S. Babu (✉)
Department of Mechanical Engineering, PSG College of Technology, Coimbatore 641004, India
e-mail: sjsham@gmail.com

E. Natarajan et al. (eds.), *Materials, Design and Manufacturing for Sustainable Environment*, Lecture Notes in Mechanical Engineering,
https://doi.org/10.1007/978-981-19-3053-9_27

When increasing the supply, current/voltage of welding the penetration depth also increasing. But, the effect of supplying current has not as much. The maximum penetration obtained at the limit level of current is 60 cm/min [1]. When the rate of feed increases, the weld metal dilution as well as its dimensions like width height also increases, however, first decreases the bead angle of contact next increases. At the same time, excluding dilution the speed of welding takes effect in opposite [2]. Materials have most likely additional inherent with in our civilization than appreciating the majority of us. Many fields such as food production, civil construction, textile, transportation are all interconnected with any of the materials [3].

However, minimum growing documentation and inputs of updated permissible energy in the field grade of steel [4]. In the deposition of welding, the effect of nitrogen attention through the input of heat as well as the composition of shielding gas exerts the microstructural effect, toughness in minimum temperature also resistance for the corrosion pitting [5]. Duplex stainless steel was weld cladded with the plate of low-alloy steel through gas metal-based arc welding at various parameters. At 28 V and 145A, favorable microstructure of the clad portion is obtained and less corrosion is observed [6]. The hardness of cladded material is established more than that of base metal. Less corrosion resistance was observed comparing with duplex stainless steel which has two-phase microstructure of austenitic and ferritic structure [7]. While AISI4140 along with AISI 316 is welded by means of filer also without filler by using GTAW-based welding and observed that tensile strength of the metals after welding is found higher than that of their candidate metals [8].

The quenching process increases the specimen strength at the same time, the tempering the tempering process is also increases its toughness [9]. Low carbon steel plates are weld cladded with AISI 308H electrodes by GMAW process. It was found that the bead width continuously increases with increase in voltage, current and electrode wire diameter. The voltage parameter during welding is observed to be the most important factor that affects bead width followed by wire diameter and current [10]. The combinational product of clad with stainless steel having thin layer and its surfaces are joined with bond of veneer integral [11]. Cladded piece also showed more hardness over the base metal of uncladded furthermore resulting in more hardness due to the presence of martensitic phases in the layers of clad [12].

From the above literature, various corrosion methods were followed for estimating clad object anti-corrosion properties. Furthermore, more studies were completed the optimization of the parameters of the process.

In our research work having steel plates with low-alloy metal clad by mean of 316 surfaces which modifying with well layering together of various materials of alloying. Also, considering a number of weld overlay techniques including their individual unique purposes as well as its applications.

2 Design Modeling of Weld Cladding

The 2D design of the weld cladding is done using AUTOCAD software using simple line, rectangle and hatch commands/tools. The front view and top view of the design are given in Fig. 1a, and 3D design of the work is carried out in SOLIDOWORKS software. The 2D dimensions are converted to 3D sketch and extruded to a solid. The isometric view of the work is given in Fig. 1b.

The base metal is a rectangular plate of 150 × 90 mm with 20 mm thickness. The metal to be deposited on top of the base plate is given the shape of another rectangle of 120 × 70 mm with 2 layers. The clad metal is deposited in 2 layers, maintaining the thickness of 5 mm each. Hence, the entire clad thickness will be of 10 mm which is decided to provide the machining allowance. This will result in loss of thickness up to 2 mm.

A finished product that resembles an overlapping coin or disc, with the beads in the middle of the disc serving as the base material of weld beat as shown in Fig. 2a and the formation of the bead of weld is depicted in Fig. 2b.

During deposition of one metal over another, it is important to keep in mind about the weld bead formation. It is very important when two or more layers are to be deposited over another. While cladding, the approach is vertical deposition of metals over the marked area. When one pass is over, for the second pass, the weld cladding/deposition is done on the intermediate positions of the first clad layer.

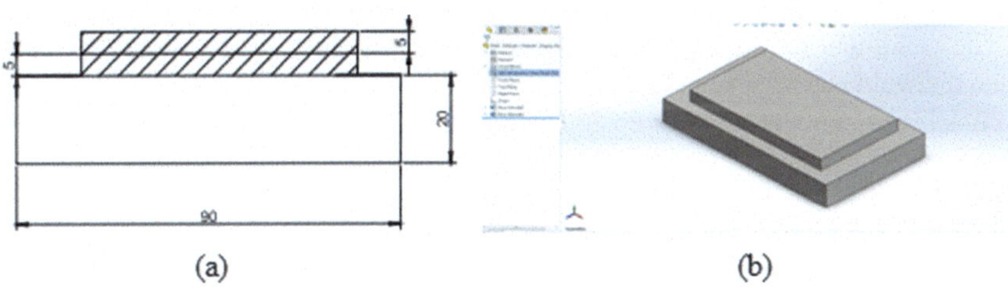

Fig. 1 Design of weld clad **a** 2D hatched portion shows the cladded area, **b** 3D isometric view

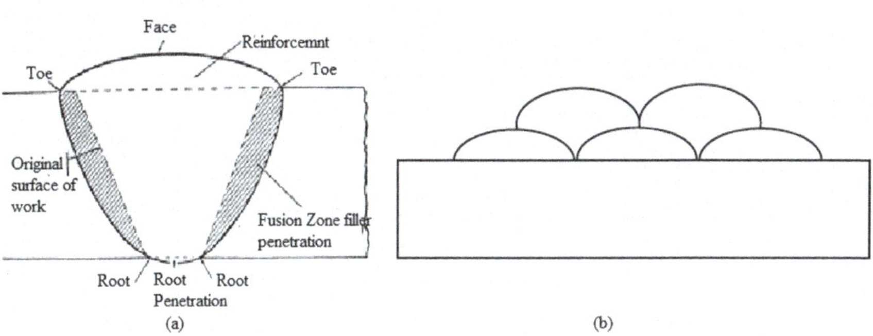

Fig. 2 **a** Various location of the beat of welding, **b** beat formation

3 Cladding Process

Cladding is a process of coating or depositing one metal over another for imparting corrosion resistant property. Weld cladding is one of the methods of cladding, where welding is used as a medium to deposit one metal over another. The metal inert gas welding is used here and proved to be efficient and provides less corrosiveness comparing others. MIG welding is one of most easily available welding methods next to arc welding. It can be used for welding thick plates and provides very less defect weldment.

In cladding of AISI 316 over EN19, the voltage and current parameters are set in the equipment. The voltage is set to 24.2 V, and the current is maintained at 139 A. The weld cladding is done on EN 19 plate with the help of MIG. By having DCRP, the electrons are emitted from the base metal which strikes the electrode producing the arc necessary to weld the wire and deposit it over the substrate. The austenitic stainless steel of one layer is cladded over the plate. The measured thickness is around 4 mm. The inter-pass temperature of the first clad layer is taken into account to minimize the risk of hydrogen-induced cracks. For reducing the risk of crack formation due to solidification/liquation in austenitic-based stainless steel risk furthermore to maintain the other mechanical properties of the metal. The second layer has 4 mm of clad thickness around it. Clad layer has 8 mm overall thickness from top face of the base plate.

Formation of the weld cladded plate with two layers and the photograph of machined EN 19 cladded stainless steel is shown in Fig. 3a, b, respectively.

The weldment requires cleaning to get rid of other dust, corrosive particles and make the weld surface even. To get a clean and even clad layer, the deposited layers are machined. This is done to ensure that the clad layer is even and there are no resemblances that would indicate that the layer was welded but resembles as a part of metal coated over another. This machining is done by shaper machine. At first, the sides and the area surrounding the clad layers are grinded. The shaper machine with HSS tool is used to clear the edges of the clad layer and make it right angled

(a) (b)

Fig. 3 **a** Weld cladded EN 19 plate, **b** machined cladded material or like a straight line. After clearing out the burrs, making the edges of the cladded area right/straight lines, the top face of the layer is machined.

4 Results and Discussion

The hardness of the materials is tested by using Rockwell hardness in the method of test general in nondestructive. When comparing the other testing method like Vickers hardness and Brinell hardness tests, the Rockwell hardness test is more user friendly as well as simple to handle. The indenter for testing the materials is made up of diamond with cone shape or else ball shape which made up of hardened steel. While testing samples, double loads are given in all time. Initially, with the small load is applied for forcing the materials to be tested through indenter, then the depth is noticed. Consequently, applying the extra load called as major load on the sample causes to develop the penetration of the depth. Now, releasing the extra load from the sample, the available force is reversed toward minor load. For estimating the value of the Rockwell hardness, the penetration depth is improved by applying as well as releasing major loads.

In view of corrosion test salt spray analysis, for obtaining the resistance of the corrosion passivity is due to a thin and invisible layer of oxides being formed on the surface. This process takes place through a reaction between the metal and the oxygen in the surrounding environment. For creating a different environmental corrosive, a test has been conducted with adjusting the salt spray in closed chamber that's called salt fog. The samples have also subjected with salt concentration at the time specified regarding the standards of industry; specifications as well as the type of product during the experimentation. For getting more atmosphere of corrosive, we have used sodium chloride 5% solution with standard in a typical test of salt spray. Furthermore, the conditions of the environmental such as humidity as well as temperature have been regulated frequently for simulating climate. With the uniform intervals, the samples have been examined for monitoring the level of corrosion and noted the level of red as well as white formation or the degradation of the coating.

The corrosion test was carried out in the salt spray analysis method. The chamber temperature was maintained from 34.5 to 35.5 °C. The pH of solutions used was 6.65–6.85 which is closer to becoming neutral yet acetic. The air pressure in the chamber was maintained at 14–18 psi. The concentration of sodium chloride solution used ranges from 4.80 to 5.30%. The collection of solution per hour is at the rate of 1.0–1.5 ml/h. The test component was loaded in the chamber at a position of 30°.

Wear test has performed using drum operating machine. The available and general test methods for wear on high-alloy carbon steel have been analyzed. The specimens for wear testing are shown in Fig. 4.

Specimen is machined in the form of pin which results in the loss of material due to abrasion and this is tabulated below. It is found that the base metal subjected to wear tends to lose more material than that of the cladded material signifying the

Fig. 4 Wear test specimen

Table 1 Wear test outcome values

Sample	Initial weight (*G*)	Final weight (*G*)	Abrasion loss (*G*)	Percentage
Cladded plate	15.3577	15.3041	0.0536	0.35
substrate (no clad layer)	14.0293	13.9621	0.06720	0.48

improved wear properties of the cladded plate. Table 1 shows the results obtain from the wear test.

By using cladding process, EN 19 weld cladded with austenitic stainless steel can improve industrial applications as such as chemical industry pipelines, oil /gas industry, gear boxes, boilers and so on. Hardness test has been carried out in the base plate (EN19) and cladded plate surface. The Rockwell hardness B scale is used to measure the hardness of the materials. The HRB designation of hardness is used. The cladded material shows lower hardness than that of the base plate.

The corrosion test has been carried out for 12 h. In these 12 h, the cladded material never showed any signs of corrosion, while the base metal got corroded easily. Figure 5 shows testing specimens (a) before (b) after corrosion.

Figure 6 shows microstructure images of the substrate clad material and weld fusion zone of (a) EN 19 (b) stainless steel (AISI 316) at hardness of 98 HRB having wear 0.48% of material loss and (c) clad layer 90 HRB having wear 0.35% of material loss.

Figure 7 shows the graphical representation of the differentiation of the hardness values (HRB) between the base metal and clad layer.

Also, it proves that the cladded plate becomes efficient in a corrosive environment. The microstructure image of the substrate, clad material and weld fusion zone is obtained. The substrate shows a very fine grain structure. The weld fusion zone is where both the metals are welded. There is a visible difference of two different grain structures, where the bottom one is that of the substrate and top is of the stainless steel. The carbon diffusion zone is visible. The clad layer showed delta ferritic (dark) and austenitic grain structures (lighter and elongated structure). Figure 8 shows the

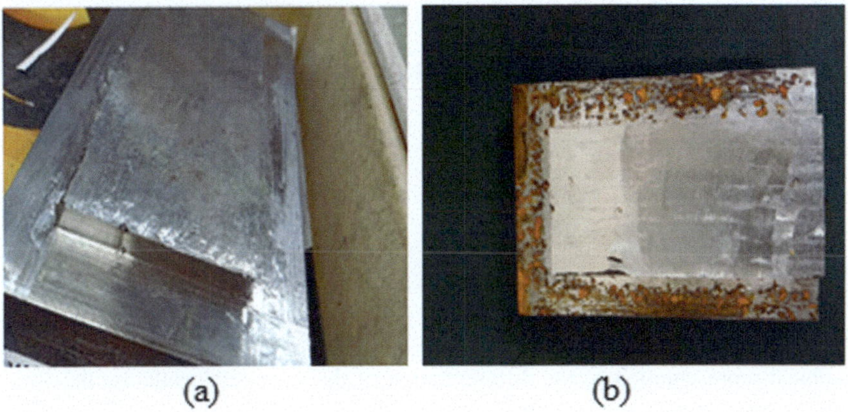

Fig. 5 Test specimens **a** before, **b** after corrosion

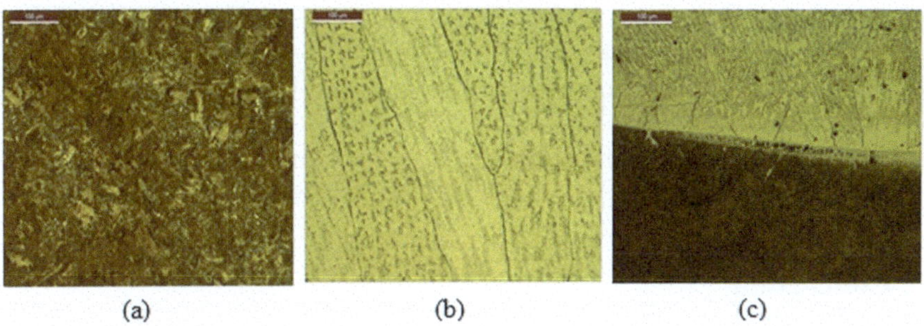

Fig. 6 Microstructure images of the substrate clad material and weld fusion zone of **a** EN 19, **b** stainless steel (AISI 316) at hardness of 98 HRB having wear 0.48% of material loss, and **c** clad layer 90 HRB having wear 0.35% of material loss

Fig. 7 Hardness values (HRB) between the base metal and clad layer

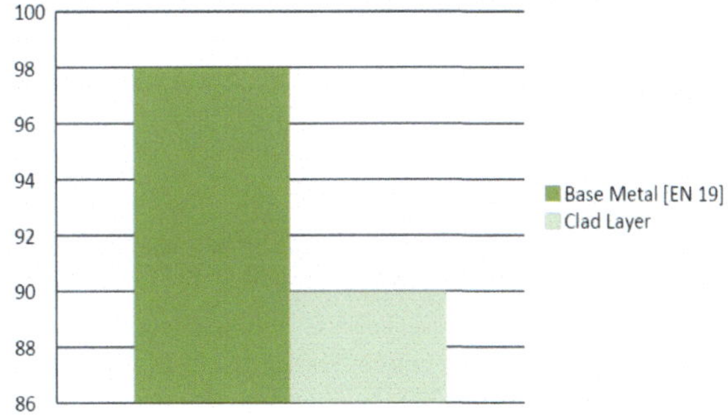

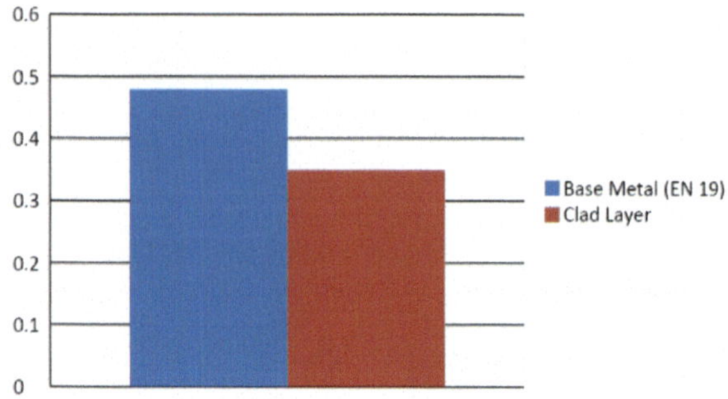

Fig. 8 Material loss % between the base metal and clad layer using wear test

graphical representation of the differentiation of the wear test results by the % of material loss between the base metal and clad layer.

Wear is nothing but the gradual removing or deformation of a material at solid surfaces. The wear test has been carried out in the drum operation machine. The cylinder size of the drum abrasive has diameter 150 mm and the length 500 mm. Abrasive sheet grade is 60. The equivalent number of revolutions of the drum is 84 times. The substrate sample of wear test had a material loss of about 0.48% while the cladded sample had 0.35%. This difference in the loss of material due to abrasion shows that the cladded plate has better wear property than the noncladded substrate.

5 Conclusion

The EN 19 cladded material has comparatively lower hardness than non-cladded material. The substrate shows fine grains under metallography which might have influenced the hardness of it. The cladded piece of base metal has high wear resistance and corrosion resistance. The components manufactured from EN 19 can be employed with this technique to get better wear and corrosion resistant property and life. The weld deposition rate has been changed the influences of weld current. Austenitic stainless steel clad layer has both austenitic and delta ferritic grain structures. The corrosion of the substrate is prevented by enriched chromium along with nickel from stainless steel electrode which was cladded to the base plate. High-strength low-alloy steel alone cannot be used in structures because of its high cost, and hence, cladding is preferred to it.

Acknowledgements The current research was supported by the Early Career Research award (Grant numbers: ECR/2018/002378) awarded by Department of Science and Technology, Science and Engineering Research Board, India. We wish to express our sincere thanks to almighty and the people who extended their help during the course of our work. We are greatly and profoundly thankful to our honorable Chairman, Col. Prof. Vel. Shri Dr. R. Rangarajan B.E. (ELEC), B.E. (MECH), M.S.(AUTO), D.Sc., and Vice Chairman, Dr. Mrs. Sakunthala Rangarajan MBBS., for facilitating us with this opportunity. We take this opportunity to extend our gratefulness to our respectable Chairperson and Managing Trustee, Smt. Mrs. Rangarajan Mahalakshmi Kishore B.E.,

M.B.A., for her continuous encouragement. Our special thanks to our cherishable Vice President Mr. K. V. D. Kishore Kumar B.E., M.B.A., for his attention toward students' community.

References

1. Karadeniz E, Ozsarac U, Yildiz C (2007) The effect of process parameters on penetration in gas metal arc welding process. Mat Des 28:649–656
2. Nouri M, Abdollah-Zadeh A, Malek F (2007) Effect of welding parameters on dilution and weld bead geometry in cladding. J Mat Sci Tech 23(6):817–822
3. William D (2010) Materials science and engineering: an introduction 8e. Wiley, Hoboken, New Jersey, ©2012. ISBN 9780470419977 (Text Book 8th edn)
4. Wessman S, Karlsson L, Pettersson R, Östberg A (2010) Study of the influence of tungsten in super duplex stainless steel welds. In: Proceedings International Conference Duplex World, Beaune, France, pp 1077–1088
5. Chakrabarti B, Hrishikesh D, Santanu D, Tapan P (2013) Effect of process parameters on clad quality of duplex stainless steel using GMAW process. Trans Indian Inst Metals 66(3)
6. Anup KV, Bidhan CB, Protap R, Samiran D (2014) The effectiveness of duplex stainless steel cladding deposit by gas metal arc welding. Indian Weld J 47(4)
7. Madduru P, Sam A, William M, Mohan P (2014) Assessment of mechanical properties of AISI 4140 and AISI 316 dissimilar weldment. Procedia Eng 75:29–33
8. Bhiswajnit K, Mithun S, Santanu D (2016) Weld cladding with austenitic stainless steel for imparting corrosion resistance. Indian Weld J 49(1):74–81
9. Rahul G, Manoj SR, Keshav O, Geethapriyan T (2018) The effect of heat treatment on microstructure and mechanical properties of EN19 alloy. Int J Mater Sci Eng 6(2)
10. Sanjit K, Dipendra K (2019) Effects of welding parameters and electrodes on weld bead using gas metal arc welding. Int J Eng Adv Technol 9(1)
11. Davis (2019) Stainless steel cladding and weld overlays. ASM specialty handbook: stainless steels, pp 108–119, 06398G J.R., Product code: Z
12. Shubrajit B, Manidipto M, Parijit S, Anish N, Viorelhn P (2020) Microstructural and wear properties of annealed medium carbon steel (EN8) cladded with martensitic steel (AISI410). Metals 10(958). https://doi.org/10.3390/met10070958

Experimental Investigations of Composite Material Using Bamboo Fiber Reinforced with Polypropylene Plastic Additives

D. Ramesh Kumar, D. Elangovan, R. Dharanidharan, Pasupuleti likhitha, and E. R. Dharanivelan

Abstract Our day-to-day life necessities of human life cannot imagine the world without composites. As necessity is the mother of invention, expectation increases day by day and the necessity also increases which ultimately gives rise for continuous research for attaining the expected result. The synthetic composite fibers are replaced by fiber-reinforced polymer composite material because of their properties and abilities. This happened due to the dominance in the properties similar to non-toxic, non-abrasive, cost efficiency and easily availability. When we come into the comparison of synthetic material with natural fiber there have been increased in mechanical properties similar to tensile strength and tensile modulus, but particular properties similar to specific gravity and specific tensile modulus will also meet the expectation for the properties needed. We mixed epoxy resin with bamboo along with polypropylene plastic and NaOH to improve its properties. The experimental characterization was studied in this research. We observed that the Composite 3 has higher tensile strength than other two composites. Significantly, Composite 2 has higher compression strength. Composite 1 has higher hardness and significantly better in both tensile and compressive strength.

Keywords Bamboo · Polypropylene · Tensile test · Impact test · Epoxy resins · Flexural test · Performance characteristics

1 Introduction

Growing competition in composites still expects more vide of variety regarding properties. Composite materials are the combination of more than two different

D. Ramesh Kumar (✉) · R. Dharanidharan · P. likhitha · E. R. Dharanivelan
Department of Mechanical Engineering, Adithya Institute of Technology, Coimbatore, Tamil Nadu 641107, India
e-mail: dramesh184@gmail.com

D. Elangovan
Department of Mechanical Engineering, PSG Institute of Technology and Applied Research, Coimbatore, Tamil Nadu 641062, India

E. Natarajan et al. (eds.), *Materials, Design and Manufacturing for Sustainable Environment*, Lecture Notes in Mechanical Engineering,
https://doi.org/10.1007/978-981-19-3053-9_28

composite material combined with dissimilar properties in the form of new material which entirely have different properties from the individual constituents [1]. Firstly, contemporary composites were used in the industrial area. It is the time; industries are looking forward for lower cost-reinforced composite material by using biodegradable. Bamboo fiber was also first examined and got a great result because they are biodegradable, recyclable, as well as sustainable. A natural fiber has been yield composite material with maximum strength-to-weight ratio. Bamboo can have a quality of industrial products. Natural glass fibers are slowly replaced by bamboo because of its high strength-weight ratio [2].

The reinforced composite materials are of two type; they are synthetic and natural fibers. To reach the expectation of the customers regarding environmental security, many natural fiber-reinforced composites were introduced. NFRCs have vast range of benefits over synthetic-based fiber-reinforced composite material. The advantages of natural fiber-reinforced composite material compared to synthetic are very high like greater strength-to-weight ratio, greater strength at temperatures, more creep resistances and maximum toughness, low weight, greater durability and design flexibility [3]. The matrices used in natural fiber-reinforced composites are either thermoset or thermoplastic. The commonly used thermoset matrices are polyester, epoxy and phenolic resin. In thermoplastic matrix, elastomers, polyethylene and polypropylenes occupied the maximum-scale position. Composite has been differentiated into three groups. They are ceramic matrix composite, metal matrix composite and polymer matrix composite.

The material we use will be selected by the application we are going to do and the result we expect. Polymer matrix composite is the most commonly used composites. This preference is mainly due to their light weight and specific properties which are so better compared to the ceramics and metals. In polymer matrix composites, low temperature and pressure can be processed. In this study, the matrix material is epoxy [4, 5]. Actually, epoxy has a very good property with glassy appearance having major advantages like good electrical insulating properties, good mechanical properties, good environmental, good adhesion to other materials and chemical resistances, etc. A fiber-reinforced polymer composite is formed when epoxy treated with natural fiber to synthesize; there is an interface happened between the matrix as well as the fiber [6, 7].

Yin Jian and Li Ming's [8] have presented that in this experimental research they combined the bamboo with different poly dendrimers for improving the internal strong of the final composite. The external morphology of the final composite was tested by scanning electron microscope (SEM) and atomic force microscope. The external characteristics are certified by X-ray photoelectron spectroscopy and Fourier transform infrared (FT-IR). This experiment proved that the resulting composite materials mechanical properties were improved. The result explained that C-PAMAM and L-PAMAM and dendrimers combined very well with the bamboo fiber's surface. It was specified that mechanical properties of the composite material were improved. They also stated that the ILLS results showed that combining g C-PAMAM onto fiber surface, increased its interfacial adhesion.

Okwuchi Smith Onyekwere et al. [9] have mentioned usage of industrial engineering in bamboo fiber production. It comprises of the processing bamboo fiber polyester composites by using experimental optimization and design techniques. The researchers used Taguchi and Grey relational analysis to find out the multiple response design and optimization of bamboo fiber polyester composites. They took samples for bamboo fiber composites at fiber contents of 10, 20, 30, 40 and 50 weight percentage. The result showed positive in mercerized-acetylated and 50 weight percentage fiber, this composition is highly preferred due to its impact on the properties. 51.1 percentage improvements are seen in the Grey relational grade, with respect to the initial parameters setting after the optimization by Grey relational analysis.

Adamu Muhammad et al. [10] have presented that about the dominant emerging material, i.e., plastic and ceramics composites. Composites have been consistent increasing in the volume and number of usages in existing markets and creating new markets. The composites are proving their weight saving property. But the problem needs to be solved is, industry needs a low-cost composite with ecofriendly nature. Bamboo fibers had properties like recyclable, biodegradable and sustainable. The practitioners and researchers have find out high quality of industrial products which can be generated from bamboo fibers. Bamboo fibers replaced natural glass fiber because of their high strength-weight ratio. New update regarding this bamboo composites are given in this article.

Guo Chen [11] have mentioned about the laminated bamboo lumber properties. They experimented LBL and saw very good response. The theoretical and experimental data matched perfectly. They made compared LBL with other composite materials about mechanical and failure modes. They also stated that further investigation is required for betterment of understanding behavior of LBL.

Alba Fernández [12] has mentioned that a researcher tells about the analysis data of the acoustic emission signals released by bamboo samples from tensile tests. Bamboos used in this research paper are increased from Bambusa vulgaris species. They conducted spectral analysis of spectral acoustic emission signal. The material suffers before breakage of the inner wall to until the final breakage of the outer wall. The research reported in this paper was increase of the center of gravity of the signal spectrum has been observed. Center of gravity more than 200 KHz has associated with fiber.

Xiaoping Li et al. [13] have researched about improving the properties of bamboo fiber and wood fiber by using pre-treatment. Due to this reason, the pre-treatment to increase the mechanical properties of bamboo fibers for the usage of woody fiber or polypropylene composite material were compared with NaOH treatment. It was found that used to reduce the amount of pectin value $p < 0.0001$ and improved tensile strength. Normally, pre-treatment there is no accurate effects on the water absorption. The research reported in this paper is stated that improvements in panel properties $p < 0.05$ and also reduced thickness swell $p < 0.05$.

2 Selection of Materials and Methods

2.1 *Methodology*

During the first step study of fiber material availability in the world. During the second step selection and preparation of the fiber and plastic material with suitable resins. During the third step preparation and generate mixing ratio for fiber, plastic and resin. During the fourth step mechanical testing like tensile test, impact test and flexural test. Finally, investigations on three samples compare with different characterizations and select suitable sample.

2.2 *Selection of Materials*

2.2.1 Bamboo Fiber Material

It is a natural fiber material used widely for all usage and applications. This is cellulose fiber extracted from the natural bamboo. It is also act as anti-bacterial and UV protected property, which is not only used in textile but very useful for better performance in composite material due to increase in mechanical properties tensile strength, durability and so on. Composition of chemical properties which are similar to the other best fibers like jute, flax, etc. The three to four years old bamboo is used for fiber preparation. Fiber is listed in during the chemical treatment process bamboo stems are treated to chemical process generated. It is softer than cotton because it has various micro gaps and it has moisture absorption (Fig. 1).

2.2.2 Epoxy Resin

Epoxy resin is used for adhesive purpose because of that strong property it adhesive the two or more different structural materials. Epoxy and resin both are having adhesive property. Where the property of epoxy is much better than the resin. The epoxy resin (LY-556) has a chemical resistivity and its medium viscosity and additionally it has a higher mechanical property. Comparing to resin epoxy is stronger in the sense of having much faster drying time (Fig. 2).

2.2.3 Polypropylene Plastic

Polypropylene plastic is one of the types of thermoplastics used commonly in industries. It is used in plastic packages, plastic machine parts and equipment material; properties of polypropylene are highly flexible, lighter and low-density properties.

Fig. 1 Bamboo fiber

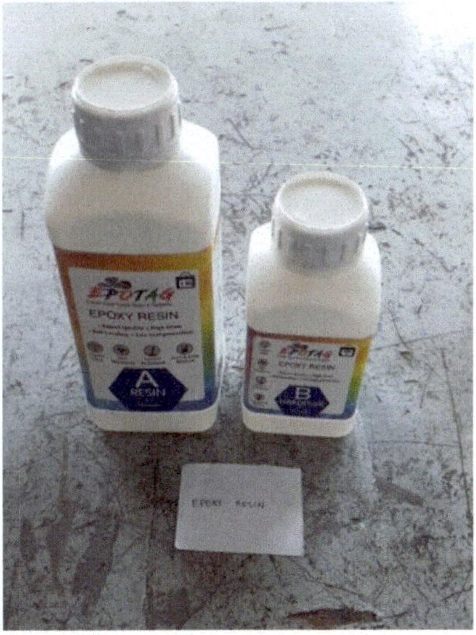

Fig. 2 Epoxy resin

The polypropylene also known as "the steel of plastic." Due to fatigue resistance, the material will regain its original shape after bending.

2.2.4 Sodium Hydroxide (NaOH)

The caustic soda or lye is commonly known as the sodium hydroxide. When it is dissolved it may produce heat. The reaction of the sodium hydroxide is tremendous on stronger acids and water solution the color of sodium hydroxide is white, and it smells odorless, and it has crystalline structure, readily available material. It plays a major role in producing the soaps, rayon, paper, exploding products, dyes and petroleum products.

2.2.5 Selection of Composition of Fiber-Reinforced Composite

See Table 1.

3 Results and Discussion

3.1 Composite 1 or Sample 1

$$\mathrm{C}-1:\mathrm{Epoxy\ resin}(10\%)+\mathrm{Polypropylene\ plastic}(30\%)+\mathrm{Bamboo\ Fiber}(60\%)$$

See Fig. 3.

Here, the Composite 1 is of the composition of Epoxy resin (10%) + Polypropylene plastic (30%) + Bamboo Fiber (60%) is mixed well and hardened with the required materials and procedure.

Table 1 Composition of fiber-reinforced composite

Composites	Composition
C-1	Epoxy resin (10%) + Polypropylene plastic (30%) + Bamboo Fiber (60%)
C-2	Epoxy resin (10%) + Polypropylene plastic (20%) + Bamboo Fiber (70%)
C-3	Epoxy resin (10%) + Polypropylene plastic (10%) + Bamboo Fiber (80%)

Fig. 3 Composite 1

Fig. 4 Composite 2

3.2 Composite 2 or Sample 2

$$C - 2 : \text{Epoxy resin}(10\%) + \text{Polypropylene plastic}(20\%) + \text{Bamboo Fiber}(70\%)$$

Here, the Composite 2 is of the composition of Epoxy resin (10%) + Polypropylene plastic (20%) + Bamboo Fiber (70%) is mixed well and hardened with the required materials and procedure (Fig. 4).

3.3 Composite3 or Sample 3

$$C - 3 : \text{Epoxy resin } (10\%) + \text{Polypropylene plastic } (10\%) + \text{Bamboo Fiber } (80\%)$$

Here, the Composite 3 is of the composition of C-3: Epoxy resin (10%) + Polypropylene plastic (10%) + Bamboo Fiber (80%) is mixed well and hardened with the required materials and procedure (Fig. 5).

Fig. 5 Composite 3

3.4 Results and Comparisons

3.4.1 Tensile Test

This above mentioned test method is determining various strength like ductility, strain hardening, yield strength, ultimate strength, Poisson's ratio and Young's modulus. The method of testing for both and measurement of force and testing of material is same.

Tensile tests were carried out as per ASTM E8 standard for bamboo composites in FIE make universal testing machine. The composites were tested and given the results in Table 2.

$$\text{Specimen dimensions : Diameter} = 12.5\,\text{mm, Gauge length} = 50\,\text{mm}$$

From the above tabulation, we can understand that the Composite 3 has ultimate tensile strength than the other Composites 1 and 2. Due to less composition of plastic, the Composite 3 has higher tensile strength compared to Composites 1 and 2 (Fig. 6).

From the above graph, we can understand that the Composite 3 has ultimate tensile strength than the other Composites 1 and 2. Due to less composition of plastic, the Composite 3 has higher tensile strength compared to Composites 1 and 2. The data clearly explains the condition of the specimen before the break even of the tensile test where the specimen in the unbroken condition. Specimen 1, 2 and 3 are taken for the tensile test. The data clearly explains the condition of the specimen after the break even of the tensile test where the specimen in the broken condition. Specimen

Table 2 Tensile test results

S. No.	Composite	Yield strength (kN)	Yield stress (MPa)	Tensile strength (MPa)	Percentage of elongation
1	Composite 1	1.52	9.52	14.54	3.96
2	Composite 2	3.75	12.35	24.78	5.22
3	Composite 3	5.26	15.43	32.25	7.14

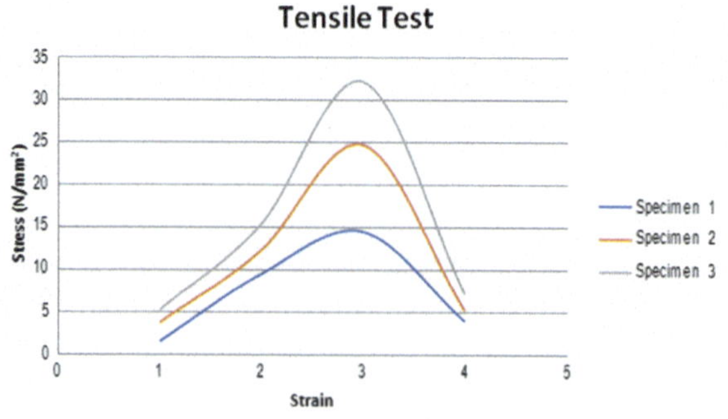

Fig. 6 Comparisons of three composite and tensile test

1, 2 and 3 are taken for the tensile test. Test results of the specimen 1, 2 and 3 are compared and calculated.

3.4.2 Impact Test

The testing facility available while the test is carried out for the Charpy impact test was given in Table 3.

Impact test was carried out as per ASTM E18 standard for composites. The samples were tested, and the results were tabulated. The results were presented in Table 4.

The Charpy impact is to evaluate the resistance of plastics to breakage by flexural shock according to method of ASTM D6110, which indicates that amount of energy required to break specimens. The data clearly explained the condition of the specimen before the break even of the Charpy impact test where the specimen in the unbroken condition. Specimen 1, 2 and 3 are taken for the Charpy impact test. Test results of the Specimen 1, 2 and 3 are compared and calculated. The data explains the condition of the specimen after the break even of the Charpy impact test where the specimen in the broken condition. Specimen 1, 2 and 3 are taken for the Charpy impact test. Test results of the Specimen 1, 2 and 3 are compared and calculated (Fig. 7).

From the above tabulation, we can understand that the Composite 3 has more impact than the other Composites 1 and 2. Due to less composition of plastic, the Composite 3 has higher impact compared to Composites 1 and 2.

Table 3 Testing facility

Machine No.	CL/ME/IMPA02
Model	AIT-300-EN
Capacity	300 J
Validity	29.11.2021
Ambient Temp. °C	25.2

Table 4 Comparisons of impact test results

S. No.	Composite	Test temp. degree (°C)	Thickness mm	Width mm	Length mm	Charpy test energy in Joules
1	Composite 1	Room temperature	10.01	10.02	55.00	6
2	Composite 2	Room temperature	10.01	10.02	55.00	8
3	Composite 3	Room temperature	10.01	10.02	55.00	12

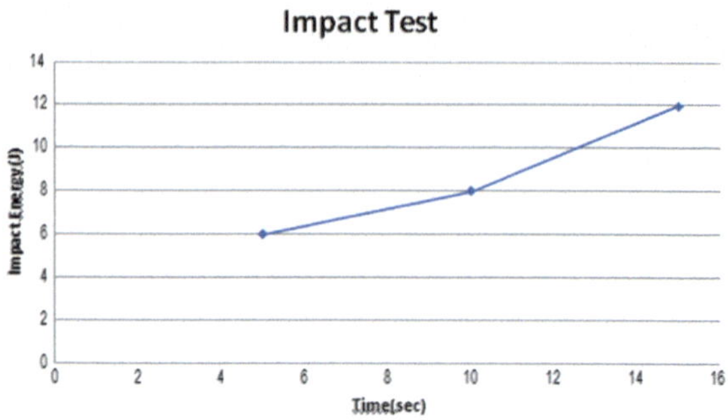

Fig. 7 Comparisons of three specimen and impact (Charpy) test

Table 5 Flexural test result

S. No.	Composite	Maximum force (kN)	Maximum displacement (mm)	Compressive strength (MPa)
1	Composite 1	2.160	0.6	3.779
2	Composite 2	2.658	0.92	5.877
3	Composite 3	3.187	1.45	6.832

3.4.3 Flexural Test

A flexure test is more inexpensive than a tensile test. The material is two points of contact, and force is applied to material through two points of contact until the sample fails. The flexural test was carried out as per ASTM standard for bamboo composite in FIE make universal testing machine. The results are analyzed and given in Table 5.

From the above tabulation, we can understand that the Composite 3 has ultimate compression strength than the other Composites 1 and 2. Due to less composition of plastic, the Composite 3 has higher composite strength compared to Composites 1 and 2.

From Fig. 8, we can understand that the Composite 3 has ultimate compression strength than the other Composites 1 and 2. Due to less composition of plastic, the Composite 3 has higher composite strength compared to Composites 1 and 2. The above figure clearly explains the condition of the specimen before the break even of the flexural test where the specimen in the unbroken condition. Specimen 1, 2 and 3 are taken for the flexural test. Test results of the Specimen 1, 2 and 3 are compared and calculated. The above figure clearly explains the condition of the specimen after the break even of the flexural test where the specimen in the broken condition. Specimen 1, 2 and 3 are taken for the flexural test. Test results of the Specimen 1, 2 and 3 are compared and calculated. Compared to these results, the Composite 3 [Epoxy resin (10%) + Polypropylene plastic (10%) + Bamboo Fiber (80%)] has higher tensile strength than other two composites. Significantly, Composite 2 [Epoxy resin (10%) + Polypropylene plastic (20%) + Bamboo Fiber (70%)] has higher compression strength. Composite 1 [Epoxy resin (10%) + Polypropylene plastic

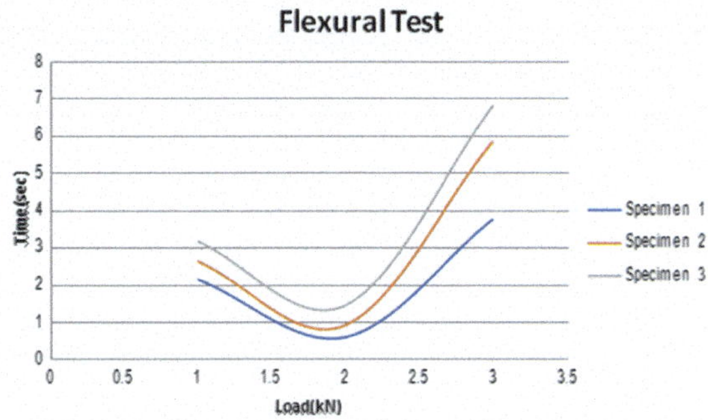

Fig. 8 Comparison of three composite material for flexural test

(30%) + Bamboo Fiber (60%)] has higher hardness and significantly better in both tensile and compressive strength.

4 Conclusion

Bamboo fiber and polypropylene plastic with epoxy resin composites were fabricated, and its mechanical behavior were studied [1, 2]. The results of experimental investigation on reinforcement with bamboo fiber materials are reported [9, 10]. It was claimed that the properties of mechanical should be increased impact strength and hardness and so on. A bamboo composite material has to increase the mechanical properties of the composite. This is due to load transfer and bonding of fibers. The various tests have been conducted in the composite material, and the results were discussed and reported in this paper. The all the test results here by concluding this bamboo fiber with epoxy resin composite have an excellent mechanical characteristic. The Composite 3 [Epoxy resin (10%) + Polypropylene plastic (10%) + Bamboo Fiber (80%)] has higher tensile strength than other two composites. Significantly, Composite 2 [Epoxy resin (10%) + Polypropylene plastic (20%) + Bamboo Fiber (70%)] has higher compression strength. Composite 1 [Epoxy resin (10%) + Polypropylene plastic (30%) + Bamboo Fiber (60%)] has higher hardness and significantly better in both tensile and compressive strength.

References

1. Tripathi P (2017) The tensile and flexural properties of bamboo hybrid composite. Int J Nat Fiber 14:1–5
2. Devi D (2016) The shear parameters of soil with and without bamboo fiber. Int J Eng Nat Fiber 5:13–15
3. Roslan SAH (2015) The tensile properties of the laminated composite. Int Eng Sci 81:111–125
4. Sing HJ, Suraj A, Suryaprakash S (2014) The natural fiber reinforced composites. Int J Eng Composite 133:34–95

5. Praveen Kumar J, Shri Hari R, Sachin Tendulkar T (2018) Natural fibers and its composites for engineering applications. Int J Sci Eng 55:87–92
6. Laemlaksakul V (2010) Physical and mechanical properties of particle board from bamboo waste. J Nano Sci Fiber 76:45–46
7. Ramachandran M (2015) Experimental study of bamboo using banana and linen fiber reinforced polymeric composites. Int J Compos Fiber Mech 76:45–46
8. Jian Y, Ming's L (2018) Enhanced mechanical properties of bamboo fiber/HDPE composites by grafting poly (amido amine) onto fiber surface. Compos Surf 25(12):1-12
9. Onyekwere et al (2020) Multi-response optimization of bamboo fiber reinforced unsaturated polyester composites using hybrid Taguchi—grey relational analysis method. J Ind Prod Eng 38(3):1–10
10. Muhammad et al (2018) Recent developments in bamboo fiber based composites: a review. Polym Bull 1–28
11. Chen et al (2018) Mechanical behavior of laminated bamboo lumber for structural application: an experimental investigation. Eur J Wood Wood Prod 78(6):1–11
12. Fernández et al (2019) Acoustic emission analysis of raw bamboo subjected to tensile tests. Mech Adv Mater Struct 28(13):1–10
13. Li X et al (2017) Improving the performance of bamboo and eucalyptus wood fiber/polypropylene composites using pectinase pre-treatments. J Wood Chem Technol 38(6):44–50

Manufacturing and Industrial Engineering

Effective Manpower Effort Reduction and Improving the Efficiency of Order Picking Process Using Class-Based Method in a Fabric Store

S. Gowtham, A. Prabukarthi, and R. Ragul

Abstract In the apparel industry, the various tasks involved start from procurement of raw material in the form of fabric and converting it into a finished product based on the requirement of the customer. The major processes involved in the fabric store namely unloading of fabric rolls on the pallets, movements of the pallets to the store, loading of rolls on the racks, retrieving rolls on the racks to the pallets, cutting of rolls to the required length, and retrieving of rolls loading to the container for delivery. The foremost process in the fabric store is retrieving rolls followed by cutting the rolled fabrics has a major impact in the context of overall processing time in connection with the various sequence of processes performed in the fabric store. The main objective of current research work is to reduce the manpower effort and redesign storage location based on class-based inventory classification which helps in improving the overall fabric picking and fabric cutting process efficiency. A major bottleneck in the picking process is systematically analyzed by regrouping fabric according to buyer, product types, and percentage of cotton. Various research references concluded that the above-said issues were predominately addressed by performing work-study followed by visual management using value stream mapping (VSM), inventory management analysis, spaghetti diagram, and simulation modeling. The expected outcome from the proposed theoretical suggestion will have an impact on increasing the order picking efficiency from 21 to 63%, and the suitable proposal was suggested to reduce the manpower effort and eliminate the waste while sizing the fabric as per the requirements of various processes.

Keywords Manpower effort reduction · Order picking efficiency · Fabric store simulation modeling · Value stream mapping · Spaghetti diagram

S. Gowtham (✉) · A. Prabukarthi · R. Ragul
Department of Mechanical Engineering, PSG College of Technology, Coimbatore, Tamil Nadu 641004, India
e-mail: gowthamcloud@gmail.com

A. Prabukarthi
e-mail: apk.mech@psgtech.ac.in

E. Natarajan et al. (eds.), *Materials, Design and Manufacturing for Sustainable Environment*, Lecture Notes in Mechanical Engineering,
https://doi.org/10.1007/978-981-19-3053-9_29

1 Introduction

Inventory is critical to a company's financial and operational success. It is, first and foremost, a financial investment for any business, and second, it is necessary for order fulfillment to customers. Because of the substantial cost of raw materials, proper inventory management is critical to an industry's success. A company may fail as a result of unnecessary surplus stocks, according to studies. As a result, inventory management optimization has a major influence on the effective maintenance of a company's financial health.

XYZ is a garment industry, which manufactures apparel products for the age group of newborn babies to infants. The raw materials like fabric rolls are stored in a fabric store. As per customer requirements, the apparel products, as well as fabric rolls, are also shipped to the customer. The fabric rolls are manufactured by their processing division and 4000 kg of fabric roll are transported by truck daily to the fabric store of the production unit. However, the placement of fabric rolls on the rack still has an issue. The fabric rolls are not well organized which leads to an increase in the picking time of the respective fabric which in turn increases the overall production lead time.

The effective utilization of manpower in the processes namely loading, unloading, and stacking plays a pivotal role in the knitting store. Manpower is not allocated appropriately in an optimized manner; there is a possibility of underutilization and overutilization of manpower. It is evident from Fig. 1 that the order picking process consumes more time (63%) compared to other processes in the store. To have effective material flow from knitting store to cutting section by minimizing the picking time and travel distance.

The main objective of current research work is to study the existing method of handling the rolled fabric from the store to the cutting section and identify a suitable proposal to reduce the efforts of workers [1] and also to redesign the storage location assignment using a class-based storage method. Many tools can help analyze the class-based method in the warehouse such as FSN analysis, ABC analysis, VED analysis, SDE analysis, and HML analysis. It was identified that "FSN analysis" is suitable for the current research because it can meet the expected goal and may reduce the number of operators while another tool concentrates on inventory management [2]. The letters F, S, and N represent fast-moving, slow-moving, and non-moving

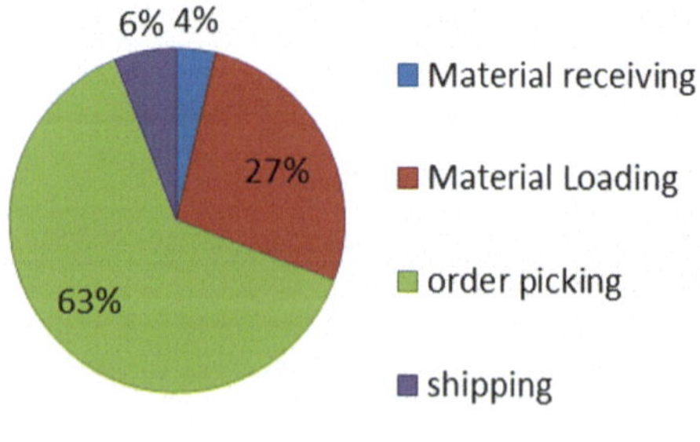

Fig. 1 Percentage of process utilization in the fabric store

items. FSN classification system classifies according to the frequency of raw materials issue. How often they are used. Fast-moving items that are issued frequently beyond inventory once in a certain period. Slow moving which in turn are infrequently exhibited items that may appear once in certain cases limit. Non-moving items are those items that are never issued from inventory at a specific time limit.

2 Methodology

The field study was performed in the warehouse of the apparel industry extensive data are collected in connection to the order picking process, and a suitable literature review was done in line with the problem statement, and a defined objective of reducing the man effort and reducing the order picking time was finalized. Initial understanding of the process was obtained using time study which helped in identifying the bottleneck process order picking time. Method study was done to understand the man effort, and a suitable proposal was suggested to reduce the man effort. VSM was used to have an overview of the processing time in a visualized manner, and discrete event simulation was used to visualize the expected outcome of the proposed kaizen. The detailed flow of the current research work is presented in Fig. 2.

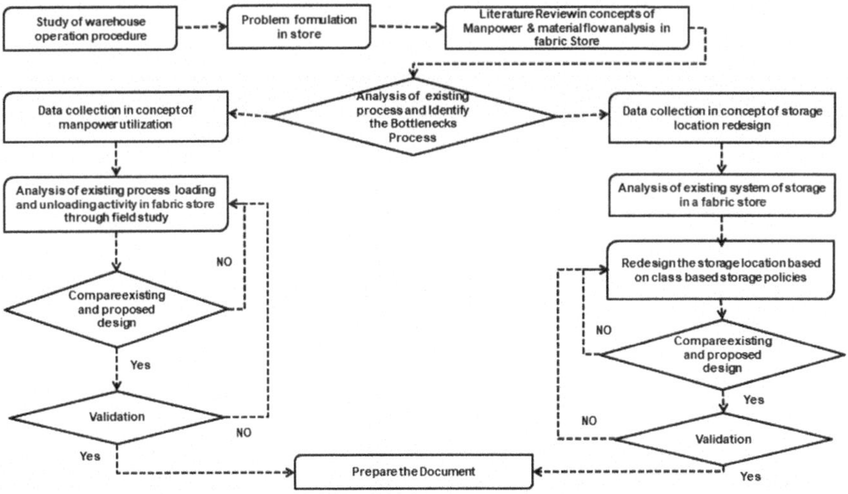

Fig. 2 Methodology

3 Analysis of Manpower in the Fabric Store

3.1 Responsibilities of the Load Man in the Fabric Store

Load man in the fabric store is involved in loading and unloading the finished fabric and transporting the fabric rolls to the cutting section. The major task performed by the load man is viz, unloading of fabric from the container on the pallets, transporting pallets, organizing rolls on racks, retrieving rolls followed by issuing it to spreaders, pintables, and manual cutters, retrieving rolls and loading to a container for delivery. They are also involved in the cutting of rolled fabrics to the required length.

3.2 Spaghetti Diagram

It is a special tool for determining the distance traveled by people or materials. Therefore, to reduce the distance traveled by parts or people, spaghetti diagrams are useful. The benefit is either faster delivery or the same delivery with less effort. This diagram is used to track product, paper or information, and people flow.

Figure 3 shows that the spaghetti diagram for the existing store, and the blue line shows the movement of the worker, and the red dots are the basic points.

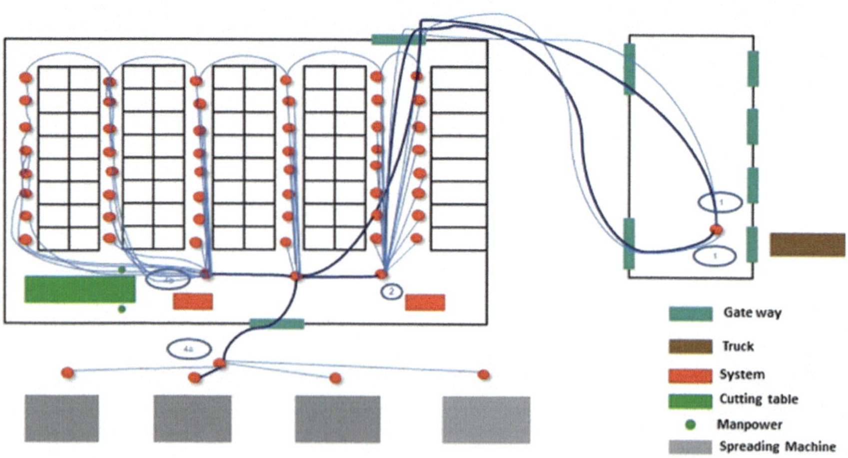

Fig. 3 Spaghetti diagram in the fabric store

PROCESS FLOW CHART		Material Type				
Chart No. 1	Sheet No. 1			SUMMARY		
				ACTIVITY		PRESENT
Subject charted: Knitting Store				Operation	○	4
Activity: Fabric rolls unloading, and loading onto racks, retrieving rolls and issuing rolls				Transport	⇨	1
				Delay	D	-
				Inspection	□	-
				Storage	▽	1
Charted by	Date :			Distance (m)		40
Approved by	Date :			Time (min)		1376

Sl.no	Description	Qty. (Rolls)	Distance (m)	Time (min)	○	⇨	D	□	▽	Remarks
1	Fabric Unloading from container on the pallet	160	-	60	●					by manual
2	Transporting pallets	160	20	16		●				by manual hand pallet
3	Organizing Rolls onto the racks	160	-	112					●	by manual
4a	Retrieving rolls & issuing rolls to spreaders, pintables and manual cutters	420	-	504	●					by manual
4b	Cutting of rolls to required length	120	-	664	●					by manual
4c	Retrieving rolls and loading to container for delivery	7	20	20	●					by manual
	Total		40	1376	4	1			1	

Fig. 4 Process flow chart

3.3 *Time Study—Manpower*

Initial analysis of the time taken for the various processes in the order picking process was analyzed by performing a time and motion study. A major outcome of the study is to identify the bottleneck process. Further, it is used for standardizing the various task involved in the order picking process. The rating factor and appropriate allowances were provided for the various task performed by the load man in the overall order picking process based on the ILO guideline. The details of process flow and time are taken for various activities and overall distance traveled as presented in Fig. 4.

4 Visual Analysis of Order Picking Process

4.1 *Existing Fabric Store Layout*

The existing fabric store is used to store raw materials like different types of fabric rolls. The fabric store manager has arranged the fabric roll at random storage method. The size of each rack is 3 m × 2.5 m × 1.5 m (length × breadth × height). The current fabric store has 216 storage locations available. The visual analysis of the existing fabric store is presented in Fig. 5.

Each rack has a capacity of 600 kg, and the average number of picklists by three shift is 40, and the average number of rolls/picklist is ten. The time is taken for the

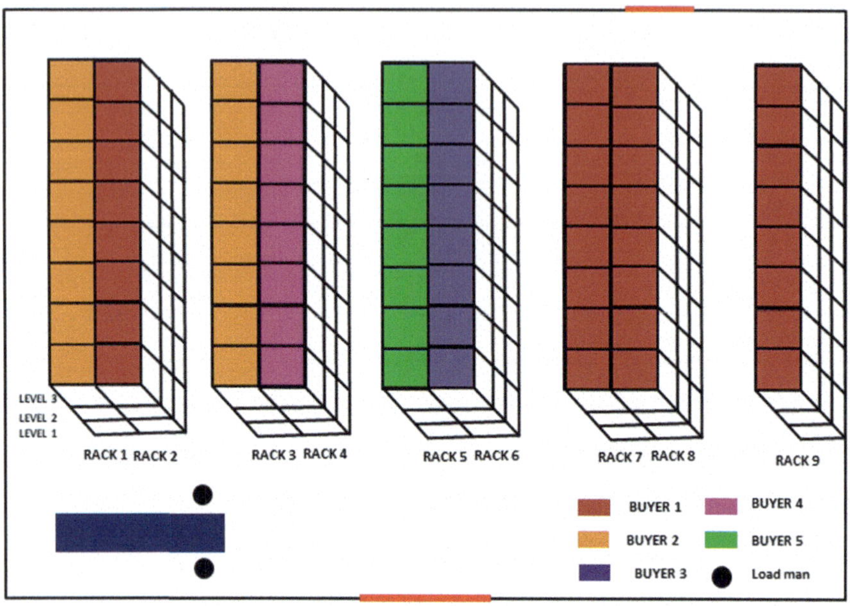

Fig. 5 Existing storage layout in the fabric store

Fig. 6 Total time taken for one picklist

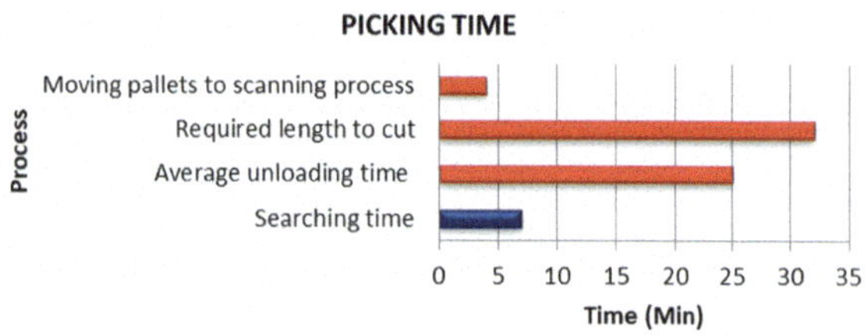

existing process in the fabric store graphically illustrated in Fig. 6. The total time taken for one picklist is 68 min (1 h 8 min). In the existing layout, the unloading time varies according to rack levels but searching time for rolls to be constant (6 min for 1picklist).

4.2 *Value Stream Mapping*

Value stream mapping is an effective tool for implementing lean systems in the industry and a very powerful tool for a visual illustration of the entire value stream-customer order, production, and shipping. The current state of VSM is drawn to identify the areas of improvement in the process fabric store [3]. The time study is carried out to calculate the lead time, inventory time, and WIP, where Fig. 7 shows that the current state VSM of fabric store is shown and the value-added time is 1388 min and the non-value-added time is 11 min. The result of the current state VSM is given in Table 1.

4.3 Cause and Effect Diagram

The cause and effect diagram is drawn to find the root cause for the problem. The cause and effect diagram is shown in Fig. 8. Mainly, the problem faced in the picking of correct fabric rolls in the racks, they are very many types of fabric rolls are available in the store.

The load man still confuses the pick the correct lot number fabric roll. After the brainstorming session carried and come up with solutions like the arrangement of fabric roll at the class-based method by balancing the problem of searching time.

4.4 Simulation

A computer model is used in simulation to better understand and enhance a real-world system. Simulation is a tried-and-true technique that allows a company to study and test its process in a virtual environment, saving time and money over actual testing. The storage location of raw material and movement of workers can be quickly introduced and adjusted in the simulation model, letting companies determine how best to fully utilize their resources and maximize efficiency. A simulation was performed for the bottleneck process to calculate manpower efficiency [4].

The assumptions made in the model are

- When the shifts start, the raw material available in source.
- Distance traveled by workers are to be mapped according to this real-world scenario likewise the results may not affect.
- Operators are to be priorities according to the process carried out.
- The store is recreated using commercial software and imported into simulation software for accurate results.
- Entities are to be sequenced manner.
- Material to be picked in rack according to the picklist.

The processes involved in the fabric store are mapped in a discrete event simulation model and presented in Fig. 9.

The simulation results showed the operators' current efficiency individually in Fig. 10. Operator 1, operator 2 operators 3, and operator 4 have to pick the pick fabric roll, and operator 5 and operator 6 doing the process of cutting of rolls to the required length. So, the overall operators' current efficiency is 95%.

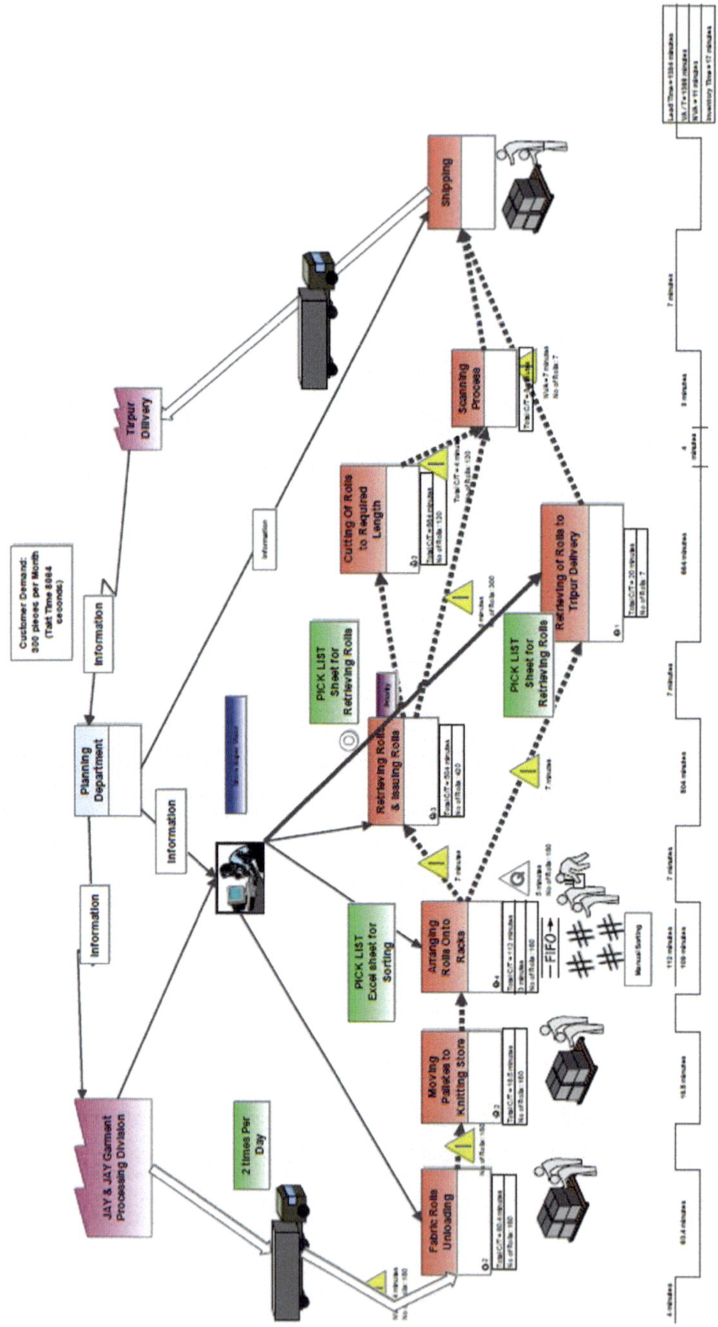

Fig. 7 Current state VSM of fabric store

Table 1 Result of current state VSM

S. No.	Bottleneck	Time (min)
1.	Cutting of rolls to required length	661
2.	Retrieving of rolls (searching + unloading)	504

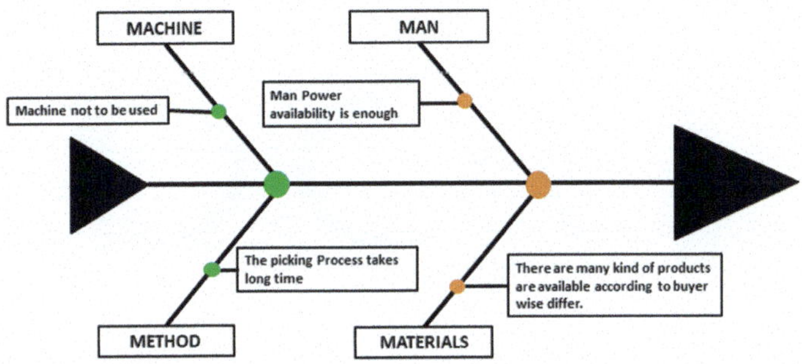

Fig. 8 Cause and effect diagram

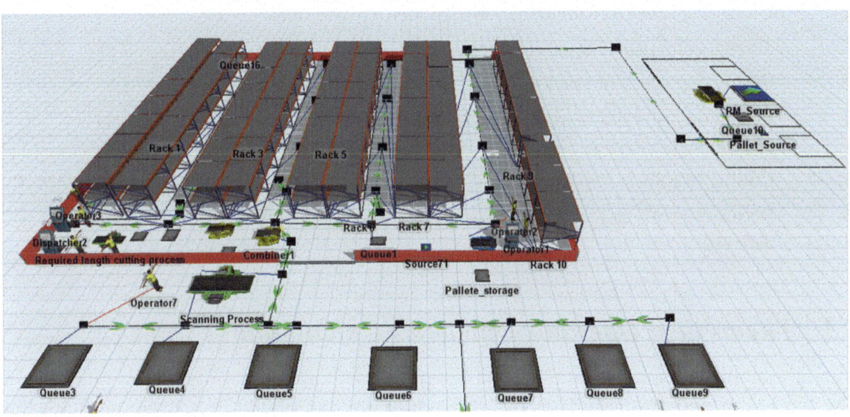

Fig. 9 Simulation model for the fabric store

5 Effort Reduction and Redesign Storage Location

5.1 Time Study in a Fabric Store

Stopwatch-based discrete-time study was performed for the six major tasks involved in the order picking process, and the details are presented in Table 2.

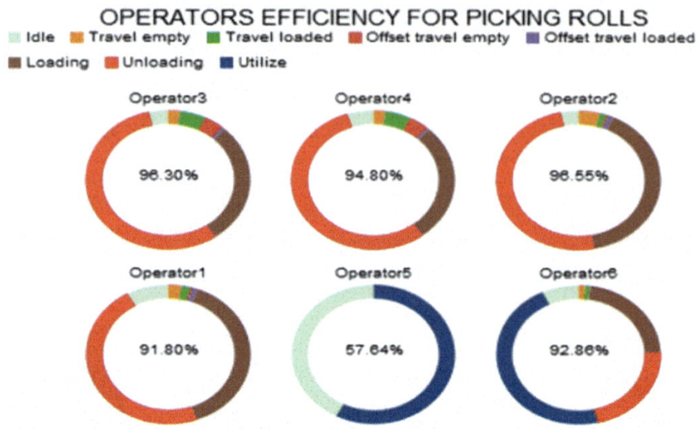

Fig. 10 Load man efficiency in store

Table 2 Overall time is taken in a fabric store

S. No.	Process	Total time (min)	Total time (hr)
1.	Unloading rolls from container onto pallets	60.4	1.01
2.	Transporting pallets to fabric store	16.51	0.28
3.	Organizing rolls onto racks	112	1.87
4.	Retrieving rolls and issuing to spreaders, pintable, and manual cutters	20	0.33
5.	Cutting of rolls to required length	664.5	11.08
6.	Retrieving rolls and loading them to a container for delivery	504.9	8.42

5.2 *Theoretical Operator Efficiency Calculation*

The load man perform activities in fabric store are calculated in percentage with the reference paper [5]. Number of work hours in individual process in fabric store

$$= 2.01 + 0.28 + 1.87 + 0.33 + 11.08 + 8.42 = 23.01\text{ hr}$$

$$\text{Load man available operating time} = 24\text{ hr}$$

$$\text{Manpower Efficiency} = \frac{\text{Number of work hours}}{\text{Total number of available hours}}$$

$$\text{Manpower Efficiency} = 95\%.$$

COMPANY LOGO	KAIZEN PROPOSAL SHEET (Idea suggestion)		KAIZEN FORM NO :	145
			DEPARTMENT:	FABRIC STORE
			SUGGESTION DATE :	13/11/2021
KAIZEN IDEA: TO FOOL PROOFING THE FABRIC CUTTING PROCESS IN REQUIRED LENGTH AT FFK STORE				
PROBLEM/PRESENT STATUS	WHY-WHY ANALYSIS & SUGGESTION FOR IMPROVEMENT		STATUS/DESIGNED CONDITION	
	WHY-WHY ANALYSIS: WHY: More time to be taken compared to other process in store WHY: It was performed manually HOW: Cutting of rolls to required length		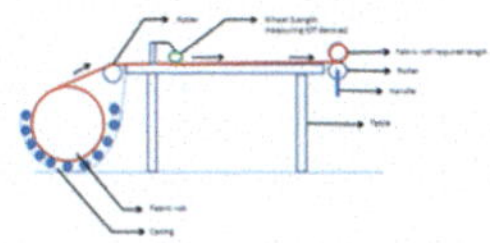	
PROBLEMS/DIFFICULTIES	TEAM LEADERS	XXXX	BENEFITS	
✓ There was difficulty for doing this job manually ✓ Approximately per day 11 hrs doing this job	TEAM MEMBERS XXXXXX		✓ This proposal may reduced the effort of worker ✓ Exact length of fabric to cut compared to manual process	
	START DATE			
	END DATE			
Approved by				

Fig. 11 Kaizen proposal sheet

5.3 *Kaizen Proposal for Manpower Effort Reduction in the Fabric Store*

The existing process of cutting of rolls to the required length consumes more time also the load man has difficulty doing this process. The kaizen proposal sheet has been developed to provide suggestions to the organization. The existing process had done manually.

The load man doing this job of cutting of rolls to required length is approx. 11 h per day. So the proposal had been taken for worker effort reduction and also existing they may not cut the fabric accurately but the proposal has been to cut the fabric accurately. The kaizen proposal sheet is illustrated in Fig. 11.

5.4 *Storage Location Redesign Based on the FSN Method*

Fabric rolls are arranged in groups based on their similarity in the type of material or substance. This group will be assigned to a specific area of the business. The similarity of materials in a group might take the form of item similarity or order list similarities [6–11]. FSN analysis is used to classify items in fabric stores based on the volume of goods moving through the store, i.e., the number of goods entering and exiting the store (consumption rate). The three types of commodities will be used to classify them, namely: non-moving goods (70%): goods that move slowly. Slow moving (20%): goods movement is not excessively fast or frequent, but neither is it excessively slow. Fast moving (10%): goods with frequent and fast movements [12].

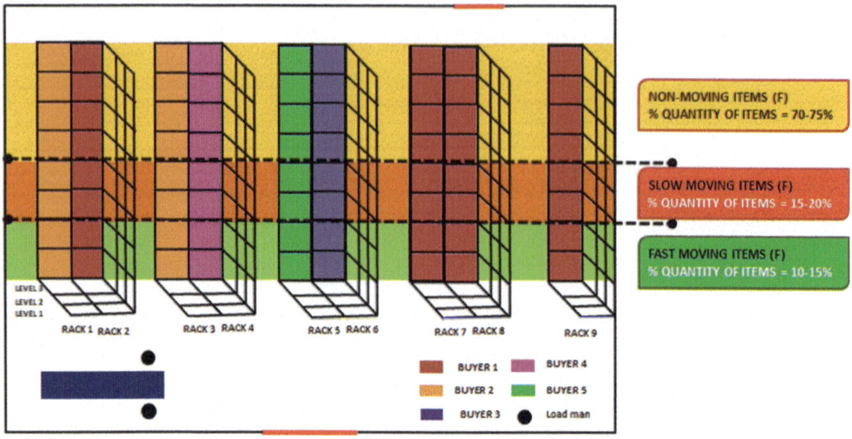

Fig. 12 FSN-based fabric roll allocation

After the data collection, the total consumption rate in the fabric store periodically based on material types was calculated. So that which material is placed in fast-moving, slow-moving, and non-moving categories based on buyer wise allocated. The FSN-based category is based on allocation for reducing the picking time like searching time is presented in Fig. 12.

5.5 Future State Value Stream Mapping

Based on the gap areas with the help of the current state value stream map, some changes were proposed and implemented in the process of the fabric store. Firstly, the process of cutting of rolls to the required length has been fully modified by portable for reducing the work effort and suitable kaizen improvements were made. Secondly, the process of retrieving rolls and issues to the cutting section has also been modified. The fabric roll is to be stored based on FSN categorization. The time study is to be calculated in the fabric store. In which the time study the future state VSM to be plotted. Figure 13 shows that the future state value stream mapping where the lead time is reduced from 1394 to 1183 min and the value-added time reduced from 1388 to 1158 min.

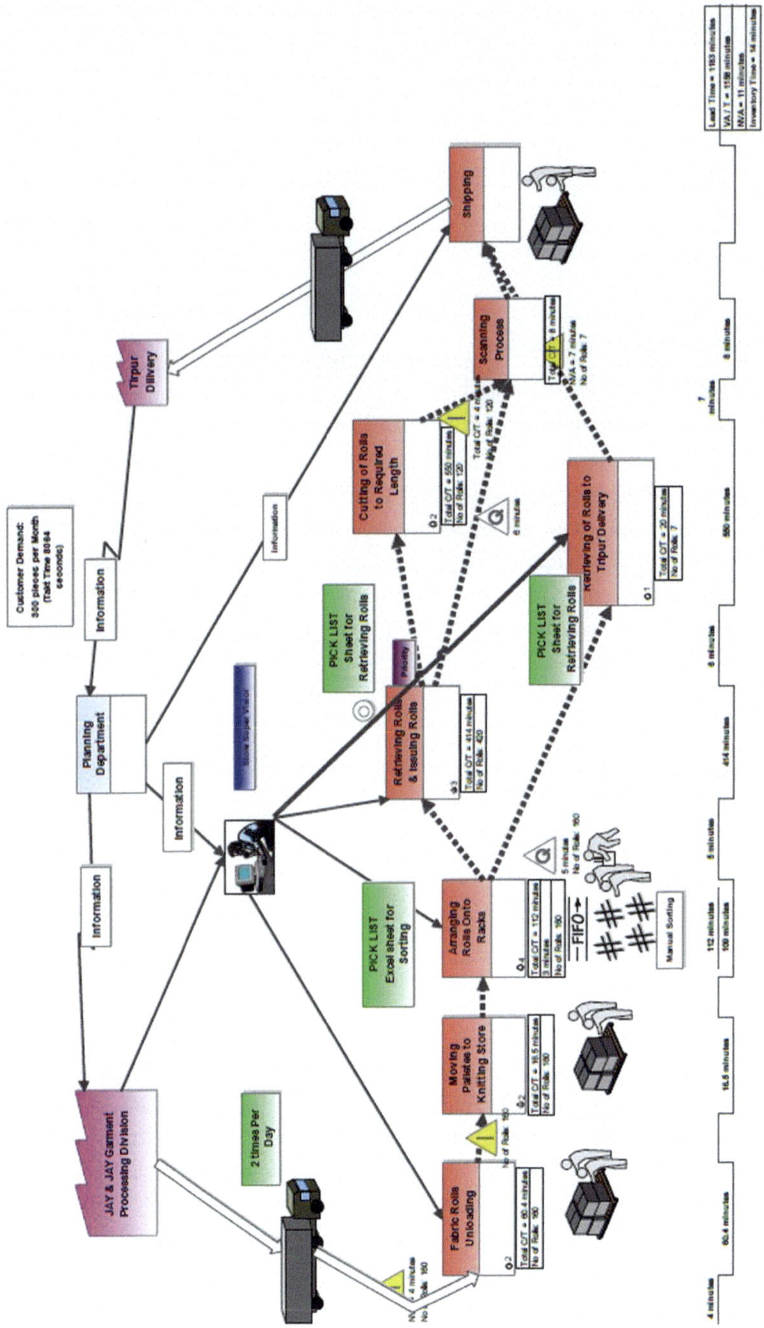

Fig. 13 Future state VSM of the fabric store

Table 3 Comparison of VSM

Process	Current state VSM		Future state VSM		Calculation	% Improvement
Retrieving rolls and issues to cutting section	504 min	8.4 h	414 min	6.9 h	$\frac{(504-414)}{414}$	21%
Value-added time	1388 min	23.1 h	1158 min	19.4 h	–	–
Cutting of rolls to required length	664 min	11.08 h	550 min	9.1 h	$\frac{(664-550)}{550}$	20%

6 Results and Discussion

6.1 Comparison of Value Stream Mapping

The value stream mapping tool was performed for analyzing the before and after in the fabric store. Among all processes, two are bottleneck processes.

The comparison table of the current state and future state VSM is listed in Table 3. The process of retrieving rolls and issuing to spreaders, pin tables, and manual cutters is improved by 21% and the cutting of rolls required length process also improved by 20%

7 Conclusion

Extensive field study was done at the fabric roll storage section of the garment industry, and initial analysis was performed by doing time study in order to identify the bottleneck processes by using industrial engineering and lean tool were identified to improve the process efficiency. Further to that the major purpose is to create a general framework for implementing lean manufacturing tools and practices in the garment industry. With the examination of the current state map, several possible improvements have been identified during this implementation in the industry. The suggested future state map depicts these potential enhancements like the process of cutting rolls to the required length has been completely modified for reducing worker effort, and another process of retrieving rolls has also been modified based on the total consumption rate, which is calculated regularly, and raw materials are placed in FSN categories based on the buyer to be allocated in the future state maps. The following results are based on the outcomes of this implementation:

- The overall lead time of the entire process is expected to get reduced of 17% while implementing the proceed kaizen.
- The existing fabric store process was simulated using discrete event simulation software, with a man effort percentage as 95%, and recommendations for kaizen

proposal were made in order to justify the improvement in the overall process efficiency.

- The current order picking efficiency is 63%, and the suggested FSN-based technique is expected to increase picking efficiency by 21%.

References

1. Tonape S, Patil K, Karandikar V (2016) Manpower optimization and method improvement for a warehouse. Int Res J Eng Technol (IRJET) 3(7):1402–1407
2. Devarajan D, Jayamohan MS (2016) Stock control in a chemical firm: combined FSN and XYZ analysis. Procedia Technol 24:562–567
3. Vasanth Kumar, Madhan Mohan, Mohanasundaram (2020) Lean supply chain management in garment industry using value stream mapping. Int J Serv Oper Manag 37(1):133–143
4. Liong C-Y, Loo CSE (2009) A simulation study of warehouse loading and unloading systems using arena. J Qual Meas Anal 5(2):45–56
5. Subramanian SK, Husin SH, Yusop Y, Hamidon AH (2008) Machine efficiency and man power utilization on production lines. In: Proceedings of the 8th WSEAS international conference, pp 70–75
6. Dixit A, Shah B, Sonwaney V (2020) Picking improvement of an FMCG warehouse: a lean perspective. Int J Logist Econ Globalisation 8(3):243–271
7. Prasetyawan Y, Ibrahim NG (2020) Warehouse improvement evaluation using lean warehousing approach and linear programming. IOP Conf Ser: Mater Sci Eng 847
8. Sooksai T (2019) The efficiency enhancement of warehouse space management with ABC analysis: a case study of ABC company limited. In: International academic research conference in Vienna
9. Tambunan MM, Syahputri K (2018) Storage design using fast-moving, slow-moving and non-moving (FSN) analysis. In: MATEC web of conferences
10. Ramaa, Subramanya (2012) Impact of warehouse management system in a supply chain. Int J Comput Appl 54(1):14–20
11. Hwang HS, Cho GS (2006) A performance evaluation model for order picking warehouse design. Comput Ind Eng 51(2):335–342
12. Candrianto FA, Gusti MA (2019) Analysis of placement maximizing planning in warehouse using FSN analysis using class-based storage method. Adv Econ, Bus Manag Res 124:682–695

Optimization of Tungsten Inert Gas Welding Process Parameters on AA6013

S. Pratheesh Kumar, K. Anand, R. Rajesh, and S. Ashwin

Abstract Inert gas tungsten arc welding (TIG) is a high-quality, high-precision welding technology that is especially well-suited for joining thin metals. Shielding gases such as helium and argon protect the weld bead from contamination by air, dust, and other contaminants. Many welding process variables, such as the heat affected zone (HAZ), weld strength, and hardness, have an impact on weld quality (HV). The pulsed-TIG welding procedure is used on the welding surface to lower the heat affected zone and hence the strength of the weld. Using optimization techniques, an attempt is made to understand the impact of TIG welding conditions on the aluminium alloy 6013. A number of parameters, such as welding current and gas flow rate, have been discovered to have an effect on response output measures such as weld tensile strength. The Taguchi method is used to determine the optimum feasible system parameters. For this investigation, the Taguchi method employs a L9 orthogonal array. First, tests were performed to determine the operating range parameters for welding aluminium 6013 alloy with a thickness of 3 mm using filler material 4043, which is frequently used for welding the 6xxx series of Al-alloys because it minimises the likelihood of stress cracking. It's time to finish up with the cutting edges of the aluminium samples. To create the designs for the welding tests, the Taguchi method was applied. The tensile test specimens were then created utilising the wire cut EDM method in accordance with ASTM requirements. Tensile testing is performed to determine the tensile strength of a weld. Based on the experimental values of the process parameters, the Taguchi algorithm computes the S/N ratio. After developing the regression models, analysis of variance was done to ensure that they were appropriate.

Keywords Tungsten inert gas welding · Optimisation · AA6013 · Welding · Design of experiments

S. Pratheesh Kumar (✉) · K. Anand · R. Rajesh · S. Ashwin
Department of Production Engineering, PSG College of Technology, Coimbatore, Tamil Nadu, India
e-mail: spratheeshkumarth@gmail.com

E. Natarajan et al. (eds.), *Materials, Design and Manufacturing for Sustainable Environment*, Lecture Notes in Mechanical Engineering,
https://doi.org/10.1007/978-981-19-3053-9_30

1 Introduction

Welding thin sheets is currently the most demanding engineering task given the current condition. Because two pass and three pass welding cannot be performed, under welding results in a significant reduction in strength, and over welding results in the production of holes in thin sheets as thin as 4 mm. Thus, perfection is the most important factor to consider throughout the sheet joining process. As a result, while combining thin sheets, it is necessary to change the weld settings. This welding operation necessitates the employment of specialised welding knowledge and methods in order to produce structurally good weldments. The quality of the welds generated by TIG welding is strongly dependent on the welding parameters selected by the welder or welding operator [1, 2]. By optimising the process parameters for butt joining, thick sheet welding on 6013 aluminium alloy of thickness 3 mm, which has outstanding corrosion resistance and stress-corrosion cracking, will be made simple and compatible. Pressure vessels and structural components will have a longer lifespan as a result of this. This project aims to make thin sheet welding simple and compatible by optimising the process parameters for butt joining on 6013 aluminium alloy of 3 mm thickness, which has excellent corrosion resistance and stress corrosion cracking resistance, by using the pulsed-TIG welding process, which reduces the heat affected zone by welding the material alternately melting and cooling [3]. As a result, the grain boundary is sharp, the chromium layer is thick, and the grain boundary is strong. Change process parameters such as current, voltage, stand-off distance, pulse on time, pulse off time, and weld speed to improve weld quality.

2 Literature Review

Messrs V. Jaya Balan and M. V. Balasubramanian conducted research in April 2008 [4] to better understand the mechanical properties of pulsed current GTA-welded titanium alloy weldments. The study's findings were released in April 2008. They used gas tungsten arc (GTA) welding to solder the titanium alloy because it is easier to operate and less expensive than other welding processes. They used four factor, five level, central composite and rotatable design matrices to optimise the number of experiments needed to reduce the number of trials. The response surface method was used to generate these mathematical models (RSM). They used the ANOVA approach to assess the models' adequacy. Their mathematical models for forecasting joint tensile characteristics were 99% accurate.

After conducting tests in November 2007 [5] to see if they could improve the mechanical properties of AA 5456 aluminium alloy welds using a magnetic arc oscillation technique, they published their findings in the journal advanced materials. To increase mechanical properties, magnetic arc oscillation welding of non-heat treatable AA 5456 aluminium alloy welds can be optimised using the Taguchi technique. The researchers discovered that the optimum scenario occurred for all of the attributes. They created regression models to help them. They investigated the connection between weld current, weld speed, amplitude, frequency and mechanical properties.

Mr. R. Dinesh Kumar, Mr. S. Elangovan and Mr. N. Siva Shanmugam carried out experiments in June 2014 [6] on the parametric optimization of the pulsed-TIG welding process in butt joining of 3041 austenitic stainless-steel sheets. They chose stainless steel because of its better fracture toughness, outstanding intergranular corrosion resistance, and lack of post-process annealing requirements. Their solution was to use pulsed-TIG welding, which minimised the heat affected zone of the weld and kept it from failing. Because to the use of stainless steel to decrease fracture growth under high pressure, pressure vessels and vehicle components became more efficient. A number of process parameters were discovered to require optimization in order to improve weld quality.

Kafali, H., and Ay, N published research on the mechanical properties of 6013-T6 aluminium alloy in the journal advanced materials in 2009 [7]. As a result, friction stir welded plate aluminium alloys were employed on the aircraft's structures, specifically the fuselage and wing fairings. The traditional riveting method was used to connect these structures, which added weight to the aircraft and increased stress in the rivet holes, resulting in fatigue cracks in the aircraft. As a result, they had to turn to welding, a unique process of combining metals. As a result of being heated past its melting point, aluminum's mechanical properties decreased. Because of these benefits, friction stir welding (FSW) was chosen as a replacement joining process.

A study published in June 2006 [8] investigated weld penetration prediction in connection to FCAW process variables such as welding current, arc voltage, distance between the tip of the nozzle and the plate, and angle between the electrode and the work. The findings were published in welding science and technology. As part of their research into the statistical approach of central composite rotatable design, they created a mathematical model to forecast weld penetration as a function of welding process parameters.

Optimising various methods of welding techniques on the aluminium alloy 6013 to produce an effective response output as weld tensile strength, the welding parameters such as welding current and gas flow rate are discovered. The heat affected zone, weld strength and hardness are having a major impact on quality of weld. Initially, some tests are conducted for identifying the operating range of welding parameters. This will increase the protrusion of the weld from the face to the root. In this research, only the process parameters of TIG welding are optimised to improve weld quality. In the realm of experimental design, the Taguchi design of experiments is an excellent strategy for achieving high quality in a small number of tests. This method has numerous advantages over traditional experimental designs, which take a long time to complete and are unworkable in many situations.

It is used in the areas where to weld a thin sheet without losing its quality of weld. Pressure vessels and structural components will have a longer lifespan as a result of this. The fuselage, wing fairings and wings are often composed of aluminium alloys such as 6061 and 7075 [9, 10]. The traditional riveting process must be utilised to join these structures. Riveting, on the other hand, adds weight to the structure of the aircraft and causes stress concentrations that may lead to fatigue cracks. Weldment is a less common way of combining two or more materials. Metal is heated to the point of melting in traditional welding procedures, diminishing its mechanical properties. Weldability is another issue with high-strength materials.

3 Design of Experiments

3.1 Taguchi Method

Experiment development is one of the most important statistical methodologies for producing high-quality systems at a minimal cost. It is used in the creation of low cost but high-quality systems. Taguchi methods are an effective and rigorous approach for developing designs that are optimised for performance, quality and cost. The optimization of process parameters via the Taguchi method is critical for achieving good quality without raising costs. Process parameter optimization has the potential to improve product quality features, and Taguchi technique-derived optimal process parameters are less sensitive to variations in environmental conditions and other noise elements.

3.1.1 Steps Followed in Taguchi Method

(i) Recognise and comprehend the impact of design factors on a process. Temperatures, pressures and other process factors that have an effect on the performance metric and may be easily changed are examples of parameter variables. It is necessary to provide the number of levels at which the parameters should be adjusted. For example, the temperature could be set between 40 and 80 °C, which are low and high temperatures, respectively. To increase the number of levels at which a parameter can be changed, the number of trials must be increased.

(ii) For the parameter design, create orthogonal arrays with the number of experiments and conditions for each experiment indicated on each row. ii. Before making a decision on orthogonal arrays, consider both the number of parameters and the degree of variation in each parameter. This topic will be discussed in greater detail further down.

(iii) Execute the tests listed in the finished array to learn more about how changes to the performance metric affect the system.

(iv) Finally, a thorough data analysis will be performed to establish how the various parameters affect this performance metric.

Table 1 Design of experiments

S. No.	Peak current	Base current	Pulse frequency	Pulse on time
	(A)	(A)	(pps)	(%)
1	160	80	2	40
2	160	85	4	50
3	160	90	6	60
4	170	80	4	60
5	170	85	6	40
6	170	90	2	50
7	180	80	6	50
8	180	85	2	60
9	180	90	4	40

3.2 Design of Experiments

To discover the welding input parameters that yield the requisite weld quality, DoE, evolutionary algorithms, and computational networks are widely utilised. Popular methods for establishing a mathematical link between welding process input parameters and weld joint output variables include DoE, evolutionary algorithms, and computational networks. Table 1 shows the weld parameters that were taken into account throughout the testing.

3.3 Selection of Filler Metal

3.3.1 Filler Metal

Filler rods are specially designed for TIG welding and usually supplied in cut lengths of 1 m in the following diameters:

(i) 1.6 mm
(ii) 2.4 mm
(iii) 3.2 mm
(iv) 4.8 mm.

Choose the filler rod compositions based on the parent metals being welded together, not the other way around. If at all possible, keep filler rods away from moisture and heat. Before using them, thoroughly clean them with stainless steel wool or an aluminium oxide cloth to eliminate any rust, scale, oil grease or moisture that could contaminate the welds. Filler rods should never be touched with bare hands after cleaning, always use supple leather or treated flame-resistant cotton.

The 4043 alloy is a great choice amongst the various types of aluminium filler metals due to the following characteristics:

(i) Less sensitive to weld cracking
(ii) Produces brighter, almost smut free welds.

3.4 *Working Range of Selected Process Parameters*

Trials were used to define the parameter working range, and the parameters were chosen based on the results of a literature review. A number of the parameters were held constant, and the key parameter ranges are illustrated in the picture below:

(i) Peak current (160–180) A.
(ii) Base current (80–90) A.
(iii) Pulse frequency (2–6) pps.
(iv) Pulse on time (40–60) % percent (%) peak time control.

4 Experimentation

From the obtained sequence, experiments were conducted with welding machine KR-315P TIG welding machine on nine pair of 6013 aluminium alloy which are of dimension 84 X 50 mm and thickness of 3 mm (Fig. 1).

Welding was employed to join the various metals. Figure 1 depicts the weld bead that resulted from the welding procedure. The welded metal is ground to flatten the weld so that the specimen may be machined properly. The tensile test profile was precisely manufactured using wire cut EDM on a CNC machine. Specimens to be utilised in experiments. The tensile specimens were created in accordance with ASTM B 557 M specifications [11]. They were manufactured with great precision, and the specimen measurements are listed below.

All the specimens are tested on the UTM machine to check the strength of each specimen. The Fig. 2 represents the tensile test specimen (before tensile test).

Fig. 1 Weld bead

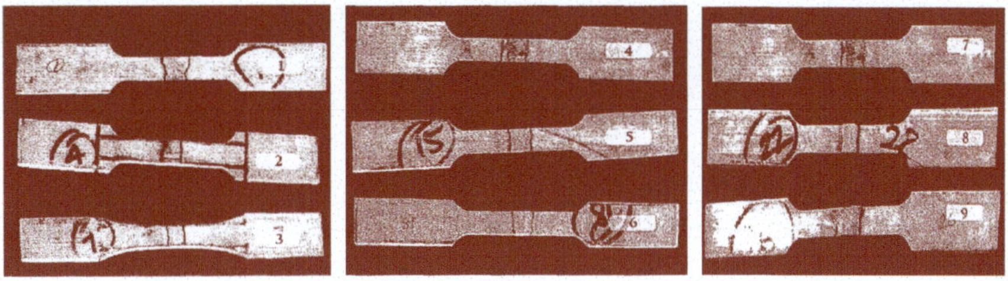

Fig. 2 Welded specimen

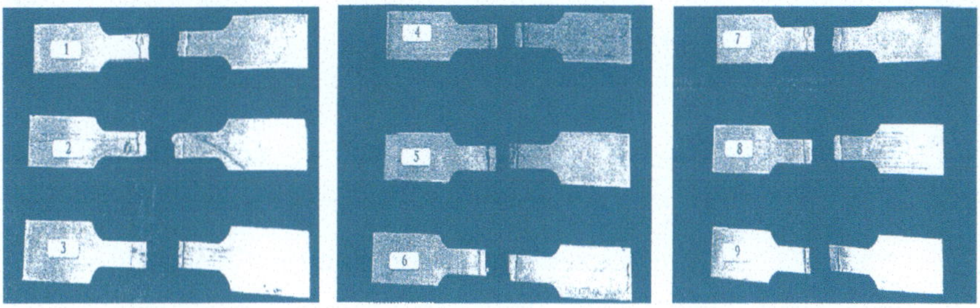

Fig. 3 Tensile test specimen

The Fig. 3 represents the tensile test specimen (after tensile test). The experimental design and the results obtained are presented in Table 2.

5 Mathematical Model

The factors and the levels used for experiments are given in Table 3. Representing the tensile properties, say tensile strength (MPa) of the response function can be expressed as,

YS = f(Peak current, base current, pulse frequency, pulse on time)

YS = f(p, b, f, t)

5.1 Regression Analysis

Regression analysis is a technique for determining the relationships between variables in a study. By far the most common method of analysis for social scientists is linear equations. The following is a representation of linear regression:

$$Y = a + b_1X_1 + b_2X_2 + b_3X_3 \ldots + b_kX_n \quad (1)$$

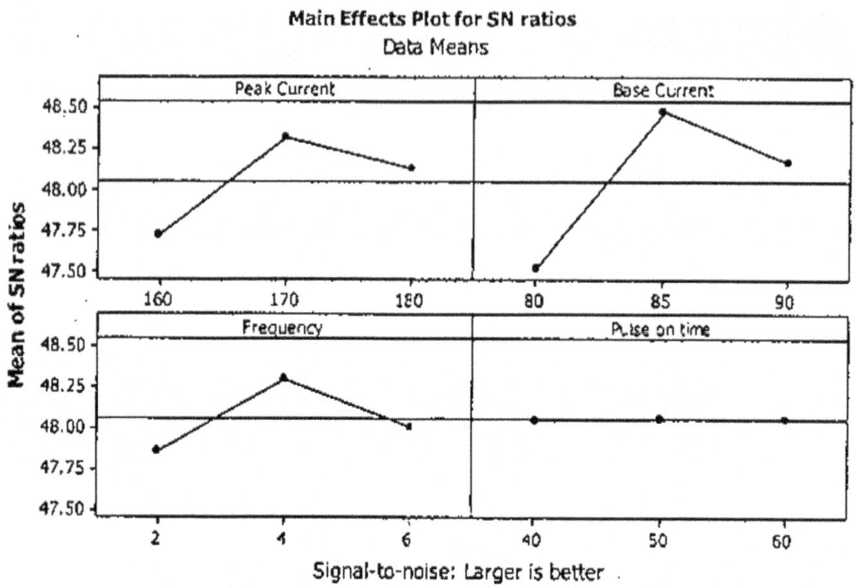

Fig. 4 SN ratio for tensile strength

Table 2 Experimental design and the results

S. No.	Peak current (A)	Base current (A)	Pulse frequency (pps)	Pulse on time (%)	Yield strength (MPa)	Tensile strength (MPa)	Elongation (%)
1	160	80	2	40	201.14	223	0.83
2	160	85	4	50	244.22	262.71	1.25
3	160	90	6	60	221.65	244.99	0.89
4	170	80	4	60	236.43	252.03	1.13
5	170	85	6	40	249.62	272.2	1.36
6	170	90	2	50	233.91	258.3	1.13
7	180	80	6	50	218.17	238.13	0.71
8	180	85	2	60	241.29	261.93	1.25
9	180	90	4	40	239.65	265.42	1.3

Table 3 Important factors and their levels

Factor	Unit	Levels	Values		
			Low	Medium	High
Peak current	Amps (A)	3	160	170	180
Base current	Amps (A)	3	80	85	90
Frequency	pulse per second (pps)	3	2	4	6
Pulse on time	%	3	40	50	60

The dependent variables bead width and weld bead hardness were investigated using regression analysis, as were the independent variables welding current, arc voltage welding speed and electrode stick out. Minitab was used to carry out the regression analysis. Tables 4, 6 and 8 demonstrate the regression analysis results for tensile, yield and elongation, respectively. Using the ANOVA statistical technique, total variation can be split down into separate causes. The overall variance in a data collection can be calculated by taking the square root of each inaccuracy and multiplying it by itself. The sum of squares of errors divided by the degree of error freedom is referred to as "error variance." It's sometimes referred to as mistake variance or simply variance. The error variance, which is a measure of the fluctuation in the results, is provided by all of the controllable parameters together. Tables 5, 7 and 9 shows the results of ANOVA analysis for component's tensile strength, yield strength and elongation percentage (Figs. 4, 5, 6, 7, 8, 9, 10, 11 and 12).

The regression equation is

$$\text{Tensile Strength} = -5 + 0.580\,\text{p} + 1.85\,\text{b} + 1.01\,\text{f} - 0.028\text{t} \quad (2)$$

The regression equation is

Table 4 Regression analysis: tensile strength

Predictor	Coefficient	SE	*T*	*P*
Constant	– 5.4	172.1	– 0.03	0.977
Peak current	0.5797	0.6983	0.83	0.453
Base current	1.852	1.397	1.33	0.256
Frequency	1.008	3.491	0.29	0.787
Pulse on time	– 0.0278	0.6983	– 0.04	0.97

Table 5 Analysis of variance: tensile strength

Source	Degree of freedom	Sum of squares	Mean squares	*F*	*P*
Regression	4	740.7	185.2	0.63	0.666
Residual error	4	1170.2	295.5		
Total	8	1910.9			

Table 6 Regression analysis: yield strength

Predictor	Coefficient	SE	*T*	*P*
Constant	17.2	188.6	0.09	0.932
Peak current	0.535	0.7651	0.7	0.523
Base current	1.316	1.53	0.86	0.438
Frequency	1.092	1.53	0.86	0.438
Pulse on time	0.1493	3.825	0.29	0.79

Table 7 Analysis of variance: yield strength

Source	Degree of Freedom	Sum of Squares	Mean squares	*F*	*P*
Regression	4	473.4	118.3	0.34	0.841
Residual error	4	1404.9	351.2		
Total	8	1878.2			

Table 8 Regression analysis: elongation

Predictor	Coefficient	SE	*T*	*P*
Constant	– 1.302	2.848	– 0.46	0.671
Peak current	0.00483	0.01155	0.42	0.697
Base current	0.02167	0.02311	0.94	0.401
Frequency	– 0.02083	0.05776	– 0.36	0.737

Table 9 Analysis of variance: elongation

Source	Degree of freedom	Sum of squares	Mean squares	F	P
Regression	4	0.10292	0.02573	0.32	0.851
Residual error	4	0.32031	0.08008		
Total	8	0.42322			

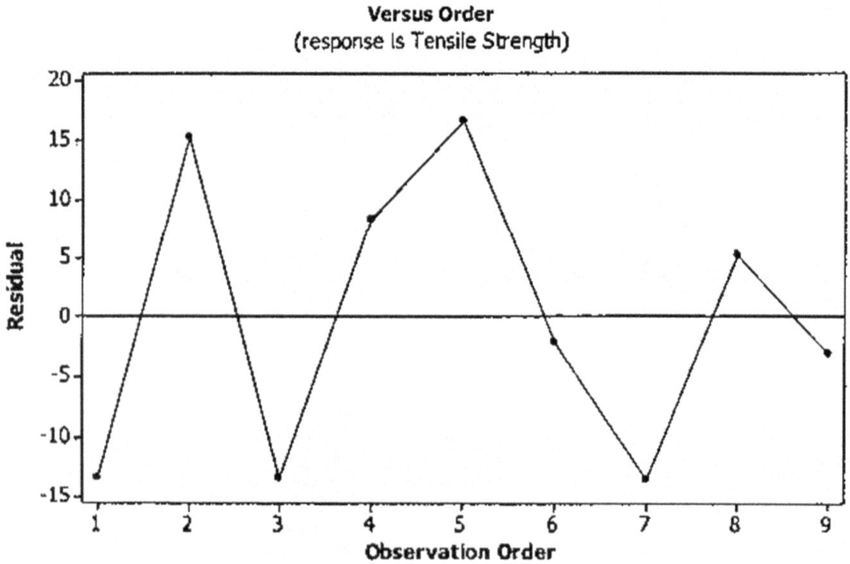

Fig. 5 Residual versus order for tensile strength

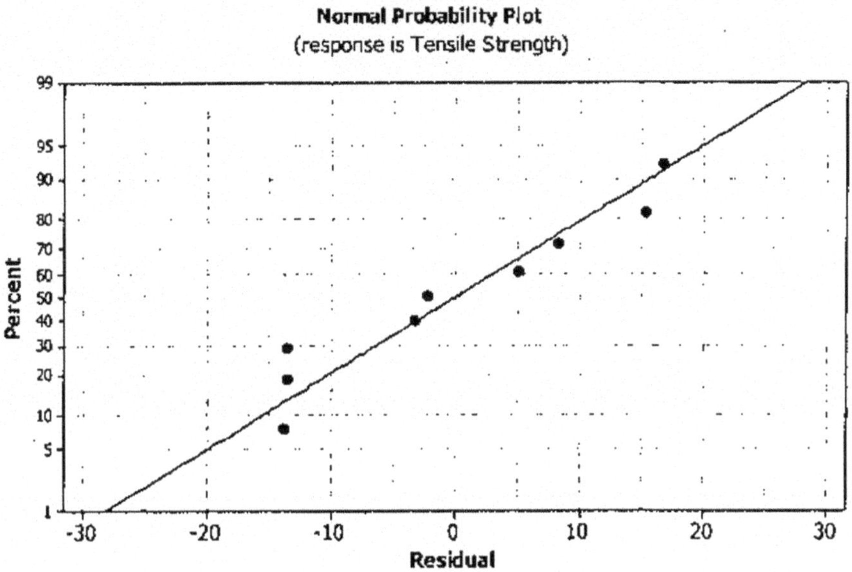

Fig. 6 Normal plot for tensile strength

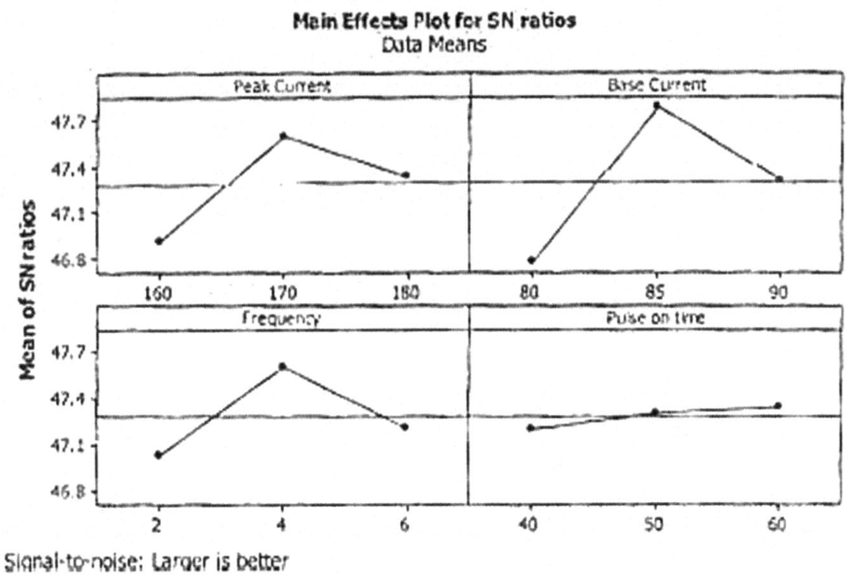

Fig. 7 SN Ratio for yield strength

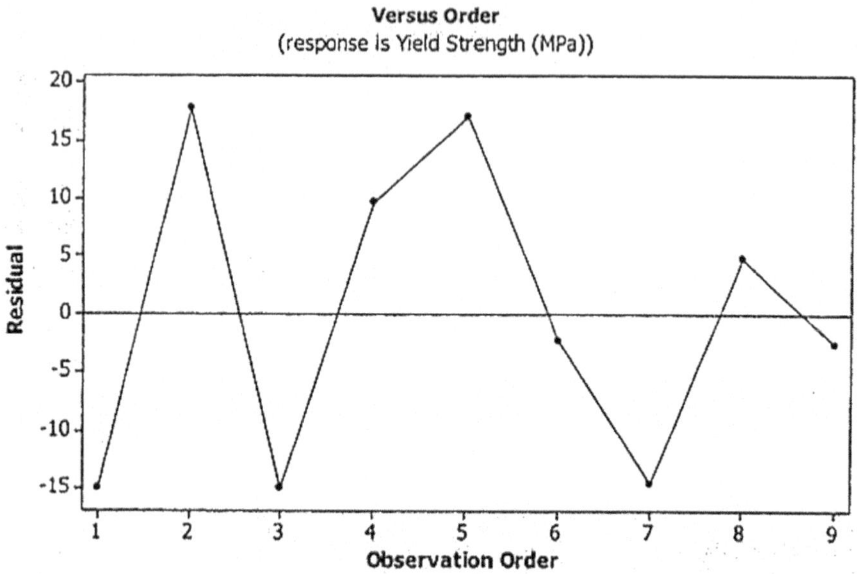

Fig.8 Residual versus order for yield strength

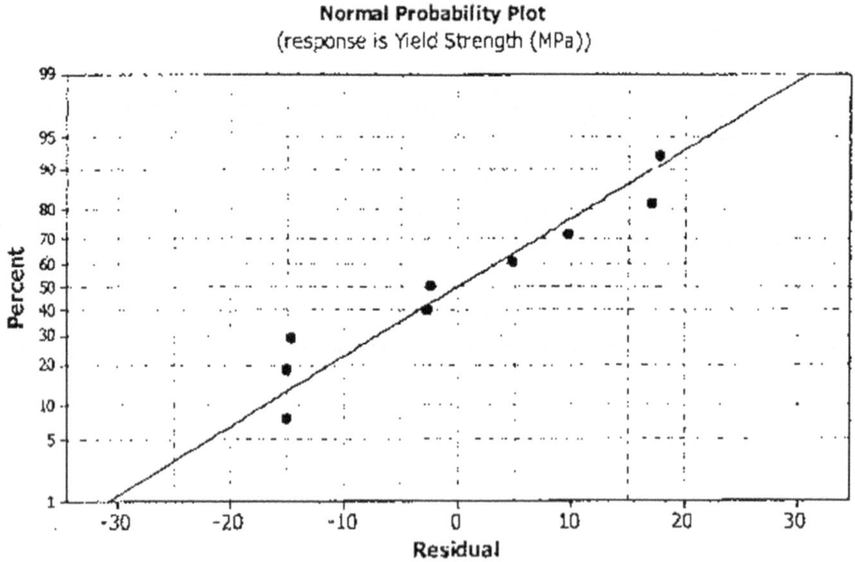

Fig. 9 Normal plot for yield strength

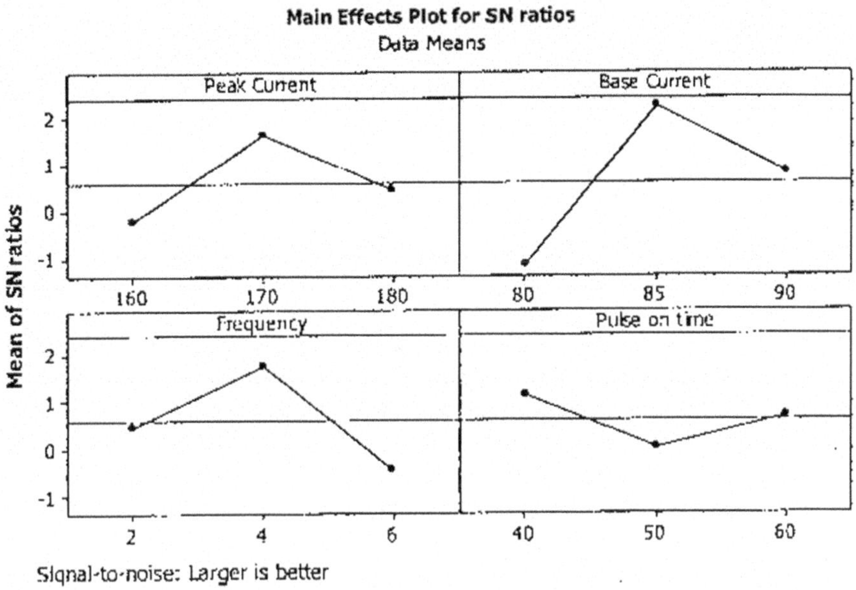

Fig. 10 SN ratio for elongation

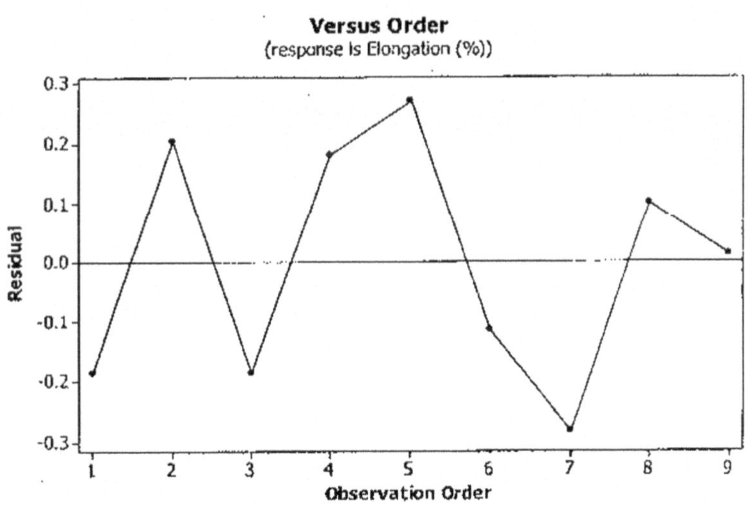

Fig. 11 Residual versus order for elongation

$$\text{Yield Strength (MPa)} = 17 + 0.535\ \text{p} + 1.32\ \text{b} + 1.09\ \text{f} + 0.149\ \text{t} \tag{3}$$

The regression equation is

$$\text{Elongation}\ (\%) = -1.30 + 0.0048\ \text{p} + 0.0217\ \text{b} - 0.0208\ \text{f} - 0.0037\ \text{t} \quad (4)$$

In regression and ANOVA, a residual plot, or graph, can be used to analyse the goodness-of-fit. Using residual plots to determine if the ordinary least squares assumptions are met is quite beneficial. If the assumptions are met, normal least squares regression coefficient estimates will have the least variance and be unbiased. Plot the residuals against the data order to confirm that the residuals are uncorrelated. The normal plot of residuals can be used to prove a critical assumption about them. To determine a design's statistical significance, compare its size and statistical significance. In other words, if the impacts were non-existent or zero, the points should fall along the fitted line. When using the normal probability plot, it is difficult to compare the extent of positive and negative impacts because negative effects show on the left side of the graph and positive effects appear on the right.

Figures 5, 8 and 11 exhibit uncorrelated residuals, but the following graphs provide unbiased coefficients with the least amount of variance as a result of regression; this can be seen by examining the previously mentioned graphs. Figures 6, 9 and 12 demonstrate that the residuals are regularly distributed.

The regression equations obtained are,

$$\text{Tensile Strength (MPa)} = -5 + 0.580\ \text{p} + 1.85\ \text{b} + 1.01\ \text{f} - 0.028\ \text{t} \quad (5)$$

$$\text{Yield Strength (MPa)} = 17 + 0.535\ \text{p} + 1.32\ \text{b} + 1.09\ \text{f} + 0.149\ \text{t} \quad (6)$$

$$\text{Elongation}\ (\%) = -1.30 + 0.0048\ \text{p} + 0.0217\ \text{b} - 0.0208\ \text{f} - 0.0037\ \text{t} \quad (7)$$

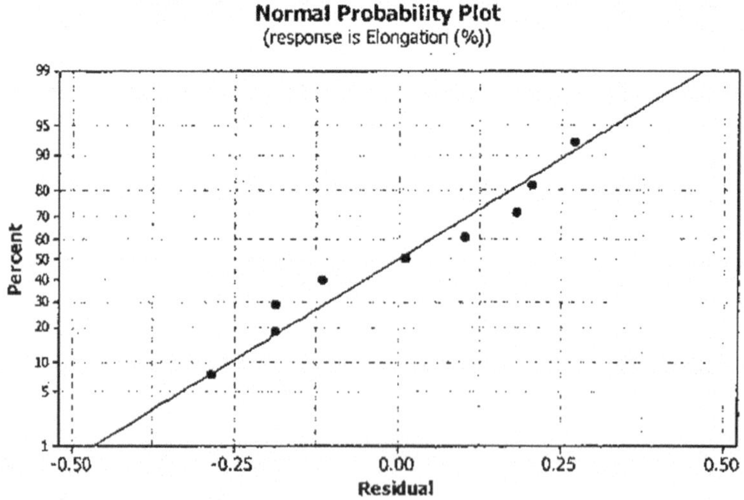

Fig. 12 Normal plot for elongation

6 Conclusion

(i) The impacts of current parameters such as peak current, base current, pulse frequency and pulse while on solid pulsed current TIG welded AA 6013 aluminium alloy materials were analysed. Taguchi experimental design for determining welding parameters has been successful. Therefore, the appropriate combination of TIG welding parameter,

(ii) Peak current of 170 A, base current of 85 A, pulse frequency of 4 Hz and the pulse on time of 50 for Tensile Strength.

(iii) Peak current of 170 A, base current of 85 A, pulse frequency of 4 Hz and the pulse on time of 60 for yield strength.

(iv) Peak current of 170 A, base current of 85 A, pulse frequency of 4 Hz and the pulse on time of 40 for elongation.

Changes in pulse on time parameters have a greater impact on ideal tensile qualities than changes in any other parameter, as evidenced by S/N ratio values comparison. The suggested levels of adjustable parameters for the pulse current TIG welding process for optimal tensile characteristics are maximum % elongation and maximum tensile strength. The investigation's eventual scope will include the following areas: When determining the study's input parameters, the peak current, base current, pulse frequency and pulse on time are all taken into account. More research can be done on the same alloy of aluminium or on different alloys of aluminium to determine other welding parameters including arc voltage, stand-off distance, gas flow rate and welding speed. To get better outcomes, the same or different materials can be treated to post-weld heat treatment.

References

1. Kumar A, Sundararajan S (2006) Selection of welding process parameters for the optimum butt-joint strength of an aluminium alloy. Mater Manuf Processes 21(8):779–782
2. Montgomery DC (1997) Design and analysis of experiments. Wiley, New York, pp 395–476
3. Aghakhani M, Mehrdad E, Hayati E (2011) Parametric optimization of gas metal arc welding process by Taguchi method on weld dilution. Int J Model Optim 1(3):216
4. Balasubramanian M, Jaya Balan V, Balasubramanian V (2008) Effect of pulsed gas tungsten arc welding on corrosion behaviour of Ti–6Al–4V titanium alloy. Mater Des 29(7):1359–1363
5. ASTM B 209M, 2004, Standard specification for aluminium and aluminium—alloy sheet and plate, pp 1–26
6. Dinesh Kumar R, Elangovan S, Siva Shanmugam N (2014) Parametric optimisation of pulsed-TIG welding process in butt joining of 304L austenitic stainless-steel sheets. Int J Res Eng Technol 6:213–219
7. Kafali H, Ay N (2009, May) Mechanical properties of 6013-T6 aluminium alloy friction stir welded plate. In: International conference on aerospace sciences and aviation technology, vol 13, No. Aerospace sciences & aviation technology, ASAT-13, May 26–28, 2009. The Military Technical College, pp 1–9
8. Faye A, Balcaen Y, Lacroix L, Alexis J (2021) Effects of welding parameters on the microstructure and mechanical properties of the AA6061 aluminium alloy joined by a Yb: YAG laser beam. J Adv Joining Process 3:100047

9. Zhang Y, Lu F, Cui H, Cai Y, Guo S, Tang X (2016) Investigation on the effects of parameters on hot cracking and tensile shear strength of overlap joint in laser welding dissimilar Al alloys. Int J Adv Manuf Technol 86(9):2895–2904
10. Tarng YS, Tsai HL, Yeh SS (1999) Modelling, optimization and classification of weld quality in tungsten inert gas welding. Int J Mach Tools Manuf 39(9):1427–1438
11. ASTM B 557M, 2004, Standard test methods of tension testing wrought and cast aluminium- and magnesium-alloy products, pp 1–18
12. Senthil Kumar T, Balasubramanian V, Sanavullah MY (2007) Influences of pulsed current tungsten inert gas welding parameter on the tensile properties of AA6061 aluminium alloy. Mater and Des 28(7):2080–2092
13. Singh L, Shah V, Singh N (2013, September) Study the influence of TIG welding parameters on weld characteristics of 5083 aluminium alloy. Int J Eng Sci Innov Technol (IJESIT) 2(5)
14. Varli AE, Gurbuz R (2007) Fatigue crack growth behaviour of 6013 aluminium alloy at different ageing conditions in two orientations. Turk J Eng Environ Sci 30(6):381–386

Experimental and Simulation Study on Deep Drawing Process to Reduce Earing

S. Pratheesh Kumar, S. Elangovan, S. Hari Chealvan, and M. Mohamed Rafeek

Abstract Deep drawing is a popular press working process because it eliminates the need for costly machining and welding steps whilst allowing for the production of components in a shorter amount of time. A prior thermo-mechanical treatment has rendered the workpiece material used in a deep drawing method anisotropic in nature, which makes it ideal for deep drawing. Earring is one of the most typical defects made when using the deep sketching approach. Understanding the process of ear formation in deep drawing allows for process change earlier in the process, which can result in a defect-free end product whilst saving money. Using experimental and modelling investigations, the purpose of this paper is to reduce ear growth during the deep sketching process. Despite the fact that various parameters such as planar anisotropy, friction between the punch and die, and blank holding force are vital for ear formation, the shape of the initial blank has a considerable impact on the final ear design. Using a modified profile of the original blank shape, the purpose of this work is to reduce the formation of ear formation. It is simulated with DEFORM 3D software, how the ears form in the original and modified blanks will look like, and the findings are validated by conducting experimental trials on the blanks.

Keywords Deep drawing · Earing · Finite element analysis · Deformation · Deform

1 Introduction

Due to the fact that sheet metal is one of the most important semi-finished materials used in the steel industry, sheet metal forming technology is an essential engineering topic. Sheet metals are characterised by a large surface area to thickness ratio compared to other types of metals. When a flat sheet of metal is formed into the appropriate shape, it is devoid of faults such as fractures or excessive localised

S. Pratheesh Kumar (✉) · S. Elangovan · S. Hari Chealvan · M. Mohamed Rafeek
Department of Production Engineering, PSG College of Technology, Coimbatore, Tamil Nadu, India
e-mail: spratheeshkumarth@gmail.com

E. Natarajan et al. (eds.), *Materials, Design and Manufacturing for Sustainable Environment*, Lecture Notes in Mechanical Engineering,
https://doi.org/10.1007/978-981-19-3053-9_31

thinning. Sheet metal shaping is one of the most common types of metal fabrication. Sheet metal is a broad phrase that refers to a variety of different materials. Metals that have been formed, on the other hand, are referred to as wrought. Sheet metal is normally rectangular in shape. Stainless steel is also rectangular in shape. Another distinguishing characteristic of a sheet's thickness is that it is far thinner than the sum of its length and width combined. Sheet metal can be created from a range of metals, including steel, copper, aluminium, gold, and brass alloys. Steel is the most common metal used in sheet metal production. The materials that are used must, however, possess specific characteristics in order to be transformed into sheet metal. Hardness and stiffness, as well as ductility, or formability, are all characteristics of metals. As a result of these regulations, it is possible that material restrictions will be imposed. Consequently, sheet metal processing is incompatible with brittle or stiff materials.

Metal sheets are semi-finished products that are lightweight, thin, stable, elastic, and flat, with the ability to be produced over a big amount of land. These features make them ideal for use as cladding and covering of any kind. Sheet metal can also be deformed in a variety of ways, such as bending, punching, or cutting. Sheet metal can be formed into a wide range of shapes, which is why it is utilised as a foundation for so many different items. Sheet metal processing encompasses a wide range of techniques, including welding, cutting, and bending in addition to punching, forming, and rolling. Forging is a subcategory of sheet metal processing and includes subcategories such as welds and glues as well as bends and punches. Other subcategories include steel building and layer forming as well as heating coils.

Sheet metal processing enables the production of both soft and hard sheets, with the differing properties being obtained using varying alloys. For instance, during metalworking, a variety of elements can be added to steel in the liquid state to alter the material qualities of the resulting sheet metal. These elements include titanium, copper, niobium, and molybdenum in addition to silicon, nickel, and chrome. Sheets can be manufactured from a variety of metals, including steel, aluminium, and others, all of which come in a variety of alloys. The different sheets are classified according to their qualities using standards that group the materials together. To make the structure of the sheets more obvious, material numbers were added.

1.1 Deep Drawing Process

To mould flat sheets of metal into cup-shaped objects such as bathtubs, shell cases, and car panels, deep drawing is utilised. In order to accomplish this, a blank of the appropriate size is placed over a shaped die, and the metal is punched into the die, as shown in Fig. 1. In order to prevent wrinkling, it is frequently necessary to apply clamping or hold-down pressure on the blank whilst pressing it against the die. In a double-action press equipped with a blank holder or hold-down ring, this is the most effective method. Deep drawing is a popular press working process because it

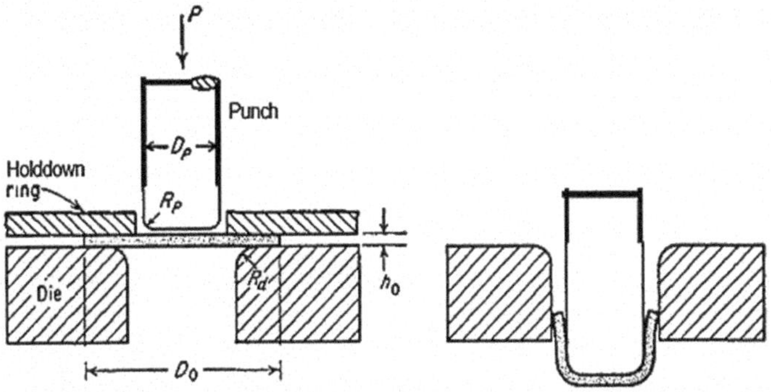

Fig. 1 Schematic representation of deep drawing process

eliminates the need for costly machining and welding steps whilst allowing for the production of components in a shorter amount of time.

1.2 Earing

During the deep drawing of anisotropic sheets, ear formations might occur, which can cause substantial problems during the manufacture of deep drawn containers. In order to reach the requisite container height and to allow the container ends to be properly attached, extra processes are required as a result of the earing. It is possible that improved control and ear reduction will reduce the necessity for these extra operations.

Earing is mostly caused by anisotropy in the sheet's planar direction. Earing is not recommended because it entails the removal of some metal from the top of the cup, which is not ideal. This increases the amount of time and money spent on the procedure. One method of avoiding this problem is to utilise a drawbead that is correctly shaped to restrict the flow of material into the die, resulting in a cup form that does not have ears. An expensive and time-consuming experimental trial and error technique was used to determine the optimal drawbead form for the production of a defect-free cup, which was ultimately unsuccessful. Numerical simulation software provides a compelling and effective alternative to traditional methods. In preparation for this inquiry, several publications on deep drawing and earing formation were gathered, and some of them are provided here as case studies.

Design of experiments and statistical analysis were utilised by Colgan and Monaghan [1] to identify the most important factors impacting the drawing process. According to the ANOVA results, the radii of the punch/die have a bigger impact on the deep drawing process than the other parameters. Detailed explanations of the causes of earing in formed cups are provided by Kishore and Kumar [2]. They experimented with several blank shapes in order to reduce earing and determined that a non-cylindrical first blank form is more effective at reducing earing.

With the help of a design hierarchy and response surface technique, Jansson and Nilsson [3] demonstrated how to improve the efficiency of sheet metal forming processes. They demonstrated how to create a forming procedure that avoids material failure whilst maintaining an acceptable through thickness strain across the thickness of the material. In their paper, Hariharan and Balaji [4] discussed how to optimise the material used in the sheet metal forming process. They demonstrated how to reduce the raw material size of the raw material used in the panel roof's construction (automobile component). They made a comparison between the experimental results and the results of finite element analysis. A 1.57% reduction in the weight of the raw material has been achieved. Kalpakjian and Schmid [5] studied the deep drawing press, the deep drawing process, deep drawing errors, and the sources of deep drawing faults in depth drawing.

Park et al. [6] investigated how to improve the formability of the elliptical deep drawing method by modifying its geometry. They looked at the impact of a variety of parameters, including corner radii, on the formability of the deep drawing process. During the deep drawing process, they ensured that the corner radius of the punch and die was optimised. The research group of Yoon et al. [7] developed a mechanism for accurately predicting earing in highly textured aluminium sheets. Generally speaking, anisotropy is not taken into consideration in finite element analysis. In this work, he quantified the anisotropy of the aluminium alloy AA5042-H2, which was previously unknown. Anisotropy and directionality were discussed in relation to the production of earings throughout the deep sketching process, according to him.

Specifically, Rao [8] investigated the deep drawing process, the impact of numerous parameters on sheet formability, and procedures and equations for determining starting blank size, the number of draws, drawing force, and the force required to hold the blank during the deep drawing process. Inal and colleagues [9] devised a method for imitating earing in textured aluminium sheets that was successful. It was for this reason that they developed a specialised finite element analysis that is limited to modelling only the sheet's flange region. Their numerical simulation of earing for two textures was then compared to see which had the best results.

Researchers Vahdat et al. [10] investigated whether the use of drawbeads may be used to minimise ear growth during the deep drawing process. Their computational technique resulted in an optimised drawbead contour that reduces the ear in deep drawing, which they were able to achieve through experimentation. They looked at a number of test problems, such as round and square cups, to see what they might learn. In the end, the best drawbead contour was achieved through an iterative design process. Drawing such cups is a time-consuming and complicated task. A topic that has recently attracted some attention is the elimination or mitigation of the earing fault during the deep drawing process.

This is expensive in terms of both money and time. In order to minimise the ear defect in deep drawing, a variety of approaches have been used to achieve this goal. However, none of them were successful in completely removing the earing. After reviewing the literature, it has been discovered that the vast bulk of the efforts in the field of ear mitigation have concentrated on optimising blank shape. The work on earring reduction in this paper has began with the use of these references as starting

Fig. 2 Deep drawn cups showing earing

points. Sheet metals are widely used in a range of metal forming sectors, including the automotive industry. In the sheet metal industry, metals such as brass, steel, and aluminium are frequently used for a wide range of purposes. Deep drawing is an important phase in the process of metal shaping [2, 11]. As shown in Fig. 2, earring is a significant flaw in deep drawing techniques. In most cases, planar anisotropy and friction between the blank and the punch are responsible for this [12–14]. Understanding the earing process in deep drawing allows for early process change, which results in a defect-free end product in the end.

2 Experiment Methods and Materials

2.1 Tooling and Equipment for Drawing

Tool steel and cast iron are the most often utilised tool and die materials, however, carbides and polymers are also occasionally employed [15]. Deep drawing is accomplished with the help of hydraulic press and mechanical press, which operate at rates ranging between 0.1 m/s and 0.3 m/s. The methodology that was employed in this investigation is depicted in Fig. 3. Figure 4 depicts the experimental setup used to study the deep sketching process.

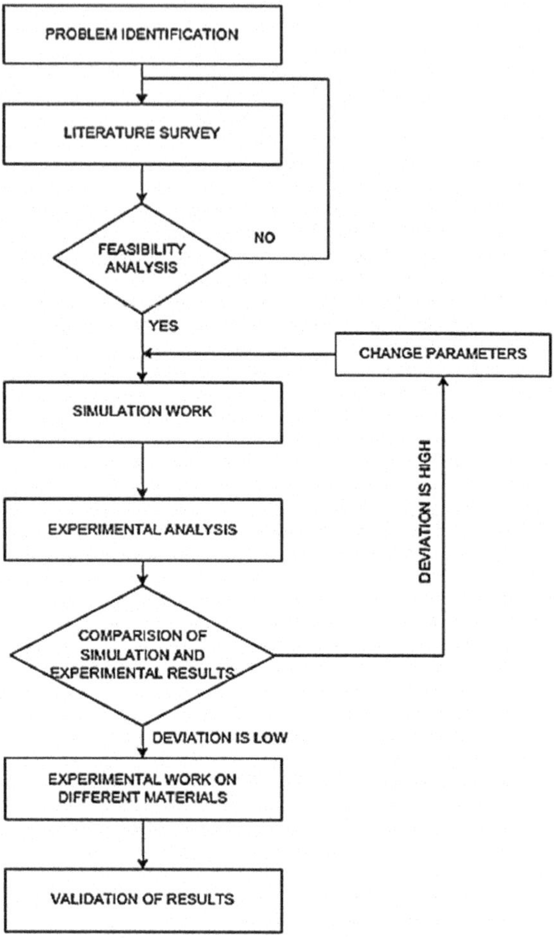

Fig. 3 Flowchart for methodology

2.2 *Parameters Considered to Reduce Earing*

(i) Directionality of sheet metal
(ii) Clearance between punch and die
(iii) Friction between blank and punch/die
(iv) Annealing of the blank
(v) Optimisation of blank shape
(vi) Use of drawbead.

Fig. 4 Deep drawing press

2.3 *Simulation Procedure*

(i) Import the geometries (blank, punch, die, and holder) and fix the required constraints.
(ii) Create mesh, enter load value for punch and die.
(iii) Enter primary die displacement value and velocity of die.
(iv) Enter the number of steps and increments for the die.
(v) Finally check and generate data.

3 Theoretical Study on Deep Drawing

The technical specification and calculation of blank size and forces are discussed in this section (Fig. 4).

3.1 *Tool Dimensions for Deep Drawing Experiment*

Blank diameter = 70 mm
Punch diameter = 36 mm
Punch profile radius = 4.2 mm
Thickness of the blank = 0.5 mm
Cup height = 25 mm

3.2 Theoretical Calculation

3.2.1 Blank Size

$$\begin{aligned}\text{Diameter } (D) &= \sqrt{(d-2r)^2 + 4d(h-r) + 2\pi r(d-0.7r)}\\ &= \sqrt{(36-2*4.2)^2 + 4*36(25-4.2) + 2\pi*4.2(36-0.7*4.2)}\\ &= 68.04 \text{ mm}\end{aligned} \tag{1}$$

$$\text{Actual blank size used} = 70 \text{ mm}$$

3.2.2 Drawing Force

$$\begin{aligned}\text{Drawing ratio } (D_R) &= \frac{\text{Blank dia}}{\text{Punch dia}}\\ &= \frac{70}{36}\\ &= 1.94\end{aligned} \tag{2}$$

$$\begin{aligned}\text{Drawing force} &= \pi t D_p Y(D_R 0.7)\\ &= \pi * 0.5 * 36 * 290(1.94 - 0.7)\\ &= 20.334 \text{ KN}\end{aligned} \tag{3}$$

3.3 Blank Holding Force

$$\begin{aligned}\text{Blank holding force} &= 1/3 * (\text{Drawing force}) \ldots\\ &= 1/3 * (26.860)\\ &= 8.953 \text{ KN}\end{aligned} \tag{4}$$

3.4 Percentage Reduction

$$\begin{aligned}\text{Percentage reduction} &= 100\left(1 - \frac{\text{shell dia}}{\text{blank dia}}\right)\\ &= 100\left(1 - \frac{36}{70}\right)\\ &= 48.57\%\end{aligned} \tag{5}$$

3.5 *Actual Calculation*

$$\text{Punch pressure} = 210\ \text{kg/cm}^2$$
$$\text{Drawing force} = P * A$$
$$= 210 * \frac{\pi}{4} * \left(\frac{36 * 36}{10 * 10}\right) * 9.81$$
$$= 20.969\ \text{KN} \quad (6)$$

The theoretical calculation shows that the drawing force is equal to the actual force applied practically in the deep drawing operation. Also, the initial blank size is slightly higher than the calculated value. The blank size and shape directly influence earing, and hence, further study will concentrate on the optimisation of initial blank shape.

4 Simulation Study

As shown in Fig. 5, circular blank ears are created at 0° and 90° to the rolling direction in a circular blank. A relationship was established between the coordinates of the sites, where the circle of the circular blank was sliced by 0° and 90° lines in the rolling direction and the corresponding coordinates of the modified blank, and the modified blank [16]. It is planned to develop a technique for optimising the shape of blanks that will consist of two stages: the first stage will involve the modification of an original circular blank, and the second stage will involve the further modification of the modified blank from the first stage (if required).

Fig. 5 Formation of ears at 0° and 90° to rolling direction

4.1 Ear Height Calculation

The blank size (D) is determined using Eq. 6

$$\begin{aligned} D &= \sqrt{(d-2r)^2 + 4d(h-r) + 2\pi r(d-0.7r)} \\ &= \sqrt{(36-2*4.2)^2 + 4*36(25-4.2) + 2\pi*4.2(36-0.7*4.2)} \\ &= 68.04 \text{ mm} \end{aligned}$$

Actual blank size that is being used = 70 mm

$$\begin{aligned} 70 &= \sqrt{(36-2*4.2)^2 + 4*36(h-4.2) + 2\pi*4.2(36-0.7*4.2)} \\ 4900 &= 761.6 + 144\text{h} - 604.8 + 872.433 \\ 144\text{h} &= 3870.607 \\ h &= 26.88 \text{ mm} \end{aligned}$$

4.2 Procedure to Modify the Blank Shape

A pair of ears were discovered in the drawn cups at 0° and 90° to the rolling direction. In general, blanks can be adjusted either by removing material from the 0° and 90° rolling directions or by adding material to the 45° rolling direction of the blank. According to the findings, the first approach of optimisation performs significantly better than the second way of optimisation [17, 18].

The Δr_1 and Δr_2 values are found using Eqs. 8 and 9

$$\Delta r_1 = r_0 - r_{45} \tag{8}$$

$$\Delta r_2 = r_{90} - r_{45} \tag{9}$$

For the purpose of changing the blank, a circle with the requisite diameter was made on graph paper and divided into eight equal halves by drawing eight lines from the centre at an angle of 45° from the centre. Equations 10 and 11 were used to get the coordinates of four new sites that were used to reduce the material in the rolling direction at 0° and 90° [19].

$$\text{Modified } X \text{ coordinate} = R - 3\Delta r_1 \tag{10}$$

$$\text{Modified } Y \text{ coordinate} = R - 3\Delta r_2 \tag{11}$$

where R is the radius of the circular blank.

Following that, four new arcs are drawn with four new centres. Specifically, each arc connects three locations, two of which are intersection points of lines drawn at 45° to the rolling direction, and the third point is the newly discovered point on the two-axis coordinate system.

4.3 Model Calculation to Find New Radius

Based on the procedure, calculations are done for both the drawn cups.

4.3.1 Model Calculation to Find for Cup 1

$$h\ (\text{at}\ r_0) = 28.31, d = 36,\ r = 4.2$$

$$\begin{aligned} D_0 &= \sqrt{(d-2r)^2 + 4d(h-r) + 2\pi r(d-0.7r)} \\ &= \sqrt{(36-2*4.2)^2 + 4*36(28.31-4.2) + 2\pi*4.2(36-0.7*4.2)} \\ &= 71.4565 \end{aligned}$$

$$r_0 = 35.7282$$

Similarly,

$$r_{45} = 34.8569$$

$$r_{90} = 35.3178$$

$$\Delta r_1 = r_0 - r_{45} = 0.8773$$

$$\Delta r_2 = r_{90} - r_{45} = 0.4669$$

Modified X coordinate $= R - 3\ \Delta r_1 = 32.3681$ mm

Modified Y coordinate $= R - 3\ \Delta r_2 = 33.5993$ mm.

4.3.2 For Cup 2

$$r_0 = 35.6475$$

$$r_{45} = 34.866$$

$$r_{90} = 35.256$$

$$\Delta r_1 = r_0 - r_{45} = 0.7815$$

$$\Delta r_2 = r_{90} - r_{45} = 0.390$$

Modified X coordinate $= R - 3\ \Delta r_1 = 32.655$ mm

Modified Y coordinate $= R - 3\ \Delta r_2 = 33.830$ mm.

4.3.3 Average Values

Modified X coordinate is 32.5 mm

Modified Y coordinate is 33.5 mm.

4.4 Modified Blank

Based on the calculated result, the modelled blank and the modified shape of the solid blank is shown in Fig. 6. With the modified blank shape, the deep drawing process is simulated in DEFORM 3D to predict the actual formation of cup and to compare the earing formation in both original and modified blank.

4.5 Simulation Steps

The positioning of die, punch, workpiece (blank), and blank holder in the simulation is shown in Fig. 7.

The initial deformation of the workpiece is shown in Fig. 8. The initial deformation starts when the upper die moves downwards along with the blank holder.

After the die travel of 25 mm downwards, the work piece is deformed fully to form a cup which is shown in Fig. 9.

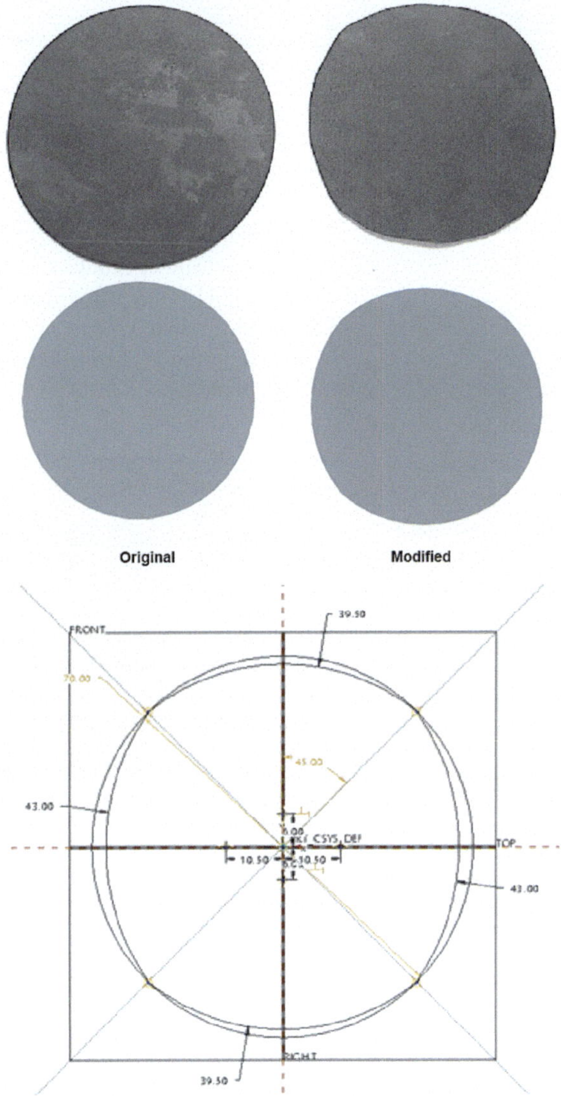

Fig. 6 Photographic and modelled view of actual blank and method to modify the blank shape

4.6 *Simulation Result*

Based on the simulated results, cup shapes formed from the original blank size and modified blank size are shown in Figs. 10 and 11, respectively. In general, performing simulation in metal forming operation is highly difficult due to process nature. In this study, simulation is conducted considering the original blank and modified blank sizes observed from the calculated results. The boundary condition of the study is selected to reflect the ideal metal forming condition in the deep drawing process.

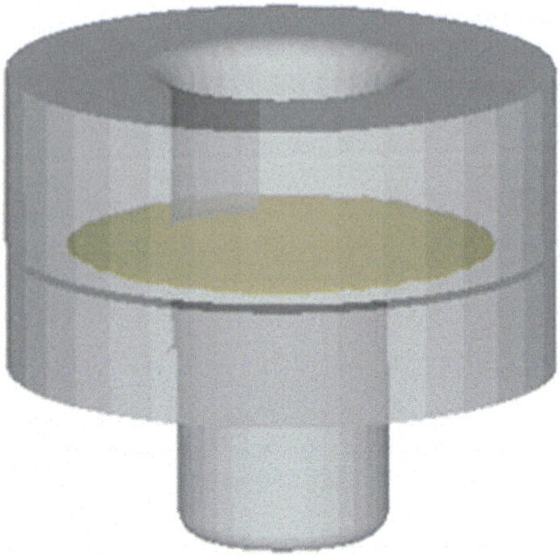

Fig. 7 Positioning of work piece, die, and punch

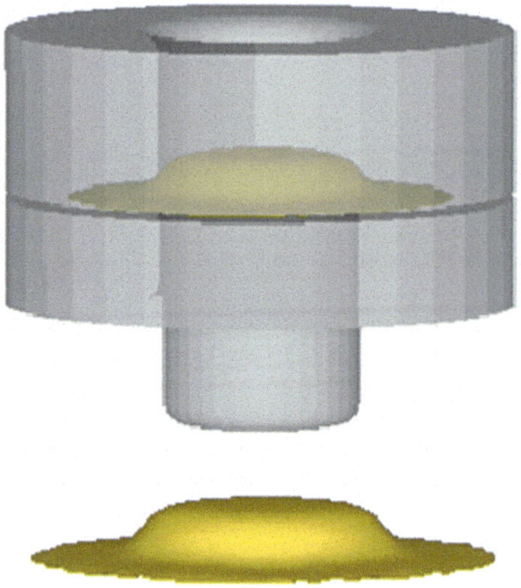

Fig. 8 Initial stage of workpiece deformation

The results obtained show the earing formation at 12% in original blank and at 8% in modified blank. In order to evaluate the simulation results, experiments are conducted, and the results are discussed in results and discussion section.

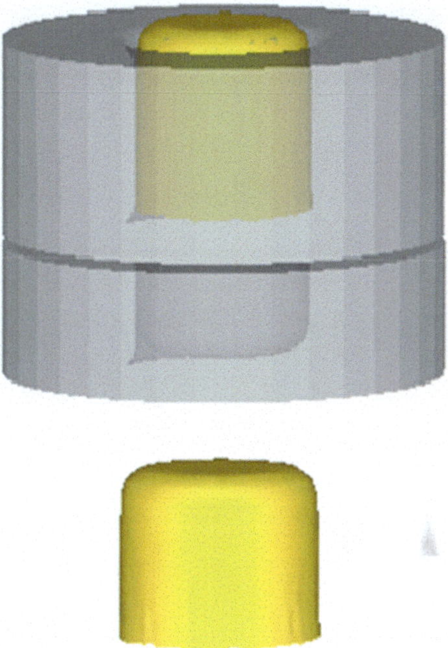

Fig. 9 Cup formation sequence

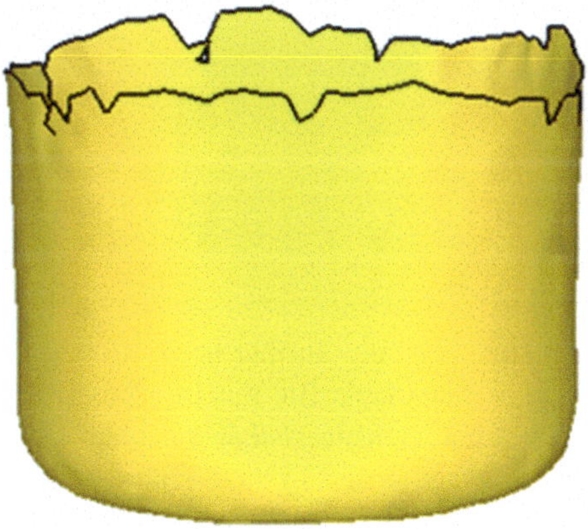

Fig. 10 Cup formed from original blank

5 Experimental Study

5.1 Initial Stage of Modification

The cup drawn from the modified blank is shown in Fig. 12 Though ears are reduced, there is irregularity in the formation of the cup. The main reason is that the blank may not be placed properly in the blank holder because of the reduction in its original

Fig. 11 Cup formed from modified blank

Fig. 12 Cup drawn from initial modified blank

diameter. This may further be controlled if the blank holder is also modified to the shape of new blank. But this is not feasible immediately and is costly.

The other reason is that the blank was cut using a snip cutter, and there are surface irregularities in the blank itself. This can be controlled if the blank will be cut using a shearing machine with the shape of the modified blank. This is also not feasible immediately and is costly. Hence, further study will be concentrated on the modification of blank shape only at 90° that is without change in diameter of the original blank.

5.2 *Second Stage of Modification*

From the simulation and experimental results, it is found that earing is mainly formed at 0°, 90°, 180°, and 360°. So, the blank diameter is reduced only at these angles.

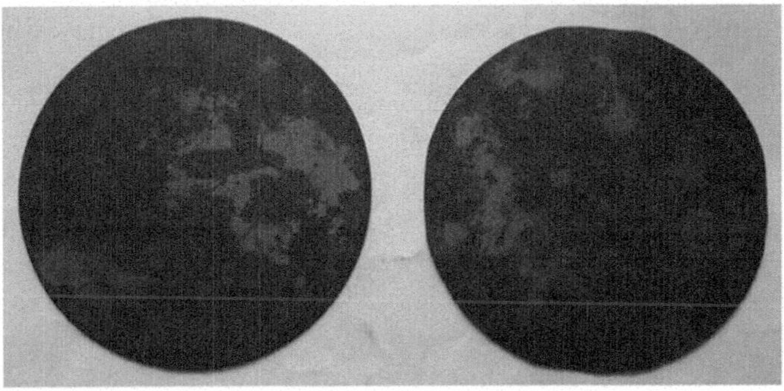

Fig. 13 Photographic view of actual blank (second stage)

Fig. 14 Cup formed from newly modified blank

Based on this, the blank shape is further modified using Pro-E and the modified shape of the solid model blank (Fig. 13).

The cup drawn from the newly modified blank is shown in Fig. 14. In the formed cup, the earring is reduced when compared to the previous modified blank.

6 Results and Discussion

6.1 Percentage of Ear Height

$$\%\ \text{Ear height} = \frac{\text{maximum cup height} - \text{minimum cup height}}{\text{minimum cup height}} * 100 \quad (12)$$

The minimum cup height after modification can be increased by using a good experimental setup to guarantee that the modified blank is properly centred in the cup chamber after modification. According to the calculation, the smallest cup height is the height of the cup that will be produced after reducing the cup's uneven top edge. The cup is deep drawn, the extent of the earing is measured, and the profile is plotted in a graph for the original and modified blanks, respectively, and the results are plotted in a graph. In the top edge of the cup, a dial gauge is used to trace the profile of the ear at 15-degree intervals, resulting in a perfect fit.

6.2 Cup Formed from Original Blank

Measured values of cup height at 15° interval on the cup drawn from original blank are shown in Table 1.

The graphs between cup height and angle are plotted using the numbers from Table 1 as a starting point. The graphs show that the maximum and minimum variances are 28.49 mm and 26.14 mm, respectively, and those the maximum and minimum variations are the same. It is necessary to calculate the minimum cup height since it is that height of the cup that will be acquired after trimming the cup's uneven top edge, which is 25 mm as shown in Fig. 15.

$$\begin{aligned}\text{\% Ear height for sample cup 1} &= \frac{28.49 - 25}{25} * 100 \\ &= 13.96\%\end{aligned}$$

$$\begin{aligned}\text{\% Ear height for sample cup 2} &= \frac{28.21 - 25}{25} * 100 \\ &= 12.84\%\end{aligned}$$

The average of percentage ear height in both samples is 13.40%, and the details are shown in Fig. 16.

6.3 Cup Formed from Modified Blank

Measured values of cup height at 15° interval on the cup drawn from modified blank are shown in Table 2.

The graphs are produced between cup height and angle based on the data obtained from Table 2. The maximum and minimum variations are 27.44 mm and 26.18 mm, respectively, according to the results of the analysis. This estimate takes into consideration the minimal cup height that will be reached after reducing the cup's uneven top edge, which is 25 mm as illustrated in Fig. 17.

Table 1 Cup height values for cup drawn from original blank

Cup 1		Cup 2	
Angle (°)	Cup height (mm)	Angle (deg)	Cup height (mm)
0	28.31	0	28.15
15	27.62	15	27.78
30	26.95	30	27.38
45	26.59	45	26.96
60	26.78	60	26.62
75	27.15	75	26.52
90	27.50	90	26.89
105	27.38	105	27.38
120	27.12	120	27.85
135	27.00	135	27.93
150	26.82	150	27.65
165	26.48	165	27.26
180	26.15	180	26.66
195	26.60	195	26.14
210	26.75	210	26.30
225	26.75	225	26.57
240	26.68	240	26.74
255	27.14	255	26.80
270	27.54	270	27.05
285	27.62	285	27.55
300	27.80	300	27.97
315	28.23	315	28.21
330	28.42	330	28.20
345	28.49	345	28.21
360	28.31	360	28.11

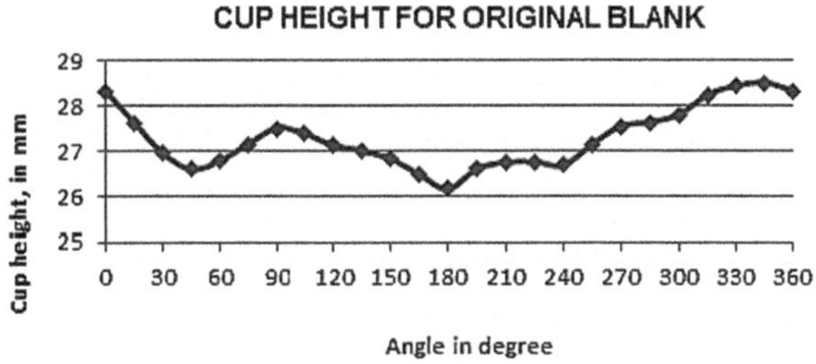

Fig. 15 Variation in height versus angle for original blank (cup 1)

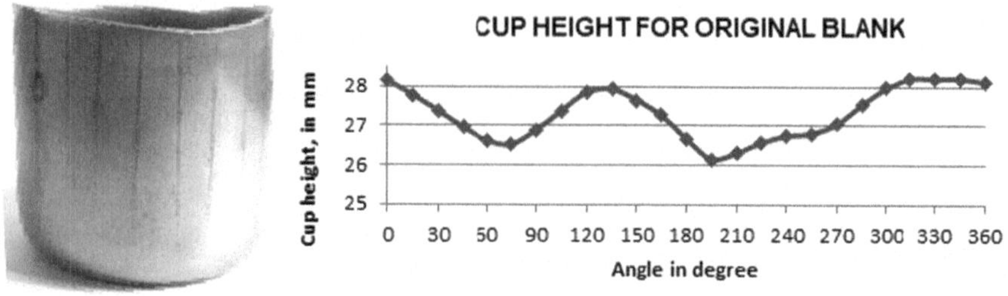

Fig. 16 Variation in height versus angle for original blank (cup 2)

Table 2 Cup height values for cup drawn from modified blank

Cup 1		Cup 2	
Angle (°)	Cup height (mm)	Angle (°)	Cup height (mm)
0	27.44	0	27.40
15	27.32	15	27.27
30	26.34	30	27.30
45	26.18	45	27.29
60	26.48	60	26.98
75	26.81	75	26.40
90	27.05	90	26.23
105	27.15	105	26.62
120	26.98	120	27.20
135	26.72	135	27.28
150	26.82	150	26.80
165	27.20	165	26.90
180	27.35	180	27.09
195	27.24	195	27.09
210	27.20	210	27.01
225	26.94	225	27.30
240	26.66	240	27.19
255	26.82	255	27.24
270	26.60	270	26.96
285	26.75	285	26.56
300	26.75	300	26.42
315	26.75	315	26.42
330	27.03	330	26.61
345	27.34	345	27.00
360	27.34	360	27.36

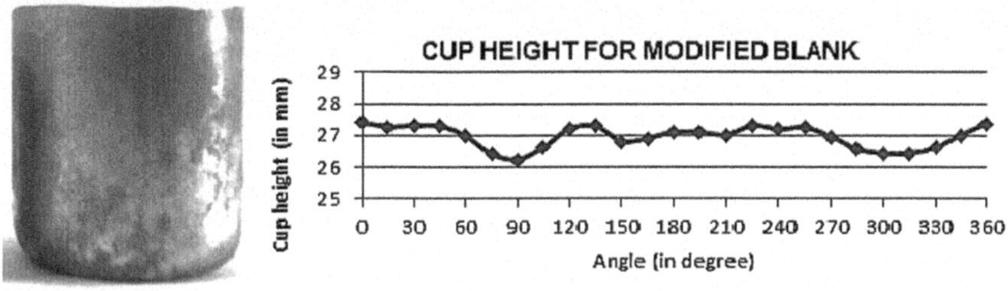

Fig. 17 Variation in height versus angle for modified blank (cup 1)

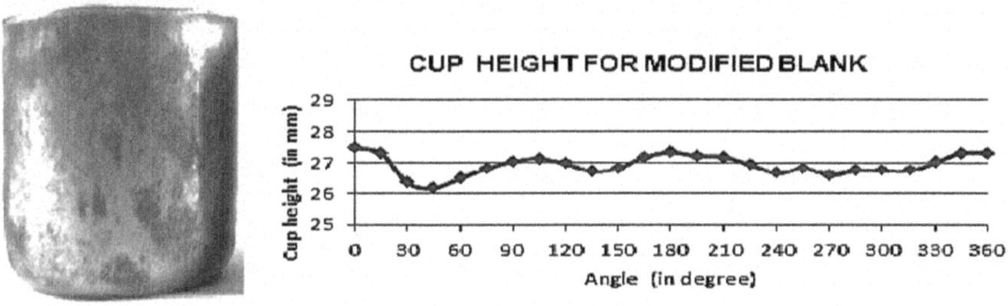

Fig. 18 Variation in height versus angle for modified blank (cup 2)

$$\begin{aligned}\% \text{ Ear height for sample cup 1} &= \frac{27.44 - 25}{25} * 100 \\ &= 9.76\%\end{aligned}$$

$$\begin{aligned}\% \text{ Ear height for sample cup 2} &= \frac{27.40 - 25}{25} * 100 \\ &= 9.60\%\end{aligned}$$

The average of percentage ear height in both samples is 9.68%, shown in Fig. 18.

6.4 Percentage Reduction in Ear Height

Based on the calculation, it is observed that the average percentage of ear height in cups drawn from original and modified blanks are 13.40% and 9.68%, respectively. Hence, the reduction in average percentage of earring for modified blank from original blank is 27.76%.

$$\text{Average of \% ear height in cup drawn from original blank} = 13.40\%$$
$$\text{Average of \% ear height in cup drawn from modified blank} = 9.68\%$$

$$\text{The maximum percentage reduction in earing} = \frac{13.40 - 9.68}{13.40} * 100$$
$$= 27.76\%$$

7 Conclusion

(i) The theoretical calculation shows that the drawing force is equal to the actual force that is being used in the deep drawing operation. But the initial blank size is slightly higher than the calculated value.
(ii) The simulation results from DEFORM 3D show that the ear formation is less in the modified blank (8%) compared to the original blank (12%) that has been already used.
(iii) The experimental results show that the percentage reduction in earing by the modification of the blank shape is 27.76% compared to the original blank shape.

Using a blank holder that is specifically designed for the modified blank, as well as cutting the new blank down using a shearing machine that is specifically designed for the changed blank, would further limit the formation of earings. Consideration of additional parameters such as blank holding force, die corner radius, clearance between punch and die, the usage of drawbead, and the qualities of the material can help to reduce the extent of earing. The understanding of anisotropy will also result in more accurate outcomes when it comes to minimising earing.

References

1. Colgan M, Monaghan J (2003) Deep drawing process analysis and experiment. J Mater Process Technol 132(1–3):35–41
2. Kishor N, Kumar DR (2002) Optimization of initial blank shape to minimize earing in deep drawing using finite element method. J Mater Process Technol 130:20–30
3. Jansson T, Nilsson L (2006) Optimizing sheet metal forming processes using a design hierarchy and response surface methodology. J Mater Process Technol 178(1–3):218–233
4. Hariharan K, Balaji C (2009) Material optimization: a case study using sheet metal-forming analysis. J Mater Process Technol 209(1):324–331
5. Jeyasimman D, Sivaprasad K, Senthil Kumar V, Narayanasamy R (2013) Carbon nanotube reinforced aluminium alloy composites-a review. J Manuf Eng 8(2):75–84
6. Park DH, Kang SS, Park SB (2001) A study on the improvement of formability for elliptical deep drawing processes. J Mater Process Technol 113(1–3):662–665
7. Yoon JH, Cazacu O, Yoon JW, Dick RE (2010) Earing predictions for strongly textured aluminium sheets. Int J Mech Sci 52(12):1563–1578
8. Rao PN (2013) Manufacturing technology, vol 1. Tata McGraw-Hill Education
9. Inal K, Wu PD, Neale KW (2000) Simulation of earing in textured aluminium sheets. Int J Plast 16(6):635–648
10. Vahdat V, Santhanam S, Chun YW (2006) A numerical investigation on the use of drawbeads to minimize ear formation in deep drawing. J Mater Process Technol 176(1–3):70–76

11. Luo L, Wei D, Wang X, Zhou C, Huang Q, Xu J, Jiang Z (2017) Effects of hydraulic pressure on wrinkling and earing in micro hydro deep drawing of SUS304 circular cups. Int J Adv Manuf Technol 90(1):189–197
12. Nagda PS, Bhatt PS, Shah MK (2017) Finite element simulation of deep drawing process to minimize earing. Int J Mech Mechatron Eng 11(2):413–416
13. Park CS, Ku TW, Kang BS, Hwang SM (2004) Process design and blank modification in the multistage rectangular deep drawing of an extreme aspect ratio. J Mater Process Technol 153:778–784
14. Tikhovskiy I, Raabe D, Roters F (2007) Simulation of earing during deep drawing of an Al–3% Mg alloy (AA 5754) using a texture component crystal plasticity FEM. J Mater Process Technol 183(2–3):169–175
15. Tran MT, Shan Z, Lee HW, Kim DK (2021) Earing reduction by varying blank holding force in deep drawing with deep neural network. Metals 11(3):395
16. Younis KM, Jaber AS (2018) Experimental and numerical study of the earing defect during square deep drawing process. Eng Technol J 36(12 Part A):1267–1275
17. Bandyopadhyay K, Panda SK, Saha P, Padmanabhan G (2015) Limiting drawing ratio and deep drawing behaviour of dual phase steel tailor welded blanks: FE simulation and experimental validation. J Mater Process Technol 217:48–64
18. Younis KM, Jaber AS (2011) Experimental and theoretical study of square deep drawing. Int J Eng Technol 29(12):2456–2467
19. Izadpanah S, Ghaderi SH, Gerdooei M (2016) Material parameters identification procedure for BBC2003 yield criterion and earing prediction in deep drawing. Int J Mech Sci 115:552–563

Design Modification of Robotic Arm for Incremental Sheet Metal Forming

S. Pratheesh Kumar, R. Mohanraj, K. Anand, and M. Mohamed Rafeek

Abstract Incremental forming, a sheet metal forming process known for its automation capability and production flexibility. The process has its own merits when compared with conventional forming process. This research aims to improve the process capability of robotic arm used for incremental sheet metal forming. Increasing the end effector's linear actuation will enhance the tool's movement range whilst also making it easier to make components with deeper depths, which is required to accomplish this goal. The study results of the robotic system developed shows the clear impact and its contribution in Incremental forming process. The simulation results reveal the credibility of robotic arm design improvements in producing precise and accurate products. The end result of the study aids in implementing robotic arm assisted part fabrication in incremental forming with linear actuator.

Keywords Robotic arm · Incremental forming · Automation · Metal forming · Rapid forming

1 Introduction

Robotic technology is becoming more prevalent across all fields of study. It is one of the most advanced production technologies available today, and it offers substantial advantages over more traditional methods of manufacturing. The use of robotics is becoming more common in academic and industrial sectors. In recent years, robotic arm impact has expanded in the sheet metal forming business because huge investments in die fabrication for metal forming are avoided. This robotic arm technology has remained viable in the sheet metal forming sector for several years because of the flexibility in shape and geometry. To understand the relevance and suitability of robotic arm in incremental forming the configuration features of robotic arm has to be understood in detail.

S. Pratheesh Kumar (✉) · R. Mohanraj · K. Anand · M. Mohamed Rafeek
Department of Production Engineering, PSG College of Technology, Coimbatore, Tamil Nadu, India
e-mail: spratheeshkumarth@gmail.com

E. Natarajan et al. (eds.), *Materials, Design and Manufacturing for Sustainable Environment*, Lecture Notes in Mechanical Engineering,
https://doi.org/10.1007/978-981-19-3053-9_32

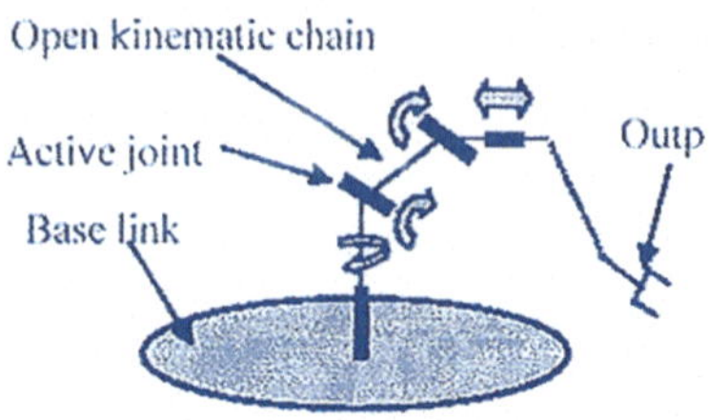

Fig. 1 Kinematic chain

1.1 Robotic Arm

A robotic arm is an electro-mechanical machine having electrical motors and electronically controllable elements that may be programmed when used in conjunction with appropriate manipulators. Robotic arms can be found in a wide variety of tasks. Manipulators of this sort have joints that allow rotational motion or translational (linear) motion between the linkages between the joints. As an example of a kinematic chain, consider how the manipulator's links are arranged as follows: (see Fig. 1). This part of a manipulator's kinematic chain comes to a halt when the manipulator is deployed, and it is specifically intended for certain purposes.

1.2 Types of Robotic Arm

Manufacturing, military surveillance, medicine, healthcare, transportation, and research are a few of the businesses that have made use of the various types of industrial robotic arms that are available. Other uses for industrial robotic arms include assembly lines and warehouses. Here are a few examples of the most commonly discussed robotic arm designs.

1.2.1 Articulated Robot Arm

Robot arms can have anywhere from two to ten or more joints, but the bulk of them have four to six articulation points or more at the minimum. As the number of joints on the robot arm increases, the number of axes does as well. The more axes the robot arm has, the more precise it may be. Robots employed in bomb disposal, for example, have appendages that resemble hands and can disarm explosive devices in a similar manner to a human hand [1]. Use this to open doors and move delicate goods like jewellery or plants. An articulating robot arm (Fig. 2a) is useful for many different tasks, such as on an assembly line or in a manufacturing process such as die casting or gas welding and spray painting.

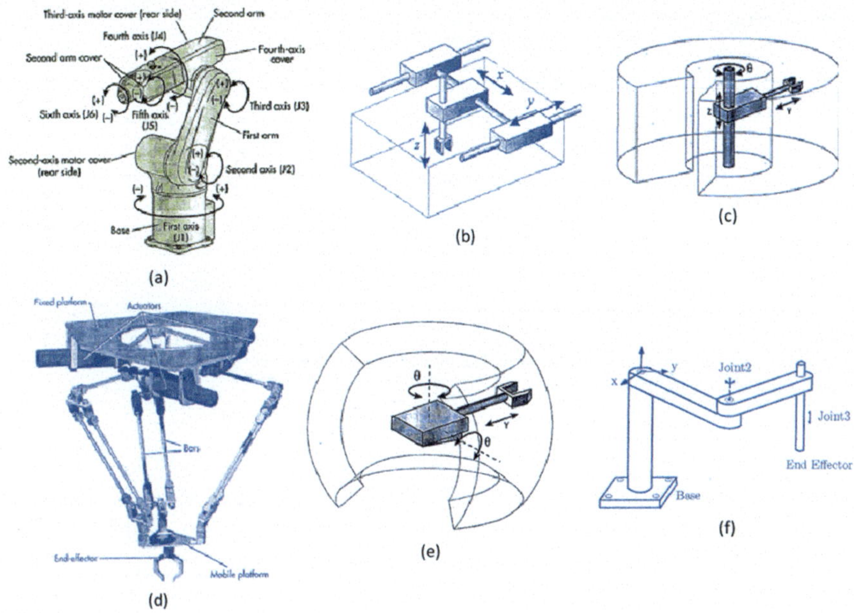

Fig. 2 **a** Articulated robot arm, **b** Cartesian robot arm, **c** cylindrical robot arm, **d** delta robot arm, **e** polar or spherical robot arm, **f** scara robot arm

1.2.2 *Cartesian Robot Arm*

Other than Cartesian, the Cartesian robots go by several other names, including rectilinear and gantry. In order to offer linear motions along each axis, these robotic arms each have three prismatic joints (see Fig. 2b). On the other hand, they could have a wrist-like attachment on one end to allow the user to rotate their wrist. Arc welding robots, assembly robots, and pick-and-place machines all use Cartesian robots.

1.2.3 Cylindrical Robot Arm

At least one rotary joint is positioned at the base of this sort of robot arm, providing rotating motion along its axis, and at least one prismatic joint is located at the top, providing linear motion and connecting the linkages between them (Fig. 2c). In most cases, there are 3. Using a cylindrical robot arm is particularly advantageous when working in industries that frequently deal with tubes and cylindrical components. Other applications include die casting, spot welding, and even measuring accretion discs in astrophysics [2, 3].

1.2.4 Delta Robot Arm

Other parallel robots can be distinguished by the existence of three parallelogram-jointed arms at the base, which are connected together at the top. A parallel robot is known as a delta robot (a type of robot with arms that have prismatic or rotational joints that converge onto a common base). This structure is in charge of ensuring that the robotic arm's end effector, where the bulk of the work is done, is always in the correct alignment (Fig. 2d). Delta robots are widely used in a variety of industries, including food processing and packaging, chemical and pharmaceutical manufacturing, and electronic manufacturing, because to their speed, precision, and sensitivity.

1.2.5 Polar or Spherical Robot Arm

To create a polar coordinate system, a robot with two rotary joints and one linear joint often uses a twisting joint to connect them to a single base. To put it another way, the work envelope and movement range will be circular in shape (Fig. 2e). Apart from gas and arc welding, polar robots are also used for environmental monitoring, planetary and deep sea research, rehabilitation and the use of physical therapy machinery [4].

1.2.6 Scara Robot Arm

According to some, SCARA stands for Selective Compliance Assembly Robot Arm, whereas to others, it stands for Selective Compliance Articulated Robot Arm (Fig. 2f). A popular application for SCARA robots is in high-speed assembly, packing, or P&P equipment. SCARA robots are cylindrical in shape with two parallel rotary joints, allowing for flexibility and suppleness (compliance) in a single plane of motion.

There are three types of robots: fully autonomous, semi-autonomous, and those controlled from a distance. A robotic arm can perform a wide range of tasks, including welding, drilling, spraying, and more. (5, 6, 7, 8) A self-sufficient robotic arm is built from various components, such as microcontrollers and motors. Thus, they can work at a higher pace whilst keeping complexity to a minimum. The extra benefit is that productivity is raised as a result.

1.3 Incremental Sheet Metal Forming

A newly developed process known as incremental sheet metal forming allows the production of asymmetric sheet metal parts without the need of a die and by the use of CNC machine tool-based path programming that induces localised plastic deformation with a spherical ending tool. Prior research has mostly focussed on the use of standalone robotic manipulators rather than the more restricted 3-axis CNC vertical machine centre or the more expensive 5 or 6-axis vertical machine centre, which are both prohibitively expensive [5, 6]. Robot-assisted incremental forming could be a viable alternative.

Forging is a metal fabrication technique used to create thin shell-like forms. Drawing a three-dimensional form is as simple as slowly pressing a blunt round tool against the surface of flat sheet metal whilst it is held in an ironing board jig (Fig. 3a, b). When compared to other sheet metal forming methods, this one requires less upfront investment because it does not call for the fabrication of expensive dies. Even so, it has not acquired general acceptance due to its slow speed and low precision. Small-scale batch production with lower precision requirements is feasible with this technique, and it is growing in popularity [7].

Being part of the international competition, which is getting more fierce as the creation and introduction of new products that are specialised to meet client needs accelerates, is critical. Continuous innovation may be achieved by bringing in more

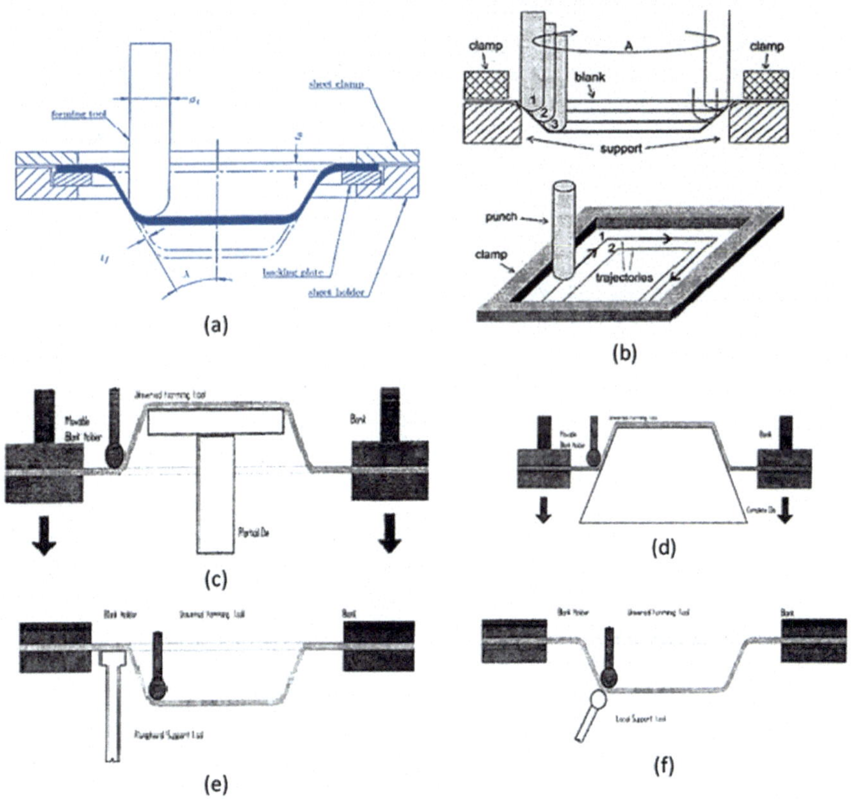

Fig. 3 **a** Single point incremental forming process (SPIF), **b** 3-dimensional strategy, **c** two point incremental forming process with partial die, **d** two point incremental forming process with complete die, **e** duplex incremental forming with peripheral supporting tool, **f** duplex incremental forming with local supporting tool

innovations whilst also cutting down on the time and money it takes to create prototypes. Because it is both economically profitable and allows for the creation of high-quality functional and flexible parts, sheet metal forming nowadays needs the use of decisive procedures. There are more complex part geometries being generated in vehicle manufacture, requiring the usage of sheet metal prototyping to allow the launch of new goods more swiftly. The need for new imaginative things that can only be produced in small quantities is also expanding. The forming capacity required is substantially lower than the traditional manner of forming because of the small plastic deformation area [8, 9]. In comparison to other standard forming techniques, the tool size is drastically reduced as a result of this innovation. Progressive deformation techniques have the benefit of allowing materials with lower ability to be deformed whilst also increasing the final result's strength. Companies rarely use the incremental forming method in spite of the numerous advantages it offers. Flexible tools are less successful at making complex shapes, hence a rigid tool is preferred. However, it was only beneficial with a simple understanding of geometry. It is still necessary to utilise a die plate to make complex shapes incrementally, but alternative modern processes are favoured since they process more quickly. For prototypes and small-scale manufacturing, the cost of technical equipment for current incremental forming procedures is excessively high.

1.4 Types of Incremental Sheet Forming

Technological advances like as computer-controlled machines, symmetric single point forming (spinning), and the creation of tool path processors in CAD software packages, amongst other things, have led to the development of new metal forming methods in recent years. One important outcome of this technology is incremental sheet forming, which enables the manufacturing of asymmetric shapes at a low cost without the need of expensive dies that are customised for each shape. This is a significant development.

The typical characteristics of incremental forming processes are:

(i) Requires a solid, small-sized forming tool
(ii) Does not need large, dedicated dies
(iii) Forming tool is in continuous contact with sheet metal
(iv) Ability of tool to move under control in three-dimensional spaces
(v) Can produce asymmetric sheet metal shapes

The incremental forming process can be totally decoupled from the spinning process once it has this ultimate property. A major development in ISF processes has been the advent of CNC milling machines and CAD software with tool path postprocessors. A patent from 1967 first described dieless formation, but no further work was done on it because the tools mentioned above were unavailable at the time. Sheet metal products are frequently made on a specialised machine or on a 3-axis

CNC mill utilising software that generates machine tool paths to generate the tool paths [10].

There are various types of asymmetric incremental sheet forming (AISF).

1.4.1 Single Point Incremental Forming (SPIF)

In recent years, SPIF has risen to the top of the list of most extensively used incremental forming methods, surpassing all others. It is gotten a lot of attention because of the different ways it can be shaped. It was possible to build the SPIF method by combining CNC technology with sheet metal spinning, shear flowing, and hammering operations. Throughout the process, the blank holder holds the flat sheet of metal tightly, allowing the forming tool to distort it. Rather than employing a die, this method forms parts with just one tool, which has a conical tip and is supported by an opposing rear plate (Fig. 3a). This minimises the effect of spring back during manufacturing. With this method, you may make prototypes or small series of sheet metal parts quickly and cheaply. You can also make a wide range of irregular-shaped, axi-symmetric components and highly customised medical goods in small batches using this method, too [11, 12].

1.4.2 Two Point Incremental Forming (TPIF) with Partial Die

Deformation occurs at the tool and the partial die in this two point incremental formation. Therefore, the die is positioned so that it is located on the other side of the sheet from where the sheet is being created. In the case of a partial die, the die's shape does not exactly match the specified geometry. Because they are pushed in opposite directions, they are capable of being moved together whilst still being driven separately (Fig. 3c). The apex of the form is more accurate with this technique than with SPIF [13], which boosts accuracy largely on the shape's periphery.

1.4.3 Two Point Incremental Forming (TPIF) with Complete Die

When a complete die is utilised on the other side of the sheet from the tool, as opposed to a partial die, the metal can be moulded according to the die's form (Fig. 3d). In spite of the increased geometric accuracy, this method has higher production costs and takes longer to make each part.

1.4.4 Duplex Incremental Forming with Peripheral Supporting Tool (DPIF-P)

A blank sheet of paper enclosed in a robust frame must have two industrial robots placed on either side of it The "master" robot holds the forming tool, whilst the "slave" robot holds the support tool (Fig. 3e). Whilst the forming tool pushes the sheet in an incremental fashion, generating the sheet in the shape of the tool path as a back plate, the peripheral tool moves along the component's boundary, giving leverage on the other side of the sheet against which the master robot can push.

1.4.5 Duplex Incremental Forming with Locally Supporting Tool (DPIF-L)

Both of these operations use robots that are equipped with universal shaping tools and housed in sturdy frames to fasten sheets together. Slave robot aids master robot in various ways. With progressive pressure, the "master" robot gradually moulds the sheet into tool path shape as depicted. Because their combined efforts result in a gap between the two tools, follower robots provide support in the opposite direction from their master robots (Fig. 3f). A robot's master and slave robots can swap roles to form concave and convex forms inside a single portion.

Single Point Incremental Forming (SPIF) uses an articulated robot arm. This project will make design changes to that arm (SPIF). In the future, we'll look more closely at the design of this robotic arm and its many parts.

Robot technology is becoming ubiquitous in all fields of manufacturing and production. It is one of the most advanced manufacturing techniques available today and offers significant advantages over more traditional manufacturing methods. The use of robotics is becoming increasingly common in a variety of industries and academic disciplines. The impact of robotic arms in the sheet metal forming business has expanded as significant investments as metal forming dies have been avoided in recent years. This robotic arm technology has been viable for many years in the sheet metal forming sector due to its flexibility in shape and geometry. These low-volume production offers greater flexibility and cost minimisation. This leads to the use of CNC milling machines and industrial robots.

The purpose of this study was to present an alternative method for manufacturing sheet metal. The part is created using a robot's step-by-step forming process. Incremental forming of sheet metal parts can be a reasonable alternative to traditional methods, especially for prototyping and small-scale. Today's production-level competition is characterised by flexibility and innovation. The traditional sheet metal forming process best used for mass production is not flexible. A significant investment is required for the production of prototypes and pre-production components. It was necessary to develop a new type of process within a reasonable time at a reasonable cost, leading to this research study. In order to better understand the incremental forming process, its feasibility and integrity with robotic arm in material deformation, a systematic literature review has been conducted and is briefed as follows.

Belchior et al. [14] investigated a coupling technique for resolving tool path variations induced by industrial robot compliance. This is accomplished by establishing two separate frustum cones, one symmetrical and one asymmetrical. By validating the FE analysis's reliability and the elastic modelling indicated in prototype applications, the final shape of manufactured components can be brought closer to the nominal requirements. Meier et al. [4] focussed on improving both the structural compliance of the accompanying serial robot structures and the springback effects of the workpiece. This study examines both an offline model-based approach and an online sensor-based approach to component accuracy enhancement. Roboforming enables the simulation of path variations due of its low stiffness. As a result, the kinematics of the robot must be changed to account for the sheet metal's shaping forces.

Meier et al. [15] invented an innovative incremental sheet metal forming process. A three-dimensional workpiece can be created without the use of a special die plate by pounding a hammer over a metal sheet positioned in a frame. Due to the low stress placed on the handling equipment in comparison to other shaping procedures, the hammering tool can be moved using a standard industrial robot. As a result, sheet metal components can be manufactured on a budget. The major beneficiaries of this new technology are small and medium-sized enterprises (SMBs). Wang et al. [16] worked on interpolated tool path used in real machining is a close approximation of the theoretical tool path, resulting in geometric divergence. This research proposes a novel technique for reducing geometric variation in order to address this issue. By adjusting the cutter location data, the geometric divergence is decreased to an acceptable level. Both empirical and computational evidence indicate that optimisation reduces the geometric deviation on the machined surface.

Meier et al. [17] studied that the precision of incremental sheet metal forming is highly dependent on the machine architecture and springback effects of the workpiece. The rigidity of the robot is critical when producing sheet metal with a robot. The FEA model calculates the forces acting on the tool's tip, whereas the MBS model calculates the route deviations caused by these forces. By combining the two models, the true path taken by the robots may be determined. By utilising this route prediction, the robot's kinematics can be corrected.

According to Liu [18] it is difficult to form materials with low overall ductility using conventional incremental sheet forming at room temperature. As a result, researchers have developed a variety of heat-assisted incremental sheet forming processes to improve the formability of these materials whilst also improving their geometric precision. His article aims to present an up-to-date history of heat-assisted incremental sheet production. Whilst a low yield strength and hardening coefficient are advantageous in terms of decreasing process pressures and springback, they can result in excessive, unwanted plastic deformation in zones adjacent to treated regions, according to Duflou al. [19] Initial study indicates that the suggested dynamic heating system can greatly reduce residual stresses. Duflou et al. [20] used a different approach, investigating the effects of localised temperature changes on material properties. A ductile zone with insufficient yield strength is formed by injecting dynamic heat directly adjacent to the stylus. Experiments are used to demonstrate how influences affect process performance. Göttmann et al. [21] proposed asymmetric incremental sheet formation as a novel manufacturing technology. A CNC-driven forming tool generates localised plastic deformation as it moves over the contour of the desired component in AISF. According to preliminary testing results, the formability of the alloy Ti Grade 5 (TiAl6V4), which is widely used in aerospace applications, can be enhanced. Bailly et al. [22] provide a unique thermocouple-integrated tool design. As a result, two distinct warm ISF methodologies were employed to examine and evaluate the developed configuration's applicability. This experiment established the capability to maintain the minimum temperatures required for Ti grade 5 formability. Mohammadi et al. [23] investigated how accumulated unwanted bulging deformation affects the generated geometry, causing it to deviate significantly from the design surface. Continuous sheet deformation beyond the contact zone promotes cone wall over formation, which then converts to cone wall under formation following laser treatment. By employing the proper laser installation process, it is possible to achieve a 42% reduction in bulge height. In many industries such as automobile, aircraft, this robotic arm for incremental sheet metal forming can be used according to recent manufacturing processes. Incremental forming of sheet metal is an extremely promising option for the production of many curved, articulated surfaces for cladding systems in architecture. Today's incremental forming processes cannot match the production speeds of stamped facade panels, nor their extreme precision and tight tolerances. However, they provide the ability to design prototypes and manufacture highly variable and economical components.

2 Design of Robot

The robotic arm was created using the solid works software; the assembled model can be shown in Fig. 4a, and more information on the individual part models can be found below.

2.1 *Base*

The robot's framework is built on top of this frame (Fig. 4b). A set of screws hold it in place on the mounting base. Bolts and screws are used to hold push-in modules and electrical installations to the base frame. Everything is gathered in a remote junction box at the rear of the car. Axis 1 motion is made possible by the servo motor and gearing mechanism inside this container. It is the part of the arm that moves the rest of the arm whilst bearing all of the weights placed on it.

2.2 *Rotating Column*

Axes 1 and 2 are driven by motors located inside a spinning column. Axis 1 is rotated because of the rotation of the spinning column. (Fig. 4c). The rotating column's electric motor powers this unit, which is attached to the base frame via axis 1's gear unit. The link arm is likewise positioned in the rotating column. All of the moving

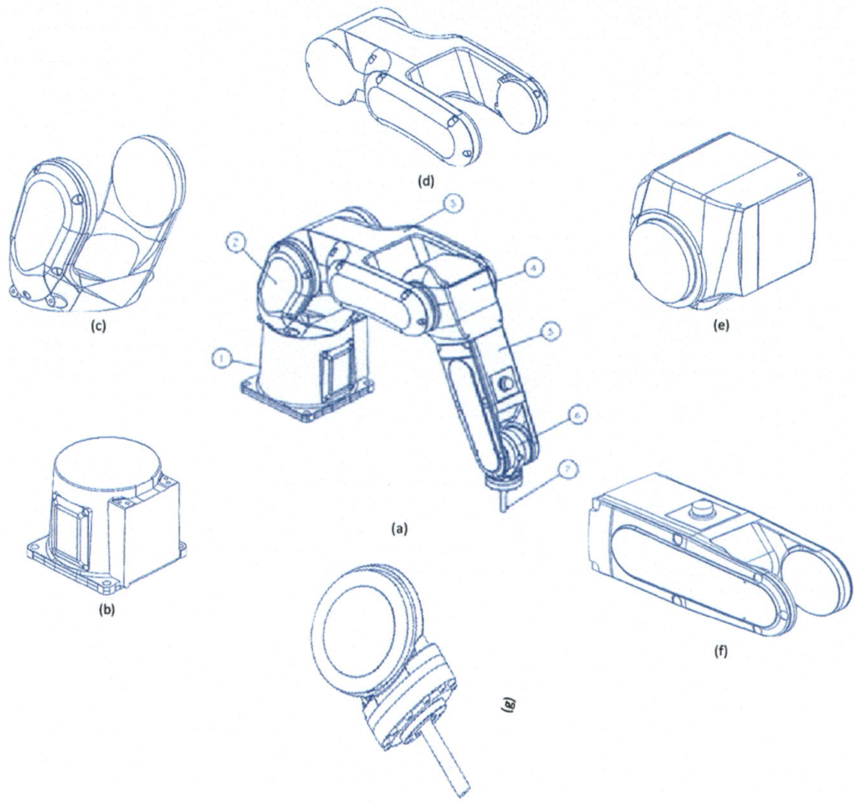

Fig. 4 **a** 6-axis industrial robotic arm, **b** base, **c** rotating column, **d** Arm Unit 1, **e** housing unit, **f** Arm Unit 2, **g** wrist unit

parts of the column are protected from the weather by the enclosure. A supply of sealing air is included in the assembly. The geometry and motion of the rotating column is a deciding factor when it comes to the working range of the robotic arm.

2.3 Arm Unit 1

Axis 2 motion is delivered by this section of the robotic arm. Axis 2 motion is critical because it dictates the vertical range and affects the robot end effector location [24, 25]. However, in some commercially available arms, the arm unit 1 also comprises motors for moving the axis 3, 4, and 5, as well as the belt drives for transmitting motion. This is not typical (Fig. 4d).

2.4 Housing Unit

Housing Unit, shown in Fig. 4e, connects Arm Units 1 and 2, as seen in Fig. 4d and 4f. The axis 3 motion of Arm unit 1 is applied to this link. A motor and gearing system send axis 4 motions to arm unit 2, which in turn passes them along to the second arm.

2.5 Arm Unit 2

During the axis 4 motion, the Arm unit 2 (Fig. 4f) works in tandem with the Wrist unit 2. Over 90-degree swivel angles can be controlled mechanically by adding or removing stops in either direction along the rotation. The arm is fitted with the corresponding buffers. Axis 5 is propelled by a motor attached to the device.

2.6 Wrist Unit

The end effector is held in position by the wrist unit (Fig. 4g). The part of the robotic arm that actually does the actual work. The Wrist unit can be used with a broad variety of tools and actuators. This component manages the movement of axes 5 and 6. The end effector can be mounted on any of the two motors that are included.

3 Design Modifications in Robotic Arm

In general, industrial robots are six-axis robots with revolute joints and motors for movement that are principally equipped. 5 axis motions have constraints in order to form the sheet metal incrementally. To make significant progress in metal forming using robotics, an additional axis must be added to the arm. Arms with up to 8 axes are commercially available [26, 27]. Adding a rotational 7th axis may lead to duplicate solutions. Because a new rotating 7th axis cannot be introduced, this study proposes a novel linear motion paradigm (Fig. 5a). There is a linear actuator sandwiched between the Arm unit 2 and the other component. This not only adds a new linear axis to the robotic arm, but it also expands the operational range of the arm. The linear motion introduced is capable of deforming complex shapes precisely by controlling the incremental motion of the form tool. This will enable better formability and precision during deformation in incremental forming. Deformation in increments is the primary feature which enhances formability in incremental forming; where the linear motion introduced gives better flexibility in case of customised part production.

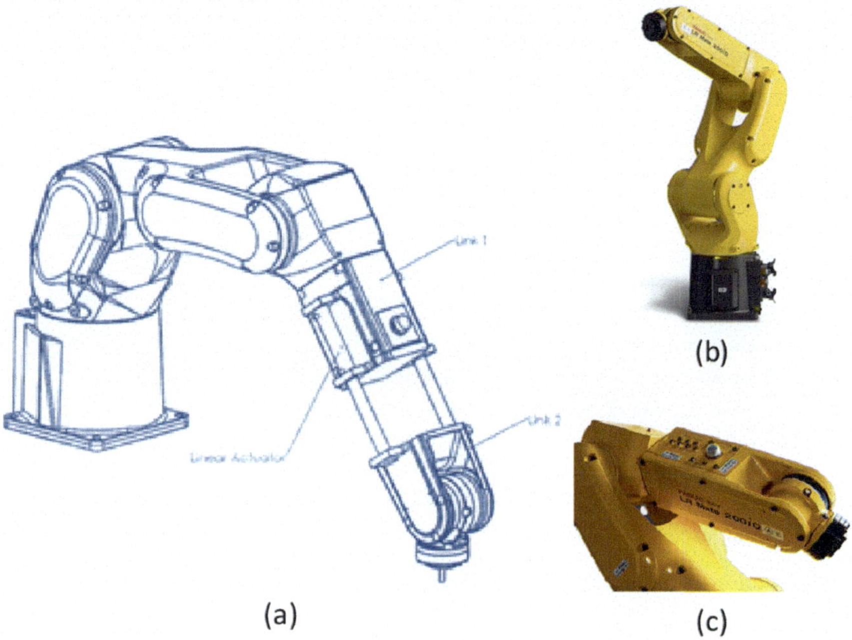

Fig. 5 **a** Arm with linear actuator, **b** Fanuc LR Mate 200id, **c** Fanuc LR Mate 200id–Robotic Arm

Since the process is highly versatile and suitable for automation, linear motion adds value to the part variability addressed during material deformation.

3.1 Design

Creating this model of the robotic arm initiate efficient metal forming and serve as a starting point for the future of automation in sheet metal forming industry. The complete arm was built in SolidWorks 2019. The arm was modelled using a FANUC LR Mate 200iD as a guide (Fig. 5b, c). A lot of people utilise this robotic arm since it is so versatile. As closely as feasible, the dimensions were matched to those of the original model. Internal components like motors and transmission systems were not researched in depth because the research is more focussed on kinematics than on the creation of a genuine robotic arm. For the linear actuation, a linear actuator with a stroke length of 100 mm was selected. To generate the required outcome in incremental sheet metal forming, the tool must follow a shape in a repeating manner. End effector was constrained to tracing the path of the 200 mm circular route established on the horizontal plane. The model's boundary conditions were specified using the SolidWorks Motion Study add-in. As an example, below are some boundary criteria to keep in mind:

Table 1 Motion range and velocity of various axes

Axis	Range	Maximum velocity
Axis 1	165°	136°/s
Axis 2	−135°/45°	95°/s
Axis 3	−65°/155°	120°/s
Axis 5	125°	258°/s
Axis 6	350°	284°/s
Linear axis	100 mm	100 mm/s

(i) The Base unit of the robot is fixed.
(ii) The steady state velocity of the end effector is given as 50 mm/s.
(iii) Axis 4 motion is restricted, since it is redundant for sheet metal forming.
(iv) The range of motion of each part of the arm is restricted to the commercially available datasheet provided by the manufacturer.
(v) The range of motion of the linear axis is 100 mm and the maximum velocity is considered to be 100 mm/s.

The motion range and maximum velocities of each axis is represented in Table 1.

3.2 *Motion Study Results*

The linear actuator's motion is primarily intended to be examined in this inquiry. Because it is a linear actuator, only linear motion results can be generated. In order to do this, a 200 mm circle was traced on the end effector and then positioned in such a way that it caused the linear actuator to expand its whole range of 100 mm. The results were shown on the y-axis versus the *x*-axis' time (measured in seconds). A summary of the motion study's conclusions may be found on this page.

(i) The linear displacement resembles a sine wave, based on the displacement data (Fig. 6).
(ii) When the actuator is fully extended and fully retracted, its maximum linear velocity is 23 mm/s, as seen on the plot of linear velocity for the actuator (Fig. 6a, b). These findings are within the expected range of the maximum permissible velocity of 100 mm/s, according to the researchers.
(iii) The acceleration map (Fig. 6c) shows a range of amplitudes between 15 and 3 mm/s^2.
(iv) To sum it up, when velocity reaches zero (or changes direction from positive to negative), acceleration is at its greatest. This can be observed in all three plots.

The use of dark black lines to illustrate the courses taken by each connection helps to make the mention of each link more understandable as the circular tour proceeds (Fig. 7a, 7b).

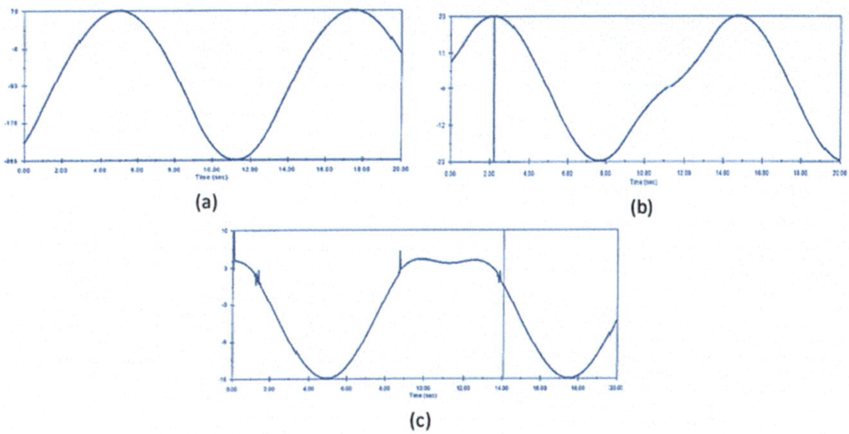

Fig. 6 **a** Linear displacement of the actuator, **b** linear velocity of the actuator, **c** linear acceleration of the actuator

One of the biggest benefits of using a linear actuator is that it increases the working range of the arm. It was determined that the arm's working range was 13.2 m^3 using the CAD model. Figure 7b, c show the volume without the linear actuator is roughly 10.47 m^3, which is an increase of about 30%. However, as opposed to using a linear actuator, using an arm with an additional rotational axis will only make movements more complex without affecting the operating range [27].

4 Design Modification in Fixture

A fixture is a clamping device used to hold metal sheet firmly in place. On the ground, the fixture is firmly secured by the straps that hold it there. Because of the novelty of robotic arms in the field of Incremental Sheet Metal Forming, there are no pre-existing fixtures or standards to adhere to when designing a fixture. Large levels of force are frequently required in sheet metal manufacturing [14]. It is feasible to save money by reducing the amount of force necessary to operate the arm by modifying the fastening. As a result, smaller robotic arms may be used during the procedure, which saves time and money.

4.1 Design of Fixture

To soften and reduce the amount of force required during the forming process, metal is frequently heated to a high temperature (annealing) in conventional metal forming processes. When designing a custom fixture for a specific technique, keep this notion

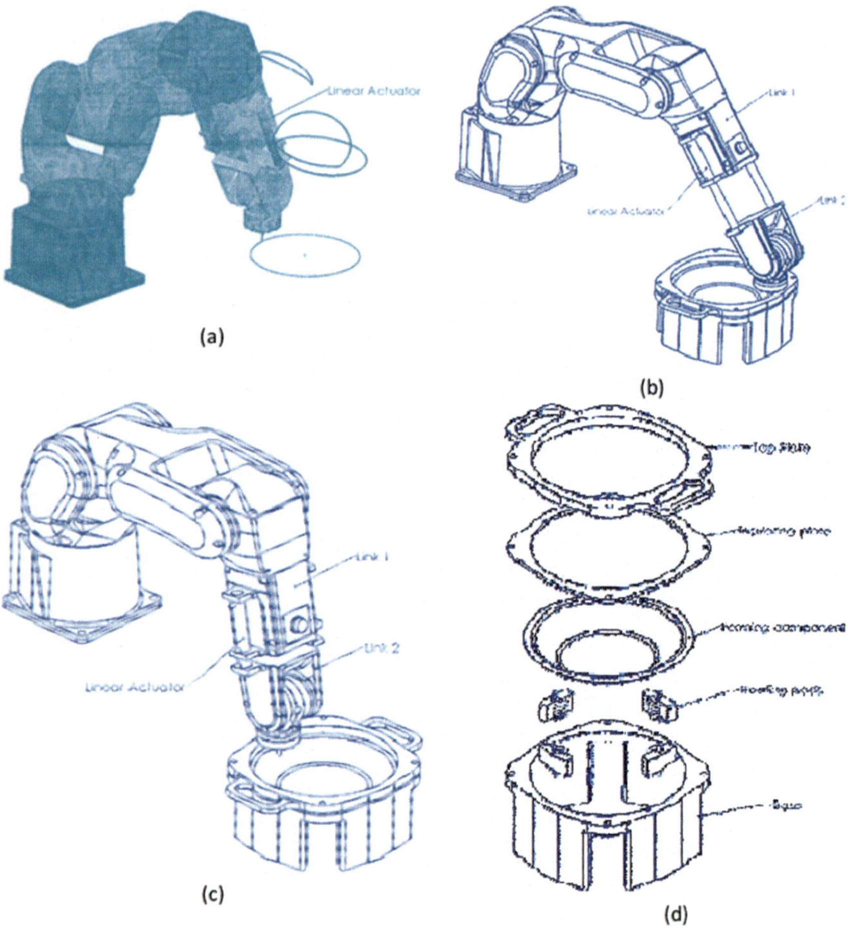

Fig. 7 **a** Paths traced by various links, **b** linear actuator in extended position, **c** linear actuator in retracted position, **d** exploded view of the modified fixture

in mind. A conventional sheet metal fixture is produced as a result, with a few design modifications. Figure 7d depicts both the redesigned fixture and the modified fixture in exploded view. The modification in high temperature forming fixture includes heating pad and insulating material which regulates the temperature developed in hot forming.

4.2 *Parts of the Fixture*

The parts of the modified fixture (Fig. 7d) are discussed below.

4.2.1 *Base*

There is a lot of weight in the fixture assembly's base. A number of large bolts hold it firmly in place. Cast iron or steel is used for the base because it has a higher strength-to-weight ratio and better thermal conductivity than other materials.

4.2.2 *Heating Pads*

Adhesive is used to adhere the heating pads to the inside of the base. When making these heating pads, copper is employed as the primary building material. Commercially available heating pads have a maximum temperature of 700 °C. In order to transport heat from the base to the sheet metal that is in contact with the sheet metal, several devices are employed.

4.2.3 *Forming Component*

In the shaping process, this piece of work will be formed. In this particular case, the metal aluminium is presumed to be used in its construction. The robotic arm's force capacity is taken into account whilst determining the sheet metal's thickness.

4.2.4 **Insulating Plate**

Sheet metal produces heat that must not be transferred to the top plate, which workers must continually handle in order to swap out work pieces. This application benefits from ceramic's low thermal conductivity and high heat resistance because of these properties.

4.2.5 *Top Plate*

The fixture's top plate holds the component firmly in place. It is attached to the workstation's base after it is been placed on the workpiece. It is possible for the top plate to remain at ambient temperature during all forming operations due to the heating pads and hot sheet metal being segregated from it.

4.3 *Thermal Analysis of Fixture*

The thermal study is carried out with the help of the SolidWorks 2019 Simulation add-in. Because the simulation results are immediately updated when the design or dimensions change, the inbuilt simulation environment is always in use. Materials

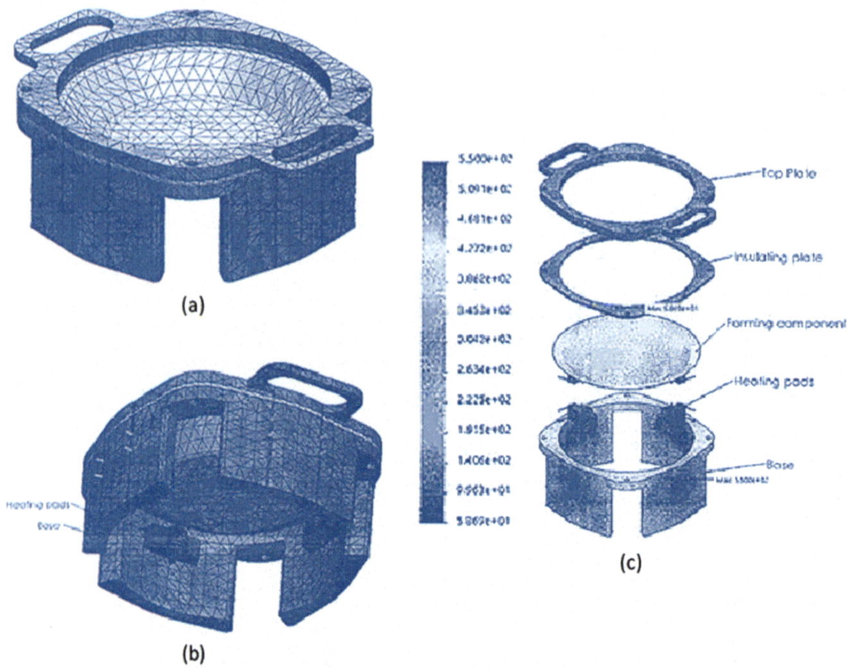

Fig. 8 **a** Meshed assembly, **b** boundary conditions, **c** temperature distribution

used in certain components can be determined by consulting the material library. The static thermal module is used because the investigation is being carried out in steady state conditions [28]. Pictured here are the meshing and boundary conditions are in Fig. 8a and b, respectively. At the input, a temperature load of 550 °C is applied to the four heating pads. Every one of the exposed surfaces has a convection coefficient of 50 W/m*, with an average on all of them. A medium meshing size is used, with the lowest element size starting at 13.5 mm and going up to 15.0 mm. All of the contact surfaces must have bonded contacts for there to be bonded contacts.

4.4 Results of Thermal Analysis

The simulation calculates and solves for thermal values across several components (Fig. 8c). When 550 °C is used as an input, the output sheet metal temperature is 350 °C, which is close to the annealing temperature of aluminium. (300–400 °C).

5 Fabrication of Prototype

Industrial robotic arms can cost anywhere from Rs.6 Lacs to Rs.3 Crores, depending on its size, power, precision, functionality, and other considerations. The cost of this undertaking much exceeds its scope. To keep costs down, a very tiny scale model of the robotic arm is built using micro servo motors instead of larger ones.

5.1 *Functioning of the Prototype*

Two arms make up the prototype: one is used to teach, whilst the other is used to learn from. After the teaching arm moves, the learning arm follows suit. The teaching arm can be moved in any direction at the user's discretion. A microcontroller transfers the teaching arm's motions to the learning arm. Potentiometers mounted on the teaching arm record the movements made by the user when the teaching arm is in operation. The microcontroller uses the user-defined programme to process the data, which is subsequently sent to the Micro servo motors for movement.

5.2 *Parts Used*

The main functional parts of the prototype are listed and explained below.

5.2.1 *Arduino UNO*

The microcontroller used in the prototype can be seen in the illustration below (Fig. 9a). A USB port on a computer can be used to communicate with the microcontroller. For data conversion from the teaching to learning arms, it is in charge [29]. With a clock frequency of 16 MHz and a memory size of 32 KB, it is fast. In all, there are 20 external component pins available.

5.2.2 *Potentiometer*

When it comes to a potentiometer, think of it as just another type of switch. Turning the knob changes the resistance, which lets us modify the resistance to get a variety of voltages. The potentiometer's resistance is 10 KCI, and its operating voltage is 5 V. The following illustration depicts this concept (Fig. 9b).

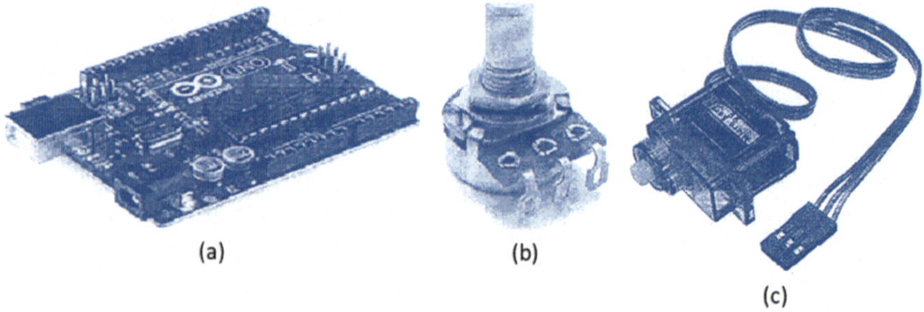

Fig. 9 **a** Arduino UNO, **b** potentiometer, **c** Servo SG90

5.2.3 *Servo SG90*

A 9-g servo motor (Fig. 9c) with an angular range of 180 degrees was used in this experiment. The motor generates a torque of about 2.5 kg cm. Operational voltage range of the motor is between 4.8 and 7.2% (voltage range).

5.3 Coding

The code developed for the prototype is given below:

```
#include
<Servo.h> Servo
one;
Servo hvo:
Servo three;
Servo four;
void setup() {
// put your setup code here, to run once:
Serial.begin(9600);
pinMode(A0,INPUT);
pinMode(A1,INPUT);
pinMode(A2,INPUT);
pinMode(A3, INPUT);
one.attach(5);
two.attach(6);
three.attach(9);
four.attach(10);
void loop() {
// put your main code here, to run repeatedly:
int a = analogRead(AO);
int s1 = map(a, 0, 1023, 180, 0);
```

```
one.write(s1);
delay(5);
int b = analogRead(A1);
int s2 = map(b, 0, 1023, 180, 0);
two.write(s2);
delay(5);
int c = analogRead(A2);
int s3 = map(c, 0, 1023, 180, 0);
three.write(s3); delay(5);
int d = analogRead(A3);
int s4 = map(d, 0, 1023, 180, 0); four.write(s4);
delay(5);
Serial.print(s1);
Serial.print(" ");
Serial.print(s2);
Serial.print(" ");
Serial.print(s3);
Serial.print(" ");
Serial.print(s4);
Serial.println(" ");
```

5.4 Cost Estimation

The cost estimation calculated for the proposed model is presented in Table 2.

Table 2 Cost sheet

Part No	Part name	Quantity (no's)	Price in (Rs.)
1	Arduino UNO	1	450
2	Potentiometer	5	250
3	Servo SG90	4	560
4	Servo MG995	2	720
5	Connecting wires	–	100
6	Battery	1	150
7	Miscellaneous	–	400
Total cost			Rs. 2630/–

6 Conclusion

Incremental forming gives a promising feature in terms of customised part production. Traditionally part fabrication in incremental forming happens by the CNC assisted form tool movement. Robotic arm is proved to be more flexible than CNC assisted incremental forming method; more economical for large parts, and easier to implement with the right tools. Because the workspace of a robot is much larger than that of a CNC machine, it is possible to create large parts with this method. This makes robotic incremental forming a very attractive method in customised part fabrication. The key research findings of this study are given below:

1. This study could be exclusively useful in the construction of an entirely new limb, specific for incremental forming.
2. Micro servo motor-powered miniature prototype was built, and cost study of the prototype was computed and presented in this study representing the feasibility of seamless integration in part fabrication.
3. The linear actuator system employed holds its upper hand in ensuring precision and accuracy in incremental forming.
4. In addition, the robotic arm programming described in this study gives the background information on the integration of linear actuator in robotic arm.

Robotic arms are already in use in nearly every industry, and there is still plenty of space for expansion and innovation in the future.

References

1. Patel K, Kalaichelvi V, Karthikeyan R, Bhattathiri S (2018) Modelling, simulation and control of incremental sheet metal forming process using CNC machine tool. Procedia Manuf 26:95–106
2. Linnemann M, Psyk V, Djakow E, Springer R, Homberg W, Landgrebe D (2019) High-speed incremental forming—new technologies for flexible production of sheet metal parts. Procedia Manuf 27:21–26
3. Gutiérrez SC, Zotovic R, Navarro MD, Meseguer MD (2017) Design and manufacturing of a prototype of a lightweight robot arm. Procedia Manuf 13:283–290
4. Meier H, Buff B, Laurischkat R, Smukala V (2009) Increasing the part accuracy in dieless robot-based incremental sheet metal forming. CIRP Ann Manuf Technol 58(1):233–238
5. Mohanty S, Regalla SP, Daseswara Rao YV (2017) Tool path planning for manufacturing of asymmetric parts by incremental sheet metal forming by means of Robotic manipulator. Mater Today 4(2):811–819
6. Mohanty S, Regalla SP, YV DR (2018) Investigation of influence of part inclination and rotation on surface quality in robot assisted incremental sheet metal forming (RAISF). CIRP J Manuf Sci Technol 22(1):37–48
7. Shah R, Pandey AB (2018) Concept for automated sorting robotic arm. Procedia Manuf 20:400–405
8. Störkle DD, Möllensiep D, Thyssen L, Kuhlenkötter B (2018) Geometry-dependent parameterization of local support in robot-based incremental sheet forming. Procedia Manuf 15:1164–1169

9. Zhang R, Shao Z, Lin J (2018) A review on modelling techniques for formability prediction of sheet metal forming. Int J Lightweight Mater Manuf 1(3):115–125
10. Kumar Y, Kumar S (2015) Incremental sheet forming (ISF). In Advances in material forming and joining. Springer India, New Delhi, pp 29–46
11. Kalo A, Newsum MJ (2014) An investigation of robotic incremental sheet metal forming as a method for prototyping parametric architectural skins. In: Robotic fabrication in architecture, art and design 2014. Springer International Publishing, Cham, pp 33–49
12. Jeswiet J, Micari F, Hirt G, Bramley A, Duflou J, Allwood J (2005) Asymmetric single point incremental forming of sheet metal. CIRP Ann Manuf Technol 54(2):88–114
13. Allwood JM, Braun D, Music O (2010) The effect of partially cut-out blanks on geometric accuracy in incremental sheet forming. J Mater Process Technol 210(11):1501–1510
14. Belchior J, Guillo M, Courteille E, Maurine P, Leotoing L, Guines D (2013) Off-line compensation of the tool path deviations on robotic machining: application to incremental sheet forming. Robot Comput Integr Manuf 29(4):58–69
15. Meier H, Zhang J, Dewald O (2007) Incremental forming of sheet metal by two industrial robots. In: AMST'05 advanced manufacturing systems and technology. Springer Vienna, Vienna, pp 437–444
16. Wang L, Li W, Si H, Yuan X, Liu Y (2020) Geometric deviation reduction method for interpolated toolpath in five-axis flank milling of the S-shaped test piece. Proc Inst Mech Eng Part B J Eng Manuf 234(5):910–919
17. Meier H, Laurischkat R, Bertsch C, Reese S (2009) Prediction of path deviation in robot based incremental sheet metal forming by means of an integrated finite element—multi body system model. Key Eng Mater 410–411:365–372
18. Liu Z (2018) Heat-assisted incremental sheet forming: a state-of-the-art review. Int J Adv Manuf Technol 98(9–12):2987–3003
19. Duflou JR, Callebaut B, Verbert J, De Baerdemaeker H (2007) Laser assisted incremental forming: formability and accuracy improvement. CIRP Ann Manuf Technol 56(1):273–276
20. Duflou JR, Callebaut B, Verbert J, De Baerdemaeker H (2008) Improved SPIF performance through dynamic local heating. Int J Mach Tools Manuf 48(5):543–549
21. Göttmann A et al (2011) Laser-assisted asymmetric incremental sheet forming of titanium sheet metal parts. Prod eng 5(3):263–271
22. Göttmann A et al (2013) A novel approach for temperature control in ISF supported by laser and resistance heating. Int J Adv Manuf Technol 67(9–12):2195–2205
23. Mohammadi A, Vanhove H, Van Bael A, Duflou JR (2016) Towards accuracy improvement in single point incremental forming of shallow parts formed under laser assisted conditions. Int J Mater Form 9(3):339–351
24. Lu HL, He K, Li JH, Wei SG, Du RX (2013) The design of a five-axis machine for water jet incremental sheet metal forming. Appl Mech Mater 395–396:949–956
25. Kwiatkowski L, Urban M, Sebastiani G, Tekkaya AE (2010) Tooling concepts to speed up incremental sheet forming. Prod eng 4(1):57–64
26. Trzepiecinski T, Lemu HG (2019) Recent developments and trends in the friction testing for conventional sheet metal forming and incremental sheet forming. Metals (Basel) 10(1):47
27. Ilangovan B (2016) Fixtureless automated incremental sheet metal forming. Loughborough University, pp 56–74
28. Hajavifard R, Maqbool F, Schmiedt-Kalenborn A, Buhl J, Bambach M, Walther F (2019) Integrated forming and surface engineering of disc springs by inducing residual stresses by incremental sheet forming. Materials (Basel) 12(10):1646
29. Stockhoff P (2017) Pinned incremental metal forming. University of North Carolina, Charlotte, pp 12–27

A Framework for Timely Delivery of Serviced Vehicles in Automotive Service Garages Using a Rough—DEMATEL Technique

Kevin Thomas, M. Uthayakumar, S. Bathrinath, M. S. Abdul Masjid, and K. Koppiahraj

Abstract The rise in global automotive utilization has increased the need for service garages. However, maintenance service providers face numerous problems at all phases of their operations. The challenges to the timely delivery of a serviced car are identified and analyzed in this study. This paper uses expert opinion, garage survey, and literature to identify issues for the prompt delivery of automotive service vehicles. The rough decision-making trial and evaluation laboratory (DEMATEL) approach establishes and assesses these barriers. This paper identifies a total of twenty-seven barriers to vehicle service through a literature review. The result indicates that misleading of the technician, lack of manpower, more demand work, lack of skilled technician, delivery time already postponed service vehicle are the most critical challenges in the timely delivery of serviced vehicles. The outcome of the work will help the stakeholders and decision-makers to make a customer-friendly service station. This study reveals the critical challenges to delivering the vehicle on time to the customer.

Keywords Automobile industry · Rough DEMATEL · Automotive garages · Engine control unit · Timely delivery

K. Thomas · M. Uthayakumar
Department of Automobile Engineering, Kalasalingam Academy of Research and Education, Krishnankovil, Tamil Nadu 626126, India

S. Bathrinath (✉) · K. Koppiahraj
Department of Mechanical Engineering, Kalasalingam Academy of Research and Education, Krishnankovil, Tamil Nadu 626126, India
e-mail: bathri@gmail.com

M. Uthayakumar · M. S. Abdul Masjid
Faculty of Mechanical Engineering Technology, Universiti Malaysia Perlis (UniMAP), Perlis, Malaysia
e-mail: shukry@unimap.edu.my

E. Natarajan et al. (eds.), *Materials, Design and Manufacturing for Sustainable Environment*, Lecture Notes in Mechanical Engineering,
https://doi.org/10.1007/978-981-19-3053-9_33

1 Introduction

According to a recent survey done in 2020, India's automobile sector is the world's fourth-largest car manufacturer and the seventh-largest commercial vehicle manufacturer. The Government of India is implementing a strategy, Automotive Mission Plan 2026, that is anticipated to place India's automotive industry among the top three global manufacturers and exporters of automobiles and components while also increasing the value of the business [1]. The automotive industry is considered one of the most important for a country's economy and trade because it shapes the industry, the cities, communal, and human life. The service sector of vehicles is equally important for the growth of the automobile industry [2]. Customer happiness and industrial progress are both feasible when the car is delivered in the best possible method.

In the Indian market, there exists a direct relationship between the production and sales of automobiles [3]. Maintaining relationships with the customers after sales is very critical in the success of business. Maintaining a vehicle in intact condition is very crucial in long-term use. Further, by establishing maintenance and service centers, the automobile industry can gain valuable customer feedback, which could be incorporated during design phases. It is also crucial to ensure the operational reliability of automobiles, which may be accomplished through regular repair and maintenance. With the growing number of vehicles on the road, it is more important than ever to have an effective service and maintenance network. Furthermore, the rapid advancement of vehicle technology and the large number of new models produced each year adds to the difficulty of delivering efficient service [4]. India's vehicle servicing and maintenance sector have unparalleled expansion potential with a sustained growth promise to keep autos in good working condition and extend their service life.

From the above information, it is evident that the service after the car delivery is very crucial in sustaining the regular business. Hence, there is a need to improvise and upgrade the quality of the service offered by the service stations. Besides knowing the importance of offering quality service to the customers, the service stations across India are facing many challenges. Henceforth, examining and evaluating the challenges faced by the service stations in offering the best service have become necessary.

Owing to increased concern on offering quality service, number of studies has been carried out. Majority of study regarding the car service are mostly review studies. Also, large proportions of the study are carried out from the developed country backdrop. Since a vast difference exists between the technical capability of the developed and developing countries, the outcome of these studies is not generalized results. So, there is a need to examine the challenges faced by the service stations located in the developing countries. In the regard, this study aims to evaluate the challenges in ensuring the timely delivery using rough DEMATEL technique. Here, DEMATEL technique is used to examine the interrelationship among the challenges. DEMATEL technique has been used in many earlier studies like challenges to blockchain [5], e-waste mitigation strategy [6], and critical success factors evaluation [7]. The major goal of this paper is to identify the obstacles to timely car delivery in a service station. Experts believe that the delay in delivery of the vehicle to the customer after service can be split into seven categories that include communication problem, delay in the service station, delay in the body shop, delay in the washing area, parts-related issues, tools-related issues, management-related issues, however, and the concerns associated to service station are discussed in this paper.

2 Barriers to Prompt Delivery of a Vehicle in a Service

Today's automotive service providers face several issues. To assimilate the barriers, this paper identifies twenty-seven challenges, namely more number of demand work, delivery time already postponed service vehicle, finding out of the extra fault, delay in final inspection, unawareness of technician about the electrical system, lack of experience in the engine control unit (ECU)-related diagnosis and on-board diagnostics (OBD), and lack of manpower [8]. The final inspector may notice some extra fault in the vehicle, mistake done by the technicians, misleading of the technician, and delay in diagnosing a fault. Other barriers include service circular from the manufacturer for replacing some parts of the vehicle, delay in tracking the engine immobilizer identification, improper bay final inspection of the technician, difficulty in dismantling the parts, difficulty in assembling the parts, and longer time to remove the rusted parts. The technician who does not know how to use the service manual procedure, lack of skilled technician, electrical or lathe works are outsourcing, delay in the road test of the vehicle, delay in the out-pass procedure to follow for the road test of the vehicle [9]. The final inspector may notice the incomplete work of the vehicle, delay in the wheel alignment or wheel balancing area, negligence of technician, supervisor, and technical advisor, overload of the technician. A traffic jam when the vehicle goes for the road test [10]. We recognized fourteen sustainability risks based on the literature and industrial expert's opinions, as shown in Table 1.

3 Methodology

The rough DEMATEL method and data collection strategy used in this work are discussed in the subsections below.

3.1 Data Collection

The research is based on information gathered exclusively from Kerala-based organizations. With the face-to-face interview method, responses are collected from each expert individually to avoid non-respondent bias. In this study, ten experts

Table 1 Barriers to prompt delivery of a vehicle in a service industry

Impediments	Description	References
More number of demand work (R1)	Workload management lacks both the conceptual foundation and the measurement tools needed to analyze it	[11, 12]
Delivery time already postponed service vehicle (R2)	Unable to complete the work, and the day before will be rerouted to the next day	[13]
Finding out of the extra fault (R3)	Technicians discover extra repairs that the customer did not indicate at the time of giving for maintenance	[13]
Delay in final inspection (R4)	Operator may take time off for personal reasons due to circumstances beyond his control	[14, 15]
Unawareness of technician about electrical system (R5)	Due to lack of appropriate training, technicians are inadequately trained in the maintenance of new cars	[11, 16]
Lack of experience in the ECU-related diagnosis and OBD (R6)	Due to the intricate integration of electro-mechanical components in automobiles, they are prone to numerous errors	[13]
Lack of manpower (R7)	Global shortage exists for technicians who were knowledgeable with the distinctive design and characteristics of modern autos	[14, 17]
Final inspector may notice some extra fault in the vehicle (R8)	Inspection process is vital to ensure client satisfaction, but it must also be efficient for the organization to save money	[13, 16]
Mistake done by the technician (R9)	Mistakes do occur if the individual performing the repair does not understand the mechanical and electrical or electronic systems' basic architecture and operation	[11, 15]
Misleading of the technician (R10)	Modern diagnostics necessitate a fundamentally different approach than that used for conventionally controlled engines	[11]
Delay in diagnosing a fault (R11)	Inexperienced automobile mechanic may take more time in diagnosing fault	[12, 14]
Service circular from the manufacturer for replacing some parts of the vehicle (R12)	Some spare parts of the vehicle may become obsolete and having a replacement for the same product is difficult	[11, 17]
Delay in tracking the engine immobilizer identification (R13)	New mechanic may take time in identifying the engine immobilizer	[14, 18]

(continued)

Table 1 (continued)

Impediments	Description	References
Improper bay final inspection of the technician (R14)	If the deviations are outside of permitted limits, the technician will need to investigate that area of the vehicle further	[11]
Difficulty in dismantling the parts (R15)	Traditional car mechanics needs to acquire knowledge on current automobile technology	[14]
Difficulty in assembling the parts (R16)	Workers working with electrical components are in demand as engines become more intricate	[11, 12]
Longer time to remove the rusted parts (R17)	Harder it gets to remove rusted and broken equipment, the lengthier the technician will have to work on the job	[14]
Technician who does not know how to use the service manual procedure (R18)	Since most the automobile mechanics have minimal education, they are not capable of understanding the service manual	[13, 15]
Lack of skilled technician (R19)	To effectively diagnose and repair/service these vehicles, highly qualified and well-trained service engineers, advisers, and technicians are required	[11]
Electrical or lathe works are outsourcing (R20)	Some electrical work is outsourced in many workshops due to a lack of infrastructure or equipment	[13]
Delay in the road test of the vehicle (R21)	Vehicles may be delayed in returning after a road test due to a traffic problem	[11, 16]
Delay in the out-pass procedure to follow for the road test of the vehicle (R22)	Technical documentation of maintenance, including the maintenance management information system, is critical for garages to achieve their maintenance goals	[14]
Final inspector may notice the incomplete work of the vehicle (R23)	When the vehicle is serviced and a final inspection is performed, it is discovered that the requisite work has not been finished	[11, 17]
Insufficient space (R24)	Apart from primary maintenance duties, appropriate space in garages is required for supporting functions such as maintenance control and analysis	[13, 18]
Negligence of technician, supervisor, and technical advisor (R25)	Service quality is subpar in terms of service quality dimensions, vehicle mechanics in the automotive repair services sector	[11]
Overload of the technician (R26)	Gap between the capacity of the information processing system required for task performance to meet expectations, and the capacity available at any one time is referred to as workload	[12, 14]

(continued)

Table 1 (continued)

Impediments	Description	References
Traffic jam when the vehicle goes for the road test (R27)	Traffic congestion that happens when the car is taken for a test drive after service might impact the quality of the service	[11]

were considered. These investigations, interestingly, are carried out using the rough DEMATEL approach. A cross-functional expert team comprises Manager (Service), Supervisor (Service), Technical Advisor, Service Advisor, and Technicians.

3.2 Rough DEMATEL Approach

The rough DEMATEL approach consists of the following steps:

Initially, m specialists are sought to select influencing parameters and the construction of m direct relation matrixes (D_{ij}') . (1) indicates the (D'_{ij}) matrix of Kth expert's I matrix, where d'_{ij} is the element of (D'_{ij}) matrix [19].

$$D_{ij}^{\prime K} = \begin{bmatrix} d_{11}^{\prime k} & \cdots & d_{1n}^{\prime k} \\ \vdots & \ddots & \vdots \\ d_{n1}^{\prime k} & \cdots & d_{nn}^{\prime k} \end{bmatrix} \tag{1}$$

The following is the steps involved in the rough DEMATEL approach.

Step 1: In this step, a combined direct relation matrix (D'_{ij}) is constructed based on the experts' D' matrices. This approach does not consider the average for construction of $(\tilde{D}_{ij})$ matrix as shown in Eq. (2).

$$\tilde{D}_{ij} = \begin{bmatrix} 1 & \cdots & \tilde{d}_{1n} \\ \vdots & \ddots & \vdots \\ \tilde{d}_{n1} & \cdots & 1 \end{bmatrix} \tag{2}$$

where $\tilde{d}_{ij}$ is the element of $(\tilde{D}_{ij})$ the matrix.

Step 2: All the matrix elements $(\tilde{D}_{ij})$ are converted into rough numbers (RN) with a lower and upper limit. Equations (3) and (4) are used for lower and upper approximation values.

$$\text{Lower approximation, } \underline{\text{App}}(d_{ij}^{\prime k}) = \cup\{X \in \cup/J(X) \le d_{ij}^{\prime k}\} \tag{3}$$

$$\text{Upper approximation, } \overline{\text{App}}(d_{ij}^{\prime k}) = \cup\{X \in U/J(X) \ge d_{ij}^{\prime k}\} \tag{4}$$

where U and X are universal that include all the objects of U.

$I = \left\{d_{ij}^{1}, \cdots d_{ij}^{m}\right\}$ for m number of expert judgment set. Equation (5) is used to find upper and lower limit of d_{ij}^{k} element.

$$\begin{cases} \underline{\text{Lim}}\left(d_{ij}^{k}\right) = \frac{\sum_{m=1}^{N_{ij}U} x_{ij}}{N_{ij}U} \\ \overline{\text{Lim}}\left(d_{ij}^{k}\right) = \frac{\sum_{m=1}^{N_{ij}U} y_{ij}}{N_{ij}U} \end{cases} \tag{5}$$

Equation (6) shows the RN for d_{ij}^{k} element.

$$\text{RN}(d_{ij}^{k}) = [\underline{\text{Lim}}(d_{ij}^{k}), \overline{\text{Lim}}(d_{ij}^{k})] \tag{6}$$

Using Eq. (7), the degree of vagueness between upper and lower number is obtained.

$$RN(d_{ij}) = \{[d_{ij}^{1L}, d_{ij}^{1U}], \ldots, [d_{ij}^{mL}, d_{ij}^{mU}]\} \tag{7}$$

Step 3: Rough total-relation matrix $\overline{(R_T)}$ is constructed using Eq. (8).

$$\overline{R_T}^{S} = \left[r_{ij}^{s}\right]_{n\times n} = \overline{R(1-\overline{R})^{-1}}, S = L, U \tag{8}$$

Step 4: Prominence (M_i) and relation (N_i) are calculated using Eqs. (9) and (10).

$$S_{R_i} = \left[s_{r_i}^{L}, s_{r_i}^{U}\right] = \left[\sum_{j=1}^{n} r_{t_{ij}}^{L}, \sum_{j=1}^{n} r_{t_{ij}}^{U}\right] \tag{9}$$

$$S_{C_j} = \left[s_{C_j}^{L}, s_{C_J}^{U}\right] = \left[\sum_{i=1}^{n} r_{t_{ij}}^{L}, \sum_{i=1}^{n} r_{t_{ij}}^{U}\right] \tag{10}$$

4 Application of the Proposed Methodology

The proposed model was applied to the automobile service station in Kerala, India. This paper uses expert opinion, garage survey, and literature to identify issues for the prompt delivery of automotive service vehicles. Initially, each of the ten experts compared the elements in pairs based on their understanding of the influence. For each expert, two initial direct relation matrices are generated in this study (one is for reliability factors and another is for recovery factors). The range of influence is 1 to 5, with 1 being no influence, 2 being a moderate influence, 3 being a medium influence, 4 being a great influence, and 5 being an extremely high influence. Because there is more than one expert, the D matrix is constructed using the average. Using the average of initial relationship matrix, the total-relationship matrix (TR) is developed, as shown in Table 2. Using Eq. (8), the total-relationship matrix was established. Next, PR and RE values are calculated using Eqs. (9) and (10). The prominence (PR) and relation (RE) values are given in Table 3. Based on the PR and RE values, the relationship diagram is plot as shown in Fig. 1. PR and RE values are normalized into crisp values and are displayed in Table 4.

5 Result and Discussions

The factors are represented using the D + R and D − R values in Fig. 1. D + R stands for prominence and indicates the importance of each element. If the value of vector D − R is positive, it signifies that the factor influences other factors in the group, and if the value is negative, it shows that other factors influence the factors in the group. The relation is referred to as D-R. In general, relative vectors are categorized into two separate groups: cause and effect groups [20]. Based on prominence and relation values, cause and effect group factors are created. Nine indicators obtained in the cause group are misleading of the technician (R 10), lack of manpower (R7), more number of demand work (R1), lack of skilled technician (R19), delivery time already postponed service vehicle (R2), negligence of technician, supervisor, and technical advisor (R25), a traffic jam when the vehicle goes for the road test (R27), overload of the technician (R26), and delay in the road test of the vehicle (R21). Eighteen indicators like delay in diagnosing a fault (R11), difficulty in dismantling the parts (R 15), improper bay final inspection of the technician (R14), the technician who does not know how to use the service manual procedure (R18), delay in final inspection (R4), lack of experience in the engine control unit (ECU)-related diagnosis and OBD (R6), the final inspector may notice some extra fault in the vehicle (R8), unawareness of technician about electrical system (R5), longer time to remove the rusted parts (R 17), the service circular from the manufacturer for replacing some parts of the vehicle (R12), delay in the out-pass procedure to follow for the road test of the vehicle (R22), electrical or lathe works are outsourcing (R20), mistake done by the technician (R9), overload of the technician (R13), the final inspector may notice the incomplete work of the vehicle (R23), insufficient space (R24), finding out of the extra fault (R3), and difficulty in assembling the parts (R16) are listed in the effect group.

Here, the challenge R10 is the most critical one. Since, the workers of the service station are not provided adequate practice on the latest technology. Hence, when there are more number of vehicles for service, being inadequate with the experience on handling latest technologies, the workers are unable to meet the demand. The second important challenge is R7. As the number of vehicles for service is not constant, there exists a fluctuation in the demand for workforce. Being an insecure job, the interest shown by the number of person required for the vehicle service is low. So, there is a need to ensure job security for vehicle service centers.

Table 2 Total-relationship matrix

		R1	R2	R3	R4	R5	R6	R7	R8	R9	R10	R11	R12	R13	R14	R15	R16	R17	R18	R19	R20	R21	R22	R23	R24	R25	R26	R27	Sum (Column)
R1	L	1	0	1.6	2.8	3.6	2.1	3	3.2	2	2.2	2	1.6	1.5	2.8	1.7	2.1	1.8	2	1.2	3.2	2.6	1.8	1.3	1.6	1.7	1.4	2.5	55.3
	U	1	0	2.3	4.3	4.5	3.8	4.2	4.1	4.2	4.0	3.8	2.3	2.3	3.9	3	3.1	3	3.8	1.6	3.9	3.2	2.4	2.2	2.3	2.4	2.2	2.8	81.8
R2	L	1.4	1	0	1.4	3.6	3.2	2.6	2.4	3.4	2.9	3.3	3.6	2.5	2.6	2.6	2.7	1.6	1.8	1.4	1.3	2.5	1.9	1.4	1.4	1.6	1.9	2	59.1
	U	1.7	1	0	1.8	4.4	4.1	3.7	3.1	4.2	4.3	4	4.3	2.8	4.1	3.9	3.9	2.4	3	1.8	2.2	3.8	3.7	2.2	1.8	2.9	3	3	82.1
R3	L	3.4	2.2	1	0	2.1	2.9	2.6	1.6	2.9	3.4	2.7	1.8	1.4	2.2	2	1.5	1.6	1.5	2	1.2	1.2	1.4	1.4	1.2	2.5	2.9	2.8	54.3
	U	4.1	4	1	0	4	3.9	3.8	2.3	4.1	4.2	4.2	3.8	3	3.9	3	2.2	2.3	2.2	2.6	1.6	1.6	1.7	1.7	2.4	3.8	4	4	80.7
R4	L	4.3	1.3	1.4	1	0	3.3	3.7	1.6	1.6	2.6	2.5	1.6	1.7	2.7	1.4	1.7	2	2.8	2.1	1.7	1.4	1.8	1.8	1.7	1.9	1.5	1.6	54.02
	U	4.7	1.7	1.9	1	0	4.2	4.5	2.3	2.3	3.8	3.7	2.3	2.4	3.8	2.3	2.9	2.6	3.9	3.2	2.4	2.2	2.5	3.7	3.7	3.7	2.9	3	79.08
R5	L	1.5	1.6	1.4	1	1	0	3.7	3.5	2.9	2.7	2.2	1.4	3	1.6	1.6	1.2	2	2.2	2.5	1.7	2.9	1.4	1.3	1.6	1.3	1.5	1.7	51.7
	U	2.9	2.3	2.3	1.5	1	0	4.6	4.2	4.1	3.9	3	2.3	4.3	2.3	2.3	1.6	3.1	3.9	3.1	2.4	4.2	2.2	2.2	2.4	1.7	2.3	3	74.4
R6	L	1.2	1.3	3.3	3.7	3.5	1	2.2	1.5	2.6	1.6	1.9	1.6	2.8	2.8	1.8	1.5	1.4	1.4	2	1.4	1.1	1.3	1.4	1.4	1.4	1.9	1.6	51.6
	U	1.6	2.2	4.2	4.6	4.4	1	3.4	2.2	3.8	2.4	3.1	2.4	4.2	4.1	2.4	2.3	1.7	2.3	3	1.7	1.6	1.8	2.3	2.3	1.8	2.6	2.3	73.09
R7	L	3.7	3.6	3.6	3.4	3.5	2.9	1	1.5	1.5	3.7	2.2	2.2	3.6	2.2	1.6	1.9	1.9	2.9	1.5	1.7	1.8	1.5	1.7	1.4	1.4	1.5	1.5	62.1
	U	4.4	4.5	4.4	4.4	4.3	4.2	1	2.3	2.3	4.5	4	4	4.3	4	3.7	2.7	3	4.1	2.2	2.4	2.4	1.8	3.1	1.9	3	2.3	2.3	88.9
R8	L	1.7	1.5	1.4	1.5	2.7	1.8	1.7	1	3.3	2.5	3.4	1.5	2.9	2.6	1.9	2.1	1.9	2.9	1.7	1.9	1.4	1.5	1.5	1.5	1.6	1.7	1.9	54.4
	U	2.4	2.3	1.7	2.3	4	2.4	2.5	1	4.1	3.1	4	2.4	4.4	3.2	3.1	4.1	3.9	4.2	3	3	1.8	2.3	2.3	2.4	2.5	2.5	3.1	79.4
R9	L	1.1	1.2	2.9	3.6	3.5	3.4	3.6	3.6	1	3.5	2.9	1.2	3	2.8	2.6	4.1	2.8	1.4	1.2	1.2	1.1	1.4	1.4	1.4	1.3	1.3	1.4	60.9
	U	1.6	1.6	4.4	4.4	4.4	4.3	4.3	4.5	1	4.1	4.4	2.2	4.3	4.1	4.1	4.3	4.1	2.3	1.6	2.2	1.6	1.7	1.7	2.3	2.2	1.7	2.2	82.7
R10	L	4.3	3.5	3.4	2.9	3.5	3.5	3.6	3.5	3.7	1	2	1.3	3.6	3.7	2.9	3.4	3.5	2.2	2.8	1.5	2.5	1.9	1.6	2	1.5	1.9	2.6	74.8
	U	4.7	4.1	4.2	4.3	4.1	4.2	4.5	4.4	4.7	1	3.7	2.2	4.3	4.5	4.1	4.2	4.2	3.9	4.1	2.2	3.7	3.3	2.4	3.7	2.3	2.6	3.8	100.8
R11	L	1.4	1.1	1.4	2.1	2.5	1.4	1.6	1.9	2	1.8	1	2.3	3.9	2	2.3	1.9	1.9	1.8	1.5	1.6	1.6	1.5	1.5	1.5	1.5	1.7	1.6	49.4

(continued)

Table 2 (continued)

		R1	R2	R3	R4	R5	R6	R7	R8	R9	R10	R11	R12	R13	R14	R15	R16	R17	R18	R19	R20	R21	R22	R23	R24	R25	R26	R27	Sum (Column)
	U	1.7	1.6	1.8	4	2.8	2.4	2.3	3.1	2.6	3	1	3.1	4.3	3.1	3.7	3.0	3	2.4	2.9	2.4	2.3	1.8	2.3	2.4	2.3	2.4	2.3	71.2
R12	L	1.2	1.3	1.1	1.3	1.3	1.4	1.6	1.6	1.5	2.9	1.5	1	1.9	2.6	2	1.4	2.5	1.8	1.8	1.8	1.4	1.9	1	1	1.2	1.2	1.2	43.3
	U	2.2	1.7	1.6	1.7	1.7	1.8	2.3	2.4	2.2	4.1	2.2	1	2.5	2.9	2.6	2.4	3.4	2.6	2.4	3	1.7	3.1	1	1.5	1.6	1.6	2.4	60.6
R13	L	3.7	2.9	3.6	2.9	2.3	2	2.1	2.6	3.2	3.5	5	2.5	1	4.3	3.5	3.5	3.4	1.7	3.4	1.6	2.1	1.6	1.5	1.6	2.6	1.8	4.3	75.3
	U	4.6	3.5	4.4	4.1	2.6	2.7	3.1	3.3	3.9	4.2	5	2.8	1	4.7	4.2	4.2	4.2	2.5	4.1	2.3	3.2	2.9	1.8	2.3	3.3	2.5	4.7	93.4
R14	L	2	1.8	1.9	2.6	1.9	1.6	1.7	1.9	2.6	2.8	2.6	2.2	1.4	1	1.4	1.8	2	1.7	1.9	1.8	1.9	2.4	1.9	1.4	1.3	1.5	1.6	51.8
	U	3.1	3	3	3.4	3	2.3	3	2.5	2.8	3.5	3.4	2.5	2.2	1	2.3	2.5	2.8	2.9	2.6	2.4	3	3.1	3	2.9	2.2	2.3	2.4	74.4
R15	L	1.4	1.9	1.8	1.8	1.7	2	1.4	1.7	1.8	1.6	2.8	1.8	2.5	1.4	1	1.4	1.4	2.1	1.6	1.6	1.4	1.6	1.4	1.5	1.7	1.7	1.3	46.3
	U	2.2	2.5	2.5	3	3.1	3.1	3	2.4	2.4	2.5	3.6	3	2.8	2.3	1	2.3	2.3	3.2	2.4	2.4	1.8	2.3	1.7	1.8	1.9	1.9	1.7	66.3
R16	L	1.7	1.4	3.7	2.9	3.7	3.1	1.8	2.1	2.7	2.6	2.1	2.1	2.5	2.8	3.7	1	3.4	1.8	1.8	1.5	1.7	1.5	1.2	1.2	1.2	1.4	1.2	58.7
	U	2.4	2.3	4.6	4.2	4.4	3.4	3.1	3.9	3.5	3.3	3.2	3.2	3.7	3.9	4.5	1	4.2	2.5	2.5	2.4	2.4	2.3	1.6	1.6	1.6	1.7	1.6	80.3
R17	L	1.6	1.8	1.4	1.5	1.6	1.6	2	2	1.6	2.9	3.6	2.4	3.5	1.8	3.7	1.8	1	3.5	1.3	1.4	1.4	1.4	1.4	1.4	1.6	1.2	1.6	53.1
	U	2.3	2.4	2.2	2.2	2.3	2.3	2.6	3.1	2.3	4	4.4	3.1	4.2	2.4	4.6	3	1	4.1	2.2	2.2	2.2	2.4	2.3	2.3	2.3	1.6	2.3	73.9
R18	L	1.4	1.8	1.5	1.4	1.4	1.4	1.2	1.4	1.7	2.3	1.5	1.6	3.4	2.2	1.1	1.4	1.5	1	1.3	1.5	1.4	1.4	1.1	1.3	1.2	1.1	1.2	41.9
	U	1.8	2.4	2.2	1.7	2.3	1.7	1.6	2.3	2.4	2.6	2.3	2.3	4	3.4	1.6	1.8	1.8	1	1.7	1.8	2.2	1.7	1.6	1.7	1.6	1.6	1.6	55.8
R19	L	2.7	2.7	3.5	3.5	2	3.2	1.9	3.7	2.8	3.4	3.5	1.6	4.3	4.3	3.3	3.2	3.3	2.4	1	1.3	1.2	1.3	1.2	1.3	1.2	1.3	1.4	67.5
	U	3.2	3.3	4.1	4.4	2.7	3.5	3	4.5	3.4	4.2	4.2	2.3	4.7	4.7	4.1	3.9	4.2	3.1	1	1.7	1.6	1.7	1.6	1.7	2.2	1.7	1.7	83.4
R20	L	1.3	1.3	1.4	1.4	1.3	1.3	1.4	1.8	1.3	2.5	2.4	1.2	2.8	2.5	1.4	1.4	1.3	1.1	1.2	1	1.4	1.2	1.2	1.3	1.2	1.4	1.2	41
	U	1.7	1.7	1.7	1.7	1.7	1.7	1.7	2.4	2.2	2.8	3.1	2.2	3.5	3.1	2.3	1.8	2.3	2.4	2.2	1	2.3	2.2	2.2	2.2	1.6	1.7	2.2	58.6
R21	L	2.8	3.6	2.6	2.9	4.2	3.6	3.6	4.2	2.9	3.5	3.7	2.6	3.5	3.7	3.7	3.5	3.4	2.4	3.4	3.4	1	2.7	3.2	3.6	3.4	4.2	3.4	89.7
	U	3.8	4.3	3.2	4.1	4.5	4.4	4.4	4.6	4.4	4.1	4.5	3.2	4.2	4.4	4.7	4.2	4.1	3.2	4	3.7	1	23.1	3.4	4.4	4.1	4.6	4.1	127.5

(continued)

Table 2 (continued)

		R1	R2	R3	R4	R5	R6	R7	R8	R9	R10	R11	R12	R13	R14	R15	R16	R17	R18	R19	R20	R21	R22	R23	R24	R25	R26	R27	Sum (Column)
R22	L	1.5	1.4	1.4	1.3	1.1	1.3	1.2	2.8	2.3	3.2	3.5	1.3	3.3	2.3	1.4	1	1.3	2.4	1.3	1.2	1.2	1	1.1	1.2	1.8	1.4	1.4	46.7
	U	2.3	2.2	2.3	1.7	1.6	1.7	2.2	3.5	2.6	3.5	4.2	2.2	3.6	2.6	2.3	1.5	1.7	3.1	1.7	1.6	1.6	1	1.6	1.6	3	2.2	2.3	62.3
R23	L	1.1	1.1	1.1	1.2	1.2	1.2	1.4	1.3	1.4	1.3	1.4	1.4	1.4	1.4	1.3	1.2	1.4	1.2	1.2	1.2	2.6	2.5	1	1	1.3	1.1	1.1	37.1
	U	1.6	1.6	1.6	1.6	1.6	1.6	1.7	2.2	1.7	2.2	2.2	2.2	2.3	2.3	2.2	2.4	2.3	2.2	2.2	2.4	3.2	3.2	1	1.5	2.3	1.6	1.6	55.7
R24	L	1.8	1.5	1.8	1.4	1.4	1.4	1.3	2.7	2.7	2	2.5	1.4	2.7	1.3	1.8	2.3	1.8	1.9	1.1	1.2	1.2	1.3	1.3	1	1.4	1.2	1.5	45.9
	U	2.4	1.8	2.4	1.7	2.3	1.8	2.2	3.3	3.4	2.6	3.2	1.8	3.3	1.7	3	2.6	2.5	2.6	1.6	1.6	1.6	1.7	1.7	1	1.7	2.2	2.4	61.2
R25	L	1.8	1.9	2.5	1.9	1.5	1.8	1.8	2	1.9	2.9	3.6	2.4	2.9	3.6	1.7	1.5	1.8	1.7	3.6	1.6	1.9	1.5	1.5	1.8	1	2.6	3.6	59.4
	U	3	2.5	3.1	3	1.8	3	2.4	2.6	2.5	4.2	4.5	3.1	4.1	4.3	2.4	2.3	3	2.4	4.3	2.4	3	2.3	2.2	3	1	3.3	4.3	81.4
R26	L	1.9	1.9	2.7	3.6	3.6	3.6	4.2	4.1	3.7	4.2	2.9	2.4	3.5	4.2	3.2	2.6	3.2	2.8	3.7	2.7	2.3	1.8	1.9	1.5	1.4	1	2.4	78.2
	U	3	2.5	3.8	4.3	4.3	4.3	4.6	4.4	4.5	4.5	3.9	3.1	4.2	4.6	3.9	3.2	3.5	3.5	4.6	3.4	2.6	2.5	3	2.4	1.7	1	2.7	95.2
R27	L	1.3	1.7	3.2	3.6	3.3	2.8	2.3	3.4	3.4	3.3	2.7	1.9	3.5	3.3	3.3	2.9	2.7	1.5	3.6	3.4	2.3	2	1.4	2.6	2.3	3.6	1	73.4
	U	2.2	2.4	3.9	4.3	3.6	3.8	2.6	4	4.1	4.1	3.9	2.6	4.1	3.6	4.1	4.1	3.4	2.3	4.3	4	2.6	3	2.3	3.2	2.6	4.3	1	91.8
Sum (Row)	L	55.4	49.2	57.8	59.6	64.1	59.9	61	65.9	65.3	73.9	72.6	51.1	74.9	71.7	61.3	57.1	59	55	53.9	47.4	47.5	45.6	40.8	42.5	44.7	47.9	51.8	
	U	73.8	66.7	75.8	80.6	82.4	78.7	83.4	86.3	87.1	95.9	98	73.2	96.3	94.1	86.6	78.8	81.2	80.9	74.2	66.0	65.5	85.0	59.3	64	64.6	65.4	72.1	

Table 3 Prominence and relation value

	X_i	Y_i	M_i	N_i
R1	1.02	1.18	2.20	−0.16
R2	0.96	1.27	2.24	−0.31
R3	1.22	1.18	2.40	0.03
R4	1.45	1.08	2.53	0.37
R5	1.37	1.09	2.46	0.28
R6	1.32	1.00	2.32	0.31
R7	1.19	1.29	2.49	−0.10
R8	1.41	1.09	2.51	0.31
R9	1.41	1.20	2.62	0.20
R10	1.55	1.58	3.13	−0.03
R11	1.52	0.96	2.49	0.56
R12	1.05	0.79	1.84	0.26
R13	1.66	1.49	3.15	0.17
R14	1.46	1.02	2.48	0.44
R15	1.32	0.87	2.19	0.45
R16	1.19	1.15	2.34	0.03
R17	1.30	1.02	2.32	0.28
R18	1.12	0.73	1.86	0.38
R19	1.09	1.28	2.37	−0.19
R20	0.96	0.74	1.71	0.21
R21	0.86	2.06	2.92	−1.19
R22	1.09	0.84	1.94	0.24
R23	0.78	0.67	1.46	0.11
R24	0.88	0.82	1.71	0.06
R25	0.88	1.19	2.07	−0.31
R26	0.94	1.53	2.48	−0.58
R27	1.06	1.44	2.51	−0.38

6 Implications of the Study

This study is an attempt to recognize and evaluate the challenges faced by the automobile vehicles service stations. Upon conducting research study under this domain, it was identified that the service stations face numerous challenges in ensuring the timely delivery of the vehicles that came for service. These problems need to be addressed in the short-run as it directly impacts the service stations' economic performance. Based on the outcome, this study suggests some implications that may help the service station overcome the above-discussed challenges. First, the service stations must adopt just-in-time (JIT) practice in the working activities [21]. With

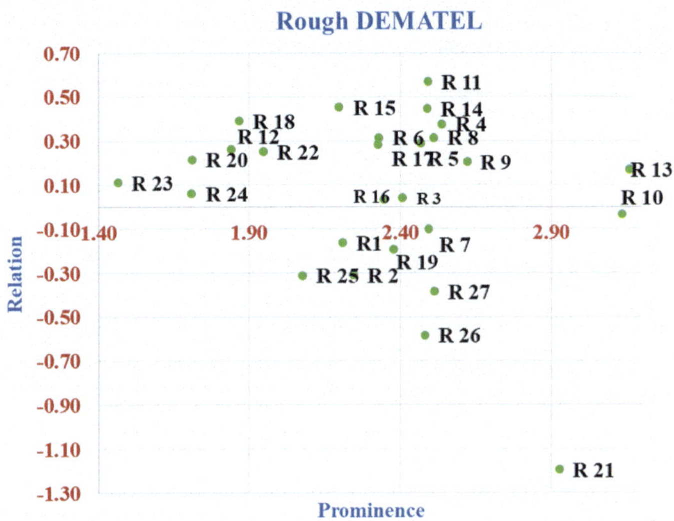

Fig. 1 Relation and prominence diagram

JIT practice, the service station may lower the time taken on each vehicle. Having an appropriate number of workers rather than short or excess will largely help witness success in JIT practice. Likewise, adopting 5 s (sort, set in order, shine, standardize, and sustain) is critical in averting the challenges. By following 5S, it is possible to eliminate time wastage in searching tools for service and ensuring timely delivery of vehicles.

7 Conclusions and Scope for Future Research

This study intended to find the barriers for the prompt delivery of vehicles in the vehicle service center; this study uses a rough DEMATEL analysis, data based on the expert suggestion that 27 factors were evaluated. Lack of skilled technician (R19), lack of manpower (R7), and negligence of technician or supervisor or technical advisor(R25) are the important factors that play a vital role in the cause of the delay in delivery of the vehicle after the service, whereas delay in diagnosing a fault (R11) and improper bay final inspection of the technician (R14) has an influential role in the delay. All the above-mentioned factors have an important effect on all other factors. So that all the twenty-seven factors considered for the data analysis play a vital role in decision-making in the service sector. The outcome of the study acts as a guide for the vehicle service centers that intends to improve their service facility. To conclude, the findings demonstrate that valuable cues are received for making sound decisions. Suppose the administration wants to achieve a high level of performance in terms of the effect group of elements, for example. In that case, they must closely monitor

Table 4 Total normalized crisp value

	∞X	∞Y
R1	0.29	0.32
R2	0.26	0.36
R3	0.43	0.32
R4	0.59	0.27
R5	0.54	0.27
R6	0.50	0.23
R7	0.41	0.37
R8	0.56	0.27
R9	0.56	0.33
R10	0.66	0.51
R11	0.64	0.21
R12	0.31	0.12
R13	0.74	0.47
R14	0.60	0.24
R15	0.50	0.16
R16	0.41	0.30
R17	0.49	0.24
R18	0.37	0.10
R19	0.34	0.37
R20	0.25	0.10
R21	0.19	0.75
R22	0.35	0.15
R23	0.13	0.07
R24	0.20	0.14
R25	0.20	0.32
R26	0.24	0.49
R27	0.32	0.45

and pay close attention to the cause group of variables. Cause indicators impact the entire system of prompt delivery of the vehicle, with ultimate goals influencing performance.

References

1. Sinha A, Mondal S, Boone T, Ganeshan R (2017) Analysis of issues controlling the feasibility of automobile remanufacturing business in India. Int J Serv Oper Manage 26(4):459–475
2. Ayala NF, Paslauski CA, Ghezzi A, Frank AG (2017) Knowledge sharing dynamics in service

suppliers' involvement for servitization of manufacturing companies. Int J Prod Econ 193:538–553
3. Bhattacharyya A, Ghosh D, Majumder A (2021) A Threadbare analysis on the state of affairs of Indian automobile sector in the aftermath of present recessional trend and its remedy. In: Productivity growth in the manufacturing sector. Emerald Publishing Limited, pp 253–267
4. Antony J, Viles E, Torres AF, de Paula TI, Fernandes MM, Cudney EA (2020) Design of experiments in the service industry: a critical literature review and future research directions. TQM J 32(6):1159–1175
5. Karuppiah K, Sankaranarayanan B, Ali SM (2021) A decision-aid model for evaluating challenges to blockchain adoption in supply chains. Int J Logist Res Appl 1–22
6. Garg CP (2021) Modeling the e-waste mitigation strategies using Grey-theory and DEMATEL framework. J Clean Prod 281:124035
7. Zhao G, Ahmed RI, Ahmad N, Yan C, Usmani MS (2021) Prioritizing critical success factors for sustainable energy sector in China: a DEMATEL approach. Energy Strat Rev 35:100635
8. Gardas BB, Raut RD, Narkhede B (2018) Reducing the exploration and production of oil: Reverse logistics in the automobile service sector. Sustain Prod Consump 16:141–153
9. Karuppiah K, Sankaranarayanan B, Ali, S. M (2021) Exploring key enablers of sustainable transportation in small-and medium-sized manufacturing enterprises. Kybernetes
10. Lamba D, Yadav DK, Barve A, Panda G (2020) Prioritizing barriers in reverse logistics of e-commerce supply chain using fuzzy-analytic hierarchy process. Electron Commer Res 20(2):381–403
11. Jain NK, Singh AK, Kaushik K (2019) Evaluating service quality in automobile maintenance and repair industry. Asia Pac J Mark Logist 32(1):117–134
12. Thakker SV, Rane SB (2018) Implementation of green supplier development process model in Indian automobile industry. Management of Environmental Quality: An International Journal 29(5):938–960
13. Yadav SK, Joseph D (2017) After-sales service quality satisfaction in Indian automobile industry. Int J Bus Inf Syst 26(3):362–377
14. Wang CN, Day JD, Farid M (2019) Service innovation model of the automobile service industry. Appl Sci 9(12):2403
15. Gupta A, Singh RK (2020) Managing operations by a logistics company for sustainable service quality: Indian perspective. Manage Environ Qual Int J 31(5):1309–1327
16. Fettermann DC, Echeveste MES, Tortorella GL (2017) The benchmarking of the use of toolkit for mass customization in the automobile industry. Benchmarking: Int J 24(6):1767–1783
17. Bhalaji RKA, Bathrinath S, Ponnambalam SG, Saravanasankar S (2021) Analyze the factors influencing human-robot interaction using MCDM method. Mater Today Proc 39:100–104
18. Bathrinath S, Mahendiran T, Ravikumar M, Karthi Shesan T, Bhalaji RKA, Koppiahraj K (2021) Analysis of risk factors in road accidents using fuzzy anp method. In: Materials, design, and manufacturing for sustainable environment, pp 739–754
19. Song W, Cao J (2017) A rough DEMATEL-based approach for evaluating interaction between requirements of product-service system. Comput Ind Eng 110:353–363
20. Karuppiah K, Sankaranarayanan B, Ali SM (2020) A fuzzy ANP–DEMATEL model on faulty behavior risks: implications for improving safety in the workplace. Int J Occup Saf Ergon 1–18
21. Zhu G, Chou MC, Tsai CW (2020) Lessons learned from the COVID-19 pandemic exposing the shortcomings of current supply chain operations: a long-term prescriptive offering. Sustainability 12(14):5858

Expert Analysis for Multi-criteria Human-in-the-Loop Input Selection for Predictive Maintenance Model

Chan Jin Yuan, Wong Yao Sheng, Jonathan Yong Chung Ee, and Wan Siu Hong

Abstract Human-in-the-Loop (HITL) concept integrating human knowledge and real-time activities into the smart and intelligent system yields an enhancement of IoT system. This paper proposes a HITL mechanism empowered with machine learning for Predictive Maintenance (PdM). A total of eight parameters were proposed as HITL inputs to integrate into the PdM system as an enhancement of the system. The parameter series are surveyed, carried out targeting Malaysia's SMEs Company for the solidification of the proposed parameter for HITL. A predictive maintenance framework is put together based on the survey result for the usage of future predictive maintenance system. In addition, the generic PdM architecture model is illustrated in a case study of a conveyor belt system with all the HITL input as proposed in earlier mentioned survey.

Keywords Human-in-the-Loop · Predictive maintenance · Machine learning

1 Introduction

The coexistence of human and engineering actors is seen through the lens of their interactions in various types of physical and digital augmentation of engineering operations. Literature shows that there are several concepts and implementations are heavily supported by human intervention for solutions in industrial system. However, its importance in sociotechnical systems and its' actual cognitive contribution of human activities to the operation of technical systems has received less attention [1].

This paper aims to study the approach for enabling Human-in-the-Loop (HITL) in Predictive Maintenance (PdM) model as well as find out what kind of a role and types of input humans can have in the PdM that can possibly enhance the overall PdM accuracy. PdM guidelines are scattered throughout literature that address parameters in either a concentrated or single-discipline manner that required a consolidation by implementation of HITL input. Thus, parameters need to be proposed and to

C. J. Yuan (✉) · W. Y. Sheng · J. Y. C. Ee · W. S. Hong
Department of Mechanical Engineering, UCSI University, Kuala Lumpur, Malaysia
e-mail: 1301539922@ucsiuniversity.edu.my

E. Natarajan et al. (eds.), *Materials, Design and Manufacturing for Sustainable Environment*, Lecture Notes in Mechanical Engineering,
https://doi.org/10.1007/978-981-19-3053-9_34

be evaluated by experts in the industries for the purpose of integration and actual implementation.

2 Literature Review

The implications for production and maintenance indicate that maintenance policies should move from a focus on short-term problems (resource usage, expense, etc.) to a consideration of longer-term goals (competitive, sustainability and strategy) [2]. Predictive Maintenance (PdM) is the method of constantly monitoring the condition of machinery to assess when a repair or replacement is needed based on the degree of deterioration. Cost savings, operating performance, product quality enhancement, and improved versatility are all advantages of this strategy [3]. For example, a study manages to build a PdM system for conveyor using few highest contributing parameters for load prediction [4].

Critical reviews on PdM related papers have highlights of PdM being beneficial, technical deployment, challenges in implementation and various conflicts of interest, as well as the reason why HITL might not be suitable in certain circumstances. Total of 35 research papers was collected to analyze and obtaining the ratio of papers that consists only predictive maintenance, and papers that consists HITL in predictive maintenance. Figure 1 presenting a pie chart to show the ratio of different industry focused PdM from 35 research papers. Keywords search in the filtering are "Predictive Maintenance" or "Maintenance" in the year range between 2016 and 2021.

Out of 35 predictive maintenance papers, only 5 of them are required human input in their proposed model. The existing predictive maintenance with human input only taking human as an accuracy improving input. Model improving input over time is main objective of involve human into the algorithm loop into machine learning

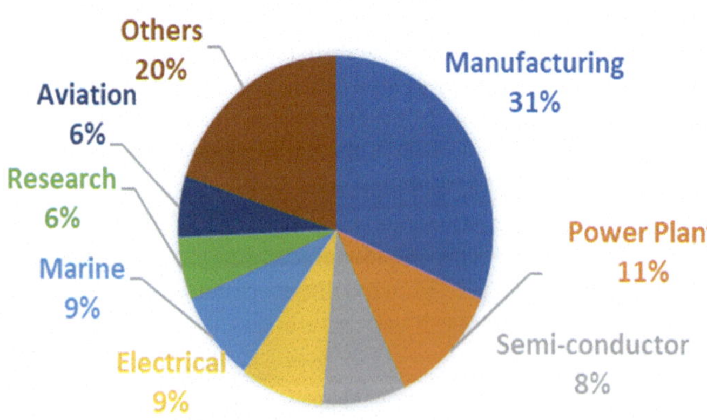

Fig. 1 Number of papers in each industry ($N = 35$)

at described by Amershi et al. [5]. After identifying all the industry, the 31% (11) of predictive maintenance research relate to manufacturing industry which none of them are related to HITL. Out of five papers with HITL, two papers contribute to power plant industry sector, while constructions, wind turbine and electricity market is mentioned in one paper each. The rest under "Others" category are related to hardware, food services, pumping station, automobile.

This review not only suggested that current trend of smart maintenance but also shows that many industries are remaining the practice of normal predictive maintenance. Also, this review means that majority researches are focused on manufacturing related industry that has not incorporate PdM and highlighting the gap that HITL can have enhancement in PdM but is unsure of the to the extent of the significance.

2.1 Human-in-the-Loop in Machine Learning

PdM models are built using machine learning in wide array of architecture of CNN, ANN, random forest, etc. HITL in machine learning can be denoted as Interactive Machine Learning (iML) in which it referring to any form of human interaction towards machine learning [5]. The algorithms with the input of agents can optimize their learning progress, where agents can be humans [6]. The ultimate purpose HITL is take advantages of human intelligence and skills to improve the quality of an algorithmic model. Thus, iML approaches can be functional when a rare or complex data set comes in. More importantly, iML enables re-traceability and explainable-AI in the medical domain [7, 8].

The general iML model normally includes the following two steps for the model. The first step is the output result will be reviewed by the human expert with experienced with the application. Generate human input as feedback to the machine learning model. The second step is the learning model will constantly be updated by integrating the human expert input. iML allow machine learning to be actively updating itself with the human expert input and this make machine learning to be more applicable to various application where human intelligence is needed for the analysis process [9].

Figure 2 shows a machine learning model with the human evaluation parameter embedded into the model. The authors introduced a human input as feedback from the monitor module into the machine learning model for condition monitoring. If the detected condition shown in the monitor is a false detection after screening by human experts. Expert will give input as dissatisfaction to the result and this particular case will be added into the training database for the future detection. Thus, this model will continuously improve with time when people are technically using it [10].

In another context, interactive machine learning can also be denoted as human-centered machine learning. Human-centered machine learning in other research emphasising visualisation for user interaction towards machine learning model [11–13].

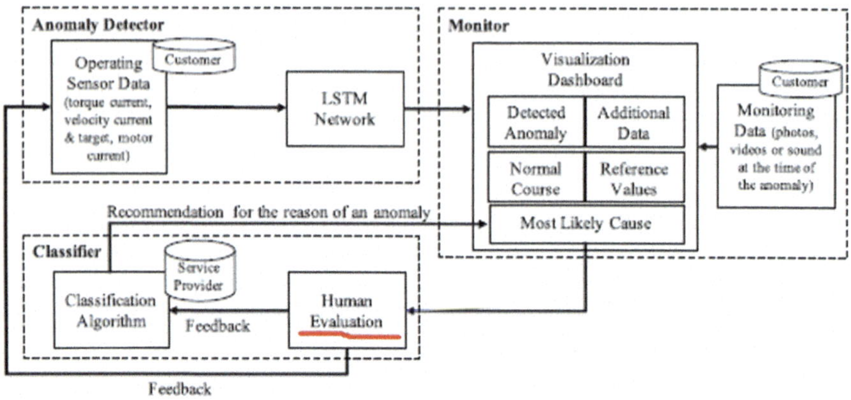

Fig. 2 Developed machine monitoring approach with human interaction [10]

2.2 *Potential Human-in-the-Loop Parameter*

Table 1 consists of 8 parameters that been proposed according to the findings in the literature.

The critical review synthesizes the 35 papers and consolidating the underpinning human factors or input that can significantly improve the maintenance system. The filtering process was identifying via the pipeline and bottleneck where decision making is done. All basis of information, processing of data, monitoring and presentation of maintenance data are crucial decision-making and hence will be listed as parameter that can be input numerically into the Machine-Learning model.

Table 1 Factors improving machine learning/predictive maintenance

Parameters	Input	References
Technical experience	(a) Scheduling data (b) Operational data	[1, 14]
Collective context sharing	Voting by stakeholders on post process data	[1]
Cumulative external data	Environmental or surrounding data (humidity, Temperature, etc.)	[15]
Dispatch maintenance rules	(a) Job due dates, (b) Scheduling rules	[16]
Data manipulations	Experts will identify high priority events to be resolve first	[17]
Cost effectiveness ratio	Cost ratio of different maintenance strategy	[18]
Computerized maintenance management system (CMMS)	Onsite real time monitoring system	[19, 20]
Remote maintenance management system (RMMS)	Real time online monitoring system that can be shared among users	[19, 21]

3 Methodology

The amalgamated set of PdM parameters will now be screened by experts to validate the relevance and significance in building the HITL PdM model. Expert survey will be conducted by sending 50 Likert scale questionnaires (Table 2) to the SMEs in Malaysia with position such as maintenance manager, maintenance engineer or other equivalent position.

After the survey, reliability test and descriptive analysis will be carried out to further analyse the collected result. Lastly, Delphi technique will be conducted by sending all the result and report to experts to review the research work.

Table 2 Questionnaire's question

No.	Questions
Q1	a. Will technical experience in the form of input data bring positive impact to increase maintenance efficiency? b. Using technical experience into estimating next maintenance, will maintenance downtime decrease?
Q2	a. Is collective agreement of maintenance team important for effective maintenance? b. Is collective agreement practical for maintenance in light industry of Malaysia?
Q3	a. Does cumulated external data bring any impact in improving estimating next maintenance? b. Does collected environment data directly relate to maintenance?
Q4	a. Will operating rule/regulations/ISO effect the interval maintenance? b. Will incorporating rules such as best practice of scheduling has any impact on the maintenance in industries?
Q5	a. Does analyzing data and visual pattern mining, including user interaction logs and equipment maintenance records by event sequence, have positive impact on maintenance? b. Does visual pattern mining allowed managers to have better decision making on maintenance?
Q6	a. Will projected cost effectiveness ratio impact maintenance strategy? b. Is cost effectiveness ratio an important consideration in managing maintenance downtime?
Q7	a. Is CMMS useful to optimize maintenance management for the purpose of saving cost and reduce downtime? b. Will computerized maintenance management system (CMMS) bring any positive impact to estimating next maintenance?
Q8	a. Is the remote maintenance and management system (RMMS) necessary towards enhancing safety in maintenance sector? b. Is the RMMS practical to reduce downtime due to machine failure?

4 Results and Discussion

The collected survey data from various maintenance personnel's by industries distribution in this survey are presented in the pie chart of Fig. 3; Fig. 4 shows mean score of each question. For the frequency of the Likert scale distribution from the survey is tabled in Fig. 5. The sampling is a total of 50 respondents in this survey.

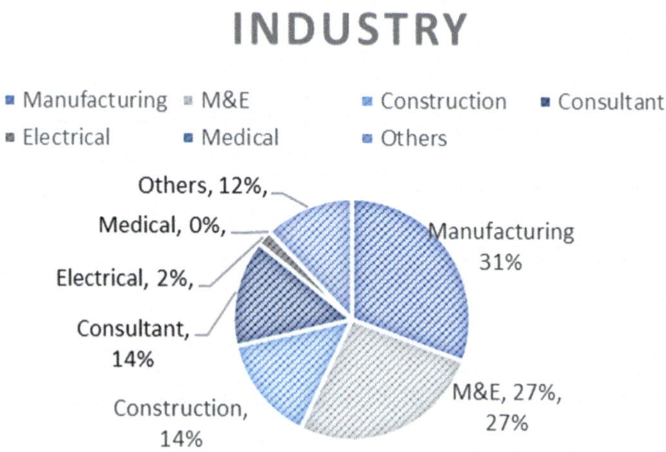

Fig. 3 Distribution of respondents in their industry area

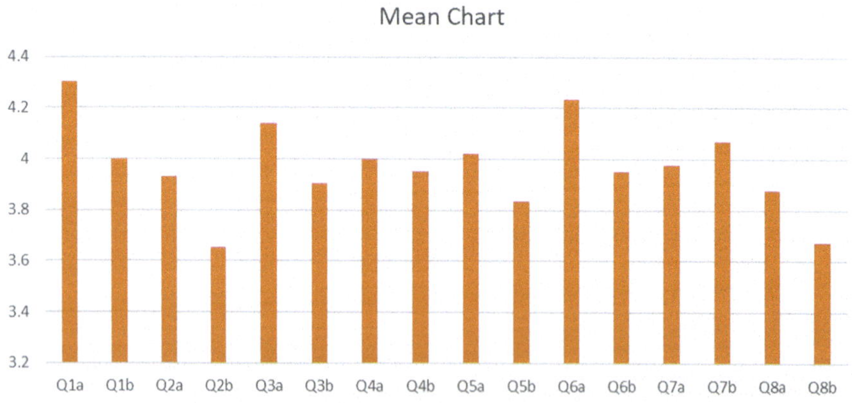

Fig. 4 Mean chart of the questionnaire response

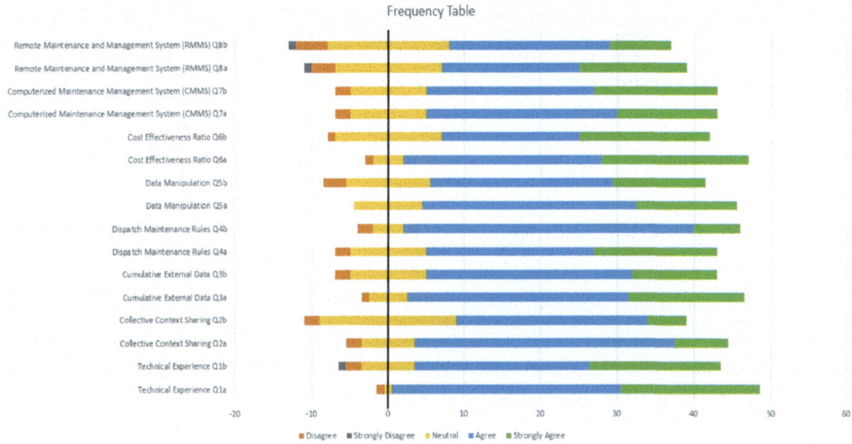

Fig. 5 Likert scale distribution chart

4.1 *Reliability Test*

Cronbach's coefficient alpha (α) was calculated for each variable in the reliability test to show the accuracy of the data collected based on the formula in Eq. 1. According to [22] coefficient alpha values greater than or equal to 0.70 are commonly accepted. The results of the reliability analysis indicate that the Cronbach's Alpha for the variables is 0.818, suggesting that the variables are reliable.

$$\alpha = \frac{K}{K-1}\left(1 - \frac{\sum_{i=1}^{K} \sigma_{Yi}^2}{\sigma_X^2}\right) \tag{1}$$

4.2 *Descriptive Analysis*

Parameter that has a lower mean score is Remote Maintenance and Management system (RMMS) that has a score of 3.62 for its question 8b. One of the possible reason is the network system structure. The communication between the product (e.g., CNC) manufacturer and their respective customer's machine tools is priority for RMMS to launch.

Technical experience is one of the high scoring inputs. The reason why is been agreed by most is because the essence linking data and knowledge is the key input of enriching the machine learning that will improve the failure prediction, decision making process and maintenance scheduling.

A slight occurrence of disagreement for remote maintenance system was one of the observances in background that's in contrast with other parameters with no strong disagreement. It's likely stemming from the unreadiness of some industries in adopting automated mechanism. Such unreadiness maybe arising from lack of confidence, complacency with the conventional or even fear of change and progression.

4.3 Delphi Technique

The results shown in the section above will be dispatched to expert in professional field for second tier verification. The second stage Delphi discloses the proposed model, parameters, and the description to close-group experts at the same time as the survey form, seeking for mutual agreements and future feedback on the survey results.

Three experts were involved in the Delphi survey, there are experts from manufacturing industry, research construction institute from Malaysia and, professional engineers institute in Malaysia. All of them agreed with the direction of the research and it is an interested topic to investigate. One of them further suggests a daily stand-up meeting around 10–15 min with maintenance team and operation team along with the presence of stakeholders, to review the performance of machine—which can be packaged, analyzed outcome and inputted for the enhancement of ML. He affirms that such meeting outcome can be made numerical for any projection and advanced analysis. Daily alignment with the relevant stakeholders will be able to achieve the best effectiveness and efficiency. On the other hand, another expert suggests more inputs related to quality of performance, time, and sustainability of business (that requires more data harvesting—as prescribed in parameters covered in Q2, Q5, Q6, Q7 and Q8) as three important elements in PdM.

4.4 PdM Model with HITL

Figure 6 shows the proposed model for the PdM with HITL. There are three main components in the model, which are Remote Maintenance and Management System (RMMS), Collective Context Sharing Centre and Physical System. Three of them are supported by a machine learning algorithm for PdM.

Physical system consists of the machine itself as well as sensors allocated around the machine to collect data required by PdM model. Figure 7a–d shows existing research project as an example of physical system. As well as the maintenance team which would be responsible for maintenance action.

Data from sensors will then be send to display dashboard (E.g., Fig. 8) as part of the Remote Maintenance and Management System (RMMS). RMMS enable workers to remotely monitor the machine at real time and can be shared among stakeholders. Computerized Maintenance and Management System (CMMS) refers to the onsite real time monitoring instead of manually monitor the system.

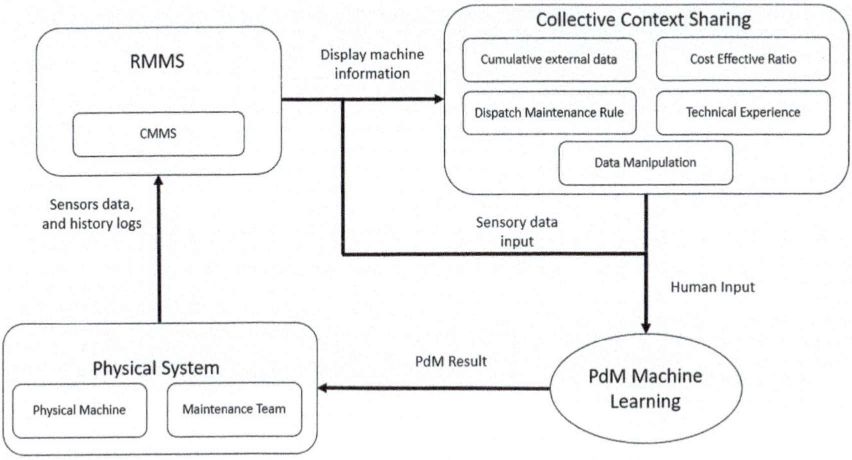

Fig. 6 Proposed PdM Model with HITL

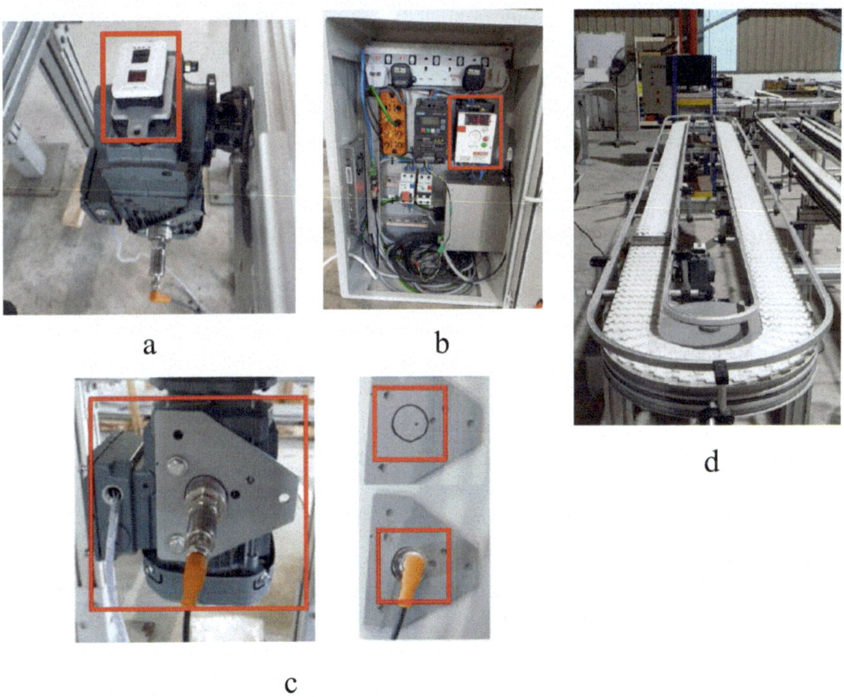

Fig. 7 **a** Bosch XDK Sensor, **b** inverter, **c** vibration sensor, **d** conveyor System Setup

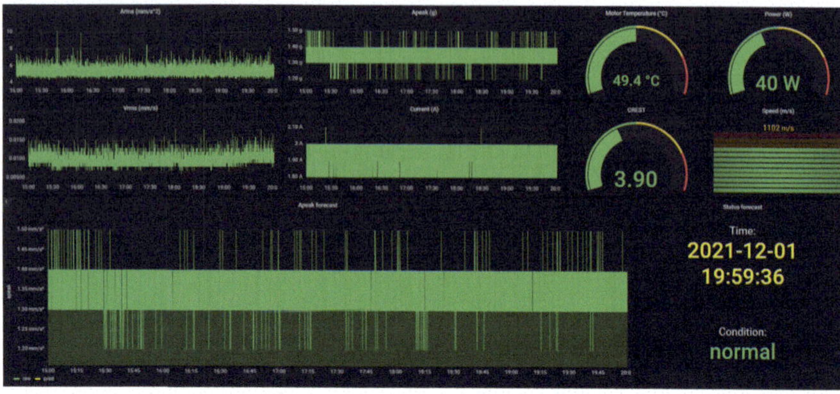

Fig. 8 Dashboard for data display

Context sharing center is a platform for exchange of knowledge for the input for HITL machine learning. Context sharing center including cumulative external data from sensors that is wearable on the operators, collect information such as ambient temperature and humidity. Cost effective ratio refers to different maintenance strategy according to machine faultiness. Dispatch maintenance rule would be the other factors that would affect the PdM, job due dates, scheduling rules are the rules that can vary over different jobs. Experts with high technical experience to the machine would also contribute their knowledge to assist in PdM model. Lastly, data manipulation represents an input that experts could identify the high priority of events to be resolve first.

Figure 9 shows a sample of existing PdM model developed by using node-red applications. This shows an idea where Machine Learning can be developed, not only in a hard coding manner but a simplified module with different function to drag and link around.

5 Conclusion

In this paper, the idea of HITL in machine learning was introduced to integrate with the machine learning technology. Eight (8) parameters were proposed in this studied as an input for HITL in predictive maintenance sector. A series of survey was carried out in Malaysia targeting Small and Medium-sized Enterprises (SMEs) with position such as maintenance manager, maintenance engineer or equivalent position. Total of 16 questions were used to test the agreement towards the proposed parameters.

From the survey, all of the proposed parameters were agreed by the experts from the SMEs with rating more than 3.6 out of 5.0. Out of the eight parameters, 'technical experiences' stand out with rating of 4.15 out of 5.0. The PdM with HITL model was proposed in this paper to be a guideline and framework incorporating all the

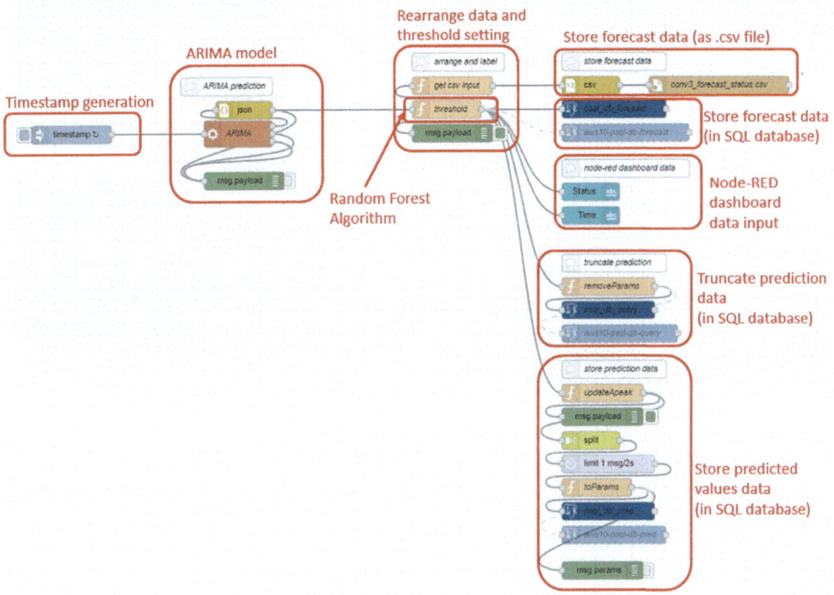

Fig. 9 PdM model on node-Red

parameters essential for building a Machine-Learning model with periodic reinforced learning. The proposed parameters may have yet to have weightage determined, this can be extension research to make the model more reliable. The rating from the expert survey could be the initial benchmark value to adjust the weightage of each parameter in the PdM model. Also, a pilot study on conveyor belt system, leveraging on the framework has been highlighted and having its platform of ML explain in fulfilling all parameters.

Moving forward, PdM model with HITL can be furthered implemented into a pilot machine to incorporate the proposed parameters for an automated decision-making prescribing maintenance schedule. At the same time, more HITL parameter can be discussed to be added into the model or to be eliminated from the existing model. There is no doubt PdM will be driven of the trend of revolution of industry 4.0 technology, each industry has to customize PdM according to each industrial-centric operation, but this proposed model will serve as a framework and checklist for a comprehensive PdM.

Acknowledgements The completion of this research work could not be possible without the support of UCSI University with Grant (REIG-FETBE-2020/011). Great thanks to the industry experts who participated in this survey, it could not be meaningful without their participation.

References

1. Emmanouilidis C, Pistofidis P, Bertoncelj L, Katsouros V, Fournaris A, Koulamas C, Ruiz-Carcel C (2019) Enabling the human in the loop: linked data and knowledge in industrial cyber-physical systems. Annu Rev Control 47:249–265. https://doi.org/10.1016/j.arcontrol.2019.03.004
2. Carnero M (2006) An evaluation system of the setting up of predictive maintenance programmes. Reliab Eng Syst Saf 91(8):945–963. https://doi.org/10.1016/j.ress.2005.09.003
3. Ruiz-Sarmiento J, Monroy J, Moreno F, Galindo C, Bonelo J, Gonzalez-Jimenez J (2020) A predictive model for the maintenance of industrial machinery in the context of industry 4.0. Eng Appl Artif Intell 87:103289. https://doi.org/10.1016/j.engappai.2019.103289
4. Goh KW, Chaw KH, Yong JC, Koh YS, Dares M, Su EL, Yeong CF (2021) Machine learning based predictive maintenance system for industrial chain conveyor system. In: RiTA 2020. Springer, Singapore, pp 251–263
5. Amershi S, Cakmak M, Knox WB, Kulesza T (2014) Power to the people: the role of humans in interactive machine learning. AI Mag 35(4):105–120
6. Holzinger A, Plass M, Holzinger K, Crişan G, Pintea C, Palade V (2016) Towards interactive machine learning (iML): applying ant colony algorithms to solve the traveling salesman problem with the Human-in-the-Loop approach. Lect Notes Comput Sci 81–95. https://doi.org/10.1007/978-3-319-45507-5_6
7. Holzinger A, Malle B, Kieseberg PM, Roth P, Muller H, Reihs R, Zatloukal K (2017) Towards the augmented pathologist: challenges of explainable-AI in digital pathology. arXiv:1712.06657v1
8. Liu M, Liu S, Zhu X, Liao Q, Wei F, Pan S (2015) An uncertainty-aware approach for exploratory microblog retrieval. arXiv:1512.04038
9. Jiang L, Liu S, Chen C (2018) Recent research advances on interactive machine learning. J Visual 22(2):401–417. https://doi.org/10.1007/s12650-018-0531-1
10. Olivotti D, Passlick J, Axjonow A, Eilers D, Breitner M (2018) Combining machine learning and domain experience: a hybrid-learning monitor approach for industrial machines. Explor Serv Sci 261–273. https://doi.org/10.1007/978-3-030-00713-3_20
11. Sacha D, Sedlmair M, Zhang L, Lee JA, Peltonen J, Weiskopf D, Keim DA (2017) What you see is what you can change: human-centered machine learning by interactive visualization. Neurocomputing 268:164–175
12. Wei Y, Wang X, Nie L, He X, Hong R, Chua TS (2019) MMGCN: Multi-modal graph convolution network for personalized recommendation of micro-video. In: Proceedings of the 27th ACM international conference on multimedia, pp 1437–1445
13. Kaluarachchi T, Reis A, Nanayakkara S (2021) A review of recent deep learning approaches in human-centered machine learning. Sensors 21(7):2514
14. Pistofidis P, Emmanouilidis C, Papadopoulos A, Botsaris PN (2016) Management of linked knowledge in industrial maintenance. Ind Manage Data Syst
15. Cimini C, Pirola F, Pinto R, Cavalieri S (2020) A human-in-the-loop manufacturing control architecture for the next generation of production systems. J Manuf Syst 54:258–271. https://doi.org/10.1016/j.jmsy.2020.01.002
16. Jimenez JJM, Schwartz S, Vingerhoeds R, Grabot B, Salaün M (2020) Towards multi-model approaches to predictive maintenance: a systematic literature survey on diagnostics and prognostics. J Manuf Syst 56:539–557
17. Zenisek J, Holzinger F, Affenzeller M (2019) Machine learning based concept drift detection for predictive maintenance. Comput Ind Eng 137:106031
18. Wang Y, Gogu C, Binaud N, Bes C, Haftka R, Kim N (2017) A cost driven predictive maintenance policy for structural airframe maintenance. Chin J Aeronaut 30(3):1242–1257. https://doi.org/10.1016/j.cja.2017.02.005
19. Selcuk S (2016) Predictive maintenance, its implementation and latest trends. Proc Inst Mech Eng Part B J Eng Manuf 231(9):1670–1679. https://doi.org/10.1177/0954405415601640

20. Labib AW (2004) A decision analysis model for maintenance policy selection using a CMMS. J Qual Maintenance Eng
21. Mori M, Fujishima M, Komatsu M, Zhao B, Liu Y (2008) Development of remote monitoring and maintenance system for machine tools. CIRP Ann 57(1):433–436
22. Sekaran U, Bougie R (2010) Research methods for business: a skill building approach, 5th edn. Wiley, Chichester

Development of Anti-siphoning Model by Automatic Identification System for Marine Security

Chan Jin Yuan, Jonathan Yong Chung Ee, Wan Siu Hong, and Siow Chee Loon

Abstract Illegal ship-to-ship (STS) fuel siphoning is a prevalent yet under-reported phenomenon due to national security confidentiality. Modeling of such scenario lacks the robustness of validating the occurrence for its own simple reason of being illegal. Nevertheless, a framework for siphoning detection is required to assist in curbing these illegal activities to preserve the rights and profit of stakeholders. This research paper is proposing a near proximity framework to identify potential illegal STS fuel siphoning outside the port region using automatic identification system (AIS) data along Straits of Malacca and Singapore. Suggested criteria on the characteristics of siphoning process are based on proximity between two vessels and total duration within the proximity. The proximity model is developed in Python to process the AIS data collected, and filters are applied according to the decide criteria with potential level of illegal siphoning. From 50 ship samples of AIS data analyzed, 18.2, 47.9 and 46.7% of the total cases are simulated with high possibility of illegal STS fuel siphoning for proximity category less than 50 m, 100 m, and 150 m, respectively, with more than 10 h duration. Enhancement techniques to the conceptual framework to identify potential illegal STS activities are also proposed in the event without on-site validation.

Keywords Ship-to-ship transfer · Fuel siphoning · Maritime safety · Automatic identification system (AIS) · Proximity tracing

C. J. Yuan (✉) · J. Y. C. Ee · W. S. Hong
Department of Mechanical Engineering, UCSI University, Kuala Lumpur, Malaysia
e-mail: 1301539922@ucsiuniversity.edu.my

S. C. Loon
Department of Aeronautics, Automotive and Ocean, University Teknology Malaysia, Johor, Malaysia

E. Natarajan et al. (eds.), *Materials, Design and Manufacturing for Sustainable Environment*, Lecture Notes in Mechanical Engineering,
https://doi.org/10.1007/978-981-19-3053-9_35

1 Introduction

1.1 Background

Status quo still shows that over 90% of the world trade is covered by international shipping industry. The global economy heavily depends on the shipping industry for import and export goods and materials as mentioned by International Chamber of Shipping in 2020 [1]. The trend of global trade by sea will continue to grow in future, due to its cost effectiveness nature of transferring goods from places to places compare to any other method. Hence, there are bounds to be illegal maritime activities amidst the booming industry.

It is reported about 3% of the fuel consumed in Southeast Asia are from illegal sources, and it is estimated the fuel value is worth up to RM 43 billion a year [2]. A report released by UNSC [3], stating various illegal STS fuel siphoning happened around the globe. Illegal STS siphoning that carried out around North Korea has been reported to be the most. Only in the duration of five months throughout 2018, there were 89 or more suspicious events for illegal STS fuel siphoning in the region and further concluded that total of 148 or more suspicious events of illegal STS fuel siphoning. Nevertheless, most siphoned reports are by triangulation via shortage and untallied oil usage, on-site detection remains a challenge.

1.2 Problem Statement

Operation of illegal STS fuel siphoning is prevalent as worldwide trend since the growth of the global shipping industries causing huge impact on economy of a country especially for oil producing country. Despite the illegal act being caught by local authorities and reported to the public, it is projected to be only a small portion of the actual illegal STS fuel siphoning volume. A series of prevention system and algorithm need to be identified to reduce or assist to eliminate this illegal activity. Constant monitoring action is required to observe and detect real-time movement of the vessel and filtering out the suspicious vessels for illegal action. In effect, there is an enormous challenge for the current industry to monitor and predict the occurrence of the illegal STS activity to validate the probing of an investigation by law. An illegal STS fuel siphoning detecting framework needs to be identified and developed. STS fuel siphoning criteria also need to be identified for the developing this framework.

2 Literature Review

2.1 Ship-to-Ship Fuel Transfer

The utilization of STS transfer was first introduced back in 1960s at Gulf of Mexico due the bustling port docking scheduling. STS transfers are referring to the transfer of cargo between two vessels that usually one of them is a large ship that cannot be loaded near the port due to limited spaces at the port [4]. The STS fuel transfer is introduced to the transfer of ship fuel at the at first. However, the STS fuel transfer had slowly extended the initial purpose of STS fuel siphoning into transfer of liquefied gases, oil, and dry cargoes such as ore [5].

The vessels involved STS transfer operation are positioned side by side. Both vessels are moving in a controllable lowest speed before there are aligned in parallel position. STS transfer operation can be either be stationary or underway depending on many different factors, security 'loopholes', and ship capability that may affect the 'non-transparent' transfer process [6].

2.2 STS Fuel Standards and Patterns

The general declared STS fuel transfer operation consists of four different phases for the safety and welfare of both mankind and sea environment during operation. The four phases that include obtaining pre-fixture information before operation. Secondly, being mooring process between two vessels once aligned in parallel position as shown in Fig. 1 and then, thirdly, followed by the transfer of cargo. Last but not least, the unmooring process of the vessels [7].

Maneuvering action is to be finished within 50–100 m proximity, and the mooring lines that can allowed to pass in between vessels are at distance around 20–30 m. The standard STS fuel operation lasts for a duration ranging from 10 to 24 h depending on the transferring quality and quantity [9].

Standard STS operation can be carried out in either at anchored (in the port) or underway (outside the port) [10]. Ships within the VTS range can be monitored by the port authorities, and STS operations outside the port range are the focused topic in this article. According to the official Website by the Malaysia Government, VTMS of Port Klang is located on the 19th floor of the Westport's commercial tower and covers 26 nautical miles (48.15 km) offshore. While VTS in other countries ranges between 200 and 400 km for ports authorities of Singapore, Shanghai, Gulf of Mexico, depending on the geographical basin [11]. Table 1 had summarized the standard and expected flow of STS transfer operation for proximity between vessels and duration of the operation.

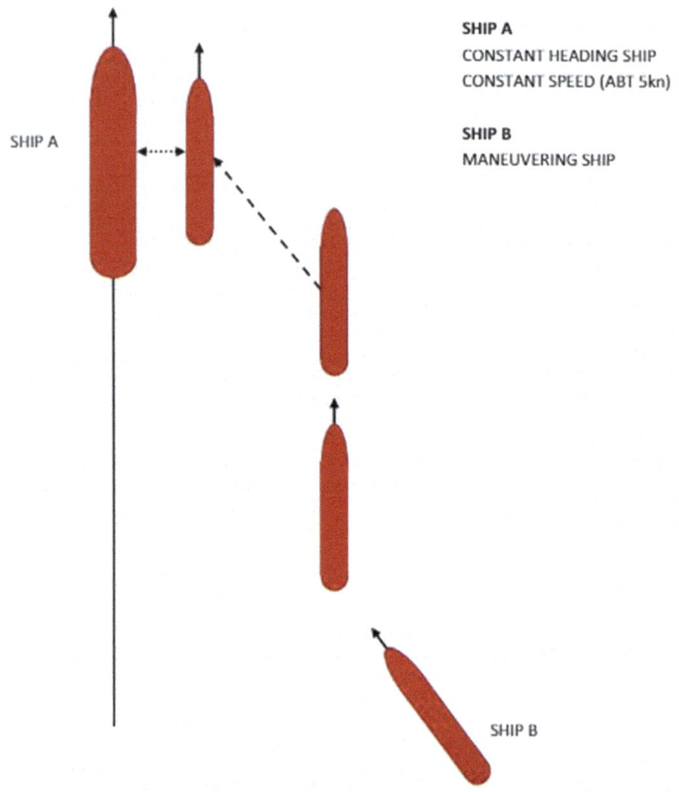

Fig. 1 Approaching route of service ship [8]

Table 1 Standard STS operation

Parameter	Range	Source
Proximity	7.62–50 m	[7, 12]
Duration	10–24 h	[12]

2.3 Illegal Activities on STS Fuel Transfer

Illegal STS fuel transfers are not the only activities that contributed to the fuel smuggling activities around the globe. In fact, fuel smuggling occurs across the supply chain of the fuel (see Fig. 2), from the wellhead, gathering station, crude oil pipeline, oil storage facility, oil bunkering station, during crude oil tanker loading, and during illegal STS transfer facilitated by oil tanker crews, the acts of robbery and piracy at sea would occurs. Figure 2 shows an illustration of fuel smuggling occurrence in the fuel supply chain, and the elements marked with in red are symbolized that the element is exposed by smuggling [13]. However, this paper only focused on illegal STS fuel transfer.

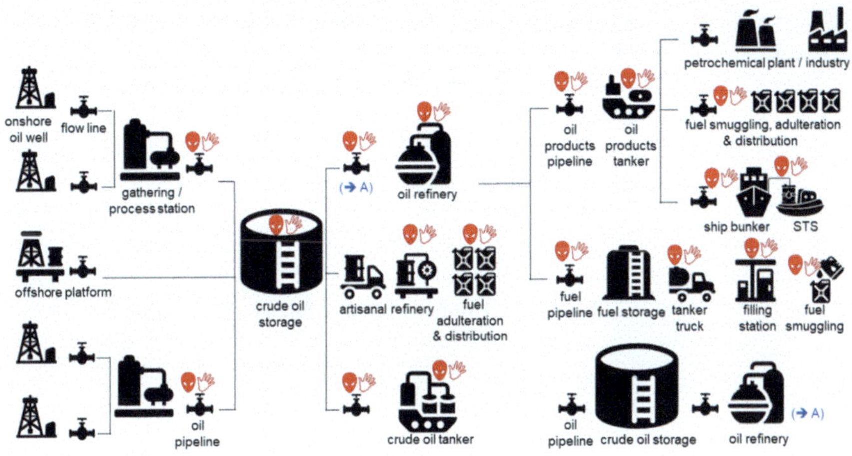

Fig. 2 Fuel theft in fuel supply chain [13]

In the year 2018, The Southern Region Marine Department had detained an oil tanker on committing illegal ship-to-ship (STS) fuel transfer at the Tanjung Pelepas Port near Gelang Patah. A vessel was arrested on 3 October 2018 at 11:30 am while carrying out siphoning activity at the port. The news mentioned the vessel which is registered at the Klang Port was transferring a total of 120 metric tons of marine gas oil (MGO) to Hong Kong-registered SafMarine Chillka [14].

Another fuel siphoning case happened in June 2019, Japanese Ministry of Foreign Affairs (MOFA) had witnessed the process of illegal STS fuel transfer between 2 unidentified ships with the North Korean flagged ship 'An San 1' which is in charge of DPRK (North Korea) sanctions. MOFA releases a photograph of 'An San 1' surrounds by 2 ships with no identification carry out illegal STS fuel transfers at the sea. STS fuel transfer at sea is one of the sources and way for North Korea to obtain ship fuel as United State of America (USA) has limited the amount of ship fuel entering or selling to North Korea as a safety measure for the world [15].

3 Methodology

3.1 General Architecture of Detecting System

The aim of this writing is to propose a conceptual framework to construct an illegal STS fuel siphoning detecting system to be apply to sea along Straits of Malacca. This research started off with collection of AIS along the Straits of Malacca and around Singapore straits. AIS data were collected by AIS receivers as a time series data and to be automatically uploaded onto the cloud. The data obtained from AIS data are further analyzed by a logic-filter to predict potential and suspicious illegal

STS fuel siphoning operation. The isolation parameters that are used to determine the occurrence of illegal STS fuel siphoning operation are (1) proximity of vessels, (2) duration in such proximity. The architecture of the logic-filter mechanism for the detection is as shown in Fig. 3.

In short, the location of each individual ship by time is tracked against all other ship by AIS data. The algorithm will first trace any two vessels with proximity closer at the set value. After which the time counter will start to time the duration of both ships' proximity, and if the duration for the proximity exceeds the set timeframe, the algorithm will record the count case for suspicious STS transfer. The algorithm does not feed information for those ship who has declared status of STS transfer.

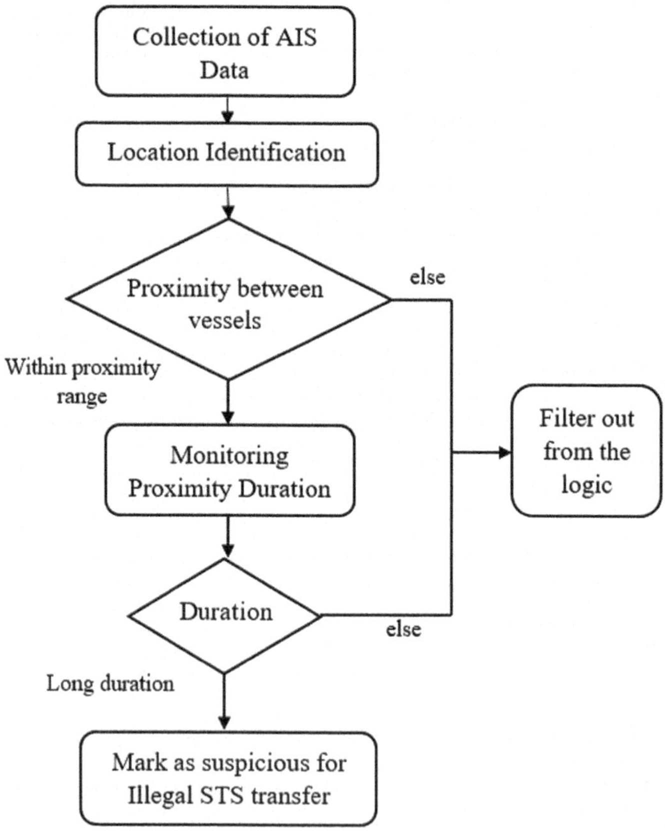

Fig. 3 Logic-filter architecture for oil-siphoning detection

3.2 AIS Data Collection

Method to collect AIS data by AIS receiver becomes part of the research due to the encryption of ship AIS data, even for the location Straits of Malacca and Singapore.

AIS receiving antennas are installed fitted with a single-board computer Raspberry Pi that act as an operating system for the integrated AIS receiver and its antenna. The built-in Wi-Fi module inside the Raspberry Pi helps to simplify the system and give the ability connect to cloud server by the Internet access. The receiver will receive AIS data sent by the vessels to AIS receiving system that had been installed along Straits of Malacca as shown in Fig. 4. The AIS receiver will received data such as date, time of message, message ID, MMSI number, navigation status, turn rate, speed over ground, position accuracy, longitude, latitude, course over ground, heading angle, time stamp, channel of AIS, and repeat indicator. One month of AIS data was collected around straits of Malacca and Singapore.

Fig. 4 Installed location of AIS receiver in Malaysia

Table 2 Table of parameters

Parameter	Level 1	Level 2	Level 3
Proximity, m (d)	$d < 150$	$d < 100$	$d < 50$
Duration, hours (t)	$1 < t < 5$	$5 < t < 10$	$t > 10$

3.3 Parameter for the Detecting System

The collected AIS data are then being processed by the logic-filter system to detect potential illegal STS fuel siphoning. The detection was done using a Python programming language to process and filter large amount of data.

The parameters that conceptually determine whether the vessels are suspicious for illegal STS fuel siphoning are includes proximity between vessels on the sea and the duration at such proximity. Three ranges of proximity are chosen based on the standard procedure of STS fuel siphoning. The purpose of setting up multiple proximity ranges and duration is to identify suspicious vessel activity and to probe the sensitivity of the parameter before detecting results for future works. It is important to include time factor for the research to eliminate vessels that is just passing by within the set ranges of proximity. The proximity between vessels and the duration are shown in Table 2. Fuel siphoning activity level will be divided into three levels, of which it represents the potential risk for illegal STS fuel siphoning. Level-1 refers to low suspicious, level-2 represents intermediate suspicious, while level-3 represents the highest suspicious to STS fuel siphoning. In short, higher level signals high probability for illegal STS fuel siphoning.

4 Result and Discussion

4.1 Development of Logic-Filter Framework

The logic-filter framework deployed in Python is separated into a two-phase filter for ease of manipulation and debugging on each filter. The first filter calculated the proximity of two vessels at one particular time by retrieving information from the raw AIS data. The computed results will be tagged with location of both vessel, proximity, date, and time.

Second filter uses results from the first phase to filter accordingly to the proximity of 150, 100, and 50 m and duration between 1–5 h, 5–10 h, and more than 10 h. After that, the pairing ships will be marked for risk of fuel siphoning potential level accordingly. Lastly, all information filtered and calculated will be written concisely as results of proximity less than 150 m, 100 m, and 50 m, respectively.

4.2 Filtered Result

This study uses 50 raw AIS data for logic framework filtering. The list excludes ships such as tug, motor hopper, and towing vessel as those are small support vessel servicing adjacent to port with no necessity for fuel siphoning. Table 3 shows the number of ships for each ship type, and Table 4 is the filtered result at second filter.

Result in Table 4 shows that total of 44 cases are detected in proximity less than 50 m with 30 cases between 1–5 h, 6 cases are between 5–10 h, and 8 cases are more than 10 h. And total of 73 and 137 cases are detected in proximity less than 100 m and 150 m.

From the results, the cases each level of proximity and duration were computed to deduce a finalized potential sequence as shown in Fig. 5. For example, close proximity less than 50 m is classified as level 3, and it is fall under level 3 for duration (more than 10 h). Thus, level 3 of proximity plus level 3 of duration equals level 6 of potential siphoning and it can express in Eq. (1).

$$\text{PL} = \text{Level of Proximity} + \text{Level for Duration} \tag{1}$$

where PL is potential level ($2 <= \text{PL} <= 6$).

Higher PL represents higher possibility for siphoning to occur, lower PL means unlikely to occur. The potential level follows a normal distribution graph's pattern from level 2 to 6 as shown in Fig. 4. However, notice that the number of detected

Table 3 Sample number of ship type

Ship type	Number of ship
Bunkering tanker	17
LPG tanker	3
Oil product tanker	25
Oil chemical tanker	1
Tanker	2
Water tanker	2
Total	50

Table 4 Overall filtered result

Proximity	$d < 50$ m (Level 3)	$d < 100$ m (Level 2)	$d < 150$ m (Level 1)
Level 1 (1–5 h)	30 (68.18%)	26 (35.62%)	52 (37.96%)
Level 2 (5–10 h)	6 (13.64%)	12 (16.44%)	21 (15.32%)
Level 3 (more than 10 h)	8 (18.18%)	35 (47.94%)	64 (46.72%)
Total cases	44	73	137

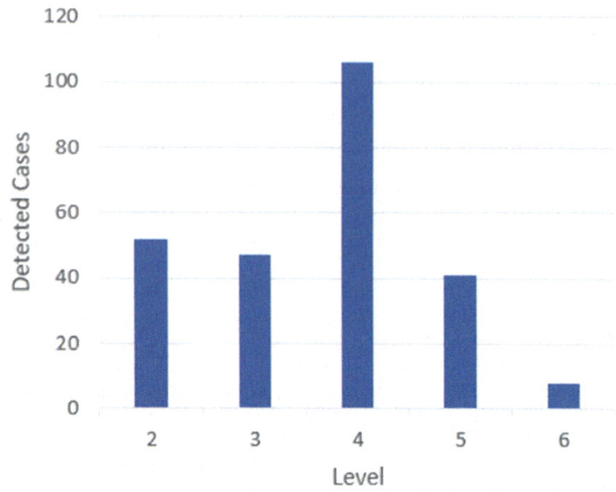

Fig. 5 Graph of detected cases against level

cases in level 2 is out of the normal distribution graph's pattern, which may stem from ships at anchorage or bunkering at sea.

4.3 Robustness of Siphoning Detection

A few parameters may be added into the framework to improve the overall detection accuracy. Filtering by speed of the vessel may be included into the framework for the reason that STS fuel siphoning required the vessels to stay stationary or sailing at a maximum speed of 5 knots to ensure the vessels line up parallelly during the siphoning process. Detection by draft line of the vessel can be also included into the framework to indirectly monitor the fuel in the vessel's given the load of the vessel is constant. However, the frequency of draft data of transmission does not sync with the normal AIS data where it is transmitted in every 6 min in normal situation, and it is subjected to vessel's crew [16]. Addition to that, the vessels distance between the ports can be a crucial parameter to consider. The AIS receiver might not receive signal that is out of receiving range of the receiver. This will require collection of AIS data far off the coast in which distances limitation can be eliminated.

There is also another uncertainty that the vessels involved in illegal activities turning off their AIS transmitter during the process. This is a 'practice' for traders or cross-border fishers to prevent getting caught by authorities who monitors by AIS information [17].

4.4 Validation Problem

The detecting systems that detect the potential STS fuel siphoning based on data have no ways to validate whether fuel siphoning really occurs without physically sending marine police or any observable unit from the sky. It is difficult to validate the occurrence of illegal siphoning without really checking it is almost impossible. Borrowing the experiment 'Schrödinger's cat', where it describes a cat that is inside a box with flask of poison can be simultaneously alive or dead until someone check on it.

What can a researcher do is to triangulate AIS data with conventional marine radar either X-band or S-band to combat the suspicion of mismatch in ship locality. X-band radar operates with frequency at 9.41 GHz [18], and Y-band radar operates at 2.94 GHz [19]. X-band radar is widely used for clearer vision while S-band radar uses in heavy weather condition or long-range detection. This may enhance the validity to predict the likelihood of illegal fuel siphoning.

Also, without custom information on whether the time-proximity detected ship pair has maritime clearance for STS, the model loops all the potential illegal ship-to-ship count without discrimination. Extension data from custom which are confidential are needed.

5 Conclusion and Future Work

This research has proposed an early concept to detect potential STS fuel siphoning using the self-collected AIS data. The framework can be further improved on as AIS data (big data) stacked on. The results suggested occurrence of illegal STS fuel siphoning at different risk level.

The current situation of the maritime world is unable to detect most of the illegal STS fuel siphoning in real time, except a direct in-situ probing on the bunkering location. Thus, this study is a conceptual framework created to launch into enhancing detection of potential illegal fuel siphoning which heaps losses on all maritime sector stakeholders.

Artificial intelligence, parameter tuning, model validation, and extensive collaboration will be needed to be done to further developed a robust STS anti-siphoning model.

Artificial intelligence such as machine learning can be applied into the siphoning detecting system. Referring other research proposing a ship detection model using convolutional neural network (CNN) algorithm for machine learning with the input by synthetic aperture radar (SAR) and AIS data to develop a robust ship detection model [20].

More parameters can be introduced such as speed over ground and draft line. Other minor improvement for better sensitivity will be to apply fuzzy logic to the existing parameter. More studies will be needed to assign the appropriate weightage on each parameter by priority or significance to model more realistically siphoning scenario that is occurring in each region. Factors that effecting the fuzzy system may include weather, wing speed, location, ship types (size).

The model may also need to be tested in other region where illegal siphoning activity is popular such as Nigeria and North Korea. It will depend on the accessibility of data.

Extensive collaboration with local authority for subject matter input or even database sharing may enhance the modeling algorithm. Given these stakeholders are responsible for the security of sea activities, they are not limited to ministry of transport, but also, maritime agencies, maritime custom, and royal navy. (Malaysia: Jabatan Kementerian Pengangkutan, MIMA, Kastam Diraja Malaysia & Tentara Laut Diraja Malaysia). To each, they uphold the maritime safety and security; giving advice and consultancy through research, training, education programs; regulate financial and maritime laws; protect the sovereignty and maritime borders of the country.

Acknowledgements The completion of this research work could not have been possible without the support of UCSI University Grant (REIG-FETBE-2020/011).

References

1. International Chamber of Shipping (2020) ICS Shipping and World Trade. https://www.ics-shipping.org/shipping-facts/shipping-and-world-trade. Retrieved 1 June 2021
2. Aqrawi-Whitcomb P, Bazilian M, Widdershoven C (2018) The world's most dangerous black markets. YaleGlobal Online. https://yaleglobal.yale.edu/content/worlds-most-dangerous-black-markets. Retrieved 7 Febr 2021
3. United Nations Security Council. (2018). S/2019/171 - E - S/2019/171. Undocs.org. Retrieved 2 June 2020, from https://www.undocs.org/S/2019/171.
4. National Academy of Sciences (NAS) (1998) Committee on oil spill risks from tank vessel lightering, Marine Board, National Academy Press, Washington DC
5. European Commission (EC) (2009) Technical support for European action to reducing GHG emissions, Annex I, Ship-to-Ship transfer. DG ENV.C3/ATA/2008/0016, Delf
6. Companies International Marine Forum (OCIMF) (2013) Ship to ship transfer guide for petroleum, chemicals, and liquefied gases, 2013
7. Stavrou D, Ventikos N (2014) Ship to ship transfer of Cargo operations: risk assessment applying a fuzzy inference system. J Risk Anal Crisis Response 4(4):214. https://doi.org/10.2991/jrarc.2014.4.4.3
8. IMO Resolution MEPC. 186(59). https://www.mavroudis.com.cy/media/cms/stsoperation_594A6B428D834.pdf. Retrieved 21 June 2021
9. Witkowska A, Smierzchalski R, Wilczynski P (2017) Approach manoeuvre during emergency Ship-to-Ship transfer operation with oil spill. Transnav. Int J Mar Navig Saf Sea Transp 11(2):145–151. https://doi.org/10.12716/1001.11.02.17
10. IMO (2009) Theta marine STS operations plan. Issuu. https://issuu.com/capt.ioannistheodorakas/docs/theta_marine_-_sts_operations_plan. Retrieved 3 June 2021
11. Segar Abdullah M (2000) VTS [vessel traffic services] in the Singapore strait: an investigation into mandatory traffic control
12. IMO (2013) Ship to ship transfer plan. Issuu. https://issuu.com/dynamarine/docs/sample_sts_plan. Retrieved 7 June 2021

13. Romsom E (2022) Global oil theft: impact and policy responses (No. wp-2022-16). World Institute for Development Economic Research (UNU-WIDER)
14. Nurul Ain Hussein I (2018) Southern region Marine department detains oil tanker ship in illegal fuel transfer. https://www.nst.com.my/news/nation/2018/10/419859/southern-region-marine-dept-detains-oil-tanker-ship-illegal-fuel-transfer. Retrieved 9 Febr 2021
15. Zwinko C (2019) Sanctioned North Korean oil tanker spotted conducting relay ship-to-ship transfers. NK News. https://www.nknews.org/2019/06/sanctioned-north-korean-oil-tanker-spotted-conducting-relay-ship-to-ship-transfers/. Retrieved 11 Febr 2021
16. Nakano T, Hasegawa K (2012) An attempt to predict manoeuvring indices using AIS data for automatic OD data acquisition. IFAC Proc Volumes 45(27):1–6. https://doi.org/10.3182/20120919-3-it-2046.00001
17. Mackenzie A (2017) Blackout: turning off vessel tracking systems. Marsh.com. https://www.marsh.com/uk/insights/risk-in-context/blackout-turning-off-vessel-tracking-systems.html. Retrieved 30 July 2021
18. Chen X, Huang W, Zhao C, Tian Y (2020) Rain detection from X-band marine radar images: a support vector machine-based approach. IEEE Trans Geosci Remote Sens 58(3):2115–2123. https://doi.org/10.1109/tgrs.2019.2953143
19. Rocadenbosch F, Barragan R, Frasier S, Waldinger J, Turner D, Tanamachi R, Dawson D (2020) Ceilometer-based rain-rate estimation: a case-study comparison with S-band radar and disdrometer retrievals in the context of VORTEX-SE. IEEE Trans Geosci Remote Sens 58(12):8268–8284. https://doi.org/10.1109/tgrs.2020.2984458
20. Song J, Kim D, Kang K (2020) Automated procurement of training data for machine learning algorithm on ship detection using AIS information. Remote Sens 12(9):1443. https://doi.org/10.3390/rs12091443

Analysing the Primary Influential Factors in Cement Manufacturing Industry Using DEMATEL Methodology

S. Bathrinath, Sai M. Nagesh, V. Dinesh, M. Sri Ram Ganesh, K. Koppiahraj, and R. K. A. Bhalaji

Abstract Cement occupies a prominent position in the construction industry. It ensures the strength and quality of the building constructed. Though the cement manufacturing industry (CMI) receives a lot of criticism from an environmental perspective, cement dependence is irreplaceable. Also, the demand for cement is increasing exponentially. Here, this paper aims to analyse the primary factor influencing the CMI regarding quality and quantity. In this regard, eighteen production factors are collected from a site survey in one of the leading southern India cement industries and a literature survey. Then, the decision-making trial and evaluation laboratory (DEMATEL) method, an MCDM method, is used to evaluate the factors. Using the DEMATEL method, the interrelationship amongst the factors is explored. Outcomes indicate that capital investment, quality control, and machinery and equipment maintenance are the influential primary factors in cement production. The result of this study may help industrial management enhance both the quality and quantity of cement.

Keywords CMG · DEMATEL · MCDM

1 Introduction

Cement is a fine powder material made by silicates of calcium, iron oxide, aluminium oxide, silica and calcium oxide [1]. India has a production capacity of 234 million tons per year, making it second place for cement manufacturing globally [2]. In concrete, cement is a key binding particle and it is the vital one for satisfying the requirements of modern infrastructure and global housing [3]. In 2006, the world

S. Bathrinath (✉) · S. M. Nagesh · V. Dinesh · M. Sri Ram Ganesh · K. Koppiahraj
Department of Mechanical Engineering, Kalasalingam Academy of Research and Education, Krishnankovil, Tamil Nadu 626126, India
e-mail: bathri@gmail.com

R. K. A. Bhalaji
Department of Mechanical Engineering, Francis Xavier Engineering College, Tirunelveli, Tamil Nadu, India

E. Natarajan et al. (eds.), *Materials, Design and Manufacturing for Sustainable Environment*, Lecture Notes in Mechanical Engineering,
https://doi.org/10.1007/978-981-19-3053-9_36

cement production was around 3.60 BT/3.30 BT (clinker) [4]. Therefore, a lot of factors are affecting cement production in the manufacturing industry. Our research factors are taken from survey methodology. Through that methodology, we have verified that the desired factors affect the production of cement.

Profit earns by the cement manufacturing industry (CMI) and the process of CMI is explained in pollution aspects [5]. ANP methodology is explained, and cement industry factor analysis is explained. Implementation of new technology helps the industry earn more profit for the long term [6]. Green gas supply chain management system's major affected factor was found using the DEMATEL methodology. DEMATEL methodology helps to identify the key factors and interrelation between the factors [7]. DEMATEL methodology is helped to identify the key factors in hospital service and find interrelation between the factors. We learned how to implement the survey results in the methodology [8]. For the past 10 years, the MCDM method has been used in different applications. Nowadays, usage of MCDM was also increased [5]. CMI has played a significant role in all phases of the cyclical growth process. India ranks the second place for the production of cement when considering the developed countries. Regarding the parameters such as financial, production and capacity, the cement industry in India is the dynamic one [9]. Below referred paper focuses on the profitability analysis of ACC Ltd, Ambuja Cements and UltraTech Cement companies in India based on operating profit ratio, PBIT ratio, gross profit ratio, cash profit ratio and net profit ratio for the period 2004–05 to 2018–19 [10]. In Battelle Memorial Institute, the DEMATEL method was originated in Geneva Research Centre. This method is used to find the configuration of complicated causal relationships through matrixes or digraphs [11]. In terms of complicated problems, there is necessary to evaluate the causal inter-relationship between the factors with graphical representation [12]. MCDM methodology was examined and explained [13].

We learn about the CMI and the DEMATEL methodology from the literature review. Some of the papers explain how to use the DEMATEL methodology for our problem and how to determine the results from the output of our methodology. Table 1 shows factors confirmed by the survey reviewer.

This research used multi-criteria decision-making (MCDM) for analysis. It is used in various domains to analyse the barriers and factors that cause its development and defects in companies or operations. MCDM has many methods as follows DEMATEL, ISM, TOPSIS, ANP, etc. Here, we used DEMATEL for analysis. Decision-making trial and evaluation laboratory (DEMATEL) is accurate and effective for analysing the prioritized factor [14].

Table 1 Factors confirmed by survey reviewer

S. No.	Factors	Notation
1	Capital investment	A
2	Location of Industry	B
3	Legalities and licenses	C
4	Material management	D
5	Quality control	E
6	Inventory management	F
7	Quality of labour	G
8	Recycle management	H
9	Data management	I
10	Supply chain management	J
11	Research and development	K
12	Pollution control	L
13	Finance and accounts department	M
14	Logistics	N
15	Health safety and environment maintenance	O
16	Customer care service	P
17	Marketing department	Q
18	Machinery and equipment maintenance	R

2 Methodology

2.1 *Site Survey*

Site survey has taken in one of the leading cement industries in southern India. Direct relationship matrix was measured with the help of officials in industry.

2.2 *DEMATEL Method*

The DEMATEL method is used to find the inter-relationships between factors with the help of a causal diagram [15]. The causal diagram clearly shows which factors are coming under the cause and effect group. It can be used in many applications such as waste management, green manufacturing, sustainable supply chain, environmental health and safety. [16–18].

Step 1: Eq. (1) is used to determine the (A) via site survey. It is calculated by rate of influence between the factors. We rated the relations between the influences based on linguistic scale (Table 2).

Table 2 Linguistic scale

Description	Scale
Very high influence	4
High influence	3
Low influence	2
Very low influence	1
No influence	0

The x_{ij} shows that the degree to which industrial experts considers i factor influences factor j. In the matrix, all the elements are set to zero in diagonal form for the condition $i = j$. For every expert, a non-negative matrix $(n \times n)$ can be fixed as $x^k = [x^k_{ij}]$, where k is the no. of respondents participated for rating the factors with $1 \leq k \leq H$ and n is the factors identified. From the experts (H) matrices are identified such as $x^1, x^2, x^3...., x^h$. Table 3 shows the direct relationship matrix. The average matrix $A = [a_{ij}]$ can be structured with the help of opinions from industry experts (H) as detailed below:

$$A = [a_{ij}] = \frac{1}{H}\sum\nolimits_{k=1}^{H} x^k_{ij} \tag{1}$$

Step 2: Eq. (2) is used to find the Matrix (D)

$$D = m \times A \tag{2}$$

where,

$$m = \min\left[\frac{1}{\max\sum\nolimits_{j-1}^{n} a_{ij}}, \frac{1}{\max\sum\nolimits_{i-1}^{n} a_{ij}}\right]$$

In matrix D falls, each element varies between 0 and 1. Table 4 shows the normalized direct relationship matrix.

Step 3: Eq. (3) is used to compute the total relation matrix in which I is an identity matrix $(n \times n)$. The t_{ij} element depicts indirectly influences i factor on j factor, and the matrix T shows the relationship amongst each and every factor.

$$T = D(I - D)^{-1} \tag{3}$$

Step 4: Compute the summation of rows and columns.

Here, r and c indicating the summation of rows and columns, respectively, which are obtained by

$$r = \left\{\sum_{j-1}^{n} t_{ij}\right\}_{n\times 1} \tag{4}$$

Table 3 Direct relationship matrix (*D*)

	A	B	C	D	E	F	G	H	I	J	K	L	M	N	O	P	Q	R
A	0	2	1	2	2	4	4	3	2	3	2	2	3	4	3	1	1	4
B	3	0	0	0	1	3	3	4	1	3	1	2	0	3	3	0	2	2
C	2	1	0	2	2	2	1	4	1	1	1	4	3	1	4	2	2	2
D	4	1	0	0	1	4	2	1	4	1	2	1	2	2	3	2	1	4
E	2	0	0	1	0	1	4	1	1	3	3	1	1	2	2	3	3	4
F	4	3	1	2	1	0	1	3	0	2	1	3	2	2	1	1	3	4
G	1	1	2	1	2	1	0	1	0	2	3	1	2	1	4	1	4	3
H	3	3	4	3	2	3	1	0	1	1	3	4	2	2	3	2	2	3
I	3	3	1	2	0	3	1	2	0	1	4	1	3	1	2	4	4	1
J	2	2	2	2	3	3	2	1	1	0	1	3	1	4	1	2	3	4
K	4	3	2	4	1	1	4	1	4	2	0	2	3	1	2	1	4	4
L	1	1	4	1	3	2	1	4	1	3	2	0	3	1	2	1	1	2
M	1	1	3	3	1	2	2	2	4	2	3	2	0	2	2	1	4	2
N	3	2	1	2	1	2	2	2	2	4	1	1	2	0	0	0	3	4
O	2	3	4	2	1	2	3	1	1	3	1	3	3	3	0	2	3	4
P	4	1	1	1	1	1	4	0	4	2	2	1	1	2	2	0	4	3
Q	1	2	1	2	3	3	4	2	3	4	3	1	4	4	1	3	0	2
R	3	4	2	3	4	2	4	3	2	4	4	3	3	4	3	3	3	0

$$c = \left\{ \sum_{i-1}^{n} t_{ij} \right\}_{1\times n} \tag{5}$$

Step 5: Eq. (6) is used to find the threshold value (α) using matrix T. This value is used to eradicate some slight problems in T matrix.

$$\alpha = \frac{\sum_{i=1}^{n} \sum_{j-1}^{n} [t_{ij}]}{N} \tag{6}$$

In Eq. (6), N is the total no. of factors in the T matrix. The causal diagram (Fig. 1) can be drawn with the help of $(r + c, r - c)$ values (Table 5).

Threshold Value: Threshold value for our calculation is 0.1355.

3 Result and Discussions

From values, we found that capital investment (A), quality control (E), recycle management (H) and machinery and equipment (R) maintenance are the major cause of factory production. $R + C$ is a value of prominence amongst the factors.

Table 4 Normalized initial direct-relation matrix (T)

	A	B	C	D	E	F	G	H	I	J	K	L	M	N	O	P	Q	R
A	0.12	0.14	0.11	0.13	0.17	0.18	0.18	0.16	0.12	0.17	0.14	0.14	0.17	0.19	0.16	0.10	0.16	0.22
B	0.14	0.08	0.07	0.07	0.12	0.14	0.14	0.15	0.08	0.14	0.10	0.12	0.09	0.14	0.13	0.06	0.13	0.15
C	0.13	0.11	0.08	0.11	0.15	0.13	0.11	0.16	0.09	0.11	0.10	0.14	0.15	0.12	0.16	0.11	0.14	0.15
D	0.16	0.10	0.07	0.07	0.13	0.15	0.11	0.10	0.08	0.11	0.12	0.10	0.12	0.13	0.13	0.10	0.12	0.18
E	0.17	0.20	0.17	0.14	0.15	0.17	0.19	0.20	0.16	0.20	0.18	0.20	0.21	0.22	0.19	0.17	0.24	0.25
F	0.14	0.14	0.09	0.11	0.13	0.09	0.11	0.14	0.07	0.13	0.10	0.14	0.13	0.13	0.10	0.09	0.16	0.18
G	0.10	0.10	0.11	0.09	0.14	0.10	0.09	0.10	0.07	0.12	0.13	0.10	0.12	0.11	0.15	0.08	0.17	0.16
H	0.16	0.15	0.16	0.14	0.15	0.16	0.13	0.11	0.10	0.13	0.15	0.17	0.14	0.15	0.16	0.12	0.16	0.19
I	0.15	0.14	0.09	0.11	0.09	0.14	0.11	0.12	0.07	0.11	0.15	0.10	0.14	0.11	0.12	0.14	0.18	0.14
J	0.13	0.13	0.11	0.11	0.16	0.15	0.13	0.11	0.09	0.10	0.11	0.14	0.12	0.17	0.11	0.11	0.17	0.19
K	0.18	0.16	0.12	0.16	0.13	0.13	0.18	0.12	0.15	0.15	0.11	0.14	0.16	0.14	0.14	0.10	0.20	0.21
L	0.11	0.10	0.14	0.09	0.14	0.12	0.10	0.15	0.08	0.14	0.12	0.09	0.14	0.11	0.12	0.09	0.12	0.15
M	0.12	0.12	0.14	0.14	0.18	0.14	0.14	0.14	0.15	0.14	0.15	0.13	0.11	0.14	0.14	0.10	0.19	0.17
N	0.14	0.12	0.09	0.11	0.15	0.13	0.13	0.12	0.10	0.16	0.10	0.10	0.13	0.10	0.09	0.07	0.16	0.19
O	0.14	0.15	0.16	0.12	0.15	0.14	0.16	0.12	0.10	0.16	0.12	0.15	0.16	0.17	0.10	0.12	0.18	0.21
P	0.13	0.10	0.08	0.09	0.10	0.10	0.14	0.08	0.12	0.12	0.11	0.09	0.10	0.12	0.11	0.06	0.17	0.16
Q	0.12	0.13	0.10	0.12	0.16	0.15	0.15	0.13	0.13	0.16	0.15	0.11	0.15	0.18	0.11	0.13	0.13	0.17
R	0.19	0.20	0.15	0.17	0.21	0.17	0.21	0.18	0.14	0.21	0.20	0.18	0.19	0.21	0.18	0.15	0.22	0.18

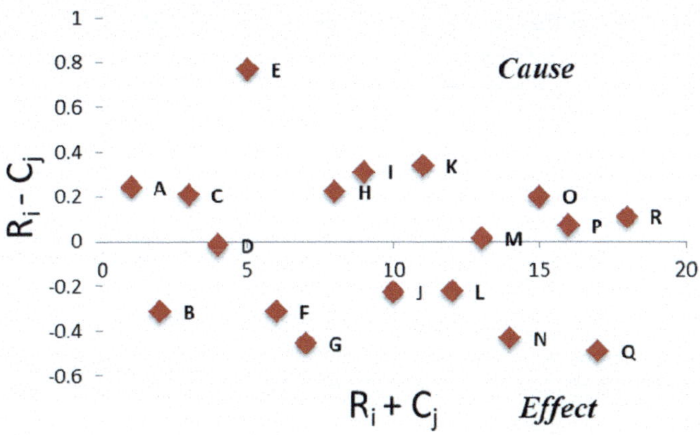

Fig. 1 Causal diagram

Table 5 $(R_i + C_j)$, $(R_i - C_j)$ matrix

S. No	$(R_i + C_j)$	$(R_i - C_j)$	Identity
Capital investment	5.31	0.24	Cause
Location of Industry	4.41	**0.32**	Effect
Legalities and licenses	4.28	0.21	Cause
Material management	4.18	**0.02**	Effect
Quality control	6.02	0.77	Cause
Inventory management	4.69	**0.32**	Effect
Quality of labour	4.53	**0.46**	Effect
Recycle management	5.03	0.22	Cause
Data management	4.13	0.31	Cause
Supply chain management	4.91	**0.23**	Effect
Research and development	5.03	0.34	Cause
Pollution Control	4.47	**0.22**	Effect
Finance and Accounts department	5.05	0.01	Cause
Logistics	4.84	**0.43**	Effect
Health safety and Environment maintenance	5.03	0.20	Cause
Customer care service	3.87	0.07	Cause
Marketing Department	5.47	0.49	Effect
Machinery and Equipment maintenance	6.57	0.11	Cause

Bold indicate effect factors

$R - C$ value denoted that the factors are either cause or effect. Since establishment of cement industry requires huge capital investement, it is always a difficult task for the industrial management in establishment. There are many reasons for the need of huge capital invetsment. As the machinery used by the cement industry are of high cost, the industrial management always needs huge capital invetsment [2]. Also, the cement manufactured by the industries need to be checked for quality. To ensure the quality and grade, the cement industry is in a situation of establishing a separate quality control department. In establishing a quality control department, the cement industry are faced with many problems like recruiting suitable person for the maintaining quality of the cement [19]. Onc of the main challenge faced by the cement industry is managing the wastes generated. Hence, the cement industry has to adopt a suitable recycle management. Although the wastes from the cement industry are effectively used in manufacturing flyash bricks, still the menace by the cement industry is a concerned one [20]. One of the main challenge faced by the cement industry regarding waste management is the particulates emitted by the industries during the production stage. The minute particles emitted by the cement industry has huge impact of the health of the local society and the workers of the cement industry.

After discussing with industry officials in factory, we have determined our output. Capital investment (A), quality control (E) as well as machinery and equipment maintenance (R) are the most important factors in cement manufacturing industry. Even though capital investment is one of the major cause of the factory output, compared to quality control and machinery and equipment maintenance, it has less involvement with other factors while running the industry.

4 Conclusion and Implications of the Study

The cement manufacturing industry has a significant impact on our country's annual income. CMI has a lot of carbon reduction and pollution control management systems. So, we are focused on product development in that company. We suggested our output to the company official to increase their product quality. Using the DEMATEL methodology gives a lot of information about various domains. We can learn the interrelation between the departments. Research about various industries and factors gives how to enlarge our view to other domains.

This study offers some value contribution to the literature on theoretical and managerial basis. Regarding the theoretical contribution, this study sheds light on the possible ways of improving the production of cement in terms of both quality and quantity. The factors discussed and evaluated in this study will act as a guide for the cement manufacturing industry. Also, the study provides some managerial implications given below for the CMI management:

- Since the establishment of any industry requires huge capital investment, increasing the number of stakeholders may lower the financial burden. Financial assistance by the government will be very influential in establishing cement industry.
- As the cement industry is criticized for adverse environmental impact, it is advisable to locate the industry on the outskirt of the residential areas.
- Cement industries in India are still mainly depended on the outdated machinery. It is essential for the cement industry to use latest machines as it offers some sort of competitive edge like reduced production time and improved production capacity.

References

1. Mafalda Matos A, Sousa-Coutinho J (2022) Municipal solid waste incineration bottom ash recycling in concrete: preliminary approach with Oporto wastes. Constr Build Mater323:126548. https://doi.org/10.1016/j.conbuildmat.2022.126548
2. Ahmed M, Bashar I, Alam ST, Wasi AI, Jerin I, Khatun S, Rahman M (2021) An overview of Asian cement industry: environmental impacts, research methodologies and mitigation measures. Sustain Prod Consum 28:1018–1039. https://doi.org/10.1016/j.spc.2021.07.024
3. Biernacki JJ, Bullard JW, Sant G, Brown K, Glasser FP, Jones S, Ley T, Livingston R, Nicoleau L, Olek J et al (2017) Cements in the 21st century: challenges, perspectives, and opportunities. J Am Ceram Soc 100:2746–2773. https://doi.org/10.1111/jace.14948
4. Gao T, Shen L, Shen M, Liu L, Chen F (2016) Analysis of material flow and consumption in cement production process. J Clean Prod 112:553–565. https://doi.org/10.1016/j.jclepro.2015.08.054
5. Zeb K, Ali Y, Khan MW (2019) Factors influencing environment and human health by cement industry. Manage Environ Qual Int J 30:751–767. https://doi.org/10.1108/MEQ-06-2018-0112
6. Kurian MC, Shalij PR, Pramod VR (2019) Maintenance strategy selection in a cement industry using analytic network process. J Qual Maint Eng 26:509–525. https://doi.org/10.1108/JQME-07-2017-0048
7. Hsu C-W, Kuo T-C, Chen S-H, Hu AH (2013) Using DEMATEL to develop a carbon management model of supplier selection in green supply chain management. J Clean Prod 56:164–172. https://doi.org/10.1016/j.jclepro.2011.09.012
8. Shieh JI, Wu HH, Huang KK (2010) A DEMATEL method in identifying key success factors of hospital service quality. Knowl-Based Syst 23:277–282. https://doi.org/10.1016/j.knosys.2010.01.013
9. Koppiahraj K, Bathrinath S, Venkatesh VG, Saravanasankar S (2019) Examination of inventory management in leather garments using ISM technique. Int J Recent Technol Eng 8:184–189. https://doi.org/10.35940/ijrte.D1040.1284S219
10. Karuppiah K, Sankaranarayanan B, Subramaniam S (2021) Evaluation of the barriers in the adoption of automated technology by the manufacturing sector: a case from India, pp 421–429
11. Susanty A, Puspitasari NB, Susanto N, Renaldi S (2020) Using a DEMATEL method to prioritize the factors contributed to the performance dairy milk supply: a comparative analysis. In: Proceedings of the 2020 IEEE 7th international conference on industrial engineering and applications (ICIEA). IEEE, pp 436–440
12. Karuppiah K, Sankaranarayanan B, Ali SM (2020) A fuzzy ANP-DEMATEL model on faulty behavior risks: implications for improving safety at the workplace. Int J Occup Saf Ergon 1–42. https://doi.org/10.1080/10803548.2020.1847486

13. Kazimieras Zavadskas E, Antucheviciene J, Chatterjee P (2018) Multiple-criteria decision-making (MCDM) techniques for business processes information management. Information 10:4. https://doi.org/10.3390/info10010004
14. Velasquez M, Hester P (2013) An analysis of multi-criteria decision making methods. Int J Oper Res 10:56–66
15. Bhalaji RKA, Bathrinath S, Ponnambalam SG, Saravanasankar S (2021) Analyze the factors influencing human-robot interaction using MCDM method. Mater Today Proc 39:100–104. https://doi.org/10.1016/j.matpr.2020.06.316
16. Karuppiah K, Sankaranarayanan B, Ali SM (2020) A fuzzy ANP-DEMATEL model on faulty behavior risks: Implications for improving safety at the workplace. Int J Occup Saf Ergon. https://doi.org/10.1080/10803548.2020.1847486
17. Bathrinath S, Charan VS, Ponnambalam SG, Saravanasankar S (2019) Identification and evaluation of criteria of agile manufacturing using DEMATEL: a case from an Indian metal fabrication industry. J Mod Manuf Syst Technol 2:61–74
18. Qarnain SS, Muthuvel S, Bathrinath S (2020) Analyzing factors necessitating conservation of energy in residential buildings of Indian subcontinent: a DEMATEL approach. Mater Today Proc. https://doi.org/10.1016/j.matpr.2020.02.039
19. Koppiahraj K, Bathrinath S, Venkatesh VG, Mani V, Shi Y (2021) Optimal sustainability assessment method selection: a practitioner perspective. Ann Oper Res. https://doi.org/10.1007/s10479-021-03946-z
20. Koppiahraj K, Bathrinath S, Narayanasamy P, Balasundar P, Parrthipan BK, Senthil S, Ramkumar T (2021) A review on the factors influencing natural fiber composite materials. In: Hybrid natural fiber composites. Elsevier, pp 185–205

Effect of Heat Treatment on Mechanical Properties of Stir-Casted Al7075-SiC Composites

P. Raghuvaran, A. Ajith, R. Arjun, S. Deepak, and N. Godwin

Abstract Composites made of aluminium are the preferred choice for most of the researches working on metal matrix composites. Some of the benefits obtained on using aluminium matrix composites are high strength with minimal weight, good resistance to wear and good stiffness. In addition, materials made from aluminium also have better electrical conductivity and thermal expansion coefficient. Amongst various methods available for manufacturing composites, stir casting is the best effective method. The reason behind using this method is achieving homogeneous distribution of reinforcement within the matrix, cost economy and availability. The impact of heat treatment on resulting properties of the composite samples was analysed in this study. Al7075 is used as reinforcement, and silicon carbide at 4, 6 and 8 wt% is used to produce composite specimens. Stir casting method is used for the manufacturing because of enormous benefits obtained by this process. Care has been taken to achieve uniform dispersion of reinforcements in the melt. The specimens are structured to ASTM standards to evaluate their properties. T6 heat treatment is performed. Tensile, hardness and wear tests were performed, and the microstructure of the samples was analysed. The properties of the heat-treated specimens were found to be superior for composite sample having 6% of SiC than other samples.

Keywords Composite · Heat treatment · Stir casting · Mechanical properties

1 Introduction

Metal matrix composites (MMCs) have the constituents of metal at its base and reinforcements in the form of organic compounds or ceramics. The objective behind the selection being improved properties than the base metal [1]. Moreover, the resulting materials will get the advantages of both matrix and reinforcement [2]. Unique properties are achieved on addition of ceramic materials like Al_2O_3 [3, 4], B_4C [4], SiC

P. Raghuvaran (✉) · A. Ajith · R. Arjun · S. Deepak · N. Godwin
Department of Mechanical Engineering, Sri Krishna College of Engineering and Technology, Coimbatore, Tamil Nadu, India
e-mail: ragumechanical@gmail.com

E. Natarajan et al. (eds.), *Materials, Design and Manufacturing for Sustainable Environment*, Lecture Notes in Mechanical Engineering,
https://doi.org/10.1007/978-981-19-3053-9_37

[5], TiC [6], TiB_2 [7, 8] and TiO_2 [9] into the metallic matrices. Properties like strength, wear resistance and reliability are improved on using particle reinforcements in MMCs. The deficiencies of the parent materials overcome on addition of reinforcements. The performance of these materials enhances as in addition to the before said mechanical properties, physical properties such as coefficient of thermal expansion, and density also increases on comparison with base metal alone [10, 11]. An increasing trend is seen on the applications of engineering material advancements. Composites are one such materials to end the search of researchers [12]. The need for lightweight materials in the globe had shifted the use of monolithic materials to composites. Reinvestigation of earlier researches revealed that the scope of MMCs got wider because of the usage of discontinuous reinforcements. As a result, MMCs find their applications in sports, aerospace and automotive industries [11].

Aluminium matrix composites (AMCs) are a form of MMCs which constitute aluminium as matrix or base metal and any one of the ceramic materials told earlier. In case of hybrid MMCs, two or more ceramic materials are mixed along with the base metal. Stir casting [8], squeeze casting [13], powder metallurgy [3] and spray deposition [14] are some conventionally available methods for manufacture of AMCs. Stir casting is the best adopted method amongst them in a conventional foundry as it is readily adoptable, cost-efficient and gives high yield. In order to achieve uniform distribution of reinforcement, casting has a prerequisite to prepare slurry in a homogeneous manner [15].

It is evident from the previous studies that the process of heat treatment is capable of increasing the resulting properties of material to a certain level [16]. It was also concluded by Smagorinski et al. [17] that additional thermal and mechanical treatments result in improving the mechanical properties of the composites. Such composites have better elongation properties on employing them to heat treatment processes. Singh et al. [18] studied the influence of T4 and T6 heat treatment process on wear behaviour of aluminium composites made by stir casting method. WC is used as reinforcement for the study. The hardness of the composites was examined for both the methods, and maximum values of hardness at specific heat treatment process were identified. The hardness and wear properties of the two processes were compared, and found that T6 heat treatment process is optimal.

Chen et al. [19] in his study on A356-SiC composites compared the wear behaviour of specimens with and without heat treatment manufactured by stir casting method. Heat-treated composites exhibited better wear resistance than that of non-heat-treated ones. Similarly, Sharma et al. [20] investigated the wear behaviour of LM30 aluminium composite reinforced with sillimanite reinforcement. Wear behaviour was superior in T6 heat-treated specimens with lesser damage on surfaces being revealed by microstructural analysis.

Kiran et al. [21] studied the wear behaviour of hybrid aluminium alloy composite having silicon carbide and graphite reinforcements. Stir-casted specimens are subjected to thermal processes. Heat treatments are done on the specimens to reduce the residual stresses. This process also helped them to overcome micro-cracks on as cast composite samples. The study concluded that the process of heat treatment and

Table 1 Chemical composition of Al7075

Elements	Zn	Cu	Mn	Mg	Fe	Cr	Ti	Si	Al
Al 7075	5.4	1.42	0.12	2.42	0.42	0.21	0.11	0.13	Remaining

formation of ceramic mixed mechanical layer during pin on disc wear test reduced the volume loss.

The review of previous literatures gives an idea that heat treatment of composites will improve the tribological and wear behaviour. Hence, with the same motto, present study is undertaken on composites having Al7075 alloy as matrix and SiC as reinforcement. To confirm the improvement of properties, wear tensile and hardness tests were undertaken.

2 Materials and Methods

2.1 Matrix Material

Aluminium 7075 is used as matrix material for the present study. Composition of Al7075 is shown in Table 1. The properties of aluminium like corrosion resistance and better strength-to-weight ratio make aluminium to choose as a replacement for other conventionally available materials copper, iron and their alloys [22]. Less cost, low density and high quality are other benefits achieved on choosing aluminium as the reinforcement.

2.2 Reinforcement

Silicon carbide is used as reinforcement in the form of particles in present study. The particle size used is around 130 μm. Three different weight proportions at 4, 6 and 8 wt% of reinforcements are used for preparing Al-SiC composites. Increasing the reinforcement content beyond a certain limit results in agglomeration of particles. This may in turn prevents homogeneous distribution of reinforcement particles and acts as a barrier to improve the strength of the composite materials manufactured [23]. Properties of Al7075 and SiC are shown in Table 2 [24].

Table 2 Properties of Al7075 and SiC

Properties	Al7075	SiC
Elastic modulus (GPa)	70–80	410
Density (g/cc)	2.81	3.1
Poisson's ratio	0.33	0.14
Hardness (HB500)	60	2800
Tensile strength (T)/compressive strength (C) (MPa)	220 (T)	3900 (C)

2.3 T6 Heat Treatment

T6 heat treatment is done on a muffle furnace shown in Fig. 1, before evaluating the mechanical properties of the samples. Heat treatment consists of the following steps [25].

1. Solution treatment—540 °C, 2 h
2. Hot water quenching
3. Artificial ageing—170 °C, 7 h.

2.4 Experimental Setup

Liquid metallurgy technique is equipped for preparing composites. Bottom pouring type stir casting machine setup as shown in Fig. 2 is used for fabricating composite specimens. Test specimens were prepared at varying volume fractions of reinforcements as mentioned earlier. Al7075 is charged into the furnace of stir casting machine, and molten melt of aluminium alloy is obtained on heating the furnace. SiC powders preheated already is introduced into the vortex of the molten liquid. A ceramic-coated steel impeller is used for stirring the particles into the melt. Stirring is performed for 10 min at a speed of 400 rpm. At a temperature of 710 °C, the molten melt is poured into cast iron mould preheated already.

3 Results and Discussion

3.1 Hardness Tests

The values of hardness of the different samples were tested, and they are plotted in Fig. 3. Composites measuring 20 mm diameter and 15 mm length are prepared from casting to perform the hardness test. Brinell hardness tests were conducted with an applied load of 500 kg on Al 7075 with 0, 4, 6 and 8 wt% of reinforcements. An improvement in hardness is found on increasing the proportion of reinforcement with

Fig. 1 Muffle furnace

the matrix. The maximum hardness of 76 BHN was for 6% of SiC, and the hardness value slightly reduces on further addition of 8% SiC reinforcement.

3.2 Tensile Tests

Tensile tests were conducted on casted specimens after machining to ASTM E8 M-04 standard [26]. The dimensions of the specimens for test as per the standard are shown in Fig. 4. Te components are machined to the required dimensions, and tensile strength is evaluated. The tensile strength (TS) variations for various combinations of SiC reinforcements in Al7075 are shown in Fig. 5.

Fig. 2 Bottom pouring type stir casting machine

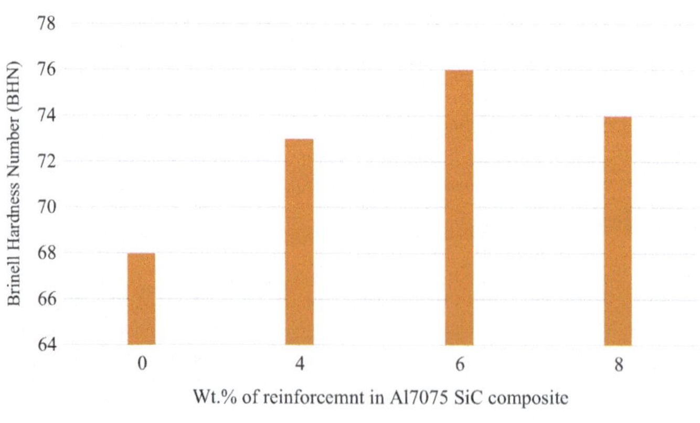

Fig. 3 Hardness of Al7075 with various SiC contents

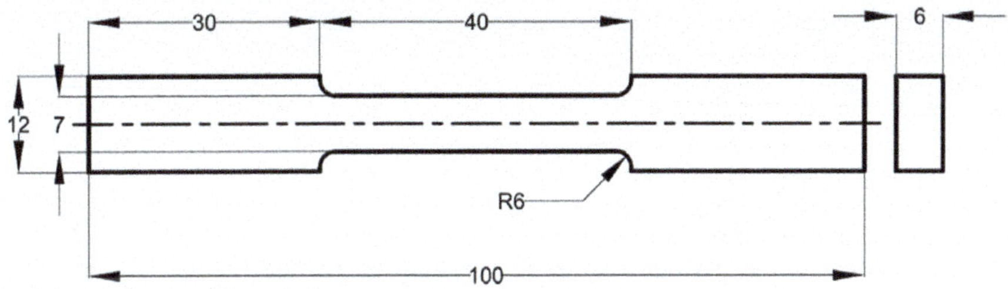

Fig. 4 Tensile specimen dimensions as per ASTM E8 M-04 standard

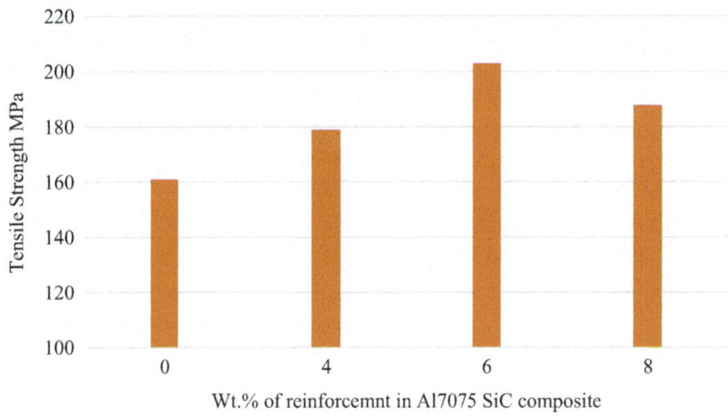

Fig. 5 Tensile strength variations for Al7075 with various SiC contents

An increase in TS value is experienced for increased content of reinforcement in the matrix. A maximum value of 203 MPa TS is achieved on Al7075 composite with 6 wt% of SiC reinforcement. Further, adding reinforcement to 8 wt% with matrix reduced TS.

3.3 *Wear Tests*

Wear tests were conducted on a computer assisted pin on disc tribometer. Tests were performed at room temperature with samples machined to ASTM G99 standards. The face of the specimens is made to contact against the rotating disc and wear loss occurred. Weight loss of specimens increased with increasing sliding distance as heat generated between the contact surfaces of the pin and rotating disc.

Wear loss for the samples was tested and is shown in Fig. 6. Wear loss is found to be minimum for composite having 6 wt% of SiC. Wear rate of the specimens was also found by dividing the weight loss with sliding distance. SiC particles carry the maximum load on uniform dispersion into the matrix, and it prevents the contact of metal to metal to a maximum extent [27]. As the ductility of composite with 8

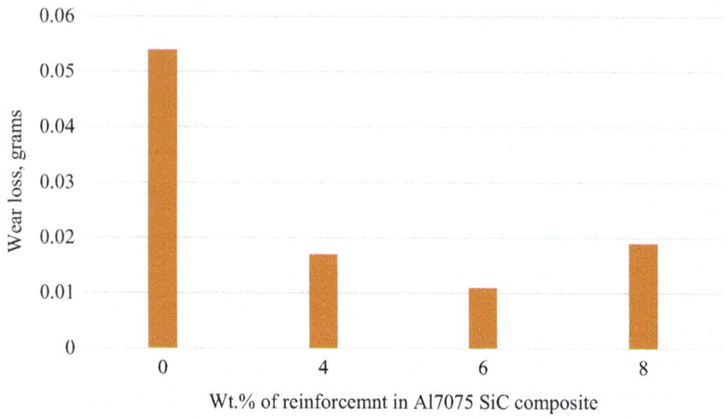

Fig. 6 Wear loss of various weight percentages of Al7075-SiC samples

wt% reinforcement increased, it resulted in having more wear loss amongst the three composite samples prepared. Wear resistance of the composites increased corresponding to increased hardness of the composites. The results are similar to those obtained by Coyal et al. [28].

3.4 *Microstructural Evaluation*

The microstructure of the Al7075-6%SiC is evaluated on an optical microscope before and after wear on Fig. 7a and b. The distribution of reinforcements within the matrix was seen on Fig. 7a, and wear patterns were seen on Fig. 7b. Homogeneous distribution of SiC particles within Al7075 matrix was seen on Fig. 7a. The figure also shows that there exists a better bonding between the interfaces of matrix and reinforcement. It is because of better stirring of reinforcements within the aluminium melt during casting process. Fine and continuous grooves in parallel to sliding direction along plastic deformations on the edges are found on the worn surface. The excellent properties of reinforcement present in the composite had attributed to the presence of fine grooves. Al7075-6%SiC experienced a lower pressure on the contact surfaces of pin and disc in addition to good resistance to ceramic particles. This in turn reduced the weight loss during wear tests. The presence of fine grooves in Fig. 7b is confirming that the presence of reinforcement played a major role in reducing the weight loss of the material subjected to wear test.

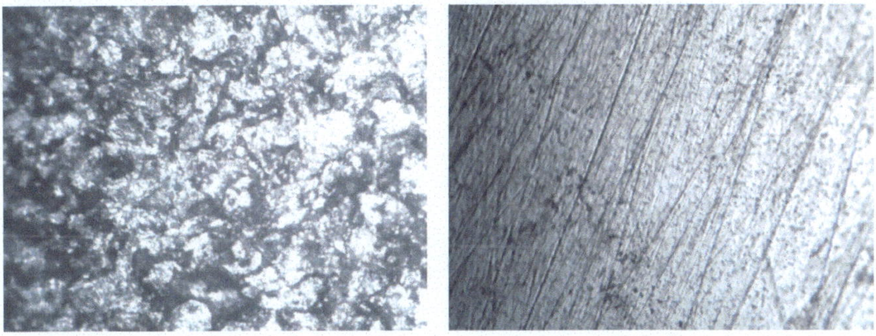

a b

Fig. 7 Optical microscopic images of Al7075-6%SiC composite, **a** Before wear and **b** after wear

4 Conclusions

After the hardness and tensile tests, the following points are concluded from this study:

1. Heat treating of composites increases the resulting properties than non-heat-treated ones.
2. Properties of composites like hardness and tensile strength increased and wear loss decreased on adding reinforcements up to 6%.
3. A maximum value of 76 BHN hardness, 203 MPa tensile strength and a minimum wear loss of 0.011 g was achieved.
4. Adding reinforcement to 8% weight reduced the resulting properties.

References

1. Sahu MK (2018) Fabrication of of aluminum aluminum matrix matrix composites composites by stir casting technique and stirring process parameters optimization. Adv Cast Tech 111–126
2. Amouri K, Kazemi Sh, Momeni A, Kazazi M (2016) Microstructure and mechanical properties of Al-nano/micro SiC composites produced by stir casting technique. Mater Sci Eng A 674:569–578
3. Fan G, Jiang Y, Tan Z, Guo Q, Xiong D-B, Yishi S, Lin R, Li H, Li Z, Zhang D (2018) Enhanced interfacial bonding and mechanical properties in CNT/Al composites fabricated by flake powder metallurgy. Carbon 130:333–339
4. Canakci A, Arslan F (2012) Abrasive wear behaviour of B4C particle reinforced Al2024 MMCs. Int J Adv Manuf Technol 63(5):785–795
5. Boobesh Nathan V, Soundararajan R, Brainard Abraham C, Vinoth E, Narayanan JK (2021) Study of mechanical and metallurgical characterization of correlated aluminium hybrid metal matrix composites. Mater Today Proc 45:1237–1242
6. Baskaran S, Anandakrishnan V, Duraiselvam M (2014) Investigations on dry sliding wear behavior of in situ casted AA7075–TiC metal matrix composites by using Taguchi technique. Mater Des 60:184–192

7. Chen F, Chen Z, Mao F, Wang T, Cao Z (2015) TiB_2 reinforced aluminum based in situ composites fabricated by stir casting. Mater Sci Eng A 625:357–368
8. Pazhouhanfar Y, Eghbali B (2018) Microstructural characterization and mechanical properties of TiB_2 reinforced Al6061 matrix composites produced using stir casting process. Mater Sci Eng, A 710:172–180
9. Shayan M, Eghbali B, Niroumand B (2020) Fabrication of AA2024−TiO_2 nanocomposites through stir casting process. Trans Nonferrous Metals Soc China 30(11):2891–2903
10. Tzamtzis S, Barekar NS, Hari Babu N, Patel J, Dhindaw BK, Fan Z (2009) Processing of advanced Al/SiC particulate metal matrix composites under intensive shearing–a novel Rheo-process. Compos Part A Appl Sci Manuf 40(2):144–151
11. Miracle DB (2005) Metal matrix composites–from science to technological significance. Compos Sci Technol 65(15–16):2526–2540
12. Bhandare RG, Sonawane PM (2013) Preparation of aluminium matrix composite by using stir casting method. Int J Eng Adv Technol (IJEAT) 3(3):61–65
13. Yuan D, Yang X, Wu S, Lü S, Hu K (2019) Development of high strength and toughness nano-SiCp/A356 composites with ultrasonic vibration and squeeze casting. J Mater Process Technol 269:1–9
14. Qiu X, Wang J-Q, Tang J-R, Gyansah L, Zhao Z-p, Xiong T-Y (2018) Microstructure, micro-hardness and tribological behavior of Al2O3 reinforced A380 aluminum alloy composite coatings prepared by cold spray technique. Surf Coat Technol 350:391–400
15. Su H, Gao W, Zhang H, Liu H, Lu J, Lu Z (2010) Optimization of stirring parameters through numerical simulation for the preparation of aluminum matrix composite by stir casting process. J Manuf Sci Eng 132(6)
16. Li RX, Li RD, Zhao YH, He LZ, Li CX, Guan HR, Hu ZQ (2004) Age-hardening behavior of cast Al–Si base alloy. Mater Lett 58(15):2096–2101
17. Smagorinski ME, Tsantrizos PG, Grenier S, Cavasin A, Brzezinski T, Kim G (1998) The properties and microstructure of Al-based composites reinforced with ceramic particles. Mater Sci Eng, A 244(1):86–90
18. Singh G, Sharma N (2021) Study on the influence of T4 and T6 heat treatment on the wear behavior of coarse and fine WC particulate reinforced LM28 Aluminium cast composites. Compos Part C: Open Access 4:100106
19. Chen R, Iwabuchi A, Shimizu T (2000) The effect of a T6 heat treatment on the fretting wear of a SiC particle-reinforced A356 aluminum alloy matrix composite. Wear 238(2):110–119
20. Sharma S, Nanda T, Pandey OP (2019) Investigation of T4 and T6 heat treatment on the wear properties of sillimanite reinforced LM30 aluminium alloy composites. Wear 426:27–36
21. Kiran TS, Prasanna Kumar M, Basavarajappa S, Viswanatha BM (2014) Dry sliding wear behavior of heat treated hybrid metal matrix composite using Taguchi techniques. Mater Des 63:294–304
22. Raghuvaran P, Baskaran J, Aagash C, Ganesh A, Gopi Krishna S (2021) Evaluation of mechanical properties of Al7075-SiC composites fabricated through stir casting technique. Mater Today Proc 45:1914–1918
23. Dhanasekaran S, Sunilraj S, Ramya G, Ravishankar S (2016) SiC and Al_2O_3 reinforced aluminum metal matrix composites for heavy vehicle clutch applications. Trans Indian Inst Metals 69(3):699–703
24. Veeresh Kumar GB, Rao CSP, Selvaraj N (2012) Mechanical and dry sliding wear behavior of Al7075 alloy-reinforced with SiC particles. J Compos Mater 46(10):1201–1209
25. Liu Z, Cheng N, Zheng Q, Wu J, Han Q, Huang Z, Xing J, Li Y, Gao Y (2018) Processing and tensile properties of A356 composites containing in situ small-sized Al3Ti particulates. Mater Sci Eng, A 710:392–399
26. Ashok Kumar B, Murugan N (2012) Metallurgical and mechanical characterization of stir cast AA6061-T6–AlNp composite. Mater Des 40:52–58

27. Rao RN, Das S (2010) Wear coefficient and reliability of sliding wear test procedure for high strength aluminium alloy and composite. Mater Des 31(7):3227–3233
28. Coyal A, Yuvaraj N, Butola R, Tyagi L (2020) An experimental analysis of tensile, hardness and wear properties of aluminium metal matrix composite through stir casting process. SN Appl Sci 2(5):1–10

Machinability Studies on Boron Carbide and Graphite Reinforced Al7029-Based Hybrid Composites

B. N. Sharath, S. Karthik, D. G. Pradeep, K. S. Madhu, and C. V. Venkatesh

Abstract Drilling characteristics in combination with machining factors such as thrust force, torque, and burr height, when drilling on Al7029+B_4C and Al7029+B_4C-Gr composites are conducted. The liquid metallurgical technique used in the production of composites. The results revealed that considerable influence on thrust force and the exit burr height was caused by the feed rate. The Al7029/10% B_4C-5% Gr hybrid composite displays reduced thrust force and burr height, while the Al7029/10% B_4C composite has the opposite effect. Graphite surface lubricating characteristics led to the decreased thrust force, torque, and burr height. Graphene-derived chips are discontinuous, allowing them to have applications that involve discontinuous chips. It has been observed that carbide tools are advantageous over the TiN-coated HSS drill bit for machining HMMCs.

Keywords Hybrid composites · Boron carbide · Graphite · Thrust force · Burr height

1 Introduction

It has been an interesting undertaking, with material science being heavily involved, in developing new lightweight and high-performance engineered materials like composites. It possesses resistance sometimes higher than that of ferrous materials.

B. N. Sharath (✉) · S. Karthik · D. G. Pradeep · K. S. Madhu · C. V. Venkatesh
Department of Mechanical Engineering, Malnad College of Engineering, VTU, Hassan, Belagavi, Karnataka, India
e-mail: sbn@mcehassan.ac.in

D. G. Pradeep
e-mail: dgp@mcehassan.ac.in

K. S. Madhu
e-mail: mks@mcehassan.ac.in

C. V. Venkatesh
e-mail: cvv@mcehassan.ac.in

E. Natarajan et al. (eds.), *Materials, Design and Manufacturing for Sustainable Environment*, Lecture Notes in Mechanical Engineering,
https://doi.org/10.1007/978-981-19-3053-9_38

This is especially important for the design of components and materials in the transportation sectors, where fuel efficiency is a key concern. Metal matrix composite has found increasing use in the engineering industry, such as aerospace, car, and marine manufacturing. These applications make use of its lightweight, high strength, stiffness, and resilience to high temperatures [1]. In addition to traditional straw-bale, non-load bearing components, modern fastened wing and, flexible wing aircraft utilise a wide range of components, such as structural integrity and dimensional stability, that previously only came into use in fixed wing aircraft. Due to the highly abrasive character of MMCs, many technical applications use MMCs as required components, although these applications remain severely hampered by their lack of machinability. Increased thrust force, and torque are all due to these consequences. However, technical advances that have been made over the last several years have called for a deeper and more systematic investigation of these composites and hybrid composites to better understand and satisfy their possible design and manufacturing applications. A wider range of machinability characteristics might be controlled while avoiding rough machining, increased tool wear, and increased cutting forces. Drilling has the potential to influence the mechanical characteristics of the workpiece via residual stresses that are extremely low around the hole opening and the fact that the surface of the hole is in an unstrained and deformed condition while new with a thick coating of heavily strained and distressed materials. Stress causes corrosion, hence, this raises the risk of corrosion on the strained area. The systematised and inexpensive machining is essential to meet the desirable dimensions and surface texture [2–4]. Basavarajappa et al. [5] performing the tests with HMMCs. In order to find out the impact of machining parameters, researchers performed tests utilising solid carbide multi-faceted drill bit ($D = 5$ mm). The findings demonstrate that feed rate influences the reliant variables, not speed. Rajmohan et al. [6], L18 orthogonal array was selected for the tests because the trials were performed on a CNC machine. The experiment findings demonstrate that feed rate and drill type are the most important variables in influencing drilling process performance. This technique may assist drilling processes by enhancing their effectiveness. Rana et al. [7], this article looks at the latest information on the way that Al-based composites are processed, microstructure, and engineered. As difficult as it is to regulate grain size and agglomeration during processing, the equally difficult challenge is to maintain a microstructure that is better than that obtained. There is further study that may help with better understanding how microstructures react to processing conditions. Chaudhary et al. [8] explore the impact of drilling parameters on aluminium 6070 alloy AMCs and HMMCs surface roughness, as well as MRR (mm^3/min) and total ruminant inventory (diversified by species) (micron). To investigate the distinct effects of each process parameter as well as their interactions, a system of multiple factorial analysis, known as ANOVA, was used. It may be concluded from the results of the ANOVA that the AMCs have better machinability than the HMMCs. Ramulu et al. [9], the drilling qualities were assessed based on how much force is used, how much wear is caused, how much formation is built up, and the overall quality of the hole. The accuracy of the drilled hole and the amount of drilling force produced were both improved when drilling with PCD drills compared to other drills. Sathia

Prathap et al. [10], AMMCs, or Al/SiC/B_4C, are being examined for machining. The use of MMCs is on the rise due to enhanced mechanical and tribological characteristics, making them better substitutes for traditional reinforced composites. The engine blocks and related component for aeroplanes are all manufactured from these many components. The hybrid MMCs machining, as well as CNC milling advancements, is covered in this article. The literature analysis clearly demonstrates that a significant amount of additional research is required to fully understand the thrust force, burr height, and chip production during composite drilling.

The purpose of this study is to look into the drilling characteristics of MMCs, with Aluminium (Al7029) as the base matrix alloy and B_4C and Gr as reinforcement, against this backdrop. Drilling aspects are studied with respect two different drill bits, namely TiN-coated HSS and carbide and by varying the machining parameters such as speed and feed as the input and thrust force (T_f), torque (T), and burr height (B_h) as the output.

2 Materials and Methods

2.1 Materials

Al7029 showing better properties both at room temperature and at higher temperatures, according to the literature. Al7029 is a high strength and age hardening alloy with a high concentration of zinc and magnesium. This composite matrix was picked as Al7029 and it is the superior one, because it combines exceptional properties such as higher strength, greater ductility, corrosion tolerance, and the ability to be heat treated. This component's excellent thermal conductivity enables it to efficiently dissipate heat. Boron carbide is an interesting reinforcing material due its stability. The most significant point to remember about B_4C is that it has lower density compared to silicon carbide and aluminium oxide. Al-boron carbide MMCs can be processed with low carbon (less time and less effort) casting techniques. In comparison with Al7029, the density of the chosen reinforcement (B_4C and Gr) was quite sparse and interfacial relationship also quite better [11, 12] (Table 1).

Table 1 Chemical compositions of Al7029

Si	Fe	Cu	Mn	Mg	Zn	Ti	V	Al
0.10	0.12	0.5–0.9	0.03	1.3–2.0	4.2–5.2	0.05	0.05	Remaining

2.2 Fabrication of Composites

The Al7029 ingot is put into a crucible, which melts the ingot until it reaches the liquid state. Then 1.5 wt% of magnesium is added to enhance the wet potential of the molten metal, and the process is repeated. It was necessary to warm the weighed B_4C reinforcement to a predetermined temperature of 450 °C to avoid the development of oxide layers surrounding the particles. It was necessary to incorporate primary reinforcement B_4C into the molten metal while maintaining consistent stirring with the mixer at 160 rpm. The preheated secondary reinforcement Gr was added to the mixture and well mixed for about 10–15 min. Hex-chloroethane (C_2Cl_6), a degassing agent, was added to the molten metal since it generates gas when heated. The mixed slury is poured into the mould at a temperature of 700 °C, which is the pouring temperature. The melt could then harden in the mould under the influence of air pressure, and the cast specimens could then be removed. According to Table 2, the test samples utilised in the current research are casted for their composition [13, 14].

Machining is performed on the casted samples to prepare the test sample which must meet the required dimensional accuracy of 140 mm × 70 mm × 10 mm. Polishing the test samples with emery sheets with grit sizes of 2000, 3000, and 4000 follows, followed by buffing.

3 Results and Discussions

The trials were carried out on Al7029+10% B_4C including Al7029+10% B_4C-Gr composite in combination with a carbide and TiN-coated HSS drill bit. Three different quantities determined through experimental study given in Table 3. In an ANOVA, many independent variables are examined for influence on the dependent variable.

Table 2 Sample composition

Sample	% Composition		
	Al7029	B_4C	Gr
Al7029+B_4C	90	10	–
Al7029+B_4C-Gr	85	10	5

Table 3 Drilling process parameters

Process parameter	Units	Levels		
		L1	L2	L3
Speed (V_c)	rpm	710	1120	1800
Feed rate (V_f)	mm/rev	0.03	0.05	0.08

3.1 Effect of Machining Parameters and Cutting Tool on Al7029+B_4C Sample and Al7029+10%B_4C–5% Gr

In several experiments (L9), varied cutting speed and feed rate were used. The thrust force (T_f), torque (T), and burr height (B_h) measurements were carried. The Al7029+B_4C composite material's data was measured and plotted in Fig. 1. Drilling mechanism depends on the reinforcing particles that are dispersed in the matrix, as well as the kind of drill and the various cutting circumstances. Due to the strength and rigidity of CNC machines, their thrust force and torque production exceeds that of conventional drilling machines in combination with greater feed rates. The removal of material from HMMCs is easier because of its lubricity. Work material, tool material, machining circumstances, and so on are all factors that influence the amount of cutting force generated.

It can be shown in Fig. 1 that the drilling performance depends on the input parameters. As the V_f is increased, both T_f and torque rises. Concerns have been raised over the fact that cutting speed has a substantial impact on the magnitude of drilling forces. Figures 1 and 2 illustrate the relationship between the amount of power applied to the drill bit and the resulting force and torque that are created while drilling with a HSS drill bit. Figure 3 shows that the V_f is the most significant component, with higher feed rates causing more T_f and torque. Presence of cutting speed has a notable impact on the degree of drilling forces.

The characteristics of the drill are defined by the ingredients of the reinforcement, the type of drill, the conditions under which it is employed, and other such parameters. Feed rate and cutting speed were both strongly connected to the force exerted when drilling, whereas T_f and T were reliant on feed rate, which results in poor surface finish. Determining how drilling works involves assessing the make-up of reinforcing particles in the matrix, the kind of drill employed, and drilling parameters including speed and feed. While V_s and V_f are equally critical to both cutting force (C_f), and T and V_f was the primary factor influencing the magnitude of drilling forces. It might

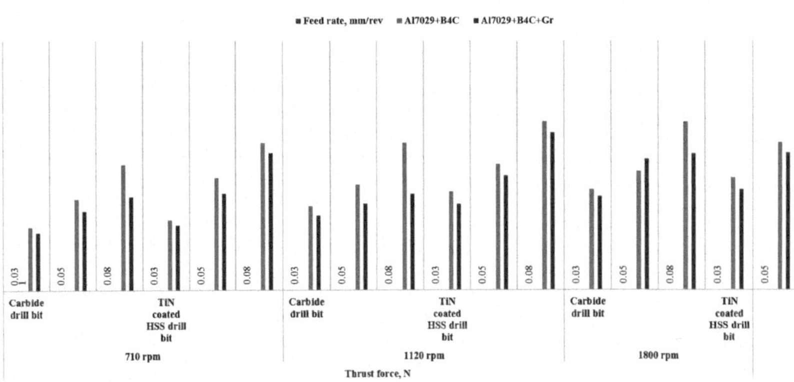

Fig. 1 Effect of speed (Vs) and feed rate on thrust force (T_f)

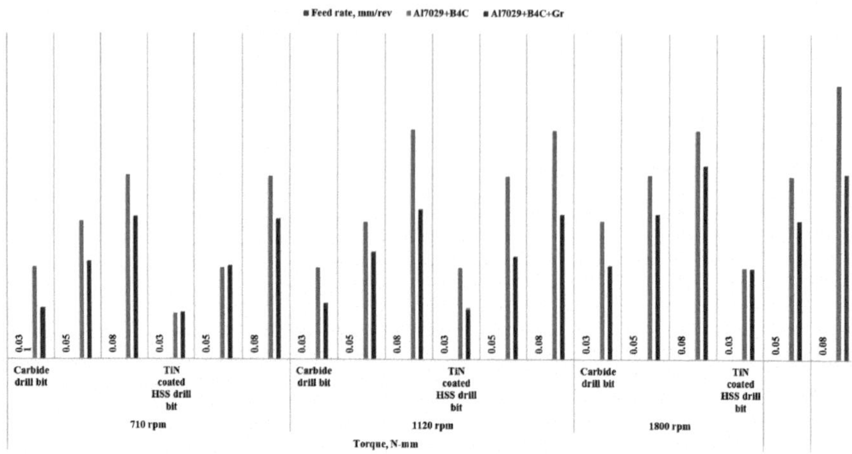

Fig. 2 Effect of speed (V_s) and feed rate (V_f) on torque

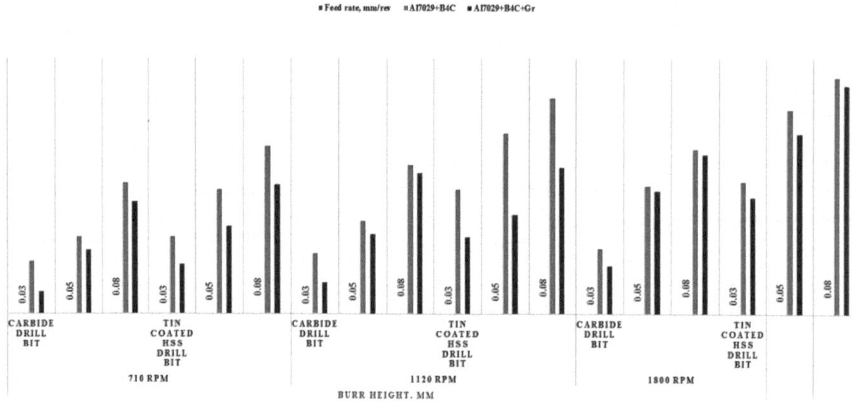

Fig. 3 Effect of speed and feed rate on burr height

result in a low-quality finish. Aluminium matrix is considerably softer than carbide; therefore, the primary obstacle to particle removal is mechanical.

To begin, the inherent abrasiveness of the reinforced particle would provide the main challenge for drilling the MMCs. In HSS drilling, it was found that Al7029 particle reinforced aluminium-based MMCs could not be cut. A large part of the difficulty in removing B_4C particles stems from the fact that aluminium matrix is softer than TiN-coated HSS. Reinforced particulate can be difficult to drill, as they are brittle and possess abrasive character. Drills using HSS could not cut Al7029 particle reinforced aluminium-based MMCs, according to the results. The major source of resistance is the mechanical removal of B_4C particles from the matrix material, which is considerably softer than TiN-coated HSS.

The influence of Vs and Vf on burr height is seen in Fig. 3. Feed rate and speed rise together, resulting in an increase in B_h. The B_h generated by the TiN-coated HSS drill bit was greater than that produced by the carbide drill bit.

Figure 3 describes the relationship between V_f and speed, which impacts the height of the burrs. The rise in burr height occurs while operating at a higher feed rate, whereas the opposite is true when operating at a higher cutting speed. Mostly because of the pushing force. In Figs. 1 and 2, it can be seen how changing the input parameters impacts thrust force and torque output because of employing a TiN-coated HSS drill bit for cutting. From the Fig. 1, the V_f influences cutting forces. Higher cutting pressures result from the HSS material's softer characteristics than those of carbide.

By going through the above statistics, V_f is the most important variable in increasing cutting forces and burr height; this is because with higher feed rates, there is an increase in the volume of chips eliminated for each revolution. Due to the side-to-side motion of the drill, the tool made significant wear on the surface and created a poor finish. With the slope become flatter, cutting forces here now function by way of friction rather than shear. This created a B_h on the exit side of the hole; thus, the exit side of the hole was not cleanly cut.

From the aforementioned information, it can be deduced that the cutting force and B_h are all proportional to the chip volume that is removed for each revolution. Poor surface finish was caused by drilling that rubbed the drill tip against the hole and generated rubbing wear. Friction, rather than shearing, caused the cutting force to be applied at this location, causing the curve to be more rounded. To produce a burr, the exit side of the hole was pushed out rather than cut neatly.

When utilising the HSS and the TiN-coated HSS drills, the surface quality is superior compared to that obtained with the hard-carbide tools. Material of the drilling instrument had a major role in the drilling forces. Drilling MMCs should not be attempted with HSS drills, since they generate more tool wear, resulting in lower-quality drilled holes and induce more drilling forces. Continuous chips were made using carbide and TiN-coated HSS drill bits that used microchip carbide particles to create their effect. Because of the damage to the sample, they are undesirable. Figure 1 demonstrates that for carbide drill bit, the maximum thrust force and torque of 186.39 N and 392.4 N-mm are produced. HSS drill bit recommends that the maximum burr height be 1.03 mm [15–19].

3.2 Optimal Combination Using ANOVA on Al7029+10%B_4C Sample

One of the main purposes of carrying ANOVA is to quantify the size of each factor's influence on the output and to calculate the error variance. A significance analysis for ANOVA results shows that the V_f and percentage of contribution are the significant aspects for material properties. 19.56% of the cutting speed is coming from

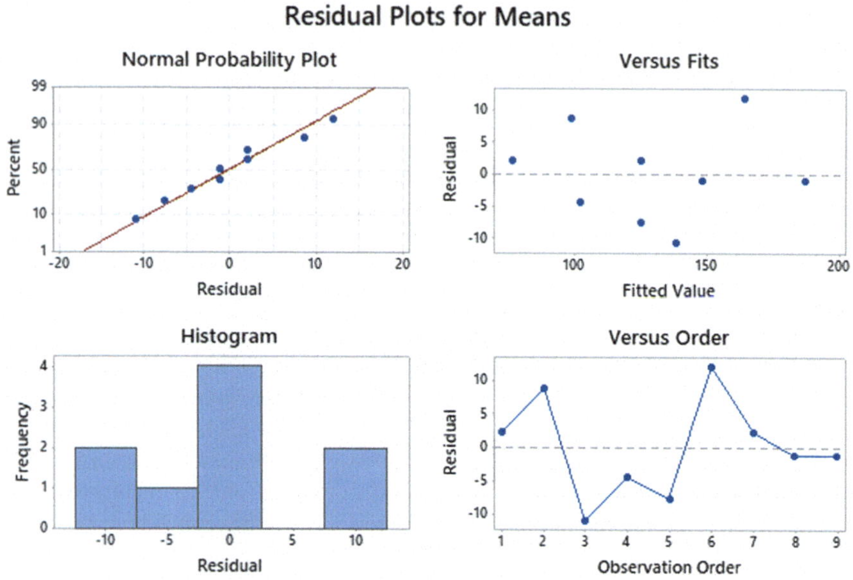

Fig. 4 Residual plots for means

contributions. According to this, when drill tools initially enter the workpiece, the thrust force will be lower. Feed rate plays a larger role than speed in generating responsiveness. 4.18% of the results are inaccurate. From Fig. 4, it can be observed that the experimental results agree with the model.

The feed rate is the most important of the parameters to understand from the main impacts graphic (Fig. 5). As the V_f increases, the T_f increases. The V_f is strongly tied to the degree of drilling forces, while the V_s had no impact on the force intensity. When feed at 0.08 mm/rev and rotating at 1800 rpm, the maximum thrust power is reached. For obtaining the maximum torque, the speed is 1800 rpm and the V_f is 0.08 mm/rev. Therefore, the cutting forces will likewise rise with increasing feed rates. Feed rate is hugely influential on burr height mostly because of the pushing force V_f of 0.08 mm/rev is higher when the burr height is set to 1800 rpm.

3.3 Optimal Combination Using ANOVA on Al7029+10%B_4C-5% Gr Sample

By observing the Fig. 6, it can be noticed that the component with the highest statistical and physical significance is the feed rate and the proportion of contribution is 62.45%. This means 35.49% of the contribution is devoted to lowering speed and the influence on this variable cannot be measured and with error of 2.04%. Increasing

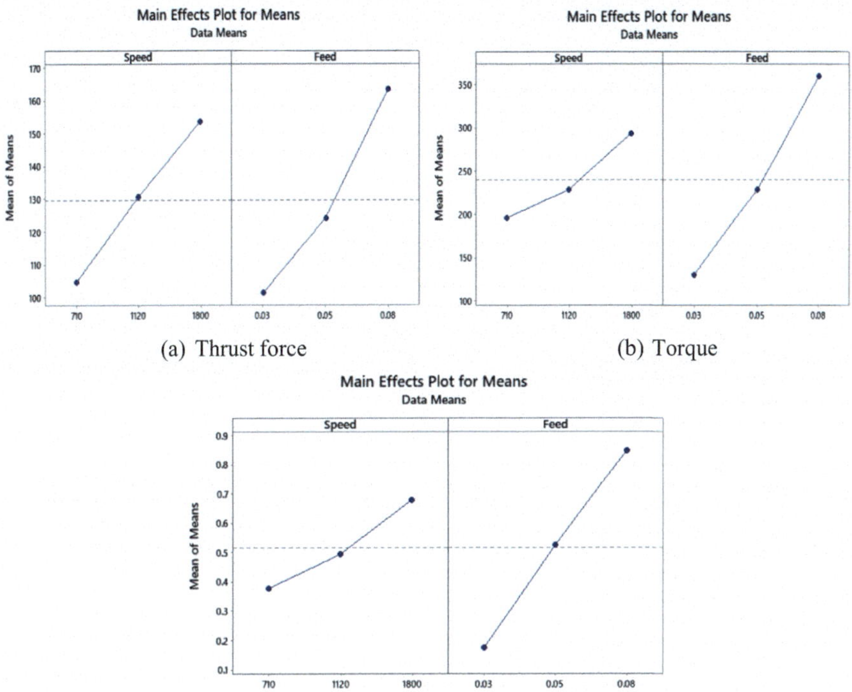

(a) Thrust force

(b) Torque

(c) Burr height

Fig. 5 Main effect plot for response variables for Al7029+10%B_4C sample

the V_f has a proportional impact on the responsive force; therefore, V_f has a greater influence on responsiveness than the speed.

Feed rate (V_f) is the crucial factor since it is shown in the main impact plot (Fig. 6). As the V_f increases and the T_f also increases. The overall cutting force depends heavily on feed rate, while the force exerted during drilling is determined to be almost minor. At V_f of 0.08, the maximum thrust force is attained. Feed rate should be increased as the cut pressure rises. Thrust force is at its highest at 1800 rpm. At an engine speed of 1120 rpm and a V_f of 0.08 mm/rev, the engine's maximum torque is achieved. It is very important to measure the burr height as it is closely linked to feed rate. The major factor behind this is found to be the thrust force. Burr height is higher at 1800 rpm and V_f of 0.08 mm/rev, at which point the burr diameter will reach a diameter of around 0.4 mm [20, 21].

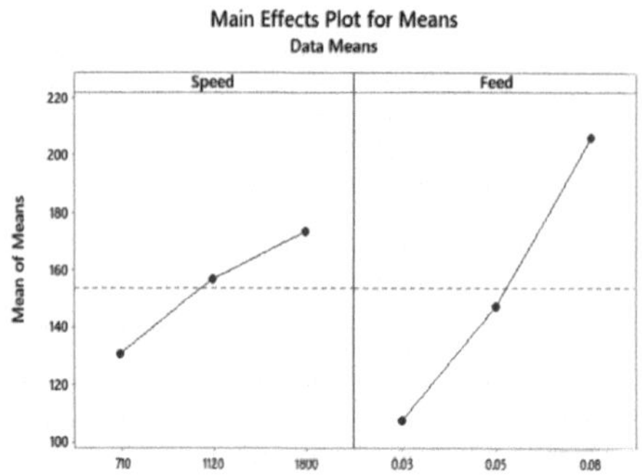

(a) Thrust force

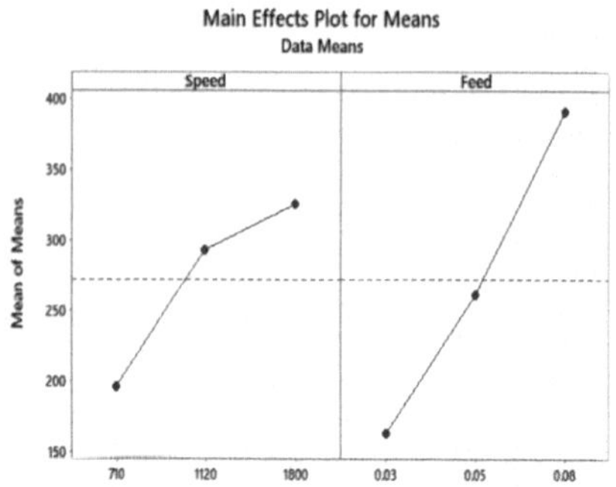

(b) Torque

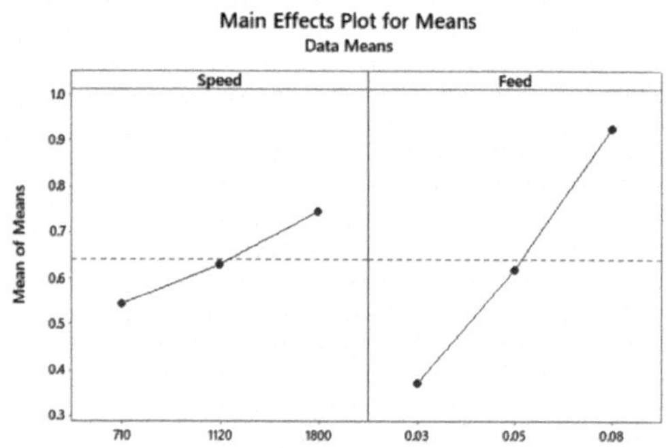

(c) Burr height

Fig. 6 Main effect plot for response variables for Al7029+10%B_4C-5% Gr sample

4 Conclusions

Following are the established findings: The L9 experiment was carried using CNC machine centre.

1. Drilling forces generated depends on the materials characteristics of the drill bit.
2. The findings reveal that carbide drill bits had the lowest output, while TiN-coated HSS drill bits generated the largest amounts.
3. As carbide drill bit was better suited for drilling of MMCs, hence carbide drill bit presents benefits over other drill bits.
4. Formation of discontinuous chip during drilling is supported by the inclusion of B_4C and Gr reinforcing composites.
5. ANOVA confirms that feed rate and speed are the most important variables and so are critical in producing surface quality, reducing the thrust force, and the burr height.

References

1. Ozben T, Kilickap E, Cakır O (2008) Investigation of mechanical and machinability properties of SiC particle reinforced Al-MMC. Mater Proc Tech 198(1–3):220–225
2. Basavarajappa S, Chandramohan G, Davim JP, Prabu M, Mukund K, Ashwin M, Prasanna Kumar M (2008) Drilling of hybrid aluminium matrix composites. Adv Manuf Technol 35(11–12):1244–1250
3. Sharath BN, Jeevan TP, Baig MA, Ashrith HS, Afzal A, Reddy AR (2021) Machinability studies on boron carbide and graphite reinforced aluminium hybrid composites. Mater Today Proc 46:8734–8741
4. Sharath BN, Venkatesh CV, Afzal A, Aslfattahi N, Aabid A, Baig M, Saleh B (2021) Multi ceramic particles inclusion in the aluminium matrix and wear characterization through experimental and response surface-artificial neural networks. Materials 14(11):2895
5. Basavarajappa S, Chandramohan G, Prabu M, Mukund K, Ashwin M (2007) Drilling of hybrid metal matrix composites—workpiece surface integrity. Mach Tools Manuf 47(1):92–96
6. Rajmohan T, Palanikumar K, Kathirvel M (2012) Optimization of machining parameters in drilling hybrid aluminium metal matrix composites. Trans Nonfer Met Soc China 22(6):1286–1297
7. Rana RS, Purohit R, Das S (2012) Review of recent studies in Al matrix composites. Sci Eng Res 3(6):1–16
8. Chaudhary G, Kumar M, Verma S, Srivastav A (2014) Optimization of drilling parameters of hybrid metal matrix composites using response surface methodology. Procedia Mater Sci 6:229–237
9. Ramulu M, Rao PN, Kao H (2002) Drilling of (Al_2O_3)p/6061 metal matrix composites. Mater Proc Tech 124(1–2):244–254
10. Sathia Prathap VSK, Vengatachalapathy K, Planiradja (2014) Machining of hybrid metal matrix composites and its further improvement-a review. Eng Adv Tech 4(1): 11–15
11. Madhu KS, Venkatesh CV, Sharath BN, Karthik S (2021) Effect of Boron Carbide on wear resistance of graphite containing Al7029 based hybrid composites and its dry sliding wear characterization through experimental, response surface method and ANOVA. Trib Finn Trib 38(3–4):48–60
12. Sharath BN, Venkatesh CV (2021) Study on effect of boron carbide, aluminium oxide and graphite on dry sliding wear behaviour of aluminium based metal matrix composite at different temperature. Trib Finn Trib 38(1–2):35–46
13. Sharath BN, Madhu KS, Venkatesh CV (2019) Experimental study on dry sliding wear behaviour of Al-B_4C-Gr metal matrix composite at different temperatures. Appl Mech Mater 895:96–101
14. Sharath BN, Venkatesh CV, Afzal A, Baig MA, Kumar AP (2021) Study on effect of ceramics on dry sliding wear behaviour of Al-Cu-Mg based metal matrix composite at different temperature. Mater Today Proc 46:8723–8733
15. Balraj US, Anitha P (2016) Machinability study on drilling of Al7075/Al_2O_{3P}/GR_P hybrid metal matrix composites. Ind J Eng 120:422–428
16. Saravanakumar A, Sasikumar P, Sivasankaran S (2014) Effect of graphite particles in drilling of hybrid aluminum matrix composite. Procedia Eng 97:495–504
17. Aldaş K, Özkul İ, Hayajneh MT (2016) Effects of machining parameters and reinforcement content on thrust force during drilling of hybrid composites. Mater Test 58(3):280–284
18. Laghari RA, Li J, Wu Y (2020) Study of machining process of SiCp/Al particle reinforced metal matrix composite using finite element analysis and experimental verification. Materials 13(23):5524
19. Tosun G, Muratoglu M (2004) The drilling of an Al/SiCp metal-matrix composites Part I: microstructure. Compos Sci Technol 64(2):299–308

20. Kumar KP, Kishore P, Laxminarayana P (2013) Prediction of thrust force and torque in drilling on aluminum 6061–T6 alloy. Int J Eng Res Technol 2(3):1–6
21. Çalışkan H, Kurbanoğlu C, Panjan P, Kramar D (2013) Investigation of the performance of carbide cutting tools with hard coatings in hard milling based on the response surface methodology. Int J Adv Manuf Technol 66(5–8):883–893

Investigation of Microscale Plastic Deformation Behavior of Copper Microgear in Forward Microextrusion Process

S. Nanthakumar, D. Rajenthirakumar, Vemula Prudhvi, S. Narendraraj, and V. P. Mowlieswaran

Abstract Microgears are critical components in the field of manufacturing complicated micro-devices for precise motion control that require lightweight, high-dimensional accuracy, and optimum performance under variable environments. Continuous demand of microcomponents has speeded up the need of producing microgears with high quality and economically from different materials. Microforming process offers a good quality of formed parts with low cost and high productivity. In this work, the deformation behavior of copper microgear is investigated under two different forming conditions. The results show that microgears extruded with high temperature resulted in good formability and improved dimensional accuracy. Microhardness measurements exhibit an inhomogeneous deformation pattern as the hardness at the center is higher when compared with the tooth. Significant reduction in microhardness and coefficient of variation were observed in the specimens formed with temperature assistance. The outcome of the current research will definitely contribute to the fundamental understanding of microextrusion of copper and enhance development microgears in a larger scale.

S. Nanthakumar (✉)
Assistant Professor, Department of Mechanical Engineering, PSG Institute of Technology and Applied Research, Neelambur, Coimbatore, Tamil Nadu, India
e-mail: kumarmechpsg@gmail.com

D. Rajenthirakumar
Professor, Department of Mechanical Engineering, PSG College of Technology, Coimbatore, Tamil Nadu, India
e-mail: drk.mech@psgtech.ac.in

V. Prudhvi · S. Narendraraj · V. P. Mowlieswaran
Department of Mechanical Engineering, PSG Institute of Technology and Applied Research, Neelambur, Coimbatore, Tamil Nadu, India
e-mail: 18me132@psgitech.ac.in

S. Narendraraj
e-mail: 18me133@psgitech.ac.in

V. P. Mowlieswaran
e-mail: 18me156@psgitech.ac.in

E. Natarajan et al. (eds.), *Materials, Design and Manufacturing for Sustainable Environment*, Lecture Notes in Mechanical Engineering,
https://doi.org/10.1007/978-981-19-3053-9_39

Keywords Microextrusion · Microgear · Temperature · Microforming · Copper

1 Introduction

The drive toward better multiple functional components for electrical and micromechanical systems continues to persist, resulting in increased demand for metallic components with the smallest dimensions and closest tolerances [1–3]. Microgears are critical components in the production of complicated microcomponents for motion control that must be ultra-lightweight, small, and compact and should have high-dimensional accuracy [4, 5].

Fabrication of these elements involves the employment of techniques that enable for the cost effective, environmentally friendly, and sustainable production of huge volumes of parts [6]. Ferrous materials are used for high-load and low-speed applications, whereas non-ferrous materials such as copper, brass, aluminum, titanium, nickel, and their constituent alloys are highly preferred for high-speed and low-load to no-load transmission applications [7, 8]. Copper is presently being concentrated because of its ability to produce complicated shapes and its widespread application in electronic parts, pharmaceutical, electromechanical, and solar panels [9, 10]. Significant cost reductions will be obtained in many industries that use microformed parts if a complete analysis of the process is developed properly. During scaling down, many researchers and industries are facing a challenging constrain named as size effect [11–13].

Irrespective of material employed in microsclae deformation size effect plays a crucial role in determining the process feasibility. Lithography technique, conventional method, and microextrusion methods of production of microgears are less explored by researchers for manufacturing microgears [14, 15]. Microgear shaft was produced through high-temperature extrusion process by employing aluminum alloy [16].

Microgear formation using bulk metallic glasses by hot embossing process was carried out, and it was found that the simulation results are with reasonable agreement with experimental results [17]. After the completion of process to eject, the microgear chemical solvents were used. 7075 aluminum alloys were hot extruded, and its microstructure and microhardness were investigated [18]. From the study, it was found that the microhardness at the center part of gear is higher when compared at the outer surface while extruding at room temperature and with proper lubricating conditions. The impact of die shapes, extrusion speed, and reduction ratio on extrusion load was studied, and it was discovered that the I-section and square section have the highest and lowest extrusion loads, respectively. Different die profiles, such as triangular, circular, I-section, square, and rectangular shapes, were used in experiments on pure aluminum and lead. The appropriate position of die openings and the choice of die land lengths were found to be key challenges in the microextrusion process [19–21]. Friction is also a very important factor in microextrusion process as it determines process feasibility and product quality. While evaluating the friction coefficient, actual process conditions should be considered rather than general assumptions. For instance in the actual topography of tool, work piece interface should be taken in to account rather than assuming it as isosceles triangle [22–24].

Proper understanding of the process and its related constraints is mandatory which will pave the way for the development of dedicated processes for manufacturing microgears. Consolidating the literature, work microextrusion process is strongly influenced by size effect, temperature, and interfacial friction. This triggers an investigation into the microextrusion process for developing microgear. In the current work, the microextrusion ability of copper is investigated.

The objective of the current study is (i) to investigate the microscale plastic deformation of copper microgear under two different forming conditions. (ii) To study the effect of temperature assistance on formability and microhardness. The outcome of this works will definitely bridge the gap between scientific issues and mass production.

2 Materials and Method

2.1 Material Processing

Copper is used as a testing specimen in the current work as it has excellent formability. Copper specimens with a diameter of 6 mm and a length of 50 mm are received. Figure 1 shows the material composition spectrum for the copper specimen, and the EDAX data are given in Table 1.

1200 grit SiC paper was used for fine polishing of samples. Then, 2.5 μm diamond paste was used for glass finish polishing. Then, the specimens are then mixed with bakelite powder and heated for 15 min and hot embossed. The specimens were etched with 10 g of ferric chloride, 60 ml hydrochloric acid, 200 ml of distilled water. After

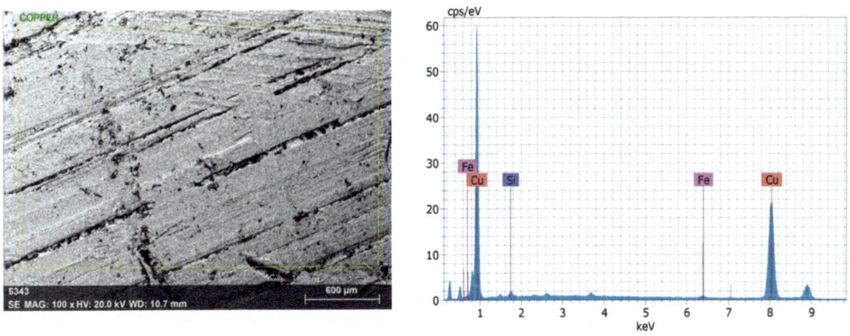

Fig. 1 Composition spectrum of copper

Table 1 EDAX data of copper

Element	Atomic weight %	Weight %
Copper	98.65	97.60
Silicon	0.79	1.77
Iron	0.56	0.63

Fig. 2 Optical microscopic image of copper (35 μm)

etching, a cotton swab is used to clean the etchant on the samples for 10–30 s. The optical microscopic image of the copper specimen is shown in Fig. 2. Line interpretation method was employed to find the grain size. The copper specimens for microextrusion trials were prepared in such a way that the surface roughness value is 1 μm (average Ra).

2.2 *Material Characterization*

Microhardness measurements were carried out using mitutoyo microhardness testing machine. The microhardness values (H_v) are measured from the midpoint in liner distance with a load of 100gf with a dwell period of 10 seconds. The measurements were made at 7–8 locations on the sample according to ASTM E348—17 standard, and the average value is considered (Figure 3). Zeiss EVO scanning electron microscope is used to carry out characterization and microstructural study.

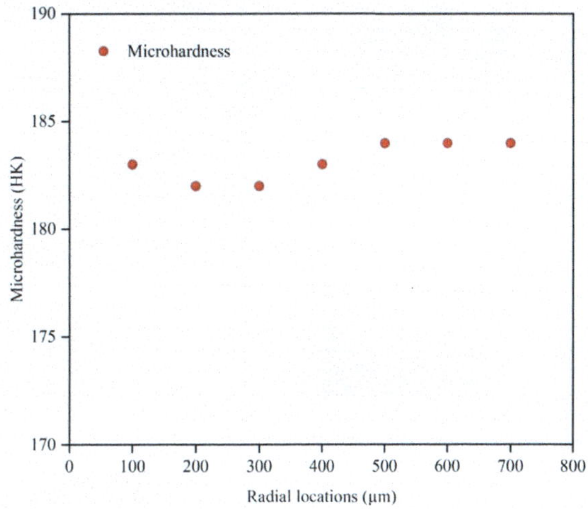

Fig. 3 Microhardness distributions of copper before extrusion

2.3 *Microextrusion System*

The microforming die assembly consists of the four primary components such as top flange, split die, bottom flange, and punch pin with punch inserts. The schematic view of the microextrusion assembly system is shown in Fig. 4. The split die is designed and fabricated using electric discharge machining process (EDM) and wire EDM process. Hot die steel is employed as the material for split die and punch pin. Necessary hardening process is carried out to achieve the hardness value of 55 HRc, and the average surface finish obtained is 0.61 μm. The die set components are designed based on the dimensions of the testing platform. The inner surface of die is coated with diamond like carbon coating (DLC) to minimize the effect of friction.

Figure 5 shows the testing platform which is equipped with a load cell (10 kN) and an LVDT (0–100 mm) to measure the forming force and the corresponding ram displacement. Punch force is measured for every 1 mm of punch travel with the speed of 0.01 mm/s. Table 2 gives the other important specifications of the machine. The prepared copper specimen is placed inside split die orifice and the setup in positioned between the die insertions at the bottom plate. The punch pin from the top pushes the copper specimen through the gear profile. The specimen undergoes deformation and forms the shape of gear profile.

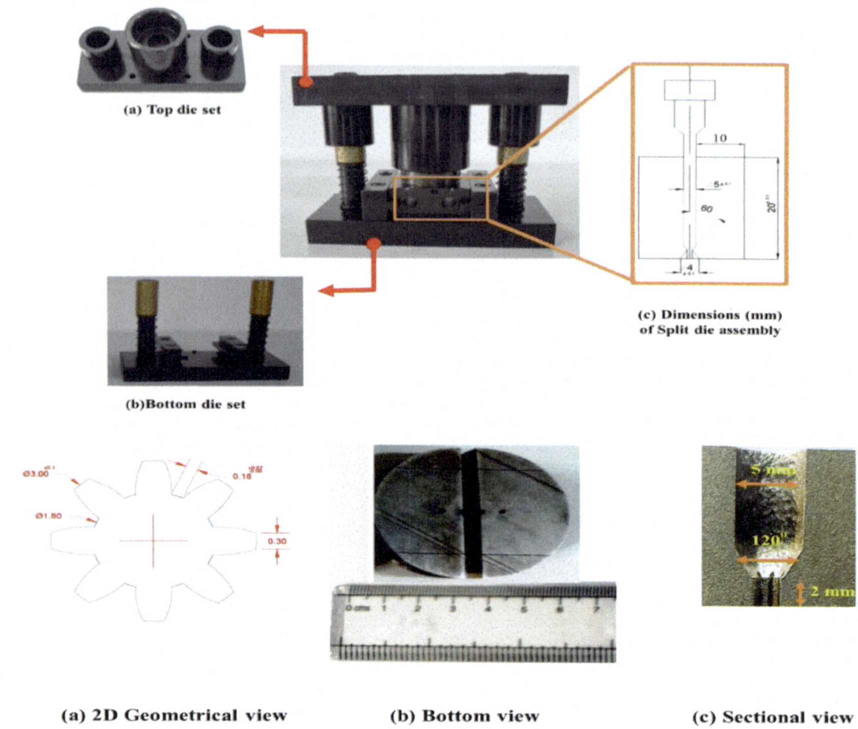

Fig. 4 Microforming system

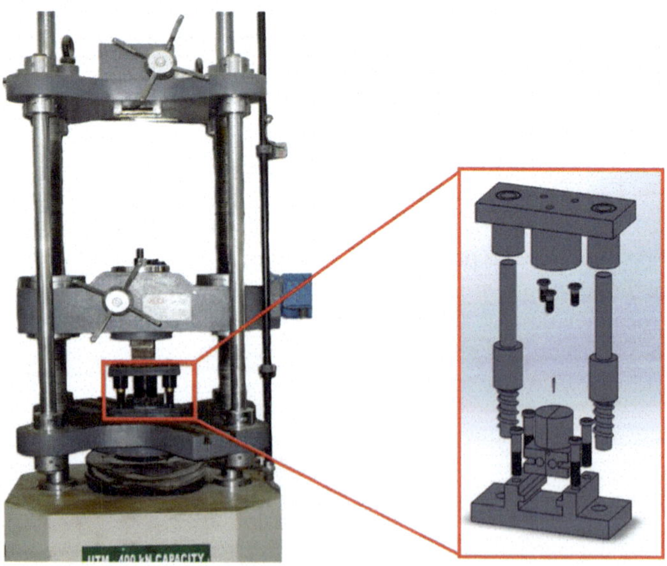

Fig. 5 Testing platform

Table 2 Microextrusion machine specifications [25]

S. No.	Parameters	Value
1	Load range	0–10 kN
2	Load measuring accuracy	± 2–100% of load cell used
3	Maximum crosshead stroke	1000 mm (without grips and load cell)
4	Clearance between columns	450 mm
5	Crosshead displacement measurement	0.01 mm
6	Power supply	230 V AC, 50 Hz single phase
7	Make/model type	FIE universal testing series Unitek 9400

3 Results and Discussion

3.1 Material Deformation Behavior

The microextrusion of copper microgear has been experimented, and the individual force displacement response had been observed for two different forming conditions (Fig. 6). Due to the complex cross section of gear profile, the flow of material inside the die orfice is improper. When the extrusion force is applied, the grains at the boundary tend to resist the deformation at a higher rate when compared to the central area. As a result, the flow of extrudate through the die orifice is not uniform, resulting in the minimum forming length. Further, it was well reported that the hardness of grain at boundaries is 30% higher when compared to interior grain region [6, 8].

The scanning electron micrographs of extruded copper microgears are shown in Fig. 7. It could be noticed that there exists both surface and edge distortions due to improper flow of material during extrusion. In addition, the center part of the gear got bulged due the resistance for shear deformation along the longitudinal surface of gear.

When the temperature is raised to 200 °C, the metal becomes softer as the dislocation density drops, and the force required for deformation lowers significantly. Further, the grain boundaries tend to widen and grains pile up along the grain boundary, reducing the force required to push the dislocation across it. There exist minor distortions found parallel to extrusion direction which are due to die roughness occurred during EDM process (Fig. 8).

Table 3 shows the forming lengths and coefficient of variation at various forming conditions. From the table, it is evident that temperature has a significant effect on forming length. In addition, temperature assistance also results in good replication of die dimensions along with good surface quality. This phenomenon is substantiated by the maximum forming length (4.5 mm) and low-coefficient of variation (3%).

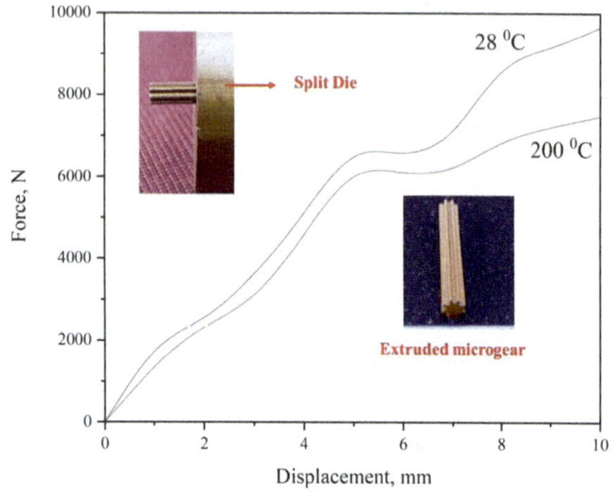

Fig. 6 Extrusion force versus displacement curve

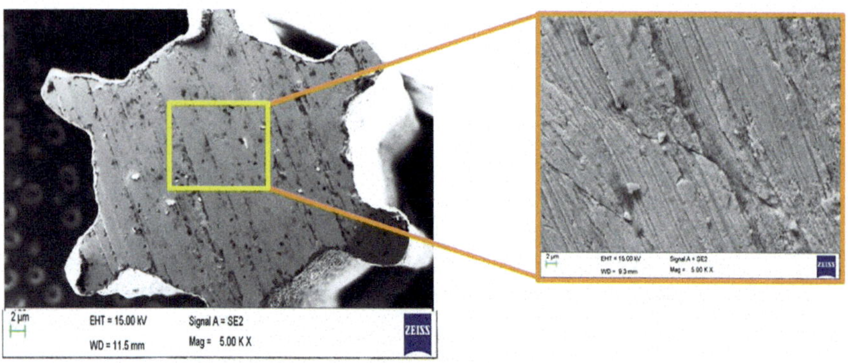

Fig. 7 SEM images of extruded copper microgears at 28 °C

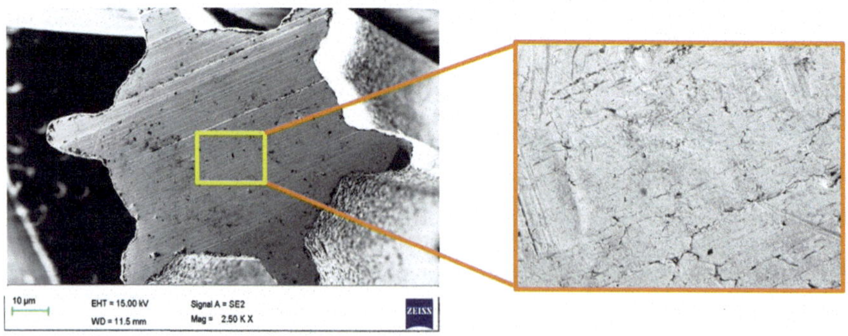

Fig. 8 SEM images of extruded copper microgears at 200 °C

Table 3 Length of extruded microgear at different friction conditions

Forming conditions (°C)	Maximum forming length (mm)	Mean length (mm)	Coefficient of variation (%)
28	3.8	3.5	6
200	4.5	4.1	3

With the application of DLC coating and better forming conditions, the deviation in gear dimensions is very minimum. Higher the temperature better the flow and formability of gear with minimum distortions.

3.2 *Microhardness Evaluations*

The average microhardness distribution of extruded copper microgear under two different forming conditions (28 and 200 °C) is shown in Fig. 9. From the figure, it could be noted that temperature has a dominant role in affecting the microhardness of microparts. It could be noted that samples extruded at room temperature have higher hardness when compared to samples extruded at 200 °C. One possible reason is when the forming temperature is raised, the metal becomes softer, and the grain boundary becomes weaker. Further, the grain size grows larger, and it makes the metal more likely to deform.

Thus, the material becomes softer, and it has excellent formability but reduced hardness. But, however, there exists inhomogeneous deformation behavior in samples extruded at room temperature as the hardness is found higher at center compared with the surface layer of extrudates [26]. But in the other case, the hardness achieved is

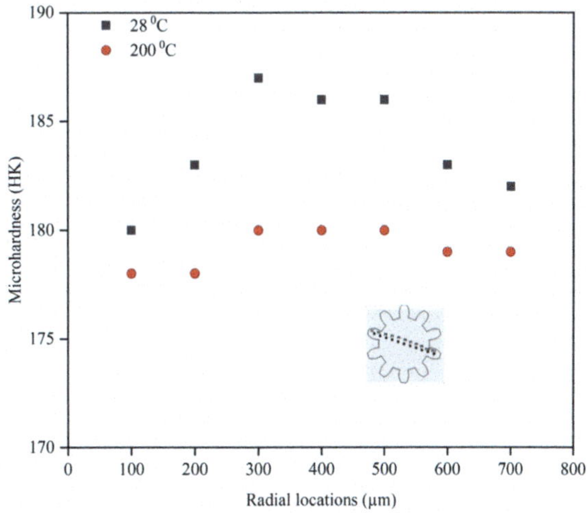

Fig. 9 Microhardness distributions of extruded copper microgear

Table 4 Average microhardness of the extruded copper microgears at different forming conditions

Forming temp (°C)	Microhardness of extruded products (H_V)							Co efficient of variation (%)
	Radial locations (μm)							
	100	200	300	400	500	600	700	
28	180	183	185	186	186	183	182	7.8
200	178	178	180	180	180	179	179	4.8

more uniform and has minimum variation. As the temperature rises, the coefficient of variation decreases, resulting in more uniform hardness (Table 4).

In their work, a 50% reduction in average hardness gradient observed by increasing the temperature from 200 to 300 °C employing CuZn15. Also adversely oriented grains can be easily plasticized by increasing the forming temperature, minimizing the material flow's reliance on grain orientation. To conclude the raise of temperature has a substantial improvement in material's formability and achieving more uniform hardness distribution.

4 Conclusions

Microgear of copper has been successfully extruded using forward microextrusion process with two different forming conditions (28 °C and 200 °C). The following conclusions are made accordingly.

- When the extrusion force is applied, the grains at the boundary tend to resist the deformation at a higher rate when compared to the central area.
- With the application of DLC coating and better forming conditions, the deviation in gear dimensions is very minimum.
- Microgear samples extruded at room temperature have higher hardness when compared to samples extruded at 200 °C.
- The raise of temperature has a substantial improvement in material's formability, and more uniform hardness with less coefficient of variation was achieved.

Declaration of Conflicting Interests The author(s) declared no potential conflicts of interest with respect to the research, authorship, and/or publication of this article.

References

1. Engel U, Eckstein R (2002) Micro-forming—from basic research to its realization. J Mater Process Technol 125:35–44
2. Fu MW, Chan WL (2013) A review on the state-of-the-art microforming technologies. Int J Adv Manuf Technol 67(9–12):2411–2437

3. Chan WL, Fu MW, Yang B (2011) Study of size effect in microextrusion process of pure copper. Mater Des 32(7):3772–3782
4. Krishnan N, Cao J, Dohda K (2007) Study of the size effects on friction conditions in microextrusion—part I: microextrusion experiments and analysis. J Manuf Sci Eng 129:669
5. Fu MW, Chan WL (2011) Geometry and grain size effects on the fracture behaviour of sheet metal in microscale plastic deformation. Mater Des 32:4738–4746. https://doi.org/10.1016/j.matdes.2011.06.039
6. Mao MY, Peng LF, Fu MW, Lai XM (2018) Co-effect of microstructure and surface constraints on plastic deformation in micro and mesoscaled forming process. Int J Adv Manuf Technol 98:1861–1886
7. Vander Voort GF, Fowler R (2012) Low-load vickers microindentation hardness testing. Adv Mater Process 170:28–33
8. Eliash T, Kazakevich M, Semenov VN, Rabkin E (2008) Nanohardness of molybdenum in the vicinity of grain boundaries and triple junctions. Acta Mater 56:5640–5652
9. Parasiz SA, Kinsey B, Krishnan N, Cao J, Li M (2007) Investigation of deformation size effects during microextrusion. J Manuf Sci Eng 129:690
10. Riahi M, Ehsanian MH, Asgari A, Djavanroodi F (2017) On a novel severe plastic deformation method: severe forward extrusion (SFE). Int J Adv Manuf Technol 93:1041–1050
11. Rane K, Strano M (2019) A comprehensive review of extrusion-based additive manufacturing processes for rapid production of metallic and ceramic parts. Adv Manuf 7(2):155–173
12. Razali AR, Qin Y (2013) A review on micro-manufacturing, micro-forming and their key issues. Procedia Eng 53:665–672
13. Imbaby MF, Jiang K (2010) Stainless steel-titania composite microgear fabricated by soft moulding and dispersing technique. Microelectron Eng 87:1650–1654. https://doi.org/10.1016/j.mee.2009.10.017
14. Tian ZQ, Tang J, Shi K et al (2005) Three-dimensional micro-fabrication on copper and nickel. J Electroanal Chem 581:153–158. https://doi.org/10.1016/j.jelechem.2004.11.041
15. Loh HT, Thian SCH, Wong YS et al (2007) Micro-mould fabrication for a micro-gear via vacuum casting. J Mater Process Technol 192–193:334–339. https://doi.org/10.1016/j.jmatprotec2007.04.098
16. Saotome Y, Iwazaki H (2001) Superplastic backward microextrusion of microparts for micro-electromechanical systems. J Mater Process Technol 119:307–311. https://doi.org/10.1016/S0924-0136(01)00957-8
17. Wang D, Shi T, Pan J, Liao G, Tang Z, Liu L (2010) Finite element simulation and experimental investigation of forming micro-gear with Zr–Cu–Ni–Al bulk metallic glass. J Mater Process Technol 210:684–688. https://doi.org/10.1016/j.jmatprotec.2009.12.005
18. Dong X, Chen F, Chen S, Liu Y, Huang Z, Chen H, Zhang X et al (2015) Microstructure and microhardness of hot extruded 7075 aluminum alloy micro-gear. J Mater Process Technol 219:199–208. https://doi.org/10.1016/j.jmatprotec.2014.12.022
19. Cheng LD, Wang CJ, Guo B, Wang ZL (2013) Size effects on plastic deformation behavior in micro radial compression of pure copper. Trans Nonferrous Met Soc China 23:2686–2691
20. Sakai T, Jonas JJ (2001) Plastic deformation: role of recovery and recrystallization. In: Encyclopedia of materials: science and technology. Elsevier, pp 7079–7084
21. Piwnik J, Mogielnicki K (2014) Experimental and FE analysis of aluminium alloy plastic flow in the forward microextrusion processes. Archiv Metall Mater 59
22. Wang C, Guo B, Shan D (2014) Friction related size-effect in microforming–a review. Manuf Rev 1:23
23. Zhen WM, Tong GQ, Chen F (2017) Deformation behaviour of materials in micro-forming with consideration of intragranular heterogeneities. Trans Nonferrous Met Soc China 27:616–626
24. Chan WL, Fu MW, Lu J, Liu JG (2010) Modeling of grain size effect on micro deformation behavior in microforming of pure copper. Mater Sci Eng A 527:6638–6648
25. Nanthakumar S, Rajenthirakumar D (2019) Influence of grain size on deformational behavior in microextrusion process. J Braz Soc Mech Sci Eng 41(3):1–12

26. Eichenhueller B, Egerer E, Engel U Micro forming at elevated temperature—forming and material behavior. Int J Adv Manuf Technol 33:119–124

Electric Discharge Drilling of Ti-6Al-4 V with O_2 as Dielectric Medium

N. Pragadish, Elango Natarajan, M. Selvam, Amares Singh, and N. Saravanakumar

Abstract Surface finish of the product refers to quality of machining operation, while production rate refers to machining performance. These two parameters are most-wanted parameters in industries, as they lead to the maximum profit. The current research utilizes oxygen (O_2) as dielectric that supports the electrical discharge machining (EDM) of Ti-6Al-4 V material and where discharge current (I), gap voltage (V), pulse-on time (T_{ON}) and oxygen (O_2) pressure (P) were selected based on Taguchi method. Experimental results were analyzed, and quadratic regression model was derived. Analysis through ANOVA revealed that the discharge current is the dominant factor in affecting the material removal, while O_2 pressure is the dominant factor on surface roughness. Validation result has confirmed that the prediction results are within the allowable limit.

Keywords Dry EDM · Taguchi method · Regression model · Material removal rate · Surface roughness

1 Introduction

Electrical discharge machining (EDM) is a non-traditional machining process which is widely applied in many industries because of its flexibility and capability of

---

N. Pragadish (✉) · M. Selvam (✉)
Department of Mechanical Engineering, Vel Tech Multitech, Avadi, Chennai, Tamilnadu, India
e-mail: pragadish@veltechmultitech.org

M. Selvam
e-mail: mselvam@veltechmultitech.org

E. Natarajan · A. Singh
Faculty of Engineering Technology and Built Environment, UCSI University, Kuala Lumpur, Malaysia

N. Saravanakumar
Department of Mechanical Engineering, PSG Institute of Technology and Applied Research, Coimbatore, Tamilnadu, India
e-mail: nsk@psgitech.ac.in

E. Natarajan et al. (eds.), *Materials, Design and Manufacturing for Sustainable Environment*, Lecture Notes in Mechanical Engineering,
https://doi.org/10.1007/978-981-19-3053-9_40

machining very hard materials as long as the materials are electrically conductive. EDM has been used for cutting hard materials like tungsten carbide, titanium, Inconel, die steel, etc., as experimented by various studies. Currently, advancements to this process allows the machining to be conducted even for non-electrically conductive materials such as ceramics [1]. Some research findings of machining of Ti alloy are found in [2]. Hourmand et al. [3] reported how EDM process parameters affect the machining of Al-Mg_2Si composite material. Aluminum powder-mixed oil-based dielectric was used while copper was used as the tool electrode. The results revealed that removal rate has increased with the current as well as voltage. But beyond a certain voltage, drop in the removal rate was observed, due to increased gap between tool and sample. Choudhary et al. [4] stated EDM as one of the most suitable processes for machining conducting materials with high hardness value. Ti-6Al-4 V, a super alloy was machined under EDM process using cryogenic-treated copper electrode and a non-treated electrode, and the performances were compared. They found narrow micro-cracks and finite recast layer on the surface machined.

Pragadish et al. [5] applied Taguchi method and gray relational analysis for machining of silicon steel. They used nickel-coated brass tool and waste cashew shell oil-mixed water for machining. The results explored the green dielectric percentage as the most influencing factor for improved conditions of material removal rate and tool wear life followed by the gap voltage. Narender et al. [6] reviewed and reported the effect of input parameters over the responses in micro-electrical discharge machining of titanium alloys. Voltage was reported as the most influencing parameter affecting the recast layer thickness. Higher material removal rate were reported for brass and zinc electrodes, whereas lower tool wear rate was observed in the copper electrodes. Jeavudeen et al. [7] studied the effect of EDM parameters and concentration of alumina powder mixed with the dielectric fluid in EDM of HSS and titanium alloys. They selected removal rate and tool wear index (TWI) as the responses. Powder concentration was observed as the main parameter in improving the TWI, whereas powder concentration and current were observed as the influencing parameters in improving the removal rate. Bhiksha et al. [8] used GRA to optimize the process parameters in electrical discharge machining of Ti-6Al-4 V alloy. Pulse-off time was observed as the major factor affecting surface roughness, layer thickness, and removal rate significantly followed by pulse-on time, powder concentration and discharge current. It was also reported that GRA can be used effectively to improve the EDM quality in modern manufacturing industries. Kunieda et al. [9] replaced liquid dielectric by gas.

Despite many advantages, drawbacks such as toxic fumes generation due to the breakdown of hydrocarbon dielectric fluid and high tool wear rate are unavoidable. Saha et al. [10] used tubular electrodes instead of solid electrode to enable the flow of gas into the machining region. Since the machining occurred only in tool region, it was not possible to drill the blind holes due to the formation of the core. Pragadish et al. [11] suggested a tool design that helps drilling the holes without formation of the core. Copper electrode was used as the tool, and oxygen was used as the dielectric medium. In their other research [12], they compared the surface hardness and microhardness of AISI D2 tool steel and LM13 aluminum alloy machined under dry EDM process and reported that AISI D2 tool steel is better to use. Liqing et al. [13] conducted dry EDM with oxygen-mixed dielectric fluid and dry EDM of cryogenically cooled work piece. They found increased removal rate in both cases and improved surface roughness in cryogenically cooled work piece. Pragadish et al. [14] conducted the drilling of LM13 aluminum alloy using a modified electrode design, where they found pulse-on time as the most influencing variable.

Numerous modifications and advancements were made in the EDM process to achieve better surface characteristics. Dry EDM is one of the green machining techniques where many drawbacks encountered by the liquid dielectric fluid were overcome. Although many experiments were conducted in the titanium alloys using EDM process, only very few results were reported in EDM of titanium alloys. In this research, blind holes were made on Ti-6Al-4 V samples using EDM with oxygen as the dielectric medium. The investigations and analysis are reported in the following sections.

2 Method of Research and Material

Die sinking EDM machine (Electronica Xpert 1) equipped with DC servo was used to conduct the experiments. An in-house made fixture as shown in Fig. 1 was attached to tool holder in order to avoid core formation in drilling blind hole [11].

Ti-6Al-4 V plate in 30 mm × 30 mm × 10 mm was considered as a sample. The experimental setup has a pure cylindrical copper rod of φ10 mm as a tool and pure oxygen (purity = 99.99%) as a dielectric medium. To enable the flow of the dielectric oxygen through the tool, some modification was made on the copper rod [11]. Three holes of dimension 2 mm were made at 180° in the place of 5, 6, and 7 mm at pitch circle diameter. Three slots of 3 mm × 2 mm were done at 180° around

Fig. 1 Dry EDM setup

the circumference as indicated in Fig. 2. This modification was done for effective flushing of debris particles.

Taguchi approach was used for experimental design of machining operations. It helped in minimizing the effect of uncontrollable factors and achieved the best response from minimum experiments [15, 16]. Current (I), pulse-on-time (T_{ON}), voltage (V), and gas pressure (P) were control factors as shown in Table 1. Duty factor and rotational speed of the tool were chosen as the fixed parameters. L_{27} orthogonal array used for conducting the experiments is shown in Table 2. Loss function was developed and further used to find signal–noise (S/N) ratio [17, 18] for lower the better (referring to surface roughness), higher the better (referring to material removal rate) [19, 20].

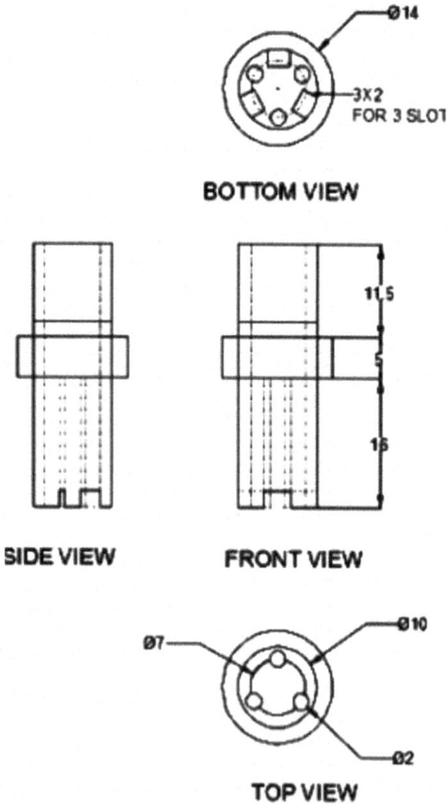

Fig. 2 Modified copper electrode

Table 1 Parameters and their levels

Parameters	1	2	3
Discharge current (I)	8	10	12
Voltage (V)	50	60	70
Pulse-on time (Ton)	100	150	200
Gas pressure (P)	1.5	2.0	2.5

Table 2 L_{27} orthogonal array from Taguchi

Experiment	I	V	Ton	P
1	8	50	100	1.5
2	8	60	100	2
3	8	70	100	2.5
4	8	50	150	2
5	8	60	150	2.5
6	8	70	150	1.5
7	8	50	200	2.5
8	8	60	200	1.5
9	8	70	200	2
10	10	50	100	2
11	10	60	100	2.5
12	10	70	100	1.5
13	10	50	150	2.5
14	10	60	150	1.5
15	10	70	150	2
16	10	50	200	1.5
17	10	60	200	2
18	10	70	200	2.5
19	12	50	100	2.5
20	12	60	100	1.5
21	12	70	100	2
22	12	50	150	1.5
23	12	60	150	2
24	12	70	150	2.5
25	12	50	200	2
26	12	60	200	2.5
27	12	70	200	1.5

3 Results and Discussion

The experimental results of removal rate and surface roughness is shown in Table 3. Regression analysis was carried out for modeling and analyzing control parameters and response variables [20]. In this study, the independent variables are I, V, T_{ON}, and P, whereas the dependent variables are removal rate and surface roughness. Equations (1) and (2) represent the prediction models in quadratic form. The coefficient of correlation R^2 values of material removal rate and surface roughness were 0.996 and 0.989, respectively.

$$\begin{aligned} \mathrm{MRR} = & -70 - 133.1I + 21.34V - 0.030T_{\mathrm{on}} + 0.8P + 8.617I^2 \\ & - 0.1645V^2 + 0.001183T_{\mathrm{on}}^2 + 8.17P^2 - 0.0810\mathrm{IV} + 0.0053\mathrm{IT}_{\mathrm{on}} \\ & + 0.81\mathrm{IP} - 0.00220\mathrm{VT}_{\mathrm{on}} - 0.472\mathrm{VP} + 0.0108T_{\mathrm{on}}P \end{aligned} \quad (1)$$

$$\begin{aligned} \mathrm{SR} = & 28.06 + 1.584I - 0.7011V - 0.0338T_{\mathrm{on}} - 2.22P - 0.0833I^2 \\ & + 0.005476V^2 + 0.000106T_{\mathrm{on}}^2 - 0.659P^2 - 0.00161\mathrm{IV} - 0.000489\mathrm{IT}_{\mathrm{on}} \\ & - 0.0170\mathrm{IP} - 0.000003\mathrm{VT}_{\mathrm{on}} + 0.0344\mathrm{VP} + 0.00255T_{\mathrm{on}}P \end{aligned} \quad (2)$$

Table 4 shows the comparison between the experimental values and predicted values from RSM models. The average absolute error for material removal rate and surface roughness were observed as 0.25% and 0.09%, respectively.

3.1 Statistical Analysis Results

The ANOVA was used to find the best combination of process characteristics, and it was done with greater precision by looking at the relative relevance of process parameters [21, 22]. The F-test statistically determines whether the parameters are significantly different. The stronger the impact on machining performance parameters, the higher the F-value. The ANOVA results for material removal rate is given in Table 5. It is observed that the current (93.81%) is the most significant parameter followed by pulse-on time (2.93%). Larger F-values of 1692.80 was observed for current. Table 6 presents the ANOVA results for surface roughness. It is observed that the pressure (75.18%) is the most significant parameter followed by current (16.52%). Larger F-values of 337.98 was observed for pressure.

Table 3 Experimental results

Experiment	I	V	TON	P	Removal rate	Surface roughness
1	8	50	100	1.5	37.13	8.63
2	8	60	100	2	57.72	6.71
3	8	70	100	2.5	48.07	6.18
4	8	50	150	2	50.21	6.88
5	8	60	150	2.5	70.93	5.16
6	8	70	150	1.5	45.14	8.13
7	8	50	200	2.5	91.56	5.59
8	8	60	200	1.5	80.00	7.81
9	8	70	200	2	66.63	7.47
10	10	50	100	2	91.29	7.25
11	10	60	100	2.5	109.29	5.14
12	10	70	100	1.5	74.82	8.08
13	10	50	150	2.5	102.84	5.19
14	10	60	150	1.5	105.74	7.55
15	10	70	150	2	89.22	7.09
16	10	50	200	1.5	105.08	8.12
17	10	60	200	2	116.84	6.27
18	10	70	200	2.5	104.36	5.71
19	12	50	100	2.5	203.63	4.46
20	12	60	100	1.5	195.88	7.05
21	12	70	100	2	176.98	6.58
22	12	50	150	1.5	197.34	7.15
23	12	60	150	2	223.52	5.11
24	12	70	150	2.5	203.61	4.56
25	12	50	200	2	221.64	6.04
26	12	60	200	2.5	245.06	4.32
27	12	70	200	1.5	204.35	6.99

3.2 *Optimization of Drilling Parameters*

Signal-to-noise (S/N) ratios representing the relationship between control variables and response variables (material removal rate and surface roughness) were computed through the experimental design for each combination of control factors. The highest value of material removal rate and lowest value of surface roughness are essential for quality improvement and cost reduction. Hence, the calculations of the S/N ratio were evaluated by the 'higher the better' for material removal rate and the 'lower the better' for surface roughness. The principal effect of control parameters on S/N ratio for material removal rate is shown in Fig. 3. The trend shows that the material removal

Table 4 Experimental and predicted values compared

Run	I	V	Ton	P	Experimental		Predicted		% Error	
					Removal rate	SR	Removal rate	SR	Removal rate	SR
1	8	50	100	1.5	37.13	8.63	37.63	8.611	−1.35	0.26
2	8	60	100	2	57.72	6.71	58.64	6.836	−1.59	-1.85
3	8	70	100	2.5	48.07	6.18	46.11	6.171	4.09	0.22
4	8	50	150	2	50.21	6.88	55.3	6.955	−10.15	-1.08
5	8	60	150	2.5	70.93	5.16	77.2	5.084	−8.84	1.56
6	8	70	150	1.5	45.14	8.13	46.63	8.12	−3.29	0.08
7	8	50	200	2.5	91.56	5.59	83.5	5.626	8.80	-0.68
8	8	60	200	1.5	80.00	7.81	79.47	7.781	0.66	0.43
9	8	70	200	2	66.63	7.47	63.56	7.398	4.60	0.95
10	10	50	100	2	91.29	7.25	84.52	7.11	7.41	1.98
11	10	60	100	2.5	109.29	5.14	106.4	5.128	2.65	0.26
12	10	70	100	1.5	74.82	8.08	74.27	8.295	0.73	-2.68
13	10	50	150	2.5	102.84	5.19	107.9	5.122	−4.92	1.35
14	10	60	150	1.5	105.74	7.55	102.3	7.408	3.26	1.84
15	10	70	150	2	89.22	7.09	86.36	6.913	3.20	2.55
16	10	50	200	1.5	105.08	8.12	105.4	8.149	−0.31	-0.32
17	10	60	200	2	116.84	6.27	124	6.45	−6.12	-2.94
18	10	70	200	2.5	104.36	5.71	109	5.86	−4.44	-2.59
19	12	50	100	2.5	203.63	4.46	206.1	4.579	−1.21	-2.71
20	12	60	100	1.5	195.88	7.05	198.8	6.996	−1.49	0.70
21	12	70	100	2	176.98	6.58	182.9	6.39	−3.34	2.83
22	12	50	150	1.5	197.34	7.15	197.1	7.189	0.12	-0.61
23	12	60	150	2	223.52	5.11	215.7	5.378	3.50	-5.28
24	12	70	150	2.5	203.61	4.56	200.7	4.677	1.43	-2.67
25	12	50	200	2	221.64	6.04	223.7	5.994	−0.93	0.71
26	12	60	200	2.5	245.06	4.32	243.1	4.088	0.80	5.43
27	12	70	200	1.5	204.35	6.99	204.4	6.998	−0.03	-0.09
Average absolute percentage error									**0.25**	**0.09**

rate increases with increase in current, pulse-on time, and pressure throughout the entire range. As for the voltage, the material removal rate has increased initially followed by a drop upon an increase in voltage. This drop may be due to increase in the distance between the tool and the workpiece, which also evidenced in poor flushing of debris particles during experiments. The debris may lead to arcing problem.

The predominant effect of control parameters on S/N ratio for surface roughness is shown in Fig. 4. The trend shows that the surface roughness increases with increase in

Table 5 ANOVA results for material removal rate

Source	Degrees of freedom	Sum of squares	Mean squares	F-value	Contribution (%)
Current	2	104,608	52,304.0	1692.80	93.81
Voltage	2	2049	1024.7	33.16	1.84
Pulse-on time	2	3271	1635.7	52.94	2.93
Pressure	2	1021	510.4	16.52	0.92
Error	18	556	30.9		0.50
Total	26	111,506			

Table 6 ANOVA results for surface roughness

Source	Degrees of freedom	Sum of squares	Mean squares	F-value	Contribution (%)
Current	2	6.5988	3.2994	74.26	16.52
Voltage	2	1.9201	0.9600	21.61	4.81
Pulse-on time	2	0.5950	0.2975	6.70	1.49
Pressure	2	30.0317	15.0159	337.98	75.18
Error	18	0.7997	0.0444		2.00
Total	26	39.9453			

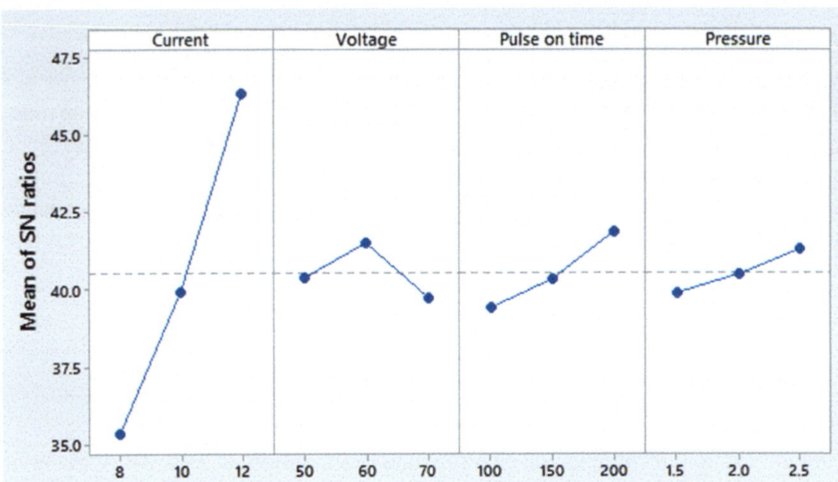

Fig. 3 Effect of process parameters on average S/N ratio for material removal rate

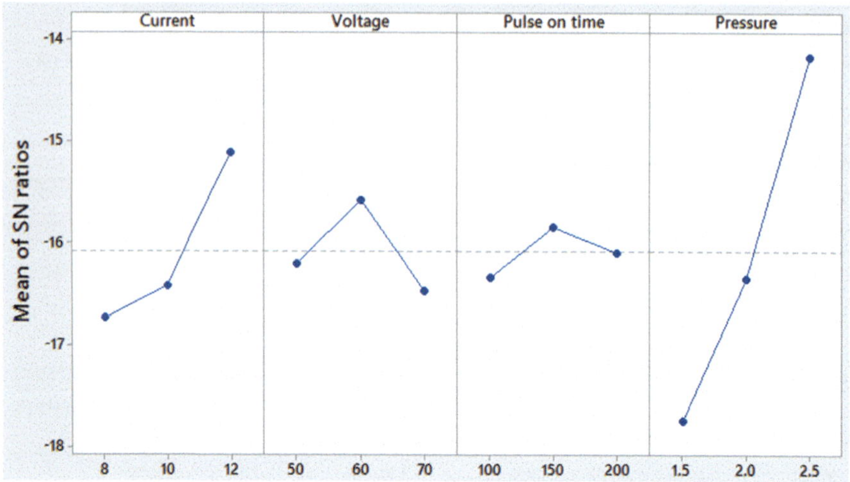

Fig. 4 Effect of process parameters on average S/N ratio for surface roughness

Table 7 Optimum factors for material removal rate and surface roughness

Current	Voltage	Pulse-on time	Pressure	Material removal rate	Surface roughness
12	57.07	200	2.5	244.56	–
12	57.88	157.576	2.5	–	3.873

current and pressure throughout the entire range. In case of voltage and pulse-on time, the surface roughness decreases up to second level before showing an increase. Since the duty factor was kept constant, the pulse-off time was dependent on the pulse-off time, and this may not have sufficient for efficient flushing of debris particles. According to the main effect plots, the optimal conditions for material removal rate and surface roughness are shown in Table 7.

4 Conclusions

Using the Taguchi technique, the experimental examination on Ti-6Al-4 V material was conducted at three levels and four factors to find the optimal values of process parameters. The following are some of the findings of the investigation:

- The coefficient of correlation R^2 values of material removal rate and surface roughness were 0.996 and 0.989, respectively. The developed models are fairly accurate and can be utilized for prediction within limits.

- The ANOVA revealed that the current is the most dominant parameter with the greatest influence on material removal rate (93.81%), and the pressure is the most significant parameter with the greatest influence on surface roughness (75.18%).
- The optimized process parameters are obtained as follows:

  Max material removal rate = 244.56 when I=12, V=57.07, T_{ON}=200 and P=2.5
  Min surface roughness = 3.873 when I=12, V=57.88, T_{ON}=157.576 and P=2.5

- Based on the findings of this study, it can be inferred that the proposed methodology may be used to solve multiple response problems in real-time industrial experiments.

References

1. Grigoriev S, Hamdy K, Volosova M, Okunkova A, Fedorov S (2021) Electrical discharge machining of oxide and nitride ceramics: a review. Mater Des 209:109965. https://doi.org/10.1016/j.matdes.2021.109965
2. S R, Palanikumar K, Boppana S, Natarajan E (2022) Analysis of chip formation and temperature measurement in machining of titanium alloy (Ti-6Al-4V). Exp Tech. https://doi.org/10.1007/s40799-021-00537-2
3. Hourmand M, Farahany S, Sarhan A, Noordin M (2014) Investigating the electrical discharge machining (EDM) parameter effects on Al-Mg2Si metal matrix composite (MMC) for high material removal rate (MRR) and less EWR–RSM approach. Int J Adv Manuf Technol 77(5–8):831–838. https://doi.org/10.1007/s00170-014-6491-2
4. Choudhary R, Kumar A, Yadav G, Yadav R, Kumar V, Akhtar J (2020) Analysis of cryogenic tool wear during electrical discharge machining of titanium alloy grade 5. Mater Today: Proc 26:864–870. https://doi.org/10.1016/j.matpr.2020.01.080
5. Pragadish N, Kaliappan S, Subramanian M et al (2022) Optimization of cardanol oil dielectric-activated EDM process parameters in machining of silicon steel. Biomass Convers Biorefin. https://doi.org/10.1007/s13399-021-02268-1
6. Singh N, Bharti P (2020) A review on micro electric discharge machining of titanium alloys. Mater Today: Proc 25:742–750. https://doi.org/10.1016/j.matpr.2019.08.235
7. Jeavudeen S, Siddhi Jailani H, Murugan M (2020) Effect of process parameters in the machining of Titanium alloy and high speed steel in powder mixed electrical discharge machining process. Mater Today: Proc 27:615–619. https://doi.org/10.1016/j.matpr.2019.12.156
8. Gugulothu B, Rao G, Bezabih M (2021) Grey relational analysis for multi-response optimization of process parameters in green electrical discharge machining of Ti-6Al-4V alloy. Mater Today: Proc 46:89–98. https://doi.org/10.1016/j.matpr.2020.06.135
9. Kunieda M, Yoshida M, Taniguchi N (1997) Electrical discharge machining in gas. CIRP Ann 46(1):143–146. https://doi.org/10.1016/s0007-8506(07)60794-x
10. Saha S, Choudhury S (2009) Experimental investigation and empirical modeling of the dry electric discharge machining process. Int J Mach Tools Manuf 49(3–4):297–308. https://doi.org/10.1016/j.ijmachtools.2008.10.012
11. Pragadish N, Kumar M (2015) Surface characteristics analysis of dry EDMed AISI D2 steel using modified tool design. J Mech Sci Technol 29(4):1737–1743. https://doi.org/10.1007/s12206-015-0347-z
12. Pragadish N, Pradeep K (2021) Comparison of surface characteristics and micro hardness of AISI D2 tool steel and LM13 aluminum alloy machined under dry EDM (green machining) process. FME Trans 49(2):488–493. https://doi.org/10.5937/fme2102488p

13. Liqing L, Yingjie S (2013) Study of Dry EDM with oxygen-mixed and cryogenic cooling approaches. Procedia CIRP 6:344–350. https://doi.org/10.1016/j.procir.2013.03.055
14. Nagarajan P, Murugesan P, Natarajan E (2019) Optimum control parameters during machining of LM13 Aluminum alloy under dry electrical discharge machining (EDM) with a modified tool design. Mater Sci 25(4):407–412. https://doi.org/10.5755/j01.ms.25.4.20899
15. Parasuraman S, Elamvazuthi I, Kanagaraj G, Natarajan E, Pugazhenthi A (2021) Assessments of process parameters on cutting force and surface roughness during drilling of AA7075/TiB2 in situ composite. Materials 14(7):1726. https://doi.org/10.3390/ma14071726
16. Kayaroganam P, Krishnan V, Natarajan E, Natarajan S, Muthusamy K (2021) Drilling parameters analysis on in-situ Al/B4C/Mica hybrid composite and an integrated optimization approach using fuzzy model and non-dominated sorting genetic algorithm. Metals (Basel) 11(12):2060. https://doi.org/10.3390/met11122060
17. Köksoy O, Muluk F (2010) Solution to the Taguchi's problem with correlated responses. Gazi Univ J Sci 17(1):59–70
18. Gupta A, Singh H, Aggarwal A (2011) Taguchi-fuzzy multi output optimization (MOO) in high speed CNC turning of AISI P-20 tool steel. Expert Syst Appl 38(6):6822–6828. https://doi.org/10.1016/j.eswa.2010.12.057
19. Ranjit KR (1990) A primer on the Taguchi method. Van Nostrand Reinhold, Competitive Manufacturing Series
20. Kaviarasan V, Venkatesan R, Natarajan E (2019) Prediction of surface quality and optimization of process parameters in drilling of Delrin using neural network. Prog Rubber Plastics Recycl Technol 35(3):149–169. https://doi.org/10.1177/1477760619855078
21. Viswanathan R, Ramesh S, Maniraj S, Subburam V (2020) Measurement and multi-response optimization of turning parameters for magnesium alloy using hybrid combination of Taguchi-GRA-PCA technique. Measurement 159:107800. https://doi.org/10.1016/j.measurement.2020.107800
22. Manoharan V, Tamilperuvalathan S, Natarajan E, Ponnusamy P (2021) Experimental investigation on electrochemical discharge machining of zirconia. Recent Adv Manuf Autom Design Energy Technol 27–34. https://doi.org/10.1007/978-981-16-4222-7_4

Optimum Tool Traverse Speed Resulting Equiaxed Recrystallized Grains and High Mechanical Strength at Swept Friction Stir Spot Welded AA7075-T6 Lap Joints

S. Suresh, Elango Natarajan, P. Vinayagamurthi, K. Venkatesan, R. Viswanathan, and S. Rajesh

Abstract Swept Friction Stir Spot Welding (SFSSW) was performed on 7075-T6 Al alloy metal and the effect of influencing tool traverse speed on weld properties were investigated. The spot weld was done in overlapping area of 35 mm × 35 mm using four different tool traverse speeds ranging between 10 and 25 mm/min, while the tool rotational speed and sweep radius were remained constant. The mechanical and metallographic tests revealed that tool traverse speed significantly affects the weld joint characterization. Worm hole defect was also noticed at lower traverse speed (10 mm/min) and higher traverse speed (25 mm/min). It must be due to insufficient plasticization of stirred material in SZ and differential thermal transformation. The mean grain size of the weld nugget was greatly reduced at the weld speed of 20 mm/min, which regarded as the optimum traverse speed to be used for SFSSW of this metal.

Keywords Swept FSSW · AA7075-T6 · Lap shear · Traverse speed · Crystallation

S. Suresh (✉)
Department of Mechanical Engineering, Muthayammal Engineering College, Rasipuram, India
e-mail: suresh.mjl@gmail.com

E. Natarajan
Faculty of Engineering, Technology and Built Environment, UCSI University, KualaLumpur, Malaysia

P. Vinayagamurthi
Department of Mechanical Engineering, Jayalakshmi Institute of Technology, Thoppur, India

K. Venkatesan
Department of Metallurgical Engineering, Government College of Engineering, Salem, India

R. Viswanathan
Department of Mechanical Engineering, AVS Engineering College, Salem, India

S. Rajesh
Department of Mechanical Engineering, Knowledge Institute of Technology, Salem, India

E. Natarajan et al. (eds.), *Materials, Design and Manufacturing for Sustainable Environment*, Lecture Notes in Mechanical Engineering,
https://doi.org/10.1007/978-981-19-3053-9_41

1 Introduction

Aluminum and its alloys have become predominant materials in the past decades in automotive, aircraft and shipbuilding due to their remarkable properties [1]. Years to go, the aluminum alloys have set marked in automobile and manufacturing sectors for weight and fuel reduction [2]. Among the available aluminum alloys (2xxx, 5xxx, 6xxx, 7xxx), aluminum 7xxx series find applications in aerospace and defense industries. It is a reliable alternative for steel due to their better mechanical and correction resistance properties [3]. To have improved aluminum structure joints, there must be a concern for selection and implementation of a suitable joining method. The Welding Institute (TWI), UK developed FSSW to overcome the difficulties arising in the conventional spot welding process like resistance spot welding (RSW) [4]. Swept FSSW, a variant of FSSW has been developed for producing spot joints on aluminum sheets, particularly to produce increased joint area. The basic principle behind the joining of materials through the FSSW process is the frictional heat and plastic deformation that soften the substrate to be welded [5]. In FSSW, a cylindrical non-consumable rotating tool is plunged into the overlapped sheets and retracted after a short period after leaving plasticized weld areas [6]. In swept FSSW, the tool is moved in a circular path after plunging, but before retraction of the tool. The weld parameters of SFSSW include rotational speed, plunge speed, travel speed and tilt angle [7, 8].

To date, several researches have been carried out in the conventional FSSW process of aluminum and alloys in terms of study of affecting weld variables [9–13]. Rotational speed and weld speed had evident effects on the material flow and heat generation during the FSSW [5, 14]. The high tool rotating speed creates a lot of heat and affects weld quality, while a lesser speed promises increased joint strength [15].

To this time a few studies have been noticed in this area of research. For example, swept FSSW was used to fabricate AA2024-T3 and AA7075-T6 to investigate the effect of significant parameters [8]. Their results show that, there would not be any further significance in weld strength once the tool plunge depth has reached a certain depth. Also reported that the tool does not create strong joints due to overheating at higher depths and longer dwell.

The FSSW process with new approach was carried out on joining Al/Fe dissimilar metals by considering process variables such as the tool rotation direction, tool feed path radius, tool rotation speed and travel speed. The tool moving path and the radius of rotation of tool are important variable that influence the shape of the Al/Fe dissimilar joint width [16]. Recently, SFSSW was conducted on AA6061-T6 with additions of nanoparticles such as SiC and Al_2O_3 to study the mechanical behavior of the joint. It was reported in their studies that the addition of ceramics in the weld spots would significantly influence the mechanical strength of joints [17, 18].

In this research, the consequence of tool traverse speed on mechanical properties and microstructure of 7075-T6 aluminum alloy joint made by SFSSW was investigated. Optical microscopic study, SEM–EDS studies and tensile tests were performed for the investigations.

2 Material and Method

7075-T6 Al alloy has a broad range of applications in aircraft structure and space structure. The chemical composition of the 7075-T6 Al alloy is listed in Table 1. Hot-rolled 7075-T6 Al alloy plates with 100 mm × 35 mm × 2 mm dimensions were used as the weld coupon, where SFSSW weld was carried out over an overlapping area of 35 mm × 35 mm.

H13 steel cylindrical tool with M5 threaded pin and machined into 3 mm long with 12 mm shoulder diameter was considered and it was hardened to 58–60 HRC for better wear resistance. Spot welded samples were prepared at four travel speeds such as 10, 15, 20 and 25 mm/min with constant parameters of 1600 rpm tool rotational speed, 0.2 mm shoulder plunge depth and 2.5 mm swept radius. The detailed process parameters are briefed in Table 2.

Weld shear tensile strength was measured using a universal tensile testing machine (Instron 4855, Norwood, MA, USA), which operated at 0.1 mm/min. Three samples were tested in each tool traverse speed condition and an average of the results was considered for the investigation. All the experiments were done in controlled environment of 20ºC. Figure 1 shows a photograph of the experimental setup and the produced samples for the lap shear test. Stereo zoom microscope (Radical–RSM 9), optical microscope (Olympus BX41M, Japan) and the scanning electron microscope (SIGMA—Carl Zeiss) were used to study the stir zone (SZ) joint characteristics.

Table 1 Composition of AA7075-T6 (wt%)

Element	Mn	Cr	Cu	Mg	Zn	Al
Wt.%	≤0.30	0.18–0.28	1.202	2.1–2.9	5.1–6.1	Rem

Table 2 SFSSW process parameters

Sample No.	Tool traverse speed (mm/min)	Tool rotational speed (rpm)	Swept radius (mm)
1	10	1600	2.5
2	15		
3	20		
4	25		

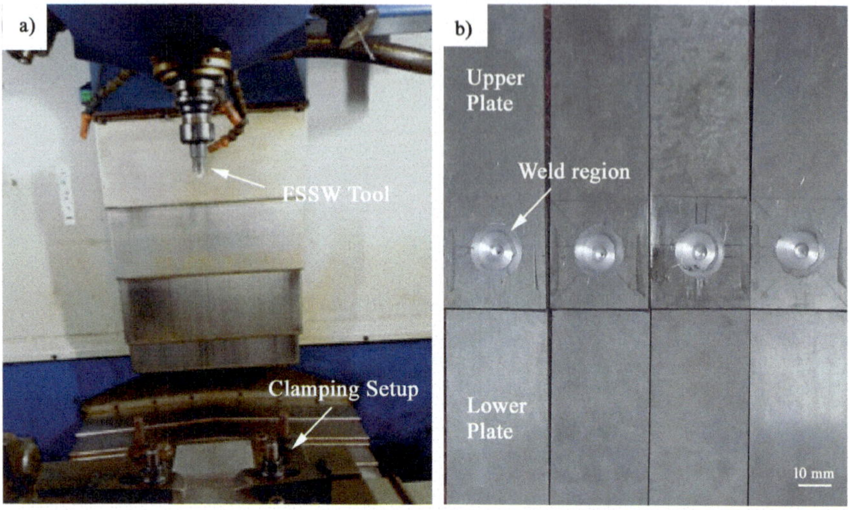

Fig. 1 Photograph of **a** FSSW setup **b** fabricated SFSS welded sample

3 Results and Discussion

3.1 Metallurgical Characterization at SZ

Figure 2 shows macroscopic images of cross-sectioned SFSSW joints prepared at various tool traverse speeds. Grain structures at SZ, TMAZ and HAZ areas [19]were examined. Wormhole defects at the end of SZ were observed at the specimen made with traverse speeds of 10 mm/min and 25 mm/min as shown in Fig. 2a, d. These were believed to be caused by applied tool traverse speed. Very low speed and high speed reduced the heat input to the weld region. As a consequence of it, the heat generated at SZ was not sufficient to plasticize the metal and material flow [20].

Microstructure of SZ of samples prepared at various tool traverse speed is shown in Fig. 3. It revealed that, the joints made at 20 mm/min has finer grains than those joints

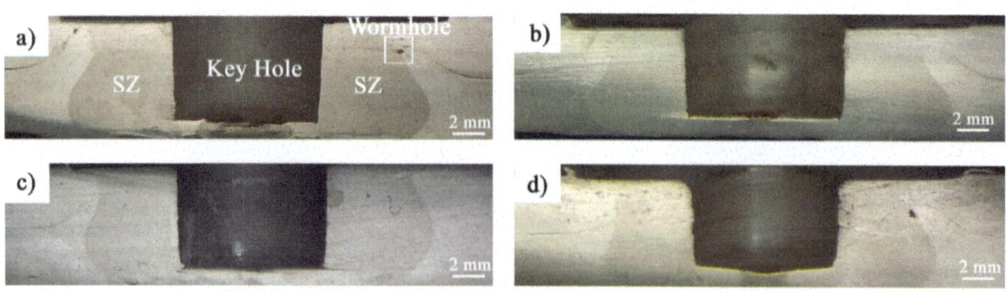

Fig. 2 Macrostructure of cross-section of samples prepared **a** at 10 mm/min **b** at 15 mm/min **c** at 20 mm/min and **d** at 25 mm/min

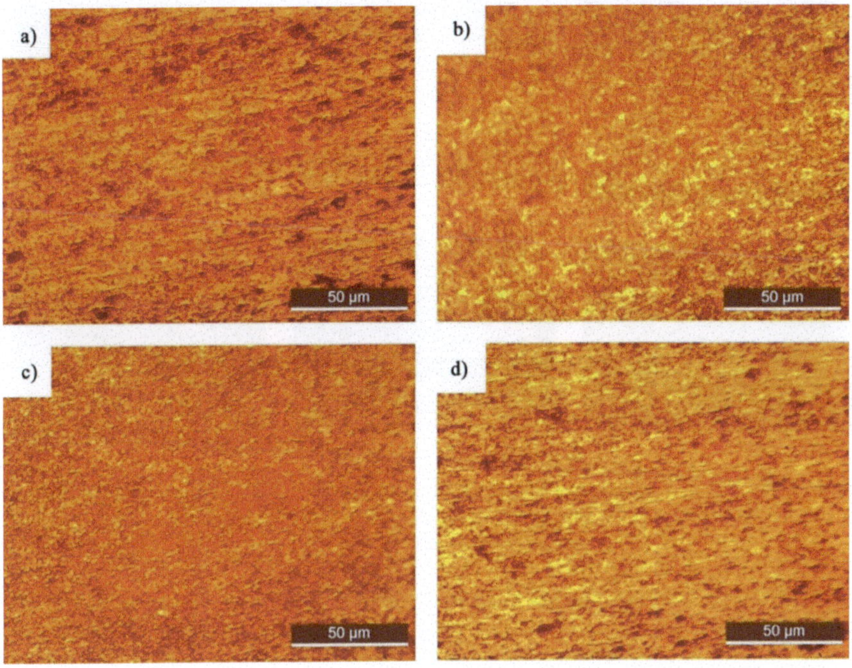

Fig. 3 microstructure of SZ of weld joints prepared **a** at 10 mm/min **b** at 15 mm/min **c** at 20 mm/min and **d** at 25 mm/min

prepared under other conditions. Greater refinement is because of dynamic recrystallization induced by appropriate frictional heat in SZ. The appropriate frictional heat has caused high plastic shear deformation and led to the fine crystallized structure as seen in Fig. 3c. Figure 3a, d show very high coarse grain structure observed in other samples.

The elemental distribution in the SZ of joint prepared at 20 mm/min is shown in Fig. 4. The evenly distributed aluminum, magnesium, zinc, copper and manganese elements is confirmed by EDS mapping, indicating refined grains at the SZ.

3.2 *Weld Strength*

Figure 5 shows the lap shear fracture load of samples prepared at 10 mm/min, 15 mm/min, 20 mm/min and 25 mm/min. A significant change in joint strength is observed while changing the traverse speed. The joint fabricated with traverse speed of 20 mm/min exhibited higher lap shear fracture load of 5430 N. As evidenced in Fig. 5, when the traverse speed is increased up to 20 mm/min, the strength of the weld joints also increases, beyond which, decreases with the further increase of tool traverse speed. Very lower traverse speed has resulted lower heat input and caused

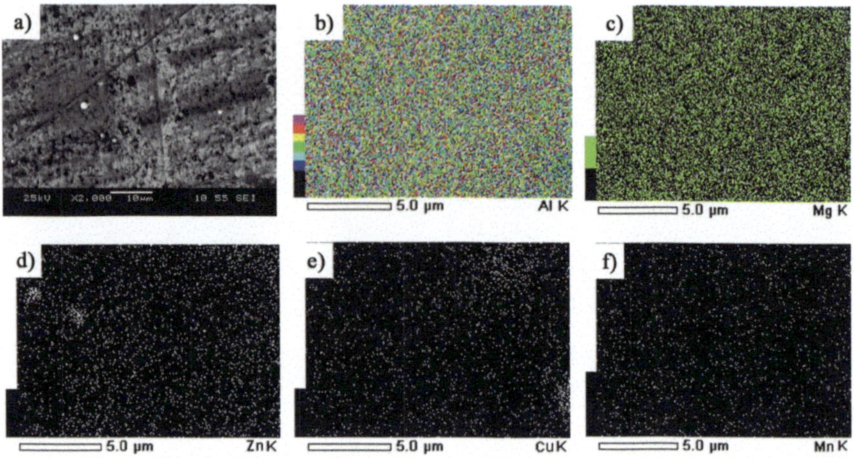

Fig. 4 EDS Elemental mapping of the SZ of the specimen under 20 mm/min **a** SEM image **b–f** the corresponding elemental mapping of Al, Mg, Zn, Cu and Mn, respectively

blowholes in the stir zone and lower strength. When very high traverse speed is used, it also ended up with poor lap shear fracture load, because of high heat generation increased grain growth. The loss in strength was attributed by lack of material flow, insufficient or very high heat output during stirring.

Figure 6 shows scanning electron microscopic images of fractured surfaces. The fracture morphology of Fig. 6 shows that the fracture was mainly occurred due to dimple fractures [21]. The fracture surface of the sample made with 20 mm/min showed many small and shallow dimples with the layered distribution as evidenced

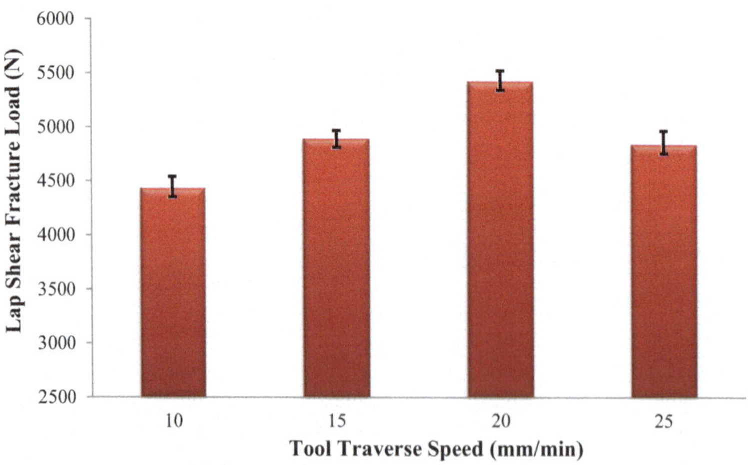

Fig. 5 Lap shear fracture load under different tool traverse speed

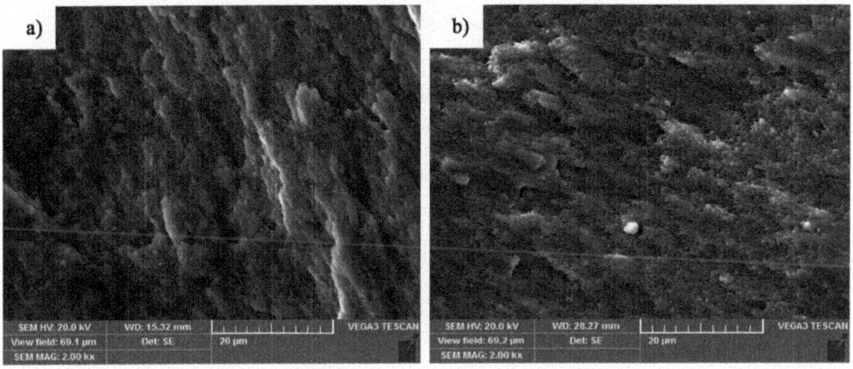

Fig. 6 Typical fracture surfaces of samples prepared at **a** 20 mm/min, **b** 25 mm/min

for ductile fracture (Fig. 6a). When the tool traverse speed was increased to 25 mm/min, the dynamic recrystallization in the SZ region was low due to high heat input [22]. As a result, the grain size was increased and joint strength was reduced. It is evidenced in the fracture morphology shown in Fig. 6b, where fracture was mainly dimple fractures with shallow and large dimples.

4 Conclusions

The conclusions of the findings can be summarized as follows.

1. Lap shear tensile load of the joint gets increased as tool traverse speed increases up to 20 mm/min, beyond which it starts decreasing. The maximum lap shear tensile load of 5430 N is recorded at 20 mm/min, which regarded as the optimum traverse speed for SFSSW 7075-T6 aluminum alloy.
2. The weld at optimum traverse speed shows much smaller equiaxed recrystallized grains and a much better uniformity of grains in the stir zone.
3. The wormhole defect was observed in the welded sample made under the tool traverse speed of 15 mm/min and 25 mm/min.
4. The lower tool traverse speed cause coarse grain microstructure and worm hole defect in weld region due to low heat input. Meanwhile, the tool traverse speed higher than 20 mm/min causes non-uniform coarse grain structure in SZ due to excessive heat generation.

References

1. Suresh S, Venkatesan K, Rajesh S (2019) Optimization of process parameters for friction stir spot welding of AA6061/Al2O3 by Taguchi method. In: AIP conference proceedings, pp 1–10. https://doi.org/10.1063/1.5117961
2. Zhong ZW (2021) Processes for environmentally friendly and/or cost-effective manufacturing. Mater Manuf Process 36:987–1009. https://doi.org/10.1080/10426914.2021.1885709
3. Madesh S, Arul TG, Perumal V (2022) Advances in welding techniques for similar and dissimilar materials. Adv Manuf Tech Eng Eng Mater. https://doi.org/10.4018/978-1-7998-9574-9.ch011
4. Bozkurt Y, Türker A, Salman S (2018) The investigation and comparison of friction stir spot welding and electrical resistance spot welding of AA2024 aluminum alloy joints. In: Proceedings of the ICENS 4th international conference on engineering and natural science, Kiev, Ukraine
5. Suresh S, Venkatesan K, Natarajan E, Rajesh S, Lim WH (2019) Evaluating weld properties of conventional and swept friction stir spot welded 6061–T6 aluminium alloy. Mater Express 9:851–860. https://doi.org/10.1166/mex.2019.1584
6. Omer MAE, Rashad M, Elsheikh AH, Showaib EA (2021) A review on friction stir welding of thermoplastic materials: recent advances and progress. Weld World. https://doi.org/10.1007/s40194-021-01178-0
7. Su ZM, He RY, Lin PC, Dong K (2014) Fatigue analyses for swept friction stir spot welds in lap-shear specimens of alclad 2024–T3 aluminum sheets. Int J Fatigue 61:129–140. https://doi.org/10.1016/j.ijfatigue.2013.11.021
8. Tweedy B M, Widener CA, Merry J D, Brown J M, Burford D A (2008) Factors affecting the properties of swept friction stir spot welds. In: SAE technical papers. https://doi.org/10.4271/2008-01-1135
9. Yang XW, Fu T, Li WY (2014) Friction stir spot welding: a review on joint macro- and microstructure, property, and process modelling. Adv Mater Sci Eng 2014:1–11. https://doi.org/10.1155/2014/697170
10. Selvamani ST (2021) Various welding processes for joining aluminium alloy with steel: effect of process parameters and observations–a review. Proc Inst Mech Eng Part C J Mech Eng Sci 095440622110596. https://doi.org/10.1177/09544062211059695
11. Neeru Singh S, Satsangi P S (2021) A review on friction stir spot welding of similar and dissimilar materials. In: Lecture notes in mechanical engineering book series (LNME) abstract, pp 531–541. https://doi.org/10.1007/978-981-33-4018-3_50
12. Jayabalakrishnan D, Balasubramanian M (2019) Friction stir weave welding (FSWW) of AA6061 aluminium alloy with a novel tool-path pattern. Aust J Mech Eng 17:133–144. https://doi.org/10.1080/14484846.2017.1373584
13. Jayabalakrishnan D, Balasubramanian M (2019) an over review of the effect of process parameters, tool geometry and joint geometry in friction stir welding and friction stir processing. Int J Appl Eng Res 10:448–455
14. Suresh S, Venkatesan K, Natarajan E, Rajesh S (2020) Influence of tool rotational speed on the properties of friction stir spot welded AA7075-T6/Al2O3 composite joint. In: Materials Today: Proceedings, pp 62–67. https://doi.org/10.1016/j.matpr.2019.08.220
15. Li Y, Sun D, Gong W (2019) Effect of tool rotational speed on the microstructure and mechanical properties of bobbin tool friction stir welded 6082-T6 aluminum alloy. Metals (Basel) 9. https://doi.org/10.3390/met9080894
16. Yoon JY, Kim C, Rhee S (2016) Effect of circumferential tool path control on friction stir spot welding of Al/Fe dissimilar metal joint. J Weld Join 34:6–11. https://doi.org/10.5781/jwj.2016.34.3.6
17. Suresh S, Venkatesan K, Natarajan E, Rajesh S (2021) Performance analysis of nano silicon carbide reinforced swept friction stir spot weld joint in AA6061-T6 alloy. SILICON 13(10):3399–3412. https://doi.org/10.1007/s12633-020-00751-4

18. Suresh S, Elango N, Venkatesan K, Lim WH, Palanikumar K, Rajesh S (2020) Sustainable friction stir spot welding of 6061–T6 aluminium alloy using improved non-dominated sorting teaching learning algorithm. J Mater Res Technol 9:11650–11674. https://doi.org/10.1016/j.jmrt.2020.08.043
19. Subramani V, Jayavel B, Sengottuvelu R, Lazar P (2019) Assessment of microstructure and mechanical properties of stir zone seam of friction stir welded magnesium AZ31B through nano-SiC. Materials (Basel) 12:1044. https://doi.org/10.3390/ma12071044
20. Sampath B, Haribalaji V (2021) Influences of welding parameters on friction stir welding of aluminum and magnesium: a review. Mater Res Proc 19:222–230. https://doi.org/10.21741/9781644901618-28
21. Suresh S, Venkatesan K, Natarajan E (2018) Influence of SiC nanoparticle reinforcement on FSS welded 6061-T6 aluminum alloy. J Nanomater. https://doi.org/10.1155/2018/7031867
22. Tao W, Shuwei D, Matsuda K, Yong Z (2022) Dynamic recrystallization and dynamic precipitation in AA6061 aluminum alloy during friction stir welding. Trans Indian Inst Met. https://doi.org/10.1007/s12666-021-02490-5

Establishing Flow and Improving Flow Velocity in Manufacturing Systems—Pointers from Theory of Constraints

Jagannathan Sekkizhar and Nagarajan Vivek

Abstract A unique aspect proposed by the theory of constraints (TOC) body of knowledge is in the process of creating flow. This approach is distinct from the current lean practices which focus on balancing of lines instead of adopting a system-driven approach. Both lean and TOC approaches have been put together after observing the Toyota production system, but each differs in the understanding and conceptualisation of the way it works. The TOC approach is far more specific on how to establish flow and improve flow in any manufacturing system and gives results much more quickly than other comparable methods. This paper tries to codify the principles of flow from the TOC approach.

Keywords Constraint · Bottleneck · Stock · Intermittent · Pacemaker process · Five focussing steps · Flow · Supermarket · Customer order schedule

1 Introduction

Lean-based manufacturing concept is derived from age old Toyota production system and most used improvement mechanism in manufacturing companies. This field of thinking originated at MIT Sloan School of Management when Womack et al. [12] wrote a book on this subject. This was also summarised by Dekier [3]. However, this did not capture all of the practices that go on inside the amazing Toyota production system. Later Rother and Shook [9] wrote a book that captured many of these principles in a better way as Shook was from the Toyota Nummi plant and a native Japanese speaker. So he understood the concepts more deeply than the earlier works. They have specified many concepts on lean that are missing in today's lean implementations. Many companies showed a good improvement in their operational performance, but

J. Sekkizhar · N. Vivek (✉)
PSG Institute of Management, PSG College of Technology, Coimbatore 641004, India
e-mail: vivek@psgim.ac.in

J. Sekkizhar
e-mail: sekkizhar@psgim.ac.in

E. Natarajan et al. (eds.), *Materials, Design and Manufacturing for Sustainable Environment*, Lecture Notes in Mechanical Engineering,
https://doi.org/10.1007/978-981-19-3053-9_42

not all. However, a significant number of companies are not able to adopt lean practices as described by the consultants and fail in their productivity too [7], (Henao et al. 2019). Though the idea of eliminating waste in all forms by focussing on the activities that create value for the customer (Womack and Jones, 1996) looks very straight forward. But in reality, it has some missing pieces, without which the results derived from the efforts are highly variable and not always useful. Production scheduling and effective utilisation of resources are the key drivers of lean management (Matt and Rauch 2013) as per many practitioners. However, these things are not properly done in the proper position (Goldratt 1994) and leads to failure in the improvement implementation. Poor goal setting and unaligned measurements with such goals are the main reason behind the lean failure [11]. These issues were addressed by Goldratt and Cox [5] where the systems concept of dealing with these issues was decoded and a standard set of rules was framed. Here there is a distinction between pull and flow very clearly as was originally intended in the Toyota production system. It has also been observed that lean gives results much more reliably when used in association with TOC framework, than on its own [8].

2 Problems in Lean (Pull versus Flow)

The entire lean world works towards a pull mindset. While pull production does have its benefits, this was not the original goal of the earliest manufacturing wizards, most notably Ford motif corp. This entire goal was achieving production flow [4]. The Ford production process was the role model around which Taichi Ohno tried to build the Toyota production process. He too tried unsuccessfully for a long time to establish the flow principles under Japanese conditions and after a lot of trial and error managed to introduce an intermediary concept called the pull process. Prior to this, manufacturing knew only two concepts, push or flow. A pull process is assumed by most people to be the Kanban system. While the Kanban is an integral part of the process, it is only a signal to withdraw materials or a signal to start production. What is overlooked here is the presence of the bins, which are a form of decoupling stock. So in reality, the pull system is serving to synchronise two processes with differing takt times to work together without any stoppages to the flow of materials in the system by adding a decoupling stock between them. Another important point which almost all lean implementers who advocate the Kanban system overlook, is the location of this decoupling stock [9].

Roser [10] has suggested that using a single larger loop requires statistically significant less inventory for the same delivery performance than two smaller loops. This implies that one should not overdo the pull system and stick to pull only in front of the pacemaker process. Too many supermarkets complicate a production system.

Any system is characterised by a bottleneck process which is termed as the pacemaker process. The entire takt of the system is decided based on the capacity of this pacemaker process. While many lean experts do acknowledge this fact, most neglect

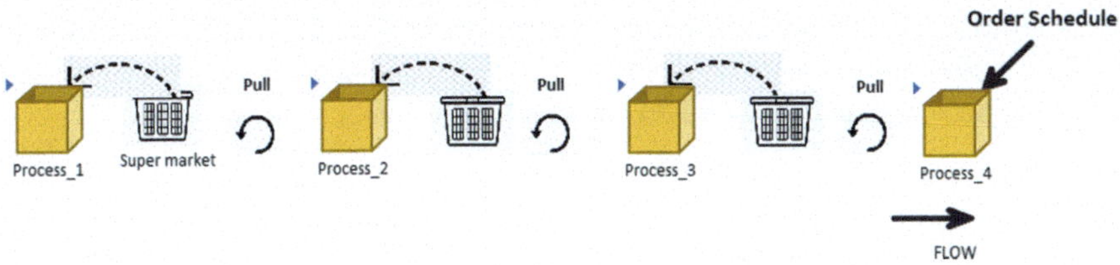

Fig. 1 Incorrect configuration in lean implementations ([9] *Source*: Rother and Shook)

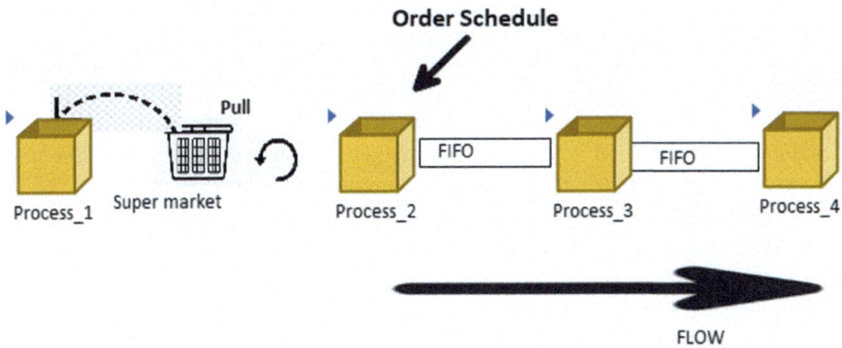

Fig. 2 Correct configuration in lean implementation ([9] *Source* Rother and Shook)

the next part of the rule, where it is mentioned that the pull should happen using decoupling stock only before the pacemaker process and not everywhere in the system. The rest of the system should be in a normal flow process. This is demonstrated in the two configurations below. Note two key points, the position of the supermarket and the process receiving the customer order schedule. Most lean implementations leave out these two critical points, which has been addressed in more detail by the TOC approach (Figs. 1 and 2).

Added to this, there is another critical rule. The customer pull should be given directly to the pacemaker process and not the last process in the system as most lean implementations strive to do. This is one point that can change the way a Kanban is used within a system. This represents a fundamental shift in the way a Kanban is used within a lean system.

Although a lean system does acknowledge the pacemaker process, it talks of balancing the different process capacities. But in most cases, this capacity is not stable and the cycle time is highly variable, leading to system flow disruptions and build-up of stock in all unlikely places. Again, the concept of waste elimination too is not focussed to benefit the system as a whole and is only aiming at local efficiencies. This was never the goal of the Toyota production system. The main aim of the Toyota production System was to minimise the material inside the system, not reduce it to zero. The system should have the minimum necessary stocks to maintain a perfect flow within the system and the entire concept of value stream mapping, originally called mojo (material and information flow map) at Toyota aimed at identifying problems with flow and eliminating them [9]. These lacunae in the lean process need to be filled, if it is to be more effective in yielding results.

3 Theory of Constraints

Goldratt and Cox [5] presented an improved approach to fill in the missing pieces in the lean body of knowledge.

TOC approach is based on four pillars [1] and extended with the fifth one by Baptista [2].

1. Inherent simplicity: Even complex systems have an underlying simplicity in them and can be understood in terms of simple, yet reliable variables. The main goal is system performance and not individual components themselves.
2. Inherent harmony: Most production systems run with conflict of interests. This is often because of assumptions on part of various people within the system, mostly thinking on local optimisation goals rather than system goals.
3. Inherent goodness of the people: This is a foundation which involves creation of respect culture within the organisation. Toyota swears by this word at each and every level. All other points disappear if this principle is neglected
4. Inherent potential: TOC is an eternal optimist. The belief is that, if other points are in place and the improvement process is properly conceptualised, any process can be improved in a short time.
5. Inherent tolerance: Don't optimise within the noise. All human-based organisations have fluctuations. Reacting to all regular fluctuations ends up amplifying variability. Planning and execution must explicitly account for them with simple mechanisms to withstand regular fluctuations (without any further actions) and react to extraordinary one.

They came up with a process of ongoing improvement (POOGI) as a set of five steps. These steps are listed below.

1. Identify the constraint resource or process
2. Exploit the constraint resource or process
3. Subordinate all other processes to the constraint resource or process
4. Elevate the constraint resource or process
5. Continue the process by going back to step 1.

The pacemaker process here is given the specific term constraint [5]. This is slightly different from the term bottleneck, but used synonymously with it for practical purposes. Here the constraint is identified by looking at the resource that has the lowest capacity. Not a process per say. Even in the TOC world, many implementations look at the process, but the unwritten rule is to look at the resources. But a process capacity is a result of the resource capacity in the first place. It is easiest to measure the capacity of a process by looking at the resources being used to complete the process. Often sharing of resources can cause disruptions in the process. This is referred to as multitasking problems in the TOC world. The most common practice is to identify a process with the lowest capacity. Usually, a process is a system and does not give us the full understanding of the major influencer of the capacity of the process. TOC approach implies a fundamental rule, "The capacity of a system is equal to the capacity of the constraint". This translates to some more implications.

1. The non-constraints are usually having surplus time
2. Improvements in non-constraints or elimination of waste in non-constraints does not result in improvement of system output
3. Any loss of productivity on the constraint leads to a loss of productivity of the entire system.
4. Similarly, any improvement in the system constraint leads to an improvement in the entire system's output.

4 Establishing Flow

Once the constraint has been identified, a huge problem has been solved with regard to system capacity. This is where TOC charts a new path. Steps 2 and 3 do something unique. In step 2, we acknowledge that the system capacity is the capacity of the constraint process. The term used here is exploit. This is not a negative term at all. It just refers to maximum utilisation of the constraint process without any waste of any kind. In most manufacturing facilities, this alone can improve the productivity of the entire system with very little need to change the processes or equipment. Once the capacity utilisation at the constraint is maximised, the next step is critical to establishing a flow. This is missed out in all lean systems. Step 3 subordinates all other non-constraint processes to the constraint. All their operating plans are subordinated to synchronise with the constraints operating plans. Even the external order coming into the system is aimed at the utilisation efficiency of the constraint [9]. This act of subordination ensures that the entire system is working to the pace of the constraint and works like an integrated system. The constraint just finishes its operations and pushes the product to the next process which is ready and waiting to process it. As a rule, the non-constraint resources will have idle time. This time is used as a time buffer to ensure smooth production flow instead of physical stocks. These points lead us to the following hypothesis.

H1: Reducing waste of resources at constraint maximises the system capacity.
H2: Subordinating the non-constraint activities to the constraint leads to the establishment of flow in the system.
H3: Customer orders, if mapped to the resource availability at the constraint, lead to easy management of the production flow.
H4: To establish flow in the system, it is not necessary to balance the capacities of all the processes. Imbalance in capacity is not a hindrance to flow. It is the improper resource loading process that leads to flow stoppages.

H5: Overloading the system, exceeding the constraint capacity, will lead to system stress and the output will not increase by any amount. The flow velocity will be equal to the capacity of the constraint in the system.

5 Increasing the Flow Velocity

Once points 2 and 3 are achieved, meaning the constraint is exploited, and the non-constraint activities are all subordinated to the constraint's production speed, any system will attain production flow which is stable and the system will not have any disruptions or stoppages. And this is not achieved using any capacity balancing process. This is capable of serving a takt that is equal to the flow velocity of the system. Now if a higher flow velocity is desired, the requirement is to increase the capacity of the constraint. It is only in this stage, that after establishing flow in the system, we strive to increase output velocity or production output. The constraint capacity increase will lead to increased system output provided two conditions are satisfied.

1. Steps 2 and 3 are ensured along with the constraint capacity increase
2. The capacity increase does not result in another process becoming the new constraint

Most improvement specialists work towards breaking a constraint by elevating it. But this step comes only after one has the constraint delivering peak performance and flow being established in the system. Elevation can be accomplished by automation, using additional resources, creating a changed process flow, etc. At the end, what matters is that the system operates on flow principles instead of just maximising productivity at every point in the system. These arguments give us the following hypothesis.

H6: TOC works to elevate flow velocity after establishing flow first.
H7: Flow velocity elevation still requires steps 2 and 3 in the TOC focussing steps to be effective.
H8: Elevation of flow velocity can very well use the principles of lean management. The effectiveness of lean work will be that much more effective if TOC steps 2 and 3 are in place first.

6 Conclusion

Flow is conceptualised in lean literature as one piece flow. But it is not about the one piece, it is about the entire process working without any interruptions. To establish flow, many of the traditional steps might not be required in the first place. All it requires is a good system study, where principles of TOC are applied to establish the flow and to improve the flow velocity. The best part is that lean can be used unchanged once the other steps in TOC are put in place. This will result in consistent improvements in the improvement efforts.

References

1. Almeida GVD, Souza FBD, Baptista HR, Gupta MC, Campos RD (2018) Fundamental principles of the toyota way and the theory of constraints: comparative analysis and synthesis. Int J Services Oper Manag 30(1):51–71
2. Baptista HR (2021) The TOC pillars—a whitepaper, https://cdn.ymaws.com/www.tocico.org/resource/resmgr/toc_bok/papers/the_toc_pillars_article_1.3.pdf
3. Dekier Ł (2012) The origins and evolution of Lean Management system. J Int Stud 5(1):46–51
4. Flanders WE (1909) A talk with flanders. Sci Am 100(3):55–57
5. Goldratt EM, Cox J (1984) The goal: excellence in manufacturing. North River Press
6. Helleno AL, de Moraes AJI, Simon AT, Helleno AL (2017) Integrating sustainability indicators and Lean manufacturing to assess manufacturing processes: application case studies in Brazilian industry. J Clean Prod 153:405–416. https://doi.org/10.1016/j.jclepro.2016.12.072
7. Jadhav JR, Mantha SB, Rane S (2014) Exploring barriers in lean implementation. Int J Lean Six Sigma 5:122–148. https://doi.org/10.1108/IJLSS-12-2012-0014
8. Moore R, Scheinkopf L (1998) Theory of constraints and lean manufacturing: friends or foes. Chesapeake Consulting Inc
9. Rother M, Shook J (2003) Learning to see: value stream mapping to add value and eliminate muda. Lean Enterprise Institute
10. Roser C (2016) Supermarket vs FIFO—what requires less inventory? April 12, 2016 https://www.allaboutlean.com/supermarket-vs-fifo-wip/
11. Rubrich, L. (2004). *How to prevent lean implementation failures: 10 reasons why failures occur*. WCM Associates.
12. Womack JP, Jones DT, Roos D (2007) The machine that changed the world: The story of lean production–toyota's secret weapon in the global car wars that is now revolutionizing world industry. Simon and Schuster

Buffer Sizing, Resizing and Monitoring in TOC DBM Approach with Respect to Seasonal Products—A Simulation Study

Jagannathan Sekkizhar and Nagarajan Vivek

Abstract Stock demand has different kinds of components. Irregular components are a highly unpredictable phenomenon in most manufacturing firms. And added to this, it rarely follows any distribution at all. Theory of constraints presents a simplified method to deal with random fluctuations in inventory demand called dynamic buffer management. This method also considers resizing buffers depending on other components of demand variation. This paper proposes a framework to deal with seasonal fluctuations in demand using the dynamic buffer management. Most current methods use highly mathematical models to deal with the stocks and safety stocks when dealing with seasonal demand patterns. This makes computations cumbersome and difficult to understand for most people. Another difficulty is setting policies for complex mathematical models which can be executed in the workplace with ease. On these counts, the model presented in this paper comes up with the answers to the problem, whilst maintaining the simplicity of policy-based implementation.

Keywords Dynamic buffer management · Demand · Lead time · Seasonal variation · Zones of inventory · Seasons · Ordering frequency · Lead time

1 Introduction

The main aim of any manufacturing operation is to meet customer demand. For this, to meet any unexpected changes in the expected demand, every firm maintains stocks (Ikezeri et al. 2019). The different stocks in use are cycle stocks to meet production needs, safety stocks (to meet changes in demand or lead times), pipeline stocks (to identify in transit stocks), Anticipatory stocks (to meet unexpected shortages or price rises), dead stock (non-moving items) [12]. Stocks make up a huge

J. Sekkizhar · N. Vivek (✉)
PSG Institute of Management, PSG College of Technology, Coimbatore 641004, India
e-mail: vivek@psgim.ac.in

J. Sekkizhar
e-mail: sekkizhar@psgim.ac.in

E. Natarajan et al. (eds.), *Materials, Design and Manufacturing for Sustainable Environment*, Lecture Notes in Mechanical Engineering,
https://doi.org/10.1007/978-981-19-3053-9_43

chunk of the invested money in a firm and hence need careful planning [1]. One of the major functions of stock is to ensure the constant availability of materials for a production process, whilst keeping the levels under control, meaning it should not be too much which implies extra investment, and at the same time it should not be too little implying stock outs. There are several approaches to dealing with demand fluctuation, but most use complex approaches like reorder points, computerised forecasting models, etc. Theory of constraints (TOC) aims at formulating a policy driven approach to dealing with this situation, which is at the same time simple enough to implement [8].

2 Components of Demand

Demand pattern has the following components if considered as a time series [1, 12].

1. Trend: This component represents the long term rise, fall or stable nature of the demand. It spans over a 10 year time frame. This paper assumes a flat trend for the computations.
2. Cycles: This is a periodic up and down movement in demand. It takes place every three or four years. It is commonly used in economic cycles like boom, recession, depression and recovery. But in more general terms, it is an up and down periodic movement of demand. This paper neglects cyclical variations in demand patterns.
3. Seasons: This component is similar to cycles, but happens within the span of a year. It is usually classified as peak season, off season and normal season. Peak season represents a highly increased demand, whilst off season represents highly reduced demand. The point about seasonality is that the periodicity can be repeating patterns within a day, within a week, within a month, within a quarter or within a year. The key point about assessing seasonal patterns is in identifying peak season and off season timings. This component is being discussed in this paper.
4. Random: This represents an unpredictable very short term variation that keeps repeating, but cannot be predicted by any identifiable pattern. It is the most difficult pattern to deal with and is usually eliminated in analysis by a process called smoothing. This paper deals with this component as well.
5. Irregular: This component refers to a rare movement in demand due to special events like an earthquake, tsunami, wars, etc. This component is excluded from the scope of discussions in this paper.

3 TOC Approach to Managing Inventory

The theory of constraints approach divides the inventory management efforts into three policies instead of trying to forecast daily demand and keep inventory levels.

Instead the stock level is measured in three easy to calculate zones. This alone is sufficient to indicate whether the stocks are adequate, too high or too low [9]. Further, it also tells us how to recompute buffer levels and also account for pipeline stocks, a topic which is neglected in most inventory management philosophies. Right inventory management is the key to proper material flows in a manufacturing system [16]. The four policies that run a TOC inventory system are,

1. Ordering policy
2. Buffer sizing policy
3. Stock level monitoring policy
4. Buffer resizing policy.

3.1 Ordering Policy

This is the simplest of the three. TOC, being based on the principles of flow and pull, uses this very same principle to manage ordering. Flow implies that no stocks are maintained in the system. But in case of asynchronous operations between two processes, there is the need for a stock. The replenishment of stocks is based on actual consumption of the stocks and not any forecast [2]. The ordering is done using a periodic ordering policy with variable order quantities rather than the usual fixed quantity batch ordering with a variable ordering frequency and reorder point reminders in the system. Here the orders placed are based on actual inventory consumption [10]. If we assume weekly demands (this will work for daily, monthly and quarterly demands as well) and three different periods for reordering, weekly, once in two weeks and once in three weeks, the orders placed would look as follows (Table 1).

Weekly ordering simply replaces what is withdrawn for that week's consumption. Once in two weeks ordering policy replaces two week consumption whilst placing the order and once in three weeks ordering policy replaces three week consumption whilst ordering. Goldratt has specified that shorter the ordering period, the lower will the

Table 1 Ordering frequency

Week	Demand	Order placed every week	Order placed, once in two weeks	Order placed once in three weeks
1	37	37		
2	35	35	72	
3	43	43		115
4	41	41	84	
5	45	45		
6	33	33	78	119
7	27	27		
8	34	34	61	
9	29	29		90

stock levels be and better the inventory turnover. This is a complete ordering policy. No other variables and rules are to be considered. Those issues will be addressed by the next two policies [3, 4].

3.2 Buffer Sizing Policy

Buffer sizing is a difficult policy under any inventory management system. TOC approach makes it easier by making certain practical assumptions. The two main determinants of buffer size are demand and lead time of replenishment [5, 6]. In reality both are variable. In this paper we consider variable demand and a fixed lead time of 2 weeks, meaning if an order is placed, then the stocks will arrive after two weeks without fail. This model can also be extended for variable lead time, but that is out of scope of this paper [11]. Most researchers assume distributions for demand and multiply the different components to generate the demand. This paper takes a different approach. The demand is assumed to be seasonal in nature. Within a season, the demand is assumed to have random variations, with only the maximum and minimum values specifying the variation limits. When the season changes, this band shifts upwards or downwards. Being random, unpredictable demand is being considered for the replenishment simulation exercise. The buffer here comprises stock on hand, in transit stocks and orders being delivered. So the only visible component is stock on hand. The buffer size is calculated by a very simple formula given below [10].

$$\text{Buffer size} = \text{Demand} \times (\text{lead time} + \text{ordering frequency})$$

Now if the demand and the lead time are variable, assuming there are no outliers in either demand or lead times, and the variation in both are random within given upper and lower limits, the formula becomes

$$\text{Buffer size} = \text{peak demand} \times (\text{maximum lead time} + \text{ordering frequency})$$

This formula removes the need to keep safety stock, which will be demonstrated in the simulation experiment being done. Some alternate suggestions use average demand and average lead times, in which case a safety stock will be required in the buffer level. But in general, the safety stock in TOC is represented by a red zone, which is explained below.

This will be plotted in a three zone band to monitor the stock levels and trigger alarm signals if the stock goes too high or drops too low. The bands in the stock level monitor are extremely simple and are generated by dividing the buffer into three zones. 0–33% representing the Red zone or the danger zone, 33–66% representing the yellow or the optimum zone, 66–100% or the overstocked zone. The red zone represents the safety stock also (Table 2).

Table 2 Stock level indicator

Buffer zone	Measure (%)
Green	66–100
Yellow	33–66
Red	0–33

3.3 *Stock Level Monitoring Policy*

There are not too many rules on how to identify overstock and understock conditions. Different papers give different rules for this. But the generally accepted rule is, if too many weeks show the inventory in the green zone continuously, the system probably has over stock. On the other hand, if too many reds are observed, it could indicate low stock levels. It is ideal if most of the weeks the stock level appears in the Yellow zone. A few greens and a few reds are not of much concern, unless there is a sharp drop or rise in stock levels (Simsit et al., 2014).

3.4 *Buffer Resizing Policy*

Goldratt [3] indicates that a number of weeks in the green zone or red zone indicate the need to resize buffer levels. Buffers should not be resized too frequently. The thumb rule is to check stock levels once in 60 days and decide on resize options. The general rule prescribed is to reduce buffer size by 33% if too many greens are discovered and increase buffer by 33% if too many reds are spotted. This approach does not consider seasonal changes [16]. The obvious reason to resize buffer levels is the changes that happen in demand and lead time values over the year. This paper tries to address this issue. Suppose the general seasonal trends are known and the bands of demand are known, the resizing approach is equally applicable without any complex formulae being used. Let us assume that in a season the demand is $D1$ with a lead time of $L1$. Assuming ordering frequency f, the buffer level to be used is $D1 \times (\text{L1} + \text{f})$. Now when the season changes and the new seasonal demand is $D2$ with a lead time of $L2$, the new buffer level will become $D2 \times (L2 + f)$. This works for both increase and decrease in seasonal demand [15]. The only question which remains is how to achieve the change in buffer levels, given that the buffer level itself is just a notional number and the physical inventory never equals this level. It is more of a monitoring graph for stock levels. The transition from demand $D1$ to demand $D2$ is to be carefully considered. This is accomplished by making certain alterations to the orders placed using the rendering policy. If $D2 > D1$, it implies a transition from slack season to peak season. Assuming $\Delta = D2 - D1$ Here Δ is the change we should bring about in buffer levels to manage the new demand levels. This is accomplished by placing an additional order of Δ along with the planned orders being placed. Sometimes, if Δ is large, it may be split over a few periods instead of placing all of it in one period itself. This way the buffer level gradually increases to a new level. By the same logic, if $D2$

$< D1$, Δ becomes negative and has to be reduced from orders being placed. Here too this reduction can be distributed over more than one time period and the buffer level gradually decreases to the desired level. The seasonal adjustments become a little more difficult if there is a sharp change in demand between seasons. For most slight changes, Δ seamlessly manages the transition. For multiple products, each having a different selling peak season, the stocks of each can be planned separately using this very same approach and monitored independently of one another.

4 Simulation Experiment Design

The simulation experiment assumes the following data (Table 3).

Three seasons have been considered. Season 2 shows an increase over season 1. Season 3 shows a decrease when compared to season 2. Lead time is kept constant as 2 weeks. Assumed ordering frequency is once a week. The demand is generated using a random function assuming uniform distribution between maximum, and minimum values in a season. Initial stock at the beginning of the simulation is assumed to be the buffer size of season 1, i.e. 600 units. But since the order lead time is 2 weeks and at the beginning of the game, for the first three weeks nothing will arrive. Seasons change every 12 weeks. The first 4 weeks are neglected to allow for stabilisation of the system. An additional stock of 450 units (150 units $\times$ 3 weeks) is added to the buffer in week 1 to allow for stabilisation of the buffer size at 600 units. This simulation is run for a period of 36 weeks, to study the transition during three types of seasons. The buffer resizing is done manually with full knowledge about demand bands in each season, unlike the automated transition in most studies. Here minor demand shifts are not considered and only big seasonal shifts are considered to address the seasonal shifts in the TOC DBM framework.

Since random numbers are being used, each simulation run will give different answers for stock positions. Here the results for 10 such simulations are demonstrated to show smooth working of the TOC buffer management system in maintaining the right quantity of stocks in the system. The software used is google sheets. A snapshot of the results for one iteration are shown below.

Table 3 Seasonal demand pattern

Season	Demand min	Demand max	Lead time (weeks)	Buffer size = max demand $\times$ (lead time + ordering frequency)	Changeover Δ
1	100	200	2	600	
2	300	400	2	1200	600
3	150	300	2	900	−300

N	O	P	Q	R	S	T	U
week	Demand	Opening stock	Order placed	Order received	Closing stock	stock status	Buffer level
1	129	600	129	129	600	G	600
2	175	600	175	175	600	G	600
3	127	600	127	127	600	G	600
4	105	600	105	129	624	G	600
5	157	624	157	175	642	G	600
6	114	642	114	127	655	G	600
7	186	655	186	105	574	G	600
8	116	574	116	157	615	G	600
9	113	615	113	114	616	C	600
10	123	616	123	186	679	G	600
11	200	679	200	116	595	G	600
12	161	595	161	113	547	G	600
13	173	547	323	123	497	C	600
14	154	497	304	200	543	G	600
15	117	543	267	161	587	G	600
16	155	587	305	323	755	G	600
17	330	547	330	304	521	Y	1200
18	339	521	339	267	449	Y	1200
19	334	449	334	305	420	Y	1200
20	347	420	347	330	403	Y	1200
21	376	403	376	339	366	R	1200
22	335	366	335	334	365	R	1200
23	336	365	336	347	376	R	1200
24	392	376	392	376	360	R	1200

er 3 ▾

Simulation Experiment results (Table 4).

A bulk of the weeks are in Yellow and Green zones, indicating that the system is working well and stock outs are under control [14]. The few reds that occur, happen at the junction of the seasonal changes. This is because of purposeful reduction in ordered quantity to adjust for the new seasonal demand. However the number of greens is slightly on the higher side, indicating that this approach is erring slightly on

Table 4 Simulation results

Simulation run	R	Y	G
1	0	20	16
2	4	20	12
3	2	20	14
4	0	16	20
5	7	15	14
6	5	18	13
7	0	17	19
8	0	14	22
9	8	14	14
10	0	17	19

Average Reds: 2.6 / 36, Average Yellows: 17.1 / 36, Average Greens: 16.3/36 the higher side. But, in spite of this, TOC DBM seasonal extension framework seems to be working without much of a problem. These observations allow us to propose the following hypothesis that could be tested more rigorously in future research studies.

H1: Random demand variations can be handled well by just following the ordering policy and stocking policy used in traditional TOC DBM framework.
H2: Inventory levels are best managed when the buffer levels are set using peak demand and peak lead times. This avoids the need to create a specific safety buffer stock.
H3: Whilst computing buffer levels, it is essential to eliminate outliers in demand and lead time data as these could distort the buffer sizes.
H4: Buffer resizing is best done manually after confirming the onset of a new season.
H5: Buffer resizing can be simplified using the standard peak demand and peak lead time knowledge about the new season.
H6: Transition to new buffer levels is best done gradually over a few periods of time, rather than a single big adjustment, especially when the buffer size is being reduced.
H7: The random patterns of demand within a season have no negative impact on the TOC DBM stock management framework.

5 Conclusion

The TOC DBM framework of stock management has been studied under different demand scenarios, but not a seasonal demand variation. Even the few studies that mention seasonality, the demand pattern shows gradual transition. Further these studies focus on automated buffer level transition and accomplish the transition abruptly every 60 days or so after checking out the number of reds and greens in the chart. However these studies totally overlook prior demand knowledge that most organisations have about peak seasons and off seasons. This study has filled this lacuna by extending the DBM model to work with gradual buffer level transition between seasons and random demand patterns within a season. This can be further tested using advanced modelling techniques. With some minor modifications, this model can be extended to account for trend and cyclical demands as well. The results obtained in this study are consistent with the expectations of the literature on theory of constraints, though not many simulation studies have been done in this area. The results demonstrate the possibility of using a policy based inventory management under seasonal demand patterns and achieve acceptable levels of inventory safety, whilst avoiding excess inventory.

References

1. Akyuz, A. O., Uysal, M., Bulbul, B. A., & Uysal, M. O. (2017). Ensemble approach for time series analysis in demand forecasting: Ensemble learning. In *2017 IEEE International Conference on INnovations in Intelligent SysTems and Applications (INISTA)* (pp. 7–12). IEEE.
2. Dalal, K., & Shah, S. (2015), Dynamic Buffer Management, International Journal of Innovative Research in Advanced Engineering (IJIRAE) ISSN: 2349–2163, 7(2)
3. Goldratt, E. M. (2002). *TOC on Distribution and Supply Chain: A Self Learning Program.*
4. Goldratt, E. M. (2009). Standing on the shoulders of giants: production concepts versus production applications. The Hitachi Tool Engineering example. *Gestão & produção, 16*(3), 333–343.
5. Goldratt EM, Cox J (2003) Production the TOC way. North river press, Great Barrington
6. Goldratt EM, Cox J (2016) The goal: a process of ongoing improvement. Routledge
7. Ikeziri LM, Souza FBD, Gupta MC, de Camargo Fiorini P (2019) Theory of constraints: review and bibliometric analysis. Int J Prod Res 57(15–16):5068–5102
8. Mabin VJ, Balderstone SJ (2020) The world of the theory of constraints: a review of the international literature. CRC Press
9. Moyo, B. (2016). *The Implications of the Theory of Constraints on Inventory Buffers.* University of Johannesburg (South Africa).
10. Reyes, J., Alvarez, K., & Vasquez, R. (2016). Dynamic buffer management for raw material supply in the footwear industry. *Journal of Industrial and Intelligent Information Vol, 4*(1).
11. Schragenheim, A (2010), Theory of Constraints Handbook, Mc Graw, Hill, New York.
12. Shah, J. (2009). *Supply chain management: text and cases.* Pearson Education India.
13. Şimşit ZT, Günay NS, Vayvay Ö (2014) Theory of constraints: a literature review. Procedia Soc Behav Sci 150:930–936
14. Stefano, G. D. S. (2020). Does the theory of constraints in supply chain management really matter? An assessment of the impacts of the TOC in the redesign of a supply chain. Master Degree Thesis, Unisinos University.
15. Wright J, King R, Goldratt EM (2006) We all fall down. North River Press, Great Barrington, MA
16. Yuan KJ, Chang SH, Li RK (2003) Enhancement of theory of constraints replenishment using a novel generic buffer management procedure. Int J Prod Res 41(4):725–740

Thermal and Fluid Sciences for Sustainability

Start/Stop Behavior of Indian Two-Wheeler Commuters in Traffic Signals: Repercussions and Propositions

J. Nagarjun, T. Prem Kumar, S. Suraj, and S. Vivek

Abstract The study was aimed at analyzing the start/stop behavior of two-wheel riders in Indian road signals and their implications on the vehicle fuel consumption and environmental pollution. The behavior was studied with the help of record of response from the two-wheel riders to the questionnaire developed. The responses were obtained by means of household interviews and Google survey from Coimbatore, Tamil Nadu, India. The minimum sample size was obtained using Cochran formula, and responses of 388 people (confidence level 95%, margin of error 5%) were recorded. The study revealed only 54.13% of the Indian public sample switch off their engine during wait time in signals. The reasons behind their start/stop response were received and analyzed. The wait time in signals at which the people respond is also recorded. Then experimentation was carried out recording the fuel consumption and CO_2 emission. It showed lower, moderate, and high-engine capacity bikes save fuel when the engine is switched off on signal timer 21.34 s, 8.34 s, and 6.08 s, respectively. When the suggested break-even time was followed, it resulted in reduction of fuel consumption and environmental pollution as much as 3.82 ml and 7.90 g, respectively, per minute of wait time in traffic signals. It was found only 8.74% of respondents have start/stop response as suggested by the study. The study revealed the people's understanding of start/stop response to the traffic signals needs major revision. The better response may lead to economic and environmental benefit to oneself and the society.

Keywords Start/stop behavior · Fuel consumption · Traffic signals · Two-wheeler · CO_2 emissions

J. Nagarjun · T. P. Kumar (✉) · S. Suraj · S. Vivek
Department of Mechanical Engineering, PSG Institute of Technology and Applied Research, Tamil Nadu, Coimbatore 641062, India
e-mail: prem50567@gmail.com

E. Natarajan et al. (eds.), *Materials, Design and Manufacturing for Sustainable Environment*, Lecture Notes in Mechanical Engineering,
https://doi.org/10.1007/978-981-19-3053-9_44

1 Introduction

Transport sector plays a major role in people's day-to-day activities. The transportation consumes lot of energy in order to operate. The primary source of energy for transportation is fossil fuel. Though the advent of electric vehicle will slowly decrease the dependence on fossil fuels, but still, petroleum vehicles are expected to stay in Indian roads as people's favorite for at least a decade to come. The major advantage is that it requires less initial investment on the vehicle. But the disadvantages are plentiful such as it incurs high-fuel expenditure and depletes natural resources which results in ever increasing demand on imported petroleum products [1]. Not only that, the fuel combustion pollutes the air, water, and soil [2]. The fossil fuel combustion generates 65% of the global greenhouse gas (GHG) emissions. The central part of cities where the traffic jam occurs, high concentration of pollutants such as CO, CO_2, VOCs, So_x, NO_x, and O_3 [3, 4] particulates may have indirect consequence on climate change and in addition may have adverse effects on human health [5]. Since the oil is majorly consumed by the road transport sector in the urban areas, the much of the effort is to be dedicated in reducing total man-made GHG emissions in those areas [6].

The absolute emission level of four wheelers has been reduced significantly in the last 20 years due to strict applicable limits on emission norms. On the other hand, the after treatment used for two-wheelers has not been efficient as cars. Hence, two wheelers are main urban source of unregulated pollutants hazardous to environment and human health [7].

Several measures have been discussed in order to reduce oil consumption by implementing regulations, vehicle efficiency enhancements, intelligent transport systems, market penetration of alternate fuel vehicles [8]. However, the fuel saving practice that could be adopted by the common commuters was not much discussed. Fuel saving issue deserves serious consideration given the ever-increasing fuel prices and its negative social and economic impacts on urban residents [9].

In order to reduce the fuel consumption and pollution, several researches have been made on driving behavior [10], emission rates [11], fuel efficiency [12], and road system [13]. While the studies focusing on commuter's response on traffic signals are rare. The commuters traveling along compact streets make frequent stops at traffic lights may contribute to higher fuel consumption on a per-mile basis [14]. The motor vehicle is considered to spend 25% of its time in the idler mode because of traffic jams and traffic signals. Further, the idler mode is considered more harmful than running mode because of its incomplete combustion of fuels. It is considered to produce high concentration of pollutant [15]. The vehicle usage in idler mode is more predominant in the traffic signals of India. Hence, it was decided to focus on the commuters' response and its repercussions in the traffic signals of an urban city in India.

In India, the road transport sector is heavily dominated by the two wheelers. For the past 3 decades, the number of two wheelers in India has grown at an average of 32% per year and thus doubling every 3 years [16]. Hence, it was decided to focus only on the response of two-wheeler commuters' practice in the traffic signals. First, the effect of start/stop response in signals as perceived by the Indian population is studied. Then, the repercussions of the response are experimentally evaluated in the laboratory environment. Following the context, the actual amount of fuel that could be saved and the amount of CO_2 that could be reduced upon performing start/stop response is recommended.

2 Methods

2.1 Study Sample

As per the motor vehicle act of India, every vehicle has to be registered with Transport Commissionerate and State Transport Authority, Government of Tamil Nadu, India before running in road. So, the statistics provided by the organization will provide detailed picture on the vehicle usage across different parts of the state. As per statistical information provided by the organization, the state has a vehicle population that amounted to 24.2 million. As of March 2018, Tamil Nadu has 21.58 million registered two wheelers across Tamil Nadu. The two-wheelers amount to 84.11% of the total vehicles registered across Tamil Nadu. In particular, Coimbatore zone has 31,26,554 two-wheeler registrations. Since the two-wheeler community is the major community, the impact it has on the environment will be much higher than the other mode of transport [17]. A common sight of two-wheeler commuters waiting in a traffic signal at Coimbatore is shown (Fig. 1).

2.2 Sample of the Study

Study exposure variables included gender, occupation, bike type, engine capacity, year of purchase, and distance covered. Study outcome variables included start/stop response (Question: Will you switch off the engine during wait time in signals? Answer: yes or no). If the answer is 'yes', then time at which the respondent switches off the engine (1–10 s, 11–30 s, 31–50 s, greater than 50 s) and motivation behind the response (monetary benefit, clutch hold difficulty, environmental concern) is also recorded. In the current study, the respondents who switch off their engines in signals are called 'yes' respondents. If the answer is 'no', the reason behind the response (no belief in fuel saving, battery life concern, maintenance issues, absence of signal timer, not interested) is recorded, and they were called 'no' respondents. A questionnaire was prepared and circulated via hard copy and soft copy. The soft copy response was obtained using the Google survey. Minimum sample size was calculated using Cochran formula [18–20]. The confidence level and margin of error were considered 95 and 5%, respectively. The corresponding Z value taken from the Z table is 1.96.

Fig. 1 Two-wheeler commuters waiting in a traffic signal at Coimbatore, Tamil Nadu, India

The expected prevalence p is considered as 0.5, and thus, q value is also 0.5.

$$n = Z^2 \frac{p(1-p)}{e^2}$$

Based on the above Cochran formula, the sample size obtained is 384.16. Then considering the total number of bikes registered in the Coimbatore region ($N =$ 3,126,554), the sample size is corrected using the formula,

$$\text{Corrected}\, n = \frac{n-1}{1+\{N\}}$$

Based on the corrected Cochran formula, the corrected sample size is 384.11. Thus, the responses of 388 participants were recorded in the survey. The outcomes were evaluated with 95% confidence intervals (CI) with $P < 0.05$ considered statistically significant. Statistical software STATA version 15.0 was used to perform all the analyzes.

3 Results and Discussion

Of the 388-response recorded in the Coimbatore region, 45.87% of the respondents never switch off their engines in traffic signals. Table 1 clearly shows the type of start whether it is self-button start or kick-start in bike plays a major role in the start/stop behavior (p <0.05). The majority of the 'no' respondents are being kick-start bike users. It can be acknowledged based on the fact that kick-start bike commuters require more physical effort during engine ignition. Hence, switching off the engine is not popular among the kick-start bike commuters. It is significant to note that gender, age, and job of the users have very little or no impact on the start/stop behavior. Table 2 lists the reason behind the start/stop response of 'yes' respondents. About 55.2% of the 'yes' respondents attributed their response toward saving fuel in order to reduce the expenditure on fuel. The second popular reason is bounded toward the environmental concern. Table 2 further gives the details of the wait time in traffic signals at which the 'yes' respondents switch off their engines. Most of the 'yes' respondents (63.3%) switch off their engine only when the signal wait is 30 s or more. Although it is possible to save the fuel when there is even lesser wait time. It was experimentally demonstrated in the later part of the paper. Table 3 relates the engine capacity of the 'yes' respondents and their wait time response. The detail is necessary for the recommendation of break-even time, since the engine ignition and fuel combustion vary in accordance with the bike capacity.

The reason behind the behavior of 'no' respondents is given in Table 4. The 'no' respondents are skeptical in switching off engines when the traffic signals lack timer. However, a few respondents suffer from bike issues whom may find it difficult to start the engine in a frequent manner. They are susceptible to run their bike in idler mode on traffic jams and traffic signals. But the primary reason is surprisingly being the belief that switching off engine would increase fuel consumption. Around 35% of 'no' respondents share same belief that igniting the engine from off condition consumes more fuel in comparison with engine running in idler condition while waiting in traffic signals. The total start/stop response of the respondents and the reason behind the response is consolidated in a bar chart (Fig. 2).

4 Experimental Study and Observations

The repercussions of the start/stop behavior were experimentally carried out in the laboratory environment. The main objective of the experiment is to identify the optimal wait time in signals for which the engine can be switched off. The fuel consumption of the engine during starting and idling condition was measured. Since the fuel consumption and CO_2 emissions of the bike mainly depend on its engine capacity, the experiment was carried out in accordance with the engine capacity as low (less than 100 cc), medium (between 100 cc and 200 cc), and high (more than 200

Table 1 Characteristics of bike respondents and their start/stop behavior ($n = 388$)

Factor recorded	Start/stop response		P value
	Yes ($n = 210$)	No ($n = 178$)	
Gender			0.677
Male($n = 346$)	186 (53.7%)	160 (46.3%)	
Female($n = 42$)	24 (57.1%)	18 (42.9%)	
Age			0.484
Below 25($n = 208$)	116 (55.7%)	92 (44.3%)	
Above 25($n = 180$)	94 (52.2%)	86 (47.8%)	
Type of start			0.002
Kick ($n = 113$)	46 (40.7%)	67 (59.3%)	
Self ($n = 105$)	65 (61.9%)	40 (38.1%)	
Both ($n = 170$)	99 (58.2%)	71 (41.8%)	
Engine capacity			0.612
High ($x >= 200$) ($n = 55$)	32 (58.2%)	23 (41.8%)	
Moderate ($100 < = x < 200$) ($n = 209$)	115 (55.0%)	94 (45.0%)	
Low (<100) ($n = 124$)	63 (50.8%)	61 (49.2%)	
Years since purchase			0.111
New ($x < = 5$) ($n = 209$)	121 (57.9%)	88 (42.1%)	
Mid ($5 < x = 10$) ($n = 112$)	60 (53.6%)	52 (46.4%)	
Old ($x > 10$) ($n = 67$)	29 (43.3%)	38 (56.7%)	
Distance covered in Km			0.131
less ($x <= 10{,}000$) ($n = 128$)	76 (59.4%)	52 (40.6%)	
Moderate ($10{,}000 < x < = 30{,}000$) ($n = 116$)	70 (60.3%)	46 (39.7%)	
High ($x > 30{,}000$) ($n = 144$)	64 (44.4%)	80 (55.6%)	
Type of people based on their job			0.937
Student ($n = 170$)	92 (54.1%)	78 (45.9%)	
Self-employed ($n = 56$)	29 (51.8%)	27 (48.2%)	
Professionals ($n = 140$)	78 (55.7%)	62 (44.3%)	
Others ($n = 22$)	11 (50.0%)	11 (50.0%)	

cc). In random five bikes of the respondents from each class were taken as samples and were analyzed, the details were given in Table 5.

The experimental data clearly demonstrate that fuel consumed during the start of bike is 6.08–21.34 times more than the fuel consumed during per second of idle time. The high-fuel consumption during the start of the engine was predominantly seen in the low-engine capacity bikes. The inferior performance during the start of the low-engine capacity bikes can be attributed to its inferior ignition technologies in

Table 2 Reasons for start/stop response and corresponding signal wait time of 'yes' respondents at which they act

Reasons for start/stop behavior of 'yes' respondents	Vehicle wait time ($n = 210$)			
	1–10 s	11–30 s	31–50 s	>50 s
	($n = 29$)	($n = 48$)	($n = 70$)	($n = 63$)
	(13.8%)	(22.9%)	(33.3%)	(30.0%)
Saving fuel ($n = 116$) (55.2%)	15 (12.9%)	28 (24.1%)	39 (33.6%)	34 (29.3%)
Clutch hold difficulty ($n = 18$) (8.6%)	03 (16.7%)	03 (16. 7%)	06 (33.3%)	06 (33.3%)
Environmental concern ($n = 72$) (34.3%)	10 (13.9%)	17 (23.6%)	23 (31.9%)	22 (30.6%)
Other reasons ($n = 4$) (1.9%)	01 (25.0%)	–	02 (50.0%)	01 (25.0%)

Table 3 The 'yes' respondent's bike engine capacity and their corresponding signal wait time at which they act

Bike engine capacity of 'yes' respondents	Vehicle wait time ($n = 210$)			
	1–10 s ($n = 29$)	11–30 s ($n = 48$)	31–50 s ($n = 70$)	>50 s ($n = 63$)
Low capacity($n = 63$) (30.0%)	06 (9.5%)	11 (17.5%)	26 (41.3%)	21 (33.3%)
Moderate capacity($n = 115$) (54.8%)	18 (15.6%)	29 (25.2%)	32 (27.8%)	36 (31.3%)
High capacity($n = 32$) (15.2%)	05 (15.6%)	08 (25.0%)	12 (37.5%)	06 (18.7%)

Table 4 Behavioral reasons for 'no' respondents during wait time in signal

Reasons for start/stop behavior of 'yes' respondents	Bike in use ($n = 178$)		
	New ($n = 88$)	Mid ($n = 52$)	Old ($n = 38$)
Lack of belief in fuel saving (35.4%)	31 (49.2%)	21 (33.3%)	11 (17.5%)
Battery life concern (11.8%)	12 (57.1%)	06 (28.6%)	03 (14.3%)
Maintenance issues (11.2%)	05 (25.0%)	06 (30.0%)	09 (45.0%)
Absence of signal timer (28.7%)	28 (54.9%)	13 (25.5%)	10 (19.6%)
Not interested (12.9%)	12 (52.2%)	06 (26.1%)	05 (21.7%)

comparison with medium and higher engine capacity bikes. Based on the difference in fuel consumption ratio between start time and idler time, the break-even time was evaluated for each class of bikes. The break-even time of low, medium, high-engine capacity bikes is found to be 21.34 s, 8.34 s, and 6.08 s, respectively. If the signal's wait timer shows above these break-even times, the study recommends the two-wheeler commuter to switch off the engine in order to save fuel and reduce environmental pollution.

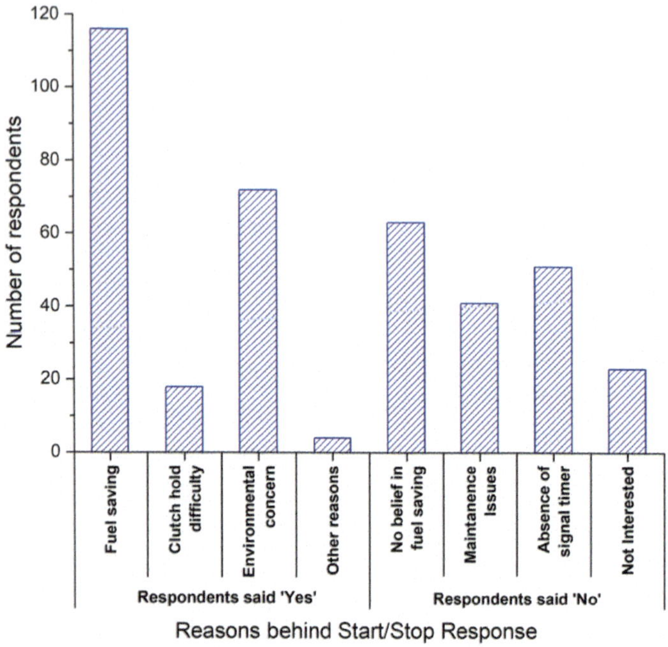

Fig. 2 Start/stop response and attributed reasons

Table 5 The bike engine capacity of the respondents and the corresponding details of fuel consumed and CO_2 emissions per minute found experimentally

Engine capacity	Mean fuel consumed per minute		Mean CO_2 emissions per minute	
	During start (ml)	During idle time (ml)	During start (g)	During idle time (g)
Low	0.720	1.938	0.91	3.78
Moderate	0.408	2.793	0.52	5.62
High	0.407	3.819	0.60	7.90

The box chart (Fig. 3) clearly demonstrates the scale of difference in break-even time between the low-engine capacity bikes and other class of bikes. The average waiting time in traffic signals in Coimbatore is found to be 60 seconds and hence that is considered for calculations of fuel consumption and emissions (Fig. 4 and Table 5).The double Y box chart (Fig. 4) represents the amount of fuel consumption and CO_2 emission that could be reduced per minute of waiting in traffic signals on switching off the engine as when following the recommended break-even time. The higher the capacity of the bikes more the fuel can be saved on following the recommendation. Since the emission of CO_2 is directly related to the fuel consumption, more the fuel saving more the CO_2 emission can be reduced.

Fig. 3 Break-even time of different class of bikes

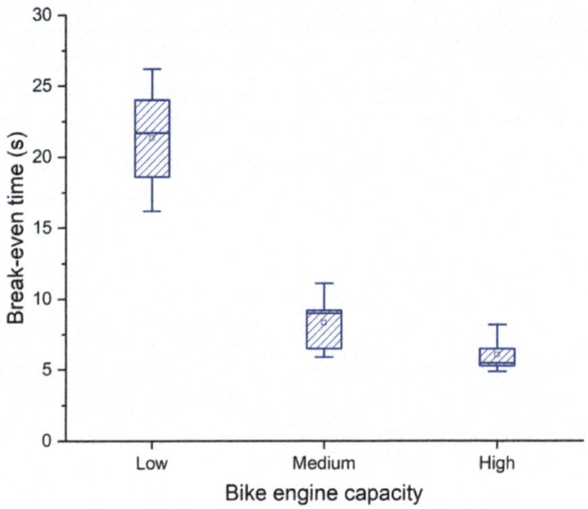

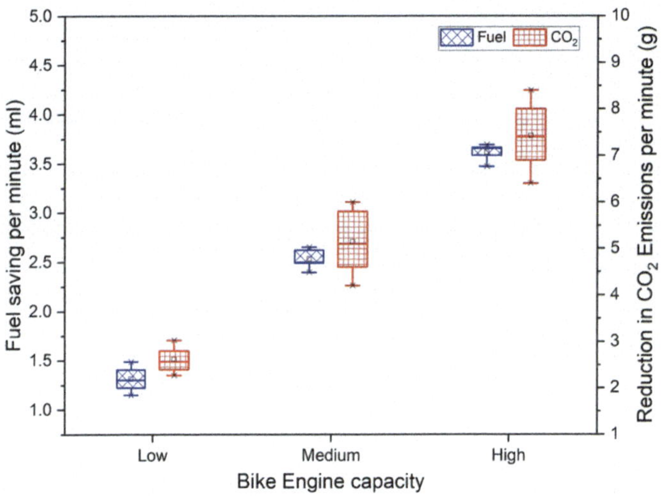

Fig. 4 Reduction in CO_2 exhaust in a bike per 60 s wait time in traffic signals on following the recommended start/stop response

5 Conclusions

When experimental result was superimposed on the respondents start/stop behavior, it was found only 8.74% of respondents have start/stop behavior in accordance to the recommended break-even time. Others have very little or no awareness toward optimal start/stop response. The experimental results suggest that people may get benefitted to one self and the society by switching off their engine if the wait time in signal is longer than 21.34 s, 8.34 s, and 6.08 s for low, medium, and high-engine capacity bikes, respectively. The benefits include fuel savings and reduction in exhaust smoke. The experimental results conclude that on following the recommended break-even time in traffic signals, it was found that on an average the low, medium, and high-engine capacity bikes can reduce the emission of CO_2 per minute by 2.6 g, 5.1 g and 7.4 g, respectively. Whereas the amount of fuel that could be saved per minute in low, medium, and high-engine capacity bikes is 1.31 ml, 2.53 ml and 3.61 ml, respectively.

From the practice and policy perspectives, public can be educated on optimal start/stop response in traffic signals. Much effort is to be made on creating awareness among the community that believes switching off engine in traffic signals increases the fuel consumption. Further, the installation of timer on each and every traffic signal is highly recommended. If the recommended practices are followed collectively, the commuters can make significant impact on reduction of fuel consumption and environmental pollution.

References

1. Alam JB, Wadud Z, Polak JW (2013) Energy demand and economic consequences of transport policy. Int J Environ Sci Technol 10:1075–1082
2. Tuzkaya UR (2009) Evaluating the environmental effects of transportation modes using an integrated methodology and an application. Int J Environ Sci Technol 6:277–290
3. Boriboonsomsin K, Barth M (2009) Impacts of road grade on fuel consumption and carbon dioxide emissions evidenced by use of advanced navigation systems. Transp Res Rec 2139:21–30
4. Venkanna R, Nikhil GN, Rao TS, Sinha PR, Swamy YV (2015) Environmental monitoring of surface ozone and other trace gases over different time scales: chemistry, transport and modeling. Int J Environ Sci Technol 12:1749–1758
5. Vasebi S, Hayeri YM, Samaras C, Hendrickson C (2018) Low-level automated light-duty vehicle technologies provide opportunities to reduce fuel consumption. Transp Res Rec 2672:60–74
6. Zahabi SA, Miranda-Moreno LF, Patterson Z, Barla P (2013) Urban transportation greenhousegas emissions and their link with urban form, transit accessibility, and emerging green technologies: Montreal, Quebec, Canada, case study. Transp Res Record 2375:45–54
7. Iodice P, Senatore A (2015) Exhaust emissions of new high-performance motorcycles in hot and coldconditions. Int J Environ Sci Technol 12:3133–3144
8. Jovanović VN, Srećković IS (2013) Action plan for the deployment of intelligent transport systems in Europe. Tehnika. 68:934–940
9. Akbari S, Habib KN (2014) Oil vulnerability in the greater Toronto area: impacts of high fuel prices on urban form and environment. Int J Environ Sci Technol 11:2347–2358
10. Shaon MRR, Qin X (2020) Crash data-based investigation into how injury severity is affected by driver errors. Transp Res Record
11. Rayle L, Pai M (2010) Scenarios for future urbanization: carbon dioxide emissions from passenger travel in three Indian cities. Transp Res Record. 2193:124–131
12. Franzese O, Lascurain MB, Loy LW (2019) Effect of heavy-truck platooning fuel-fficiency gains on overall fuel efficiency. Transp Res Rec 2673:188–196
13. Zhang N, Alipour A (2019) Integrated framework for risk and resilience assessment of the road network under Inland flooding. Transp Res Record. 0361198119855975.

14. Wang X, Liu C, Kostyniuk L, Shen Q, Bao S (2014) The influence of street environments on fuel efficiency: insights from naturalistic driving. Int J Environ Sci Technol 11:2291–2306
15. Ning Z, Cheung CS, Lu Y, Liu MA, Hung WT (2005) Experimental and numerical study of the dispersion of motor vehicle pollutants under idle condition. Atmos Environ 39:7880–7893
16. Badami MG, Iyer NV (2006) Motorized two-wheeled vehicle emissions in India: behavioral and institutional issues. Transp Res Rec 1954:22–28
17. Tamil Nadu State transport authority. https://tnsta.gov.in/pdf/g5.pdf
18. Cochran W (2007) Sampling techniques. Wiley, United Kingdom
19. Israel G Sampling the evidence of extension program impact . https://edis.ifas.ufl.edu/pd005
20. Israel GD Determining sample size. http://www.giimt.ac.in/web/wp-content/uploads/2017/10/2_Glenn-D.-Israel_Determining-Sample-Size.pdf

Design and Development of River Water Trash Collector for a Sustainable Environment

Chidambaram Vigneswaran, M. Afifah Zaynab, J. Keerthana, R. Harish Krishna, and R. Hariharan

Abstract Every year, nearly 10–20 million tons of plastic end up in the oceans. According to a recent study, there are currently 5.25 trillion plastic particles floating in the world's oceans, weighing a total of 268,940 ton. This work aims at designing a system that could eliminate floating trash from water to alleviate the issue of water pollution. Currently, the costlier autonomous river cleaning robots make use of AI technologies that complicate the control system. In order to minimize the complication of the system, a prototype has been developed for easy manufacturing and implementation to produce effective results. The prototype uses one 220 V motor to create suction, removing debris out of the water using a catch bag that is inside the bin. The clean water is then forced back out. The motor is powered with the help of eco-friendly flow generators. The major parameters that come into play are the pressure gradient and the flow rate. The weight of the entire structure has been taken into account as it directly affects the flotation characteristics of the river water trash collector (RWTC). Each RWTC unit could capture an average of 12 kg marine litter per day.

Keywords Environment · Computer aided design · Water pollution · Eco-friendly · Trash collector

C. Vigneswaran (✉) · M. Afifah Zaynab · J. Keerthana · R. Harish Krishna · R. Hariharan
Department of Production Engineering, PSG College of Technology, Coimbatore, India
e-mail: cvn.prod@psgtech.ac.in

M. Afifah Zaynab
e-mail: 18p101@psgtech.ac.in

J. Keerthana
e-mail: 18p108@psgtech.ac.in

R. Harish Krishna
e-mail: 18p107@psgtech.ac.in

E. Natarajan et al. (eds.), *Materials, Design and Manufacturing for Sustainable Environment*, Lecture Notes in Mechanical Engineering,
https://doi.org/10.1007/978-981-19-3053-9_45

1 Introduction

About 80% of India's water is heavily polluted because people dump raw sewage, sediment, and garbage into the country's rivers and lakes. This makes the water undrinkable, and people have to rely on illegal and expensive water sources. Every year more than 1.5 million Indian children die from diarrhea. Experts predict that of the entire Indian population, 40% of the population may not have access to clean water by 2030. Water pollution in India not only endangers people's health and food safety, but also leads to a decline in India's GDP and economic stagnation [1]. When the country's water pollution exceeds a certain limit, not only will GDP growth be reduced by one third, but also agricultural income from areas close to industrial areas will be reduced by 9%. The financial cost of ocean plastic pollution is also high. According to conservative forecasts from March 2020, the direct damage to the blue economy of the Association of Southeast Asian Nations is $2.1 billion a year. Residents of coastal areas suffer from the adverse health effects of plastic pollution and tidal waste. Finding strategies to prevent plastic and other garbage from polluting water sources and cleaning the mup is also crucial [2]. Exclusive cleaning boats are now used in various locations for cleaning. This cleaning boat has a gap with a net in the front to trap floating waste. This process requires only one boat operator and another collector. Staffs collect enormous waste in containers to avoid congestion. The garbage will be collected once a day and the whole process takes about 3 h for the whole river [3]. The use of river water trash collector (RWTC) can simplify this process to a great extent. The general idea of the RWTC comes from the seabin concept which acts as a trash collector in the sea. The seabin smart technology (SST) as shown in Fig. 1 merges the concept of a pool skimmer and a garbage bin to create a litter capture system for the marine environment. SST filters the top 20 mm of the water column for marine litter, as well as surface pollutants including micro plastics, plastic fibers, oils, and fuels. Seabins use surface tension to capture floating marine litter in the catch net. The surface tension is enhanced by a submersible pump that drives water through the unit at a rate of 25000 L/hour and up to 6 m from the unit. SST has a great impact on reducing marine plastic pollution in calm environments such as water ways, marines, and ports and harbors [4]. A significant proportion of litter captured by seabins originated on land and is transported into the marine setting by physical environmental factors such as rain or wind [5]. From this existing design, several improvements are made in its cost and functions. The seabin remains attached to a pole, whereas the RWTC can move freely along the river water [6]. The concept of a ballast tank is used in the floating mechanism. In RWTC, the submersible pump is powered with the help of flow generators which makes the technology greener [7]. Initially, the load that can be carried by RWTC was fixed as 12 kg since weight estimation plays a crucial role in designing the floating mechanism [8]. The buoyancy force calculations were done to determine the diameter of the system. Then, CAD modeling was done using the dimensions obtained from the calculations. The hydrostatic pressure analysis was done using

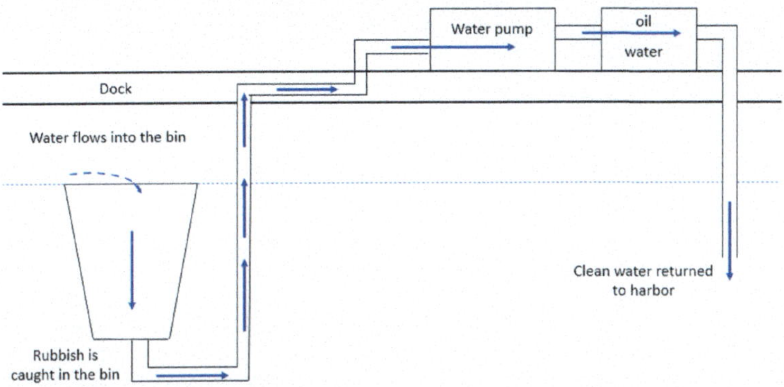

Fig. 1 Schematic diagram of the seabin

ANSYS to observe the deformation of the structure. Then, the stability analysis of the RWTC in presence of water stream was performed using Open Foam software.

2 Methods

The general idea for the prototype was inspired from the carburettor design as shown in Fig. 2. It works on the principle that when air flows through the throat of the venturi, due to decrease in cross-section, there will be a pressure drop at that point. Since the water on the sea level will be at atmospheric pressure, the water will try to move toward the pressure drop region. Once the water enters into the venturi, it will start flowing downwards and the trash will be collected at the bottom using a catch bag. Implementation of this method can result in significant reduction of the cost of the product. Since a pump is not required and suction happens due to pressure drop created by the converging cross-section. But since water is sucked in through pipes, it will not be possible to suck in wastes which are large in size and there are high chances of blockage.

A. **Working mechanism**

The conceptual ideas have been developed and finalized to overcome the limitations discussed earlier. The primary objective is to achieve buoyancy using a ballast tank. The RWTC has a compartment under the main hull which is used to store water, which then stabilizes the vessel. Based on this concept, the RWTC will easily descend to the exact level above the water level instead of using rocks or iron ballast. The ballast tank is also used to temporarily pump water when entering shallow water. The operation of ballast tanks is as follows; initially, four of the solenoid valves were placed above and below the tank. This would allow water to enter and move the air. Once all the air is sucked in and the RWTC is on par with the water level, all four

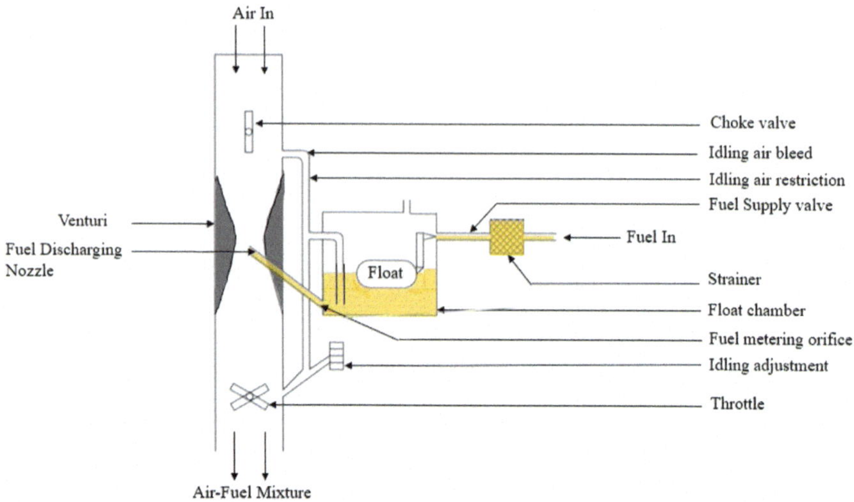

Fig. 2 Carburetor design

solenoid valves will be closed and the RWTC is fully submerged. Since, the developed model has to float at the water level; the weight force must be equal to the force of the buoyancy. The RWTC consists of two layers, the outer layer acting as a ballast tank, while the inner layer will acts as a waste collection tank, as shown in Figure 3. A sand pack is inserted between the outer and inner vessels to help RWTC sink. The vessel inside consists of a submersible electric water pump powered by a 12V water generator mounted inside the RWTC's side-mounted float. A bag is present on the inner bin to capture waste floating in the water. The major advantage of the RWTC is that it uses a green power source which is a 12V flow generator.

B. **Concept selection**

Using morphological chart and PUGH matrix, the concept selection and scoring were performed and a suitable solution has been selected as shown in Fig. 4.

C. Material selection

Ashby chart has been used for selecting materials for each component. The materials assigned for the different components of RWTC are listed in Table.1.

3 Analysis of the Product

Since RWTC is made up of many different components, it is critical to calculate the weight of each one. By equating weight force with the buoyant force, vessel's appropriate diameter has been determined.

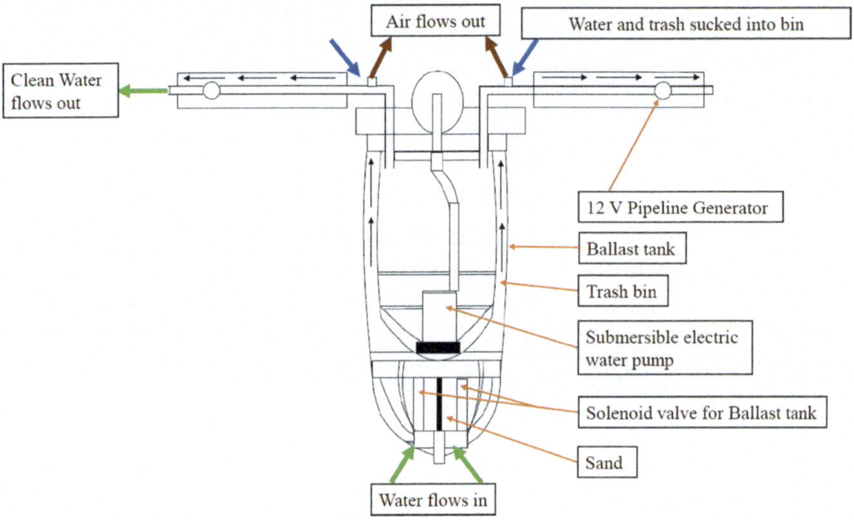

Fig. 3 Conceptual design of RWTC

Fig. 4 Developed CAD model of the RWTC

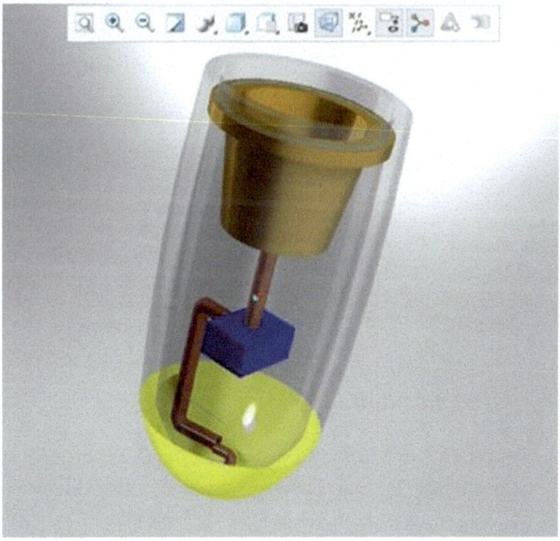

D. Input parameters for design calculations

Volumetric flow rate, Q =400 lpm =0.0066 m^3/s= 6.6lps
Discharge level=20cm=0.65ft
Suction level = 70 cm = 2.69 ft
Density $\rho = 1000$ kg/m^3
Overall efficiency, $\eta_0 = 90\%$

Table 1 Material selection for the developed prototype model

S. No.	Component	Objective	Constraints	Material
1	Outer vessel	Maximum strength	Minimum weight	PVC
2	L structure	Maximum yield strength (stiffness)	Minimum weight	PVC
3	Trash bin	Maximum strength	Minimum cost	Al 6061
4	Transmission pipes	Maximum yield strength (stiffness)	Minimum cost	Stainless steel

Rotational speed of the impeller = 2880 rpm
Diameter of inlet pipe = 30mm (0.098 ft)
Diameter' of outlet = 25mm (0.082 ft)
Atmospheric pressure, P_{atm} = 101.325 kPa

I. **Fluid Velocity at Inlet**

It is the velocity with which the fluid flows

$$v_1 = Q/A = 0.0066/3.14 * 0.015^2$$

$$v_1 = 9.34\text{m/s}$$

II. **Fluid velocity at outlet**

$$\begin{aligned} v_2 &= \frac{Q}{A} \\ &= \frac{0.0066}{3.14 * 0.0125^2} \\ v_2 &= 13.45\ \text{m/s} \end{aligned}$$

III. **Head calculation**

Total head = Static head friction head + velocity head.
Static head (ft) = Discharge level- Suction level = −0.65−2.29= −2.94 ft
Friction head (discharge) = $f * \left(\frac{\text{length}}{\text{dia}}\right) * \left(\frac{v^2}{2g}\right) * \left(\frac{v^2}{2g}\right)$.
Length of pipe= 0.65+0.4=1.05 ft
Friction head (discharge) = $0.03 * \left(\frac{1.05}{0.082}\right) * \left(\frac{44.12^2}{2*32}\right) = 11.69$ ft.
Friction head (suction) = $0.03 * \left(\frac{2.29}{0.098}\right) * \left(\frac{30.6^2}{2*32}\right) = 10.25$ ft.
Total friction head= 11.69−10.25 = 1.44 ft
Velocity head (discharge) = $\frac{v^2}{2g} = \frac{44.12^2}{2*32} = 30.41$ ft.
Velocity head (suction) = $\frac{30.62^2}{2*32} = 14.63$ ft.
Total velocity head= 30.41−14.63 = 15.78 ft
Total head (H) = −2.94+1.44+15.78 = 14.28 ft. = 4.35 m

IV. **Inlet pressure**

It is the pressure produced by the pump at the inlet side.

$$P_{\text{in}} = \frac{H*\text{specific gravity}}{2.31} = \frac{10.61*1}{2.31} = 4.59\,\text{psi} = 31.64\,\text{kPa}$$

Suction pressure = 31.64 kPa.
So, the pressure gradient due to pumping action = $P_{\text{atm}} - P_{\text{in}}$ = 69.71 kPa.
Pressure gradient = 69.71 kPa.

V. **Input power of pump**

$$\text{Input power} = \frac{(\text{Density}*\text{Volumetric flow}*\text{Head})}{1000*102*\text{efficiency}} = \frac{1000*6.6*4.65}{1000*102*0.8} = 0.376\,\text{kW} = 376\,\text{W}$$

VI. **Reynolds number**

The Reynolds number is the ratio of inertial forces to viscous forces.

$$\text{Re} = \frac{D*v*\rho}{\mu} = \frac{0.025*13.45*1000}{8.9*10^{-4}} = 3808.98$$

Since Re >3500, it is a turbulent flow.

VII. **Weight analysis**

In order for the body to float, the weight force must be balanced by the buoyant force. The total mass was calculated and the weight force was determined. The weight force is then equated with the buoyant force to determine the diameter of the outer vessel. The height of the outer vessel was assumed to be 90 cm.

$$\text{Total weight} = 7 + 3 + (1.68*2) + 0.3 + 0.4 + 2.5 = 16.5\,\text{kg}$$

VIII. **Buoyancy force analysis**

Any particle immersed in a fluid, exerts an upward force known as buoyancy. The buoyant force is caused by the fluid's pressure on the item. Since, pressure rises with depth, the pressure at the bottom of an object is always greater than the pressure at the top, resulting in a net upward force. The weight force must be equal to or less than the buoyancy force exerted on the RWTC. The load that can be submerged in water is about 20 kg which equals to 396.2 N. This load was is equal to the expected trash that can be collected inside the trash bag.

Weight force, $F_w = \text{mg} = (16.5 + 20)9.81 = 358.06$ N.
Buoyant force, $F_b = \rho\text{gV} = 1000*9.81(\pi r^2 h)$.
Equating weight force with the buoyant force.
$358.06 = 30{,}803.4*h*r^2$.
$27{,}723.06*r^2 = 358.06$.
$r^2 = 0.01291$.
$r = 0.1136\ \text{m} = 11.36\ \text{cm}$.
Diameter of the outer vessel = 23 cm.
Height of the outer vessel = 90 cm.

4 CAD Modeling

The conceptual model was generated and the design has been finalized and modeled. The input data was provided from the values obtained from the design calculations presented in Sect. 3. Creo 5.0.2 software was used to model the parts. Figures 4, 5 and 6 are the CAD models that were modeled in the Creo software.

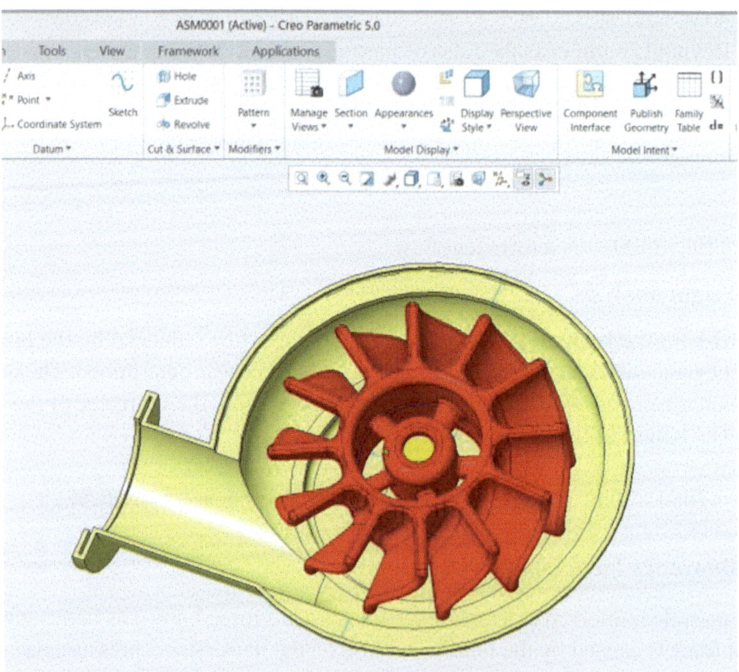

Fig. 5 Developed CAD model of the centrifugal pump

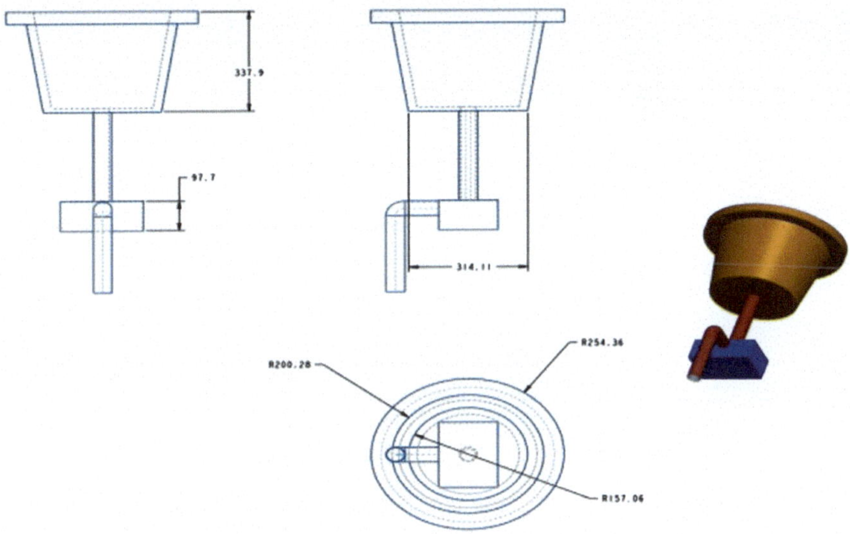

Fig. 6 Drafting of the inner vessel of RWTC

5 Simulation and Analysis

The center of buoyancy is calculated as

X Coordinate = −272.79 mm.

Y Coordinate = −29.14 mm.

Z Coordinate = −69.61 mm.

A. **Stability analysis**

Generally, the velocity of water streams in rivers is in the range of 0–3 m/s. We have used Open Foam software to simulate the RWTC in the presence of a water stream. From that, it can be inferred that the RWTC can maintain its stability if the velocity of the water stream reaches 3 m/s.

B. **Hydrostatic pressure analysis**

The formula used to calculate the hydrostatic pressure: $p = \rho gh$.

Height of the RWTC = 1 m.

Density of water = 1000 kg/m^3.

Max hydrostatic pressure acting on the RWTC = 9810 Pa.

The deformation due to the hydrostatic pressure has been found by simulation through ANSYS static structural solver as shown in Fig. 7. Since the deformation is in the range of 10^{-6} mm, the hydrostatic pressure does not produce any significant deformation of the RWTC.

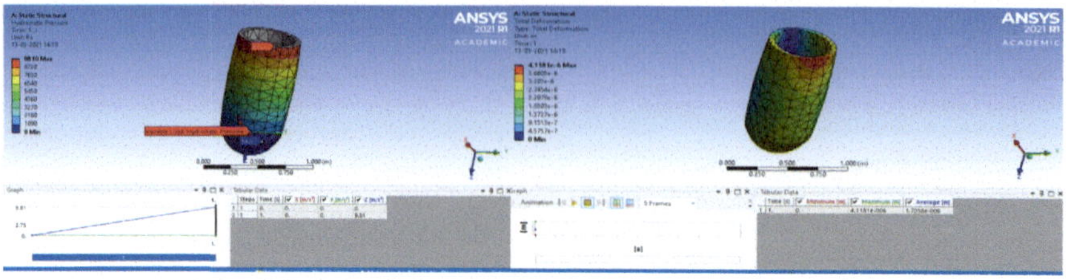

Fig. 7 Static structural analysis of RWTC

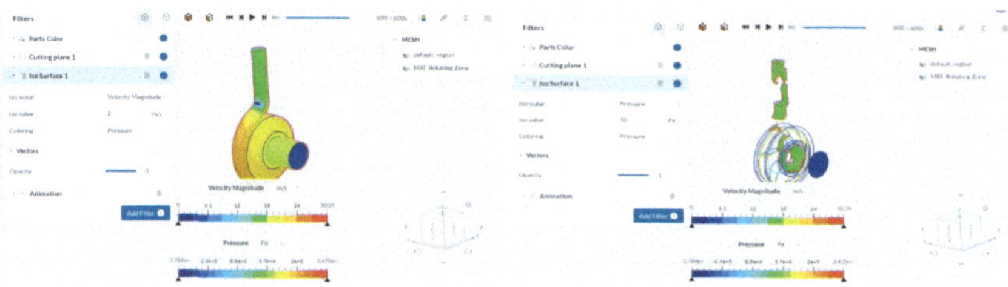

Fig. 8 Velocity and pressure analysis of RWTC

C. **Velocity and pressure variation in the pump**

The velocity and pressure variations within the pump were simulated using Simscale software as shown in Fig. 8. It could be inferred that the maximum velocity reached is 30.39 m/s, and the peak pressure within the pump is 0.347 MPa.

6 Cost Estimation

The direct cost and indirect costs associated with the manufacture of the product are estimated and listed in Table 2. The existing design of the seabin costs roughly around Rs. 50,000, while the newly developed RWTC design costs around Rs. 6000. As a result, the savings would be roughly Rs.44,000 in total.

7 Conclusion

The material for each component was selected based on the required objective and constraints. Following that, design calculations were carried out based on the strength of the selected materials. Values obtained from the calculations were used to model the components of the RWTC in the CAD software. Static structural analysis was performed on the created model to analyze the total deformation due to hydrostatic

Table 2 Cost estimation of items for the developed prototype model

S. No.	Item of expenditure		Total cost for the prototype (in ₹)
1	Direct material cost per component		4335
	a. Pump	₹700	
	b. Battery	₹1300	
	c. Stainless steel pipes	₹165	
	d. Solenoid valve	₹1200	
	e. Trash bin net	₹20	
	f. Outer vessel	₹450	
	g. Inner vessel	₹500	
2	Direct labor cost per day		500
3	Administrative expenses		200
Total cost of the product developed			5035

pressure. Further, work includes performing stability analysis. Once the analysis is successful, it provides proof that the RWTC design is stable enough to withstand the velocity of water flow. Based on the simulation, the static structural analysis proves that the RWTC does not undergo any significant deformation. Also, the Simscale simulation helps in determining the inlet and outlet pressure of the centrifugal pump.

8 Future Scope

The current work makes use of flow generators to power the motors but these flow generators have limited power generation capacity. Therefore, high capacity pumps cannot be powered in this way. The best alternative would be to use a solar power system along with the flow generators. As a result, there would be sufficient power to run high capacity pumps which in turn improves the efficiency of RWTC.

References

1. Helinski OK, Poor CJ, Wolfand JM (2021) Ridding our rivers of plastic: A framework for plastic pollution capture device selection. Mar Pollut Bull 165:112095
2. Zhang Y, Zhang Y, Gao Y, Zhang H, Cao J, Cai J, Kong X (2011) Water pollution control technology and strategy for river–lake systems: a case study in Gehu Lake and Taige Canal. Ecotoxicology 20(5):1154–1159
3. Fisher-Vanden K, Olmstead S (2013) Moving pollution trading from air to water: potential, problems, and prognosis. J Econ Perspect 27(1):147–172
4. Nikiema J, Asiedu Z, Mateo-Sagasta J, Saad D, Lamizana B (2020) Catalogue of technologies to address the risks of contamination of water bodies with plastics and microplastics
5. Mohazzab P (2017) Archimedes' principle revisited. J Appl Math Phys 5(04):836
6. Simões AS, Carrion R, Martins ACG, Costa RB, Schvarcz AF, Buzo VR, Felipe MH, Ferrari BA (2006) Autonomous mobile robots designing for the medical trash collector task. In: 2006 IEEE 3rd Latin American robotics symposium, IEEE, pp 234–239
7. Mukhtar MF, Rosley MIF, Lubis AMHS, Tamaldin N, Hussin MSF, Damanhuri AAM, Azlan KA, Hanizat NH (2020) Development of river trash collector system. J Phys: Conf Series 1529(4):042029. IOP Publishing
8. Anagnostopoulou C, Kagemoto H, Sao K et al (2016) Concept design and dynamic analyses of a floating vertical-axis wind turbine: case study of power supply to offshore Greek islands. J Ocean Eng Mar Energy 2:85–104

Comparison of Sensor-Controlled Hybrid Electric Vehicle Controlled by PID-P&O MPPT and PID-INC MPPT

Srikanth Ravipati, M. Venkatesan, and Y. Srinivasa Rao

Abstract The growth in the automobile industry has been rising due to the improvements in the design with the employment of renewable energy resources and energy storage devices. This manuscript provides a novel solar fuel cell-based hybrid electric vehicle with maximum power point tracking system. Also two tracking algorithms named as perturb and observe method and incremental conductance have been analyzed with the combination of proportional + integral + derivative controller. The proposed vehicle has been controlled with the sensors generating hall signals and speed form the vehicle. The advantage of integrating renewable energy sources is reduces the climatic changes, scattered energy production and superior energy security. It concludes with the comparison of performance of two proposed trackers for the proposed electric vehicle using MATLAB Simulink software.

Keywords Hybrid electric vehicle · Solar fuel hybrid system · Perturb and observe MPPT · Incremental conductance technique

1 Introduction

The conventional vehicles exploiting the life of entire world due to huge amount of gases that are being spitted out of the vehicles. If this goes on increasing, the survival may become a complicated issue [1]. So there has to be an alternative that overcomes the problems of fuel-based vehicle and replaces with renewable-based vehicle. So, among the numerous types of sources, solar energy has been [2, 3] opted due to its huge advantages and also fuel cell has been chosen as another source for driving the vehicle.

S. Ravipati · Y. Srinivasa Rao
Department of Electrical and Electronics Engineering, Vignan's Foundation for Science, Technology and Research, Guntur, Andhra Pradesh, India

M. Venkatesan (✉)
Department of Electrical and Electronics Engineering, Vignan's Lara Institute of Technology and Science, Guntur, Andhra Pradesh, India
e-mail: drmvenkatesan_vlits@vignan.ac.in

E. Natarajan et al. (eds.), *Materials, Design and Manufacturing for Sustainable Environment*, Lecture Notes in Mechanical Engineering,
https://doi.org/10.1007/978-981-19-3053-9_46

There have been number of renewable energy sources like solar, wind and tidal of which solar energy is ahead among the numerous types of renewable sources in the aspect of reduced noise pollution and production of clean energy. As a single source cannot provide continuous power all the times, hydrogen-based fuel cell has been chosen as another source as it possess certain advantages like reduced emissions, higher reliability and production of water as an outcome. Pure electric vehicles reduce global warming effect and environmental pollution.

But the renewable sources suffer from problem of variable power values for varying irradiance and temperature conditions. So, to extract the maximum power from the solar panel, tracking algorithm has been opted which [4–6] tracks the maximum power for variable weather conditions. Among the available tracking techniques, like hill climbing, fixed duty cycle, constant voltage method and perturb and observe, incremental conductance has been chosen for tracking the solar energy. Perturb and observe tracking algorithm finds the peak power point with certain perturbations and observations with a step size. Incremental [7] conductance algorithm finds the peak power point with the change in both power of the solar system and also with the change in voltage. Also, this tracking algorithm suffers with the problems of transient variations in the output parameters. So, to attain stability in the transient [8] and steady state performance, the trackers have been analyzed with the PID controller where the proportional gain reduces the steady state error, integral gain nullifies the steady state error and derivative gain reduces the overshoot problem.

The paper is organized as follows: Sect. 1 presents introduction to the proposed PID fed MPPT trackers-based electric vehicle, Sect. 2 deals with the block diagram explanation of proposed MPPT trackers-based hybrid electric vehicle, Sect. 3 deals with perturb observe and incremental conductance MPPT techniques, Sect. 4 deals with the simulation of the proposed vehicle and results and Sect. 5 deals with the conclusions drawn from the projected work.

2 Block Diagram Explanation of Proposed Vehicle

The block diagram model of the MPPT controlled vehicle is shown in Fig. 1 which consists of photovoltaic system and fuel cell system generating electricity with two types of MPPT techniques fed PID controller.

When the solar rays fall up on the panel, it generates DC energy and when the hydrogen is being pumped into the fuel cell, it also generates DC energy. The DC energy from both the sources is integrated, and the hybrid energy will be directed to high gain interleaved boost converter which enlarges the magnitude of voltage at the output of the converter.

The boosted output from the converter will be passed to inverter which converts the DC output from the converter into AC input to drive the [10–13] BLDC motor. Also, the integrated output will be passed to [14] maximum power point tracker for attaining the maximum power all the times from the hybrid system.

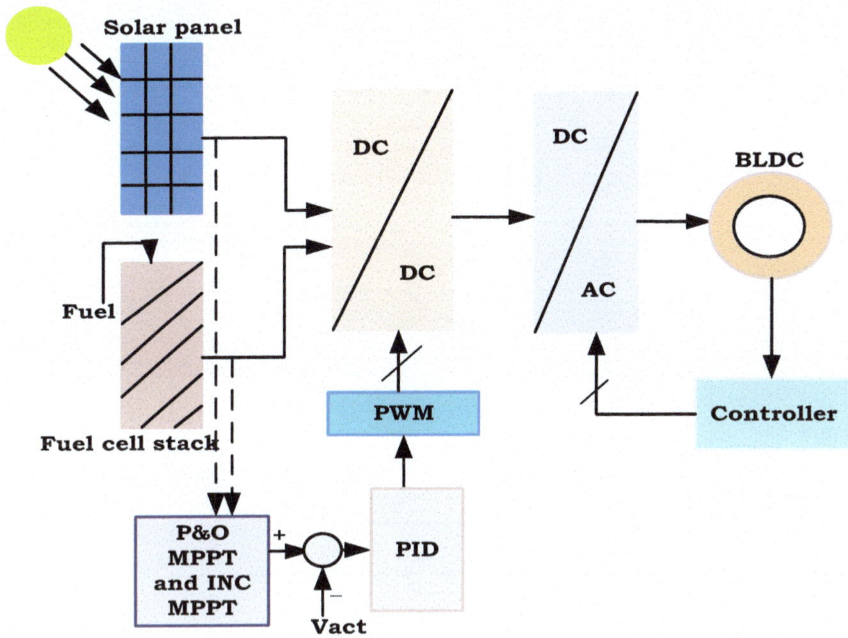

Fig. 1 Block diagram of the proposed vehicle

The maximum voltage tracked by the tracker is fed to PID controller where it compares the actual voltage with the peak voltages and generates processed error. The processed error will be compared with the carrier signal there by generating gate pulse for the high gain interleaved boost converter.

The expression for the voltage at the output of the solar cell can be given as

$$V_{\mathrm{pv}} = I_{\mathrm{pv}}\, R \tag{1}$$

The expression for the current through the solar cell can be given as

$$I_{\mathrm{pv}} = I_{\mathrm{sc}}\, e^{\frac{V}{\eta V_t}-1} \tag{2}$$

The expression for the filter inductances of the converter can be given as

$$L = \frac{K\, V_{\mathrm{in}}}{\Lambda I\, f_s} \tag{3}$$

The expression for the filter capacitances of the converter can be given as

$$C = \frac{K\, V_0}{R\Lambda V\, f_s} \tag{4}$$

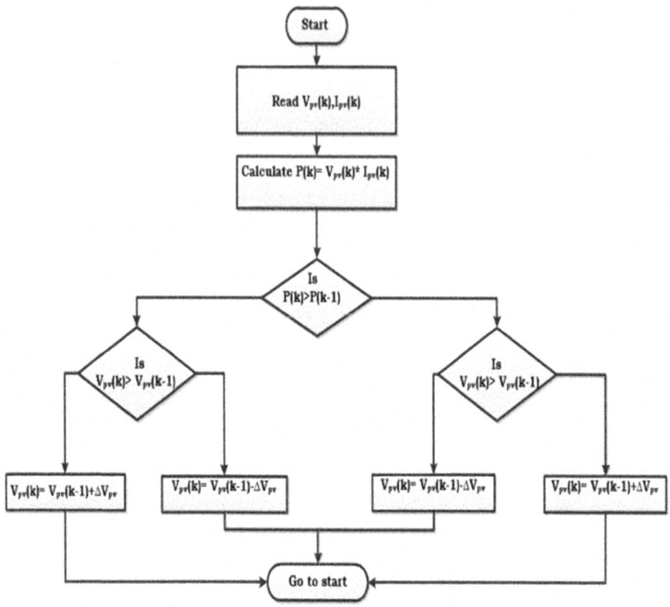

Fig. 2 P&O MPPT flowchart

3 PID Controller Fed MPPT Trackers

3.1 Perturb and Observe MPPT

The flowchart for perturb and observe MPPT technique is given in Fig. 2 where the inputs for the tracker are voltage and current of the hybrid system. Then it calculates change in power and change in voltage. The changes are being compared with their past change values, and the difference between past value and present value will be calculated. If there occurs greater than zero, then perturbation [9, 10] will be in the similar path. If there exists lesser than zero then perturbation gets reversed and moves in reverse direction.

3.2 Incremental Conductance MPPT

Perturb and observe will suffer from the problem of swinging around the maximum power point which can be overcome with incremental conductance MPPT technique anywhere it achieves the peak power and prevents the perturbations if maximum point is reached. If maximum power is not achieved, then lane of perturbation will be achieved [9, 10] using $\mathrm{d}I/\mathrm{d}V$ and $-I/V$. It is attained through the verity that $\mathrm{d}P/\mathrm{d}V$ is negative, if the tracking is to the right side of maximum power and that $\mathrm{d}P/\mathrm{d}V$ is positive, if the tracking is to the left side of maximum power. It calculates at what point peak power will be attained instead of fluctuating around maximum power point like as perturb and observe method does (Fig. 2).

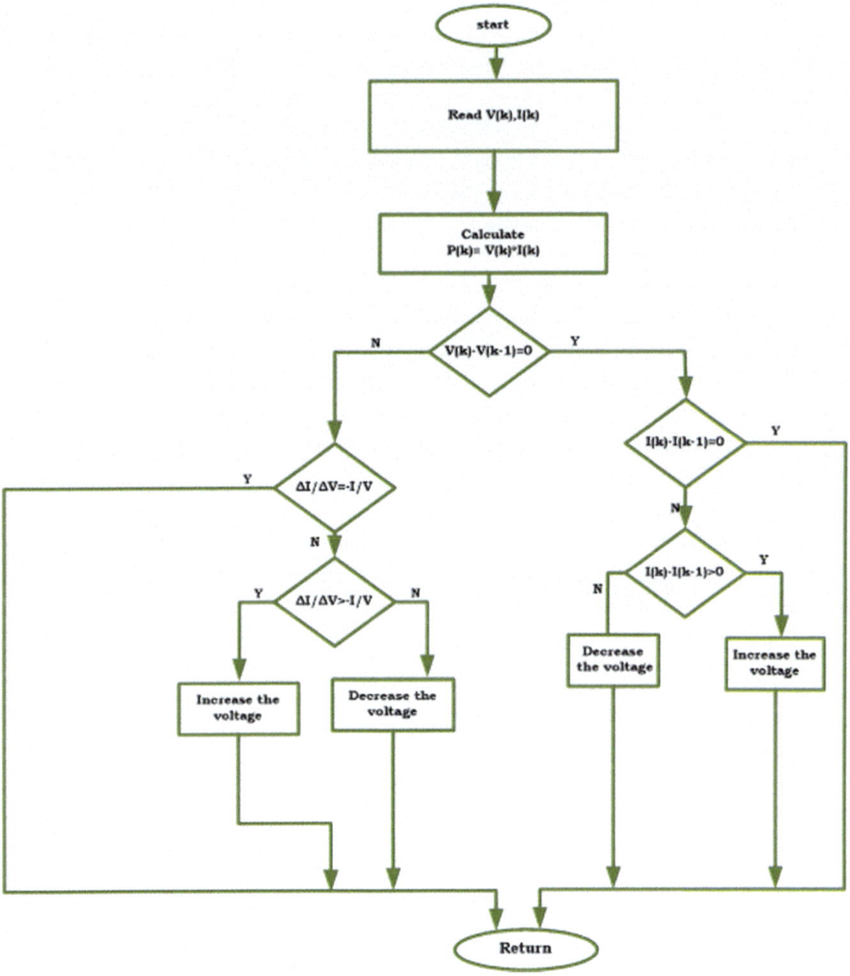

Fig. 3 Incremental conductance MPPT flowchart

4 Simulation of Proposed Electric Vehicle

The proposed PID controller-based electric vehicle has been simulated using MATLAB Simulink, and the following is the results obtained.

The solar and fuel cell systems are integrated and the integrated voltage of hybrid system is 45 V is shown in Fig. 4.

The integrated output has been fed to the converter which is controlled with PID controller-based MPPT controller whose voltage, current and power are shown Fig. 5. An output voltage of 360 V has been produced from the HGIBC whose peak value is 385 V and settled at $t = 0.21$ s with PID-P&O controller as revealed in Fig. 5 and also output voltage of 372 V has been produced from the HGIBC whose peak value is 379 V and settled at $t = 0.1$ s with PID-INC controller. The current of 6.1A has been obtained from the boost converter whose peak value is 6.5A and settled at $t = 0.21$ s with PID-P&O controller as revealed and also current of 6.28A has been obtained from the boost converter whose peak value is 6.1A and settled at $t = 0.1$ s with PID-INC controller. The output power of the converter has been observed to be 2280 W whose peak value is 2500 W and settled at $t = 0.21$ s with PID-P&O controller as revealed and also output power of the converter has been observed to be 2336 W whose peak value is 2450 W and settled at $t = 0.1$ s with PID-INC controller.

The output of the converter has been directed into the inverter which generates AC output for driving the BLDC motor and Fig. 6 shows the voltage generated from the inverter with PID-P&O and PID-INC controller.

The output from the HGIB converter is passed to the three-phase inverter to produce AC output for the BLDC motor whose value is 360 V, and the performance characteristics (speed, torque, EMFs' and currents) of the motor are as shown in Figs. 7 and 9 whose speed of BLDC motor is 1450RPM.

The torque obtained is 0.65 N m, three-phase generated EMF whose magnitude is 85 V, three-phase currents whose magnitude is 19A with PID-P&O controller.

Fig. 4 Potential of integrated system

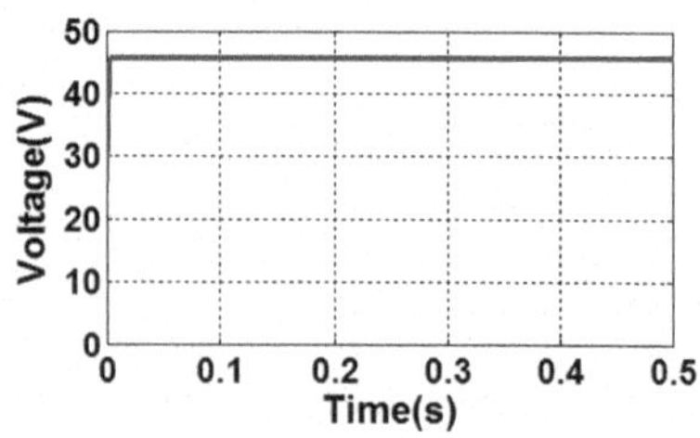

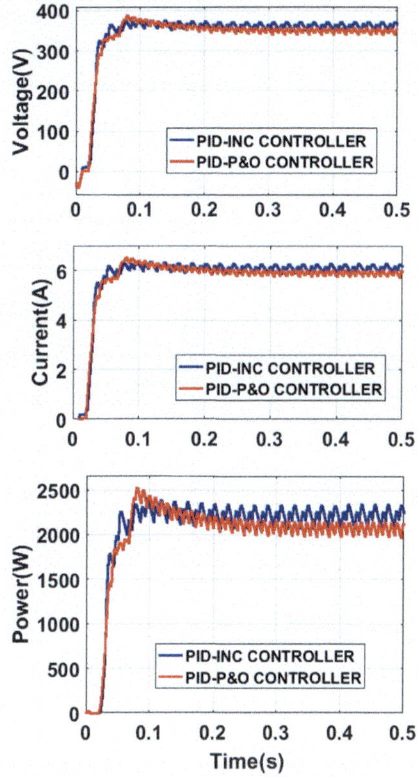

Fig. 5 HGIBC output voltage, current and power with PID-INC and PID-P&O controller

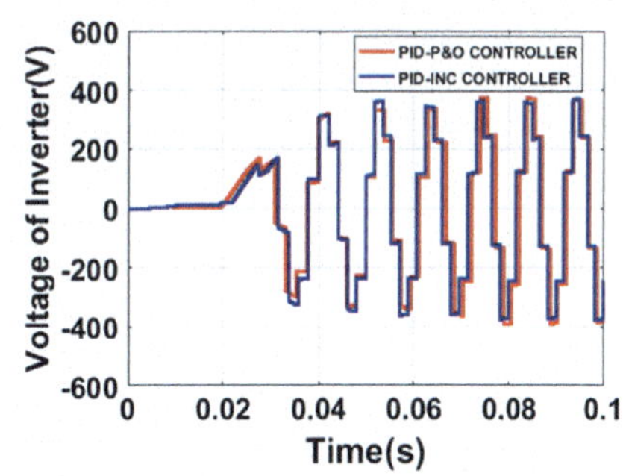

Fig. 6 Voltage of inverter with PID-INC and PID-P&O controller

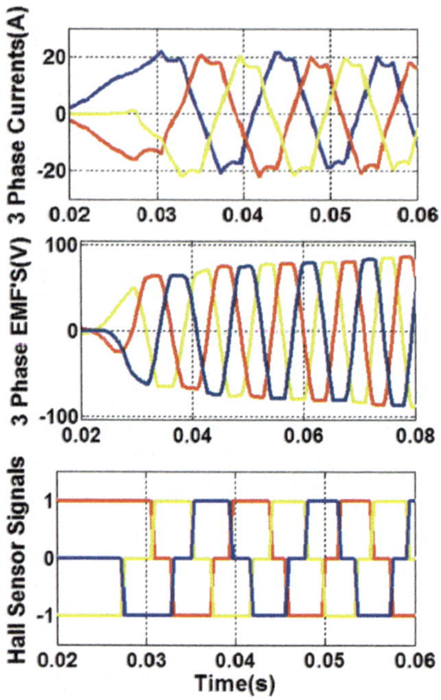

Fig. 7 Currents, EMF's and hall sensor signals of BLDC motor with PID-P&O controller

The output from the HGIB converter is passed to the three-phase inverter to produce AC output for the BLDC motor whose value is 372 V and the performance characteristics (speed, torque, EMFs' and currents) of the motor are as shown in Figs. 8 and 9 whose speed of BLDC motor is 1520RPM and torque obtained is 0.705 N m, three-phase generated EMF whose magnitude is 87 V, three-phase currents whose magnitude is 20.3A with PID-INC controller (Tables 1 and 2).

Fig. 8 Currents, EMF's and hall sensor signals of BLDC motor with PID-INC controller

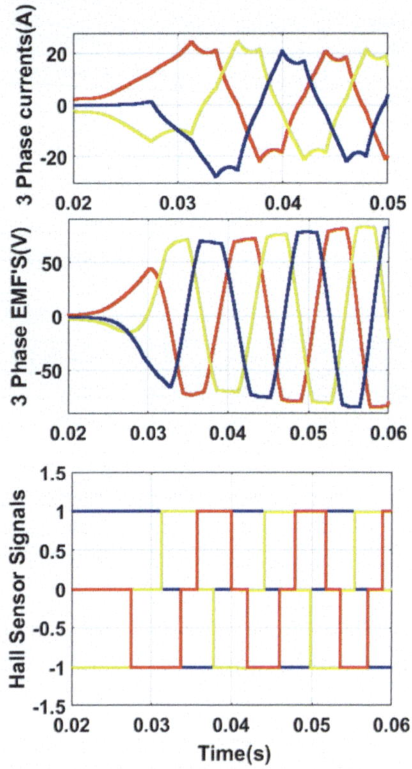

Fig. 9 Speed and torque of BLDC motor with PID-INC and PID-P&O controller

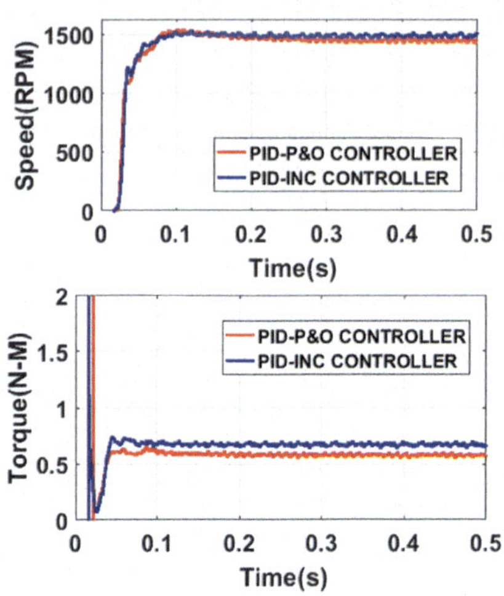

Table 1 Assessment of DC link voltage, current and power of HGIBC with PID-INC and PID-P&O MPPT

Parameter	PID-INC MPPT	PID-P&O MPPT
DC voltage	372 V	360 V
DC current	6.28A	6.1A
DC power	2336 W	2280 W
Speed	1520RPM	1490RPM
Torque	0.705 N m	0.69 Nm
Torque ripples	0.11 N m	0.15 Nm

Table 2 Comparison table in terms of static and dynamic performances with PID-INC and PID-P&O MPPT

Performance	Parameter	PID-INC MPPT	PID-P&O MPPT
Dynamic	Response time (ms)	0.1	0.21
Dynamic	Overshoot (W)	150	300
Static	Ripple (W)	180	200
Efficiency	Output power avg value (W)	2336	2280

5 Conclusions

The design of renewable energy-based electric vehicle with the use of the solar and fuel cell as the supply resources. Also, PID-based MPPT tracking system has been designed in this paper with two trackers named as incremental conductance and perturb and observe method. The assessment of the performance of controller-based maximum power point technique in terms of voltage, current, power, response time, torque ripples and efficiency of the system and also comparison has been made in between incremental conductance-based MPPT system and also perturb and observe-based MPPT system with designing the system using MATLAB Simulink.

References

1. Chowdhury MSA, Al Mamun KA, Rahman AM (2016) Modelling and simulation of power system of battery, solar and fuel cell powered Hybrid Electric vehicle. In: 2016 3rd international conference on electrical engineering and information communication technology (ICEEICT). IEEE, pp 1–6
2. Bhadra S, Mukhopadhyay P, Bhattacharya S, Debnath S, Jhampati S, Chandra A (2020) Design and development of solar power hybrid electric vehicles charging station. In: 2020 IEEE 1st international conference for convergence in engineering (ICCE). IEEE, pp 285–289
3. Ali M, Mohammad S, Rahman MM (2019) Modelling a solar charge station for electric vehicle with storage backup. In: 2019 1st international conference on advances in science, engineering and robotics technology (ICASERT). IEEE, pp 1–4

4. Petrusic A, Janjic A (2021) Renewable energy tracking and optimization in a hybrid electric vehicle charging station. Appl Sci 11(1):245
5. Chowdhury MSA, Rahman AM, Samrat NH (2015) A comprehensive study on green technologies used in the vehicle. In: 2015 3rd international conference on green energy and technology (ICGET). IEEE, pp 1–5
6. Azeem I, Baig MMA, Uddin MH (2018) A strategy to evaluate MPPT techniques. In: 2018 Asian conference on energy, power and transportation electrification (ACEPT). IEEE, pp 1–4
7. Kchaou A, Naamane A, Koubaa Y, M'Sirdi NK (2016) Comparative study of different MPPT techniques for a stand-alone PV system. In: 2016 17th international conference on sciences and techniques of automatic control and computer engineering (STA). IEEE, pp 629–634
8. Selmi T, Abdul-Niby M, Devis L, Davis A (2014) P&O mppt implementation using matlab/simulink. In: 2014 ninth international conference on ecological vehicles and renewable energies (EVER). IEEE, pp 1–4
9. Hamidon FZ, Aziz PA, Yunus NM (2012) Photovoltaic array modelling with P&O MPPT algorithm in MATLAB. In: 2012 international conference on statistics in science, business and engineering (ICSSBE). IEEE, pp 1–5
10. Chung TM, Daniyal H, Sulaiman MH, Bakar MS (2016) Comparative study of P&O and modified incremental conductance algorithm in solar maximum power point tracking
11. Apatya YA, Subiantoro A, Yusivar F (2017) Design and prototyping of 3-phase BLDC motor. In: 2017 15th international conference on quality in research (qir): international symposium on electrical and computer engineering. IEEE, pp 209–214
12. Suganthi P, Nagapavithra S, Umamaheswari S (2017) Modeling and simulation of closed loop speed control for BLDC motor. In: 2017 conference on emerging devices and smart systems (ICEDSS). IEEE, pp 229–233
13. Sarala P, Kodad SF, Sarvesh B (2016) Analysis of closed loop current controlled BLDC motor drive. In: 2016 international conference on electrical, electronics, and optimization techniques (ICEEOT). IEEE, pp 1464–1468
14. Sreeram K (2018) Design of fuzzy logic controller for speed control of sensorless BLDC motor drive. In: 2018 international conference on control, power, communication and computing technologies (ICCPCCT). IEEE, pp 18–24

ANFIS Prediction Using Neuro-Fuzzy Model of Experimental Study on Concentric Tube Heat Pipe Heat Exchanger Using Acetone

P. Ramkumar, A. Kajavali, S. Ramasamy, C. M. Vivek, and M. Sivasubramanian

Abstract To safeguard the electronic cooling systems and also to utilize the excess heat in the components, the heat pipe is implemented. The study deals with the design of a concentric tube heat pipe heat exchanger (CTHPHE) using acetone and water. The investigation is carried out for (0° and 90°) and further with inclination angles (10°–80°). The result shows that the higher values are obtained for 0° than 90°, similarly in case of inclination angles, the 60º possess maximum while relating with 10°. The increment in effectiveness, heat transfer coefficient, observed for 60° than 10° as 36.3%, 58.49%. The observed average experimental value for effectiveness as 50.84% and predicted value as 49.52%. The experiment it is warranted for acetone demonstrated enhanced results, and results are compared with numerical analysis using neuro-fuzzy system. The results indicated that the heat pipe heat exchanger appropriateness for application involving heat dissipation in waste heat recovery systems.

Keywords Heat pipe heat exchanger · ANFIS · Effectiveness · Electronic cooling system

---

P. Ramkumar (✉) · M. Sivasubramanian
Department of Mechanical Engineering, Kalasalingam Academy of Research and Education, Virudhunagar, Tamil Nadu, India
e-mail: rkmailmech@gmail.com

A. Kajavali
Department of Mechanical Engineering, Government College of Engineering, Bargur, Tamil Nadu 635104, India

S. Ramasamy
Department of Mechanical Engineering, Government Polytechnic College, Melur, Tamil Nadu, India

C. M. Vivek
Department of Mechanical Engineering, Periyar Maniammai Institute of Technology, Thanjavur, Tamil Nadu, India

E. Natarajan et al. (eds.), *Materials, Design and Manufacturing for Sustainable Environment*, Lecture Notes in Mechanical Engineering,
https://doi.org/10.1007/978-981-19-3053-9_47

1 Introduction

The heat pipe implied in various applications at industries and domestically. The heat pipe used for effective heat recovery system. The heat pipes were implemented in industries and processing plants to enhance the performance of the system [1, 2]. In exhaust gas system of automobiles, HPHE was used [3]. In HPHE, the refrigerants were used its environmental friendly [4]. In spacecraft, the LHP was experimented [5]. The HP used in insulated gate bipolar transistor module shows efficient operation [6]. The studies made on helicoidally DPHE by neuro-fuzzy module shows acceptable and compelling outputs on relating with the results [7]. The lower thermal resistance obtained for de-ionized water as fill ratio of 50% for pulsating heat pipe [8]. The CLPHP with fill ratios, number of turns, gravitational effect and heat input, shows the maximum performance for maximum number of turns in wick [9, 10].

The less thermal resistance was obtained for ammonia HP in satellite applications [11]. Methane cryogenic LHP had maximum performance at various heat loads [12]. The pulsating heat pipe was investigated using different refrigerants for R134a was greater than R404A and R600a [13–15]. The RHP using conical condenser, and numerical study using Navier–Stokes equations [16]. The heat pipe and PCM were used for solar thermal electricity production and energy storage [17]. The heat pipes were studied with various angles, wick construction, and operational fluid [18, 19]. The PHP using DI water at gravitational angle possesses lesser heat resistance of 0.077 K/W [20].

The DPHE with EG water with the ratio of 20:80 at 0.08% shows the resultant performance of 1.148 [21]. The CTHPHE with acetone in the waste heat recuperation system for 60° has 1600 W at 30°C [22]. The performance of CTHPHE was analyzed for different conditions are working and heat carrying fluids, geometry, and angles [23, 24]. The literatures possess the earlier studies for limited constraints of the HPHE. In this analysis, the HP is manufactured and placed inside a CTHE for investigating the performance of electronic cooling system.

2 Experimentation Details and Procedure

2.1 Design Details

In Fig. 1, it shows the fabricated HPHE. The copper HP of 19 and 17 mm as OD and ID is kept inside the GI heat exchanger shell at evaporator and condenser regions of 50 and 35 mm. The red and blue arrow denote the hot and cold fluid given in Fig. 2. The HPHE is fabricated for total 1000 mm, with evaporator and condenser section as 700 and 300 mm. HP with SS wick of mess size as 50 holes/inch. Acetone and water as working fluid and heat transport fluid. The 5 L bath of hot and cold fluid has 3 LPM rotameters with heater of 2 kW. An 0.5 HP pump for circulating the hot

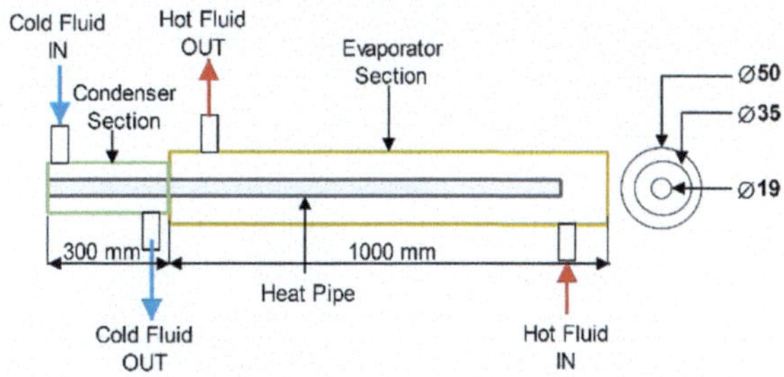

Fig. 1 Schematic view of a CTHPHE

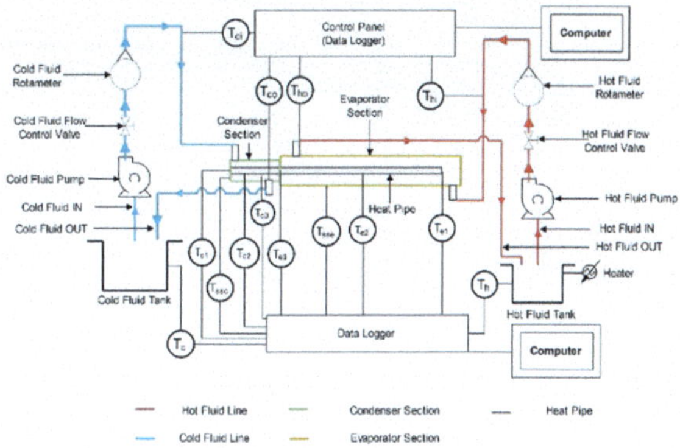

Fig. 2 Schematic sketch of CTHPHE

water inside the evaporator, similarly the rotameters are used to regulate the flow rates. The K—type thermocouples are used to check the temperature.

2.2 *Experimental Procedure*

The Fig. 3 depicts the experimentation of HPHE. The gases inside the HP are evacuated with the vacuum pump and warranted using 30 inches of mercury vacuum gage. The 50% fill ratio of acetone is charged [25]. HPHE is initially positioned for 0° and 90° angles and extended for other angles of 10°–80°.

The inlet flow rates of hot and cold fluid are 40 and 20 LPH, and temperatures are 50 and 30.5 °C. Continued for various angles of 10 to 80° with 60 to 120 LPH. The outer circumference of the CTHPHE is isolated with the foam of 20 mm.

2.3 *Uncertainty Study*

The Kline and McClintock's technique are used for uncertainty analysis [26]. The regular calibration is carried out to minimize the instrumental uncertainty. The

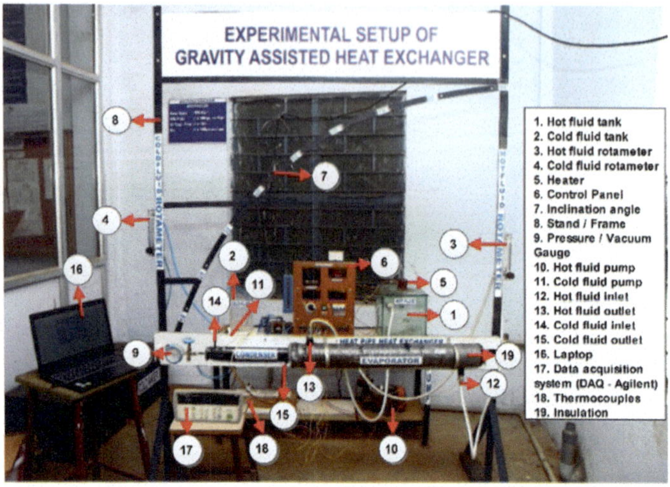

Fig. 3 Photographic view of CTHPHE

uncertainty for effectiveness and heat transfer coefficient are 7.47 and 5.63%. The instrumental uncertainty are analyzed to minimize the errors.

3 Results and Discussion

The investigation is done using above parameters, and the results are produced. In Fig. 4, it depicts the effectiveness for 0° and 90° as 60 °C and 100 LPH for water as HTF. The observed values are 50.84 and 45.76% with the increment in effectiveness of 11.10%.

In Fig. 5, the study is done for 10°–80° the effectiveness as 29.8–40.6%, for 10°–60° and correspondingly for 70° and 80° as 32.5 and 34.2%. In case of 10°–60°, there is increase in effectiveness is happens by wick capillary movement and quick circulation of acetone from hot to cold zone, happens by the driving force of water to the hot zone, thus heat provided and vapor formation are even, hence high heat transfer observed, this leads to occurrence of high effectiveness of 36.3% for 60° than 10°. Similarly, beyond 60° results reduced to 24.98% for 70° and 18.81% for 80°, this happens by the liquid film formation at condenser zone. The effectiveness increases from 40 to 100 LPH, beyond that it decreases for 120 LPH [27] and happens by the small heat absorption and ejection of the acetone at either zones.

Figure 6, the heat transfer coefficient for 0° and 90°, 60 °C and 100 LPH for water as HTF. The values are 1584 W/m^2 °C and 1324 W/m^2 °C with the increment in effectiveness of 19.6%. In Fig. 7, for 10°–80° angles, the heat transfer coefficient achieved is 674 W/m^2 °C and 1068.5 W/m^2 °C for 10°–60° with increment of 58.4%, and similarly for 70° and 80° as 787 W/m^2 °C and 836.5 W/m^2 °C, respectively.

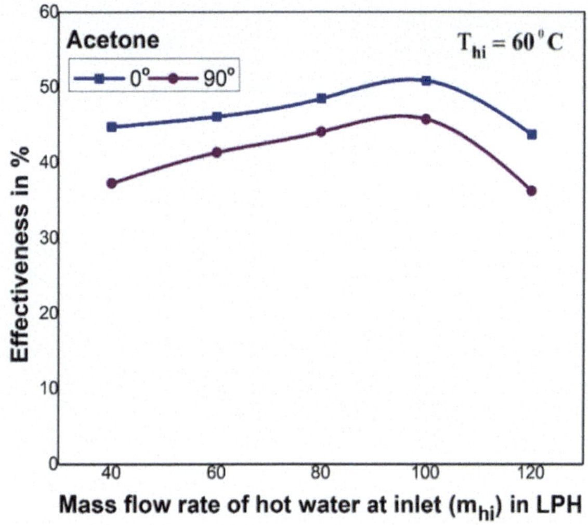

Fig. 4 Effectiveness (0° and 90°)

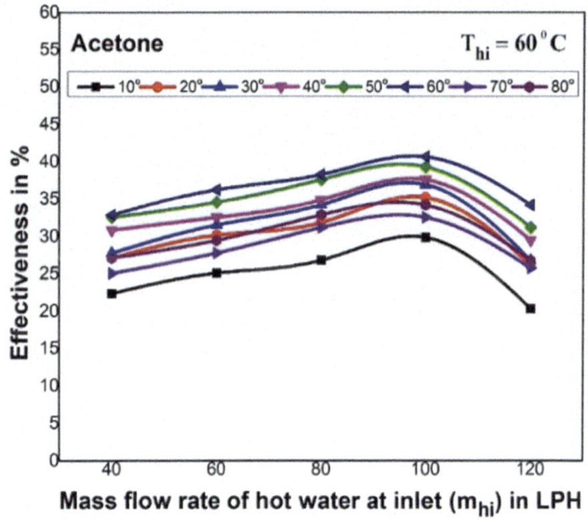

Fig. 5 Effectiveness (10°–80°)

3.1 Neuro-Fuzzy Model

An ANFIS technique is used to make a prediction technique using the experiment data. Neuro-fuzzy interpretation tool (MATLAB 2020) is analyzed by prediction model. The Fig. 8 shows experiment data allied using the input module of the plot. The complete experimental data are separated as 70, 20, and 10% for the training, testing, and checking method.

In case of training procedure, the module along the experimentation data and forms the neuro-fuzzy network, this makes the connection for input and output features.

Fig. 6 Heat transfer coefficient (0° and 90°)

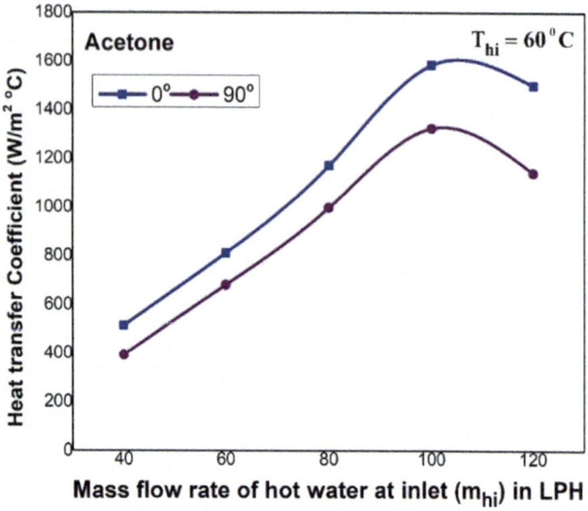

Fig. 7 Heat transfer coefficient (10°–80°)

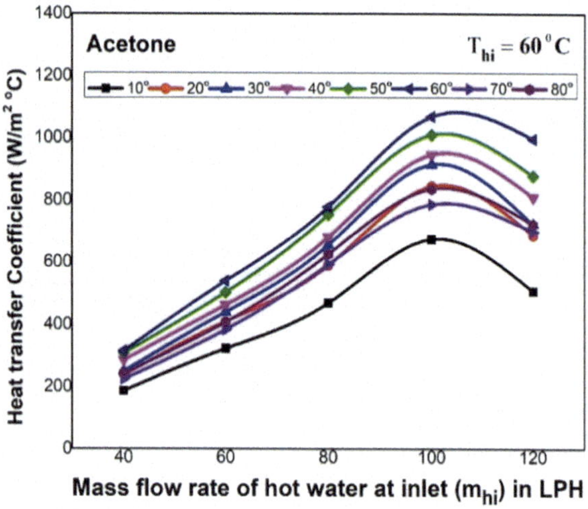

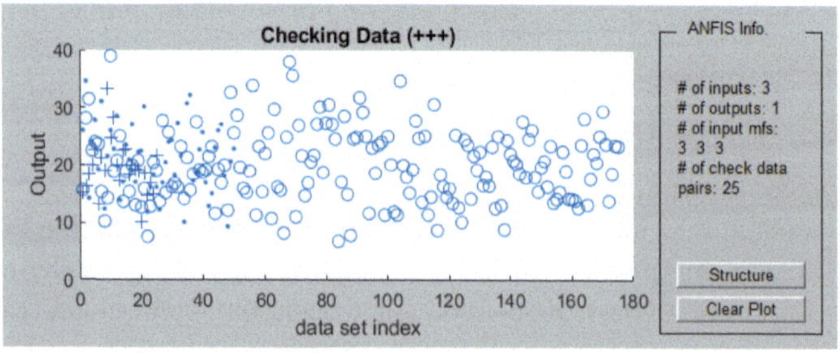

Fig. 8 Experimentation data for the process input module

The testing and checking data check the performance along the established prediction model. The ANFIS established prediction model creates outcome matching of the testing data; hence, the output is related among the experiment data along the stage of accurateness of the prediction model.

The Fig. 9 shows the network for the neuro-fuzzy structure. The 03 input features are angles, mass flow rates, and temperatures, and the outcome is effectiveness. The Sugeno created neuro-fuzzy system as created and 03 functions for each input constraint are defined as low, medium, and high. Membership purpose demonstrates as bell MF depicts in Fig. 10. The study is steered for additional membership functions as triangular, gauss, trapezoidal, but bell function is measured suitable for improved outcomes with less quantity of processing time.

The Fig. 11 depicts the features after every input is related with a association along the outcomes are attained. The procedure recurrences along link is creating with minimum quantity of error. Membership purpose of the model is skilled by the experimentation data. Numerous epochs are determined the 500 epochs provide good training system, this attributes with the minimum testing error as 5.5459. Checking

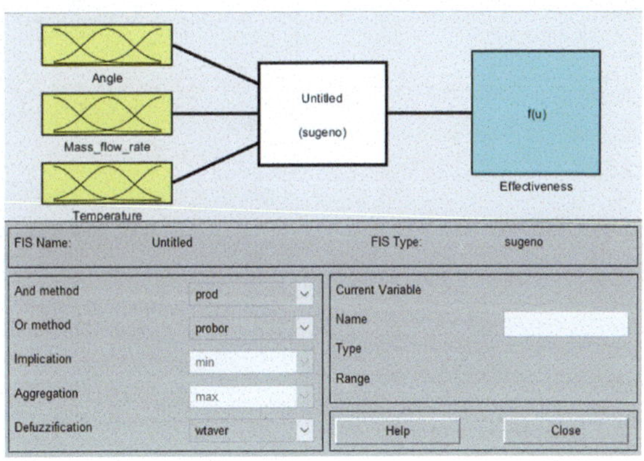

Fig. 9 Network—neuro-fuzzy system

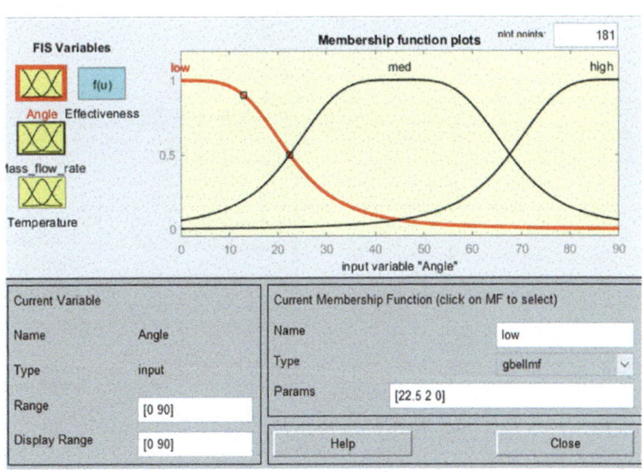

Fig. 10 Membership—bell MF

error is exposed as blue line, and this specifies the checking error nearer to the testing error (5.7620) shown in Fig. 12. The testing and checking data denoted along the experiment data. The testing vs predicted data are given in Fig. 13. This determines that both the plots of established prediction model are stable among experimental and predicted results are closer, while the training is finished a rule viewer permits to comprehend the difference of output while the input is changes. The prediction of data input is shown in Fig. 14. The average angle as 45°, m_{hi} as 80 LPH, T_{hi} as 60 °C, and obtained average effectiveness as 36% (Figs. 15 and 16).

The rule observer the prediction model is studied by locating the predicted results among equivalent test data. Error percentage is considered using the Eq. 1,

$$\text{Error\%} = \frac{|\text{Predicted value} - \text{Experimental Value}|}{\text{Experimental Value}} \times 100 \quad (1)$$

Fig. 11 Neuro-fuzzy inference architecture—experiment data

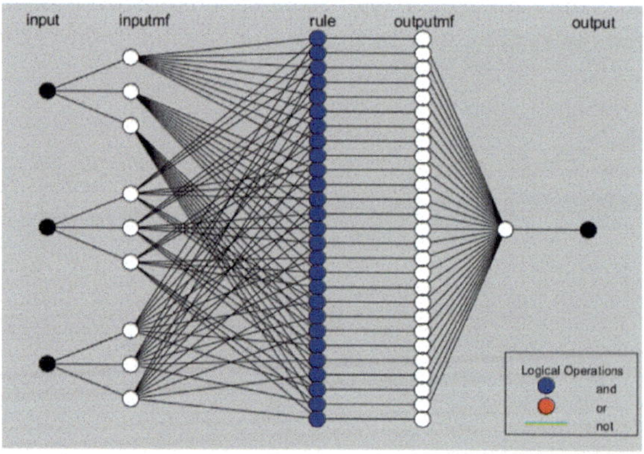

Fig. 12 Difference of error using epochs

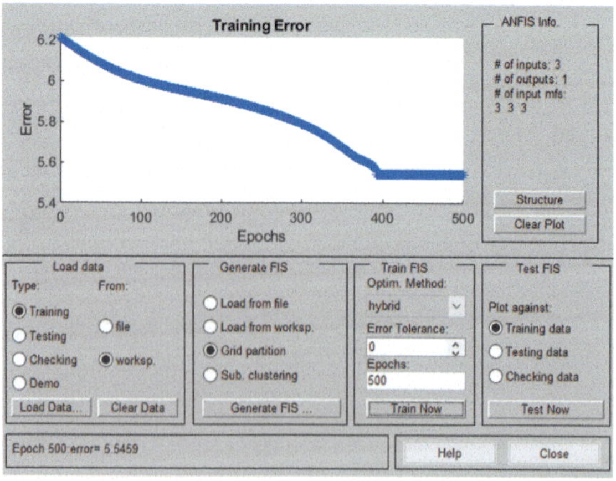

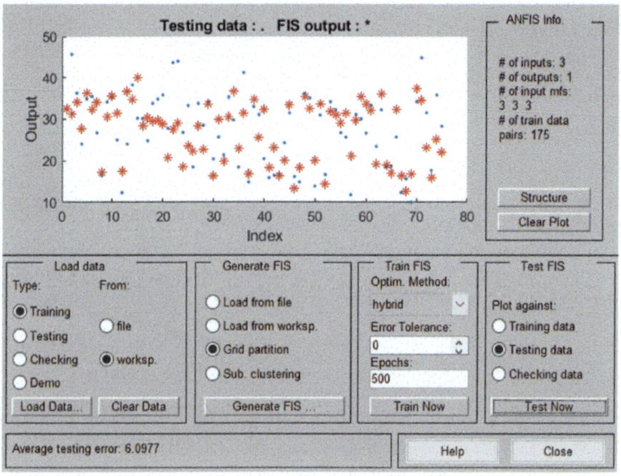

Fig. 13 Testing versus predicted data

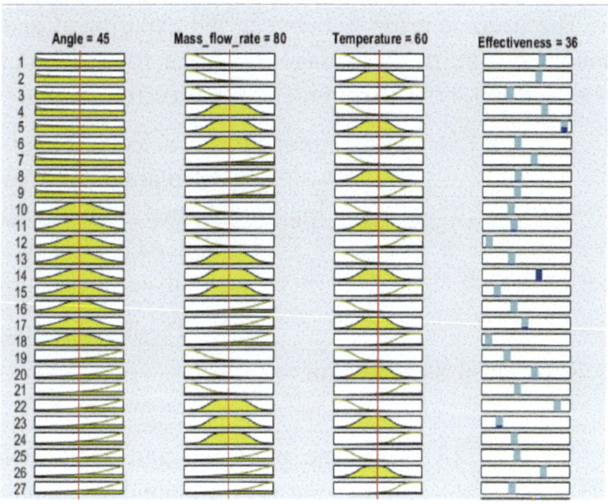

Fig. 14 Variance for output by input data

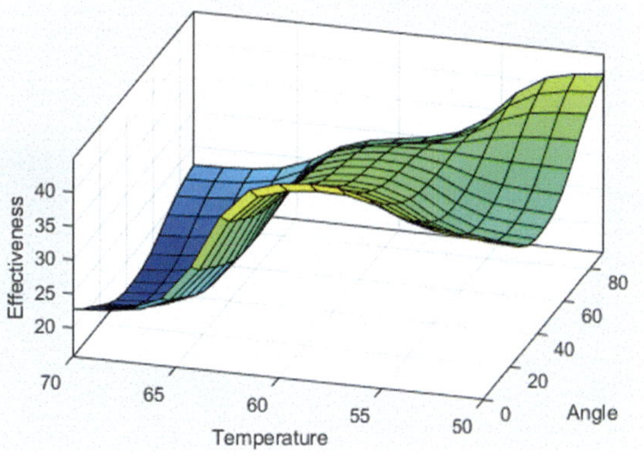

Fig. 15 Effectiveness with m_{hi} for angles

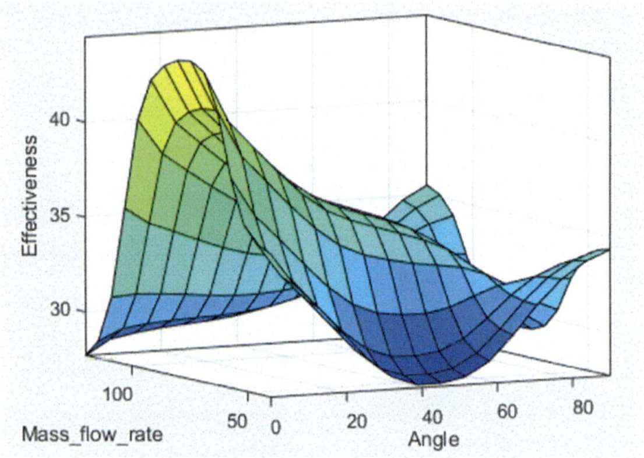

Fig. 16 Effectiveness with T_{hi} angles

The average error % between the experiment and predicted value as 6.70% is adequate. The mean average deviation for the experimental and predicted result obtained is 1.47. The root mean square error result is 1.85 given in Eq. 2, (Fig. 17).

$$\text{Mean Average Deviation} = \frac{\sum_{i=1}^{n} |\text{Experimental Value} = -\text{Predicted Value}|}{n}$$

$$\text{Root Mean Square Error} = \sqrt{\frac{\sum_{i=1}^{n} (\text{Predicted Value} - \text{Experimental Value})}{n}} \quad (2)$$

The Fig. 18 shows the predicted and experimental results using the testing data. The graph signifies the very minute discrepancy among the experimental and predicted results. The given established prediction model is consistent.

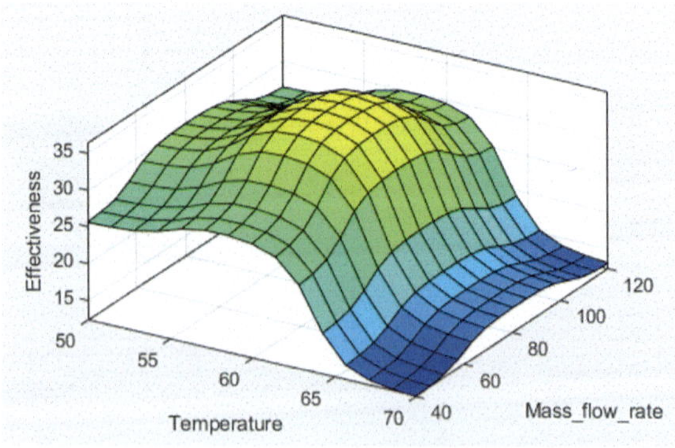

Fig. 17 Effectiveness with T_{hi} for M_{hi}

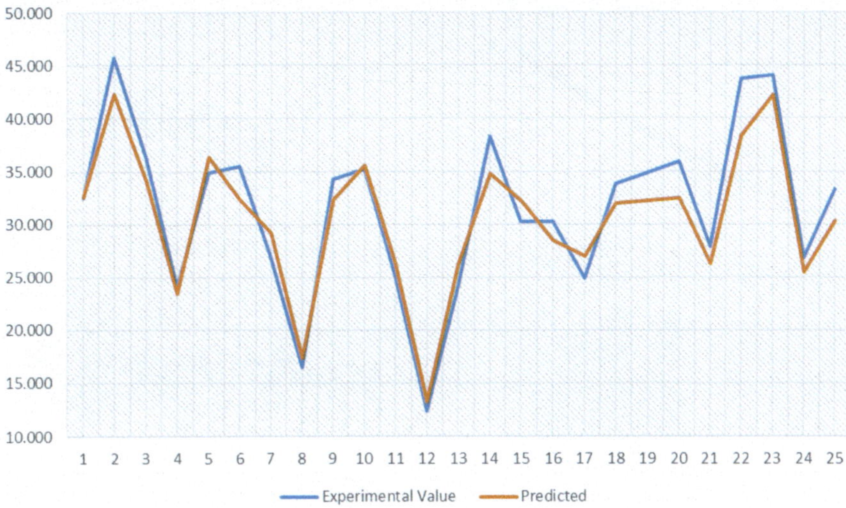

Fig. 18 Predicted and experimental value for the testing data

3.2 *Analysis of Variance*

Data are used to analyze the effectiveness for 03 input constraints. The normal probability plot describes the results are associated among the mean line, by representing the robustness of the demonstrated experiments. The Fig. 19 gives the histogram discloses bell shape curve for confirming the experiment interpretations are consistent and no inconsistency in the data.

The outcomes are presented in Table 1, the F-value clearly shows maximum inducing factor disturbing effectiveness is temperature and further the inclination angle and the least as mass flow rate. Analysis of variance system is attained as R^2 results are 97%, this specifies analysis technique is effective.

The Fig. 20 shows the tendency of the input limitations over the output is clearly given along the plot. The plot of the input features shows the maximum effectiveness for 0° and 90°. The effectiveness rises till 100 LPH and shows a decline curve. Similarly for temperature, the effectiveness peaks at 60 °C and then starts to decrease for increment in temperature.

At angle 0°, for 100 LPH and 60°C, the effectiveness obtained is maximum at 50.847. The readings are given as the neuro-fuzzy, and predicted results are attained for similar inputs. Experiment and predicted results are stated in the Table 2. This shows that the error % are adequate for additional confirming the neuro-fuzzy model as consistent. The main things plot the input constraints at the minimum effectiveness are attained. The minimum effectiveness of 9.367% for 10° angle, 120 LPH and 60 °C. At this constraint set, the predicted value is 10.91. This clearly shows that the neuro-fuzzy model forecasts the results using tolerable error of margin.

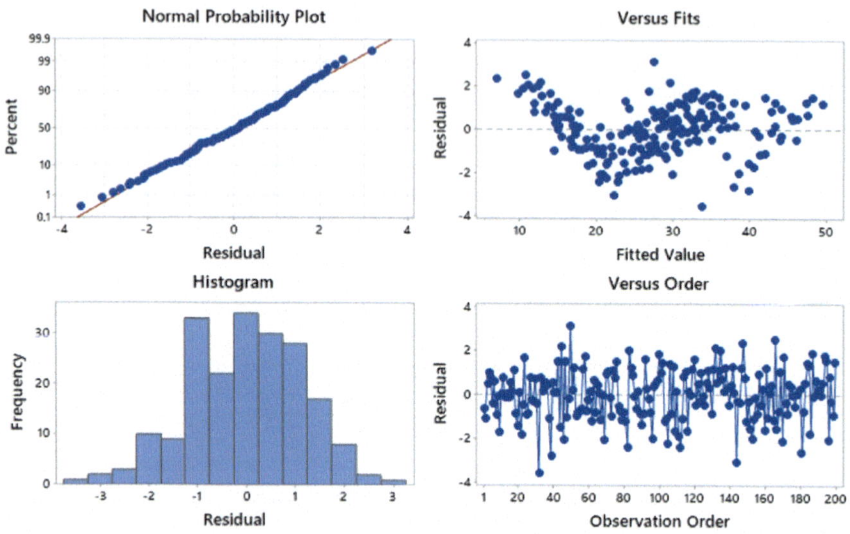

Fig. 19 Analysis of variance for given data

Table 1 Variance of analysis of F-value

Parameter	Degree of freedom	Sum of square	Mean of square	F-value
Angles	9	7158	795	527
Mass flow rates	4	1265	316	209
Temperatures	4	7859	1964	1303

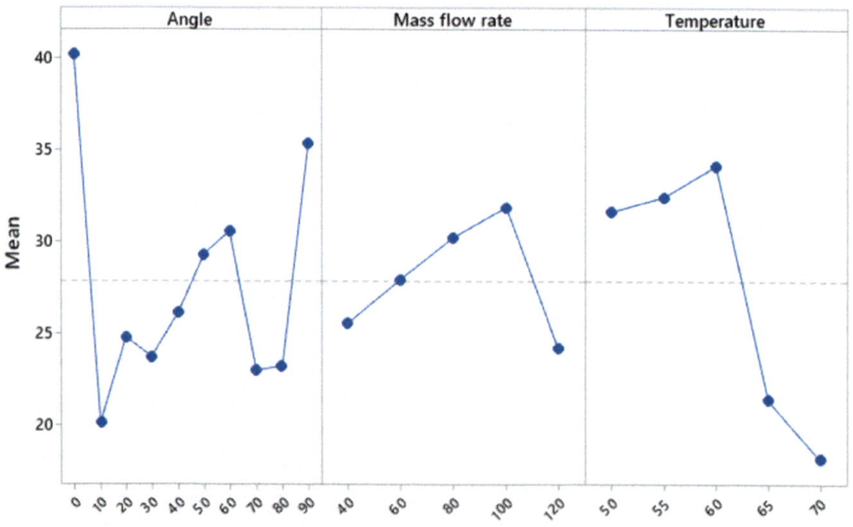

Fig. 20 Plots represent the experimental data

Table 2 Experimental and predicted results

S.No.	Angle (°)	Mass flow rate (LPH)	Temperature (°C)	Experimental value (%)	Predicted value (%)
Maximum effectiveness					
1	0	100	60	50.847	49.52
Minimum effectiveness					
2	10	120	70	9.367	10.91

4 Conclusions

The study shows the experimental and numerical study of neuro-fuzzy interpolation for acetone, water, and gravitational angles. The results show higher for 0° than 90° angle. Correspondingly, while related with 10–80°, the higher results are obtained for 60° angle. This investigation shows that for optimum situations, increment in effectiveness and heat transfer coefficient obtained is 11.10% and 19.61%, respectively, for 0° than 90°. In case of inclination angles 36.3% and 58.49%, respectively, for the inclined angles of 60° while comparing with 10°. The experimentation values are closer with predicted value, thus neuro-fuzzy interpolation shows better outcomes and robust.

References

1. Noie Baghban SH, Majideian GR (2000) Waste heat recovery using heat pipe heat exchanger (HPHE) for surgery rooms in hospitals. Appl Therm Eng 20(14):1271–1282
2. Vasiliev LL (2005) Heat pipes in modern heat exchangers. Appl Therm Eng 25(1):1–19
3. Feng Y, Yuan X, Lin G (2003) Waste heat recovery using heat pipe heat exchanger for heating automobile using exhaust gas. Appl Therm Eng 23(3):367–372
4. Longo GA, Righetti G, Zilio C, Bertolo F (2014) Experimental and theoretical analysis of a heat pipe heat exchanger operating with a low global warming potential refrigerant. Appl Therm Eng 65(1–2):361–368
5. Wang L, Miao J, Gong M, Zhou Q, Liu C, Zhang H, Fan H (2019) Research on the heat transfer characteristics of a loop heat pipe used as mainline heat transfer mode for spacecraft. J Therm Sci 28(4):736–744
6. Lu J, Shen L, Huang Q, Sun D, Li B, Tan Y (2019) Investigation of a rectangular heat pipe radiator with parallel heat flow structure for cooling high-power IGBT modules. Inter J Therm Sci 135:83–93
7. Mehrabi M, Pesteei SM (2011) Modeling of heat transfer and fluid flow characteristics of helicoidal double-pipe heat exchangers using adaptive neuro-fuzzy inference system (ANFIS). Inter Comm Heat Mass Transf 38(4):525–532
8. Durga B, Zhang H, Cai W, Li F (2018) An experimental investigation of thermal performance of pulsating heat pipe with alcohols and surfactant solutions. Inter J Heat Mass Transf 117:1032–1040
9. Vipul PM, Hemant kumar B. Mehta, (2019) Experimental investigations on the effect of influencing parameters on operating regime of a closed loop pulsating heat pipe. J Enhan Heat Transf 26(4):333–344

10. Jia H, Jia L, Tan Z (2013) An experimental investigation on heat transfer performance of nanofluid pulsating heat pipe. J Therm Sci 22(5):484–490
11. Vivek Patel K (2018) An efficient optimization and comparative analysis of ammonia and methanol heat pipe for satellite application. Ener Conver Manag 165:382–395
12. Yuandong G, Lin G, Zhang H, Miao J (2018) Investigation on thermal behaviours of a methane charged cryogenic loop heat pipe. Energy 157:516–525
13. Wang X, Jia L (2016) Experimental study on heat transfer performance of pulsating heat pipe with refrigerants. J of Therm Sci 25(5):449–453
14. Chaoling H, Zou L (2015) Study on the heat transfer characteristics of a moderate-temperature heat pipe heat exchanger. Inter J Heat Mass Transf 91:302–310
15. Didi Z, Li G, Liu Y, Tian X (2018) Simulation and experimental studies of R134a flow condensation characteristics in a pump-assisted separate heat pipe. Inter J Heat Mass Transf 126:1020–1030
16. Lian W, Han T (2019) Flow and heat transfer in a rotating heat pipe with a conical condenser. Inter Comm Heat Mass Transf 101:70–75
17. Shabgard H, Bergman TL, Sharifi N, Faghri A (2010) High temperature latent heat thermal energy storage using heat pipes. Inter J Heat Mass Transf 53(15–16):2979–2988
18. Masoud R, Asgary K, Jesri S (2010) Thermal characteristics of a resurfaced condenser and evaporator closed two-phase thermosiphon. Inter Comm Heat Mass Transf 37(6):703–710
19. Venkatachalapathy S, Kumaresan G, Suresh S (2015) Performance analysis of cylindrical heat pipe using nanofluids–an experimental study. Inter J Multiph Flow 72:188–197
20. Shang F, Fan S, Yang Q, Liu J (2020) An experimental investigation on heat transfer performance of pulsating heat pipe. J Mech Sci Tech 34(1):425–433
21. Tumuluri K, Panitapu Bhramara G, Abhiram, (2020) Experimental investigations on thermal performance of double pipe heat exchanger using EG-water-based sic nanofluid. J Enhan Heat Transf 27(3):249–266
22. Xiaoxing H, Wang Y (2018) Experimental investigation of the thermal performance of a novel concentric tube heat pipe heat exchanger. Inter J Heat Mass Transf 127:1338–1342
23. Ramkumar P, Sivasubramanian M, Rajesh Kanna P, Raveendiran P (2021) Heat transfer behaviour on influence of an adiabatic section on concentric tube shell assisted heat pipe heat exchanger. Inter J Amb Ener 42(06):672–681
24. Ramkumar P, Sivasubramanian M, Rajesh Kanna P, Raveendiran P (2021) An experimental inquisition of waste heat recovery in electronic component system using concentric tube heat pipe heat exchanger with different working fluids under gravity assistance. Micropro Microsyst 83:104033
25. Washburn EW (1930) International critical tables of numerical data, physics, chemistry and technology. National research council. The National Academies Press, Washington: D.C
26. Holman JP (2007) Experimental methods for engineers, 7th edn. McGraw-Hill, New York
27. Raveendiran P, Sivaraman B (2015) Heat transfer coefficient and friction factor characteristics of a gravity assisted baffled shell and heat-pipe heat exchanger. J Eng Sci Tech 10(6):802–810

Experimental Investigation on the Effect of DEE Addition in a Biogas-Biodiesel and Biogas-Diesel Fueled Dual-Fuel Engine

V. Kishorre Annanth, M. Abinash, M. Sreekanth, and M. Feroskhan

Abstract Nowadays, biodiesel is an encouraging perspective that attracts researchers to utilize it as an alternative fuel in this fast-growing society. It is also a cause of engine performance collapsing due to less energy per unit volume. The present study compares the effect of DEE addition on the emission and performance characteristics of biogas-biodiesel and biogas-diesel fueled dual-fuel engines. Here, a typical single cylinder CI engine is used in a dual-fuel mode, where the primary fuel utilized is biogas, and diesel and biodiesel (palm oil) are used as secondary fuel with a DEE blending ratio of 5%. Effect of DEE addition on volumetric efficiency, brake-specific energy consumption, brake thermal efficiency, CO, HC, and NO_x emissions is studied for diesel and biodiesel blends. The AVL 5-gas emission analyzer was used to measure the exhaust gas emissions; while for few samples, density and calorific value have also been measured. A high-brake thermal efficiency increase of 3.46% was observed in the experiments when 5% DEE used with 12 lpm biogas flow rate in both fuels. Also, a decrease in load decreases NO_x emission and escalates HC emission in both fuels. The optimal emission characteristics and engine performance were built up to a greater extent due to DEE being used as an ignition improver during the experiment.

Keywords Biogas · Biodiesel · DEE · Dual fuel

Nomenclature

DEE	Di-ethyl ether
CI	Compression ignition
BTE	Brake thermal efficiency
BSEC	Brake-specific energy consumption

V. Kishorre Annanth · M. Abinash · M. Sreekanth · M. Feroskhan (✉)
School of Mechanical Engineering, Vellore Institute of Technology (VIT), Chennai, Tamil Nadu, India
e-mail: feroskhan.m@vit.ac.in

E. Natarajan et al. (eds.), *Materials, Design and Manufacturing for Sustainable Environment*, Lecture Notes in Mechanical Engineering,
https://doi.org/10.1007/978-981-19-3053-9_48

VE	Volumetric efficiency
HC	Hydrocarbon emissions
CO	Carbon monoxide emissions
NO_x	Nitrogen dioxide emissions
CH_4	Methane
CO_2	Carbon dioxide
BD	Bio-diesel
D	Diesel
BG	Biogas
R^2	Coefficient of regression
RMSE	Root mean square error

1 Introduction

Since the existence of fuel, it has played a vital role in our daily lives with great growing demand and importance. Due to this, it has encouraged various researchers around the globe to look for an alternative to these conventional resources as there is a fear of depletion. One among the few industries that uses a large quantity of fuel is the transportation industry, and its increasing demand for cars necessitates accelerating research into alternative fuels. Numerous studies with variations in primary and pilot fuels have been made to understand the performance and emission characteristics of dual-fuel diesel engines and their purposes [1–4]. It is noticed that to obtain better engine stability and fuel conversion efficiencies, low-load running necessitates significantly lower boost pressures and higher intake temperatures [5]. The characteristics of DEE that influence NO_x production are high-cetane number, oxygen content, high-latent heat of vaporization, and low-calorific value. DEE reduces flash point, viscosity, pour point, fire point, density, heating value, and cloud point of different fuels, and fuel blends while increasing cetane number. As the cetane number increases, it decreases the ignition delay period, resulting in a shorter combustion duration, thus helping to lower the No_x emissions [6, 7]. The wood pyrolysis oil of 5, 10, and 15% by volume is mixed with 95, 90, and 85% by volume of Jatropha methyl ester resulting in Jatropha oil emulsion, X1JOE5, X1JOE10, and X1JOE15, where at full load, the brake thermal efficiency is found out to be 0.5%, 4.75%, and 7.4%, respectively, proving to be significantly better compared to that of diesel at full load. When fueled with X1JOE15, a remarkable decline of 2.5% in NO emissions was also observed [8]. Fuel fumigation technique was executed in a direct injection diesel engine by fueling with bioethanol, and when compared with bioethanol fumigation to diesel operation at full load, the smoke emissions and maximum brake-specific nitric oxide were determined to be 25% and 24.2% lower, respectively [9]. When dual-fuel mode was compared with the diesel-only method, significant unidentified losses were found in the dual-fuel mode at low loads. In contrast, at higher loads exhaust gas and coolant are significantly better. Apart from the air–fuel ratio and exhaust gas temperature, which shows an unexceptional rise with methane enrichment, biogas composition has no effect on any measures [10]. Concerning specific gas consumption, thermal efficiency, and brake power output, methane enriched biogas nearly performed identically to compressed natural gas [11]. Due to air displacement, the volumetric efficiency of dual-fuel mode and manifold heating is also reduced. Significant reductions in air density and knocking index can be achieved through the dual-fuel mode [12, 13]. The biodiesel biogas dual-fuel mode with injection timing of 24.5 and DEE injection of 4% is a dual-fuel operation that outperformed other DEE injection conditions in terms of performance, emission, and combustion characteristics [14]. The biodiesel blends' emission characteristics and engine performance in a dual-fuel engine were compared using statistical regression analysis by changing the compression ratio, revealing that all parameters except HC and CO emissions are directly proportional to the compression ratio [15]. Using correlations provided by other researchers, the authors [16] estimated the air–fuel ratio and particulate matter from exhaust emission and smoke data using semi-analytical methods. Researchers used modeling and experiment on a standard rail engine to establish a clear correlation between biodiesel blend ratio and energy, economic, and environmental indices [17]. Multiple mathematical models and regression analysis were used to predict the brake-specific fuel consumption and emissions of a CI engine in terms of various factors [18, 19]. DEE being utilized as a fuel additive alongside diesel and other alternative fuels has been proved in numerous studies due to its top-notch qualities such as low-self-ignition temperature, high-cetane number, flammability, oxygen content, and high miscibility with diesel fuel [20].

According to the available open literature, no studies have been conducted to compare DEE blends' inclusion in a biogas-biodiesel and biogas-diesel dual-fueled CI engine. In the present study, various comparisons related to emission and performance characteristics of DEE blending ratio and constant biogas flow rate in a biogas-diesel and biogas-biodiesel fueled dual-fuel engine are carried out, and studies have been analyzed and compared experientially, giving an enhanced and effective performance overall.

2 Methodology

Figure 1a, b exhibits the schematic portrayal of the pilot setup utilized in this research. One cylinder 4-stroke CI engine, a mixing chamber, a smoke sensor, an emission analyzer, a dynamometer, and CH_4 and CO_2 cylinders are all part of the system. Here, an eddy current dynamometer acted as both a starter and a load controller on the engine.

In this study, biogas is made from 3:2 combinations of CH_4 and CO_2 gases, with the flow rate regulated by pressure regulators. The intake manifold injects biogas and ambient air into the engine, while a separate pipeline introduces modified fuel into the intake manifold. A smoke sensor is used to measure the amount of smoke, and an emission analyzer is used to measure the amount of CO, NO_x, and HC emissions.

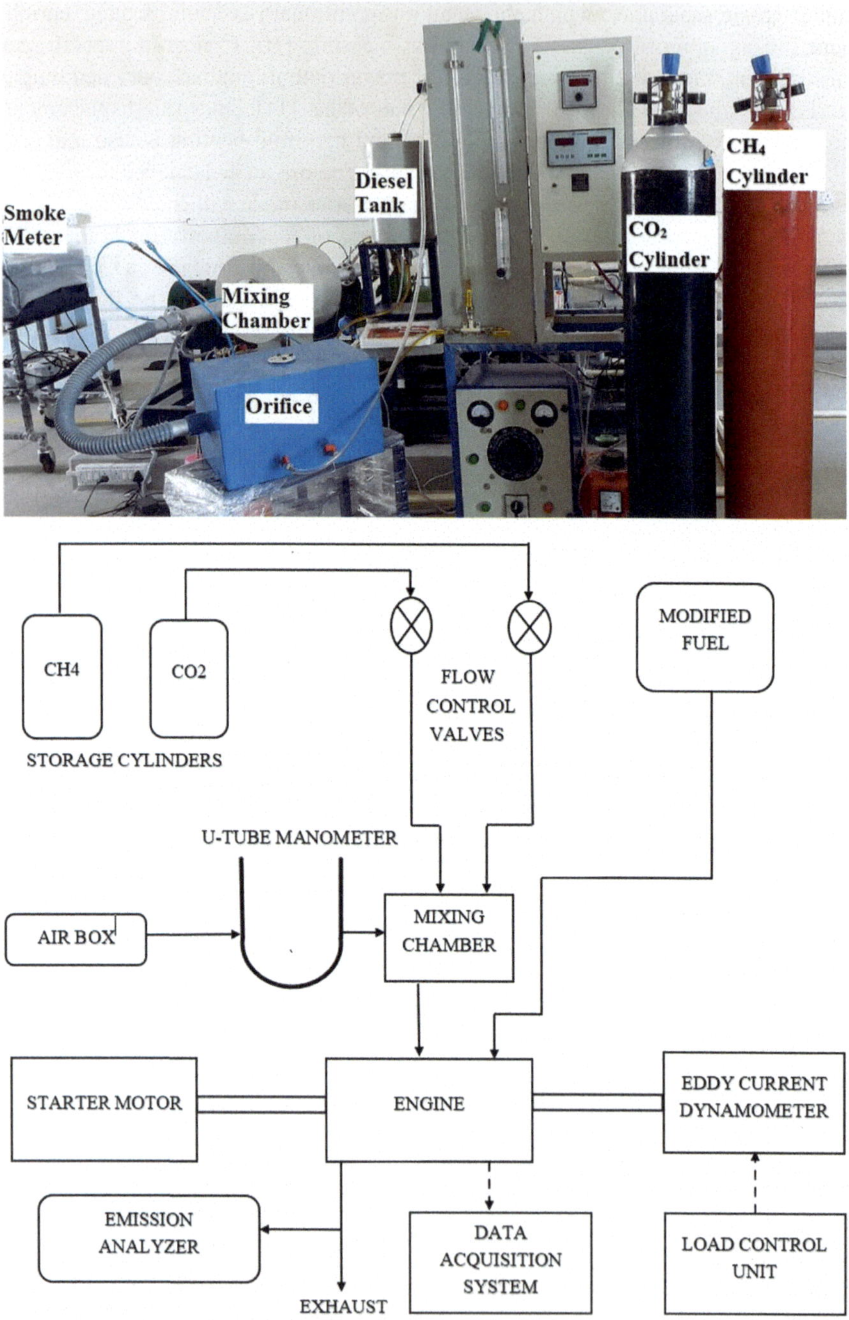

Fig. 1 **a** portrait and **b** schematic layout of the experimental setup

Table 1 Specifications of engine

Parameters	Values
Cubic capacity	481 cm^3
Horsepower	5 hp
Working principle	4 stroke, CI
Bore and Stroke	87.5 mm and 80 mm
Compression ratio	17:1
Nozzle opening pressure	200 bar
Peak pressure	75 bar
Fuel injection pressure	200 MPa

These sensors are connected to the exhaust system's tailpipe, where they trap exhaust gas and provide signals to the analyzer.

Table 1 contains the engine specifications, whereas Table 2 details the input parameters. Biogas is now the principal fuel, with biodiesel (palm oil) and diesel having a DEE blend of fraction 5% by volume serving as supplementary fuels. At constant biogas flow rates of 12 lpm, the influence of fuel mixes is investigated at four different engine loads (5, 10, 15, and 20 N m). The experimental findings map the impact of varying fuel mixes on the engine's performance and emissions in dual-fuel mode.

The fuel's terminology and real nature are explained in the following section, referenced throughout the paper. B20 biodiesel is made by combining biodiesel (20% palm oil by volume) with regular fuel (80% by volume). Biodiesel blend ($F1$) and diesel blend ($F2$) are formed by adding 5% DEE blend with 950 ml of B20 biodiesel and diesel, respectively. In addition, diesel-only fuel (1000 ml of diesel) is considered for comparative studies concerning the fuel blends. These two fuel blends ($F1$, $F2$)—are made by combining DEE with B20 biodiesel and diesel, respectively, in the proportions shown below, with density and calorific values listed in Table 3.

Table 2 Input parameters

Input parameters	Values
DEE blend ratio (% vol.)	5
Methane fraction (%)	60
Load (N m)	5, 10, 15, and 20
Intake temperature (°C)	30
Biogas flow rate (lpm)	12

Table 3 Fuel properties

Properties	B20	DEE	$F1$	$F2$
Density (kg/m^3)	845.4	713.4	827.36	825
Calorific value (MJ/kg)	39.3	33.9	38.1	40.9

Biodiesel blend ($F1$): 50 ml of pure DEE + 950 ml of B20 biodiesel.
Diesel blend ($F2$): 50 ml of pure DEE + 950 ml of diesel.

3 Analysis of Engine Performance and Emission Characteristics

We know that a given fuel's performance and emission characteristics can be roughly approximated by knowing the ratio of engine parameters or the chemical formula. Predicting these characteristics for fuel blends like biogas-biodiesel and biogas-diesel is challenging, but it may be accomplished using simple correlations. After obtaining a clear picture of the difficulty of the engine parameters measuring system, we have developed correlations that use load, DEE, BD, D, BG, R^2, and RMSE in calculating brake thermal efficiency, brake-specific energy consumption, volumetric efficiency, hydrocarbon emissions, carbon monoxide emissions, and nitrogen dioxide emissions.

Brake thermal efficiency (η_{BTE}) is expressed with R^2 (0.9) and RMSE (2.26) values as

$$\begin{aligned}\eta_{\mathrm{BTE}} =& 1.186 * \mathrm{load} - 1.187 * 10^{14} * \mathrm{DEE} - 0.095 * \mathrm{BD} \\ &+ 0.134 * D + 7.82 * 10^{13}\mathrm{BG}\end{aligned} \tag{1}$$

Brake-specific energy consumption (BSEC) is expressed with R^2 (0.76) and RMSE (0.82) values as

$$\begin{aligned}\mathrm{BSEC} =& - 0.24 * \mathrm{load} + 8.135 * 10^{13} * \mathrm{DEE} + 0.117 * \mathrm{BD} \\ &+ 0.067 * D - 3.39 * 10^{13}\mathrm{BG}\end{aligned} \tag{2}$$

Volumetric efficiency (η_{vol}) is expressed with R^2 (0.9) and RMSE (1.145) values as

$$\begin{aligned}\eta_{\mathrm{vol}} =& 0.054 * \mathrm{load} + 1.015 * 10^{14} * \mathrm{DEE} + 0.712 * \mathrm{BD} \\ &+ 0.912 * D - 4.23 * 10^{13}\mathrm{BG}\end{aligned} \tag{3}$$

Hydrocarbon emissions (HC) are expressed with R^2 (0.9) and RMSE (64.3) values as

$$\begin{aligned}\mathrm{HC} =& - 9.9 * \mathrm{load} + 1.014 * 10^{16} * \mathrm{DEE} + 1.94 * \mathrm{BD} \\ &+ 1.8 * D - 4.22 * 10^{15}\mathrm{BG}\end{aligned} \tag{4}$$

Carbon monoxide emissions (CO) are expressed with R^2 (0.6) and RMSE (0.028) values as

$$\begin{aligned} \text{CO} =& 0.0002 * \text{load} + 2.11 * \text{DEE} - 0.0025 * \text{BD} \\ &+ 0.0011 * D - 8.8 * 10^{11}\text{BG} \end{aligned} \tag{5}$$

Nitrogen dioxide emissions (NO_x) are expressed with R^2 (0.84) and RMSE (144.1) values as

$$\begin{aligned} \text{NO}_\text{x} =& 34.67 * \text{load} + 1.85 * 10^{16} * \text{DEE} - 9.7 * \text{BD} \\ &+ 0.487 * D - 7.69 * 10^{15}\text{BG} \end{aligned} \tag{6}$$

4 Results and Discussion

4.1 Brake Thermal Efficiency (BTE)

The influence of fuel blends on brake thermal efficiency (BTE) at different engine loads with constant biogas flow rate and DEE at 12 lpm and 5%, respectively, can be found in Fig. 2. Despite higher heat value, diesel-only contributes to the elevated BTE at small loads of fuel blends used in this investigation. From the graph (Fig. 2), it is noticed that the BTE escalated due to an increase in load. This inclination is predominantly a result of a rising cylinder temperature and superior combustion process at large loads; it releases a tremendous amount of heat due to the effect of exothermic reactions. With the constant biogas flow rate of 12 lpm, BTE reaches a highest of 35.29% at full load, which is excess of diesel by 2.54%. This leaning in BTE likely contributes to the complete fuel combustion. The ascent of volumetric efficiency increases the biogas introduction via the inlet manifold as outlined in the prior work [1].

BTE drops to 11.68% at the engine load of 5 N m for 12 lpm biogas flow rate and 5% DEE blend for biodiesel dual-fuel mode. It extends to a maximum of 31.84% at 20 N m engine load, 12lpm biogas flow rate, and 5% DEE blend for diesel dual-fuel mode. Major concern for the bond deterioration between reinforcing bars and concrete from past decades is due to the main reinforcement corrosion.

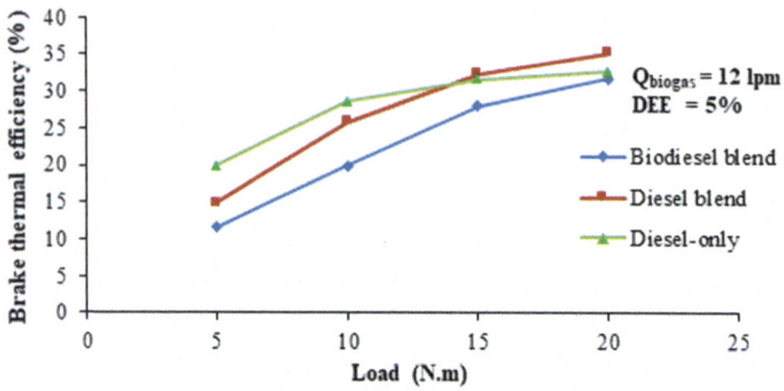

Fig. 2 Influence of load and fuel blend on brake thermal efficiency

4.2 *Brake-Specific Energy Consumption (BSEC)*

Figure 3 represents the variations of brake-specific energy consumption (BSEC) at different loads with constant biogas flow rate and DEE at 12 lpm and 5%, respectively. As shown in Fig. 3, because of the escalation of engine load, BSEC gradually decreases during the experiment. At small loads, compared to the dual-fuel mode, BSEC is less for diesel-only mode because of the small change of gaseous fuel to work. It is also observed that at full load, compared to diesel-only mode, lower BSEC is obtained in diesel dual-fuel mode due to supplementing biogas with methane for enhanced combustion [10].

BSEC attained a maximum of 8.55 at the engine load of 5 N m for 12 lpm biogas flow rate and 5% DEE blend for biodiesel blend. It reaches the lowest value of 2.83 at 20 N m engine load at the same operating conditions for diesel blend.

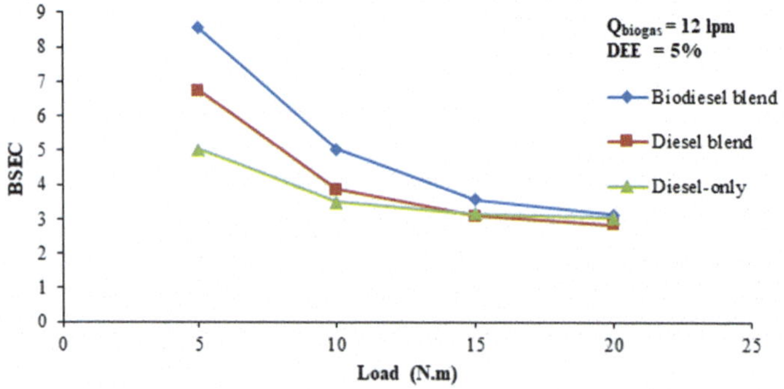

Fig. 3 Influence of load and fuel blend on brake-specific energy consumption

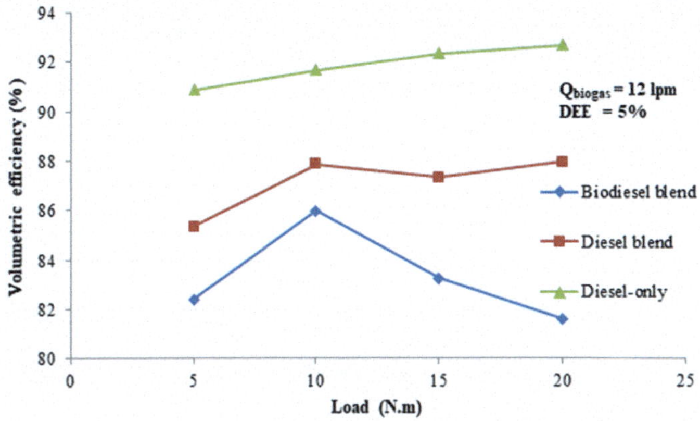

Fig. 4 Influence of load and fuel blend on volumetric efficiency

4.3 Volumetric Efficiency (VE)

The dissimilarity of volumetric efficiency (VE) at various loads with constant biogas flow rate and DEE at 12 lpm and 5%, respectively, is portrayed in Fig. 4. With an escalation in engine load, VE descends regardless of the fuel blends. The possible reason is that the cylinder wall is hot, which lessens the instigated air density, and the outcome is the fall in volumetric efficiency [9]. At 12 lpm, the biodiesel blend drastically decreased at 20 N m load, and this depletion in VE is perhaps due to the induction of biogas [11]. The VE of a diesel-only engine is increased from 5 to 20 N m when compared to a dual-fuel engine.

VE of engine dropped to 81.63% at the engine load of 20 N m for 12 lpm biogas flow rate, and 5% DEE blend for biodiesel blend. It reaches a maximum of 87.96% at 20 N m engine load at the same operating conditions for diesel blend.

4.4 Hydrocarbon Emissions (HC)

Figure 5 exhibits the outline of hydrocarbon (HC) emissions with the applied loads at different fuel blends with constant biogas flow rate and DEE blend at 12 lpm and 5%, respectively. Diesel's thermos-physical properties insist its induction happen at high pressure; subsequently, HC emission from diesel combustion is considered lower across different loads in contrast to other fuels used in this study. From Fig. 5, the increase in load resulted in the decrease of HC emissions due to oxygen molecules, which essentially provide carbon dioxide and water that assisted in combustion. At 12 lpm, HC emission decreases drastically to 220 ppm for 20 N m engine load, where the depletion with diesel is 70 ppm. The reduced flame velocity at the induction of biogas flow rate may have resulted in the integration of inhibitor gases by heat

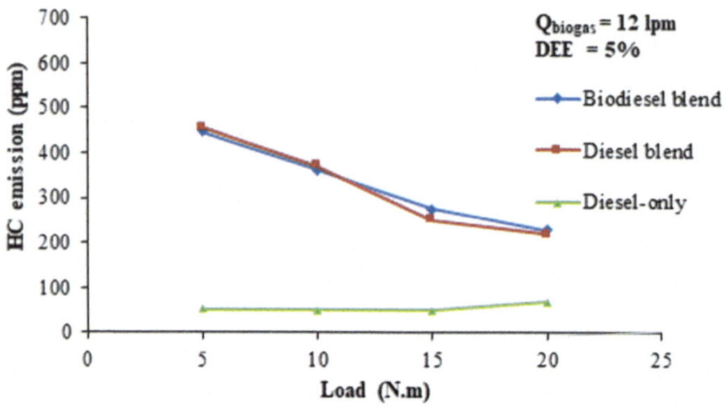

Fig. 5 Influence of load and fuel blend on hydrocarbon emissions

production, resulting in a small flame temperature. In addition, induction of biogas through the inlet manifold reduces the air-induced capacity, establishing a rich fuel blend area that sequentially escalates the incomplete combustion [2, 3]. A maximum of 457 ppm of HC emission is witnessed at 5 N m loads with 12 lpm of biogas flow rate and 5% DEE blend.

4.5 *Carbon Monoxide Emissions (CO)*

Figure 6 conveys the carbon monoxide (CO) emissions contrast variation with the applied loads for the fuel blends at constant biogas flow rates and DEE blend, 12 lpm, and 5%, respectively. In diesel combustion, lower CO emission is due to the result of the lean mixture, and with an escalation in load, a decrease in CO emission reasonably occurs. The existence of low-engine temperature and lean mixture at part loads, partial combustion of the total air–fuel mixture, and barely any of it goes into the exhaust. This may also raise the combustion rate due to the large cylinder gas temperature [8]. The lowest and highest value of CO emissions was found to be 0.13 and 0.23% at 20 N m for fuel blends, constant biogas flow rate with constant DEE blend ratio.

4.6 *Nitrogen Dioxide Emissions (NO_x)*

Figure 7 illustrates the nitrogen dioxide (NO_x) emissions trend with the applied load for fuel blends at constant biogas flow rate and DEE blend, 12 lpm and 5%, respectively. At all engine loads, NO_x emissions are rated to be higher with diesel operation. It can be examined from Fig. 7, the NOx emissions showed an inclination

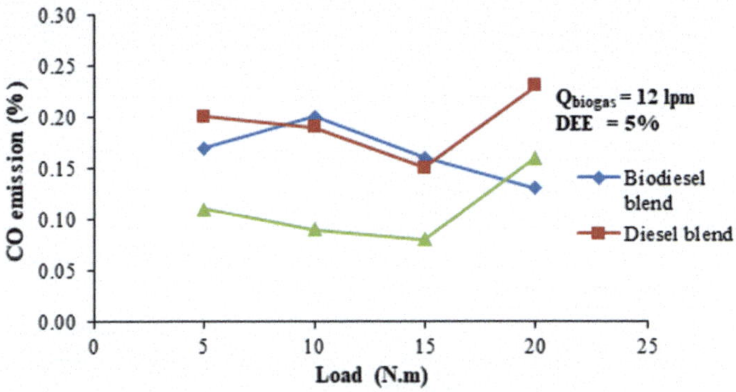

Fig. 6 Influence of load and fuel blend on carbon monoxide emissions

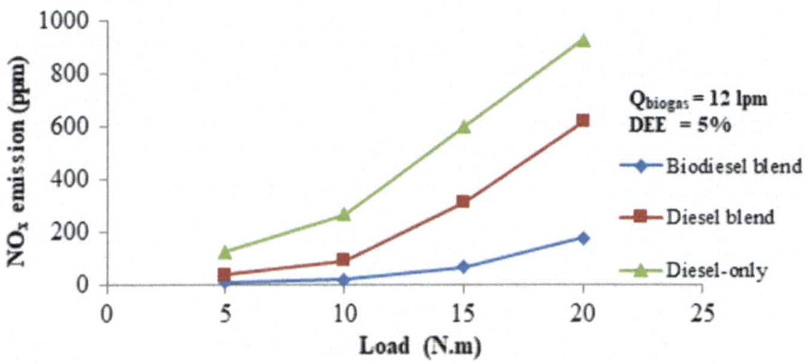

Fig. 7 Influence of load and fuel blend on nitrogen dioxide emissions

with an escalation of the engine load. At full load, blended biodiesel fuel has a high of 618 ppm, whereas 929 ppm is found with diesel operation.

At elevated load, ignition of additional gaseous fuel, the liberation of extra energy, the inclination in combustion temperature, and the inclusion of other oxygen molecules in combustion resulted in the advancement of more NOx. Due to an increase in NOx emissions, progressive injection occurs, increases the temperature, and cylinder pressure escalates because additional fuel burns near the top dead center [4, 5]. The lowest noted NO_x emission in blended fuels is 10 ppm at 5 N m engine load with 12 lpm of biogas flow rate at 5% DEE blend. The most negligible NO_x emission in diesel operation is determined to be 127 ppm at the same operative state.

5 Conclusions

The present investigation has been carried out to examine the performance and emission characteristics of a CI engine that experiences a dual-fuel mode of combustion.

From experimental studies, the subsequent conclusions have come up and are as follows:

- Brake thermal efficiency (BTE) enhanced to 35% at high loads because of the injection of DEE blend at 5% with 12 lpm of biogas in diesel dual-fuel mixture, hence bringing down the overall pollution.
- Varying the load in increasing order led to a substantial decrease in brake-specific energy consumption in the engine, having the most negligible value of 2.83 at 12 lpm and 5% DEE blend in diesel dual-fuel mode.
- At high loads, the engine's volumetric efficiency, which when used in diesel dual-fuel mode, obtained an increase of 6.33% compared to biodiesel blend at 12 lpm biogas flow rate and DEE blend of 5%.
- A sharp decline was seen in HC and CO emissions, 220 ppm and 0.13%, diesel, and biodiesel dual-fuel mode, respectively, at 20 N m, 12 lpm due to improvement of combustion by DEE blend at 5%.
- The addition of load displayed a significant increase in the NOx concentrations of 618 ppm at 20 N m, 12 lpm having 5% DEE blend in diesel possibly due to the rise in combustion temperature.
- Formulated on the results obtained through our experiments, it is confirmed that the engine performance is enhanced in the diesel dual-fuel mode. In contrast, the emission characteristics are improved in biodiesel blend/dual-fuel mode because of the introduction of DEE addition. The ideal experimental results for superior performance and emission characteristics obtained are at 5% DEE having a load of 20 N m with a biogas flow rate of 12 lpm in both the dual-fuel modes. Paper gave an overall view about the reasons for degradation of bond, and the pull-out test through which the strength of the bond is determined and also covered the basic concept of bond-slip relationship with an example of lightweight aggregate concrete.

References

1. Barik D, Murugan S (2014) Simultaneous reduction of NOx and smoke in a dual fuel DI diesel engine. Energy Convers Manage 84(2):217–226
2. Papagiannakis RG, Hountalas DT, Rakopoulos CD (2007) Theoretical study of the effects of pilot fuel quantity and its injection timing on the performance and emissions of a dual fuel diesel engine. Energy Convers Manage 48(11):2951–2961
3. Kyunghyun R (2013) Effects of pilot injection pressure on the combustion and emissions characteristics in a diesel engine using biodiesel–CNG dual fuel. Energy Convers Manage 76:506–516
4. Sahoo BB, Sahoo N, Saha UK (2009) Effect of engine parameters and type of gaseous fuel on the performance of dual-fuel gas diesel engines – a critical review. Renew Sustain Energy Rev 13(6–7):1151–1184
5. Krishnan SR, Srinivasan KK, Singh S, Bell SR, Midkiff KC, Gong W, Fiveland SB, Willi M (2004) Strategies for reduced NO_x emissions in pilot-ignited natural gas engines. ASME J Eng Gas Turb Power Trans 126(3):665–71

6. Sezer İ (2018) A review study on the using of diethyl ether in diesel engines: effects on NOx emissions. Int J Automot Eng Technol 7(4):164–183
7. Sezer İ (2020) A review study on using diethyl ether in diesel engines: effects on fuel properties, injection, and combustion characteristics. Energy Environ 31(2):179–214
8. Prakash R, Singh RK, Murugan S (2013) Experimental investigation on a diesel engine fueled with bio-oil derived from waste wood–biodiesel emulsions. Energy 55:610–618
9. Dulari H, Murugan S (2014) Bioethanol fumigation in a DI diesel engine. Fuel 130:324–333
10. Feroskhan M, Ismail S (2016) Investigation of the effects of biogas composition on the performance of a biogas–diesel dual fuel CI engine. Biofuels 7(6):593–601
11. Chandra R, Vijay VK, Subbarao PMV, Khura TK (2011) Performance evaluation of a constant speed IC engine on CNG, methane enriched biogas and biogas. Appl Energy 88(11):3969–3977
12. Feroskhan M, Ismail S (2020) Evaluating the effect of intake parameters on the performance of a biogas–diesel dual-fuel engine using the Taguchi method. Biofuels 11(4):441–449
13. Da Costa RBR, Valle RM, Hernández JJ, Malaquias ACT, Coronado CJR, Pujatti FJP (2020) Experimental investigation on the potential of biogas/ethanol dual-fuel spark-ignition engine for power generation: Combustion, performance and pollutant emission analysis. Appl Energy 261:114438
14. Barik D, Murugan S (2016) Effects of diethyl ether (DEE) injection on combustion performance and emission characteristics of Karanja methyl ester (KME)–biogas fueled dual fuel diesel engine. Fuel 164:286–296
15. Navdeep SD, Gyanendra SG, Ajay S (2021) Experimental investigations on the performance and emissions characteristics of dual biodiesel blends on a varying compression ratio diesel engine. SN Appl Sci 3(622)
16. Donepudi J, Puli RK, Madhu MK (2011) Performance and emission characteristics of diesel engine run on biofuels based on experimental and semi analytical methods. Int J Energy Environ (IJEE) 2(5):899–908
17. Tuan NN, Nguyen XK, Le AT (2021) The correlation of biodiesel blends with the common rail diesel engine's performance and emission characteristics. Energies 14:2986
18. Nasrin SS, Abbas R, Abdulali F, Mohamad HA (2016) Modeling of specific fuel consumption and emission parameters of compression ignition engine using nanofluid combustion experimental data. Fuel Process Technol 154:37–43
19. Shashank M, Anand K (2017) A composition based approach for predicting performance and emission characteristics of biodiesel fuelled engine. SAE Technical Paper 01–2340
20. Miller Jothi NK, Nagarajan G, Renganarayanan S (2008) LPG fueled diesel engine using diethyl ether with exhaust gas recirculation. Int J Therm Sci 47(4):450–457

A Comprehensive Thermodynamic Evaluation of a Geothermal Power Plant Coupled with Organic Rankine Cycles at Full and Part Loads

M. Sreekanth and M. Feroskhan

Abstract A geothermal power plant coupled with a two-stage organic Rankine cycle has been studied using energy and exergy analysis. For the first time, the refrigeration load needed to provide the coolant for the condensers of the Rankine cycles has been considered for analysis. Also, for the first time, a part load performance study has been carried out. It has been found that the refrigeration load of the condensers' coolant has a significant influence on the net power produced as the coolant is required at very low temperatures. The coolant refrigeration load caused a drop of 6.2% points of energy efficiency and 11.9% points of exergy efficiency. At part load (80%), the energy and exergy efficiencies have dropped by 3% points and 2% points, respectively. From the present study, it could be concluded that it is prudent to consider the coolant load for a realistic performance analysis based both on energy and exergy. From the study, it turns out that the organic Rankine cycles are parasitical in nature and actually consume the power produced by the main geothermal plant turbine.

Nomenclature

Symbol	Name
$\dot{E}$	Rate of energy transfer, kW
h	Specific enthalpy, kJ/kg
h_0	Dead state specific enthalpy, kJ/kg
$\dot{m}$	Mass flow rate, kg/s

M. Sreekanth (✉) · M. Feroskhan
School of Mechanical Engineering, Vellore Institute of Technology Chennai, Vandalur-Kelambakkam Road, Chennai 600127, India
e-mail: manavalla.sreekanth@vit.ac.in

M. Feroskhan
e-mail: feroskhan.m@vit.ac.in

M. Sreekanth
Electric Vehicle Incubation, Testing and Research Centre, Vellore Institute of Technology Chennai, Vandalur-Kelambakkam Road, Chennai 600127, India

E. Natarajan et al. (eds.), *Materials, Design and Manufacturing for Sustainable Environment*, Lecture Notes in Mechanical Engineering,
https://doi.org/10.1007/978-981-19-3053-9_49

$\dot{Q}$	Rate of heat transfer, kW
S	Specific entropy, kJ/kg-K
S_0	Dead state specific entropy, kJ/kg-K
S_{gen}	Entropy generated, kJ/K
T	Absolute temperature, K
T_0	Dead state temperature, K
$\dot{W}$	Rate of work, kW
$\dot{W}_{net}$	Net work, kW
$\dot{X}$	Rate of exergy transfer, kW
$X_{\text{destroyed}}$	Exergy destroyed, kW
η_{en}	Energy efficiency
η_{ex}	Exergy efficiency

1 Introduction

The world has realized that using fossil fuels to meet the energy demands is not environmentally sustainable, and the only way out is to wean away from them. At the same time, alternative sources of energy need to be explored and developed to fill in the giant shoes of fossil fuels [1]. Renewable sources of energy offer another way to meet the energy demands. The advantage of renewable sources of energy is that they are either free (solar, wing, ocean energies) or are available at low cost (biomass, agricultural residue) [2]. However, each source of energy has its pros and cons, and hence, the choice of the source of energy needs to be carefully made. As of now, solar and wind energies are becoming competitive with fossil fuel energy on the basis of price [3]. However, solar and wind energy are not continuously available, and hence, provision to store large amount of energy must be made [4].

One source of renewable is geothermal energy which is continuously available for a long duration. However, the distribution of geothermal energy is not uniform all over the world, and a few locations are fortunate to have this source. This form of energy is attractive in locations that are remotely located and are difficult to connect to the grid. Power plants based on geothermal energy can be installed as captive power plants and meet the local energy needs [5]. Since the source of hot water stream usually has low-flow rate, geothermal thermal power plants have to be small in size, usually a few mega Watts [6]. Moreover, since the source of energy is of low grade (hot water at moderate temperature and low pressure), its thermodynamic conversion to useful energy involves poor energy and exergy efficiency [7]. This calls for implementing energy conservation measures in the power generation process to improve the overall efficiency including both power and heat.

One benefit of geothermal energy is that the method of utilization using a steam power plant is well established at all magnitudes. Hence, setting up of a power plant using geothermal energy can be carried out modularly available components which can decrease the overall capital cost [8]. Several innovative ways of enhancing the power from geothermal energy sources have been suggested. A recent study suggested a hybrid power plant combining geothermal as well as solar energies [9]. Another study by Liu et al. [10] proposed a hybrid power plant involving geothermal energy as well as fossil fuel power plant.

Opting for hybrid power plants is one way of complimenting the available geothermal energy and it comes with the associated increase in cost. One option that does not involve hybrid energy sources but utilizes the available geothermal to the maximum possible extent is the involvement of organic Rankine cycle [11]. In an organic Rankine cycle (ORC), the commonly used working fluid, i.e. steam is replaced by another fluid which evaporates at low temperatures and pressures. The generated vapour is expanded in a turbine, and power generation can take place in more than one stage [12]. Several studies involving energy, exergy, environmental, and economic analysis of geothermal plants (single as well as hybrid) have been carried out of late [13]. Optimization studies also have been carried out along with 3-E analysis [14]. Assad et al. [15] studied a single flash geothermal plant by optimizing the separator pressure. They have not considered inclusion of ORC. Fan et al. [16] have studied a single flash geothermal plant and combined it with a two-stage ORC using different working fluids, viz. R227ea and R116. Two-stage ORC combined with geothermal plants are commonly suggested instead of three stage ones as the heat energy available decreases rapidly towards lower stages. It becomes expensive to extract useful energy from the lower stages.

From the above literature study, it is understood that many studies considered hybrid as well as ORC to analyze the energy, exergy, environment and economic performance of geothermal power plants. However, no study considered the energy needed to provide the coolant necessary to condense the working fluid in the ORC. Since the condensation temperatures can be very low (often sub-zero), it could require considerable amount of energy to provide a coolant to enable condensation. The energy quantity can be high enough to offset the energy, exergy, and economics projected by ignoring it. Also, there was no study carried out considering the part load performance of a single flash geothermal power plant with two-stage ORC. Part load performance is important to evaluate the functioning when the load is lower than the design load or the hot water supply rate drops. Hence, the present study sets out to carry out a comprehensive study by.

i. Including the energy required to provide the coolant for condensation and
ii. Part load performance of the geothermal power plant with two-stage ORC

To conduct the study, a recently published work by Fan et al. [16] will be modified by including vapour compression refrigeration systems for each of the ORCs. The full load calculations will be carried out to produce 10 MW turbine power and part load at 80% of the full load, i.e. 8 MW of turbine power. Also, performance evaluation will be carried out using energy and exergy analysis.

2 Methodology

In the present work, a flow sheeting software named Cycle-Tempo has been used to simulate the system. All the necessary components are incorporated into the system and are connected using appropriate fluid pipes. In the current system, components like steam drum, turbines, condensers, heat exchangers, compressors, and pumps are required. The working fluids include water, steam, air, and refrigerants. Necessary inputs like pressure, temperature, mass flow rate, isentropic efficiency, pressure drop, etc., are needed, and the software internally builds the necessary mass balance, energy balance, and exergy balance equations and solves them for the unknown quantities. The software incorporates sub-routines which compute the fluid properties at the required conditions. The equations are solved until convergence is achieved.

3 System Description

The system, shown in Fig. 1, broadly consists of a source of pressurized hot water which is utilized in the geothermal plant after converting the water to steam using a flash evaporator. The steam is expanded in the steam turbine, and the condensate is directly mixed with the liquid fraction separated in the flash evaporator. The energy contained in this mixed stream is utilized in a topping organic Rankine cycle to evaporate a refrigerant R227ea, which in turn is expanded in a turbine which can handle evaporated refrigerant. Then, the refrigerant is condensed in a condenser and returned to the evaporator. Before condensing, the heat in the expanded vapour is transferred to the evaporator of the bottoming organic Rankine cycle. Similar events occur in the bottoming cycle using another refrigerant R410A at different pressures and temperatures. In the refrigeration systems, the evaporator exit is taken as saturated vapour, whilst the condenser exit is taken as saturated liquid, for simplicity. All the heat exchangers are counter flow type. In both the topping and bottoming cycles, the condensate needs a coolant fluid which is obtained by incorporating a vapour compression refrigeration system in each of the condensers. The refrigeration systems in turn have their own condensers which use ambient air as the coolant. Since the bottoming cycle operates at a very low-refrigerant temperature, it needs a multistage refrigeration system to efficiently operate the compressors. After complete utilization, the geothermal water is pumped back into the reinjection well at 104 °C.

4 Model Formulation

The following assumptions are made in the model formulation:

i. The system operates at steady-state conditions.

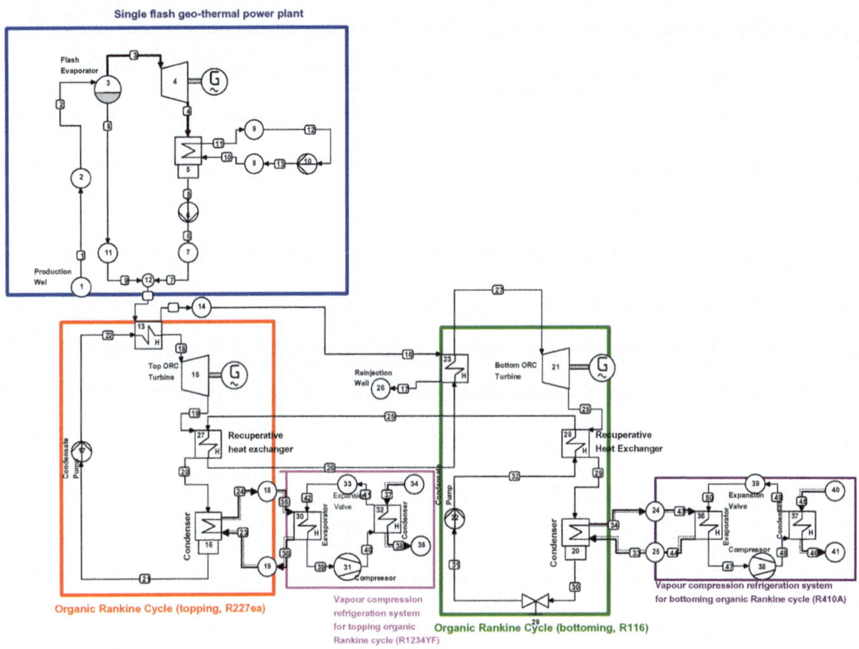

Fig. 1 Single flash geothermal power plant with a two-stage ORC and refrigeration system for condenser coolant

ii. The turbines have an isentropic efficiency of 78%, pumps 80%, and compressors 85%.
iii. At the evaporator and condenser exit in the two vapour compression system, the refrigerant is dry saturated and saturated liquid, respectively.
iv. Changes in the kinetic and potential energy are neglected throughout.
v. Dead state is taken as 1 bar and 25 °C.

The following equations are framed and solved for all the components in the steady flow system:

Mass Balance:

$$\sum_{in} \dot{m} = \sum_{out} \dot{m} \tag{1}$$

Energy Balance:

$$\sum_{in} \dot{E} = \sum_{out} \dot{E} \tag{2}$$

For multiple streams:

$$\dot{Q} - \dot{W} = \sum_{\text{out}} \dot{m}h - \sum_{\text{in}} \dot{m}h \tag{3}$$

For single stream:

$$\dot{Q} - \dot{W} = \dot{m}[h_2 - h_1] \tag{4}$$

Exergy Balance:

$$\underbrace{\dot{X}_{\text{in}} - \dot{X}_{\text{out}}}_{\substack{\text{Rate of net exergy}\\ \text{transfer by heat,}\\ \text{work, and mass}}} - \dot{X}_{\text{destroyed}} = 0 \tag{5}$$

$$\dot{X}_{\text{heat}} = \left(1 - \frac{T_0}{T}\right)\dot{Q} \tag{6}$$

$$\dot{X}_{\text{work}} = \dot{W}_{\text{useful}} \tag{7}$$

$$\dot{X}_{\text{mass}} = \dot{m}\psi \tag{8}$$

$$\psi = (h - h_0) - T_0(s - s_0) \tag{9}$$

$$\dot{X}_{\text{destroyed}} = T_0\dot{S}_{\text{gen}} \tag{10}$$

$$\dot{W}_{\text{net}} = \sum \dot{W}_{\text{turbine}} - \sum \dot{W}_{\text{pump}} - \dot{W}_{\text{compressor}} \tag{11}$$

$$\eta_{\text{energy}} = \frac{\text{Net work}}{\text{Rate of heat carried by geothermal water}} \tag{12}$$

$$\eta_{\text{exergy}} = \frac{\text{Net work}}{\text{Exergy drop between production well and reinjection well}} \tag{13}$$

5 Model Validation

From the arrangement shown in Fig. 1, the two vapour compression refrigeration systems have been removed which makes the arrangement same as that studied by Fan et al. [16], as shown in Fig. 2. The inputs given by them have been used, and their results have been reproduced by the existing model. Table 1 summarizes and juxtaposes the results obtained by the present model and by Fan et al. [16]. Pressure, temperature, enthalpy, mass flow rate, and exergy in each pipe have been compared. It can be seen that excepting for the enthalpies computed for the two refrigerants (R227ea and R116) shown with bold face in italics, all other parameters and values are very closely produced by the existing model. As for the enthalpies of the refrigerants, the huge difference is due to the method of computing the enthalpy by Cycle-Tempo (used for the present work) and Engineering Equation Solver (EES, used by fan et al. [16]). Cycle-Tempo uses 25 °C as the reference temperature for computing the enthalpy whilst EES uses 0 °C. The steady flow energy equation used in these studies uses enthalpy differences only, and hence, the reference temperature used is insignificant as far as the end result is concerned. For example, since exergy is computed using differences in enthalpy and entropy, there should be no difference in its values in each pipe. This can be observed from the exergy values shown in Table 1, where the difference is less than 1% maximum. With this result, it can be concluded that the present model is capable of accurately simulating the work of Fan et al. [16].

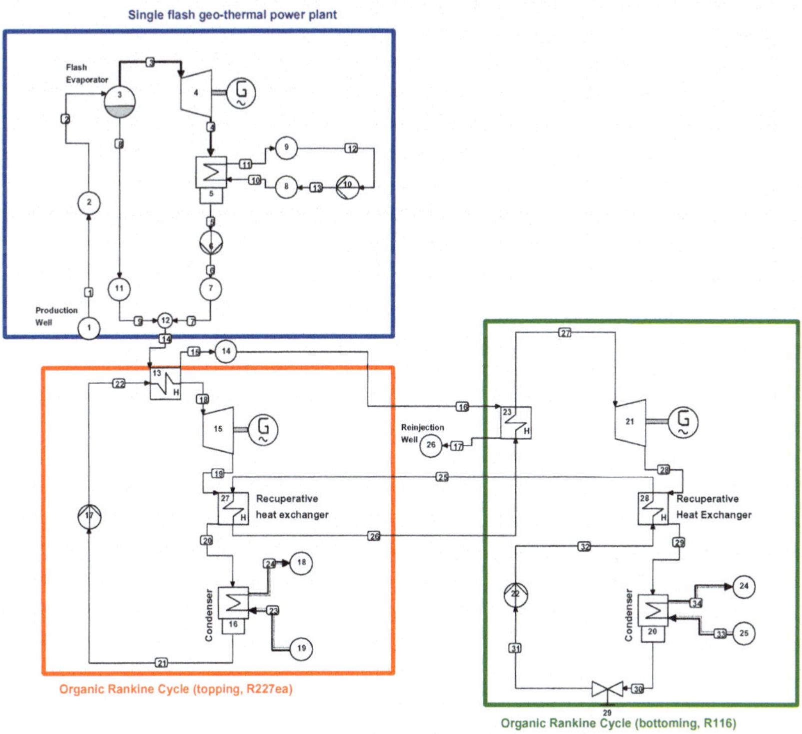

Fig. 2 Single flash geothermal power plant with two-stage ORC [16]

Table 1 Validation of the present work with Fan et al. [16]

Pipe No	Pressure, bar		Temperature, (°C)		Enthalpy, (kJ/kg)		Mass flow rate, (kg/s)		Exergy, (kW)		Working fluid
	Present work	Fan et al.	Present work	Fan et al.	Present work	Fan et al.	Present work	Fan et al.	Present work	Fan et al.	
1	85.84	85.84	300	300	1344.6	1344	50	50	18,942	18,931	Water/steam
2	8.918	8.918	174.97	175	1343.81	1344	50	50	16,199.58	16,205	
3	8.918	8.918	174.97	175	2772.68	2773	14.835	14.83	11,900.3	11,903	
4	0.1234	0.1234	49.98	50.1	2252.51	2253	14.835	14.83	2403.54	2405	
5	0.1234	0.1234	49.98	50.2	209.26	209.3	14.835	14.83	60.15	60.28	
6	15	–	50.2	–	211.14	–	14.835	–	82.92	–	
7	15	15	50.2	50.3	211.14	211.2	14.835	14.83	82.92	83.04	
8	8.918	8.918	174.97	175.2	741.02	741.2	35.165	35.17	4299.27	4302	
9	15	15	174.9	175.2	741.02	741.2	35.165	35.17	4315.21	4302	
10	3	–	25	–	105.11	–	1456.555	–	290.22	–	
11	2	–	30	–	125.92	–	1456.555	–	396.61	–	
12	2	–	30	–	125.92	–	1456.555	–	396.61	–	
13	3	–	30.01	–	126.05	–	1456.555	–	533.49	–	
14	3.8	3.489	138.74	138.9	583.8	584	50	50	3685.48	3687	
15	3.489	3.489	108	108.4	453.06	454.3	50	50	2064.51	2078	
16	3.489	3.489	108	108.4	453.06	454.3	50	50	2064.51	2078	
17	3.489	3.489	104	104	436.16	435.4	50	50	1884	1875	
18	24	24	133.9	133.9	***425.16***	***243.9***	27.062	27	1430.54	1420	R227ea
19	1	1	70.6	70	***384.87***	***203.8***	27.062	27	68.28	65.86	
20	1	1	−16.65	−16.4	***275.43***	***82.84***	27.062	27	230.17	283	

(continued)

Table 1 (continued)

Pipe No	Pressure, bar		Temperature, (°C)		Enthalpy, (kJ/kg)		Mass flow rate, (kg/s)		Exergy, (kW)		Working fluid
	Present work	Fan et al.	Present work	Fan et al.	Present work	Fan et al.	Present work	Fan et al.	Present work	Fan et al.	
21	1	1	−16.65	−16.4	***181.74***	***1.753***	27.062	27	641.86	636.5	
22	24	24	−15.52	−16.3	***183.59***	***3.638***	27.062	27	680.46	675.5	
23	1.5	–	−25	–	−154.69	–	436.545	–	17,898.9	–	Air
24	1.2	–	−20	–	−148.89	–	436.545	–	9229.41	–	
25	28	28	−7.78	−7.4	***188.86***	***−95.44***	22.695	25.2	1453	1451	R116
26	28	28	64.8	64.8	***319.35***	***34.12***	22.695	25.2	1419.6	1419	
27	28	28	103.2	103.2	***356.57***	***71.56***	22.695	25.2	1573.7	1574	
28	1	1	31.01	31.2	***307.82***	***22.83***	22.695	25.2	−4.72	−3.874	
29	1	1	−71.9	−72	***235.41***	***−50.15***	22.695	25.2	336.43	342.4	
30	1	1	−78.34	−78.3	***114.35***	***−170.5***	25.2	25.2	1952.17	1951	
31	1	1	−78.34	−78.3	***114.35***	***−170.5***	25.2	25.2	1952.17	1951	
32	28	28	−76.94	−76.9	***116.45***	***−168.4***	22.695	25.2	1988.92	1988	
33	1.5	–	−90	–	−222.05	–	593.192	–	39,823.16	–	Air
34	1.2	–	−85	–	−216.91	–	593.192	–	26,717	–	

6 Results and Discussion

The flowsheet is executed at full load and at 80% load. Major results at full load are shown in Fig. 3a–c. Each of these figures are taken from the Fig. 1 and shown with the numerical results.

In Fig. 3a, the geothermal plant is shown with its flash evaporator, turbine, condenser, and condensate pump along with the coolant circuit which uses water as the working fluid. Full load conditions imply that the three turbines together produce 10 MW of power. It can be seen that the turbine produces a power of 7.7 MW using a generator having 98% efficiency. In Fig. 3b, the topping cycle is shown. The turbine in this case produces little over 1 MW of electric power, and the refrigeration system consumes 0.86 MW. Figure 3c shown the bottoming cycle and the turbine here produces 1.2 MW whilst the refrigeration system consumes 1.6 MW of power. The power consumed by the pumps in the geothermal power plant, topping cycle, and bottoming cycle is 0.2, 0.03, 0.058, and 0.06 MW, which is negligible compared to the power produced by the turbines and that consumed by compressors.

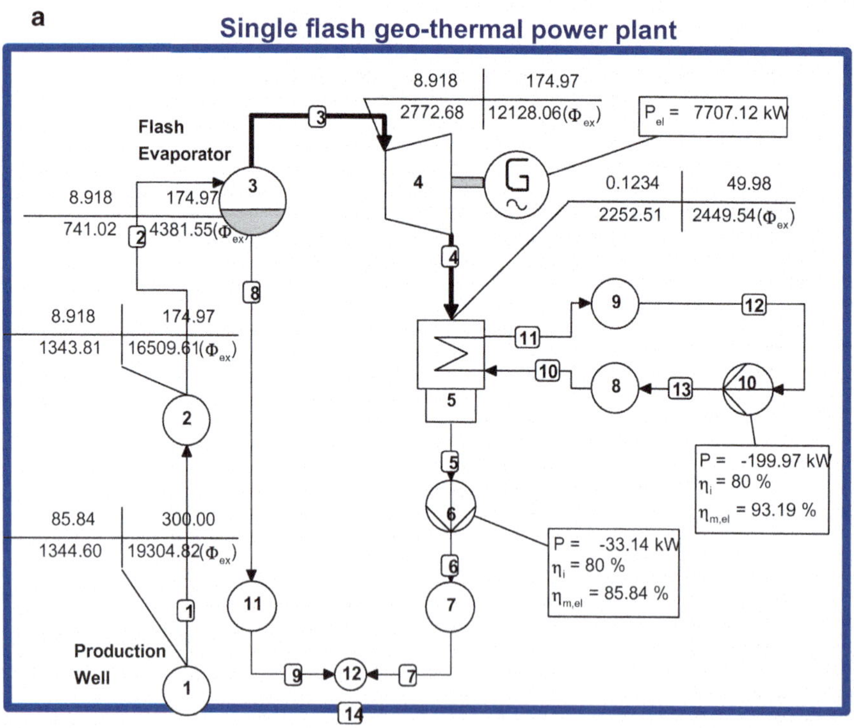

Fig. 3 **a** Single flash geothermal power plant at full load, **b** topping organic Rankine cycle using R227ea as the working fluid at full load, **c** bottoming organic Rankine cycle using R410A as the working fluid at full load

b

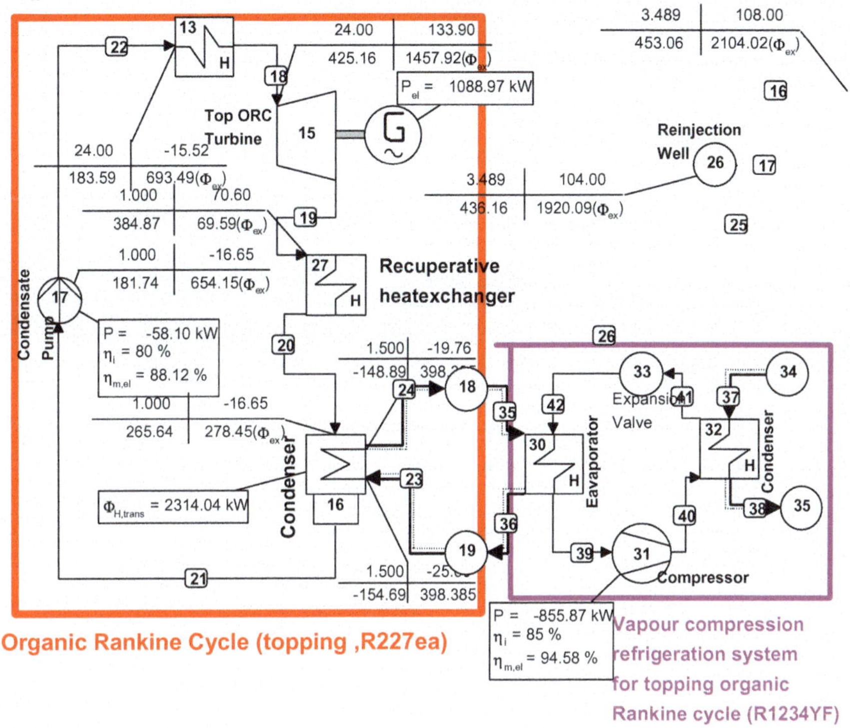

Fig. 3 (continued)

The refrigerants used in topping and bottoming cycles are same as those used by Fan et al. [16] whilst those used in the vapour compression systems are chosen based on the condenser coolant temperature requirement. Since the coolant in both ORCs is required at very low temperatures, refrigerant R1234YF is chosen for the topping cycle and R410A for the bottoming cycle.

Table 2 shows the power produced, power consumed, and energy and exergy efficiencies (as computed using Eqs. 12 and 13) for various components used in the system. It can be seen that the turbine power produced as estimated by the present work and by Fan et al. [16] is very close and is around 9.8 MW. Some information is not available as indicated in the table. Also, Fan et al. [16] did not consider the vapour compression refrigeration system, and hence, the net power shown in the present work and by them is different. The refrigeration system is very much power intensive as very low temperatures need to be produced, and hence, it can be seen that in both the topping and bottoming cycles, the compressors in the refrigeration systems consume power very much close to that generated by the respective turbines. In case of the bottoming cycle, the refrigeration system consumes 1.6 MW whilst the corresponding turbine produces only 1.2 MW. This reflects in the drop in the energy and exergy efficiencies in the present work. The energy and exergy efficiencies of

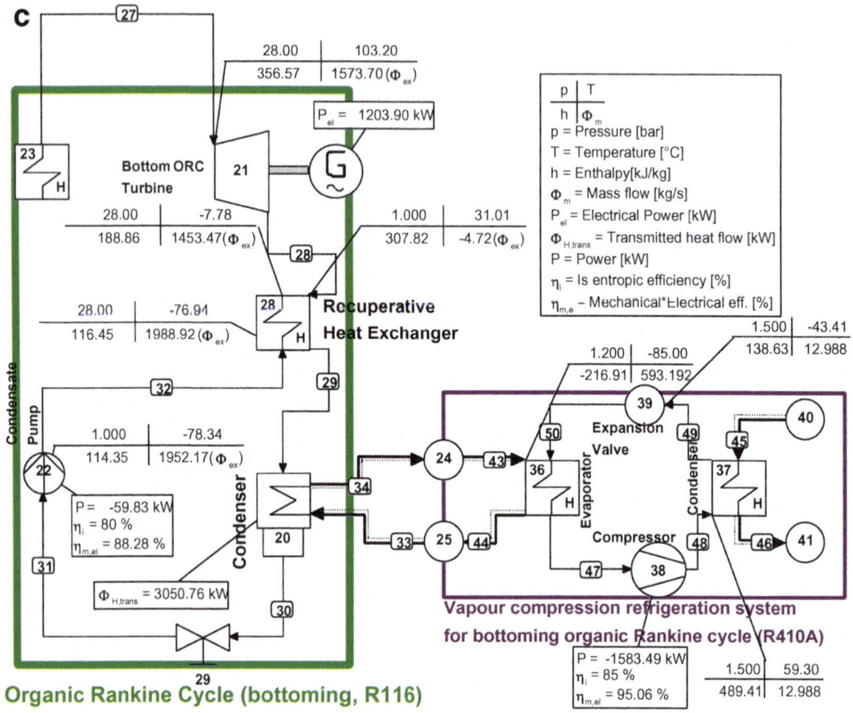

Fig. 3 (continued)

the present work are found to be 15.6 and 46.1% whilst that reported by Fan et al. [16] are 21.78 and 58%, respectively. This is because they did not consider the power consumed to provide a coolant fluid for the ORCs whilst the present work did.

Overall, it is observed that including the ORCs actually results in a drop in the overall power produced as the refrigeration systems consume some power produced by the main geothermal power plant turbine. Hence, the present choice of the refrigerant and operating conditions of the ORCs is inadequate and needs to be changed. Else, the exhaust steam of the main geothermal turbine can be used for process heat.

At part load, the turbines are set to produce a total power of 8 MW (80% full load), and the system is executed. It is observed that at part load, the energy and exergy efficiencies have dropped by 3% points and 2% points, respectively. This is understandable as the flowrates of all the fluids decrease as well as the operating conditions deviate from the design values. This decreases the power produced by the turbines, whilst the accessories consume almost the same amount of power. Moreover, the irreversibilities at part loads would be more than those at full load.

Any improvement in the energy utilization can be achieved by understanding where and how the exergy in the components is destroyed. Exergy destroyed is computed using Eq. 10 and is shown for full as well as part load. Figure 4 shown the exergy destroyed in major components at full and part loads. In general, exergy

Table 2 Summary and comparison of power produced/consumed by various components [16]

Component	Power produced/consumed, kW, present work	Power produced/consumed, kW, Fan et al.
Geothermal power plant turbine	+7562.39	–
Organic Rankine cycle (topping) turbine	+1068.52	–
Organic Rankine cycle (bottoming) turbine	+1203.9	–
Total turbine power	+9834.8	+9898
Geothermal plant pump	−32.54	–
Geothermal plant condenser cooling water pump	−196.23	–
Topping cycle feed pump	−57.07	–
Bottoming cycle pump	−59.83	–
Total pumping power	−345.67	
Refrigeration system compressor power (topping cycle)	−817.67	–
Refrigeration system compressor power (bottoming cycle)	−1583.39	–
Refrigeration system total power consumption	−2401.1	–
Net power produced	7088	9898
Energy of geothermal water source	45,422	45,438
Net exergy of geothermal water source	15,257	15,257
Energy efficiency	7088/45422 = 15.6%	21.78%
Exergy efficiency	7038/15257 = 46.1%	58.03%

destroyed is greater at full load than at part load which could be due to the greater mass flow rates. The geothermal plant condenser destroys the greatest amount of exergy closely followed by tits turbine. The node where the geothermal plant condensate mixes with the saturated liquid separated in the flash evaporator also destroys exergy to an amount greater that the ORC turbines and heat exchangers. In the geothermal plant condenser, the temperature difference between the hot and cold fluid is more leading to greater irreversibility. Similarly, in the node, where mixing of two fluids takes place, exergy destruction is higher than in other components. As for the geothermal power plant turbine, it operates at low pressure and temperature (8.9 bar, 175 °C) which according to the 2nd law of thermodynamics is not desirable. Also, the isentropic efficiency of the turbines is taken as 78% which could also be a reason for the high amount of exergy destruction.

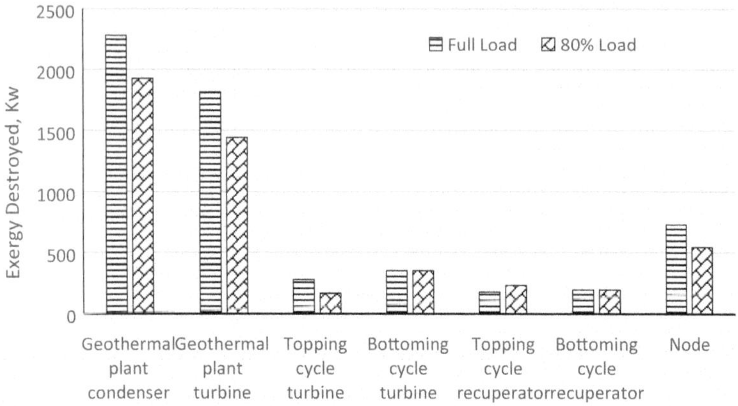

Fig. 4 Exergy destruction in various components at full load and 80% load

7 Conclusions

The following conclusions can be drawn from the present study:

i. At full load, the energy and exergy efficiencies are 15.6 and 41.5%, respectively, whilst at 80% load, they are 12.58 and 39.5%. Hence, it is always preferable to operate the plant at full load conditions.
ii. When the refrigeration system is included in the process, the net power is lower than when it is not considered. However, since it is required to include a refrigeration system, it would be sensible to not include the two-stage ORCs at all, especially with the refrigerants used in the present study.
iii. Since the coolant of the topping and bottoming cycle condensers is at very low temperature, the refrigeration load and hence power consumed by the compressor is very high, offsetting the power produced by the two ORC turbines.
iv. The exergy destructed during the part load operation is lower than during the full load condition. The exergy destruction decreased from 15–40% for various components.

References

1. Biernat K (eds) (2016) Alternative fuels: technical and environmental conditions. BoD–Books on Demand
2. Nelson VC, Starcher KL (2015) Introduction to renewable energy. CRC Press
3. Hirth L (2013) The market value of variable renewables: The effect of solar wind power variability on their relative price. Energy Econ 38:218–236
4. Sivaram V, Dabiri JO, Hart DM (2018) The need for continued innovation in solar, wind, and energy storage. Joule 2(9):1639–1642

5. Moya D, Aldás C, Kaparaju P (2018) Geothermal energy: Power plant technology and direct heat applications. Renew Sustain Energy Rev 94:889–901
6. Mburu M (2009) Geothermal energy utilization. Explor Geoth Resour 1:1–1
7. Çengel YA, Boles MA (2015) Thermodynamics: an engineering approach, 8th edn. The McGraw-Hill Companies, Inc., New York
8. Fridleifsson IB (2001) Geothermal energy for the benefit of the people. Renew Sustain Energy Rev 5(3):299–312
9. El Haj AM, Ahmadi MH, Sadeghzadeh M, Yassin A, Issakhov A (2021) Renewable hybrid energy systems using geothermal energy: hybrid solar thermal–geothermal power plant. Int J Low-Carbon Technol 16(2):518–530
10. Liu Q, Shang L, Duan Y (2016) Performance analyses of a hybrid geothermal–fossil power generation system using low-enthalpy geothermal resources. Appl Energy 162:149–162
11. Kolahi M, Yari M, Mahmoudi SM, Mohammadkhani F (2016) Thermodynamic and economic performance improvement of ORCs through using zeotropic mixtures: case of waste heat recovery in an offshore platform. Case Stud Therm Eng 8:51–70
12. Abdolalipouradl M, Mohammadkhani F, Khalilarya S (2020) A comparative analysis of novel combined flash-binary cycles for Sabalan geothermal wells: Thermodynamic and exergoeconomic viewpoints. Energy 209:118235
13. Kazemi H, Ehyaei MA (2018) Energy, exergy, and economic analysis of a geothermal power plant. Adv Geo-Energ Res 2(2):190–209
14. Behzadi A, Gholamian E, Ahmadi P, Habibollahzade A, Ashjaee M (2018) Energy, exergy and exergoeconomic (3E) analyses and multi-objective optimization of a solar and geothermal based integrated energy system. Appl Therm Eng 143:1011–1022
15. Assad ME, Aryanfar Y, Radman S, Yousef B, Pakatchian M (2021) Energy and exergy analyses of single flash geothermal power plant at optimum separator temperature. Int J Low Carbon Technol 16(3):873–881
16. Fan G, Gao Y, Ayed H, Marzouki R, Aryanfar Y, Jarad F, Guo P (2021) Energy and exergy and economic (3E) analysis of a two-stage organic Rankine cycle for single flash geothermal power plant exhaust exergy recovery. Case Stud Therm Eng 28:101554

Numerical Investigations on the Flow Characteristics of Dual Cavity in a Strut-Based Scramjet Combustors

N. Maheswaran and S. Jeyakumar

Abstract This paper emphasizes the implications of dual cavity in a DLR supersonic combustion ramjet combustor. The two-dimensional reacting supersonic flow is computationally analyzed using ANSYS Fluent 18.0. The RANS equation is used in association with the k-ω SST turbulence model and a global one-step reaction mechanism to predict hydrogen-air combustion. The numerical outcomes have been validated using the experimental values. The simulation results of the cavity configurations and the standard DLR supersonic combustion ramjet combustor are compared. The wall static pressure distribution increases with the incorporation of cavities in the combustor compared to the DLR scramjet due to the additional shock wave generation from the edges of the cavity. The complete combustion is achieved within a short combustor length for cavities compared to the DLR scramjet model.

Keywords Scramjet · Cavity flow · Dual cavity · Strut injection · Combustion efficiency

1 Introduction

The fuel mixing with incoming supersonic air and the subsequent scramjet combustor combustion process is one of the challenging tasks [1–4]. Scramjets appear to be relatively simple and have no mechanical components, but designing a new scramjet engine is a difficult task since it involves transitioning from a ramjet to a scramjet mode [5], supersonic air with fuel mixing [6–8], and combustion process [9]. The proper blending of fuel and air, as well as combustion stability, is critical to the scramjet combustor's performance. The fuel injection systems and flame holding devices have a big impact on these two characteristics. As a result, an enhanced fuel

N. Maheswaran (✉)
Aeronautical Engineering, Hindusthan Institute of Technology, Coimbatore, Tamilnadu, India
e-mail: leomahesn@gmail.com

S. Jeyakumar
Aeronautical Engineering, Kalasalingam Academy of Research & Education, Krishnankoil, Tamilnadu, India

E. Natarajan et al. (eds.), *Materials, Design and Manufacturing for Sustainable Environment*, Lecture Notes in Mechanical Engineering,
https://doi.org/10.1007/978-981-19-3053-9_50

injection approach with a cavity-based flame holder is required in order to improve the scramjet performance.

Several studies [10–20] have explored at cavity flame holder models with various limiting factors aiming at seeing the cavity flow field and analyze the influence of the cavity on scramjet combustor in enhancing the mixing, and combustion processes. The residence time of supersonic air is a fundamental metric that many researchers address since it is a function of all performance characteristics, and noted that it is enhanced by implementation of a cavity flame holder. Huang et al. [21] examined a number of fuel injection approaches including strut injection, cantilever ramp injection, and wall injection, as well as a cavity flame holder in a scramjet combustor. All the fuel injection approaches are analyzed, and a potential fuel injection strategy is predicted using a fuel injection scheme based on a cavity flame holder.

The supersonic combustion ramjet engine performance with the influence of cavity is explored by Xing et al. [22]. The aim of the investigation is to examine the impact of a junction plate put at the cavity holder's entry. They claimed that locating fuel-rich locations and achieving flame stability in the cavity flow field disrupts the flow field significantly, improves early ignition, lowers ignition delay, and reduces scramjet weight by shortening the scramjet engine. Manna et al. [23] found that changing strut positions and fuel injection strategies enhanced the scramjet combustor's combustion efficiency. The slope injection of fuel into a scramjet combustion chamber is identified statistically by Chandra Murty and Chakraborty [24]. The primary goal of the study is to look at the effect of chemical kinetics on the internal flow field of a scramjet combustor. The k-ω turbulence model outperforms the k-ε turbulence model in a comparison of two turbulence models.

Gao [25] experimented the impact of the cavity on supersonic flow field mixing and flame stability. They found that the flame stabilization is improved on the basis of the shape and size of the cavity. Chang et al. [26] did investigation about the collaborative configuration of strut and cavity flame holder, as well as multi-staged fuel injection, and determined that multi-staged fuel injection increased the combustor's performance. Furthermore, combining the strut with the cavity flame holder considerably improves the combustor's effectiveness. In a scramjet combustor, Yang et al. [27] investigated the effects of parallel and tandem dual-cavity configurations. They claimed that flame stability has improved for both configurations. The modification of the cavity's aspect ratio affects the combustion phenomena, according to Navinkumar et al. [28]. The transverse fuel injection has a notable effect on mixing and effectiveness of combustion, which is thoroughly examined by Huang [29] in the context of single and multiport injection designs.

The open literature reveals the influence of cavities in a strut-based combustor. The result of aft wall cavity with double inclination in a strut injected supersonic combustion ramjet combustor is not reported in the literature and is discussed in the present investigation. This paper reveals the influence of double at the rear wall slanting cavities having a parallel injection within a reacting supersonic field. The flow parameters are investigated based on the flow factors like wall pressure and centerline pressure distribution, pressure contours with streamlines, and combustion efficiency.

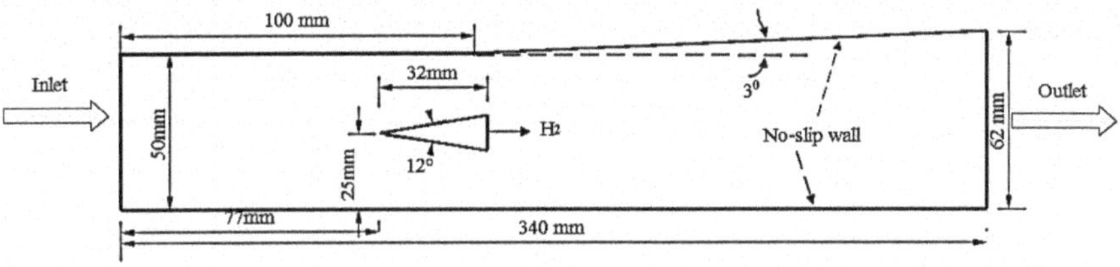

Fig. 1 Schematic view of the DLR combustor

2 Configuration of DLR Supersonic Combustion Ramjet Combustor

The DLR supersonic combustion ramjet combustor has been experimented with by Waidaman et al. [ref] is considered as the baseline model for the study, and the experimental values are used for the validation of the computational results; the layout is revealed in Fig. 1. The combustor has geometrical dimensions of 40 mm × 50 mm × 340 mm. The top wall of the combustor is kept constant for about 100 mm from the inlet and is diverged to 3 degrees toward the outlet. The incoming supersonic air is allowed into the combustor at Mach 2.0. A strut injector is positioned at the centerline of the combustor, at a distance of 77 mm from the inlet. The strut injector is provided with 15 holes each with 1 mm diameter through which hydrogen fuel is injected at Mach 1.0.

3 Numerical Methodology

3.1 Computational Domain Modeling

The computational domain for the baseline DLR scramjet model is identical to the experimental diagram (Fig. 1), and the dual cavity positioned at the top and bottom walls of the combustor is exhibited in Fig. 2. The cavities are located at a distance of 120 mm from the inlet. The 90° aft wall of the cavity is incorporated at the top most wall of the combustor and the bottom wall, and double aft wall inclined cavity is attached. The deepness of the cavity is kept constant as 10 mm, and the geometrical features of the lower wall cavity configuration are presented in Fig. 3. The various cavity rear wall angles used for the study are detailed in Table 1.

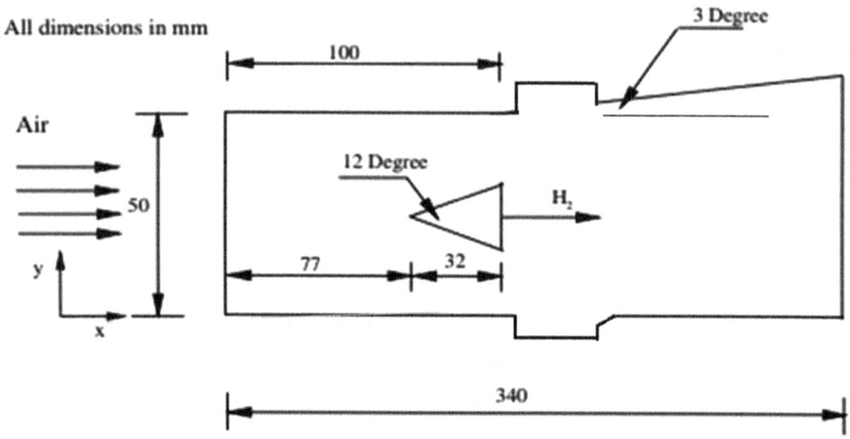

Fig. 2 Schematic view of the modified new DLR combustor design

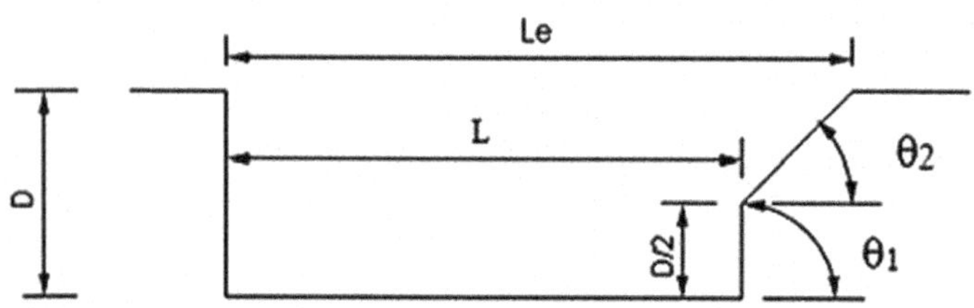

Fig. 3 Schematic view of the cavity

Table 1 Geometrical detail of bottom wall cavity configurations

Case	Cavity length/depth	Effective length/depth	Major angle, θ_1 (°)	Minor angle, θ_2 (°)
1	7	7.5	90	45
2	7	7.8	90	30
3	7	8.8	90	15

3.2 *Mathematical and Numerical Modeling*

The governing equations play an important role to solve and minimize the complexness of the numerical problems because the flow behavior of internal flow extremely depends on the selection of governing equations. The flow in a scramjet combustor is considered to be steady, compressible, and turbulent. Hence, the 2D RANS equation is to be used as governing equation. With the help of above-mentioned governing equation, we can find the exact location of shock waves and their features. Turbulence is demonstrated with the aid of the density-based shear stress transport standard $k - \omega$ model. Combustion is simulated by the help of the finite rate/eddy dissipation

Table 2 Inflow boundary conditions of air and hydrogen

Variable	Air	Hydrogen
Mach number	2	1
Axial velocity (m/s)	730	1200
Static pressure (bar)	1	1
Static temperature (K)	340	150
Density (kg/m^3)	1.002	0.097
Mass fraction of O_2	0.232	0
Mass fraction of N_2	0.736	0
Mass fraction of H_2O	0.032	0
Mass fraction of H_2	0	1

equation. The mass fraction of each species could be calculated with the aid of the convection–diffusion equation.

3.3 *Boundary Conditions*

The numerical governing equations are solved using boundary conditions, and these conditions have a substantial influence on the accuracy of the results. The inflow conditions are defined with pressure inlet and pressure outlet for the outlet of the computational domain. For the supersonic conditions, the air and fuel inlets are defined using the variables listed in Table 2. The combustor walls, as well as the strut walls, are all described as no-slip and adiabatic.

3.4 *Combustion Modeling*

For the supersonic combustion modeling technique [40–43], the finite rate chemical turbulence model with eddy dissipation is employed. The scramjet combustor's overall performance characteristics can be predicted with substantially reduced computing cost using a global one-step chemical reaction of hydrogen-air combustion.

3.5 *Validation*

Figure 4 depicts the experimental and numerical shadowgraph pictures of the DLR supersonic combustion ramjet model in a reactive flow field. Both experimental and numerical pictures appear to show identical shock generation at the strut leading and

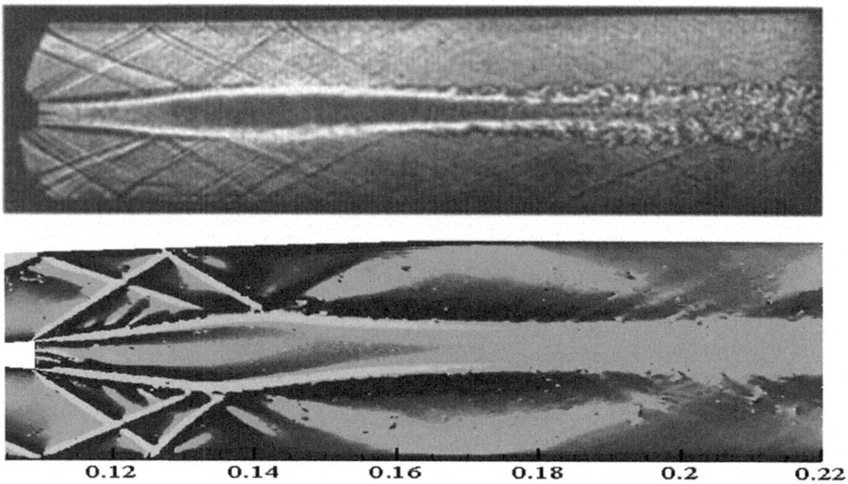

Fig. 4 Comparison of DLR experimental (top) and numerical (bottom) shadowgraph pictures

trailing edges, as well as shock reflections off the combustor's bottom and top walls and shock to shock interactions. The present computational methodology could be used to continue the current cavity study.

4 Results and Discussion

The reacting flow field attributes of a strut-based scramjet combustor with a dual cavity are reported in this section. The formation of pre-combustion shocks and recirculation region plays an important role to increase the performance of scramjet combustors. Implementation of the dual cavity in the combustor generates more vortices inside the cavity and enhances the recirculation region. Moreover, the pre-combustion shock waves generated from the leading and trailing edges of the cavity create a compressive zone in the cavity region. The static pressure distribution at the bottom wall of the cavity is depicted in Fig. 5.

Peak pressure is noted at the axial distance of 0.135 m on the combustor inlet for the DLR scramjet model. This is due to the interaction of the oblique shock from the leading and trailing edges of the strut with the wall boundary layer creating a compressive region. With the addition of a cavity in the combustor, which creates a subsonic area within the cavity, the pressure peak region widens along the flow direction. The shock waves that arise from the cavity that reduces the incoming air velocity create a compressive zone in the cavity region downstream of the strut, which can be noticed from the centerline pressure distribution plotted, Fig. 6.

Figure 7 depicts the formation of recirculation zones in the combustor for different with and without cavity configurations. From the plot, it is noticed that a large

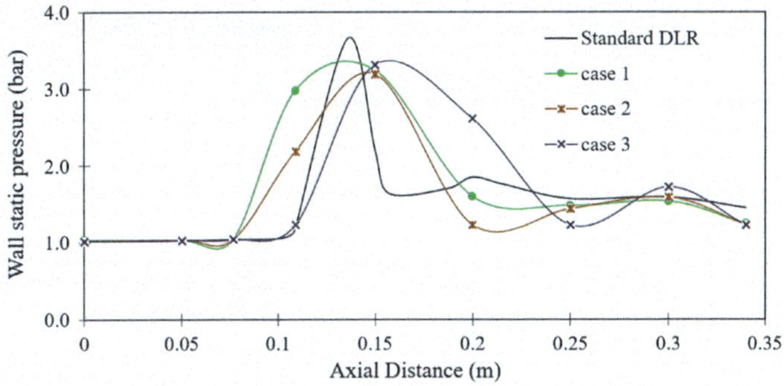

Fig. 5 Static pressure at combustor bottom wall

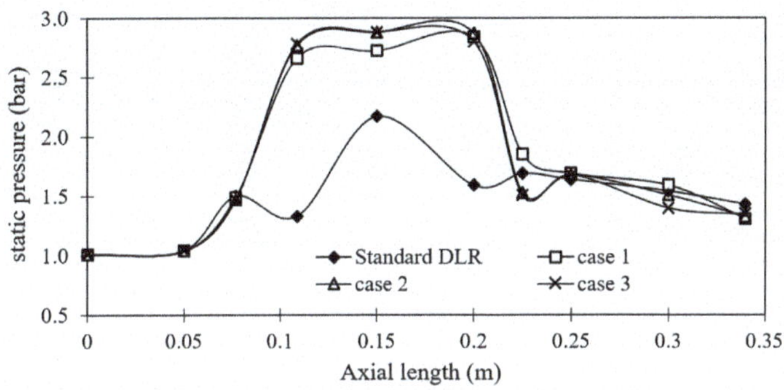

Fig. 6 Static pressure at the middle of the combustor

recirculation zone is observed downstream of the strut base which enhances fuel-air mixing. As the incorporation of the cavity at the top and bottom walls of the combustor, additional recirculation zones are developed inside the cavity region. Moreover, the recirculation region widens downstream of the strut with a decrease in the bottom aft wall inclination. This indicates an increase in the mixing and combustion in the cavity region with the variation of the cavity aft wall inclination.

The combustion efficiency is a vital parameter in the performance evaluation of the scramjet combustor. The combustion efficiency can be calculated by the expression [ref],

$$\eta_c(x) = 1 - \frac{\int A(x)\rho_{gas} u Y_{H_2} dA}{\dot{m}_{H_{2(inj)}}} = 1 - \frac{\dot{m}_{H_{2(x)}}}{\dot{m}_{H_{2(inj)}}}$$

where $\dot{m}_{H_{2(x)}}$ is the mass flow rate of hydrogen at the location "x" and $\dot{m}_{H_{2(inj)}}$ is the mass flow rate of the hydrogen injected from the strut injector, and ρ and u are the density and velocity of the stream. The combustion efficiency for the cavity cases and standard DLR scramjet model is given in Fig. 8. From the plots, the study observed that the total combustion is observed at the axial distance of 0.25 m from

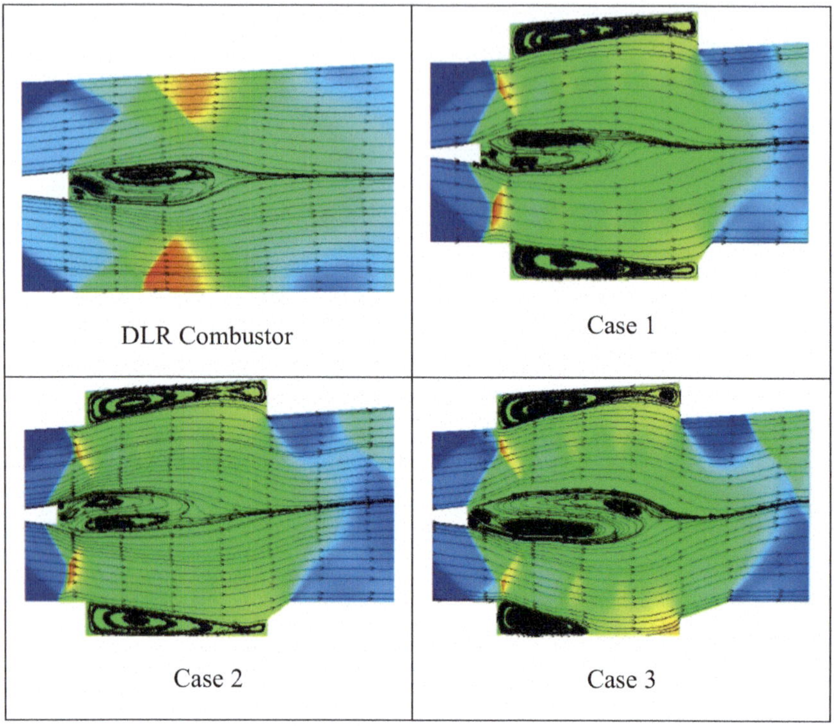

Fig. 7 Pressure contour with streamline for different cavity configurations compared to the DLR scramjet model

the combustor inlet for the DLR scramjet model. However, complete combustion is achieved at 0.185 mm with the incorporation of cavities in the combustor. Moreover, the reduction in the aft wall minor angle from 45° to 15° improves the combustion efficiency.

5 Conclusions

The implication of dual cavity with rear wall inclination in a DLR supersonic combustion ramjet combustor is numerically investigated using a two-dimensional steady flow RANS equation. The efficiency of the combustor is analyzed based on the wall pressures, streamlines, and combustion efficiency. The peak pressures and the axial length of the combustor are noticed with the dual-cavity configurations related to the DLR scramjet combustor due to the shock interactions generated from the cavity region. Moreover, the compressive regime above the cavity region decelerates the incoming supersonic air which enhances hydrogen-air mixing and achieves complete combustion within a short axial distance related to the DLR supersonic combustion

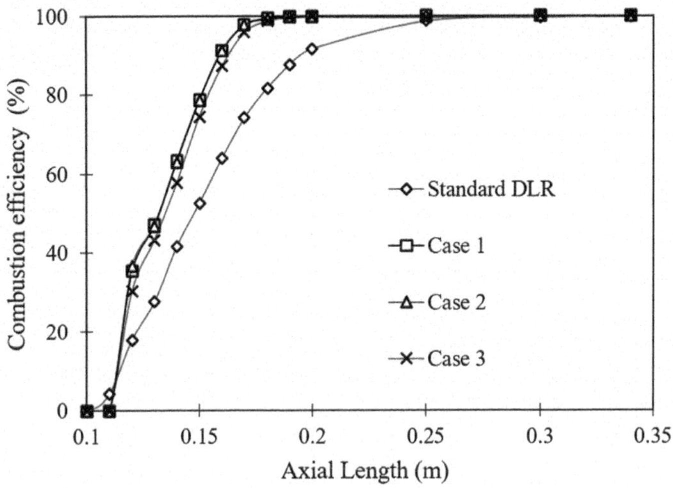

Fig. 8 Combustion efficiency for DLR scramjet model compared with various cavity configurations

ramjet model. The cavity secondary aft wall angle of 15° widens the recirculation zone in the lateral direction of the combustor and provides complete combustion within the shortened axial distance compared to the other configurations.

References

1. Curran ET (2001) Scramjet engines: the first forty years. J Propuls Power 17(6):1138–1148. https://doi.org/10.2514/2.5875
2. Voland RT, Huebner LD, McClinton CR (2006) X-43a hypersonic vehicle technology development. Acta Astronaut 59(1):181–191. https://doi.org/10.1016/j.actaastro.2006.02.021
3. Smirnov NN, Betelin VB, Nikitin VF, Phylippov YuG, Koo J (2014) Detonation engine fed by acetylene oxygen mixture. Acta Astronaut 104(1) 134–146. https://doi.org/10.1016/j.actaastro.2014.07.019
4. Fulton JA, Edwards JR, Cutler A, McDaniel J, Goyne C (2016) Turbulence/chemistry interactions in a ramp-stabilized supersonic hydrogen air diffusion flame, Combust Flame 174:152–165. https://doi.org/10.1016/j.combustflame.2016.09.017
5. Yan Z, Bing C, Gang L, Baoxi W, Xu X (2014) Influencing factors on the mode transition in a dual-mode scramjet, Acta Astronaut 103:1–15. https://doi.org/10.1016/j.actaastro.2014.06.006
6. Chang J, Hu Q, Yu D, Bao W (2011) Classifier utility modeling and analysis of hypersonic inlet start/unstart considering training data costs. Acta Astronau 69(9):841–847. https://doi.org/10.1016/j.actaastro.2011.05.035
7. Dharavath M, Manna P, Chakraborty D (2015) Numerical exploration of mixing and combustion in ethylene fueled scramjet combustor. Acta Astronaut 117:305–318. https://doi.org/10.1016/j.actaastro.2015.08.014
8. Srikrishnan AR, Kurian J, Sriramulu V (1996) An experimental investigation of thermal mixing and combustion in supersonic flows. Combust Flame 420 107(4):464–474. https://doi.org/10.1016/S0010-2180(96)00084-3

9. Tian Y, Xiao B, Zhang S, Xing J (2015) Experimental and computational study on combustion performance of a kerosene fueled dual-mode scramjet engine. Aerosp Sci Technol 46:451–458. https://doi.org/10.1016/j.ast.2015.09.002
10. Li J, Song W, Luo F, Shi D (2014) Experimental investigation of vitiation effects on supersonic combustor performance. Acta Astronaut 96:296–302. https://doi.org/10.1016/j.actaastro.2013.11.016
11. Wang H, Wang Z, Sun M, Qin N (2015) Large eddy simulation of a hydrogen-fueled scramjet combustor with dual cavity. Acta Astronaut 108:119–128. https://doi.org/10.1016/j.actaastro.2014.12.008
12. Wang H, Li P, Sun M, Wei J (2017) Entrainment 480 characteristics of cavity shear layers in supersonic flows. Acta Astronaut 137:214–221
13. Cai Z, Liu X, Gong C, Sun M, Wang Z, Bai X-S (2016) Large eddy simulation of the fuel transport and mixing process in a scramjet combustor with rearwall-expansion cavity. Acta Astronaut 126:375–381
14. Hugangand W, Wang Z-g, Yan L, Liu W-d (2012) Numerical validation and parametric investigation on the cold flow field of a typical cavity-based scramjet combustor. Acta Astronautica 80:132–140
15. Huang W, Luo S-b, Pourkashanian M, Ma L, Ingham DB, Liu J, Wang Z-g (2010) Numerical simulations of a typical hydrogen fueled scramjet combustor with a cavity flame holder. In: Proceedings of the world congress on engineering 2010 vol II WCE 2010, June 30–July 2
16. Lu W, Zhansen Q, Liangjie G (2015) Numerical study of the combustion field in dual-cavity scramjet combustor. Procedia Eng 99:313–319. In: 2014 Asia-Pacific international symposium on aerospace technology, APISAT2014 24–26 Sept 2014 Shanghai, China
17. Zong Y, Bao W, Chang J, Hu J, Yang Q, Song J, Wu M (2015) Effect of fuel injection allocation on the combustion characteristics of a cavity-strut model scramjet. J Aerosp Eng 28
18. Bao W, Zong Y, Chang J, Hu J, Yang Q, Song J, Wu M (2014) Effects of upstream strut on the combustion of liquid kerosene in a model cavity scramjet. Proc Inst Mech Eng Part G: J Aerosp Eng 228:2323–2328
19. Zhang C, Chang J, Zhang Y, Wang Y, Bao W (2017) Flow field characteristics analysis and combustion modes classification for a strut/cavity dual-mode combustor, Acta Astronaut 515 137:44–51
20. Pourkashanian M, Ma L, Ingham DB, Luo S-B, Wang Z-G (2012) Effect of geometric parameters on the drag of the cavity flame holder based on the variance analysis method. Aerosp Sci Technol 21(1):24–30
21. Huang W et al (2011) Overview of fuel injection techniques for scramjet engines. In: Proceedings of ASME Turbo Expo 2011, 6–10 June 2011. Vancouver, British Columbia, Canada
22. Xing F, Zhao MM, Zhang S (2012) Simulations of a cavity based two dimensional scramjet model. In: 18th Australasian fluid mechanics conference, Launceston, Dec 2012
23. Manna P, Dharavath M, Sinha PK, Chakraborty D (2013) Optimization of a fight-worthy scramjet combustor through CFD. Aerosp Sci Technol 21(1):138–146
24. Chandra Murty MSR, Chakraborty D (2012) Numerical simulation of angular injection of hydrogen fuel in scramjet combustor. Proc IMechE Part G: J. Aerosp Eng 226 (7):861–872
25. Gao P et al (2012) The numerical research on the scramjet combustion with different Structure cavity. Adv Mater Res 468–471:1444–1447
26. Chang J, Shi W, Bao W. Influence factor analysis of performance parameter for a strut/cavity supersonic combustor. In: 51st AIAA/SAE/ASEE joint propulsion conference, AIAA propulsion and energy forum, (AIAA 2015-3944)
27. Yang Y, Wang Z, Sun M, Wang H, Li L (2015) Numerical and experimental study on ame structure characteristics in a supersonic combustor with dual-cavity. Acta Astronaut 117:376–389
28. Mahto NK, Choubey G, Suneetha L, Pandey KM (2016) Effect of variation of length-to-depth ratio and mach number on the performance of a typical double cavity scramjet combustor. Acta Astronaut 128:540–550

29. Huang W (2016) Transverse jet in supersonic crossows. Aerosp Sci Technol 50:183–195
30. Pandey KM (2016) Investigation on the effects of operating variables on the performance of two-strut scramjet combustor. Int J Hydrogen Energy 41(45):20753–20770
31. Choubey G, Pandey KM (2017) Effect of different strut-wall injection techniques on the performance of two-strut scramjet combustor. Int J Hydrogen Energy 42(18):13259–13275
32. Choubey G, Pandey KM (2016) Effect of variation of angle of attack on the performance of two-strut scramjet combustor. Int J Hydrogen Energy 41 (26) (2016) 11455 -11470.
33. Raul R, Gilreath H, Sullins G (1992) Numerical and experimental investigation of mixing enhancement in scramjets. In: AIAA fourth international aerospace planes conference, Orlando, FL, 1–4 Dec 1992
34. Obula Reddy K (2016) Numerical analysis of hydrogen fuel scramjet combustor with turbulence development inserts and with different turbulence models. Int J Hydrogen Energy
35. Wang H, Wang Z, Sun M, Wu H (2013) Combustion modes of hydrogen jet combustion in a cavity-based supersonic combustor. Int J Hydrogen Energy 38(27):12078–12089
36. Huang W, Yan L (2016) Numerical investigation on the ram scram transition mechanism in a strut-based dual-mode scramjet combustor. Int J Hydrogen Energy 41(8):4799–4807
37. Choubey G, Pandey KM (2017) Effect of different strut plus wall injection techniques on the performance of two-strut scramjet combustor. Int J Hydrogen Energy 42(18):13259–13275
38. Choubey G, Pandey KM (2016) Effect of variation of angle of attack on the performance of two-strut scramjet combustor. Int J Hydrogen Energy 41(26):11455–11470
39. Huang W (2015) Investigation on the effect of strut configurations and locations on the combustion performance of a typical scramjet combustor. J Mech Sci Technol 29:5485–5496
40. Huang W, Wang ZG, Luo SB, Liu J (2011) Parametric effects on the combustion flow field of a typical strut-based scramjet combustor. Chin Sci Bull 56(35):3871–3877
41. Gerlinger P, Nold K, Aigner M (2010) Inuence of reaction mechanisms, grid spacing, and inflow conditions on the numerical simulation of lifted supersonic flames. Int J Numer Meth Fluids 62(12):1357–1380
42. Strhle J, Myhrvold T (2007) An evaluation of detailed reaction mechanisms for hydrogen combustion under gas turbine conditions. Int J Hydrogen Energy 32(1):125–135
43. Kumaran K, Babu V (2009) Investigation of the effect of chemistry models on the numerical predictions of the supersonic combustion of hydrogen. Combust Flame 156(4):826–841
44. Moses Devaprasanna M, Maheswaran N (2016) A CFD based numerical analysis of scramjet combustor. Int J Eng Res Technol 5(12)

Performance Analysis of Vortex Tube Refrigeration System by Experimental Method

B. Aravinth, S. Manivannan, C. Rameshkannan, M. Subramaniyan, and V. Raj Kumar

Abstract The vortex tube may be an uncommon cooling device, without a transferring components which could create bloodless air and warm air from suppressed compressed fueloline while now no longer touching the encompassing area. While the competitive air is tangentially injected into the vortex chamber, a robust vortex glide is created that may be divided into two air streams, one warm movement at the rims and any other bloodless in the backbone at every end. It is used economically consisting of cooling of slicing equipment, heating gadget, fridges, etc. It is far weightless and desires a small toilet. Also, the preliminary fee is low, and its running expenses degree a rectangular anywhere suppressed fueloline is provided quickly. The cutting-edge mission is to research the overall performance of the vortex tube with the aid of using dynamical size of air in water and in bloodless climates. The vortex air-cooled tube cooling gadget turned into evolved with the assist of equipment consisting of rotameter, strain transmitter, and RTD temperature sensors. In the course of this experiment, an alittle vortex tube made from stainless steel turned into used. The vortex tube is nicely prepared with ideal insulation. The study agrees with the multiple thermo-physical properties of the compressed gasoline utilised as the active fluid. Strain, air temperature, and air temperature are determined at utterly unique locations for utterly unique pressures of water performance by adjusting the modern terminal control of the vortex tube.

Keywords Vortex tube refrigeration system · Rotameter · Pressure transmitter and RTD temperature sensors

B. Aravinth · C. Rameshkannan · M. Subramaniyan
Department of Mechanical Engineering, Dr. Navalar Nedunchezhiyan College of Engineering, Tholudur, India

S. Manivannan (✉)
Department of Mechanical Engineering, Center for Material Science, Karpagam Academy of Higher Education, Coimbatore, India
e-mail: manivannan.s@kahedu.edu.in

V. Raj Kumar
Department of Mechanical Engineering, PSG Institute of Technology and Applied Research, Coimbatore, India

E. Natarajan et al. (eds.), *Materials, Design and Manufacturing for Sustainable Environment*, Lecture Notes in Mechanical Engineering,
https://doi.org/10.1007/978-981-19-3053-9_51

1 Introduction

The ortex tube may be an uncommon cooling tool, and not using a shifting elements able to generating bloodless air and warm air from a compressed fueloline deliver while now no longer shifting the atmosphere. Excessive fueloline is compelled via the era chamber, and because of the excessive and restricted volume, the compressed air deliver head is reborn in a kinetic head that produces relevant air stream to the internal partitions of the tube. It is miles clean that 1/2 of the cooling unit does now no longer encompass any shifting part [1–5]. As the compressed fuel is multiplied to a higher degree of stream and aggressively injected into a rotating chamber. The compressed fueloline inserted into the vortex tube passes via pipes via the tangent unit to the closed counter bore. These bombs positioned air into the vortex motion. This circulating airflow rotates at a particular perspective and passes via a brand new tube in the rotating shell type, nearly like a whirlwind. Way to the nozzle above the tube, best the outer shell of the propellant is authorized to break out at that moment. The ultimate fueloline is compelled to include an partner diploma vortex inner of the decreased width among the outer vortex. Part of the fresh, excessive-pace air is authorized to go out the manipulate valve. A few gradual air waft is compelled to counteract the waft upward among excessive-pace air [6, 7]. The provided fueloline deliver is taken from the device and added via the gap of the manipulate valve. Pressurized fueloline is managed with the aid of using putting it at decided on selective stress. It passes via the rotameter via the fueloline tubes. The air is then transferred to a stress transmitter that is constantly related to the water withinside the vortex tube. The frame of the water vortex tube has holes, air from the device this is disbursed to the vortex tube. Therefore, the swirl motion is achieved with the aid of using injecting air into the frame of the water pipe. High-pace air actions to a brand new a part of the tube. While managed with the aid of using a transportable button, the pinnacle is made. It reasons the air temperature to upward thrust and withinside the center of the middle, a part of the air expands freely withinside the depressed region on the opposite aspect of the vortex tube. In the middle, the air grows, and bloodless air is taken at the other stop of the tube. In bloodless weather, any other stress transmitter is used to compress the stress, and in addition, a rotameter is used to stabilize air temperature withinside the bloodless. By putting RTD into the water, warm garage region, bloodless termination region, and above the brand new tube, the unit temperature region is measured. In the brand new completing phase, an RTD probe detector is used [8–11]. To keep away from warmth loss in the vortex tube region, the tube is nicely insulated with an appropriate insulant. For excessive air stress, a controller is used. By controlling the manipulate valve of the device tool, the extent is measured in the rotameter. The compressor is used for the mistaken managing capacity. The gauge related to the tank of the device tool suggests readings (Fig. 1).

In this section, a few important terms commonly used in vortex tube are defined.

Fig. 1 Experimental setup of vortex tube refrigeration system

1.1 Cold Mass Fraction

The cold part of the mass is the most important parameter that shows the function of the vortex tube and the temperature or power separation within the vortex tube. The cold component is defined as the rate of cold air flow rate and the air flow rate.

$$\mu = \frac{m_c}{m_i} \tag{1.1}$$

The cold mass fraction can be controlled by the cone valve, which is placed at the hot tube end, and it is expressed in Eq. 1.1.

1.2 Cold Air Temperature Drops

Cold air temperature drop is defined as the difference in temperature between entry air temperature, and cold air temperature is expressed in Eq. 1.2

$$\Delta T_c = T_i - T_c \tag{1.2}$$

in which T_i is the entry air temperature and T_c is the cold air temperature.

1.3 Hot Air Temperature Rise

Hot air temperature rise is defined as the difference in temperature between entry air temperature and hot air temperature is expressed in the Eq. 1.3

$$\Delta T_h = T_h - T_i \tag{1.3}$$

in which T_i is the entry air temperature and T_h is the hot air temperature.

2 Artificial Neural Networks

In this study, synthetic nerve networks typically model three parameters of the most important parameters, body water pressure, flow fee, and partial cold weight withinside the cooling operation of the counter flow vortex tube. During this process, the test information typically trains and validates the neural network model via MATLAB (R2013a) with the case software program software for the neural networks that embody the tool. During this model, a decrease in temperature withinside the cold region has been considered because of the cooling function of the vortex tube. Based on verified experimental data, the proper operation of the vortex tube cooling system identified three water parameters in advance. During this study, the neural network model has become advanced to diploma the link most of the three most important parameters and therefore the decrease in cold temperature of the vortex tube. In general, a good way to create size issues related to networks that weigh up to a few hundred, Levenberg–Marquardt (LM) law achieves faster integration and therefore a far higher accuracy. Therefore, the luminous flux unit has become decided on because of the network education method.

3 Results and Discussion

In this study, ANN has been used to investigate the effects of three important parameters like inlet pressure, orifice diameter, and cold mass fraction (p_i, $D_{0,}$ and μ_c) on the cooling performance of the vortex tube. Figure 2 shows the structure of neural network.

Figure 3 shows the working curve of the neural network. The best training performance is 1.9516 for 5000 epochs. Figure 4 shows the neural curve of the neural network. Overall, the value of R is 0.9877 indicating that this ANN model is equal to the test value. Figure 4 shows the ANN predictive test values that ensure that the ANN values are close to the test values (Figs. 5 and 6).

Figures 7, 8, 9, 10, and 11 show the difference between the cold and cold temperature differences between the various pressures and widths of the orifice. From the

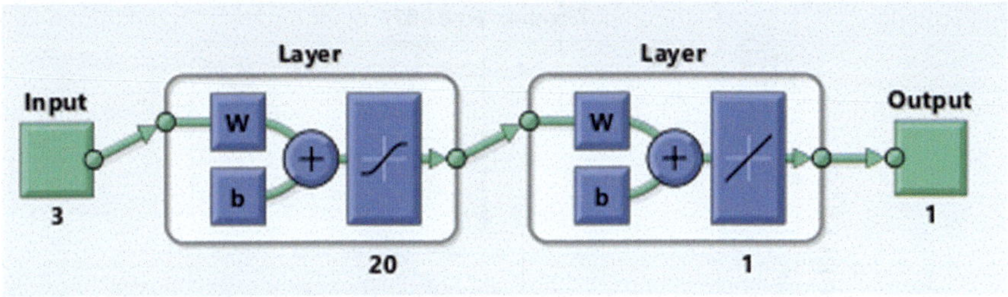

Fig. 2 Structure of neural network

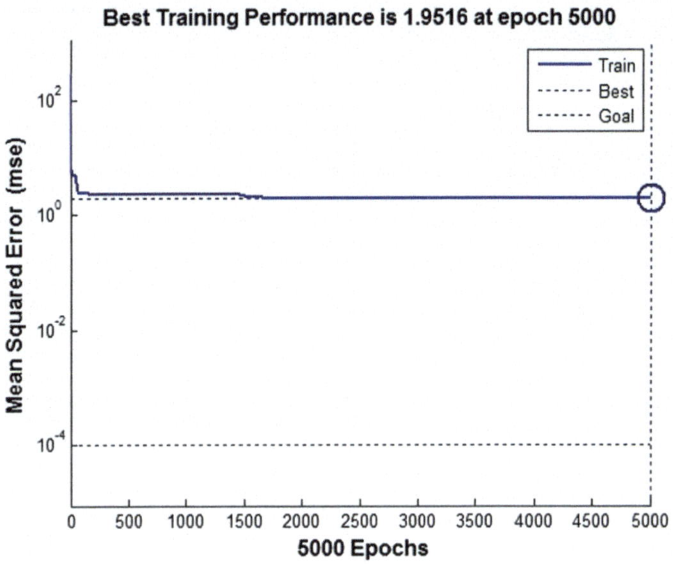

Fig. 3 Performance curve of the neural network

above graphs, it is noted that ANN and test values are close. High temperatures are found when the frost fraction is 0.4–0.6 in all cases.

The maximum temperature difference is obtained when

Inlet pressure	= 6 bar
Orifice diameter	= 4 mm
Cold mass fraction	= 0.58.

4 Conclusion

The artificial neural network version was utilized to examine the effect of the enter parameter on the cooling characteristic of the vortex tube. Comparisons between

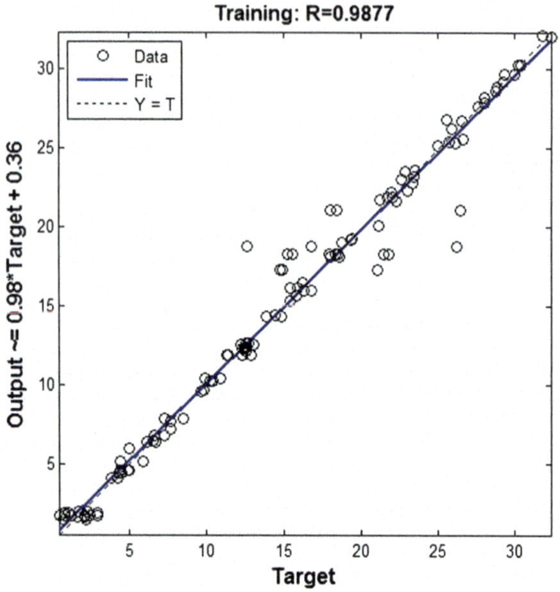

Fig. 4 Linear regression of the neural network

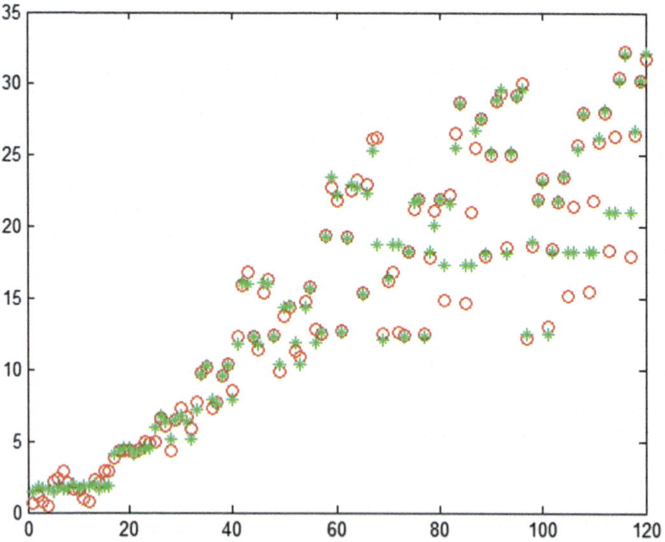

Fig. 5 Experiment versus ANN

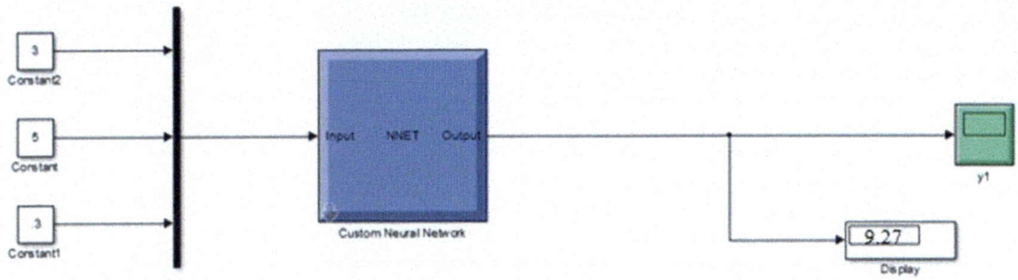

Fig. 6 Output functions

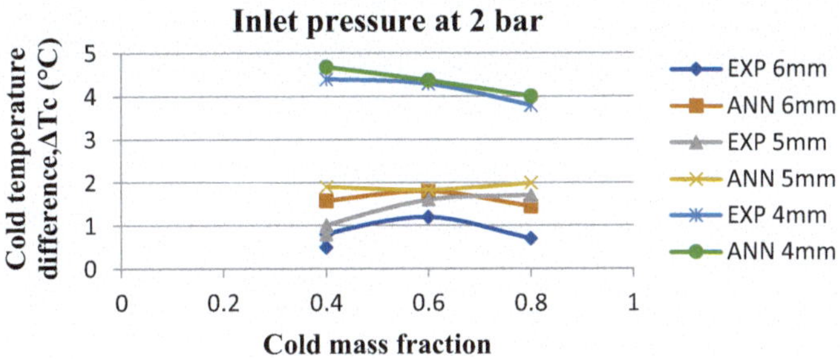

Fig. 7 Variation temperatures with respect to cold mass fraction at 2 bar

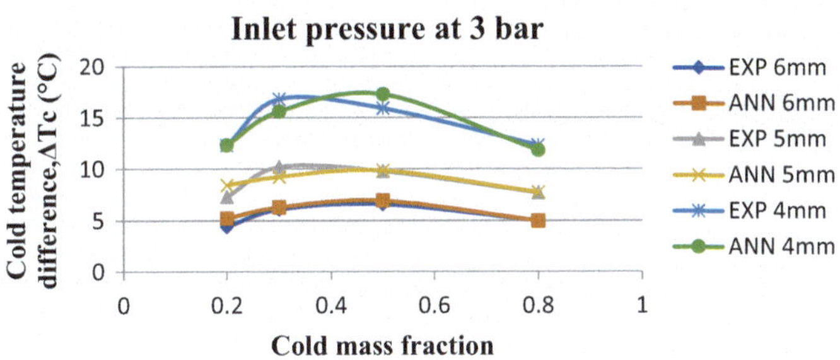

Fig. 8 Variation temperatures with respect to cold mass fraction at 3 bar

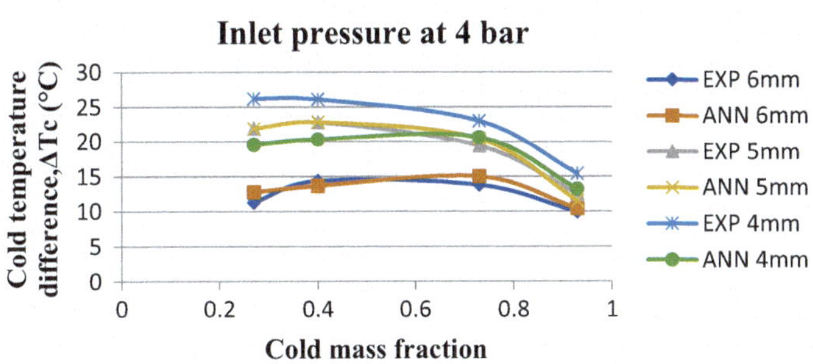

Fig. 9 Variation temperatures with respect to cold mass fraction at 4 bar

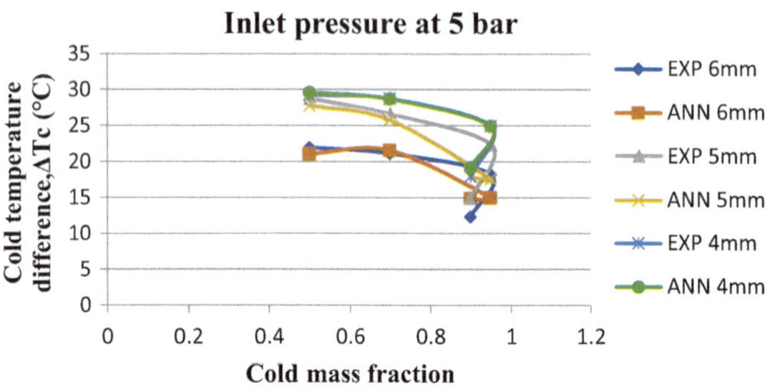

Fig. 10 Variation temperatures with respect to cold mass fraction at 5 bar

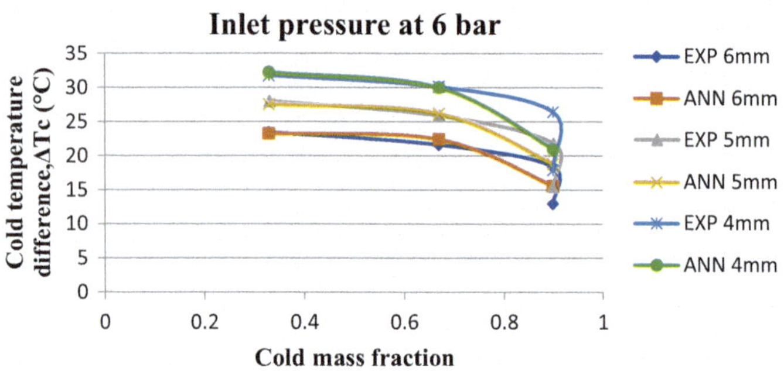

Fig. 11 Variation temperatures with respect to cold mass fraction at 6 bar

ANN predictions and records revealed that this version had a high level of accuracy. According to this, when the temperature is lower, the bloodless a portion of 0.4 to 0.6 occurs in all circumstances. The current results show that an increase in intake stress increases the vortex tube's overall cooling performance. The cooling performance of the vortex tube decreases as the orifice width increases. The four-mm vortex generator produces a most temperature distinction of 32.3 °C, and the ANN value with a most temperature distinction of 31.93 °C turned into acquired because of the running stress of 6 bar withinside the bloodless segment of 0.57.

References

1. Ahlborn B, Gordon J (2000) The vortex tube as a classical thermodynamic refrigeration cycle. J Appl Phys 88:3645–3653
2. Amitani T, Adachi T, Kato T (2004) A study on temperature separation in a large vortex tube. Jpn Soc Mech Eng 49:877–884
3. Arbuzov DYN, Lebedev AV, Pravdina MK, Yavorskii NI (1997) Observation of large-scale hydrodynamic structures in a vortex tube and the Ranque effect. Tech Phys Lett 23(12):938–940

4. Deissler RG, Perlmutter M (1960) Analysis of the flow and energy separation in a vortex tube. Int J Heat Mass Transfer 1:173–191
5. Eiamsa S, Promvonge P (2007) Numerical investigation of the thermal separation in Ranque-Hilsch vortex tube. Int J Heat Mass Trans 50:821–832
6. Gutsol AF (1997) The ranque effect. Phys Uspekhi 40:639658
7. Harnett JP, Eckert E (1957) Experimental study of the velocity and temperature distribution in a high velocity vortex-type flow. Trans ASME 79:751–758
8. Hilsch R (1947) The use of the expansion of gases in a centrifugal field as cooling process. Rev Sci Instrum 1:208–214
9. Prabakaran J, Vaidyanathan S (2010) Effect of orifice and pressure of counter flow vortex tube. Ind J Sci Technol
10. Kurosaka M (2010) Acoustic streaming in swirling flows. J Fluid Mech 124:139–172
11. Lay JE (1959) An experimental and analytical study of vortex flow temperature separation by superposition of spiral and axial flows. Part I Trans ASME J Heat Transfer 81(4):202–212

Investigational on Biodiesel Exploitation Orange Peel Oil After CI Engine

B. Aravinth, C. Rameshkannan, S. Manivannan, V. Raj Kumar, and N. Rajasekar

Abstract Improved attentions of the reduction of fossil fuels then the herbal aids of biodiesel gasoline have increased nowadays. Its essential rewards are that it is miles one of the most renewable fuels to be had now and is non-toxic and decaying. It can also be used right away on most diesel engines without the want for huge engine modification. Even though, the fee of biodiesel is a first-rate barrier to its income likened to petroleum-based totally diesel gasoline. Castoff cooking oil is one of the most assets of biodiesel production. Biodiesel production from crude vegetable oil affords a triple-aspect answer: financial, environmental and waste management. New system era advanced through the years has made it feasible to supply biodiesel from recycled fossil fuels which include the notable of biodiesel vegetable oil without a different appealing price drop. Therefore, biodiesel made from recycled frying oil has the equal ability to be used. Many preceding evaluations estimate the rate of biodiesel manufacturing based at the assumptions made with the aid of their authors, concerning production volume, feedstock and chemical technology. From a waste management mind-set, the manufacturing of biodiesel through used frying oils is useful to the environment, as it provides a cleanser way to put off that merchandise.

Keywords Diesel · Orange peel oil · Chemical compounds

B. Aravinth · C. Rameshkannan · N. Rajasekar
Department of Mechanical Engineering, Dr. Navalar Nedunchezhiyan College of Engineering, Tholudur, India

S. Manivannan (✉)
Centre for Material Science, Department of Mechanical Engineering, Karpagam Academy of Higher Education, Coimbatore, India
e-mail: manivannan.s@kahedu.edu.in

V. Raj Kumar
Department of Mechanical Engineering, PSG Institute of Technology and Applied Research, Coimbatore, India

E. Natarajan et al. (eds.), *Materials, Design and Manufacturing for Sustainable Environment*, Lecture Notes in Mechanical Engineering,
https://doi.org/10.1007/978-981-19-3053-9_52

1 Introduction

Over the last 100 years, mineral intake has multiplied because of modernization and the dramatic increase in international population. This will increase the demand for energy, that's driven by gas consumption, which has caused gasoline shortages, rising costs and negative environmental consequences which include international warming, ozone depletion, deforestation, acid reflux disease and photochemical emissions. considering that fossil fuels are a confined supply of energy, this creates a need for one of a kind electricity assets that can be similarly social, low cost and environmentally identical. the two have played a major role in looking for strength in the shipping and business sectors, turning into predominant customers of energy. The kinetic version is designed for steam immersion in lemongrass (*Cymbopogon* spp.). This fashionable version has been tested on driver scales, smoke extraction units and clusters of lemon to 70 to one thousand kg. For closer contact with lemongrass and steam, two advanced additives are designed for those devices. It is far noteworthy, however that the loose packaging of plant fabric inside the current distillation and the extent of steam injection, will increase the oil yield [1]. The conduct of the immediately line and the temporary axis of the upper stage version method that oil production in the distillation unit is not fast, but some time is needed to irrigate the oil filler grass and osmosis. Vegetable oil-like oils were around for a hundred years while diesel engine founder Rudolph Diesel first examined nutmeg oil in his urgent engine [2]. The consequences of the three stages of ripening on lemongrass harvesting in vital oils, chemical composition and intermediate content material. The lemongrass plant became planted using a 4-dimensional blockchain layout, at college Agriculture Park, university Putra Malaysia [3]. It changed into reported that experimental research have been done to assess the performance and launch of a right away injection diesel engine combined with Jatropa biodiesel prepared with methanol to come across jatropa oil methyl ester [4, 5]. Biodiesel produced sixteen with lemon grass using the Trans-esterification method used as gasoline. Overall performance checks were executed and as compared with diesel [6]. Gasoline emissions have been analyzed for distinctive load concentrations of diesel and biodiesel. A brand new take a look at designed for the satisfactory aggregate, namely, the crank pressure drawing diagram drawn for the best blend and the pressure variations for distinctive masses of the composite element is plotted in the p-v diagram [7]. Test effects display that the thermal performance of all compounds is slightly decrease or almost identical compared to diesel. But NO_x emissions are barely higher in comparison to diesel gas [8, 9].

He noted that the development of different fuels for internal combustion engines is frequently an evolutionary technique in which petroleum-related issues are encountered and crucial gasoline traits and their specific hassle-solving limitations are recognized. In this regard, this take a look at gives an overview of the symptoms of lemongrass fat burning [10]. It has been studied that supplemental biodiesel is regularly desired to enhance the overall performance and release of diesel engine features, and excessive gas injection stress works to enhance overall performance and decrease carbon emissions. Within the present paintings, Isobutanol and ethanol as additives of diesel–biodiesel compounds were investigated with the aid of testing an instantaneous injection diesel engine, and may be upgraded to 250 bars, because of which the brakes' performance and fuel economic system are progressed [11]. In this work solvent extraction strategies and extraction methods had been used to extract the vital oils from the lemongrass leaf. The solvent extraction approach produced 1.85% and the extraction method produced zero.86% of vital oils, respectively. From the analysis the solvent extraction method has yielded better yields because of less publicity to air and heat [12]. It investigated whether or not ethanol turned into blended with lemongrass blend of natural diesel oil and investigated the impact of ethanol concentration on the heat, emissions and performance of a specific diesel engine injection. The awareness of low ethanol changed into investigated with a aggregate of natural lemongrass oil, take a look at effects acquired from these compounds are akin to the ones located in diesel, tested mixtures found out one of a kind overall performance, exhaust homes and combustion capabilities as compared to diesel gasoline [13].

In short describe the production base of Biodiesel numerous issues—assets, opportunities, demanding situations, crop formation, testing and many others mentioned the production of Biodiesel. An vital a part of the paintings presented to speak about vital troubles associated with the layout of the Biodiesel manufacturing facility, gives crucial info needed for the layout of the Biodiesel plant layout problem, and affords a likely mathematical model for evaluating the Biodiesel enterprise. design [14]. Measuring microalgae oil emissions is a major step inside the complete biodiesel manufacturing system. These days, greater powerful carbon dioxide has been proposed to update common solvent extraction methods due to the fact they are reliable, innocent, chemically solid and less expensive. It makes use of locally customary solvent, which may be effortlessly separated from products. In addition to the usage of supercritical carbon dioxide it permits easy separation of the product. On this paper, the normal production of biodiesel with the aid of first-generation feedstock, the usage of chemical catalysts and solvent extraction [15].

2 Methodology

Orange peel oil, a long-lasting hot plant that produces aromatic oils. The term lemongrass is derived from the lemon-like aroma of vital oils. This plant is local to Asia and Australia. Lemongrass turned into one of the herbs that carried the spice line from Asia to Europe. Limemongrass industrial oil is satisfactory called Cochin oil for international alternate, with 90% of it being exported from Cochin port. The state of Kerala in India turned into accountable for the manufacturing and export of lemongrass oil. Worldwide annual production of lemongrass oil is about 1000 t from an area of 16,000 ha. In India, it is cultivated on an area of 4000 ha and the yearly production is 250 t. This plant is broadly grown in bad, remote and polluted regions and close to ties as a living cover. A well-tired root device of the plant helps to preserve soil and water.

Orange peel leaves are gathered inside the garden. A sample of the plant was lately reduce, 10 cm from the foundation, in the morning at the day of series. Orange peel oil, a percentage of the vital oil yield of slightly dried leaves is found to be higher than that of sparkling leaves. Consequently, as soon as amassed, the plant is dried at room temperature for 4 days, after which saved in a plastic bag sealed in ambient temperature and protected from mild. The extraction yield increases via reducing the particle size due to the excessive amount of fats released as the destroy cells are destroyed with the aid of digestion. To improve the efficiency of the collection, the plant material was immersed in its distilled water 30 min before being extracted [16].

150 g of dried lemongrass pattern turned into weighed with a pattern of sliced lemon grass and placed in a lower flask with 1 L easy. 500 ml of N-hexane solvent became poured right into a flask. The flask and contents had been allowed to stand for 36 h. this is done to extract all of the fats from the lemongrass and to extract it completely. After that the output changed into decreased to some other 1 L beaker. 200 ml of ethanol is brought to extract vital oils because the essential oil is dissolved in ethanol. The mixture is then transferred to a 500 ml separator and separated by means of a method called a liquid/liquid separation technique. The contents of the setting apart panel were allowed and allowed to reach equilibrium, which had been divided into layers (relying on their one of a kind densities). The decrease ethanol extracted and the top hexane layer have been accrued into separate 250 ml beakers and placed in a water bath. With the aid of 78% that is finished on the way to extract ethanol 24 leaving most effective natural vital oils. The oil yield became decided by way of measuring the output on the electronic scale. The distinction between the final weight of the extracted beaker and the preliminary weight of the empty beaker gave the burden of the crucial oil.

A 150 g pattern of fresh lemongrass is placed in a 1-L circular backside flask containing 250 ml of distilled water. The flask turned into outfitted with a rubber cap attached to a condenser and heated. The room temperature water flows over the counter in the intervening time with a 25-degree condenser to tighten the confirmation gadget. While the water reaches a hundred it starts off evolved to boil freeing crucial oils from lemongrass. While lemongrass is hot, critical oils are extracted from the leaf and mixed with water vapor. Both pass thru the condenser and the vapor is secreted into the liquid. With the use of ice block, cooling changed into made possible and the conversion of critical oils became averted. The condensate is accumulated immediately using a 500 ml beaker and poured right into a separator. This creates layers of oil and water. The setting apart tap faucet was turned on to drain the water while the oil was quickly accrued in a 100 ml sealed bottle. Biodiesel was released through the trans-esterification process. The physical properties and biodiesel properties of orange peel oil were tested by ASTM standards and compared with diesel fuel. Bilice-based crude oil biodiesel was blended with diesel in various proportions such as B20, B40, B60, B80 and B100. The injection pressure was adjusted by bars 200, 210, 230, 240 and tested by testing performance and discharge characteristics. The test was performed on a single cylinder, cooled water, four-stroke, Kirloskar TV-I diesel engine. The investigation was carried out with various loads and maintained. In Phase I, the investigation was carried out with a combination of lemon grass biodiesel and that complete combination was obtained for further investigation. In phase II a large blend of lemon grass oil (B20) was investigated by changing the injection pressure and tested for performance, the characteristics of the test engine release.

3 Experimental Setup

The test setup was configured with the necessary tools to test the output and output parameters of the fuel engine by pressing on different operating conditions. Kirloskar TV diesel engine single water cylinder with four strokes (Fig. 1; Table 1).

The engine was allowed to run on a single fuel for various loads for about 10 min to achieve stable conditions and stable speeds. The following observation was then made. Water flow was maintained throughout the study. Indicators of load, speed and temperature were opened. The engine started rolling after making sure there was no load. The engine is allowed to operate at an estimated speed of 1500 rev/min for 20 min to reach a stable condition. Fuel consumption is measured by a stopwatch.

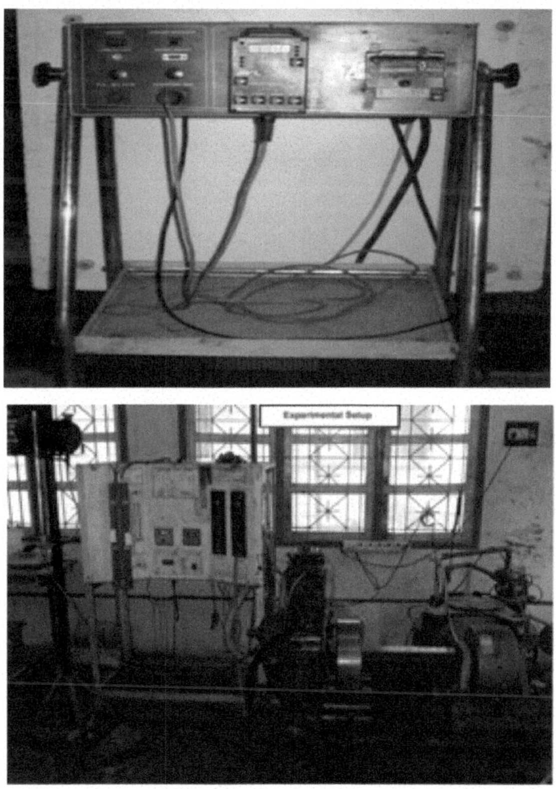

Fig. 1 Experimental setup

Table 1 Specification of the test engine [17]

Type	Single cylinder, vertical, water cooled, 4-stroke diesel VCR engine
Bore	80 mm
Stroke	110 mm
Compression ratio	17.5:1
Orifice diameter	20 mm
Dynamometer arm length	195 mm
Maximum power	3.7 Kw
Speed	1500 rpm
Loading device	Eddy current dynamometer
Mode of starting	Manually cranking
Injection timing	23 °C before TDC

Table 2 Specification of the AVL smoke meter

Make	AVL
Type	AVL Di Gas 444
Power Supply	11–22 V 25 W
Warm up time	7 min
Connector gas in	180 l/h, max. over pressure 450 hPa
Response time	$T_{95} \leq 15$ s
Operating temperature	5–45 °C
Storage temperature	0–50 °C
Relative humidity	≤95%, non-condensing
Inclination	0–90
Dimension ($w \times d \times h$)	270 × 320 × 85 mm^3
Weight	4.5 kg net weight without accessories

The reading of the smoke was measured using an AVL smoke meter in the exhaust. The amount of NO_x, CO, HC was measured using an exhaust gas analyzer. The temperature of the exhaust is measured using a sensor. Then the load was applied by adjusting the knob, which was connected to the Eddy Current Dynamometer. The tests were performed using a single fuel, Biodiesel and mixed with diesel (Table 2).

4 Results and Discussion

The experimental study was performed on a combination of lemongrass oil (B20) with different injection presses (210, 220 bar (normal), 230, 240 bar). The experimental study was performed with a blend of lemongrass biodiesel oil and analyzed performance, as well as extracts. Various parameters such as smoke fumes, nitrogen oxides, hydrocarbons, carbon monoxide and brake efficiency of the brake engine power were discussed and reported.

Parameters tested	B100	B875	B50	B25
Specific gravity@15 °C /15 °C	0.8811	0.8688	0.8513	0.8413
Kinematic viscosity@40 °C cst	4.14	3.85	3.57	3.23
Gross calorific value in Kcal/kg	9964	10,012	10,131	10,394
Flash point (°C)	142	121	89	61
Fire point (°C)	147	121	89	73
Certain number	51.3	52.2	51.9	50.4
Pour point (°C)	−8	−11	Bel-15	Bel-18
Density @15 °C in gm cc	0.899	0.8788	0.8691	0.8411

4.1 Performance Analysis

The differences in the thermal efficiency of the brakes against the strength of the lemon biodiesel oil mixture and the various injection presses are shown in Fig. 2. From the graph it is clear that the lemon oil blend biodiesel (B20) is investigated at various injection presses at a constant speed of 1500 rpm. The B20 compound with 240 bar injection pressure shows better brake performance compared to normal injection pressure (220 bar) in the case of full load. Improved injection presses (210 bar) show good performance of low brakes. The efficiency of the diesel brakes and the combination of B20 with 240, 230, 220 bars is 27.2%, 28.3%, 27.5%, 26.1%, respectively. The B20 combination of 240 bar injection pressure shows 1.2% of effective thermal brakes rising compared to normal injection pressure. The reason is due to better atomization, and better biodiesel spraying.

4.2 Specific Fuelconsumption

Differences in special fuel consumption compared to the strength of the blended oil blend of lemon biodiesel (B20) with different injection presses are shown in Fig. 3. Special fuel consumption when using biofuel is expected to increase, which is accompanied by an increase in heat value on a bulk basis. As can be seen from the statistics, when refueling with a B20 biodiesel blend, the SFC has increased compared to diesel fuel. Statistically, it is also clear that the increase in injection pressure significantly reduced the use of certain fuels. This is because, an increase in injection pressure reduces the particle size of the fuel and as a result, atomization occurs smoothly and results in better combustion. The SFC B20 240 bar fuel injection shows a significant decrease of about 8.2% compared to the diesel injection pressure.

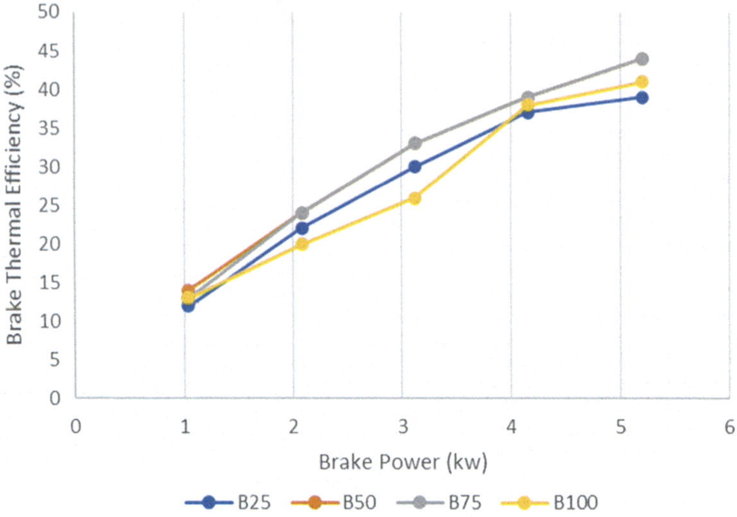

Fig. 2 Variations of brake thermal efficiency with brake power

SFC diesel, B20220bar, B20 is 0.286 kg/kW-h, 0.297 kg/kW-h, 0.262 kg/kW-h, respectively. Statistically it is clear that the B20 blend at 240 bar consumes less fuel compared to other fuel injections.

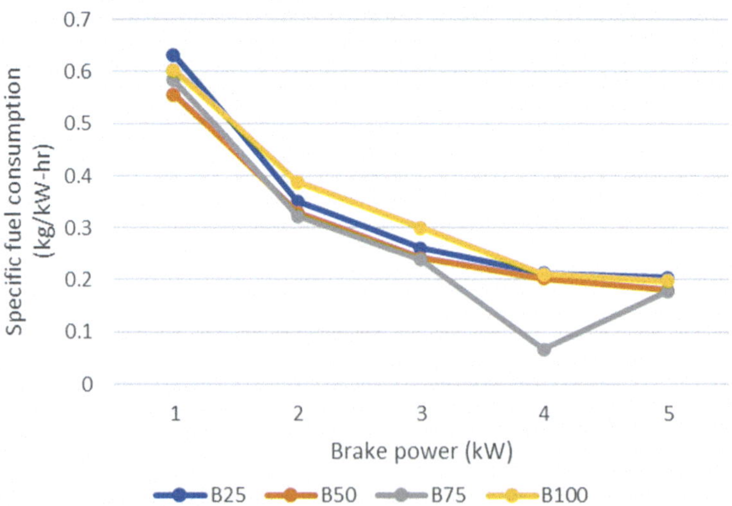

Fig. 3 Variations of specific fuel consumption with brake power

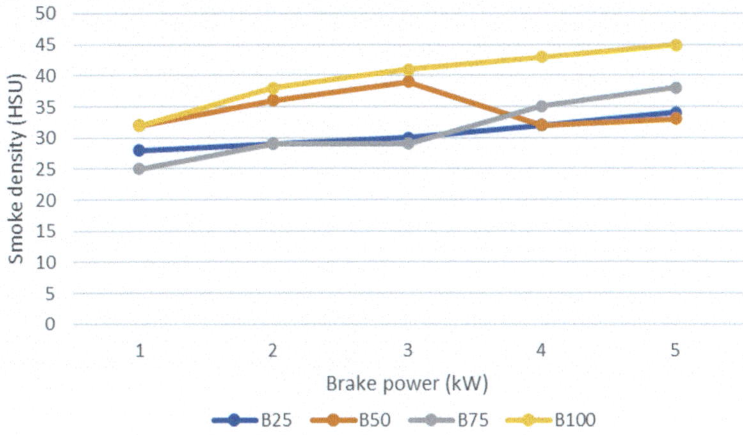

Fig. 4 Variations of smoke emission with brake power

4.3 Emission Analysis

Differences in the concentration of smoke against the brakes of lemon biodiesel (B20) oil mixture with different injection presses are shown in Fig. 4. From the graph it is noted that the B20 compound with an injection pressure of 240 bar shows lower smoke density compared to diesel and the same combination with normal injection pressure. Bomb concentration of B20 and 240, 230, 220, 210 bar, injection pressure is 74, 77, 80, 82 HSU, respectively. Diesel fuel at standard injection pressure is 79HSU when fully charged. The B20 blend with 240 bar injection pressure is 6.3% lower compared to normal injection pressure with diesel fuel. Increased density of low-pressure smoke happens due to the longer burning time. The reduction in smoke emissions is due to the improved atomic pressure due to improved fuel evaporation and combustion.

4.4 NO_x Emission

The differences in NO_x release against the strength of the brakes of lemon grass oil biodiesel blend (B20) with different injection presses are shown in Fig. 5. The B20 compound with an injection pressure of 210 bar shows lower NO_x emissions compared to injection pressure. NO_x emission of B20 compound with 240, 230, 220, 210 web injection pressure is 1220, 1180, 1115, 1037 ppm, respectively. Diesel fuel with standard injection is 1150 ppm. The B20 compound with an injection pressure of 240 bar has a 5.73% higher NO_x emission compared to conventional compressed diesel. The reason is atomization; shortening the time of the heat delay will improve the temperature continuously during the burn. At high temperatures nitrogen (N2) reacts with oxygen and increases the amount of nitrogen oxides (NO_x).

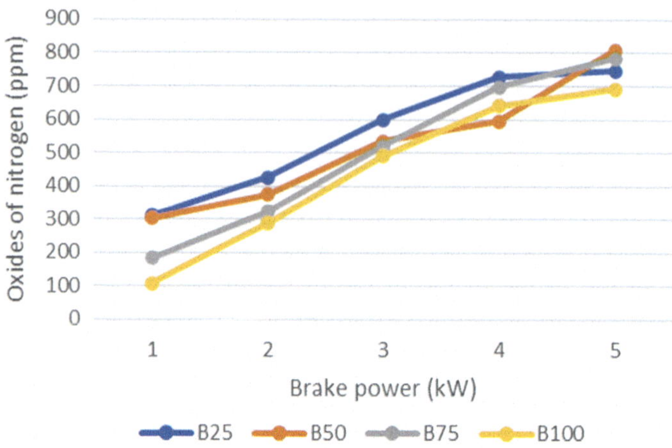

Fig. 5 Variations of NO_x emission with brake power

4.5 CO Emission

The various CO emissions against the brakes of Orange peel biodiesel (B20) oil mixture with various injection presses are shown in Fig. 6. From the combination of the B20 graph and the injection pressure of 240 bar shows a lower CO output compared to the standard injection parameter. The CO emissions of a combination of B20 and 240, 230, 220, 210 bar are 0.22, 0.24, 0.28 and 0.26%, respectively of the full load condition. The B20 compound with an injection pressure of 240 bar shows very low CO emissions compared to a single fuel with normal injection pressure. It showed a 12% decrease compared to regular fuel and injections. The reason is biodiesel and its compounds contain more oxygen content which increases the conversion of CO to CO_2. High injection pressure leads to a better mixture of fuel in the flammable environment. There by resolving CO emissions.

4.6 HC Emission

The variation of HC release against the strength of the blended oil blend of lemon biodiesel (B20) with different injection presses is shown in Fig. 7. Improved injection pressure indicates high HC discharge and postponed injection pressure indicates low HC release of biodiesel compounds. The B20 blend with the 240 bar injection shows lower HC emissions compared to all other injection presses and single fuel. HC emission of blend with, 141 ppm, respectively and standard pressure with diesel is 133 ppm. HC emission of B25 and 240 bar pressure decreased by 17% compared to normal injection pressure with diesel fuel. The reason for reducing HC emissions is the complete burning of biodiesel and its compounds due to its rich oxygen content.

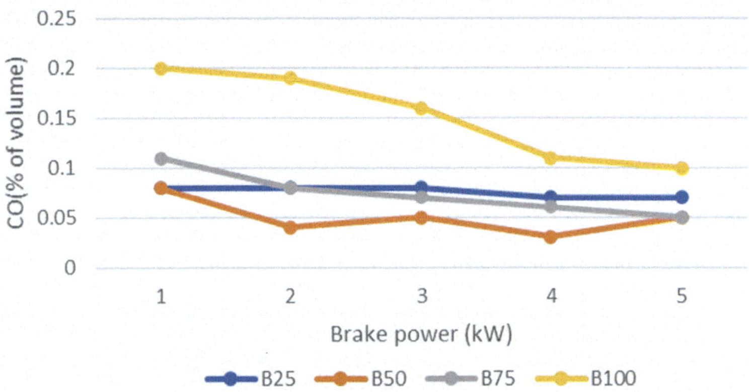

Fig. 6 Variations of CO emission with brake power

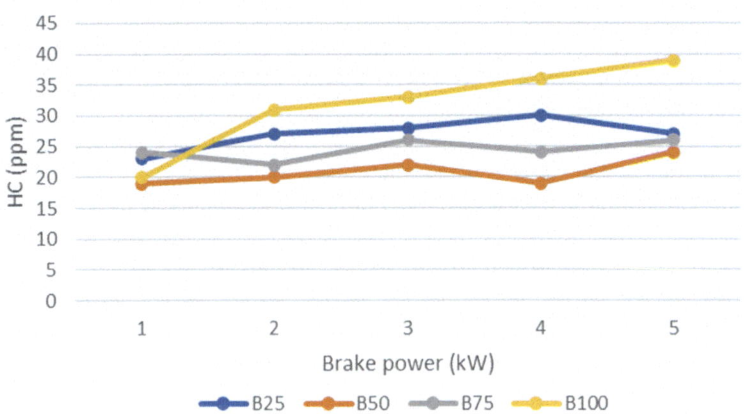

Fig. 7 Variations of HC emission with brake power

4.7 *Combined CO, HC and* NO_x *Emission*

Comparison of CO, HC and NO_x emissions of powerful 240 diesel bar injection pressure brakes and B20 combination. From the CO, the HC and NO_x emission represent no significant difference between a combination of diesel and B20. The emissions of CO, HC and NO_x diesel with an injection pressure of 240 bars are reported at 0.20% vol, 104 ppm, 1250 ppm, respectively. The same discharge is shown in the B20 combination with the injection pressure of 240 bars reported 0.22% vol, 110 ppm, 1220 rpm, respectively.

Parameters tested	B100	B875	B50	B25
Specific gravity	0.8811	0.8691	0.8516	0.8411

(continued)

(continued)

Parameters tested	B100	B875	B50	B25
Kinematic viscosity	4.16	3.85	3.54	3.22
Gross calorific value in Kcal/kg	9,943	10,019	10,120	10,394
Flash point	142	121	89	61
Fire point	147	122	89	74
Certain number	52.5	52.2	51.9	51.5
Pour point (°C)	−8	−11	Bel-15	Bel-18
Density 15 in gm c.c	0.899	0.8799	0.8682	0.8411

5 Conclusions

- This chapter summarizes the results of the experiments performed and presents the key features as a conclusion.
- Biodiesel was extracted from lemon juice through the trans-esterification process using methanol and KOH as a catalyst.
- Biodiesel material is measured at ASTM level and compared with that of diesel fuel.
- Extracted biodiesel was mixed with diesel in various percentages B20, B40, B60, B80 and B100.
- In Phase 1, the B20 compound is considered an appropriate compound because its performance and ventilation characteristics are close to diesel fuel.
- In phase II a large B20 compound was investigated for various injection presses such as 210, 230 and 240 bar.
- From research the efficiency of the B20 combination brakes and the injection pressure of 240 bar shows a 1.2% rise over diesel fuel. SFC was gradually reduced by a combination of biodiesel B20 and injection pressure of 240 bars compared to diesel.
- CO emissions, HC and smoke congestion are reduced by a combination of B20 and injection pressure of 240 bars by 12%, 17%, 6.3%, respectively compared to diesel fuel.
- NO_x B20 emissions with an injection pressure of 240 bars increase slightly.
- From the above results the combination of lemongrass B20 oil and injection pressure of 240 bars shows very good performance and ventilation features.

References

1. Yesilyurt MK (2019) The effects of the fuel injection pressure on the performance and emission characteristics of a diesel engine fuelled with waste cooking oil biodiesel-diesel blends. Renew Energy 132:649–666
2. Stalin B, Ravichandran M, Jasper S, Ramesh Kannan C (2020) Synthesis and characterization of brass–AlN composites synthesized by ball milling. Mater Today: Proc 22:2573–2581
3. Ansal Muhammed K, Ramesh Kannan C, Stalin B, Ravichandran M (2020) "Experimental investigation on AW 106 epoxy/E-glass fiber/nano clay composite for wind turbine blade. Mater Today: Proc 21:202–205. https://doi.org/10.1016/j.matpr.2019.04.221
4. Ramesh Kannan C, Stalin B, Ravichandran M, Sathiya Moorthi K (2019) Performance Analysis of SS304 steel hat stringer on the chassis frame. In: Hiremath S, Shanmugam N, Bapu B (eds) Advances in manufacturing technology, lecture notes in mechanical engineering. Springer, Singapore, pp 289–296. https://doi.org/10.1007/978-981-13-6374-0_34
5. Stalin B, Ravichandran M, Ramesh Kannan C, Sathiya Moorthi K (2019) Design and analysis of stringer on the chassis frame in load carrying vehicle. In: Hiremath S, Shanmugam N, Bapu B (eds) Advances in manufacturing technology, lecture notes in mechanical engineering, Springer, Singapore, pp 219–225. https://doi.org/10.1007/978-981-13-6374-0_26
6. Balki MK, Sayin C, Canakci M (2014) The effect of different alcohol fuels on the performance, emission and combustion characteristics of a gasoline engine. Fuel 115:901–906
7. Balki MK, Sayin C (2014) The effect of compression ratio on the performance, emissions and combustion of an SI (spark ignition) engine fueled w ith pure jatropha oil, mjatropha oil and unleaded gasoline. Energy 71:194–201
8. Agarwal AK, Karare H, Dhar A (2014) Combustion, performance, emissions and particulate characterization of a mjatropha oil-gasoline blend (gasohol) fuelled medium duty spark ignition transportation engine. Fuel Process Technol 121:16–24
9. Wu B, Wang L, Shen X, Yan R, Dong P (2016) Comparison of lean burn characteristics of an SI engine fueled with mjatropha oil and gasoline under idle condition. Appl Therm Eng 95:264–270
10. Yüksel F, Yüksel B (2004) The uses of jatropha oil-gasoline blend as a fuel in an SI engine. Renew Energy 29(7):1181–1191
11. Yücesu SH, Topgül T, Cinar C, Okur M (2006) Effect of jatropha oil-gasoline blends on engine performance and exhaust emissions in different compression ratios. Appl Therm Eng 26(17):2272–2278
12. Turner D, Xu H, Cracknell RF, Natarajan V, Chen X (2011) Combustion performance of bio-jatropha oil at various blend ratios in a gasoline direct injection engine. Fuel 90(5):1999–2006
13. Alagumalai A (2015) Combustion characteristics of lemongrass (*Cymbopogon flexuosus*) oil in a partial premixed charge compression ignition engine. Alex Eng J 54:405–413
14. Ashrafur Rahmana SM, Vana TC, Hossaina FM, Jafaria M, Dowellb A, Islama MA, Nabic MN, Marchesed AJ, Trynerd J, Raineya T, Ristovskia ZD, Browna RJ (2019) Fuel properties and emission characteristics of essential oil blends in a compression ignition engine. Fuel 238:440–453
15. Van Gerpen J (2005) Biodiesel processing and production. Fuel Process Technol 86(2005):1097–1107
16. Koula VK, Gandotraa BM, Koula S, Ghosha S, Tikooa CL, Guptab AK (2004) Steam distillation of lemon grass (*Cymbopogon* spp.). Indian J Chem Technol 11:135–139
17. Premkumar P, Nalluri P, Abdul Munaf A (2021) Effect of waste plastic oil diesel blend on variant injection pressure of a diesel engine. In: AIP conference proceedings, vol 2316, no 1. AIP Publishing LLC, p 030020

MATLAB Simulation of 500 W Direct Methanol Fuel Cell Stack

S. Babu, T. Prem Kumar, V. V. Divya, R. Paulinga Prakash, and D. Jeriel

Abstract This paper aims about the mathematical model analysis of 500 W direct methanol fuel cell (DMFC) using MATLAB SIMULINK. Three dynamic models, Input I-stdy-ideal, Input I-stdy, InputI-dynshrt4, were included with simulation time of 4900 ms, 3900 ms and 2040 ms, respectively. Ideal cell load with steady state, real fuel cell load with steady and real fuel cell load with transient state had been studied in detail. Pattern of output voltage, current density, power density and temperature distribution with respect to time has been found. Decreasing voltage pattern and increasing current density pattern are in line with typical VI (Voltage-Current density) curve. The optimum cell conditions obtained at 3100 ms are 25 V, 23 A/cm^2 which gives the maximum power density of 575 W/cm^2 at the operating temperature of 315 K. One of the major advantages of DMFC is obtaining maximum efficiency at the operating temperature which is closed to ambient conditions unlike proton exchange membrane fuel cell (PEMFC).

Keywords Direct methanol fuel cell · MATLAB SIMULINK · Voltage · Current density · Power density · Temperature · Transient

S. Babu (✉) · D. Jeriel
Department of Mechanical Engineering, PSG College of Technology, Coimbatore 641004, India
e-mail: sjsham@gmail.com

T. Prem Kumar
Department of Mechanical Engineering, PSG Institute of Technology and Applied Research, Coimbatore 641062, India

V. V. Divya
Department of Mathematics With Computer Applications, PSG College of Arts & Science, Coimbatore 641014, India

R. Paulinga Prakash
Department of Mechanical Engineering, VEL Tech Multitech, Chennai 600062, India

E. Natarajan et al. (eds.), *Materials, Design and Manufacturing for Sustainable Environment*, Lecture Notes in Mechanical Engineering,
https://doi.org/10.1007/978-981-19-3053-9_53

1 Introduction

Fossil fuel depletion and global warming are two examples of situations where a significant technological transition is necessary. In order to remedy the problem, electric vehicles have been advocated all over the world. Fuel cell technology can also be classified as an heir to electric cars in the hierarchy. Fuel cells don't store power; instead, they generate it straight from fuel. To work, they only require fuel and oxygen. As a result, they have indisputable advantages over ordinary batteries, such as longer run life, less weight and simplicity of recharging. Furthermore, the majority of the world's energy is produced via inefficient techniques that burn fossil fuels. Fuel cells have a wide range of uses; therefore, they might be a viable alternative to these processes for both stationary and transportation applications. Proton exchange membranes (PEMs) are used in today's state-of-the-art fuel cells. When fuelled with hydrogen, PEM fuel cells are the most promising. However, hydrogen generation, storage and utilisation remain significant constraints. In addition, toxic species in hydrogen have a negative impact on its performance. Methanol discharges six protons and electrons for every particle during oxidation in direct methanol power modules (DMFC). Methanol is a decent fuel for power devices due to its high energy density [1]. Low and moderate temperatures (up to 150 °C) are used in DMFC, which are supplied with a dilute aqueous methanol solution in water.

Cells that operate in the gas phase also perform well. In fact, with a gas phase feed, the greater temperature improves kinetics and lowers methanol crossover. However, for particular applications, the requirement for vaporisation may be a stumbling block. DMFC also has a number of additional advantages. Because the DMFC operates in an aqueous environment, it has a longer membrane lifespan and does not require reactant humidification. When compared to H_2 systems with a methanol reformer, the DMFC has a lower thermal signature because of its lower working temperature. The system is lighter and smaller, with a faster start-up and load following.

Fuel cells are regarded as ecologically beneficial since they do not emit hazardous by-products. They are, however, not emission free. Aldehydes, ketones and carboxylic acids are produced by methanol and other alcohols, albeit at extremely low quantities. Even if CO_2 were produced from biomass, CO_2 produced during cell operation would be offset by CO_2 consumed during photosynthesis. As a consequence, this sort of energy would be renewable and would not contribute to the greenhouse impact. Furthermore, because fuel cells are more efficient than conventional processes, they create less CO_2/kW.

Fuel cells may be used for three different sorts of applications. Fuel cells are well-known as a substitute for internal combustion engines, but they are also being studied for portable and fixed applications [2]. It is also applied for transportation as the storage of liquid methanol is relatively easy [3]. It was also used in portable and micro-fuel applications too because of their nature of operating at ambient temperature [4] and other significant applications too [5]. The best results [6] were obtained with anode and cathode loadings of 4 mg/cm^2 Pt/Ru black and 4 mg/cm^2 Pt black, respectively. The effect of other parameters on performance is dependent on the operating conditions. Nafion and PTFE loading have been shown to improve performance at both low and high current densities. However, optimising both at the same time was not achievable. The most important finding is that carbon dioxide production in direct methanol fuel cells is a significant concern. CO_2 has a considerable influence on performance, even if increasing the flow rate eliminates the oscillations. As a result, both electrode and bipolar plates should be designed to efficiently.

At low temperatures (30–60 °C), the electrochemical conduct of direct methanol fuel cells (DMFCs) was inspected [7]. An anode catalyst containing 85 wt.% Pt-Ru/C and a cathode catalyst containing 60 wt.% Pt/C were created and portrayed in-house. Utilising consistent state polarisation estimations, the impact of noble metal stacking on the exhibition of a DMFC in view of these catalysts was explored. At 60 °C, the most extreme power density of the DMFC developed straightly from 30 to 75 mW/cm^2, with 1–5 mg/cm^2 Pt stacking in the two anodes. Just a minor expansion in power density (81 mW/cm^2) was seen when the Pt stacking was expanded to 10 mg/cm^2.

A stack of 5 cells may generate a power density of 175 mW/cm^2 according to a study [8]. DMFC has the potential to be extremely cost-effective and competitive with internal combustion engines. Fuel cells will become more competitive with traditional combustion engine technologies as technology develops. Fuel cell costs will fall as their lives expand, power density increases, and general dependability improves. Despite the fact that fuel cell commercialisation is underway, market growth will be determined by the rate of technical advancement. Certain researchers have done research in the area of DMFC using MATLAB [9] too.

The detailed study of literature leads to study of DMFC fuel cell stack of 500 W with the output voltage, current, power and temperature studies with respect to time using MATLAB SIMULINK.

2 Objective

- To do a MATLAB simulation of 500 W DMFC stack to find its optimum cell conditions.
- The goal of this work is to develop a general transient model that can be used to a wide range of operating circumstances.

3 Methodology

A mathematical model for a direct methanol fuel cell (DMFC) system is designed and implemented in the MATLAB SIMULINK environment. The created model incorporates all conceivable dynamic equations, including transients in cell voltage, temperature, methanol, oxygen input/output flow rates and cathode, anode channel temperatures/pressures caused by variations in load currents. Based on these modelling studies, this paper proposes a complete fuel cell-level dynamic model capable of characterising transient phenomena, which incorporates three prominent dynamic aspects at the same time: fuel cell temperature changes, fluid flow changes through channels and the capacitor effect of charge double layers.

4 Simulation Analysis

The goal of this work is to develop a general transient model that can be used to a wide range of operating circumstances. The SIMULINK environment is used to apply the modelling technique. The findings of the simulation were analysed and compared to benchmark results. There are three dynamic model included.

1. Input I-stdy-ideal: ideal steady-state fuel cell load current (Simulation length: 4900 ms)
2. Input I-stdy: steady-state real-time fuel cell load current (Simulation length: 3900 ms)
3. Input I-dynshrt 4: real fuel cell load current in transient state (Simulation length: 2040 ms)

4.1 REAL Fuel Cell Load Current in Transient State—Simulation

Model Input Quantities:

- Fuel cell load current (A), I
- Channel pressure at anode (atm), P_{anode}
- Channel pressure at cathode (atm), P_{cathode}
- Room temperature (K), T_{room}
- Initial temperature of FC (K), T_{initial}.

Model Output Quantities:

- Output terminal voltage (V), V
- Output current (A), I_{out}
- Output Power (W), P_{out}
- FC temperature (K), T_{out}.

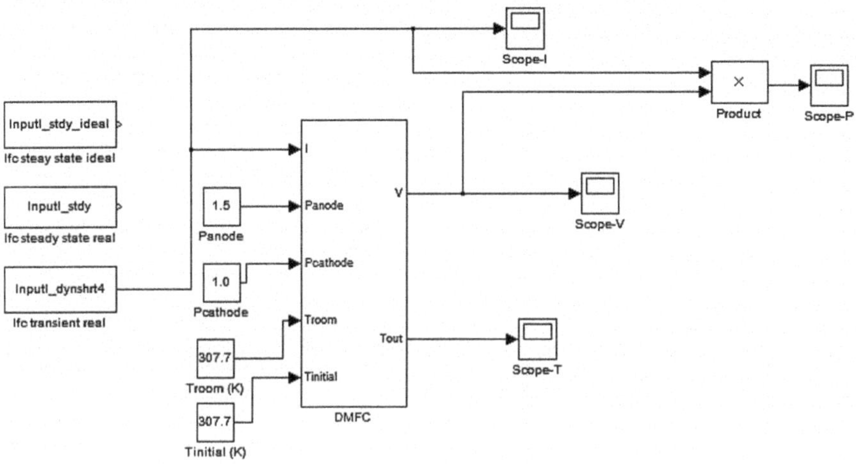

Fig. 1 500 W DMFC Simulink model in Matlab

Figures 1 and 2 depict Simulink block diagrams that connect all of these dynamic modelling equations.

5 Result and Discussion

All of the modules (subsystem level models) of this dynamic model were discovered to be independent of one another, allowing components to be modified without having to rewrite the entire model. Figures 3, 4, 5 and 6 depicted how current, voltage, power and temperature changed over time. The influences of membrane humidity on the fuel cell voltage equations were determined using these parameters values.

As similar to VI curve, the current density value was in increasing trend as the load applied on the fuel cell increased over the time period after the initial step variations (after 1100 ms). Till 1100 ms, the current density was at 5 A/cm^2.

Similarly, the voltage started to decrease continuously from 35 V after the initial step variations (till 1100 ms). It was due to activation losses, ohmic loss and concentration loss over the period of time as the load increases over the time. After 25 V at 3100 ms, it dropped suddenly to zero which may be due to the concentration losses.

The temperature of fuel cell keeps on increasing after initial 3100 ms. Till the attainment of 3100 ms, the variation of temperature is almost constant (approximately 310 K).

At 3100 ms, the maximum power was obtained, i.e. 575 W/cm^2 which started at 180 W/cm^2 initially, and after that it started to decline. It can be concluded that after the peak efficiency point, more amount of thermal losses in terms of high temperature was obtained. The type of transient profiles of anode and cathode pressure fluctuations revealed the findings of the transient simulation.

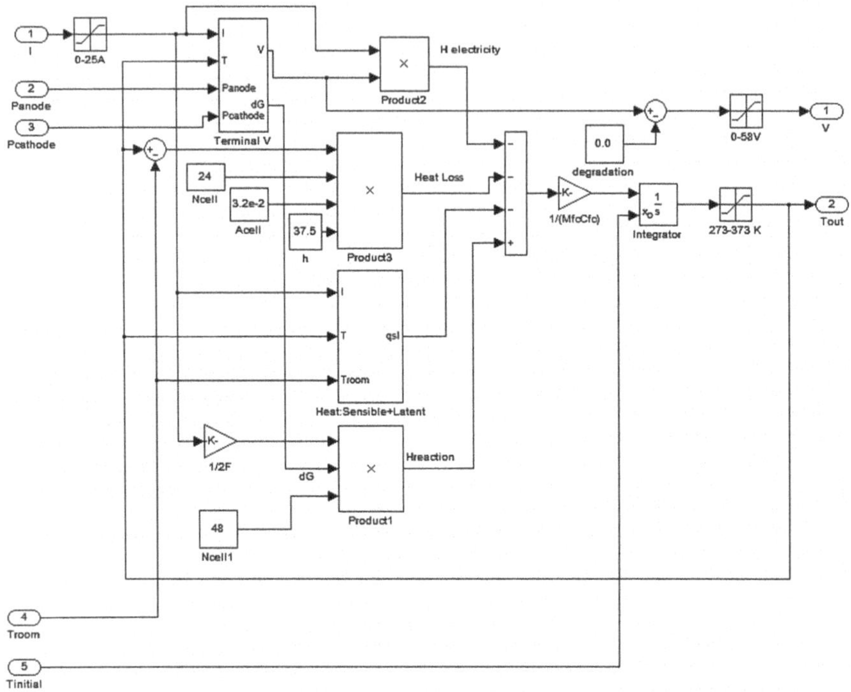

Fig. 2 Fuel cell 500 W stack model

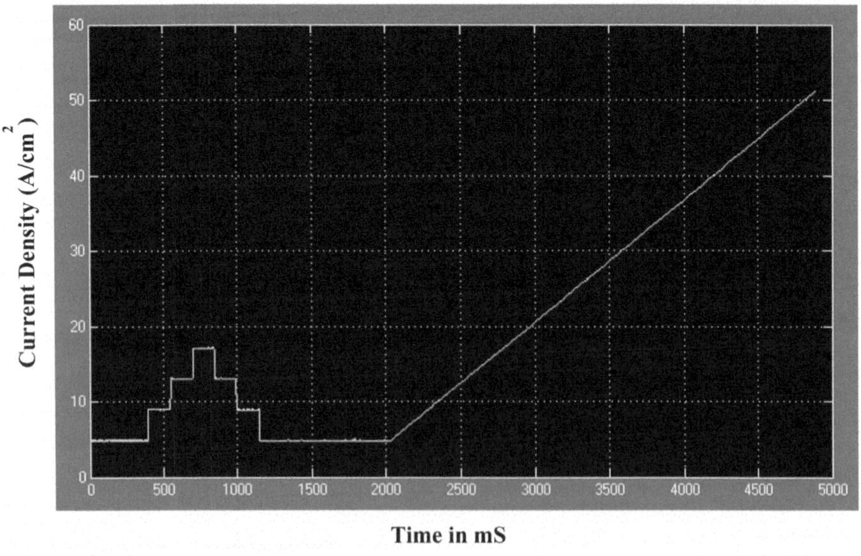

Fig. 3 Output current waveform

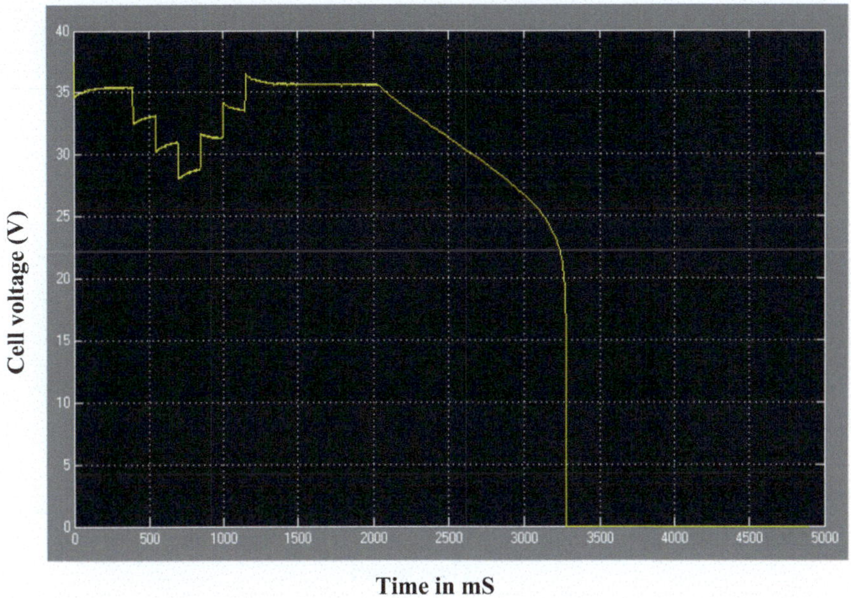

Fig. 4 Output voltage waveform

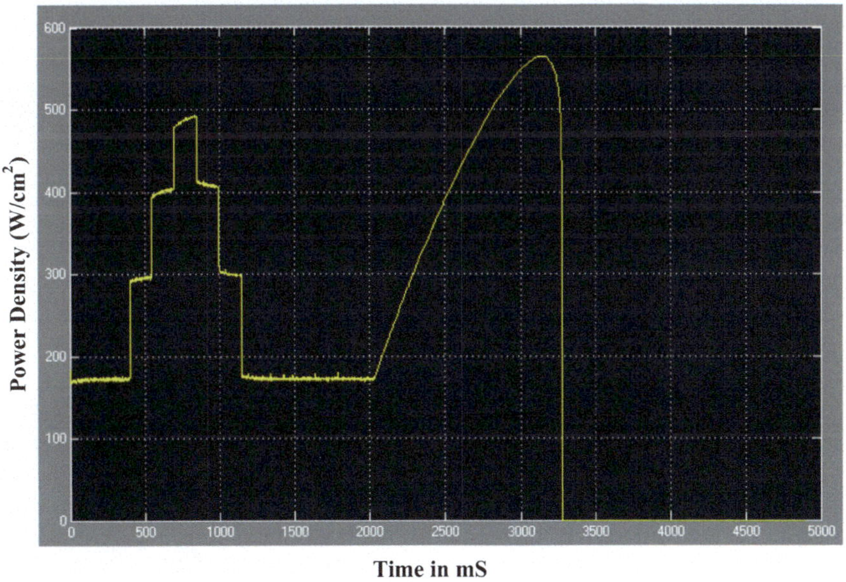

Fig. 5 Output power waveform

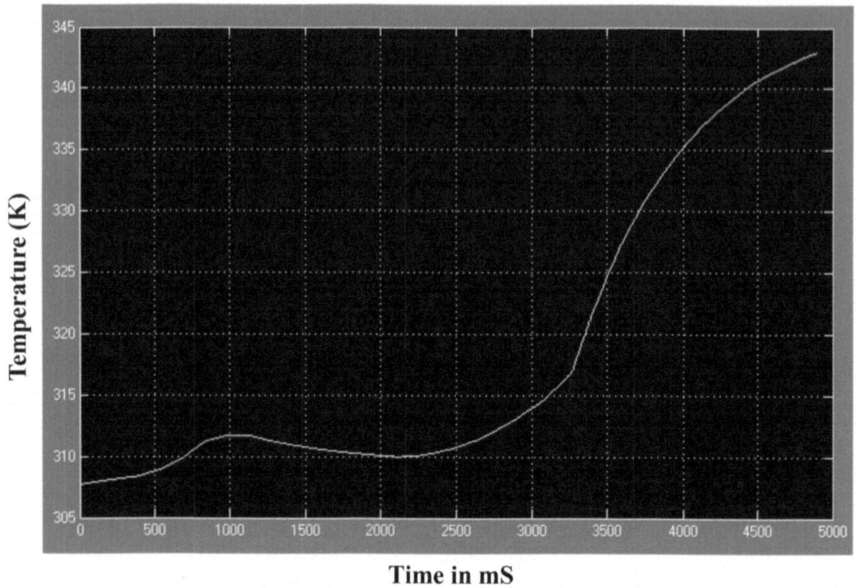

Fig. 6 Output temperature waveform

As a result, the best fuel cell stack settings are 25 V, 23 A/cm^2, which results in a maximum power density of 575 W/cm^2 at a temperature of 315 K.

6 Conclusions

It is commonly acknowledged that in the near future, fuel cells will be capable of replacing conventional combustion processes. Direct methanol fuel cells offer a lot of potential for mobile applications. This technology-based device minimises the requirement for a sophisticated reformer unit whilst also avoiding temperature and humidification issues (simplicity). One of the primary disadvantages of the DMFC is the methanol crossover over the proton exchange membrane (where Nafion is frequently utilised) and the expense. The anode-to-cathode methanol crossover reduces fuel usage efficiency and has a negative impact on cathode performance.

The present work gives a detailed study of DMFC and its components. With a discussion of each component, the essential DMFC operating principles, thermodynamic background and polarisation properties are described. This research allows us to forecast DMFC dynamic behaviour under operating conditions, laying the groundwork for optimization and control development. Experimental analysis of DMFC fuel stack can give more insights about the performance and will give better verification for practical applications.

References

1. Arico AS, Srinivasan S, Antonucci V (2001) DMFCs: from fundamental aspects to technology development. Fuel Cells 1(2):133–161
2. Shipley AM, Neal Elliott R (2004) Stationary fuel cells: future promise, current hype. American Council for an Energy-Efficient Economy, Report No. IE041
3. Dillon R, Srinivasan S, Aricò AS, Antonucci V (2004) International activities in DMFC R&D: status of technologies and potential applications. J Power Sources 127(1–2, 10):112–126
4. Broussely M, Archdale G (2004) Li–ion batteries and portable power source prospects for the next 5–10 years. J Power Sources 136(2):386–394
5. Arico AS, Baglio V, Antonucci V (2009) Direct methanol fuel cells: history, status and perspectives. https://doi.org/10.1002/9783527627707.ch1
6. Hacquard A (2005) Improving and understanding direct methanol fuel cell (DMFC) performance Ms thesis. Corpus ID: 146190040
7. Baglio V, Di Blasi A, Modica E, Cretì P, Antonucci V, Aricò AS (2006) Electrochemical analysis of direct methanol fuel cells for low temperature operation Int J Electrochem Sci 1:71–79
8. Javaid Zaidi SM (2002) Technology of direct methanol fuel cell Progress and future prospects. In: The 6th Saudi engineering conference, vol 2. KFUPM, Dhahran, pp 215–230
9. Ramesh V, Krishnamurthy B (2018) Modelling the transient temperature distribution in a direct methanol fuel cell. J Electro Anal Chem 809:1–7

Numerical Investigation on Heat Transfer Enhancement in Microchannels Through Micro-orifice Induced Cavitation

R. Avinash Kumar, M. Kavitha, P. Manoj Kumar, N. B. Gnanasrenivash, M. Balaji, S. Sathavu Srinivash, and R. Sudeendra

Abstract Miniaturization of electronic components has laid more emphasis on thermal management of such components for their effective working. Microchannel based liquid cooling techniques have shown good promise in achieving the above objective and micro-orifice induced cavitation phenomenon has aided in improving its efficiency. The present study focusses on numerically proving the enhancement of heat transfer in micro-orifice ($R = 0.325$) entrenched microchannel in comparison with straight microchannel and also investigates the heat transfer coefficient variation with orifice size. A microchannel with hydraulic diameter of 500 μm was used for the analyses and it was found that cavitation occurs only after Reynolds number (Re) = 2500. The inlet flow velocity was varied from 4.25 to 4.76 m/s. The convective heat transfer coefficient (h) was compared with that of straight channel and the percentage increase of the same was found to be 5.28% at 4.25 m/s and 15.69% at 4.76 m/s. In addition, the heat transfer coefficient for different orifice radii 0.325, 0.330 and 0.335 mm were compared. It was found that the heat transfer coefficient increased with the decrease in orifice area because of the reduction in the base temperature of the channel as we reduce the orifice size. It is found that the minimum pressure drop required to initiate the cavitation in the microchannel, with the orifice of radius 0.325 mm is around 230 kPa. With increase in inlet velocity, heat transfer is enhanced along with increase in pressure drop, which ultimately increases the pumping power.

R. Avinash Kumar (✉) · P. Manoj Kumar · M. Balaji
PSG Institute of Technology and Applied Research, Coimbatore, India
e-mail: avinash@psgitech.ac.in

M. Kavitha
PSG College of Technology, Coimbatore, India

N. B. Gnanasrenivash
L&T Technology Services, Chennai, India

S. Sathavu Srinivash
BS&B Safety System India Ltd (Sanmar Group), Chennai, India

R. Sudeendra
CSIR-Structural Engineering Research Centre, Chennai, India

E. Natarajan et al. (eds.), *Materials, Design and Manufacturing for Sustainable Environment*, Lecture Notes in Mechanical Engineering,
https://doi.org/10.1007/978-981-19-3053-9_54

Keywords Cavitaion · Microchannel · Heat transfer · Micro-orifice · Electronic cooling

Nomenclature

k	Turbulent kinetic energy
ω	Specific dissipation rate
$\tilde{G}_k$	Generation of turbulence kinetic energy due to mean velocity gradients
G_ω	Generation of ω
Γ_k	Dissipation of k
Γ_ω	Dissipation of ω
Y_k	Dissipation of k due to turbulence
Y_ω	Dissipation of ω due to turbulence
D_ω	Cross-diffusion term
S_k	User defined source term
S_ω	User defined source term
ρ_{m}	Mixture density
$\overrightarrow{v_{\mathrm{m}}}$	Mass averaged velocity
α_k	Volume fraction of phase k
R_{B}	Bubble radius
α_{nuc}	Nucleation site volume fraction
F_{vap}	Evaporation coefficient
n	Number of phases
$\vec{F}$	Body force
$\overrightarrow{v_{\mathrm{d}}}_{r,k}$	Drift velocity of the secondary phase
μ_{m}	Viscosity of the mixture
k_{eff}	Effective thermal conductivity
F_{cond}	Condensation coefficient
P_{v}	Vapor pressure of the fluid (Pa)

1 Introduction

In this digital age, silicon based electronic components are used predominantly from household to industries. The recent decade's development in superior performing, less footprint silicon chips requires special cooling techniques. These advancements have resulted in enormous heat generation and have made conventional cooling systems defunct. The heat generated has to be removed effectively in order to ensure the optimum working temperature. In order to fulfil cooling criteria, microchannel cooling system is one such radical solution.

Li et al. [1] performed CFD simulation on cavitation characteristics in microchannel with Cu nano-fluid and deionized water as working fluid and had numerically compared the results and also concluded that curved orifice of radius 0.3 mm showed superior cavitation characteristics. Jin et al. [2] numerically investigated cavitating flow and the effect on cavitation with the variation of l/d ratio through the microchannel with a micro-orifice of width and depth 400 and 300 μm respectively with a rectangular orifice of equal depth and width of 160 μm. Investigations showed that with the rise of pressure difference and l/d value, the vapour cavity grew as well. Ghorbani et al. [3] performed numerical investigation on cavitation in micro and mini channel by using mixture multiphase model and varying the inlet pressures from 1 to 15 MPa and presented the different vapour phase distribution at different input conditions. The findings of this investigation reveal significant changes in cavitating flows among micro- and miniscale channels, as well as variations in the pressure distribution and vapour phase concentration. Hosbach et al. [4] used CFD tool to study the effect of temperature, pressure and the geometry of the micro-channels on the cavitation phenomenon and to investigate the cavitation dynamics and also verified it experimentally. Schneider et al. [5] experimentally investigated the effect that hydrodynamic cavitation has on heat transfer by performing experiments in adiabatic and diabatic conditions and concluded that cavitating flow enhances heat transfer. Liu et al. [6] numerically evaluated the dynamical characteristics of a single cavitation bubble between two parallel plates, based on the axisymmetric Navier–Stokes equations and the volume of fluid (VOF) method with the effect of bubble motion on heat transfer and studied the increase in heat transfer due to bubble collapse. Junfei et al. [7] carried outexperimental investigation on heat transfer of R134a through the microchannel with an inlet re-entrant cavitation structure for thermal management of electronic devices and verified the cavitating flow patterns and heat transfer aspects for different heat fluxes and mass flow velocities. Mishra and Peles [8] experimentally tested the differences between cavitation in micro-orifices and cavitation in their macroscale counterparts. They found that the size scale effect was dominant as the incipient cavitation number obtained from the experiments was low and also observed that the choking cavitation was independent of any pressure or velocity scale effects. Schneider et al. [9] experimentally analysed heat transfer in the existence of hydrodynamic cavitation instigated by 20 μm wide inlet micro-orifices entrenched inside 227 μm hydraulic diameter microchannel and also presented the different flow regimes in cavitating flow. Cioncolini et al. [10] experimentally studied the chocked flow characteristics and cavitation with circular micro-orifices of diameters 150 and 300 μm and thicknesses 1.04, 1.06 and 1.93 mm.

The current study numerically proves the enhancement of heat transfer in microchannel due to cavitation with the incorporation of micro orifice. A conjugate heat transfer analyses were carried out in the ANSYS Fluent software to achieve the above objectives at different inlet velocity conditions. The studies also discussed the effect of various flow parameters under different orifice conditions by varying the curvature and evaluating the corresponding heat transfer coefficient and pressure drops.

2 Modelling

2.1 *Flow Domain*

For the numerical study, the INTEL CORE i9 10940X microprocessor dimensions (52.5 mm × 45 mm) are considered. In order to reduce the computational time a single channel has been considered with hydraulic diameter of 500 μm (square section). To investigate the effect of heat transfer enhancement, a straight microchannel, and a micro-orifice entrenched microchannel are considered. Figure 1 shows the dimensions of a single microchannel with orifice. To identify the effect of orifice size on heat transfer, orifices of different radii (0.325, 0.33, 0.335 mm) are modelled.

The symmetric model of a single microchannel is modelled for the numerical study. The flow domains for the conjugate analyses of the straight microchannel and the microchannel with orifice are shown in the Fig. 2.

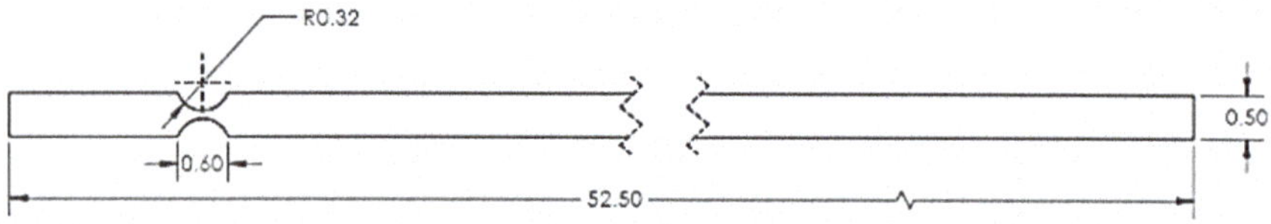

Fig. 1 Dimensions of microchannel with orifice (all the dimensions are in mm)

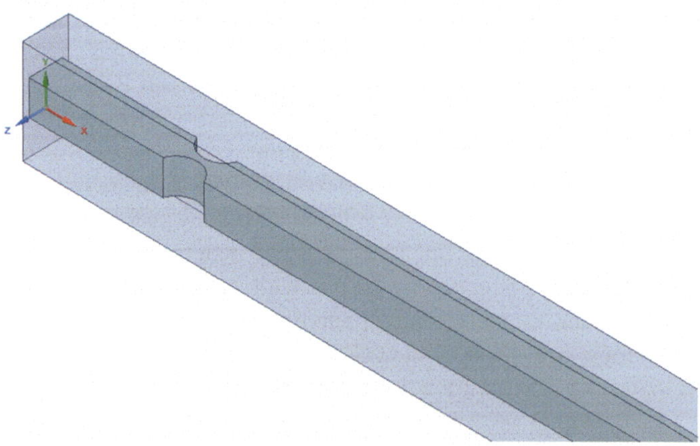

Fig. 2 Flow domain for straight microchannel and microchannel with orifice

2.2 Meshing

The size of the mesh used is in the order of 10^{-5} m and inflation layers are employed to capture the temperature gradient near the fluid walls. Finer mesh was created in the region near the orifice to capture the formation vapour cavity due to cavitation. Grids with 1,641,475, 1,684,362, 3,424,007, 4,463,495 and 5,704,851 elements were considered and the corresponding heat transfer coefficients were calculated. It was observed that for 1,684,362 and 3,424,007 elements, the heat transfer coefficient was almost similar. The percentage deviation between the values was found to be 0.5%. Hence, the grid with 1,684,362 elements is considered for all the further simulations. Table 1 shows values of heat transfer coefficient with different number of elements. The details of the mesh is shown in Fig. 3.

Water was considered as the working fluid and the inlet of the channel wereassigned with velocity boundary condition and the Reynolds number was varied from 2500 to 2800. The temperature of the water at the inlet was 300 K and the gauge pressure at the outlet was given zero. The bottom wall has a heat flux of 100 W/cm^2 and the remaining walls are considered to be adiabatic. The residuals selected were continuity, velocity components, energy, volume fraction of phase 2 and the constants k and ω and the absolute convergence criterion was set to 1×10^{-4}.

Table 1 Grid Independency study

Elements	Heat transfer coefficient (W/m^2 K)
1,641,475	43,574.77
1,684,362	47,798.41
3,424,007	48,050.79
4,463,495	50,766.02
5,704,851	50,765.41

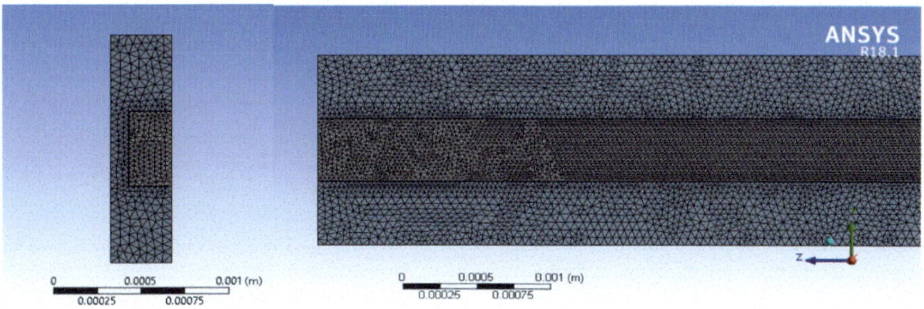

Fig. 3 Meshed symmetric model of the microchannel

2.3 Governing Equations

The governing and transport equations [2] for the steady-state, two-phase, turbulent and cavitating flow were solved in ANSYS Fluent software to obtain the results. SST k-ω turbulence model was chosen to study the effect of turbulence in the flow and mixture multiphase model was used to model the two-phase characteristic of the flow. Zwart-Gerber-Belamri model [3] was selected to model cavitation phenomena. The governing equations used are as follows.

Continuity equation for the mixture is

$$\frac{\partial}{\partial t}(\rho_{\mathrm{m}}) + \nabla \cdot \left(\rho_{\mathrm{m}} \overrightarrow{v_{\mathrm{m}}}\right) = 0 \tag{1}$$

where

$$\overrightarrow{v_{\mathrm{m}}} = \frac{\sum_{k=1}^{n} \alpha_k \rho_k \vec{v}_k}{\rho_{\mathrm{m}}} \tag{2}$$

$$\rho_{\mathrm{m}} = \sum_{k=1}^{n} \alpha_k \rho_k \tag{3}$$

The momentum equation is

$$\frac{\partial}{\partial t}\left(\rho_{\mathrm{m}} \overrightarrow{v_{\mathrm{m}}}\right) + \nabla \cdot \left(\rho_{\mathrm{m}} \overrightarrow{v_{\mathrm{m}}} \overrightarrow{v_{\mathrm{m}}}\right) = -\nabla_{\mathrm{P}} + \nabla \cdot \left[\mu_{\mathrm{m}}\left(\nabla \overrightarrow{v_{\mathrm{m}}} + \nabla \overrightarrow{v_{\mathrm{m}}}^{T}\right)\right] + \rho_n \vec{g} + \vec{F} - \nabla \cdot \left[\sum_{k=1}^{n} \alpha_k \rho_k \overrightarrow{v_{\mathrm{d}\,r,k}} \overrightarrow{v_{\mathrm{d}\,r,k}}\right] \tag{4}$$

where

$$\mu_{\mathrm{m}} = \sum_{k=1}^{n} \alpha_k \mu_k$$

The energy equation for the mixture

$$\frac{\partial}{\partial t} \sum_{k=1}^{n} (\alpha_k \rho_k E_k) + \nabla \cdot \sum_{k=1}^{n} \left(\alpha_k \overrightarrow{v_k} (\rho_k E_k + p)\right) = \nabla \cdot (k_{\mathrm{eff}} \nabla T) + S_{\mathrm{E}} \tag{5}$$

The SST k-ω model has a similar form to the standard $k - \omega$ model:

$$\frac{\partial}{\partial t}(\rho k) + \frac{\partial}{\partial x_i}(p k u_i) = \frac{\partial}{\partial x_j}\left(\Gamma_k \frac{\partial k}{\partial x_j}\right) + \tilde{G}_k - Y_k + S_k \tag{6}$$

and

$$\frac{\partial}{\partial t}(\rho\omega) + \frac{\partial}{\partial x_i}(\rho\omega u_i) = \frac{\partial}{\partial x_j}\left(\Gamma_\omega \frac{\partial \omega}{\partial x_j}\right) + G_\omega - Y_\omega + D_{\mathrm{w}} + S_\omega \tag{7}$$

Zwart-Gerber-Belamri Model

$$R = n\left[4\pi R_{\mathrm{B}}^2 \rho_v \frac{DR_{\mathrm{B}}}{Dt}\right] \tag{8}$$

If $P \leq P_{\mathrm{v}}$

$$R_{\mathrm{e}} = F_{\mathrm{vap}} \cdot \frac{3\alpha_{\mathrm{nuc}}(1-\alpha_{\mathrm{v}})\rho_{\mathrm{v}}}{R_{\mathrm{B}}} \sqrt{\frac{2}{3}\left(\frac{P_{\mathrm{v}} - P}{\rho_{\mathrm{l}}}\right)} \tag{9}$$

If $P \geq P_{\mathrm{v}}$

$$R_{\mathrm{e}} = F_{\mathrm{cond}} \cdot \frac{3\alpha_{\mathrm{nuc}}(\alpha_{\mathrm{v}})\rho_{\mathrm{v}}}{R_{\mathrm{B}}} \sqrt{\frac{2}{3}\left(\frac{P_{\mathrm{v}} - P}{\rho_{\mathrm{l}}}\right)} \tag{10}$$

where,

R_{B} bubble radius $= 10^{-6}$ m
α_{nuc} nucleation site volume fraction $= 5 \times 10^{-4}$
F_{vap} evaporation coefficient $= 50$
F_{cond} condensation coefficient $= 0.01$.

2.4 Validation

Figure 4 shows the variation of velocity along the centreline of the microchannel and Fig. 5 shows the vapour volume fraction formed in the orifice due to cavitation. The same model dimensions as mentioned in the existing literature [2] were used and the velocity variation and vapour volume fraction values were obtained. The obtained values and trend show close agreement with that of the values and trend mentioned in the above literature. The deviation of the obtained results with respect to those presented in the literature was about 0.47%. The maximum velocity obtained in the present study is 23.1 m/s whereas the maximum velocity given in the paper is 23.209 m/s.

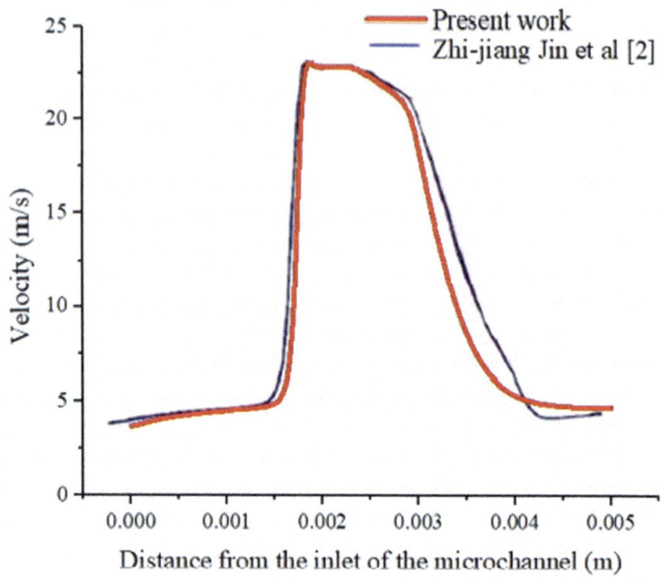

Fig. 4 Velocity along the centreline

<table>
<tr><th>Results in the literature</th><th>Results obtained in the validation</th></tr>
<tr><td>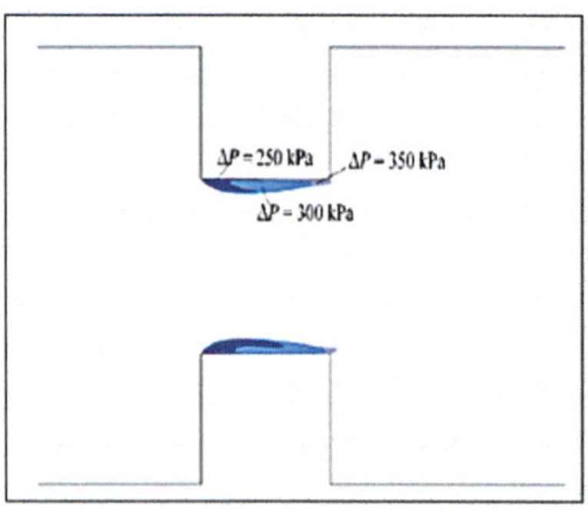

Vapor volume fraction on the symmetry plane</td><td>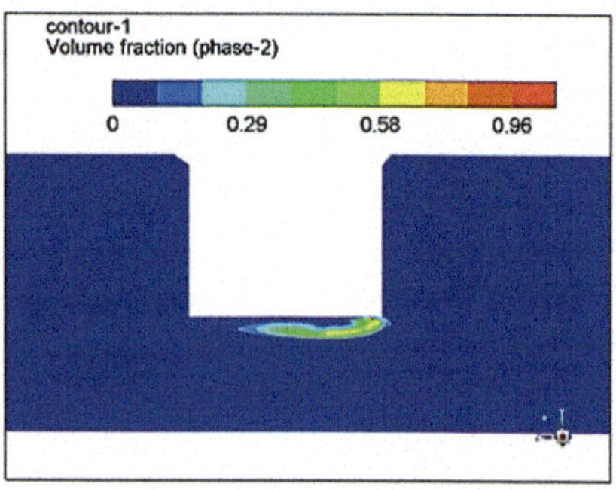

Vapor volume fraction on the symmetry plane</td></tr>
<tr><td>Maximum vapor volume fraction under different grids
<table>
<tr><td>Grid</td><td>Coarse</td><td>Regular</td><td>Fine</td></tr>
<tr><td>α_{max}</td><td>0.937</td><td>0.963</td><td>0.965</td></tr>
</table></td><td>Maximum vapor volume fraction
<table>
<tr><td>α_{max}</td><td>0.9627</td></tr>
</table></td></tr>
</table>

Fig. 5 Vapour volume fraction in the orifice

3 Results and Discussion

The results obtained from numerical analysis for both straight and micro orifice entrenched microchannel are discussed below. This study primarily focusseson cavitation induced heat transfer enhancement with the incorporation of microorifice. The performance was studied by comparingthe simulation results of a straight channel and a micro orifice ($R = 0.325$ mm) entrenched microchannel at same conditions. Further the effect of different orifice sizes were analysed and compared. The pressure drop required across the orifice of radius 0.325 mm to initiate the cavitation is around 230 kPa. Figure 6 shows the vapour volume formation near the orifice.

The heat transfer coefficient was calculated in the straight microchannel for different inlet velocity conditions corresponding to Reynolds number from 2500 to 2800 (4.25, 4.42, 4.59 and 4.76 m/s). Under similar conditions, the heat transfer coefficient was calculated for microchannel with orifice of radius = 0.325 mm (Orifice 1). Figures 7, 8 and 9 show the variation of heat transfer coefficient, base temperature, pressure drop respectively with inlet velocity for cavitating and non cavitating flow conditions. It is noticed that with increase in velocity, the heat transfer coefficient increases, base temperature decreases and pressure drop increases. Also, the cavitating flow in the microchannel with orifice has increased heat transfer coefficient, reduced base temperature and increased pressure drop when compared to the non cavitating flow in the straight microchannel. Table 2 shows the percentage increase in heat transfer in cavitating flow when compared to non cavitating flow.

Because of the bubble collapse due to cavitation, the heat transfer coefficient increased when micro orifice is incorporated. The fluid is forced to move against the walls of the microchannel during bubble collapse, which enables in heat transfer augmentation [6]. However, the pressure drop in the curved orifice was considerably greater than in the straight channel. This is caused by the sudden expansion of the fluid and the establishment of eddies at the orifice's exit. The energy necessary to

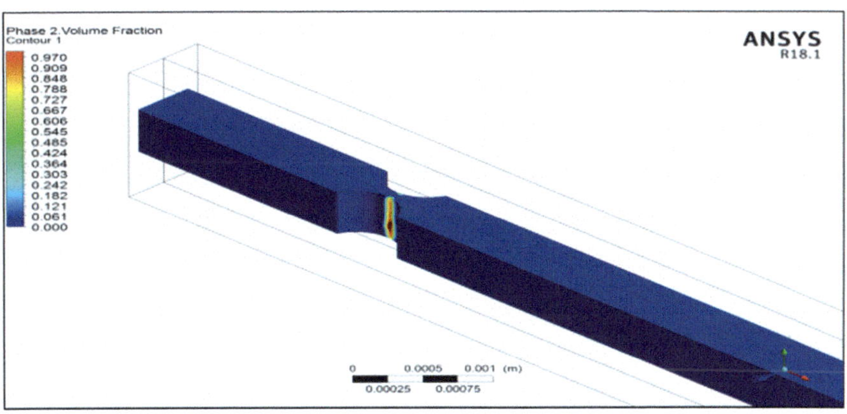

Fig. 6 Vapour volume fraction contour obtained in the present study

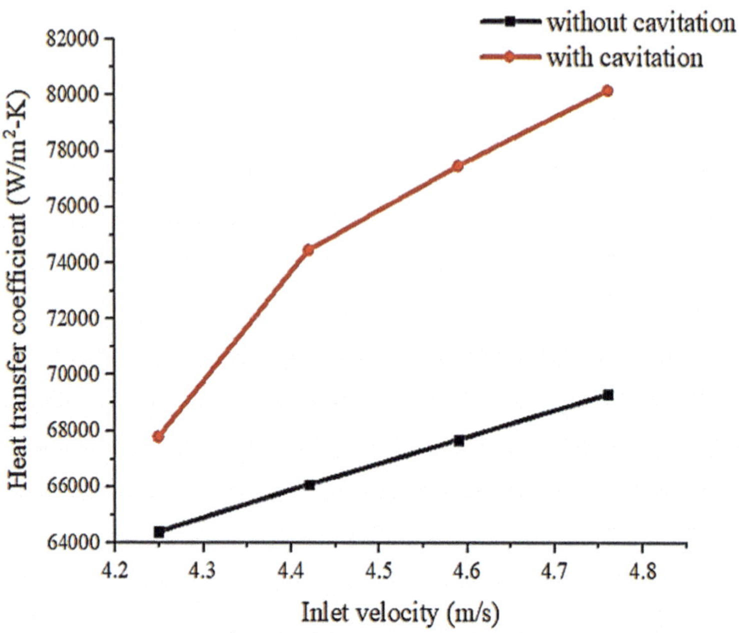

Fig. 7 Heat transfer coefficient versus inlet velocity

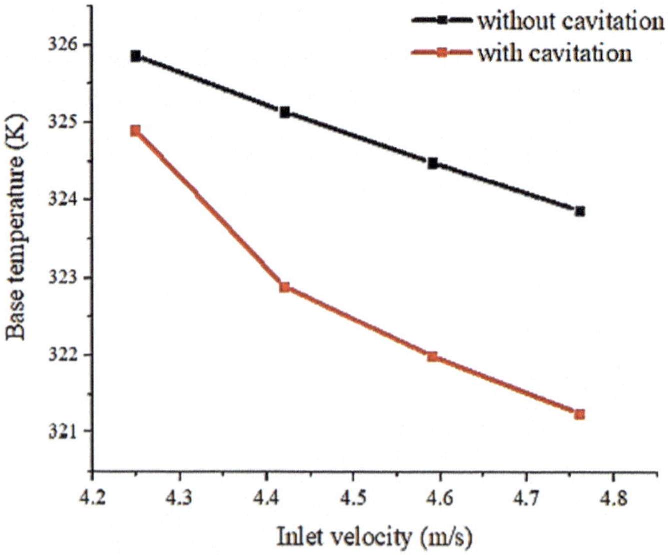

Fig. 8 Base temperature versus inlet velocity

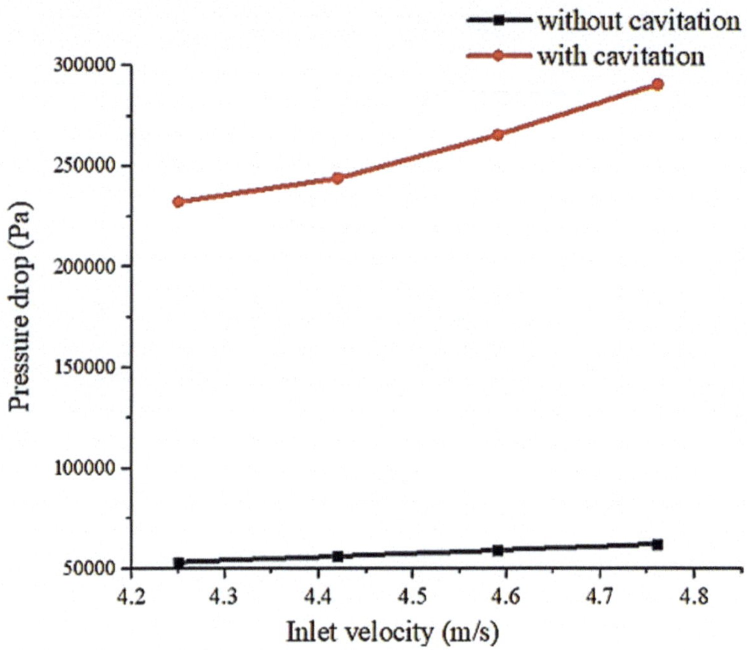

Fig. 9 Pressure drop versus inlet velocity

Table 2 Variation of heat transfer coefficient and pressure drop in cavitating and non-cavitating flow

Inlet velocity (m/s)	Percentage increase in heat transfer	Percentage increase in pressure drop
4.25	5.28	335.73
4.42	12.68	333.96
4.59	14.48	348.77
4.76	15.69	367.38

generate the eddies is extracted from the fluid's potential energy, thus raising the pressure drop [9]. Because the base temperature was found to be slightly lower, the heat transfer coefficient increased.

In order to study the effect of orifice size, the studies were further extended for different orifice sizes of radii 0.33 mm (Orifice 2) and 0.335 mm (Orifice 3). Increasing the orifice radius decreases the minimum area of the flow, resulting in more cavitation and pressure drop across the orifice. Figures 10, 11 and 12 show the variation of heat transfer coefficient, base temperature, pressure drop respectively with inlet velocity for different orifice sizes.

It was observed that decreasing the orifice area increases velocity and decreases the pressure of the fluid. This makes the flow more turbulent and creates more cavitation in the orifice which increases the heat transfer coefficient and reduces the base temperature. Table 3 shows the percentage increase in heat transfer coefficient for

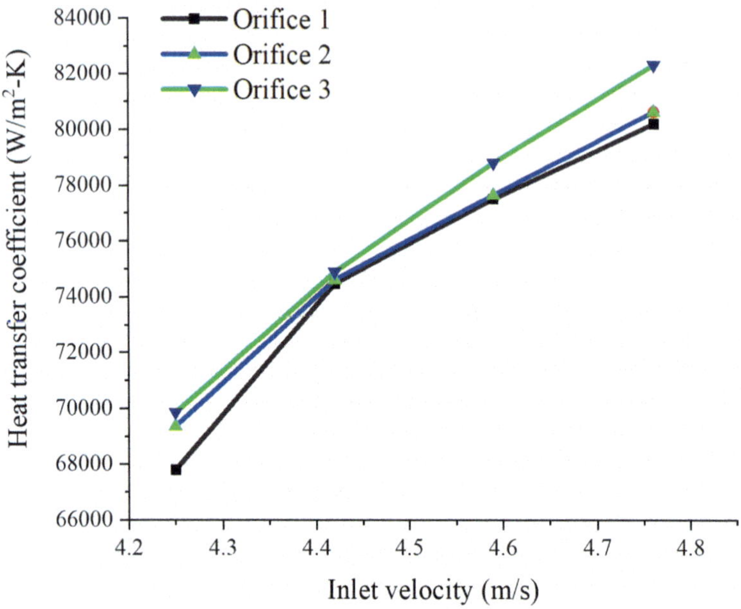

Fig. 10 Heat transfer coefficient versus inlet velocity

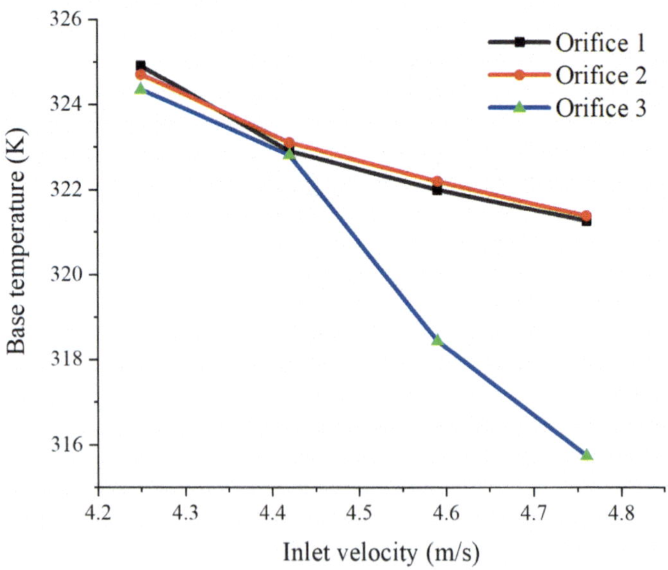

Fig. 11 Base temperature versus inlet velocity

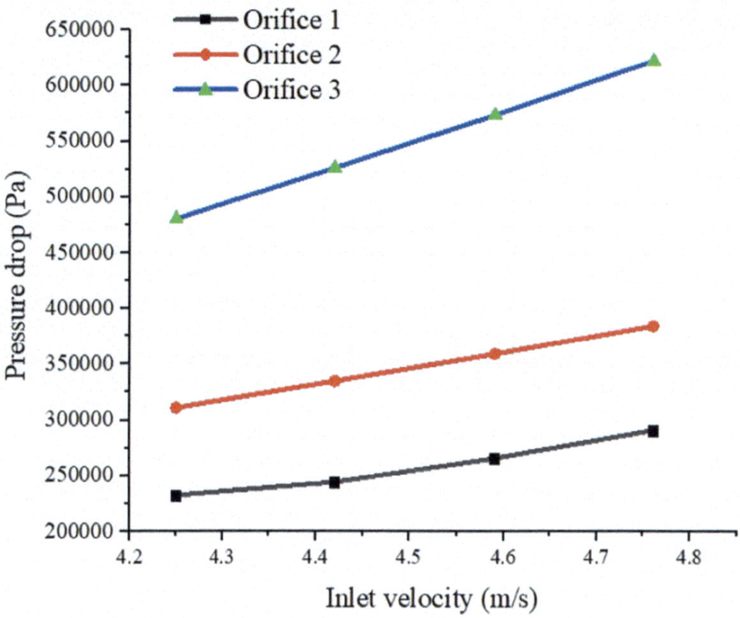

Fig. 12 Pressure drop versus inlet velocity

different orifice sizes when compared with non cavitating flow. As the orifice size is reduced the pressure drop increased significantly high which may lead to rise the pumping power required to pump the fluid through the channels.

Table 3 Percentage increase in heat transfer coefficient in channels with orifice compared to straight channel

Inlet velocity (m/s)	Orifice 1 ($R = 0.325$ mm) Percentage increase in		Orifice 2 ($R = 0.33$ mm) Percentage increase in		Orifice 3 ($R = 0.335$ mm) Percentage increase in	
	Heat transfer coefficient	Pressure drop	Heat transfer coefficient	Pressure drop	Heat transfer coefficient	Pressure drop
4.25	5.28	335.73	7.71	483.57	8.49	800.75
4.42	12.68	333.96	12.89	495.34	13.32	835.59
4.59	14.48	348.77	14.7	506.84	16.43	869.04
4.76	15.69	367.38	16.33	518.01	18.75	900.7

4 Conclusions

The present study investigates the effect of cavitation on enhancement of heat transfer in themicrochannel with orifice when compared with the straight microchannel. Thefollowing conclusions were arrived.

- The pressure drop required to initiate the cavitation in the microchannel (Dh = 500 μm) with the orifice of radius 0.325 mm was found to be around 230 kPa.
- An increase in heat transfer coefficient was obtained in the cavitating flowconditions when compared to the non-cavitating flow, in addition to theincrease in pressure drop.
- During the flow with inlet velocity of 4.76 m/s, the increase in heat transfer coefficient was around 15.69, 16.33 and 18.75% in cavitating flow conditions when comparedwith the non-cavitating flow conditions.
- The increase in the heat transfer was observed with the decrease in theorifice area. There is an increment of 2.64% in the heat transfer coefficient inthe microchannel with the orifice radius 0.335 mm when compared with the orifice radius 0.325 mm at 4.76 m/s.

Considering the results obtained from the numerical studies, it may be concluded that the cavitation induced by micro orifice in a microchannel enhances the heat transfer significantly with the penalty of pressure drop. This method may be recommended for the cases where the pumping power consumption is not an important criteria.

References

1. Li T, Liu B, Zhou J, Xi W, Huai X, Zhang H (2020) A comparative study of cavitation characteristics of nano-fluid and deionized water in micro-channels. Micro Mach 11(3):310
2. Zhi-jiangJin Z-j, Gao Z-x, Li X-j, Qian J-y (2019) Cavitating flow through a micro-orifice. Micro Mach 10(3):191
3. Ghorbani M, Yildiz M, Gozuacik D, Kosar A (2016) Cavitating nozzle flows in micro and minichannelsunder the effect of turbulence. J Mech Sci Technol 30(6):2565–2581
4. Hosbach M, Gitau S, Sander T, Leuteritz U, Pfitzner M (2019) Effect of taper, pressure and temperature on cavitation extent and dynamics in micro-channels. Exp Thermal Fluid Sci 108:25–38
5. Schneider B, Kosar A, Peles Y (2007) Hydrodynamic cavitation and boiling in refrigerant (R-123) flow inside microchannels. Int J Heat Mass Transf 50(13–14):2838–2854
6. Liu B, Cai J, Huai X (2014) Heat transfer with the growth and collapse of cavitation bubble between two parallel heated walls. Int J Heat Mass Transf 78:830–838
7. Junfei Y, Lin W, Zhanwei W, Yingying T (2019) Experimental investigation of heat transfer in microchannel with inlet cavitation structure. J Therm Sci 30:294–301
8. Mishra C, Peles Y (2005) Cavitation in flow through a microorifice inside a silicon microchannel. Am Inst Phys 17(1):013601-1–013601-15

9. Schneider B, Kosar A, Kuo C-J, Mishra C, Cole GS, Scaringe RP, Peles Y (2006) Cavitation enhanced heat transfer in microchannels. J Heat Transfer 128(12):1293–1301
10. Cioncolini A, Scenini F, Duff J, Szolcek M, Curioni M (2016) Choked cavitation in micro-orifices: an experimental study. Exp Thermal Fluid Sci 74:49–57

Impact of Surface Roughness on the Aerodynamic and Aeroacoustic Performance of the Darrieus Wind Turbine

J. Sarathkumar Sebastin, B. Madhan Kumar, M. Shreedharan, Ajay Kumar Javadala, V. Manoj, and C. Haribabu

Abstract The recent advancements in small wind turbines show an increased demand for the Darrieus wind turbine in the urban environment. The present work is devoted to the numerical analysis of the aeroacoustic sound emission from a straight-bladed Darrieus wind turbine with NACA0018. Computations are performed at Reynolds number of 28,000 with a tip speed ratio of 0.4 using the unsteady Reynolds averaged Navier Stokes equations with Ffowcs Williams-Hawking's (FW-H) acoustic analogy. The FW-H method is based on a free-field Green's function where the scheme uses a porous integration surface and implements an advanced time formulation. The study aims to obtain a numerical methodology to predict the effect of surface roughness on wind turbines' aerodynamic power and sound emission.

Keywords Darrieus wind turbine · Aeroacoustic · Aerodynamic power · Sound emission

1 Introduction

Wind energy has become essential due to ecologically viable power generation requirements. This vision is due to the adverse effects of combusting fossil fuels for power creation. Wind power usage leads to new wind turbines setting up near city zones. Hence, it is crucial to advance noise emission performance to the sound contamination in inhabited zones. The word noise defines the individual sensitivity

J. Sarathkumar Sebastin (✉)
CFD Center, Kalasalingam Academy of Research and Education, Virudhunagar, India
e-mail: sarathkumar.j@klu.ac.in

B. Madhan Kumar
IIT-Madras, Chennai, India

M. Shreedharan · A. K. Javadala · V. Manoj · C. Haribabu
Department of Aeronautical Engineering, Kalasalingam Academy of Research and Education, Virudhunagar, India
e-mail: 9921149003@klu.ac.in

E. Natarajan et al. (eds.), *Materials, Design and Manufacturing for Sustainable Environment*, Lecture Notes in Mechanical Engineering,
https://doi.org/10.1007/978-981-19-3053-9_55

of unnecessary sound waves in the human ear. The average human ear can able to receive the sound in the range of 20 Hz to 20 kHz. The Small wind turbines are well-advised an answer for wind power reaping, particularly on small scales in city zones. Inquiries in this space have concentrated on vertical axis wind turbines (VAWT). The VAWT does not need a yaw system. It can be applied in turbulent streams at low construction costs.

The utmost broadly known scheme of VAWTs is the Darrieus turbine [1], as shown in Figs. 1 and 2, established by the Scientist Darrieus in the twentieth century. Converting lift force is the supreme effective tactic to transform wind power into mechanical power [2]. In contrast, a representative drag-type method is a Savonius wind turbine. With a maximum power coefficient of 30%, the Savonius turbine does not apply to profitable wind power usage. Owing to wind turbines' growing energy and sound quality over the past decades, researching wind turbines' sound emissions is still a vital research area (Fig. 3). In common, wind turbines have different sound sources. They can be distributed into two collections.

The first kind is mechanical sound, e.g. sound generated by auxiliary equipment such as gearboxes, generators, yaw units, cooling systems, and fluid power devices. The sound conduction path of mechanical sound is an airborne or structural sound. Another sound source is aerodynamic sound. Both sounds strongly depend on the rotor's geometry, airfoil profiles, and the adjacent flow situations. The sound conduction path of mechanical sound is an airborne or structural sound. Another sound source is aerodynamic sound. The sound conduction path of mechanical sound is an airborne or structural sound. Another sound source is aerodynamic sound. Both sounds strongly depend on the rotor's geometry, airfoil profiles, and the adjacent flow situations. The flow of air usually produces aerodynamic sound through the blades. Low-frequency sound is considered noise when revolving blades impede localised flow variations due to wind speed. We know the default frequency range for the low-frequency sound is not less than 10 Hz to 200 Hz. Low-frequency sound

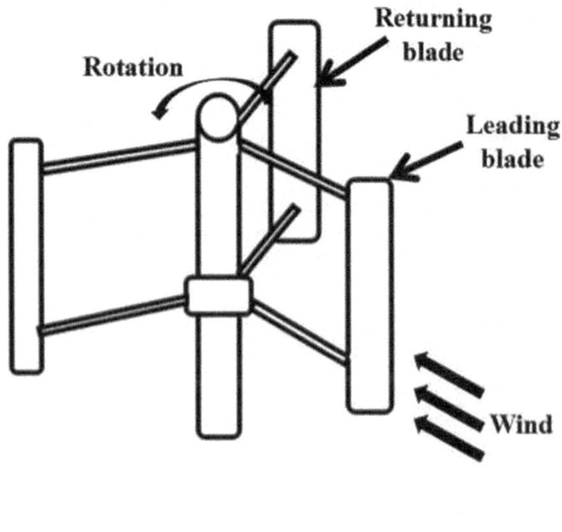

Fig. 1 Three-bladed Darrieus wind turbine

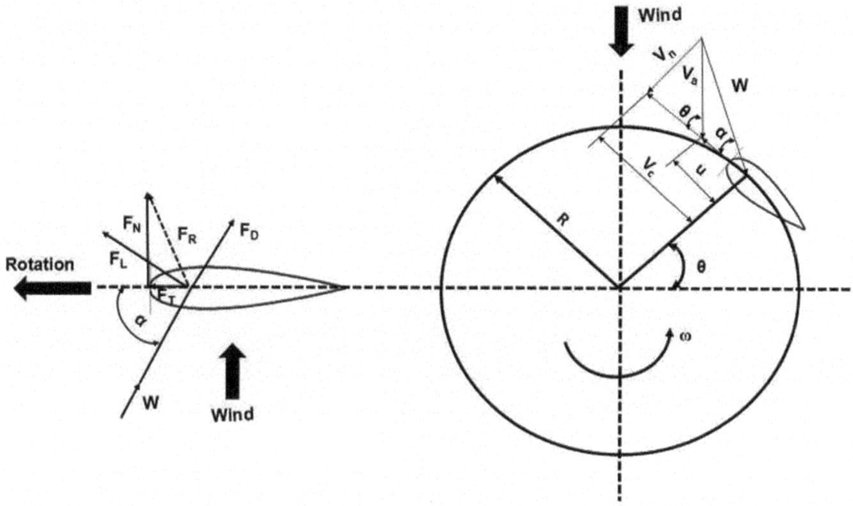

Fig. 2 Loads and speed dissemination on Darrieus rotor airfoil

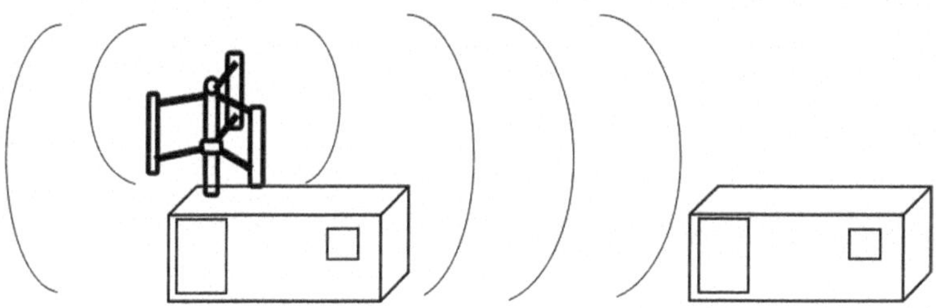

Fig. 3 Wind turbine sound propagation

is primarily linked to the wind blade running frequency and high harmonics. The airfoil profile of blades can create aerodynamic lift when exposed to the incident wind. This aerodynamic lift provides a moment along the blade axis, which allows the wind turbine's main shaft to rotate. Three-bladed VAWTs with straight blades are well suited for small-scale power production due to their blade design simplicity. In addition, VAWTs have certain benefits over HAWTs. Surface roughness is the one promising parameter to increase the aerodynamic power. But still, the studies need to analyse the sound Signals.

2 Problem Definition

In current years, experimental and numerical studies focussing on aerodynamic behaviour have been conducted in the field of VAWTs. Compared to traditional

HAWTs, straight-bladed VAWTs are considered a viable solution in the municipal location. Therefore, it is essential to advance rotor performance and control noise discharge from the turbine blade to surrounding habitats. Based on the previous literature (Table 1), the main features for achieving high performance of the rotor are (1) Solidity [3–20]; (2) the number of blades [21–23]; (3) blade profile [24–31]; (4) surface roughness [32–35]; (5) Government Effect [36–44]; (6) Pitch Control Strategy [45–54]; (7) Reynolds Number Effect [55–60]. Due to the recent progress of the CFD, the numerical investigation of the Darrieus wind turbines has attracted attention. The researchers found that the wind turbine's aeroacoustic characteristics could be calculated with reasonable accuracy using the CFD algorithm. Although these findings have revealed that the straight blade Darrieus wind turbine can be used to maximise aerodynamic performance, little consideration has been given to the aerodynamic noise issues. In this paper, attempts are being made to obtain a valid numerical method for finding a wind turbine's noise emission. Ffowcs Williams-Hawkings (FW-H) acoustic analogy is used to study sound emission. The study aims to analyse the effect of surface roughness on the Darrieus turbine. Weber has experimentally and numerically investigated the Darrieus wind turbine with three blades. It is hoped that the current work will lead to more research in the aeroacoustics of the Darrieus wind turbine.

3 Methodology

3.1 Computational Domain and Mesh Details

The current study uses an unstructured mesh for the three-blade model. The computational domain consists of two areas, the stationary and the rotating regions (Fig. 3). The overlapping grid method has been applied, permitting high-quality meshes specifically for every grid constituent. The rotor computational domain is circular with a radius R. The domain size is in the x-direction is $24R$ (from $-8R$ to $16R$) and in the Y $16R$ (from $8R$ to $-8R$) (Fig. 3). The receiver is situated at 1 m from the rotor centre to receive the acoustic signal (Fig. 3). The grid system consists of 1.05 lakh cells with a near-wall distance $y+ <1$.

3.2 CFD Settings

Aerodynamic Study: A comparison of published experimental findings and CFD for an H-rotor Darrieus turbine was used to evaluate the numerical turbulence model. When utilising the SST K-ω turbulence model, these results tell a good agreement between numerical and experimental functions of dB and C_p. Other investigations involving rotating blades and airfoils have shown a similar trend, demonstrating the

Table 1 Outline of previous studies

Author	Method	Re(×10)	Key features
Castelli et al. [10]	Num. 2D	5.2	Correlation between rotor power and torque concerning flow characteristics
Mohamed [30]	Num. 2D	N/A	Performance study of a turbine with different blade shapes
Mohamed [31]	Num. 2D	N/A	Aeroacoustic study on a three-bladed turbine
Rossetti and Paves [61]	Num. 2D and 3D	3.4	Start-up characteristics of turbine
Weber et al. [62]	Exp./Num	2.8	Validation of numerical model with noise measurement of NACA0018 profile
Eboibi et al. [19]	Exp	1.2–2.1	Performance study of a turbine concerning solidity effects
Howell et al. [9]	Exp	0.67, 1.00	Solidity effect and Surface roughness
Li et al. [13]	Exp	0.8–1.1	Effect of blade's pitch angle, solidity and wind velocity
Du et al. [63]	Exp	1.0, 0.81, 0.67	Effects of various design aspects concerning turbine start-up
Mazarbhuiya et al. [64]	Exp	1.3–2.0	Effects of asymmetric blades with turbine performance
Desskosky et al. [65]	Num. 2D	2.8	Aerodynamic and Aeroacoustic study with a three-bladed rotor

Re = Reynolds number; Exp. = experimental measurement; Num. = numerical simulation

SST K-ω use models for quick CFD calculations. For spinning bodies, the SST K-ω model is commonly preferred. A novel transport equation for the turbulent dissipation rate is included in this model. Compared to the standard $k-\omega$ model, the SST K-ω model usually produces better results for whirling and separation flows. Near all walls, the $y+$ values are about one, within the required range for best-practise CFD. Depicts the torque coefficient variation for the rotor. After completing the first three revolutions, the statically stabilised state is reached. For the power curve computation, the last six revolutions are considered. A three-bladed NACA0021 profile with a chord length of 0.0858 m and a Rotor radius of 0.515 m makes up the computational domain of the rotating region. The blades are joined with 0.25c spokes. The computational domain with blade profiles is shown in Fig. 3. In this simulation, the spoke connection and shaft are disregarded to avoid unnecessary calculation time.

Aeroacoustic Study: The unsteady Reynolds averaged Navier Stokes (URANS) equations resolve the H-rotor Darrieus wind turbine's flow field. The noise emission is obtained using the Ffowcs Williams-Hawkings acoustic analogy using the "correlated length" method. The technique assumes that the vortex detaching correlates along a blade's firm length. The two-dimensional way can acquire major aerodynamic characteristics at a low cost. The rotating velocity of the airfoil is 83.78 rad/s. The finite volume method with the second-order upwind spatial discretisation scheme and second-order implicit temporal discretisation systems is used in the paper. The SIMPLE algorithm for pressure-velocity coupling and a fixed time-stepping (0.0001 s) method is adopted. The prior studies can establish the fluid-governing equations, correlation method, and the acoustic analogy. The CFD simulations were implemented in the ANSYS Fluent and developed a MATLAB code for acoustic signal treating. The NACA0018 airfoils are evenly distributed circumferentially in the flow field, and the link rod is not included in the study. The computational domain contains a rotating and stationary area. To characterise the blade aerodynamics and the relevant noise mechanisms, in this paper, the numerical analysis is carried out at Re 28,000.

3.3 *Grid Independence Test*

The total number of cells in the computational domain can influence the results due to the numerical approximations in RANS equations. However, the increase in grid size also necessitates more computational resources. Hence, a grid independence test is conducted in this paper to obtain a suitable mesh. Table 1 shows the grid independence tests with the acoustic parameter SPLT (Eq. 1), where SPLT is the tonal SPL. The case-3 with 1.05 lakhs cells is selected in this paper. Beyond 1.05 lakhs, the acoustic property changes are found to be insignificant (Table 2).

$$\mathrm{SPLT} = 20 \log (p/p_o) \tag{1}$$

where p is the sound pressure and p_o is the reference sound pressure of 0.00002 pascals = 0 dB.

Table 2 Grid independence test

Case	No. of cells (Lakhs)	SPLT (dB)
1	0.51	70.7
2	0.89	76.5
3	1.05	78.2
4	1.24	78.2

4 Result and Discussion

A comparison of published experimental findings and CFD for an H-rotor Darrieus turbine was used to evaluate the numerical turbulence model. When utilising the SST K-ω turbulence model, these results tell a good agreement between numerical and experimental functions of dB and C_p. Other investigations involving rotating blades and airfoils have shown a similar trend, demonstrating the SST K-ω use models for quick CFD calculations. For spinning bodies, the SST K-ω model is commonly preferred. A novel transport equation for the turbulent dissipation rate is included in this model. Compared to the standard K-ω model, the SST K-ω model usually produces better results for whirling and separation flows. Near all walls, the $y+$ values are about one, within the required range for best-practise CFD. Depicts the torque coefficient variation for the rotor. After completing the first three revolutions, the statically stabilised state is reached. For the power curve computation, the last six revolutions are considered. The results show a good agreement between experimental and numerical results regarding prediction—validation of the experimental result in Table 3.

The turbine's self-noise depends on the blades' force variation and airflow through the spin. Thickness noise is continuously connected with loading sound, but not vice versa. Low Mac numbers generate only monotonous sound (thickness sound) and dipole sound (loading sound). However, sound pressure level is an important characteristic that indicates the sound an observer wants. Total sound emission depends on sound source power, spatial distribution, and temporal events. Radiation sound waves are superposed, displaying stout direction due to different radiations and phase offsets.

The NACA0018 shape turbine's chord span is 0.05 m, and it has 0.4 m in diameter and 0.2 m tall. To measure acoustic pressure, free-field microphones were placed at a distance of 1 m. The blade passing frequency for the setup is 40 Hz. Periodic signals observed for six complete revolutions of physical simulation time: 0.45 s. Acoustic results show the best fit in the critical frequency range. The blade passing frequency is the most vital regime, which is also accurately predicted. The comparison to the whole noise spectrum is made from 30 Hz because Weber reported that frequencies lesser than 29–30 Hz are wind tunnel noise.

Table 3 Validation of the experimental result

Author	Method	Parameter	Validation	
			Author's result	Present study
Marco Raciti Castelli	Experimental	Power coefficient	0.3	0.272
Weber	Experimental	Overall sound pressure level (dB)	0.82	0.79

Table 4 Effect of surface roughness

Surface roughness (Wall: Airfoil)	Power coefficient (C_p)	Overall sound pressure level (dB)
0	0.272	93.6683
0.0001	0.2598	93.4931
0.0002	0.265676034	93.403
0.0003	0.281709	93.2098
0.0004	0.345873718	93.3816
0.0005	0.39	93.2694
0.0006	0.481133964	93.4691
0.0007	0.536282	93.3856

5 Effects of Surface Roughness

A turbulent boundary layer will be giving high resistant to the air separation from the rotor blade aboveground, the profile drag could be minimised where there was previously an air separation in the presence of the laminar flow. This is possible despite the improvised skin drag, the summary of the drag on the aerofoil is reduced. Skin friction drag will get increased if the surface is too frictional, and speedy variations in the surface chart will make locations for decentralised air separation, increasing the entire drag. The blades' surface accomplishment was visibly rough due to the manufacturing process. Because the blades were made of foam, it was impossible to utilise a surface profiler to identify the exact magnitude of the surface roughness. However, the enormous roughness scales were determined to be 0.3, 0.5, and 0.7 mm in height, as result in a non-dimensional roughness of (x/C) equal to 0.00015 to 0.00065. The turbine rotor blades were first put through their paces in their as-built state. The results were simulated in Fluent 19 and tabulated below. Table 4, will explain the effect of surface roughness in power coefficient and sound pressure level. The effective surface roughness range lies between 0.00015 and 0.0003. Because further increasing surface roughness will improve the power coefficient and increase the level of sound.

6 Conclusion

Darrieus VAWTs have the excellence of operating under small wind velocity and city zone circumstances. However, it writhes from deprived aerodynamic performance compared to HAWT. This paper's outcomes display the effect of using the H-rotor Darrieus VAWT on noise radiation. Numerical inquiries of an H-Darrieus wind blade noise generation have been made to conform with the experimental data. There is a noble agreement amongst the numerical acoustic and experimental measurements. The examined operating point is characterised by the low tip speed ratio of $\lambda = 0.4$

and Re = 28,000. Forthcoming studies will include an emphasis on the acoustics of various pitch angles. The proposed method can be utilised for aeroacoustic optimisation to create VAWTs. The results clearly explain the effects of surface roughness; it is only effective on the below 0.0003 roughness range. Future investigations will emphasise the acoustics of various pitch angles incorporated with surface roughness. The proposed method can be used to improve the aeroacoustic performance of VAWTs.

References

1. Du L, Ingram G, Dominy RG (2019) A review of H-Darrieus wind turbine aerodynamic research. Proc Inst Mech Eng C J Mech Eng Sci 233(23–24):7590–7616. https://doi.org/10.1177/0954406219885962
2. Hashem I, Mohamed MH (2017) Aerodynamic performance enhancements of H-rotor Darrieus wind turbine. Energy. https://doi.org/10.1016/j.energy.2017.10.036
3. Templin RJ (1974) Aerodynamic performance theory for the NRC vertical-axis wind turbine. Laboratory technical report, LTR-LA-160. National Research Council of Canada, Ottawa
4. Blackwell BF, Sheldahl RE and Feltz LV (1977) Wind tunnel performance data for the Darrieus wind turbine with NACA0012 blades. SAND7-0130. Sandia Laboratories, Albuquerque, NM
5. Consul CA, Willden RHJ, Ferrer E et al (2009) Influence of solidity on the performance of a cross-flow turbine. In: Proceedings of the 8th European wave and tidal energy conference, Uppsala, Sweden, January 2009
6. Gosselin R, Dumas G, Boudreau M (2016) Parametric study of H-Darrieus vertical-axis turbines using CFD simulations. Renew Sustain Energy 8
7. Vassberg JC, Gopinath AK, Jameson A (2005) Revisiting the vertical-axis wind turbine design using advanced computational fluid dynamics. In: The 43rd AIAA aerospace sciences meeting and exhibit, Reno, Nevada, USA, 10–13 Jan 2005
8. Worstell MH (1982) Aerodynamic performance of the 17-m-diameter Darrieus wind turbine in the three-bladed configuration: an addendum. SAND-79-1753. Sandia Laboratories, Albuquerque, NM
9. Howell R, Qin N, Edwards J et al (2010) Wind tunnel and numerical study of a small vertical axis wind turbine. Renew Energy 35:412–422
10. Castelli MR, Englaro A, Benini E (2011) The Darrieus wind turbine: proposal for a new performance prediction model based on CFD. Energy 36:4919–4934
11. Subramanian A, Yogesh SA, Sivanandan H et al (2017) Effect of airfoil and solidity on performance of small scale vertical axis wind turbine using three dimensional CFD model. Energy 133:179–190
12. Rezaeiha A, Montazeri H, Blocken B (2018) Towards optimal aerodynamic design of vertical axis wind turbines: impact of solidity and number of blades. Energy 165:1129–1148
13. Li Q, Maeda T, Kamada Y et al (2016) Effect of solidity on aerodynamic forces around straight-bladed vertical axis wind turbine by wind tunnel experiments (depending on number of blades). Renew Energy 96:928–939
14. Sagharichi A, Zamani M, Ghasemi A (2018) Effect of solidity on the performance of variable-pitch vertical axis wind turbine. Energy 161:753–775
15. Kirke BK, Lazauskas L (1991) Enhancing the performance of a vertical axis wind turbine using a simple variable pitch system. Wind Eng 15:187–195
16. Musgrove PJ, Mays ID (1978) Development of the variable geometry vertical axis windmill. In: International symposium on wind energy systems, Amsterdam, Netherlands, 3–6 Oct 1978
17. Roh SC, Kang SH (2013) Effects of a blade profile, the Reynolds number, and the solidity on the performance of a straight bladed vertical axis wind turbine. J Mech Sci Technol 27:3299–3307

18. Simhan K (1984) A review of calculation methods for the determination of performance characteristics of vertical axis wind energy converters with special reference to the influence of solidity and starting characteristics. Fachberichte Physik. Report no. 12. Bremen University, Bremen
19. Eboibi O, Danao LAM, Howell RJ (2016) Experimental investigation of the influence of solidity on the performance and flow field aerodynamics of vertical axis wind turbines at low Reynolds numbers. Renew Energy 92:474–483
20. Singh MA, Biswas A, Misra RD (2015) Investigation of self-starting and high rotor solidity on the performance of a three S1210 blade H-type Darrieus rotor. Renew Energy 76:381–387
21. Sabaeifard P, Razzaghi H, Forouzandeh A Determination of vertical axis wind turbines optimal configuration through CFD simulations. In: 2012 International conference on future environment and energy, Singapore.
22. Berg DE (1990) Customised airfoils and their impact on VAWT cost of energy. SAND90-1148C. Sandia Laboratories, Albuquerque, NM
23. Klimas PC (1992) Tailored airfoils for vertical axis wind turbines. SAND84-1062. Sandia Laboratories, Albuquerque, NM
24. Masson C, Leclerc C, Paraschivoiu I (1998) Appropriate dynamic-stall models for performance predictions of VAWTs with NLF blades. Int J Rotating Mach 4:129–139
25. Angell RK, Musgrove PJ, Galbraith RAMCD (1988) Unsteady wind tunnel testing of thick section aerofoils for use on large scale vertical axis wind turbine. In: Wind energy conversion, proceeding of 10th BWEA conference, London, UK, 22–24 Mar 1988, pp 195–203
26. Islam M, Ting DSK, Fartaj A (2007) Desirable airfoil features for smaller capacity straight bladed VAWT. Wind Eng 31:165–196
27. Bianchini A, Ferrari L, Magnani S (2011) Start-up behavior of a three-bladed H-Darrieus vawt: experimental and numerical analysis. ASME Turbo Expo 6:811–820
28. Bausas MD, Danao LAM (2015) The aerodynamics of a camber-bladed vertical axis wind turbine in unsteady wind. Energy 93:1155–1164
29. Chen C-C, Kuo C-H (2013) Effects of pitch angle and blade camber on flow characteristics and performance of small-size Darrieus VAWT. J Vis 16:65–74
30. Mohamed MH (2012) Performance investigation of H-rotor Darrieus turbine with new airfoil shapes. Energy 47:522–530
31. Mohamed MH (2014) Aero-acoustics noise evaluation of H-rotor Darrieus wind turbines. Energy 65:596–604
32. Ashwill TD (1992) Measured data for the Sandia 34 m vertical axis wind turbine. SAND91-2228. Sandia Laboratories, Albuquerque, NM
33. Li Y, Tagawa K, Liu W (2010) Performance effects of attachment on blade on a straight-bladed vertical axis wind turbine. Curr Appl Phys 10:S335–S338
34. Priegue L, Stoesser T (2017) The influence of blade roughness on the performance of a vertical axis tidal turbine. Int J Mar Energy 17:136–146
35. Edwards JM ()2012 The influence of aerodynamic stall on the performance of vertical axis wind turbines. Ph.D. thesis, Sheffield University
36. Ma N, Lei H, Han Z et al (2018) Airfoil optimisation to improve power performance of a high-solidity vertical axis wind turbine at a moderate tip speed ratio. Energy 150:236–252
37. Migliore PG, Wolfe WP, Fanucci JB (1980) Flow curvature effects on Darrieus turbine blade aerodynamics. Energy 4:49–55
38. Rainbird JM, Bianchini A, Balduzzi F et al (2015) On the influence of virtual camber effect on airfoil polars for use in simulations of Darrieus wind turbines. Energy Convers Manag 106:373–384
39. Sharpe DJ (1984) Refinements and developments of the multiple streamtube theory for the aerodynamic performance of vertical axis wind turbines. In: Proceedings of the BWEA wind energy conference
40. Paraschivoiu I, Delclaux F (1983) Double-multiple streamtube model with recent improvements. AIAA J 7:250–255

41. Paraschivoiu I, Frauni P, Beguier C (1985) Streamtube expansion effects on the Darrieus wind turbine. AIAA J 1:150–155
42. Paraschivoiu I (1987) Double-multiple streamtube model for studying vertical axis wind turbine. AIAA J 4:370–377
43. Soraghan CE, Leithead WE, Feuchtwang J et al Double multiple streamtube model for variable pitch vertical axis wind turbines. In: The 31st AIAA applied aerodynamics conference, San Diego, USA, 24–27 June 2013
44. Du L, Imgram G, Dominy RG (2019) Time-accurate blade surface static pressure behaviour on a rotating HDarrieus wind turbine. Wind Energy 22:563–575
45. Bossanyi EA (2003) Wind turbine control for load reduction. Wind Energy 6:229–244
46. Kosaku T, Sano M, Nakatani K (2002) Optimum pitch control for variable-pitch vertical-axis wind turbines by a single stage model on the momentum theory. In: IEEE international conference on systems
47. Vandenberghe D, Dick E (1987) Optimum pitch control for vertical axis wind turbines. Wind Eng 11:237–247
48. Chougule P, Nielsen S (2014) Overview and design of selfacting pitch control mechanism for vertical axis wind turbine using multi body simulation approach. J Phys Conf Ser 524
49. Chen B, Su S, Viola IM et al (2018) Numerical investigation of vertical-axis tidal turbines with sinusoidal pitching blades. Ocean Eng 155:75–87
50. Abdalrahman G, Melek W, Lien F (2017) Pitch angle control for a small-scale Darrieus vertical axis wind turbine with straight blades (H-type VAWT). Renew Energy 114:1353–1362
51. Vandenberghe D, Dick E (1987) A theoretical and experimental investigation into straight bladed vertical axis wind turbine with second order harmonic pitch control. Wind Eng 11:237–247
52. Bhutta M, Hayat N, Farooq AU et al (2012) Vertical axis wind turbine—A review of various configurations and design techniques. Renew Sustain Energy Rev 16:1926–1939
53. Rezaeiha A, Kalkman I, Blocken B (2017) Effect of pitch angle on power performance and aerodynamics of a vertical axis wind turbine. Appl Energy 197:132–150
54. Parra-Santos T, Trullen DJP, Gallegos A et al (2016) Influence of fixed pitch angle on the performance of small scale. In: Proceedings of the ASME 2016 fluids engineering division summer meeting, Washington, DC, USA, 10–14 July 2016
55. Meana-Fernandez A, Solis-Gallego I, Oro JMF et al (2018) Parametrical evaluation of the aerodynamic performance of vertical axis wind turbines for the proposal of optimised designs. Energy 147:504–517
56. Amstrong S, Fiedler A, Tullis S (2012) Flow separation on a high Reynolds number, high solidity vertical axis wind turbine with straight and canted blades and canted blades with fences. Renew Energy 41:13–22
57. Zanforlin S, Deluca S (2018) Effects of the Reynolds number and the tip losses on the optimal aspect ratio of straight-bladed vertical axis wind turbines. Energy 148:179–195
58. Sheldahl RE, Klimas PC, Feltz LV (1980) Aerodynamic performance of a 5-m diameter Darrieus turbine with extruded NACA-0015 blades. SAND80-0179. Sandia Laboratories, Albuquerque, NM
59. Danao LA, Eboibi O, Howell R (2014) An experimental investigation into the influence of unsteady wind on the performance of a vertical axis wind turbine. Appl Energy 116:111–124
60. Lam HF, Liu YM, Peng HY, Lee CF, Liu HJ (2018) Assessment of solidity effect on the power performance of H-rotor vertical axis wind turbines in turbulent flows. J Renew Sustain Energy 10:023304. https://doi.org/10.1063/1.5023120
61. Rossetti A, Pavesi G (2013) Comparison of different numerical approaches to the study of the H-Darrieus turbines start-up. Renew Energy 50:7–19
62. Weber J, Becker S, Scheit C, Grabinger J, Kaltenbacher M (2015) Aeroacoustics of Darrieus wind turbine. Int J Aeroacoust 14(5–6):883–902. https://doi.org/10.1260/1475-472X.14.5-6.883
63. Du L (2016) Numerical and experimental investigations of Darrieus wind turbine start-up and operation. Ph.D. thesis, University of Durham, UK

64. Mazarbhuiya H, Biswas A, Sharma KK (2018) Performance investigations of modified asymmetric blade H-Darrieus VAWT rotors. Renew Sustain Energy 10
65. Dessoky* A, Bangga G, Lutz T, Kramer E (2019) Aerodynamic and aeroacoustic performance assessment of H-rotor Darrieus VAWT equipped with wind-lens technology. Energy 175:76–97

Parametric Studies on Screen-Splitted Air-Conditioned Room for Reduced Energy Consumption

J. Abhishek, S. Moulieswaran, S. Nirmal Raj, S. Pradeep, and P. Manoj Kumar

Abstract Energy plays a vital role in modern society. Conventional fuel resources are depleting at a faster pace at present. In order to meet the future energy demands, energy should be mainly produced from renewable resources and also by conserving the usage of energy. There are many household appliances which consume more amount of energy. Air conditioner is one of the most energy consuming appliances. It consumes almost half the energy in the household energy consumption. Many techniques such as cooling system using desiccant cooling, amount of refrigerant charge to be added, using multiple sensors, and control mechanism to ensure optimal energy savings are employed in air conditioner to increase the overall efficiency. Also, various other techniques have been experimented to reduce the overall power consumption. In this study, an effort is taken to reduce the power consumed by the air conditioner by making an enclosure around the bed. To ensure human comfort and safety, we used different materials to arrive at the minimum power consumption. Variation of relative humidity, temperature, and oxygen level inside the enclosure setup with different materials has been studied in comparison with the full room and to check whether the various parameters are in safe limit for human comfort.

Keywords Energy conservation · Air-conditioning · Enclosure · Human comfort

1 Introduction

Energy is the quantitative measure of all kinds of work done by human beings and nature. Energy plays a vital role in the modern world. Energy has become the most important thing for the human beings, industries, etc. The amount of energy consumption has been increasing continuously due to increase in population and rapid industrialization. Energy can be produced from both renewable and non-renewable resources. Conservation of energy is also important in the present scenario. The

J. Abhishek · S. Moulieswaran · S. Nirmal Raj · S. Pradeep · P. Manoj Kumar (✉)
Department of Mechanical Engineering, PSG Institute of Technology and Applied Research, Neelambur, Coimbatore, Tamil Nadu, India
e-mail: manoj@psgitech.ac.in

E. Natarajan et al. (eds.), *Materials, Design and Manufacturing for Sustainable Environment*, Lecture Notes in Mechanical Engineering,
https://doi.org/10.1007/978-981-19-3053-9_56

energy conservation is attained through efficient energy usage; in this case, energy use is decreased at the same time getting the same outcome as a result. Energy conservation is an inexpensive solution to energy shortages, and it is more environmentally friendly alternative to energy production. Air-conditioning is one of the major consumers of electrical energy in many parts of the world today. Air-conditioning is unavoidable in today's world. With increasing global warming and pollution, air-conditioning is a necessary for human comfort. The comfort doesn't come without a cost or without any side effect. Already today, air-conditioning started causing energy shortage in the countries. For example, China, which has the world's largest population, is having a shortage in electricity in which air-conditioning is one of the major contributors to it. The demand can be expected to increase because of changing working times, increased comfort expectations and global warming. In every country, air-conditioning plays a vital role in domestic as well as industrial applications. Hence, a small saving in day-to-day use can result in a large saving of electricity when accounted annually. Teng et al. [1] used an energy saving device to improve the performance of a split air conditioner. A moisture transferring a quick drying textile material was used on the shell surface of the compressor. Cloud-based artificial intelligence was used by Lee and Tsai [2] to control the air conditioners remotely. The model based predictive control improved the performance of a slit type air conditioner by 9.12%. In order to reduce the load on air conditioner during hot and dry weather, Dhamneya et al. [3] theoretically studied the evaporative cooling technique to improve the performance using cooling pads. Deng et al. [4] studied the impact of air conditioner operation on the energy consumption of a model building with different types of exterior wall thermal insulation. Phase change material was used in the form of packed bed by Loem et al. [5] to reduce the power consumption of a 1TR air conditioner. They used a set of spherical balls and integrated with the air conditioner for reducing the air temperature before entering the evaporator coil. Yang et al. [6] made hybrid air-conditioning system combining fresh air ventilator and direct evaporative cooling water spray system. Energy savings using inverted air conditioners under real-time varying climatic conditions were studied by Lim et al. [7]. Jinming et al. [8] studied numerically and experimentally the savings in energy of an air conditioner for heat and cooling loads by using PCM as floor material in office. Nethaji et al. [9] used the condensate from evaporator coils to circulate along the inner walls of the room which resulted in 6 to 8% decrease in air conditioner power consumption. Mainil et al. [10] used an evaporative cooling module before the condenser to reduce the air temperature to the condenser thereby increasing the COP of the air-conditioning system.

The present study is about improving the efficiency of domestic air-conditioning systems by covering the area of the bed with a screen so that it is sufficient to cool a smaller volume than the whole bedroom. Nowadays, air conditioners have become an integral part of the domestic appliances. Major use of the air-conditioning system for the domestic purpose is during sleeping time. An average sleeping time of human is around 6–8 h. Thus, improving efficiency in this specific application contributes to a major saving of electricity.

2 Experiment Details

For conducting the experiment, a room with 16 × 10 × 10 feet dimensions was chosen. The cot was positioned near to the wall to which the air conditioner was fixed. The layout of the room is shown in Fig. 1.

During the experiment, parameters such as temperature, humidity, power consumed by the air conditioner, and the oxygen level inside the enclosure were measured. Oxygen is one of the important factors because within enclosure there should be optimum level of oxygen that the human needs to survive. DHT11 temperature and humidity sensor was used to measure the temperature and humidity levels

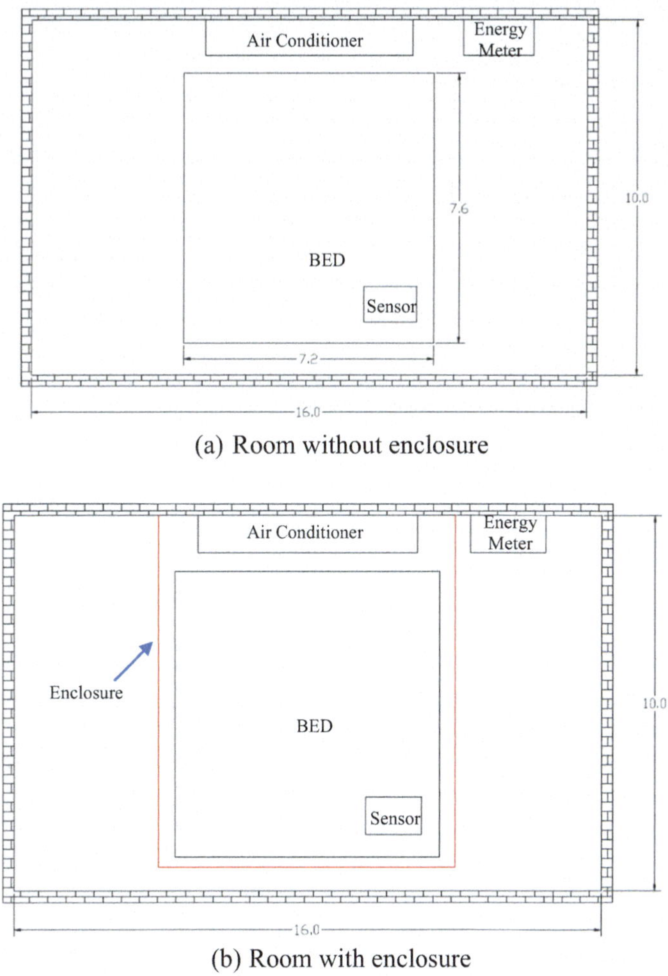

Fig. 1 Schematic of the experimental setup

near to the bed. It is combined with a high-performance 8-bit microcontroller. It uses a capacitive humidity sensor and a thermistor to measure the temperature and humidity. The enclosure material is made to cover the perimeter of the bed, so it allows the air conditioner to cool the particular volume than the whole room. Polyster and cotton were chosen as the enclosure material. The following points were considered while selecting the material:

1. The material should be breathable since it should re circulate oxygen so that it does not cause any discomfort to human.
2. The enclosure arrangement should be easily portable and collapsible.
3. The material should be durable.
4. The material should be as light as possible.

3 Results and Discussion

Various experiments were conducted to determine the variation in temperature, humidity, and oxygen level in full room as well as confined size of the room. They are used to make a comparative study over the full room with the confined enclosure setup. Also, the energy consumption was measured over the period the experiment was conducted. Initially, the room temperature and relative humidity are noted down before the start of the experiment. Then, the air conditioner is turned on, and the Arduino is programmed and connected to the DHT 11 sensor which determines and displays the room temperature and relative humidity for every 15 min. The experiments were conducted from 10:00PM to 6:00 AM, and the various parameters are noted during this period.

The variation of temperature and humidity was measured for three different AC temperatures 25, 21, 18 °C. The energy meter initial reading is noted down before the experiment is started, and it is noted again after the experiment is ended. The difference of the energy meter readings gives the energy consumed by the air conditioner. Oxygen content readings are measured for the confined enclosure setup using a handheld oxygen sensor for every one hour. Oxygen levels are noted down for both cotton and polyester during the AC set temperature of 21 °C. The oxygen level variations are also noted down for a single person sleeping inside the setup as well as for two persons sleeping inside the setup.

Figure 2 shows the variation in temperature and relative humidity near to be bed for a period of 8 h when people are occupying the bed for three different air conditioners set temperature and without any enclosure. The temperature continuously decreases as the air conditioner is working and later increases because of the addition of heat because of the metabolic activity. The relative humidity remains the same for 21 and 18 °C set temperature, whereas RH is found to increase continuously when the set temperature is 25 °C. This may be due to the fact that air can accommodate more water vapor in it at higher temperatures.

Figure 3 shows the variation of temperature near to the bed when enclosure is used to reduce the cooling space. It can be seen that when polyester is used as the

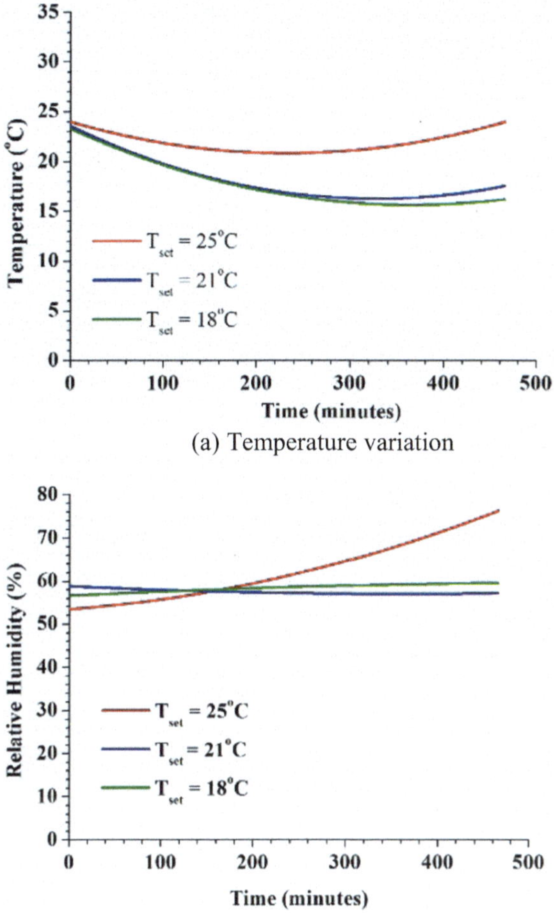

(a) Temperature variation

(b) RH variation

Fig. 2 Temperature and RH variation in the room without enclosure

screen material the temperatures are high inside the enclosed area. Cotton seems to be a better material for making enclosure than polyester.

Figure 4 shows the variation in oxygen concentration inside the enclosure when a single person is occupying the bed and when two persons are occupying. It can be seen that the oxygen levels are getting reduced initially as more oxygen is consumed for breathing, and later, the concentration is increasing may be because of oxygen diffusing in through the gaps in the screen arrangement. It can also be observed that the oxygen levels are not going below the critical levels in the 8 h of testing.

Figure 5 shows the energy consumed by the air conditioner during the 8 h operation. It can be seen that energy consumption is minimum when polyester enclosure is used. This shows that polyester-based enclosure is ideal for energy consumption reduction in air conditioners.

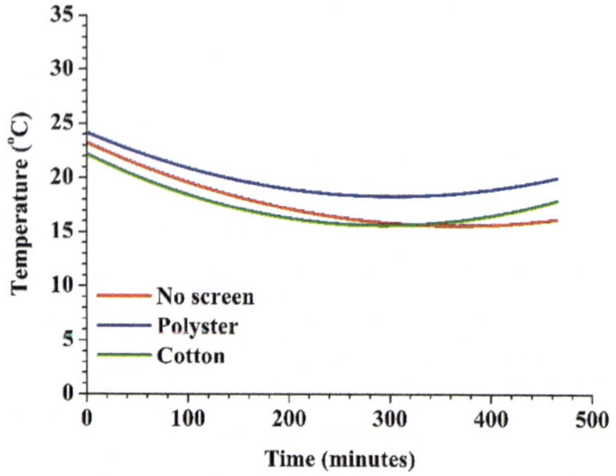

Fig. 3 Temperature variation near to the bed

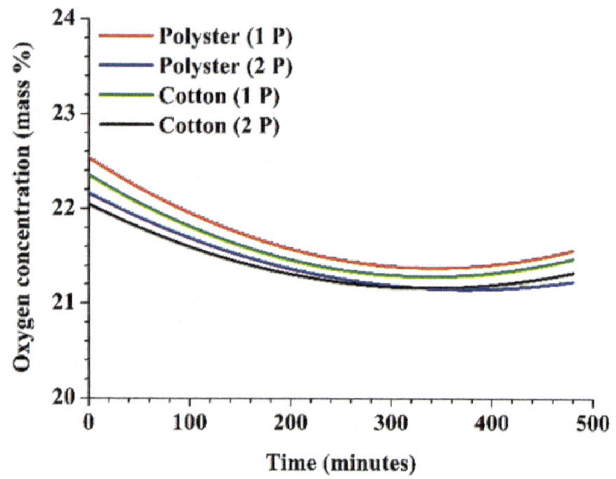

Fig. 4 Oxygen concentration (mass %) near to the bed

4 Conclusions

The suitability of using enclosure near to the bed for energy consumption reduction from an air conditioner was studied experimentally. It was observed that there is quite a lot of savings in energy when enclosure is used without much drop in oxygen levels. Polyester seems to be ideal material for screen-splitting purpose.

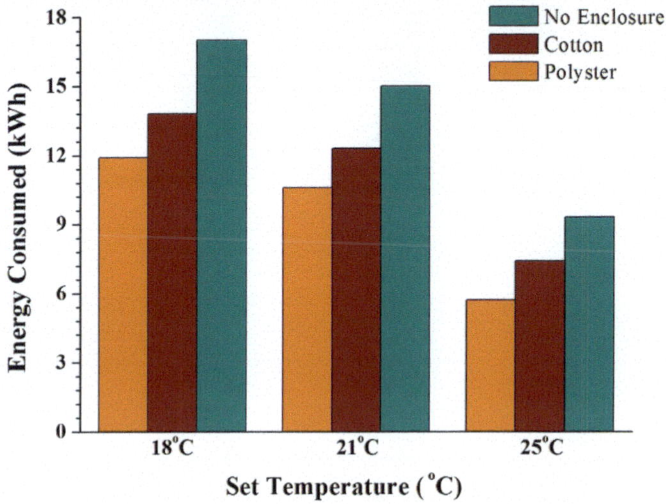

Fig. 5 Comparison of energy consumed with and without enclosure

References

1. Chen W-H, Mo H-E, Teng T-P (2018) Performance improvement of a split air conditioner by using an energy saving device. Energy Build 174:380–387
2. Lee D, Tsai F-P (2020) Air conditioning energy saving from cloud-based artificial intelligence: case study of a split-type air conditioner. Energies 13(8):2001
3. Dhamneya AK, Rajput SPS, Singh A (2018) Theoretical performance analysis of window air conditioner combined with evaporative cooling for better indoor thermal comfort and energy saving. J Build Eng 17:52–64
4. Deng Q, Liu D, Zhou Z, Ma S (2021) The effect of air-conditioner operation modes on the energy-saving capacity of external wall insulation in residential buildings. Energy Explor Exploit 39(2):1–17
5. Loem S, Deethayat T, Asanakham A, Kiatsiriroat T (2020) Study on phase change material thermal characteristics during air charging/discharging for energy saving of air-conditioner. Heat Mass Transf 56:2121–2133
6. Yang Z, Zhao J, Wang B, Zhuang R, Li X, Xiao H, Shi W (2021) Experimental performance analysis of hybrid air conditioner in cooling season. Build Environ 204:108160
7. Lim J, Yoon MS, Al-Qahtani T, Nam Y (2019) Feasibility study on variable-speed air conditioner under hot climate based on real-scale experiment and energy simulation. Energies 12(8):1489
8. Jinming S, Xinyu H, Huichuan G, Xiaodong S, Zhilong X, Xiaona Z, Zhili S, Waseem A, Chong Q, Ruimin Y, Ruqiang Z (2020) Experimental investigation and numerical validation on the energy-saving performance of a passive PCM floor for a real scale building. ES Energy Environ 8:21–28
9. Nethaji N, Mohideen T, Nethaji M (2019) Energy conservation in room air conditioner unit by recovering cold energy from condensate. Int J Refrig 104:95–102
10. Mainil RI, Sulaiman AW, Mainil AK, Aziz A (2021) Performance enhancement of split air conditioner during evaporative cooling application. J Adv Res Fluid Mech Thermal Sci 86(2):147–156

The manufacturer's authorised representative in the EU is Springer Nature Customer Service Centre GmbH, Europaplatz 3, 69115 Heidelberg, Germany. If you have any concerns regarding our products, please contact ProductSafety@springernature.com

Printed and bound by CPI Group (UK) Ltd, Croydon, CR0 4YY
30/09/2026
02225315-0020